Library of
VOID
Davidson College

*Methods of
Experimental Physics*

VOLUME 11

SOLID STATE PHYSICS

METHODS OF EXPERIMENTAL PHYSICS:

L. Marton, *Editor-in-Chief*

Claire Marton, *Assistant Editor*

1. Classical Methods
   Edited by Immanuel Estermann
2. Electronic Methods, Second Edition (in two parts)
   Edited by E. Bleuler and R. O. Haxby
3. Molecular Physics, Second Edition (in two parts)
   Edited by Dudley Williams
4. Atomic and Electron Physics—Part A: Atomic Sources and Detectors, Part B: Free Atoms
   Edited by Vernon W. Hughes and Howard L. Schultz
5. Nuclear Physics (in two parts)
   Edited by Luke C. L. Yuan and Chien-Shiung Wu
6. Solid State Physics (in two parts)
   Edited by K. Lark-Horovitz and Vivian A. Johnson
7. Atomic and Electron Physics—Atomic Interactions (in two parts)
   Edited by Benjamin Bederson and Wade L. Fite
8. Problems and Solutions for Students
   Edited by L. Marton and W. F. Hornyak
9. Plasma Physics (in two parts)
   Edited by Hans R. Griem and Ralph H. Lovberg
10. Physical Principles of Far-Infrared Radiation
    L. C. Robinson
11. Solid State Physics
    Edited by R. V. Coleman
12. Astrophysics—Part A: Optical and Infrared
    Edited by N. Carleton

Volume 11

Solid State Physics

Edited by

R. V. COLEMAN

Department of Physics
University of Virginia
Charlottesville, Virginia

1974

ACADEMIC PRESS · New York and London
A Subsidiary of Harcourt Brace Jovanovich, Publishers

COPYRIGHT © 1974, BY ACADEMIC PRESS, INC.
ALL RIGHTS RESERVED.
NO PART OF THIS PUBLICATION MAY BE REPRODUCED OR
TRANSMITTED IN ANY FORM OR BY ANY MEANS, ELECTRONIC
OR MECHANICAL, INCLUDING PHOTOCOPY, RECORDING, OR ANY
INFORMATION STORAGE AND RETRIEVAL SYSTEM, WITHOUT
PERMISSION IN WRITING FROM THE PUBLISHER.

ACADEMIC PRESS, INC.
111 Fifth Avenue, New York, New York 10003

United Kingdom Edition published by
ACADEMIC PRESS, INC. (LONDON) LTD.
24/28 Oval Road, London NW1

Library of Congress Cataloging in Publication Data

Coleman, Robert Vincent.
    Solid state physics.

    (Methods of experimental physics, v. 11)
    Supplement to Solid state physics, edited by K. Lark-
Horovitz and V. A. Johnson.
    Includes bibliographical references.
    1. Solids.   I.   Lark-Horovitz, Karl, 1891-1958, ed.
Solid state physics.   II.   Title.   III.   Series.
QC176.C57   530.4'1   74-5190
ISBN 0–12–475911–8

PRINTED IN THE UNITED STATES OF AMERICA

CONTENTS

CONTRIBUTORS xv
FOREWORD xvii
PREFACE xix

1. Experimental Methods of Measuring High-Field Magnetoresistance in Metals

   by W. A. REED

   1.1. Introduction 1
       1.1.1. Summary of the Monotonic Effects 1
       1.1.2. Magnetic Breakdown 8
       1.1.3. Quantum Oscillations 9
   1.2. Sample Preparation and Evaluation 10
       1.2.1. Methods for Cutting Samples 10
       1.2.2. Methods for Attaching Leads 12
       1.2.3. Sample Mounting 14
       1.2.4. Sample Evaluation 15
   1.3. Sample Holders 15
   1.4. Magnetic Fields and Low Temperatures 20
   1.5. Measurement Techniques 21
       1.5.1. DC Method 21
       1.5.2. AC Method 23
       1.5.3. Pulsed Field Method 25
       1.5.4. Inductive (Helicon) Method 26
       1.5.5. Induced Torque Method 27
   1.6. Data Collection and Processing 29
   1.7. Special Topics 29
       1.7.1. High-Pressure Measurements 29
       1.7.2. Low-Field Effects 30
       1.7.3. Anomalous Longitudinal Magnetoresistance . 30

2. Experimental Methods for the de Haas–van Alphen Effect

by J. R. ANDERSON and D. R. STONE

    2.1. The de Haas–van Alphen Effect 33
    2.2. dHvA Expressions 35
        2.2.1. Frequencies 35
        2.2.2. Amplitudes 37
        2.2.3. **B** versus **H** 40
    2.3. Field Modulation Technique 44
        2.3.1. dHvA Frequency Selectivity 46
        2.3.2. Block Diagram 47
        2.3.3. Frequency Measurements 51
        2.3.4. Amplitude Measurements 54
        2.3.5. Computer Analysis 55
    2.4. Examples of dHvA Experiments 56
        2.4.1. dHvA Studies in Lead 56
        2.4.2. Cyclotron Mass Measurements in Indium . . 60
        2.4.3. Ferromagnetic Metals 61
    2.5. Summary 64

3. Experimental Techniques for Visible and Ultraviolet Photoemission

by G. F. DERBENWICK, D. T. PIERCE, and W. E. SPICER

    3.1. Introduction 67
        3.1.1. Photoemission Measurements Past and Present   67
        3.1.2. Physics of the Photoemission Process 74
    3.2. Photoelectric Quantum Yield 81
        3.2.1. Definition and Measurement 81
        3.2.2. Calibration of Reference Phototube 82
    3.3. Measurement of Energy Distribution Curves . . . 84
        3.3.1. The Retarding Field Analyzer 84
        3.3.2. Electronics for the Retarding Field Analyzer   89
        3.3.3. Other Types of Energy Analyzers 99
        3.3.4. Angular EDC Measurements 100
        3.3.5. EDC Measurements As a Function of Temperature 101

3.4.	Sample Preparation	102
	3.4.1. Cleavage	103
	3.4.2. Evaporation	104
	3.4.3. Heat Cleaning	105
	3.4.4. Ion Bombardment Cleaning	107
	3.4.5. Lowering Electron Affinity by Applying Surface Layers	110
3.5	High Vacuum Photoemission Chambers	114
	3.5.1. Vacuum Pumps	114
	3.5.2. Photoemission Chambers	116
3.6.	Monochromator and Light Sources	120
3.7.	Directions for Future Research	121

4. Experiments on Electron Tunneling in Solids

by R. V. Coleman, R. C. Morris, and J. E. Christopher

4.1.	Introduction	123
4.2.	Tunneling in Superconductors	126
	4.2.1. Summary of Superconducting Tunnel Characteristics	126
	4.2.2. Tunneling between Normal Metal and a Superconductor	128
	4.2.3. Tunneling between Two Superconductors	132
	4.2.4. Deviations from Ideal Tunneling Behavior	133
	4.2.5. Experimental Determination of the Energy Gap	135
	4.2.6. Tunneling into Single Crystals	140
	4.2.7. Tunneling Measurements of Density of States and Phonon Spectra	144
	4.2.8. Tunneling Used as a Probe of the Electron–Phonon Interaction	149
	4.2.9. Phonon Generation and Detection	150
	4.2.10. Geometrical Resonances in Tunneling	152
	4.2.11. Superconducting Tunneling in High Magnetic Fields	153
4.3.	Normal-Metal Tunneling	157
	4.3.1. Introduction to Normal-Metal Tunneling	157
	4.3.2. Elastic Normal-Metal Tunneling	157
	4.3.3. Dispersion Relations in the Barrier	161

4.3.4.	High-Voltage Tunneling	161
4.3.5.	Inelastic Tunneling	163
4.3.6.	Molecular Excitations in Barriers	165
4.3.7.	Barrier Excitations in Normal-Metal Tunneling	167
4.3.8.	Electrode Excitations in Normal-Metal Tunneling	169
4.3.9.	Zero-Bias Anomalies	170

4.4. Semiconductors in Tunnel Junctions 177
    4.4.1. Summary 177
    4.4.2. Metal–Insulator–Semiconductor Tunnel Junctions 178
    4.4.3. Metal–Semiconductor Tunnel Junctions . . . 181
    4.4.4. p-n Tunnel Junctions 184
    4.4.5. Electrode–Semiconductor–Electrode Tunnel Junctions 187

4.5. Special Topics on Experimental Techniques 187
    4.5.1. Junction Fabrication 187
    4.5.2. Junction Testing 192
    4.5.3. Measurement Circuits 193

5. Experiments Using Weakly Linked Superconductors

by B. S. DEAVER, Jr., and D. A. VINCENT

5.1. Introduction 199
5.2. Experiments That Study the Josephson Effects . . . 201
    5.2.1. The Josephson Effects 201
    5.2.2. General Comments on Experiments 208
    5.2.3. The dc Supercurrent 212
    5.2.4. Microwave-Induced Steps on the I–V Curve 217
    5.2.5. Frequency Dependence of the Josephson Current and the Riedel Singularity 225
    5.2.6. Self-Induced Supercurrent Steps—Cavity Modes 229
    5.2.7. Radiation from the Oscillating Supercurrent . 233
    5.2.8. Plasma Resonance 238
    5.2.9. Subharmonic Structure on the I–V Curve . . 239
    5.2.10. Phonon Generation by the Josephson Effect . 243
    5.2.11. Effects of Thermodynamic Fluctuations and Noise 245

5.3. Characteristics of Various Types of Weakly Linked
     Superconductors 251
     5.3.1. Types of Weak Links 251
     5.3.2. Current–Phase Relation and Various Phenom-
            enological Descriptions of Weak Links ... 262
     5.3.3. Equivalent Circuit Models 268
5.4. Applications of Weakly Linked Superconductors .. 273
     5.4.1. Measurements of e/h 273
     5.4.2. Voltage Standard 274
     5.4.3. Superconducting Rings Containing a Single
            Weak Link 275
     5.4.4. Superconducting Rings Containing Two Weak
            Links 289
     5.4.5. The Clarke Slug 296
     5.4.6. Detection, Mixing, Harmonic Generation, and
            Parametric Amplification 299
     5.4.7. Digital Devices 304

6. Experimental Methods in Mössbauer Spectroscopy

by R. L. Cohen and G. K. Wertheim

6.1. Introduction to Mössbauer Spectroscopy 307
     6.1.1. The Measurables 311
6.2. Mössbauer Spectrometers 316
     6.2.1. Drives and Data Collection 316
     6.2.2. Gamma-Ray Detection 346
     6.2.3. Radioactive Sources for Mössbauer Experi-
            ments 350
6.3. Auxiliary Equipment 352
     6.3.1. Low-Temperature Techniques 352
     6.3.2. Furnaces 359
     6.3.3. Magnets 361
     6.3.4. High-Pressure Mössbauer Experiment ... 363
     6.3.5. Data Handling 365
Appendix A. Data Analysis 366
Appendix B. Useful Graphs 367
Bibliography 368

7. Ultrasonic Studies of the Properties of Solids

by E. R. Fuller, Jr., A. V. Granato, J. Holder, and E. R. Naimon

    7.1. Introduction 371
    7.2. Ultrasonic Waves in Solids 374
        7.2.1. Longitudinal Waves in an Isotropic Solid . . 375
        7.2.2. Small-Amplitude Waves in Cyrstals 376
        7.2.3. Finitely Strained Solids 381
        7.2.4. Attenuation and Dispersion 389
    7.3. Experimental Techniques 391
        7.3.1. Generation of Ultrasonic Waves 391
        7.3.2. Sample Preparation 396
        7.3.3. Velocity and Attenuation Measurements . . 397
    7.4. Applications of Ultrasonic Waves to Measuring Physical Properties 410
        7.4.1. Elastic Constants 410
        7.4.2. Real Materials 421
    Appendix A. Thermoelasticity Theory and Related Elastic Coefficients 433
        A.1. Thermoelasticity 433

8. The Use of Ions in the Study of Quantum Liquids

by Frank E. Moss

    8.1. Review of the Structure of Ions in Liquid Helium . 443
        8.1.1. The Bubble Model for the Negative Ion and the Positronium Atom 443
        8.1.2. The Electrostriction Model for the Positive Ion 445
    8.2. Production of Ions in Liquid Helium 446
        8.2.1. Radioactive Sources 446
        8.2.2. Photoelectric Injection of Electrons 448
        8.2.3. Injection of Hot Electrons by Tunnel Diodes 449
        8.2.4. Injection of Electrons from Thermionic Cathodes 451
        8.2.5. Field Emission 452
        8.2.6. Gaseous Discharges and Laser Breakdown . 453

8.3. Methods for Measurement of the Ionic Drift Velocity 454
    8.3.1. The Velocity Spectrometer 454
    8.3.2. The Cunsolo Method 456
    8.3.3. Velocity Measurements Using Signal Averagers 458
    8.3.4. The Space–Charge Limited Diode 461

8.4. The Use of Ion Mobility Measurements in the Study of Microscopic Excitations in Liquid Helium 463
    8.4.1. Introduction 463
    8.4.2. The Roton Region 465
    8.4.3. The Phonon Region 467
    8.4.4. He^3–He^4 Solutions 468
    8.4.5. Pure He^3 468

8.5. Ion Techniques for Studying Macroscopic Quantum Excitations 469
    8.5.1. Rotating Superfluid He^4 469
    8.5.2. Capture of Ions by Quantized Vortex Lines . 470
    8.5.3. Escape of Ions from Quantized Vortex Lines 474
    8.5.4. Creation of Vortex Rings with Ions 476
    8.5.5. The Use of Ions in Studies on the Structure of Turbulent Superfluid Helium 479
    8.5.6. Mobility of Ions along Linear Vortex Lines . 481

8.6. Studies of Superfluid Surfaces and Films 481
    8.6.1. Interaction of Ions with the Free Liquid Surface 481
    8.6.2. Motion of Ions in Superfluid Films 483

9. Thermometry at Ultralow Temperatures
by WALTER WEYHMANN

9.1. Introduction 485

9.2. Resistance Thermometers 489

9.3. Susceptibility Thermometers 502
    9.3.1. Paramagnetic Susceptibility of Localized Atomic Moments 504
    9.3.2. Paramagnetic Susceptibility of Nuclear Moments 511

9.4. Miscellaneous Thermometers 533
  9.4.1. Thermocouples 534
  9.4.2. ³He Melting Curve 535
  9.4.3. Superconductors 536
  9.4.4. Capacitors 538
9.5. Conclusions 538

10. Superconducting Microwave Resonators
by JOHN M. PIERCE

10.1. Introduction 541
10.2. Microwave Properties of Superconductors 544
  10.2.1. The Surface Impedance of Normal and Superconducting Metals 544
  10.2.2. The Superconducting Surface Resistance . . 550
  10.2.3. The Superconducting Surface Reactance . . 552
  10.2.4. The Residual Surface Resistance 553
  10.2.5. Dependence of the Surface Impedance on rf Field Level 560
  10.2.6. The rf Critical Field 562
10.3. Methods of Fabricating High-Q Superconducting Resonators and Measuring Their Properties 567
  10.3.1. Design and Fabrication of High Q Resonators 567
  10.3.2. Materials for High-Q Superconducting Resonators 570
  10.3.3. Measurement Techniques for High-Q Resonators 575
  10.3.4. Frequency Measurement and Control 580
  10.3.5. Coupling Networks for Superconducting Resonators 581
  10.3.6. Special Cryogenic Microwave Techniques . . 584
10.4. Brief Review of Experiments Using Superconducting Resonators 584
  10.4.1. Frequency Stability in Superconducting Resonators 585
  10.4.2. Tuning Superconducting Resonators . . . 587
  10.4.3. Frequency Standards Referenced to Superconducting Resonators 588

	10.4.4.	High-Field Applications	590
	10.4.5.	Material Property Studies	591
	10.4.6.	Miscellaneous Applications	592

11. Superconducting Device Technology

11.1. Superconducting Magnets ... 595
by C. D. Graham, Jr.

11.1.1.	Introduction	595
11.1.2.	General Considerations	596
11.1.3.	Materials	598
11.1.4.	Stabilization	600
11.1.5.	Field Uniformity	600
11.1.6.	Field Measurement	602
11.1.7.	Power Supply Considerations	603
11.1.8.	Operation in Swept Field	604
11.1.9.	Reversed Field	605
11.1.10.	Liquid Helium Consumption	605
11.1.11.	Operating Procedure	606
11.1.12.	Costs	607

11.2. Superconducting Shielding ... 609
by W. O. Hamilton

11.2.1.	Introduction	609
11.2.2.	Shielding against Time-Varying Fields	610
11.2.3.	Shielding against dc Fields	612
11.2.4.	Conclusions	617

12. Experimental Methods in the Preparation and Measurement of Thin Films
by. D. C. Larson

12.1. Preparation of Thin Films ... 620

12.1.1.	Introduction	620
12.1.2.	Thermal Evaporation	622
12.1.3.	Sputtering	637
12.1.4.	Chemical Deposition	645

12.2. Thin Film Measurements 652
    12.2.1. Introduction 652
    12.2.2. Thickness Measurements 653
    12.2.3. Mechanical Measurements 660
    12.2.4. Optical Measurements 665

13. The Observation of Magnetic Domains
by D. J. CRAIK

13.1. Introduction 675
13.2. Principles Governing Domain Formations 676
13.3. Survey of Methods 683
    13.3.1. Specimen and Specimen Surface Preparation 684
13.4. Powder Pattern (Colloid) Technique 687
13.5. Specialized Techniques and Pattern Formation . . . 692
    13.5.1. Replica Method for Electron Microscopy . . 692
    13.5.2. Vapor Condensation Method 693
    13.5.3. The Formation of Powder Patterns 695
13.6. Optical and Magnetooptical Properties 697
    13.6.1. Faraday and Birefringence Methods 707
    13.6.2. Magnetic Birefringence 715
    13.6.3. The Polar Kerr Method 718
    13.6.4. Longitudinal Kerr Method 721
    13.6.5. Transverse Kerr Method 726
13.7. Lorentz (Electron) Microscopy 726
13.8. Scanning Electron Microscopy 734
13.9. X-Ray Method 736
13.10. Methods for Antiferromagnetic Domains 740
13.11. Further Methods 743

AUTHOR INDEX 745
SUBJECT INDEX 768

CONTRIBUTORS

Numbers in parentheses indicate the pages on which the authors' contributions begin.

J. R. ANDERSON, *Department of Physics and Astronomy, University of Maryland, College Park, Maryland* (33)

J. E. CHRISTOPHER, *Department of Physics and Astronomy, University of Kentucky, Lexington, Kentucky* (123)

R. L. COHEN, *Bell Telephone Laboratories, Inc., Murray Hill, New Jersey* (307)

R. V. COLEMAN, *Department of Physics, University of Virginia, Charlottesville, Virginia* (123)

D. J. CRAIK, *Department of Chemistry, University of Nottingham, University Park, Nottingham, England* (675)

B. S. DEAVER, Jr., *Department of Physics, University of Virginia, Charlottesville, Virginia* (199)

G. F. DERBENWICK,* *Stanford University, Stanford, California* (67)

E. R. FULLER, Jr.,† *Department of Physics, Materials Research Laboratory, University of Illinois, Urbana, Illinois* (371)

C. D. GRAHAM, Jr., *Department of Metallurgy and Materials Science, University of Pennsylvania, Philadelphia, Pennsylvania* (595)

A. V. GRANATO, *Department of Physics, Materials Research Laboratory, University of Illinois, Urbana, Illinois* (371)

W. O. HAMILTON, *Department of Physics, Louisiana State University, Baton Rouge, Louisiana* (609)

---

* Present address: Bell Telephone Laboratories, Inc., Murray Hill, New Jersey 07974.

† Present address: National Bureau of Standards, Institute for Materials Research, Washington, D. C. 20234.

J. HOLDER, *Department of Geology, Materials Research Laboratory, University of Illinois, Urbana, Illinois* (371)

D. C. LARSON, *Department of Physics, Drexel University, Philadelphia, Pennsylvania* (619)

R. C. MORRIS,* *Department of Physics, University of Virginia, Charlottesville, Virginia* (123)

FRANK E. MOSS, *Department of Physics, University of Missouri, St. Louis, Missouri* (443)

E. R. NAIMON, *Department of Physics, Materials Research Laboratory, University of Illinois, Urbana, Illinois* (371)

D. T. PIERCE, *Stanford University, Stanford, California* (67)

JOHN M. PIERCE,[†] *Department of Physics, University of Virginia, Charlottesville, Virginia* (541)

W. A. REED, *Bell Telephone Laboratories, Inc., Murray Hill, New Jersey* (1)

W. E. SPICER, *Stanford University, Stanford, California* (67)

D. R. STONE, *Laboratory for Physical Sciences, College Park, Maryland* (33)

D. A. VINCENT, *Department of Physics, University of Virginia, Charlottesville, Virginia* (199)

G. K. WERTHEIM, *Bell Telephone Laboratories, Inc., Murray Hill, New Jersey* (307)

WALTER WEYHMANN, *Tate Laboratory of Physics, University of Minnesota, Minneapolis, Minnesota* (485)

---

* Present address: Department of Physics, Florida State University, Tallahassee, Florida 32306.

† Present address: Develco, Inc., 530 Logue Avenue, Mountainview, California 94040.

FOREWORD

A few months ago, in writing a foreword to the second edition of the Molecular Physics volume (Volume 3) of our series, I discussed the reasons for issuing a revised edition of one of the earlier books. At the time that decision was made, we considered the revision of the Solid State Physics volumes (Volumes 6A and 6B) published in 1959. As Professor Coleman explains in his preface to this volume, it appeared more advisable to supplement the earlier volumes rather than revise their contents completely.

With the tremendous growth of solid state physics a supplement of the present size cannot be complete. It is to be hoped that within a few years it shall be possible to add further material to that presented here.

Professor Coleman and the contributors of this volume had a challenging task before them to live up to the standards set by the editors and authors of Volumes 6A and 6B. I hope our readers will agree with me that they succeeded in their efforts; I for one am very grateful for their untiring work and collaboration.

L. Marton

PREFACE

The present volume on solid state physics gives a detailed and comprehensive treatment of a number of current research areas that have developed extensively since the original volumes in this series devoted to solid state physics were published in 1959 (Volume 6, Parts A and B). The current volume is intended to supplement and expand many of the subjects covered in the original volumes and demonstrates the tremendous growth and sophistication that have developed in the experimental techniques of solid state physics. The original volumes contain excellent introductions to many of the basic experimental techniques used in solid state physics and it is the purpose of the present volume to provide a more specialized treatment of some of the topics of great current interest. The volume contains complete descriptions of the experimental methods as well as discussions of what physical quantities can be measured and how they relate to the basic problems of solid state physics. The volume has been written in the hope of providing a useful source book and guide for researchers who wish to do experimental work in the various fields covered. It should appeal to graduate students as well as to more senior researchers who wish to start research in areas outside their specialty. The volume emphasizes the experimental background and technique of the subjects covered, but it also provides an overall view of the physical principles involved and should provide stimulating reading for anyone who wishes to become more familiar with advanced solid state subjects.

The volume covers a variety of subjects representing modern solid state physics and should give the reader a fairly comprehensive view, but due to the great scope of current research in solid state no attempt has been made to cover all subjects.

I wish to take this opportunity to express my great appreciation and thanks to all of the contributors who made this volume possible. Their cooperation and hard work have been essential to success. I also wish to thank the publisher and Dr. L. Marton for continued support and encouragement in this project.

ROBERT V. COLEMAN

1. EXPERIMENTAL METHODS OF MEASURING HIGH-FIELD MAGNETORESISTANCE IN METALS*

1.1. Introduction

1.1.1. Summary of the Monotonic Effects

Although it had been known for many years that the galvanomagnetic properties of pure metals at low temperatures were highly anisotropic with respect to the field direction, it was not until Lifshitz and his co-workers[1] developed their high-field theory that the galvanomagnetic effects could be related to the topology of the Fermi surface. The high-field condition means that for all carriers

$$\omega_c \bar{\tau} = (eB/m^*)\bar{\tau} \gg 1, \qquad (1.1.1)$$

where ω_c is the cyclotron frequency, $\bar{\tau}$ is the relaxation time averaged around each orbit, and e, B, and m^* are the electron charge, magnetic field, and effective mass, respectively, measured in electromagnetic units. Physically this condition means that each carrier will repeat its periodic motion many times before scattering. One of the beauties of the Lifshitz theory is that in the high-field regime the details of the scattering processes are relatively unimportant and the major features of the galvanomagnetic properties are governed by the geometry of the Fermi surface.

Of all the galvanomagnetic effects,[2] the two most often measured are the magnetoresistance and Hall effect. The magnetoresistance, which is measured on pair 1 of the sample shown in Fig. 1, is defined as $[\varrho(B) - \varrho(0)]/\varrho(0) \equiv \Delta\varrho/\varrho$ and may be either of order one or greater

---

[1] I. M. Lifshitz, M. Ya. Azbel, and M. I. Kaganov, *J. Exp. Theor. Phys.* **30**, 220 (1955) [*English transl.*: *Sov. Phys. JETP* **3**, 143 (1956)]; I. M. Lifshitz and V. G. Peschanskii, *J. Exp. Theor. Phys.* **35**, 1251 (1958) [*English transl.*: *Sov. Phys. JETP* **8**, 875 (1958)].

[2] A description of all the galvanomagnetic effects can be found in J. P. Jan, *Solid State Phys.* **5**, 1 (1957).

---

* Part 1 is by W. A. Reed.

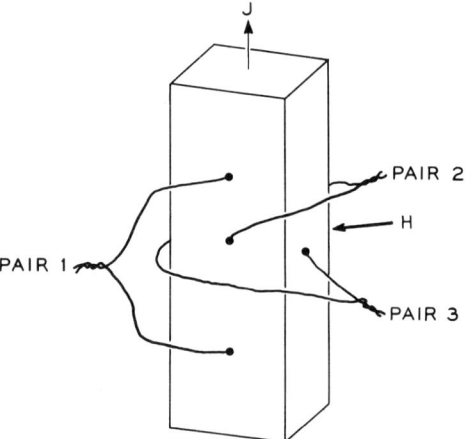

FIG. 1. Schematic diagram of sample geometry and placement of potential leads for galvanomagnetic measurements. Pair 1: resistance leads; pairs 2 and 3: Hall leads.

than 10^5, depending upon the metal and magnetic field direction. The Hall field E_H, whose components are measured on pairs 2 and 3, is normal to both the current and magnetic field. It also depends upon the metal and field direction and is related to the current density J and the magnetic field B through the Hall coefficient R by $\mathbf{E}_\mathrm{H} = R\mathbf{J} \times \mathbf{B}$.

The motion of an electron in a magnetic field is periodic, and, when viewed in k space, describes an orbit on the Fermi surface in a plane normal to the magnetic field. The basic types of orbits viewed in the extended zone scheme are illustrated in Fig. 2 on a Fermi surface which might be found in a cubic metal. In part (a) of this figure the field is parallel to a $\langle 100 \rangle$ axis and only closed orbits are permitted. An orbit is considered closed if the carrier executing the orbit returns to its starting point, and it has an electron/hole character[3] if it encloses a region of occupied/unoccupied states. This figure also demonstrates how a Fermi surface of one character can support orbits with the opposite character for special directions of the field. In parts (b) and (c) of this figure the magnetic field has been rotated away from the $\langle 100 \rangle$ axis to illustrate the two types of open orbits. An orbit is considered open when a carrier on that orbit does not return to its original position in the extended zone scheme. Periodic open orbits occur only when the field is in a crystallographic plane, whereas the aperiodic open orbits occur for the field within a solid

---

[3] For a precise definition of the character of a sheet of Fermi surface see E. Fawcett and W. A. Reed, *Phys. Rev.* **131**, 2463 (1963).

1.1. INTRODUCTION

angle around high-symmetry crystallographic axes. The terms *periodic* and *aperiodic* refer to whether the motion along the orbit is repetitive, but in each case a net direction of the open orbit is defined.

The "state of compensation" of a metal is also an important concept in terms of the galvanomagnetic properties. A metal is compensated if the number of electrons equals the number of holes ($n_e = n_h$) and uncompensated if the numbers of electrons and holes are unequal ($n_e \neq n_h$). It is easy to understand physically how compensation can occur. Each Brillouin zone can hold two electrons, one spin up and one spin down, and the total number of electrons available to fill the zones is equal to the atomic number of the metal Z times the number of atoms per unit cell s. Thus if sZ is an even number, there are just enough electrons to fill $sZ/2$ Brillouin zones. However, the crystal is a conductor and not an insulator, so that some electrons must be promoted to a higher

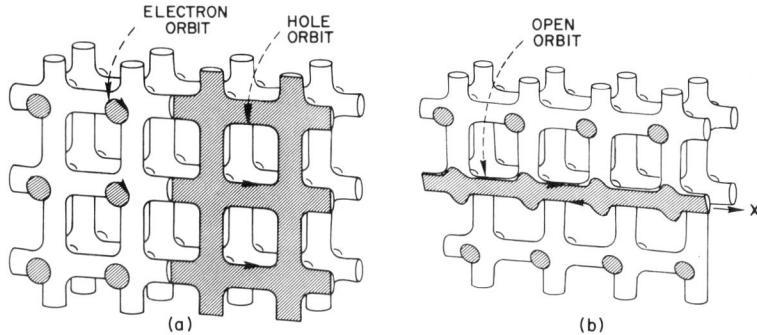

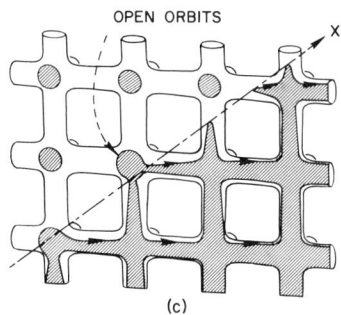

FIG. 2. Model Fermi surface illustrating possible types of open and closed orbits: (a) $B \parallel \langle 100 \rangle$ axis, closed electron and hole orbits; (b) B in (001) plane; (c) B near $\langle 100 \rangle$ axis [adapted from R. G. Chambers, *in* "The Fermi Surface" (W. A. Harrison and M. B. Webb, eds.), p. 100. Wiley, New York, 1960].

zone leaving an equal number of holes behind in the lower zones. This description is true only for nonmagnetic metals and a rigorous statement of the conditions for compensation in both magnetic and nonmagnetic metals can be found in Ref. 3.

The theory of Lifshitz et al.[1] relates in a simple way the field dependence of the magnetoresistance and Hall effect to the state of compensation and the closed or open nature of the orbits. The most important results of this theory are summarized as follows:

I. When only closed orbits exist and $n_e \neq n_h$, the magnetoresistance saturates for all directions of the magnetic field and current. The Hall coefficient is related to the Fermi surface by the expression $R = (\Delta n\, e)^{-1}$, where $\Delta n = n_e - n_h$.

II. When only closed orbits exist and $n_e = n_h$, the magnetoresistance rises quadratically with the magnetic field for all directions of field and current except for $B \parallel J$, where it saturates. The Hall coefficient is a constant but is not related to the Fermi surface in a simple way.

III. For magnetic field directions where the Fermi surface permits open orbits with a single average direction, the magnetoresistance is quadratic ($\sim B^2$) in the field and depends upon current direction as

$$\Delta\varrho/\varrho = a + bB^2 \cos^2 \alpha, \tag{1.1.2}$$

where a and b are constants and α is the angle between the current direction and the open orbit direction. This is true regardless of the state of compensation. The Hall coefficient is a constant but cannot be directly related to the shape of the Fermi surface. Physically this means that an open orbit behaves like a two-dimensional conductor, conducting only in the plane normal to the open orbit direction.

IV. For magnetic field directions where the Fermi surface permits more than one direction of open orbits, the magnetoresistance saturates for all directions of current as in case I, but now the Hall coefficient is proportional to B^{-2}.

Magnetoresistance data that illustrate the various situations are shown in Fig. 3. Figure 3a is typical of an uncompensated metal (Cu) with a Fermi surface which permits open orbits. In region a only electron orbits exist, the magnetoresistance saturates at a low value ($\Delta\varrho/\varrho \approx 1$), and the Hall coefficient corresponds to one electron/atom. Point b is also an example of case I, but for this field direction both electron and hole orbits exist similar to the example in Fig. 2a. For a singular field direction

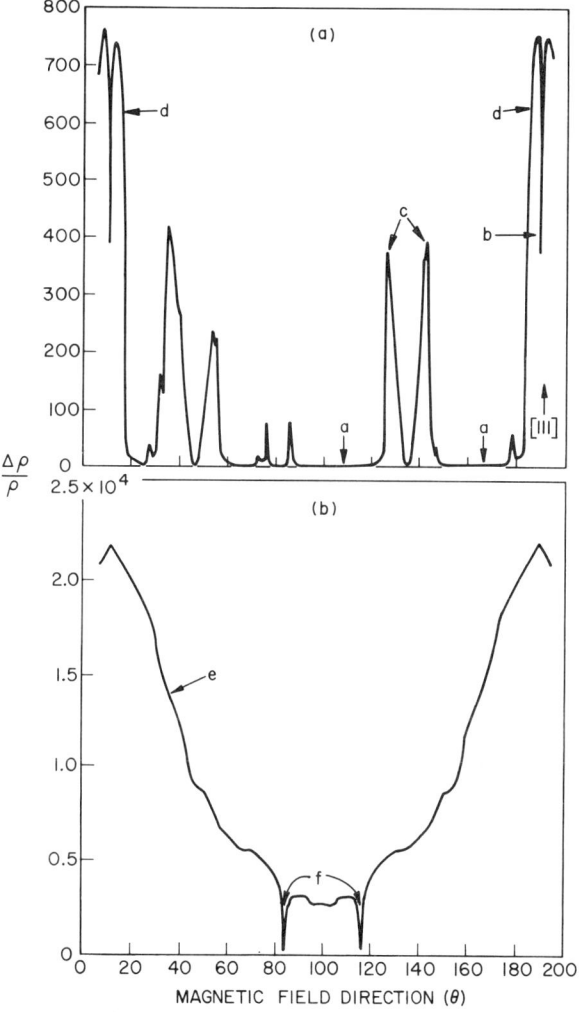

FIG. 3. Magnetoresistance rotation curve (a) uncompensated metal, Cu; (b) compensated metal, Zn (see text for details); [after W. A. Reed and E. Fawcett, *Science* **146**, 603 (1964). Copyright 1964 by the Amer. Ass. for the Advancement of Science].

such as this, the number of hole orbits allowed is directly related to the thickness of the arms, so that the Hall coefficient provides a method of measuring the arm diameters.[4] The spikes in the data (points c), which

[4] J. R. Klauder, W. A. Reed, G. F. Brennert, and J. E. Kunzler, *Phys. Rev.* **141**, 592 (1966).

occur when the field lies in certain crystallographic planes, are due to periodic open orbits, whereas the region labeled d is due to aperiodic open orbits. These are both examples of case III behavior. Figure 3b is typical of a compensated metal (Zn) whose Fermi surface also supports open orbits. The large background magnetoresistance (region e) is due to $n_e = n_h$ as in case II, and the sharp minima (points f) are a result of open orbits whose directions are normal to the current. It is easy to see from this figure that open orbits are more easily detected in an uncompensated metal since they appear as large spikes against a low background magnetoresistance. Only then the angle α between the current and the open orbit direction is $\pi/2$ [see Eq. (1.1.2)] will the existence of an open orbit fail to appear as a spike. The converse, however, is true for a compensated metal. Since the background is high the effect of an open orbit is easily seen only when $\alpha \approx \pi/2$.

To investigate the galvanomagnetic properties of a metal, the procedure is to measure the field dependence of the magnetoresistance for a large number of field directions and for at least two nonequivalent current directions. If the magnetoresistance is quadratic for most field directions, the metal is considered compensated, whereas, if the magnetoresistance saturates, the metal is considered uncompensated. The field directions at which open orbits are observed are then plotted on a stereogram as shown in Fig. 4. The lines represent field directions where open orbits appear as spikes in the data similar to points c in Fig. 3a. The shaded areas are called "two-dimensional regions" and are the field directions that correspond to data labeled d in Fig. 3a. It is inside these two-dimensional regions where the aperiodic open orbits occur. The centers of these regions represent singular field directions where both electron and hole orbits occur. The Hall constant should be measured for these singular directions to obtain geometric information about the multiply connected surfaces. Having plotted the stereogram, an attempt then is made to construct a Fermi surface with the proper topology. Often band structure calculations or data from other types of Fermi surface measurements are useful in deciding on the proper topology. It was this type of data analysis that allowed Klauder and Kunzler[5] to deduce the topology of the Fermi surface shown in Fig. 5 from the stereogram of Fig. 4. It should be noted that the galvanomagnetic data provided information about the diameter of the necks and the directions of open orbits but did

---

[5] J. R. Klauder and J. E. Kunzler, "The Fermi Surface" (W. A. Harrison and M. B. Webb, eds.), p. 125. Wiley, New York, 1960.

1.1. INTRODUCTION

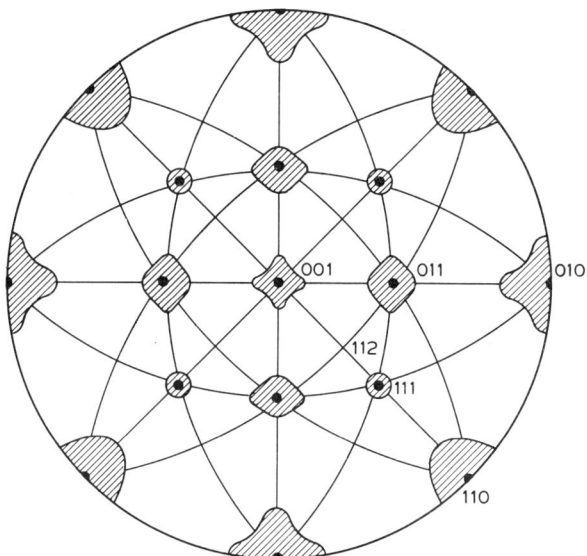

Fig. 4. Stereogram of magnetoresistance of copper [after W. A. Reed and E. Fawcett, *Science* **146**, 603 (1964). Copyright 1964 by the Amer. Ass. for the Advancement of Science].

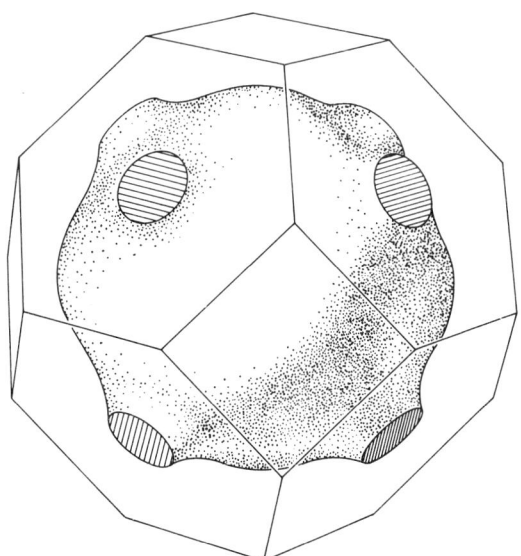

Fig. 5. Fermi surface of copper [after W. A. Reed and E. Fawcett, *Science* **146**, 603 (1964). Copyright 1964 by the Amer. Ass. for the Advancement of Science].

not provide the information needed to determine the undulations on the body. That information was obtained from de Haas–van Alphen and magnetoacoustic measurements.

In the preceding discussion, the internal field **B** is used rather than the applied field **H**. Although in diamagnetic materials the distinction is usually minor,[6] in ferromagnetic materials it is essential to interpret the data in terms of **B**. This was demonstrated in a study of the galvanomagnetic properties of Ni, where Reed and Fawcett[7] showed that **B** and not **H** was required to calculate the correct value of the Hall coefficient. A more exotic example is provided by Gold's study of the de Haas–van Alphen effect in iron.[8] He found that to interpret the data he had to account for the rotation of **B** within the sample as a function of **H** for **H** applied along a direction other than the easy axis of magnetization.

The preceding has been brief and is only intended to give the reader the "flavor" of the galvanomagnetic effects. Several articles have been published which deal completely or in part with the theory and the reader is directed to them. Two articles devoted completely to the galvanomagnetic effects are by Fawcett,[9] which is very comprehensive, and by Reed and Fawcett,[10] which is more an introductory article. Comprehensive articles that are partially devoted to the galvanomagnetic effects have been published by Pippard[11] and introductory articles devoted to the general topic of Fermi surfaces have been published by Harrison[12] and Mackintosh.[13]

1.1.2. Magnetic Breakdown

At high fields, there is a possibility that an electron will have enough energy to tunnel through an energy gap and appear on a new sheet of

---

[6] For an example of when the distinction between B and H is important in diamagnetic materials see J. H. Condon, *Phys. Rev.* **145**, 526 (1966).

[7] W. A. Reed and E. Fawcett, *J. Appl. Phys.* **35**, 754 (1964).

[8] A. V. Gold, *Proc. Int. Conf. Magn., Nottingham, 1964*, p. 124. Inst. of Phys. and the Phys. Soc., London.

[9] E. Fawcett, *Advan. Phys.* **13**, 139 (1964).

[10] W. A. Reed and E. Fawcett, *Science* **146**, 603 (1964).

[11] A. B. Pippard, "The Dynamics of Conduction of Electrons." Gordon and Breach, New York, 1965; A. B. Pippard, *Rep. Progr. Phys.* **23**, 176 (1960) (The Physical Society, London).

[12] W. A. Harrison, *Science* **134**, 915 (1961).

[13] A. R. Mackintosh, *Sci. Amer.* **209**, 110 (1963).

1.1. INTRODUCTION

Fermi surface instead of being Bragg diffracted. This phenomenon, called *magnetic breakdown*, can greatly complicate the interpretation of the galvanomagnetic properties, since orbits can change from hole to electron-like or from closed to open simply by increasing the magnetic field. The probability that magnetic breakdown will occur is given by

$$P = \exp(-B/B_0), \qquad (1.1.3)$$

where $B_0 \approx m^* \Delta^2/e\hbar\varepsilon_F$, Δ is the energy gap, and ε_F is Fermi energy.

Since energy gaps of 0.1 eV or less are common in metals, the effects of magnetic breakdown are easily seen in fields of 10 to 100 kOe. It is therefore necessary in studying the galvanomagnetic properties to watch for possible changes in the field dependence over the range of fields available. If magnetic breakdown does occur, it can be used to estimate the size the energy gaps breaking down. Since there is an excellent and recent review of magnetic breakdown by Stark and Falicov,[14] it is not necessary to give a detailed account of the phenomenon here. The important point to remember is that even with magnetic breakdown, it is the types of orbits at a given field that determine the galvanomagnetic properties. The types of the orbits may change with field, but the theory by Lifshitz *et al.*[1] remains valid at each field as long as $\omega_c \bar{\tau} \gg 1$ for all orbits.

1.1.3. Quantum Oscillations

Theoretically quantum oscillations, whose origin is the same as the de Haas–van Alphen effect, should be superimposed on the monotonic component of the galvanomagnetic effects. Except for the semimetals Bi, Sb, and As, these oscillations are generally not observed due to their extremely small amplitudes. However, under the appropriate conditions the amplitudes become very large and even dominate the magneto-resistance.[14,15] These large amplitudes are due to magnetic breakdown from orbits that give a quadratic magnetoresistance at low fields to orbits that give a saturating magnetoresistance at high fields (or vice versa). As the Landau levels pass through the Fermi level, the energy gap oscillates, which causes the breakdown probability to oscillate. As the probability oscillates the magnetoresistance switches between a large and small value, which amplifies the normal Shubnikov–de Haas oscillations.

---

[14] R. W. Stark and L. M. Falicov, "Progress in Low Temperature Physics" (C. J Gorter, ed.), p. 235. Wiley, New York, 1967.
[15] W. A. Reed and J. H. Condon, *Phys. Rev. B* **1**, 3504 (1970).

Since the amplitudes of these magnetoresistance oscillations are increasing at the fields where the de Haas–van Alphen amplitudes are decreasing, a study of the magnetoresistive oscillations at high fields can give additional information about the extremal areas of the Fermi surface.

1.2. Sample Preparation and Evaluation

1.2.1. Methods for Cutting Samples

The methods used to grow, cut, orient, and mount the single-crystal samples must be selected so as to minimize sample strain and contamination. Most metals are very soft when purified to better than 10 ppm total impurities and must be treated with great care at all stages of the experiment. If large crystals are available, it is usually desirable to prepare several samples with the current axes parallel to the major crystallographic axes. To help ensure a uniform current density the samples should be long and thin with a length-to-width ratio greater than 10 : 1. From a practical point of view, this means a width ranging from 0.5 to 2 mm wide and a length of 5 to 30 mm. Samples with smaller dimensions can and have been used but the problems of attaching leads without introducing strain becomes more difficult. A restriction on the minimum sample size is that the mean free path of the electron l at a few degrees Kelvin should be less than the sample width d. If $l \approx d$, the sample can still provide information about the qualitative feature of the Fermi surface but quantitative information should be considered suspect until demonstrated otherwise.

One method of cutting a sample to the desired shape in a relatively strain-free manner is by acid-etching using an acid-string saw. This instrument runs an acid-laden Saran string lightly over the sample until the acid etches out a cut. The method, however, is limited to materials for which corrosive acids are available. After cutting, the surfaces are quite ragged and must be acid-lapped smooth. Descriptions and designs for typical string saws may be found in Madden and Asher[16] and Young and Wilson.[17] A variation of the acid-string saw is the electrolytic saw.[18] It is similar in design but uses electrolysis for the cutting action, which extends its use to more metals.

---

[16] Robert Maddin and W. R. Asher, *Rev. Sci. Instrum.* **21**, 881 (1950).
[17] F. W. Young, Jr. and T. R. Wilson, *Rev. Sci. Instrum.* **32**, 559 (1961).
[18] U. Bonse, E. te Kaat, and E. Kappler, *J. Sci. Instrum.* **42**, 631 (1965).

1.2. SAMPLE PREPARATION AND EVALUATION

A second method is spark erosion or spark cutting. This method is based on the fact that when a spark jumps between two conductors it erodes away part of both conductors. A schematic diagram of a spark cutter is shown in Fig. 6. The power supply charges the capacitor to about 300 V and the servo mechanism advances the tool toward the sample until an arc occurs. The servo keeps the tool from touching the sample but still close enough for the arching to continue. The tool may be a thin metal sheet, but the smoothest and most accurate cuts are obtained from a 0.013 to 0.015 cm diameter copper wire that is continuously fed through the cut. The speed of the cut is determined by the values of

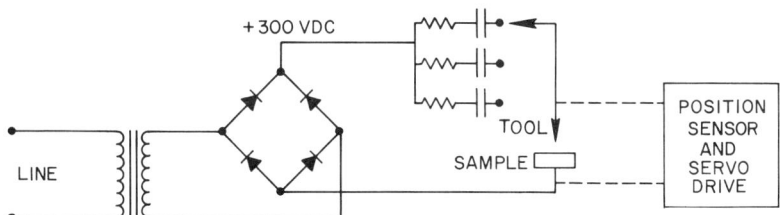

FIG. 6. Schematic diagram for a spark cutter (see text for discussion).

the resistors and capacitors. This method may inflict severe damage to a sample but such damage is usually localized at the surface of the cut and does not penetrate very far into the crystal. The damaged layer, which for most materials can be kept to less than 0.025 cm with the proper selection of spark energies, can be removed either chemically or electrochemically. To facilitate the orientation of the samples it is convenient to build a sample holder which fits both the X-ray equipment as well as the spark cutter. With care the sample and crystallographic axes can be aligned to better than one degree of arc. A variation of the spark cutter is the spark planer, which is used to produce surfaces that are smoother than the wire cut surface. The wire is replaced by a rotating disk and the arcing takes place over the entire sample surface. For both cutting and planing the sample may be mechanically fastened to the holder or glued down with either a conducting epoxy or a one-to-one mixture of a silver paste and Duco cement.[†] With a conducting adhesive the cut can be made through the adhesive and still keep the sample secure. Although spark cutters

---

[†] Both a silver paste grade number 7548 and Duco Cement are manufactured by E. I. DuPont de Nemours & Co., Wilmington, Delaware.

suitable for laboratory use are available commercially,[†] they are simple enough to be build in the lab.[19]

Methods of cutting which use some form of mechanical abrasion such as a diamond-wheel saw or polishing paper have been used on occasion with limited success. These techniques usually introduce strains which penetrate far into the sample and should be avoided if possible.

1.2.2. Methods for Attaching Leads

The usual methods of measuring the magnetoresistance and Hall effect require the attaching of leads to the sample after it has been cut to the desired dimensions and etched. The general position of the leads is shown in Fig. 1. The current leads should cover the ends of the sample in a manner that ensures a uniform current density in the sample. This is a nontrivial problem (see Section 1.7.3) and it seems almost impossible to guarantee uniform current distribution in crystals with extremely anisotropic conductivity tensors. However, in most cases the current distribution can be made sufficiently uniform for meaningful measurements. Small blocks with a cross section the same as the crystal or ribbons which cover the end of the crystal have generally worked well as long as the connection is made over the entire end. The potential leads are placed near the center of the sample, where the current should be most uniform. A rule of thumb is that the resistance leads (pair 1, Fig. 1) be spaced apart no more than $\frac{1}{3}$ the sample length. The Hall leads (pairs 2 and 3) should be placed in the center and perpendicular to each other. This arrangement of potential leads permits the direction and magnitude of the electric field to be calculated from the potentials measured on the three pairs. The contact area of the potential leads should be kept as small as possible to minimize the distortion of the current density by the leads and each pair of leads should be twisted to minimize induced voltages when the magnetic field is sweep or rotated. It is also important that the Hall leads be accurately placed in a plane normal to the current direction. This precaution will minimize the amount of the resistive component measured on the Hall leads and thus make the Hall measurements more precise. Since the samples are relatively small, the lead attachment is

---

[19] J. R. Merrill, D. Giovanielli, and A. A. Bright, *Rev. Sci. Instrum.* **41**, 31 (1970).

[†] Servomet, manufactured by Metals Research Ltd., Cambridge, England.

1.2. SAMPLE PREPARATION AND EVALUATION

most easily done under a microscope, a 7–30 × zoom binocular microscope being ideal.

There are a number of ways to attach the leads to a sample, the most appropriate method depending on the particular metal to be measured. The most common method is soldering. There is a wide variety of solder compositions and fluxes and it is often simply a case of "cut and try" until one works.[20] The combinations which usually work are (a) 60 Sn/40 Pb with rosin or acid flux, (b) pure In with no flux on a clean freshly etched surface, (c) 26 Sn/54 Bi/20 Cd with acid flux, and (d) pure In with an ultrasonic soldering iron.[†] Another method, which works well on metals like Be which form a tough oxide coat, is to heat the sample to about 200°C, draw some solid 60 Sn/40 Pb solder down to a fine wire, dip the solder in a stainless steel flux,[20] and then touch the sample with the tip of the solder. The corrosive flux removes the oxide as the solder melts and wets the metal underneath. After the solder dots are placed on the sample, the leads may be attached by soldering to the dots with an iron in a conventional way. If the sample can be heated to 500°C, another method is to put the initial "dots" on the sample with colloidal silver.[‡] These dots can be made very small if applied with a fine capillary tube. After the dots are fired about 30 min at 400°C, the leads can be attached to them using a silver saturated solder (97.5 Pb/1.0 Sn/1.5 Ag).

Other methods of attaching leads are spot welding, conducting paste or epoxy, and pressure. Spot welding works well in many cases but if not done carefully it may damage the sample. It is also difficult to spot weld current leads and have the contact area cover the end of the sample. Conducting pastes and epoxies work but the contacts generally have low mechanical strength (pastes) and/or high contact resistance (epoxies).[§] Pressure contacts have also been used with some success but tend to be noisy and may damage the sample if it is soft. These last three methods should be avoided unless the other methods fail.

---

[20] For a list of solders and fluxes see W. H. Warren, Jr., and W. G. Bader, *Rev. Sci. Instrum.* **40**, 180 (1969).

[†] A typical unit is manufactured by Sonobond Corp., West Chester, Pennsylvania.

[‡] Engelhard Industries Inc., Hanovia Liquid Gold Division, East Newark, New Jersey.

[§] A typical conducting epoxy is Epo-tek 417 manufactured by Epoxy Technology, Inc., Watertown, Massachusetts. Silver paste can be removed with acetone and many epoxies can be stripped with warm dimethylformamide. Dimethylformamide should be used with care in a hood since the vapors are harmful.

1.2.3. Sample Mounting

After the leads are attached the author has found it convenient, although not necessary, to mount the sample on a subassembly such as shown in Fig. 7. The advantage of using subassemblies is that once the leads are attached to a sample they never need be disturbed again. The subassembly mounts on the holder and leads, which are permanently mounted on the holder, are soldered to the back of the pins on the subassembly.

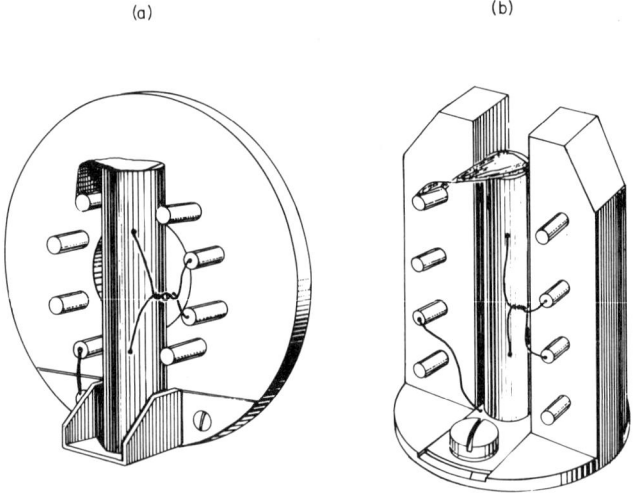

Fig. 7. Sample mount subassemblies: (a) used with holder shown in Fig. 8; (b) used with holder shown in Fig. 11.

This arrangement permits the subassembly to be mounted, removed, and remounted without affecting the sample. The soldering of the leads to the pins has not produced any problem with thermoelectric voltages, presumably because these effects are small at low temperatures, and all parts of the holder immersed in the liquid He are sufficiently isothermal. The subassemblies shown in Fig. 7 are designed for samples large enough to be supported at one end by soldering to the copper strip. For smaller samples such as whiskers it may be necessary to support them by laying them on an insulating strip. When mounting the sample it is necessary to provide for the differential thermal expansions of the various materials by holding the sample fixed at only one point and using flexible leads at all other points.

1.2.4. Sample Evaluation

There are various techniques available for evaluating the sample to be measured. Impurities can be measured by a spark or mass-spectrographic analysis and the crystal perfection can be judged from X-ray photographs. However, the most usual and useful method is the residual resistance ratio. This method is based on the assumption that at room temperature the electron mean free path (and hence the resistance) is completely dominated by thermal scattering, whereas at the lowest temperatures the mean free path is restricted primarily by the electrons scattering from impurities and crystal imperfections. Thus the residual resistance ratio, defined as $R_{300\,\mathrm{K}}/R_{4.2\,\mathrm{K}}$, is a relative measure of the low-temperature mean free path of the electrons between samples of the same material.† Since the high-field condition is $\omega_c \bar{\tau} \gg 1$, it is desirable to use crystals with large resistance ratios, which imply large $\bar{\tau}$ at low temperatures. Ratios in Sn and Pb have been measured in excess of 10^5, whereas high-field galvanomagnetic effects have been measured in some transition metals with ratios as low as 100. It should, however, be clearly understood what a resistance ratio means. This ratio is a good way of comparing the "high-field quality" of samples of the same material, since it does measure the relative electrons relaxation times at low temperatures. It is not a particularly good means of evaluating samples of different materials. It is also not a measure *per se* of sample purity. Different impurities will affect the ratio by different amounts and if the impurity has segregated or precipitated in the sample, it will have little effect on the ratio. This is because the ratio only measures the "most conducting path." Another example of how the resistance must be carefully interpreted is in ferromagnetic materials. It was found in iron that the resistance taken at 4.2 K in zero magnetic field was too large, due to magnetoresistance from the internal field. A correct ratio was estimated by measuring the resistance in low magnetic fields and extrapolating back to zero field.[21]

1.3. Sample Holders

Since the high-field galvanomagnetic effects are very sensitive to the direction of the magnetic field, it is necessary to design a holder capable

---

[21] W. A. Reed and E. Fawcett, *Phys. Rev.* **136**, A422 (1964); *Proc. Int. Conf. Magn. Nottingham 1964*, p. 120. Inst. of Phys. and the Phys. Soc., London.

† In some very pure samples, the resistance at 4.2 K has not reached the residual limit.

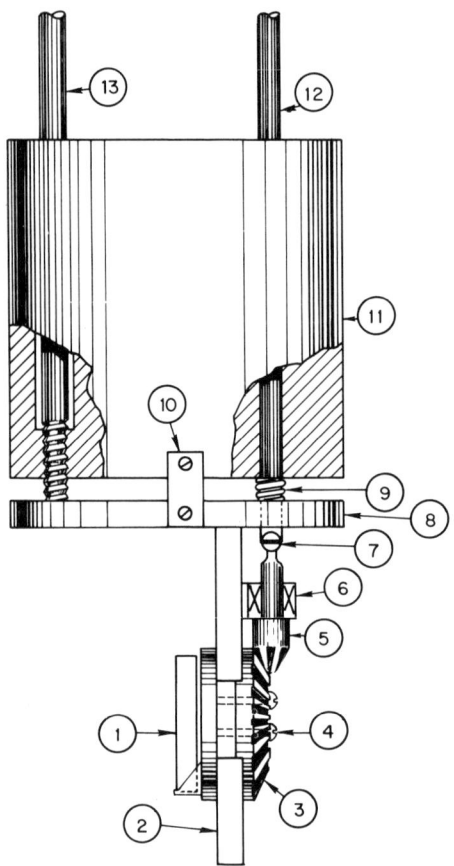

FIG. 8. Sample holder with one axis of rotation: (1) sample subassembly, (2) brass plate, (3) gear, (4) screws, (5) pinion gear, (6) bushing, (7) ball joint, (8) brass disk, (9) spring, (10) hinge, (11) housing, (12) rotation drive rod, (13) tip drive rod.

of both precisely aligning the sample relative to the magnetic field and rotating it while immersed in liquid helium. If the measurements are to be made in a magnet that can rotate about a vertical axis, then rotation of the sample about one horizontal axis is sufficient. Holders with only one axis of rotation are relatively easy to design and typical examples have been published by Milliken and Young[22] and by Datars and Dixon.[23] Another scheme, used by the author and shown in Fig. 8, permits tilting

[22] J. C. Milliken and R. C. Young, *Phys. Rev.* **148**, 558 (1966).
[23] W. R. Datars and A. E. Dixon, *Phys. Rev.* **154**, 576 (1967).

the sample $\pm 5°$ about one horizontal axis and complete rotation about a second horizontal axis. The sample subassembly of Fig. 7a is slid into the hole in the brass plate (2) and screwed to a gear (3). This gear is meshed with a pinion gear (5), which is rotated with rod (12). The pinion gear is held by the bushing (6) and connected to the rod at a ball joint (7). The rotation about this axis is limited only by the wires attached to the pins on the subassembly through holes in the gear. The sample is tilted by the rod (13), which is threaded into the brass block (11) and acts as a pusher on the disk (8). The disk is hinged at (10) and pressed against the rod by the spring (9). Sample alignment of better than 0.1° can be obtained with this holder.

If the measurements are to be made in a solenoid, then the holder should rotate the sample about two different axes normal to the solenoid axis so that the magnetoresistance can be measured for all magnetic field directions without remounting the sample. It is considerably more difficult to design a holder that keeps motion about one axis independent of motion about the other. One solution of the problem is given by Milliken and Young[22] (Fig. 9), who used two pulleys to provide the rota-

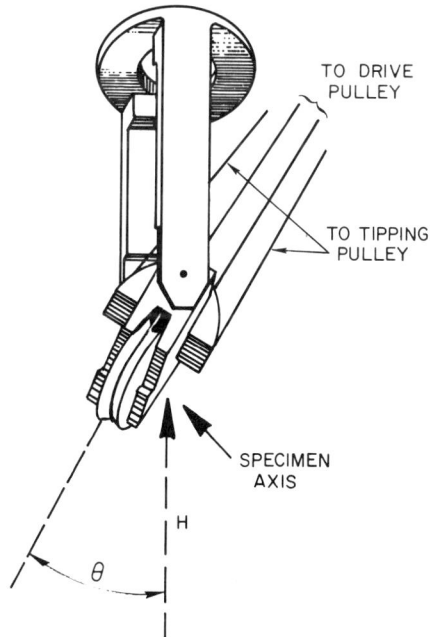

FIG. 9. Sample holder with two axes of rotation [after J. C. Milliken and R. C. Young, *Phys. Rev.* **148**, 558 (1968)].

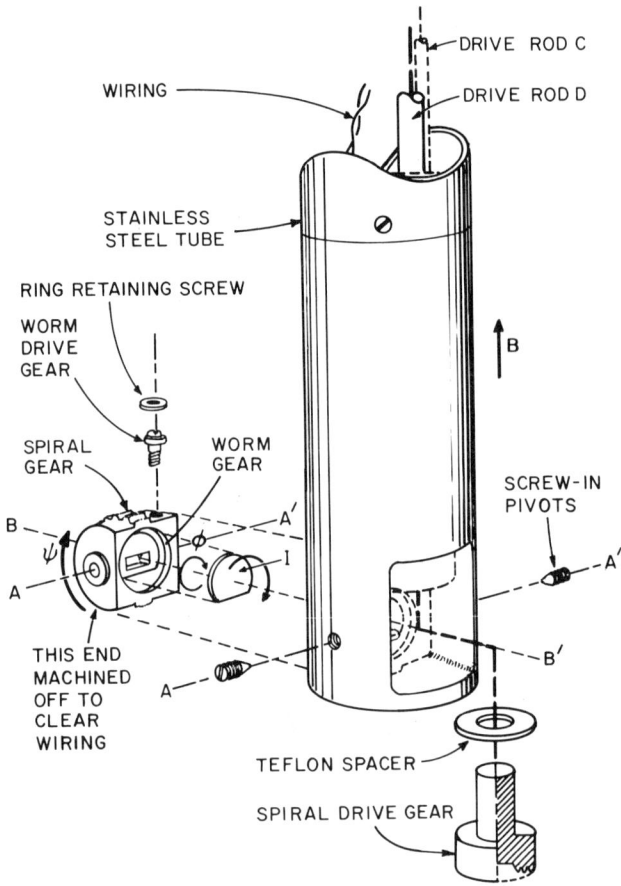

Fig. 10. Sample holder with two axes of rotation [after D. J. Sellmyer, *Rev. Sci. Instrum.* **38**, 434 (1967)].

tion about the different axes. The wires are connected at the top of the holder to protractors, which record the orientation of the sample, and to motors for rotating or tipping the sample at a uniform rate. This holder is relatively simple and inexpensive to construct and is in principle free from backlash and interaction of the axes. However, some care is required in arranging the pulley wires so they do not foul each other.

Another solution, reported by Sellmyer,[24] is shown in Fig. 10. The sample is mounted on the disk I, which is tilted about axis B by the

[24] D. J. Sellmyer, *Rev. Sci. Instrum.* **38**, 434 (1967).

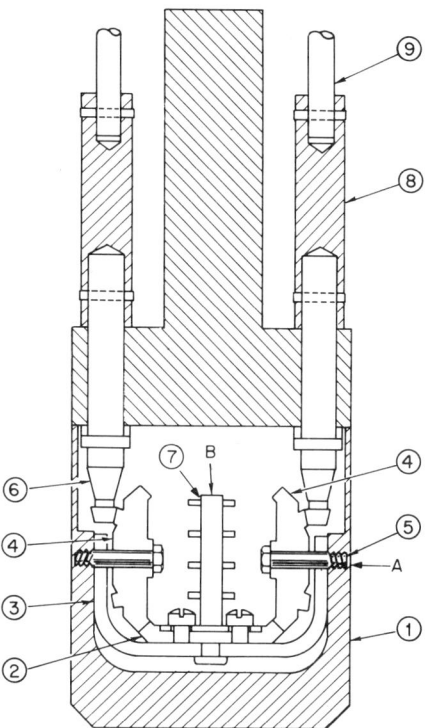

FIG. 11. Sample holder with two axes of rotation [after G. F. Brennert, W. A. Reed, and E. Fawcett, *Rev. Sci. Instrum.* **36**, 1267 (1965)]: (1) housing, (2) gear, (3) yoke, (4) gears, (5) gear bearing, (6) pinion drive gear, (7) sample subassembly, (8) collar, (9) control rod.

worm gear attached to rod C. After the sample has been tilted to the desired ψ angle, the rod C is raised and locked. Rotation about the A axis is then provided by the spiral gear which is motor driven. The advantages of this holder are that it can be made quite small and is also relatively inexpensive to build. Its major drawback is that the rotation must be returned to $\psi = 0°$ each time the angle ϕ needs to be changed. The backlash in the drive rods is not a serious problem since it is reproducible and can be accounted for by always approaching the desired angle from the same direction of rotation.

A more complicated but highly versatile holder has been reported by Brennert et al.[25] and is shown in Fig. 11. It is based upon the principle

---

[25] G. F. Brennert, W. A. Reed, and E. Fawcett, *Rev. Sci. Instrum.* **36**, 1267 (1965).

of a differential gear. If the gears (4) are driven in the same direction, a sample (7) mounted on gear (2) will tip about axis A. If, however, the gears (4) are driven in opposite directions, the sample rotates about axis B. The control rods (9) are driven with motors mounted on the top of the holder and the sample orientation is recorded by a dummy sample mounted on a second differential assembly driven by the same motors. To ensure the tipping and rotating motions are kept separate, the control rods must rotate at the same speed. This is accomplished with stepping motors run from the same variable oscillator. This not only keeps the rotation speed the same but also allows for easy variation of rotation speeds. Although Brennert *et al.* report a housing diameter of 3.8 cm, a scaled-down version with a diameter of 2.61 cm has been built.[26] The advantages of this holder are its reproducibility ($\pm 0.1°$), versatility, and ability to accommodate long samples [up to about $\frac{3}{4}$ the diameter fo the housing (1)]. Its major disadvantage is cost compared to other designs.

1.4. Magnetic Fields and Low Temperatures

It has been stated earlier that the high-field conditions are only achieved at low temperatures and in large magnetic fields. Low temperatures mean the boiling point of liquid helium (4.2 K) or below. Designs for dewars capable of holding liquid He and also fitting into an electromagnet or solenoid are easily found in *Review of Scientific Instruments*[27] or *Journal of Scientific Instruments*.[28]

Magnetic fields up to about 30 to 35 kOe are most conveniently generated by iron-core electromagnets. For higher fields, superconducting solenoids are capable up to about 140 kOe and a Bitter-type solenoid at the Francis Bitter National Magnet Laboratory in Cambridge, Massachusetts reaches a field of slightly over 200 kOe. For fields above 200 kOe, pulsed magnets are required. For every pure samples, fields under 30 kOe are adequate, but for less pure samples and for studies of magnetic breakdown, fields of 100 to 150 kOe may be required. The homogeneity of the field becomes most important in studying the oscillatory effects. A general rule is that the variation of the field over the sample should be no greater than one half the separation of the oscillations in oersteds.

---

[26] A. J. Arko, Private communication (1970).
[27] *Rev. Sci. Instrum.* Amer. Inst. of Phys., New York.
[28] *J. Sci. Instrum.* Inst. of Phys. and the Phys. Soc., London.

1.5. Measurement Techniques

1.5.1. DC Method

The experimental system for measuring and recording galvanomagnetic data by the dc method is very simple and is shown in Fig. 12, A direct current is supplied to the sample and the potential drop across either the magnetoresistance or Hall leads is first amplified, and then recorded on the y axis of an X–Y recorder. Either the field intensity or field direction is recorded on the x axis of the recorder. If the data is complicated, the output can, in addition, be digitized and recorded on either paper or magnetic tape and then processed by a computer. The current source should be

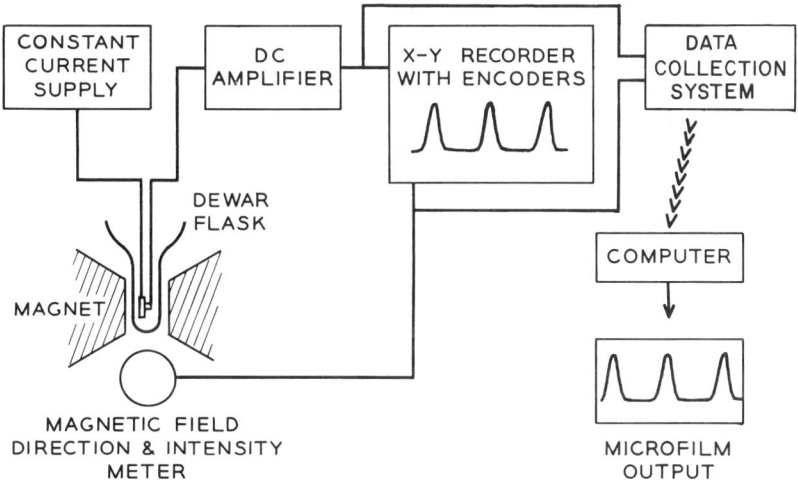

FIG. 12. Schematic diagram of experimental systems for measuring galvanomagnetic data by the dc method [adapted from W. A. Reed and E. Fawcett, *Science* **146**, 603 (1964). Copyright 1964 by the Amer. Ass. for the Advancement of Science].

capable of supplying from about 10^{-5} to 10 A and be regulated to better than 0.01%. Batteries, when stabilized, work well, but excellent electronic supplies are commercially available and are more convenient. The amplifier generally needs a sensitivity of 10^{-9} V and a range of 10^{-8} to 10^{-2} V. Several suitable amplifiers are commercially available. Occasionally sensitivities greater than 10^{-9} V are needed. Ries and Satterthwaite[29] and Clark and Fickett[30] have published papers which contain designs for

---

[29] R. P. Ries and C. B. Satterthwaite, *Rev. Sci. Instrum.* **38**, 1203 (1967).
[30] A. F. Clark and F. R. Fickett, *Rev. Sci. Instrum.* **40**, 465 (1969).

amplifiers with sensitivities of $\sim 10^{-12}$ V and also references to other designs. The rest of the components in the measuring system are all standard units and commercially available.

The connection of the current supply and amplifiers to the sample should be done with great care, since noise and drifts of 10^{-6} V can obscure the signal. All leads should be twisted pairs with all connections electrically and thermally shielded. If switches are used on the potential leads, they should have gold plated contacts. If an electronic current supply is used, its power should come through a high-quality isolation transformer, and the ground side of the supply's output should be connected to the chassis ground. The author has found that without the isolation transformer there is usually enough ac leakage to ground through the transformer of the power supply, to give sizable offset voltages and noise.

Each potential measured during the course of an experiment has four components, which can be identified both by their origin and by their parity with respect to the reversal of current and magnetic field. The first component is due to the resistance of the sample V_R and it reverses sign with current but not with field. The second component is due to the Hall effect V_H and it reverses sign with both current and field. The next component is due to contact potentials and thermoelectric effects in the measuring circuit V_C, and is assumed to be independent of current and field. The last component is induced by the changing magnetic flux through the loops in the leads V_I, and reverses with field but is independent of current.[31] Thus a field sweep or rotation curve should be made for the four combinations of forward and reverse current and forward and reverse field. Each component can then be separated out by combining the potentials as

$$V_R = \tfrac{1}{4}[V(++) + V(+-) - V(-+) - V(--)],$$
$$V_H = \tfrac{1}{4}[V(++) - V(+-) - V(-+) + V(--)],$$
$$V_C = \tfrac{1}{4}[V(++) + V(+-) + V(-+) + V(--)],$$
$$V_I = \tfrac{1}{4}[V(++) - V(+-) + V(-+) - V(--)],$$

where $V(I, B) = V(+-)$ means the voltage measured with the current in the forward direction and the field in the reverse direction. Since

---

[31] There has been reported a voltage which reverses sign with field but not with current and is proportional to I^2. The origin of this effect is unknown. See W. A. Reed and J. A. Marcus, *Phys. Rev.* **126**, 1298 (1962); Klauder *et al.* Ref. 4.

1.5. MEASUREMENT TECHNIQUES

these four components compose the potentials measured on each pair, it is clear how the precise placement of the potential leads on the crystal can minimize an unwanted component. If, however, the desired component is large and the other components can be shown to be negligible, then the procedure may be simplified. An example of this simplification is the measurement of only $V(++)$ to obtain V_R if V_H, V_C, and V_I are small, and only $V(++)$ and $V(+-)$ to obtain V_H if V_I is small.

1.5.2. AC Method

The ac techniques commonly use either (a) alternating current with a static magnetic field, or (b) field modulation with a constant current. Methods which alternate the field (actually reverse it) are not practical and will not be considered. Since the methods to be discussed in this section use the same samples and holders described in the previous section, it is easy to switch from one method to the other.

A schematic diagram of a circuit for both method (a) and (b) is shown in Fig. 13. When the switches are in position A the bipolar current supply is connected to the sample, which is in a static magnetic field. The potentials are detected and amplified by a phase-sensitive detector and recorded on an X–Y recorder or by a data collection system. The method records the resistance R as a function of either the field intensity of field direction. The oscillatory signal to drive the bipolar power supply is usually provided by the reference signal of the phase-sensitive detector but may be provided by an external oscillator. When the switches are in position B, the bipolar supply is connected to the field modulation coils (designed to give ± 1 kOe) and a dc power supply is connected to the sample. The data is recorded in the same manner except now dR/dH is recorded rather than R. In both cases the detector should have a sensitivity of 10^{-9} V.

A description of method (a) and experimental data using this method have been published by Woolam.[32] The advantages of this method are increased signal-to-noise through the use of phase-sensitive detecting, a direct measurement of R (if the capacitance and inductance are small), and effective elimination of errors due to thermoelectric and magnetic field-induced voltages. The disadvantages are that absolute calibration is a bit difficult and, when the Shubnikov–de Haas oscillations are the main interest, it does not discriminate against the monotonic background and

---

[32] J. A. Woolam, *Cryogenics* **8**, 312 (1968).

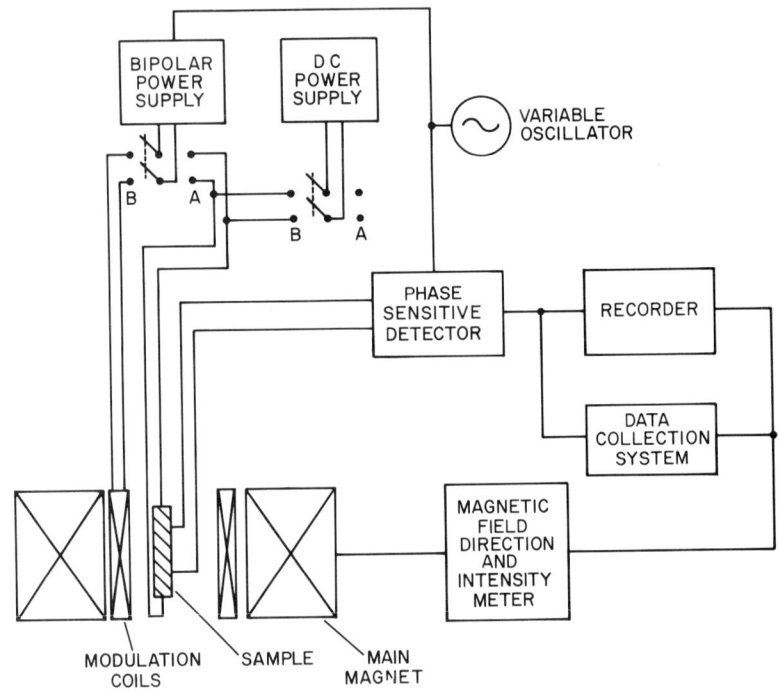

FIG. 13. Schematic diagram of an experimental system for measuring galvanomagnetic data by the ac method.

enhance the low-amplitude, high-frequency oscillations [as does method (b)]. This method has also a possible pit fall when measurements are made on very pure uncompensated metals. Reed and Meincke[33] have shown that the potential leads act as single-turn coils and will detect any helicon waves excited by the current.

Method (b) is identical to the field modulation method used to measure the de Haas–van Alphen effect except the potential leads to the sample replace the pickup coil. A discussion of both the theory and technology of this method has been published by Stark and Windmiller.[34] This is an excellent way to measure the oscillatory components of the galvanomagnetic effects since discrimination for or against different frequencies is easily accomplished through the use of filters and/or detection at higher harmonics. To detect the harmonics and to drive the modulation coils, and external oscillator is required as a reference. The phase sensitive

[33] W. A. Reed and P. P. M. Meincke, *Phys. Rev.* **163**, 664 (1967).
[34] R. W. Stark and L. R. Windmiller, *Cryogenics* **8**, 272 (1968).

detector is then tuned to a frequency 2, 3, 4, etc. times the reference. The monotonic components cannot be measured by this method and the absolute amplitudes of the oscillatory components are obtained only with difficulty.

1.5.3. Pulsed Field Method

If there is a need to make measurements above 150 to 200 kOe, pulsed fields are required. A schematic circuit for measuring the galvanomagnetic effects in a pulsed field is shown in Fig. 14. A direct current is supplied to the sample and the voltage drop across the potential leads is fed into a wide-band amplifier whose output is recorded on one beam

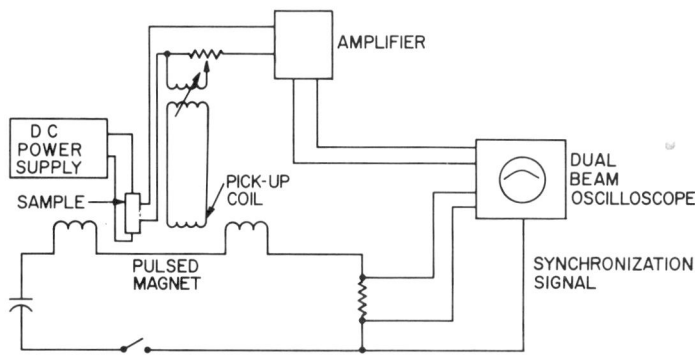

FIG. 14. Schematic diagram of experimental system for measuring galvanomagnetic data by the pulsed-field method.

of a dual beam oscilloscope. The amplifier needs a sensitivity of about 10^{-7} V and a bandwidth of about 1 kHz. To balance the large emf's induced when the field is pulsed, a fraction of the potential from a pickup coil is also fed into the amplifier. The variable inductance coil in the pickup circuit is used to adjust the phase of the balancing signal. The field intensity is recorded by the second beam of the oscilloscope by measuring the potential drop across a standard resistor. A variation of this circuit has been published by Alekseevskii and Egorov,[35] who used an alternating current in the sample and resonant detection of the signal. Their method eliminates the need of a pickup coil but requires balancing out the signal induced by the alternating current.

[35] N. E. Aleksevskii and V. S. Egorov, *J. Exp. Theoret. Phys.* **45**, 448 (1968) [*English transl.: Sov. Phys. JETP* **18**, 309 (1964)].

The pulsed field method does present special difficulties such as dynamic stresses on the sample, induced emf's, and eddy-current heating. However, these problems can be solved, and measurements up to 300 to 400 kOe are possible.

1.5.4. Inductive (Helicon) Method

This method is quite distinct from the methods previously discussed in that no leads are attached to the sample. A coil is used to excite electromagnetic waves in the crystal and the frequencies and fields that permit standing wave resonances are measured. The resonant frequencies and fields for a sample of rectangular geometry are then related to the magnetoresistance and Hall coefficient by[36]

$$\varrho(B) = 8d^2 \nu_r/(1 + u^2)^{1/2} \times 10^{-7} \quad \Omega\text{m} \qquad (1.5.1)$$

and

$$R = 8d^2 \nu_r u/B(1 + u^2)^{1/2}, \qquad (1.5.2)$$

where ν_r is the resonant frequency, d is the sample thickness, and $u = RB/\varrho(B)$ is the Hall angle. These resonances are known as helicon resonances.

A schematic diagram of the circuit used in this method is shown in Fig. 15. The sample is usually cut into a plate with typical dimension of $10 \times 10 \times 2$ mm. A static magnetic field is placed normal to the largest face and two perpendicular coils are wrapped around the sample normal to the smaller faces. An oscillatory signal is supplied to the drive coil and the signal from the pickup coil is measured with a phase-sensitive detector. The signals both in and out-of-phase with the driving signal are plotted versus frequency on a recorder.

The expressions relating the resonant frequency and field to the resistivity tensor are a result of solving a boundary-value problem for the electromagnetic waves in the metal. Legendy[37] has derived a solution of the boundary-value problem for infinite slabs and cylinders and has calculated the corrections needed when samples with finite dimensions are used. Amundsen[38] has applied the results of Legendy to his data on indium and his calculated Hall coefficient is consistent with the predicted high-field value. For a more detailed discussion on the use of helicons in

---

[36] R. G. Chambers and B. K. Jones, *Proc. Roy. Soc. (London)* **A270**, 417 (1962).
[37] C. R. Legendy, *Phys. Rev.* **135**, A1713 (1964).
[38] T. Amundsen, *Proc. Roy. Soc.* **88**, 757 (1966).

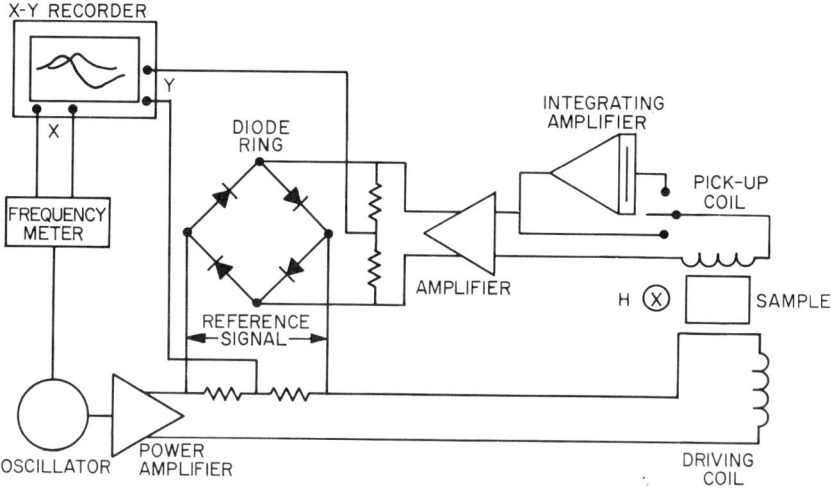

Fig. 15. Schematic diagram of experimental system for measuring galvanomagnetic data by inductive method [after M. T. Taylor, J. R. Merrill, and R. Bowers, *Phys. Rev.* **129**, 2525 (1963)].

galvanomagnetic studies, which includes a description of the apparatus, experimental methods, and data, the reader is directed to the articles by Merrill *et al.*[39]

The greatest advantage of the inductive technique is that no leads need be attached to the sample. This can be important for particular materials like the alkali metals, but, in general, does not represent much of an advantage. The major disadvantage is that only uncompensated metals or singular field directions in compensated metals may be studied, which severely limits the metals available. In addition it must still be demonstrated that this method can give the detailed information provided by either the ac or dc methods. The inductive method is therefore a useful technique in special situations but is not the most appropriate method for a broad study of galvanomagnetic properties.

1.5.5. Induced Torque Method

This method is also quite different from the traditional probe methods. A spherical or ellipsoidal sample is hung from a torque magnetometer[40]

---

[39] John R. Merrill, *Phys. Rev.* **166**, 716 (1968); D. V. Giovanilli and J. R. Merrill, *Phys. Rev.* **2**, 812 (1970); D. V. Giovanilli and J. R. Merrill (to be published).

[40] A suitable magnetometer has been described by J. H. Condon and J. A. Marcus, *Phys. Rev.* **134**, A446 (1964).

in a uniform magnetic field which rotates slowly about the axis of the magnetometer. The rotating field generates eddy currents in the sample which produce a torque. Since the magnitude and anisotropy of the eddy currents are related to the high-field magnetoconductivity tensor, a measurement of the torque is a means of studying the galvanomagnetic properties. Datars and Cook[41] have used this technique to study the open orbits in Cd, Ga, and Cu, and have found their results from this method agreed with the standard four-probe method. Lass and Pippard[42] have not only measured the induced torque in potassium but also have derived expressions relating the torque to the conductivity tensor. Similar expressions have also been derived by Visscher and Falicov[43] and by Condon.[44]

The primary advantages of this technique are that no leads are attached to the sample, the same sample can be used to map out the entire stereogram of the magnetoresistance, and the sensitivity is high when the conductivity is high. One problem with this method is that open orbits whose directions are not perpendicular to the axis of rotation can generate torques about axes other than the magnetometer axis and force the sample to swing away from its equilibrium position. Since most magnetometers are sensitive to lateral motion, this will produce an extraneous signal. A second problem has been pointed out by Condon.[44] In his derivation of the induced torque he calculated all the components of the torque and not just the component parallel to the magnetometer. He found that in the case of an uncompensated metal there is a torque, proportional to $n_e - n_h$ and parallel to dB/dt, which dominates in the limit of high fields. This torque will also force the sample to swing away from its equilibrium position and, if the rotation axis is not colinear with the magnetometer axis, will produce an extraneous signal. Condon actually used this effect to measure the Hall constant of copper for a general field direction and his results agree with Klauder et al.[4] to within 3%.

Although these extra torques may make accurate numerical measurements difficult, the results of Datars and Cook[41] show that qualitative information is possible. The torque technique is new and is probably

---

[41] W. R. Datars and J. R. Cook, *Phys. Rev.* **187**, 769 (1969); J. R. Cook and W. R. Datars, *ibid.* **1**, 1415 (1970); W. R. Datars, *Can. J. Phys.* **48**, 1806 (1970).

[42] J. S. Lass and A. B. Pippard, *J. Phys. E Sci. Instrum.* **3**, 137 (1970).

[43] P. B. Visscher and L. M. Falicov, *Phys. Rev.* **2**, 1518 (1970).

[44] J. H. Condon, Private communication (1970).

not yet developed to its full potential. However, unless it can produce results comparable in detail to studies done by the dc method, it will remain a method used primarily in those situations where probes cannot be attached to the sample.

1.6. Data Collection and Processing

Detailed studies of the monotonic galvanomagnetic properties such as in Cu[4] or of the oscillatory properties as in IrO_2[45] require the collection and processing of large quantities of data. It is therefore convenient and at times necessary to use some form of automated data collection and computerized processing of the data. For many experiments an automated system is a luxury which trades the experimenters' time for the computers' time. However, there are cases, such as analyzing oscillatory data containing ten or more separate frequencies, where automatic recording and computer analysis is required. Generally the choice between recording the data on punched paper tape or on magnetic tape is based on economics. The hardware for recording on paper tape is less expensive than for recording on magnetic tape, but the input of paper tape to the computer is slower and generally more expensive. It is also possible to have on-line data processing by small dedicated computers but the costs of such systems is very high at present. It is not particularly useful to discuss specific data recording systems, since the technology and hardware of these systems are changing so rapidly and the types of systems are so varied. It is therefore suggested that at the time a system is needed, the instrument journals and the various manufacturers be consulted.

1.7. Special Topics

1.7.1. High-Pressure Measurements

A field which has received relatively little attention is the measurement of the galvanomagnetic effects under pressure. Hydrostatic or uniaxial pressure in some materials can change the topology of the Fermi surface or change the energy gaps involved in magnetic breakdown which should be easily detected by the magnetoresistance. A description of a pressure cell capable of 60 kbars has been published by Vais-

---

[45] W. D. Ryden, W. A. Reed, and E. Greiner, *Phys. Rev.* B **6**, 2089 (1972).

nys and Kirk,[46] and both a description of apparatus and data on Zn and Cd have been published by Gaidukov and Itskevich.[47] Additional references can be found in these papers. There are a number of experimental problems associated with this technique such as the pressures may not be hydrostatic, a fair number of leads must be passed into the pressure cell, and it is difficult to achieve high pressures, high fields, and low temperatures all at the same time. However, the development of superconducting magnets and high-strength materials have made the third problem less formidable.

1.7.2. Low-Field Effects

The question is often raised whether the low-field galvanomagnetic coefficients are useful in determining the Fermi surface of metals. The answer in general is "no." The theories that treat the low-field coefficients[48–50] first assume a Fermi surface consisting of several ellipsoids and then calculate the coefficients characteristic of this surface. While any set of ellipsoids has a unique set of galvanomagnetic coefficients, the converse is not true. A case in point is the data published on gallium by Reed and Marcus.[51] Although low-field coefficients could be calculated from galvanomagnetic data, no set of ellipsoids is consistent with all of the coefficients. This is not too surprising since the complete Fermi surface of gallium cannot be approximated by a set of ellipsoids.[52] With the possible exception of the semimetals Bi, Sb, and As, the low-field effects are not a profitable way to study the Fermi surface of metals.

1.7.3. Anomalous Longitudinal Magnetoresistance

Although longitudinal magnetoresistance $(B \parallel J)$ does not provide information about the topology of the Fermi surface, it is often measured and almost as often reported to be "anomalous." "Normal" longitudinal magnetoresistance[1] should saturate at $\Delta\varrho/\varrho \approx 1$ and unless size effects

---

[46] J. R. Vaisnys and R. S. Kirk, *Rev. Sci. Instrum.* **36**, 1799 (1965).

[47] Yu. P. Gaidukov and E. S. Itskevich, *J. Exp. Theoret. Phys.* **45**, 71 (1963) [*English transl.: Sov. Phys. JETP* **18**, 51 (1964)].

[48] C. Herring, *Bell System Tech. J.* **34**, 237 (1955).

[49] B. Abeles and S. Meiboom, *Phys. Rev.* **101**, 544 (1956).

[50] J. R. Drabble and R. Wolfe, *Proc. Phys. Soc. (London)* **69**, 1101 (1956).

[51] W. A. Reed and J. A. Marcus, *Phys. Rev.* **130**, 957 (1963).

[52] W. A. Reed, *Phys. Rev.* **188**, 1184 (1969).

or magnetic impurity effects are present the resistance should not drop below its zero-field value. Recent measurements on chromium[53] have shown that the anomaly observed in this metal is due to nonuniform injection of the current into the sample coupled with a highly anisotropic conductivity tensor. The current, in effect, is constrained to flow on one side of the sample and not on the other. An experimental and theoretical study by Reed et al.[54] confirms this idea. It is quite likely that other anomalous results such as those reported for bismuth[55] and gallium[56] can be explained on this model. The work of Reed et al.[54] simply reaffirms the fact that great care must be taken in the preparation and mounting of samples; and even then problems may occur.

---

[53] A. J. Arko, J. A. Marcus and W. A. Reed, *Phys. Rev.* **176**, 671 (1968).
[54] W. A. Reed, E. Blount, J. A. Marcus, and A. J. Arko, *J. Appl. Phys.* **42**, 5453 (1971).
[55] Suichi Otake and Shigetoshi Koite, *J. Phys. Soc. Japan* **24**, 1176 (1968).
[56] J. Yahia, C. C. Lee and E. Fournier, *Phys. Rev. Lett.* **23**, 293 (1969).

2. EXPERIMENTAL METHODS FOR THE DE HAAS–VAN ALPHEN EFFECT*

2.1. The de Haas–van Alphen Effect

During the last two decades major advances have been made in our knowledge of the electronic structure of conducting solids. This increased understanding has resulted partially from improved techniques for band structure calculations and partially from the availability of detailed Fermi surface information, especially as provided by de Haas–van Alphen (dHvA) experiments.

The dHvA effect consists of magnetization oscillations periodic in the inverse magnetic induction, which are observed in pure metals at low temperatures. These once curious oscillations, which were first discovered in 1929 as force oscillations in a bismuth crystal cooled to liquid hydrogen temperatures,[1] now provide a powerful probe of the quantum properties of degenerate electron systems.

The force balance, originally used by de Haas and van Alphen, is of historical interest only, since large magnetic field gradients are required. These gradients cause a drastic overall reduction of dHvA amplitudes due to the variation of the phase of the oscillations over the specimen volume.[2] Recognizing this, Shoenberg developed the torque method,[3] which has been used successfully to study the dHvA effect in a variety of metal[4] and semimetal crystals. The applicability of this method has been restricted, however, because a magnetic field transverse to the torque

---

[1] W. J. de Haas and P. M. van Alphen, *Leiden Commun.* **208d, 212a** (1930).

[2] D. Shoenberg, *Phil. Trans. Roy. Soc. (London)* **A255**, 85 (1962).

[3] D. Shoenberg and M. Z. Uddin, *Proc. Roy. Soc.* **A156**, 687 (1936).

[4] The first torque experiment on a true metal was performed on zinc by J. A. Marcus, *Phys. Rev.* **71**, 559 (1947).

---

* Part 2 is by J. R. Anderson and D. R. Stone.

suspension axis is required and this has been obtained conveniently only with iron-core electromagnets, which produce relatively low magnetic fields.[5]

Another major advance was Shoenberg's introduction of impulsive fields.[6] For those experiments, a pulsed magnetic field was used to produce a time varying magnetization which was sensed by means of a small pickup coil surrounding the sample. Fields up to about 250 kOe of 10 to 20 msec duration have been obtained in this manner, greatly extending dHvA measurement capabilities. A recent improvement has been the use of a digital data acquisition system for storage of the dHvA data from a single field pulse so that the data can then be analyzed with a high-speed digital computer. Evidence of the success of this scheme is demonstrated in the dHvA studies in iron by Gold et al.[7]

Although the pulsed field technique has been developed to a high level of sophistication, the use of slowly varying fields in conjunction with the field modulation technique[8] has all but superseded the older methods. The advantages gained by circumventing the noise-bandwidth and eddy-current problems inherent in the pulsed field technique appear to more than compensate for the more modest magnetic fields that obtain from superconducting and iron magnets. Therefore, we shall confine our discussion to the details of this field modulation approach. For a description of the torque and pulsed field systems see Condon and Marcus[9] and Panousis and Gold,[10] respectively, and the references therein. A rather comprehensive review of the dHvA technique has already been given by Gold,[11] and experimental details have been presented by Windmiller and Ketterson[12] and Stark and Windmiller.[13]

---

[5] J. Vanderkooy, *J. Sci. Instrum.* **2**, 718 (1969), recently has adapted the torque method to the awkward geometry of a superconducting solenoid.

[6] D. Shoenberg, *Physica* **19**, 791 (1953).

[7] A. V. Gold, L. Hodges, P. T. Panousis, and D. R. Stone, *Int. J. Magn.* **2**, 357 (1971).

[8] D. Shoenberg and P. Stiles, *Phys. Lett.* **4**, 274 (1963).

[9] J. H. Condon and J. A. Marcus, *Phys. Rev.* **134**, A446 (1964).

[10] P. T. Panousis and A. V. Gold, *Rev. Sci. Instrum.* **40**, 120 (1969). See also J. R. Anderson, USAEC Rep. IS-762 (1962).

[11] A. V. Gold, *in* "Solid State Physics, Vol. 1: Electrons in Metals" (J. F. Cochran and R. R. Haering, eds.), pp. 39–126. Gordon and Breach, New York, 1968.

[12] Windmiller and Ketterson, *Rev. Sci. Instrum.* **39**, 1672 (1968).

[13] R. W. Stark and L. R. Windmiller, *Cryogenics* **8**, 272 (1968).

2.2. dHvA Expressions

2.2.1. Frequencies

As a prelude to the discussion of the field modulation technique, we give the theoretical expression for the oscillatory component of magnetization M_{osc} due to the conduction electrons. A detailed justification of these results is beyond the scope of this review; appropriate references can be found in the article by Gold.[11]

For the case of an electron gas under isothermal conditions, $\mathbf{M}_{osc}$ is given by

$$\mathbf{M}_{osc} = \sum_{j,l} \frac{\mathbf{M}_j \cos}{l^{1/2}} \left(\frac{l\pi g_j \mu_j}{2} \right) \frac{\exp(-l\gamma \mu_j T_D{}^j/|\mathbf{B}|)}{\sinh(l\lambda \mu_j T/|\mathbf{B}|)} \sin\left(\frac{\hbar c \mathscr{A}_j(\mathbf{B})}{e|\mathbf{B}|} + \gamma_{lj} \right), \tag{2.2.1}$$

where

$$\mathbf{M}_j = \frac{2kT}{(2\pi)^{1/2}} \left(\frac{e}{\hbar c} \right)^{3/2} \frac{\hbar c \mathscr{A}_j(\mathbf{B})}{2\pi e} |\mathbf{B}|^{-1/2} \left| \frac{\partial^2 \mathscr{A}_j(\mathbf{B})}{\partial k_B{}^2} \right|^{-1/2}$$
$$\times \left(\hat{B} + \frac{\partial \ln \mathscr{A}_j(\mathbf{B})}{\partial \theta} \hat{\theta} + \frac{1}{\sin \theta} \frac{\partial \ln \mathscr{A}_j(\mathbf{B})}{\partial \phi} \hat{\phi} \right).$$

The sum on j is taken over all extremal sections of the Fermi surface in planes perpendicular to $\mathbf{B}$ and the sum on l is over all positive integers. The various quantities entering these expressions are defined in Table I.

Equation (2.2.1) was first derived for the case of a noninteracting gas in an external magnetic field $\mathbf{H}$, under the assumption of an arbitrary dispersion relation.[14] For an interacting electron system, Pippard[15] has used thermodynamic arguments to show that the magnetic induction $\mathbf{B}$ rather than the magnetic field $\mathbf{H}$ belongs in the expression for the oscillatory magnetization. A similar conclusion has also been obtained both experimentally[16-19] and theoretically[20] for ferromagnetic metals.

First, we consider the argument of the sine in Eq. (2.2.1). The phase factor γ_{lj} is constant and will be neglected for the remainder of

---

[14] I. M. Lifshitz and A. M. Kosevich, *Sov. Phys. JETP* **2**, 636 (1956).
[15] A. B. Pippard, *Proc. Roy. Soc. (London)* **A272**, 192 (1963).
[16] J. R. Anderson and A. V. Gold, *Phys. Rev. Lett.* **10**, 227 (1962).
[17] L. Hodges, D. R. Stone, and A. V. Gold, *Phys. Rev. Lett.* **19**, 655 (1967).
[18] D. C. Tsui, *Phys. Rev.* **164**, 669 (1967).
[19] J. R. Anderson, J. J. Hudak, and D. R. Stone, *AIP Conf. Proc.* (C. D. Graham and J. J. Rhyne, eds.) No. 5, 477. AIP, New York, 1972.
[20] C. Kittel, *Phys. Rev. Lett.* **10**, 339 (1963).

TABLE I. Notation Used in Eq. (2.2.1)

Symbol	Description
e	Electron charge (4.803×10^{-10} esu)
$\hbar$	(Planck's constant)/2π (1.0545×10^{-27} erg sec)
m_0	Free electron mass (9.11×10^{-28} grams)
k_b	Boltzmann's constant (1.38×10^{-16} ergs/K)
c	Velocity of light (2.998×10^{10} cm/sec)
$\mathscr{A}_j(\mathbf{B})$	jth extremal Fermi surface cross section in a plane perpendicular to $\mathbf{B}$ (cm^{-2})
$\mathbf{B}(\hat{B}, \hat{\theta}, \hat{\phi})$	The total magnetic induction within a specimen (G). $\hat{B}$ is in the direction of $\mathbf{B}$. $\hat{B}, \hat{\theta}, \hat{\phi}$ define an orthogonal set of unit vectors for a spherical coordinate system.
$\mathbf{k}_B$	Wave vector parallel to $\mathbf{B}$ (cm^{-1})
$T_D{}^j$	Dingle temperature for orbit j (K)
T	Absolute temperature (K)
g_j	Effective spin-splitting factor for jth extremal orbit on the Fermi surface (see text)
μ_j	Cyclotron mass ratio (m^*/m_0)
E_F	Fermi energy (ergs)
γ_{lj}	Phase factor for jth orbit and lth harmonic
$\partial^2 \mathscr{A}_j(B)/\partial k_B{}^2$	Curvature factor (2π for a spherical Fermi surface)
λ	$2\pi^2 m_0 c k_b / e\hbar = 147$ kG.

this Chapter. The argument $\hbar c \mathscr{A}_j(\mathbf{B})/(e \mid \mathbf{B} \mid)$ is large, greater than $1000 \times 2\pi$ for the usual experimental situation. (For example, for the second zone hole surface in aluminum for $\mathbf{B} = 50$ kG, $\hbar c \mathscr{A}_j(\mathbf{B})/(e \mid \mathbf{B} \mid)$ is about $6000 \times 2\pi$).. Thus the major variation of the magnetization with magnetic field results from the sine in Eq. (2.2.1.); compared to this, the other factors in Eq. (2.2.1) change slowly with $\mathbf{B}$. Consequently, each component of the magnetization oscillates periodically in $\mid B \mid^{-1}$ to a very good approximation and a dHvA frequency F_j may be defined by $F_j = \hbar c \mathscr{A}_j(B)/2\pi e$.

From a measurement of the oscillation frequency F_j, one can obtain the jth extremal cross section of the Fermi surface in a plane perpendicular to $\mathbf{B}$. The frequency F and area $\mathscr{A}$ are related by

$$F \text{ (G)} = 1.04728 \mathscr{A} \text{ (Å}^{-2}) = 3.741 \times 10^8 \mathscr{A} \text{ (au}^{-2}).$$

In order to establish the shape of a Fermi surface one must study the dHvA frequencies for many orientations of magnetic field. Although, in general, it is not possible to obtain a unique Fermi surface from such frequency data, correlations with band structure calculations have proved highly successful for a number of metals.[21] In addition, it has been shown that if the extremal areas are obtained at all orientations for a closed centrosymmetric surface, a unique inversion from areas to radii is possible.[22] Mueller[23] has used inversion schemes to obtain radii with only a limited amount of area data.

2.2.2. Amplitudes

A study of the amplitude factor in Eq. (2.2.1),

$$\mathbf{D}_{jl} = \frac{\mathbf{M}_j \cos(l\pi\mu_j g_j/2) \exp(-l\lambda\mu_j T_D{}^j/|\mathbf{B}|)}{\sinh(l\lambda\mu_j T/|\mathbf{B}|)}, \qquad (2.2.2)$$

provides important information about electronic structure and carrier scattering in a metal. From a measurement of the temperature dependence of the amplitude at constant $\mathbf{B}$ the cyclotron mass ratio μ_j is obtained. It can be easily shown that

$$\mu_j = (\hbar^2/2\pi m_0) \, (\partial \mathscr{A}_j(\mathbf{B})/\partial E)_{E_F}, \qquad (2.2.3)$$

where the derivative with respect to the energy E is evaluated at E_F, and therefore, from a measurement of μ_j, the dependence upon energy of an extremal cross section of the Fermi surface can be found. In practice the argument of the sinh in Eq. (2.2.2) is usually greater than one and $\sinh^{-1}(l\lambda\mu_j T/|\mathbf{B}|) \approx 2\exp(-l\lambda\mu_j T/|\mathbf{B}|)$ (Less than a 1% error is made if this argument is greater than 2.2.) In this case μ_j can be obtained directly from the slope of a plot of $\ln(D_{jl}/T)$ versus T at constant $|\mathbf{B}|$. As an example, Fig. 1 shows the amplitude of the cobalt $[0001]\beta$

---

[21] A. P. Cracknell, *Advan. Phys.* **18**, 681 (1969); **20**, 1 (1971).
[22] I. M. Lifshitz and A. V. Pogorelov, *Dokl. Akad. Nauk SSSR* **96**, 1143 (1954).
[23] F. M. Mueller, *Phys. Rev.* **148**, 636 (1966).

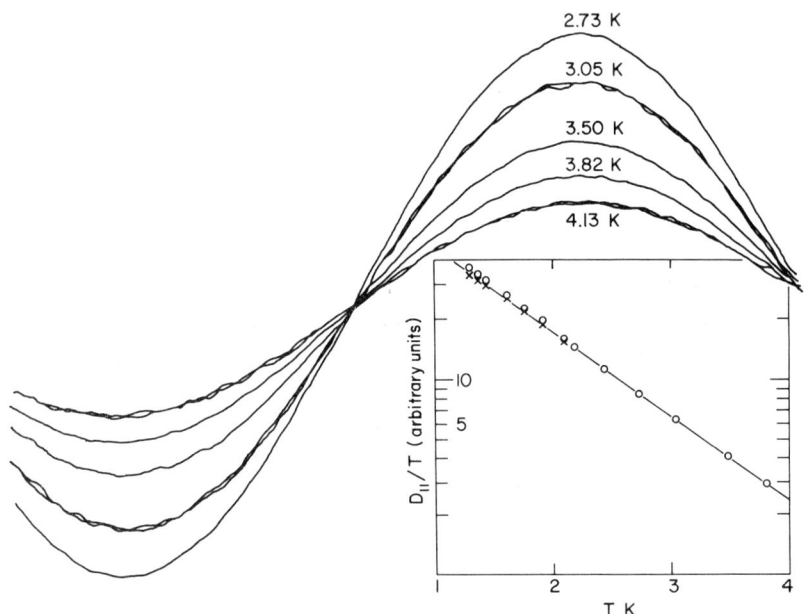

FIG. 1. Temperature dependence of the amplitude of the β oscillations ($F = 3.53 \times 10^6$ MG) in cobalt for $H_{ap} \parallel [0001]$. Approximately one-half of a DHvA oscillation cycle is shown for different temperatures at a value of magnetic induction $\mathbf{B} = 34.89$ kG. The field separation between the maximum and minimum amplitude is about 172 G. The insert at the lower right is a logarithmic plot from which a cyclotron mass ratio, $\mu = 0.23$, has been determined. The ○ points are the measured amplitudes divided by the temperature, while × points have been corrected for the sinh term using the value $\mu = 0.23$.

oscillations at temperatures between 4.13 and 2.73 K. These data were obtained by sweeping over one-half of a dHvA cycle in a reproducible manner and displaying the data on an X–Y recorder. In the insert in the lower right corner of Fig. 1, $\ln(D_{11}/T)$ has been plotted for temperatures between 1.3 and 4.13 K. At the lowest temperatures, the sinh correction must be made as discussed in Section 2.3.4.

After the cyclotron mass has been determined, from the magnetic field dependence of the dHvA amplitude at constant temperature one can obtain a carrier relaxation time τ, which is related to the Dingle temperature T_D by

$$T_D{}^j = (\hbar/2\pi k_b) \langle 1/\tau_j \rangle. \tag{2.2.4}$$

Here $\langle 1/\tau_j \rangle$ is the reciprocal of the relaxation time averaged around the *j*th orbit. Phillips and Gold[24] have shown that for Pb specimens of high perfection, T_D^j is less than 0.1 K. We find similar results for Nb. In poorer crystals T_D^j can be as much as several degrees.

In some cases, information about the Lande splitting factor g_j which is 2.0023 for free electrons, also can be obtained from a study of dHvA amplitudes. If $g_j\mu_j$ is an odd integer, the factor $\cos(\pi g_j\mu_j/2)$ in Eq. (2.2.1)

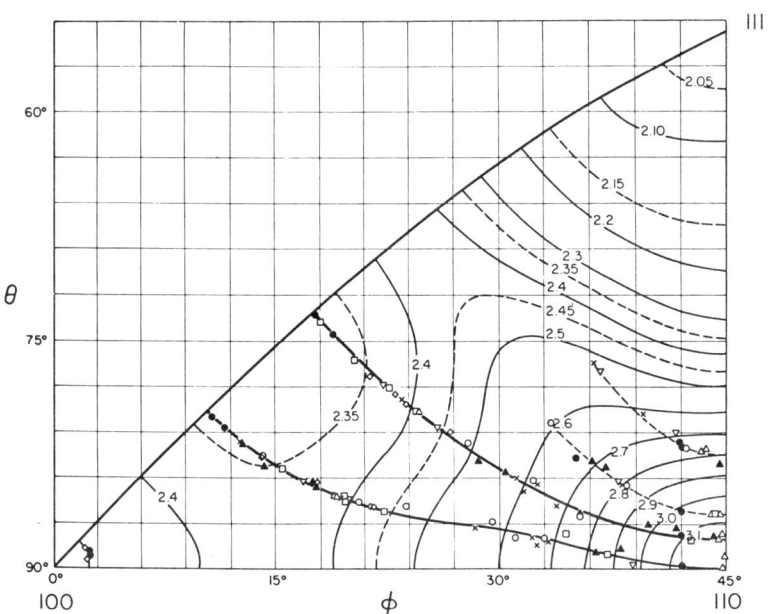

FIG. 2. Spin-splitting zeros in platinum. Lines have been drawn through the data points ($\triangle$, $\bigcirc$, $\times$, $\triangledown$, $\bullet$, $\blacktriangle$, and $\square$) to indicate contours at which the dHvA amplitude vanished because the spin-splitting term became zero. The remaining lines with numbers ranging from 2.05 to 3.1 represent contours of constant effective mass [from J. B. Ketterson and L. R. Windmiller, *Phys. Rev.* **B2**, 4813 (1970)].

vanishes. A rather elegant study of these spin-splitting zeros has been made in platinum by Ketterson and Windmiller,[25] and Fig. 2 shows the directions of **B** for spin-splitting zeros for one sheet of this Fermi surface superimposed upon a contour map of accurate values of μ. It should also be pointed out that the spin-splitting factor $\cos(\pi g_j\mu_j/2)$ is absent in ferromagnetic materials.

[24] R. A. Phillips and A. V. Gold, *Phys. Rev.* **178**, 932 (1969).
[25] J. B. Ketterson and L. R. Windmiller, *Phys. Rev.* **B2**, 4813 (1970).

2.2.3. B versus H

In Eq. (2.2.1) the magnetization was given as a function of the magnetic induction **B**, where **B** can be expressed in terms of the applied field $\mathbf{H}_{ap}$ as

$$\mathbf{B} = \mathbf{H}_{ap} + 4\pi\mathbf{M} + \mathbf{H}_d. \qquad (2.2.5)$$

Here **M** is the total magnetization and $\mathbf{H}_d$ is the demagnetizing field. Although it is usual to replace **B** by $\mathbf{H}_{ap}$ in the dHvA espressions, we now address briefly the complications which arise when the other terms in (2.2.2) must be included. Two cases will be considered: (1) use of ferromagnetic samples and (2) large amplitudes for the oscillatory component of magnetization.

For a ferromagnetic sample it is important to include both the average magnetization $\mathbf{M}_{av}$ and the demagnetizing field due to sample shape. (We assume here that the measurements are made on a single magnetic domain. To our knowledge no dHvA oscillations have been observed at an applied field less than that necessary to achieve technical saturation.) If the bounding surface of a specimen is of second degree, then $\mathbf{M}_{av}$ and $\mathbf{H}_d$ will be uniform throughout the sample and $\mathbf{H}_d$ can be expressed as

$$\mathbf{H}_d = -\mathbf{D} \cdot \mathbf{M}_{av}, \qquad (2.2.6)$$

where **D** is the demagnetization tensor which depends upon only the sample shape. As a result, the quantity that replaces **B** in Eq. (2.2.1) is $\mathbf{B}_{fe}$, where

$$\mathbf{B}_{fe} = \mathbf{H}_{ap} + (4\pi\mathbf{I} - \mathbf{D}) \cdot \mathbf{M}_{av}. \qquad (2.2.7)$$

For a general second-degree sample shape, $\mathbf{B}_{fe}$ will not be parallel to $\mathbf{H}_{ap}$ and must be calculated from a knowledge of **D** (Fig. 3.). The choice of a sphere is very convenient since in this case $\mathbf{D} = (4\pi/3)\mathbf{I}$, where **I** is a unit tensor, and

$$\mathbf{B}_{fe} = \mathbf{H}_{ap} + (8\pi/3)\mathbf{M}_{av}. \qquad (2.2.8)$$

An additional difficulty arises if it is necessary to include magnetocrystalline anisotropy, since even for a spherically shaped sample $\mathbf{M}_{av}$ may not be parallel to $\mathbf{H}_{ap}$ and $\mathbf{B}_{fe}$ must be calculated taking the anisotropy constants into account. However, when these corrections are made, it is possible to determine $|\mathbf{M}_{av}|$ from the periodicity of the dHvA oscillations. This has been done for iron and nickel,[16-18] where the mag-

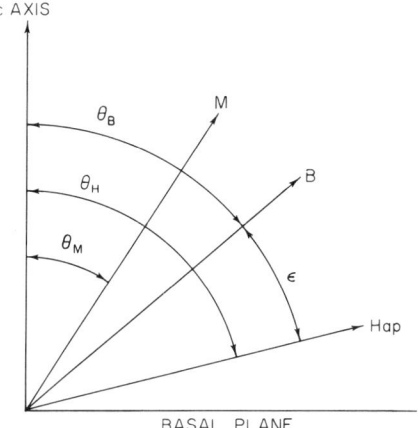

Fig. 3. Relation between $\mathbf{H}_{ap}$ and the directions of $\mathbf{M}$ and $\mathbf{B}$ for a cobalt single crystal shown schematically. The c axis is the easy direction of magnetization.

netocrystalline anisotropy can be neglected, and for cobalt,[19] where the contribution from the magnetocrystalline anisotropy is large.

The second important complication we consider is magnetic interaction, that is the situation in which the oscillatory magnetization is large and must be included in $\mathbf{B}$ and in Eq. (2.2.1). We will consider only nonferromagnetic samples for which $\mathbf{M}_{av}$ can be neglected. The expression for $\mathbf{B}$ then becomes

$$\mathbf{B}_{MI} = \mathbf{H}_{ap} + (4\pi\mathbf{I} - \mathbf{D}) \cdot \mathbf{M}_{osc} \qquad (2.2.9)$$

and as a result $\mathbf{M}_{osc}$ is given implicity rather than explicitly by Eq. (2.2.1). Shoenberg[26] and Condon[27] have shown that the magnetic interaction effect is especially important when $|dM_{osc}/dB| > 1/4\pi$. Under these conditions the line shape of the dHvA oscillations is seriously influenced and, in addition, if more than one dHvA frequency is present, extensive mixing will occur.

In order to exhibit this mixing, we will summarize a discussion given by Shoenberg.[26] (An iterative approach has been used by Phillips and Gold[24] to study the same problem.) To simplify this discussion, we shall omit $\mathbf{D}$ by treating the case of a very long, thin, cylindrical sample with $\mathbf{H}_{ap}$ and $\mathbf{M}_{osc}$ oriented along the sample axis. We will therefore

---

[26] D. Shoenberg, *Can. J. Phys.* **46**, 1915 (1968).
[27] J. H. Condon, *Phys. Rev.* **145**, 526 (1966).

drop the vector notation and write

$$B_{MI} = H_{ap} + 4\pi M,$$

so that Eq. (2.2.1) can be simplified as

$$M_{osc} = D \sin[2\pi F/(H_{ap} + 4\pi M_{osc})]. \quad (2.2.10)$$

Now let $x = 1/H_{ap}$ and then Eq. (2.2.10) becomes

$$M = D \sin[2\pi Fx/(1 + 4\pi Mx)]. \quad (2.2.11)$$

Even though dM/dH_{ap} can be of the order of 1, $M/H_{ap} \ll 1$ and therefore we expand the denominator of the argument in Eq. (2.2.11) to obtain

$$M = D \sin[2\pi Fx - 8\pi^2 Fx_0^2 M]. \quad (2.2.12)$$

Because x is slowly varying under normal experimental conditions compared to M, which is oscillatory, we have set x equal to a constant value x_0 in the second term in the argument of the sine in Eq. (2.2.12). The magnetization M as expressed by Eq. (2.2.8) is clearly periodic in x. Letting $z = 2\pi Fx$, $y = 8\pi^2 x_0^2 FM$, and $a = 4\pi x_0^2 D$, we obtain, finally,

$$y = a \sin(z - y). \quad (2.2.13)$$

For $a < 1$, the solution of Eq. (2.2.13) can be obtained as a Fourier series

$$y = \sum_{n=1}^{\infty} (-1)^{n+1}(2/n) J_n(na) \sin nz. \quad (2.2.14)$$

This expression shows that harmonics of the fundamental are introduced by magnetic interactions. For $a > 1$, the solution to Eq. (2.2.13), which becomes multivalued, is discussed in some detail by Shoenberg.[26]

To show the mixing that occurs we again follow Shoenberg and include two frequencies, F_1 and F_2, so that Eq. (2.2.11)

$$M = D_1 \sin \frac{2\pi F_1 x}{1 + 4\pi Mx} + D_2 \sin \frac{2\pi F_2 x}{1 + 4\pi Mx}. \quad (2.2.15)$$

Here we have assumed that the origin of x has been chosen such that both oscillations are in phase. Letting

$$a = 4\pi x_0^2 D_1, \quad \beta a = 4\pi x_0^2 D_2, \quad z = 2\pi F_1 x, \quad \alpha = F_2/F_1, \quad (2.2.16)$$

we can rewrite Eq. (2.2.15) as

$$y = a \sin(z - y) + \beta a \sin \alpha(z - y). \qquad (2.2.17)$$

If $a(1 + \alpha\beta) < 1$, Eq. (2.2.17) can be inverted to give y as a Fourier series

$$y = \sum_{n=1}^{\infty} (-1)^{n+1}\left(\frac{2}{n}\right)$$

$$\times \left[J_n(na) J_0(n\beta a) \sin(nz) + \frac{1}{\alpha} J_0(n\alpha a) J_n(n\alpha\beta a) \sin(n\alpha z) \right]$$

$$+ \sum_{n=1}^{\infty} \sum_{m=-\infty}^{\infty}{}' (-1)^{n+m+1}\left(\frac{2}{n + m + \alpha}\right)$$

$$\times J_n(na + m\alpha a) J_m(n\beta a + m\alpha\beta a) \sin(n + m\alpha)z, \qquad (2.2.18)$$

where the prime means that $m = 0$ is excluded from the summation.

The mixing of the two frequencies is demonstrated by the occurrence of $\sin(n + m\alpha)z$ in Eq. (2.2.18); this equation can be used to estimate the relative amplitudes of sum and difference frequencies produced by magnetic interactions. For example, consider the terms for $n = 1$ and $m = \pm 1$. The relative amplitudes become

$$\frac{A^+}{A^-} = \frac{(1 - \alpha) J_1[a(1 + \alpha)] J_1[\beta a(1 + \alpha)]}{(1 + \alpha) J_1[a(1 - \alpha)] J_1[\beta a(1 - \alpha)]}, \qquad (2.2.19)$$

where A^+ is the amplitude of the sum frequency and A^- is the amplitude of the difference frequency.

If the arguments of the Bessel functions are small, we may expand the Bessel functions, keeping only the first terms, to obtain

$$A^+/A^- \approx (1 + \alpha)/(1 - \alpha) = (F_1 + F_2)/(F_1 - F_2). \qquad (2.2.20)$$

We see that to this approximation the temperature and field dependences are the same for both terms and in fact that amplitudes will be nearly the same if $F_2 \ll F_1$. This result has been demonstrated by Phillips and Gold[24] for sum and difference frequencies in lead. As an example of these effects, we show high-frequency F_1, dHvA oscillations in indium mixed with a second frequency F_2 such that $F_1/F_2 \approx 65$ (Fig. 4). The Fourier transform (Fig. 4) shows the sum and difference frequencies that result from the two indium frequencies because of magnetic interactions.

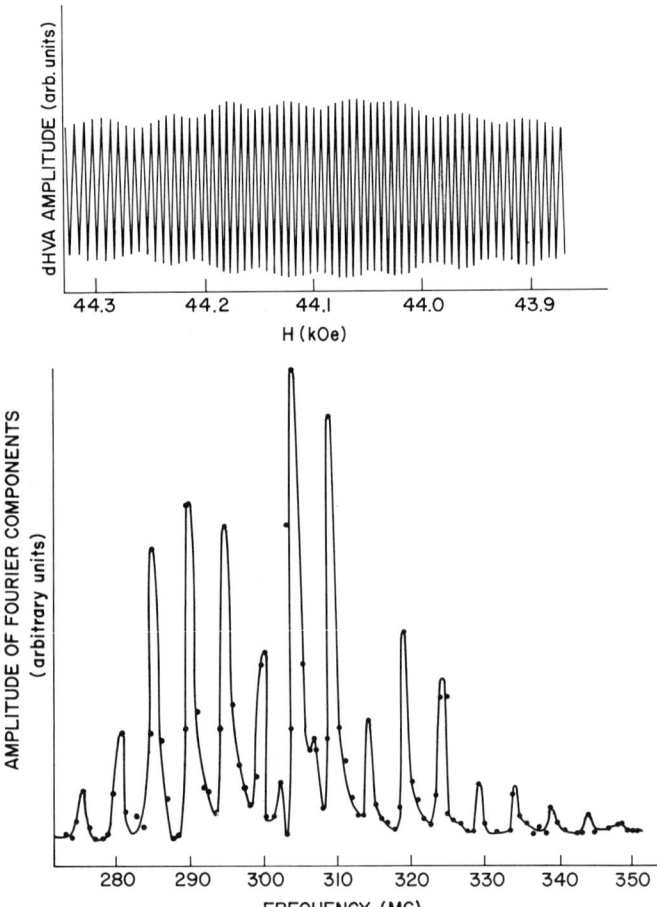

Fig. 4. dHvA oscillations in In for $H \parallel [110]$ indicating mixing effects. The fundamental frequency component of the oscillations, shown in the upper portion of the figure, is 304 MG. This is the largest peak in the Fourier transform, shown in the lower part of the figure. There is a lower frequency in indium of about 4.6 MG. All the other peaks in the Fourier transform appear to be sums and differences of the 304-MG frequency and multiples of the 4.6-MG frequency.

2.3. Field Modulation Technique

The field modulation technique is probably the simplest and most versatile approach for dHvA studies. We will limit our discussion to the experimental system in which the sample can be thought of as the core of a transformer with the modulation impressed upon the primary

2.3. FIELD MODULATION TECHNIQUE

and the signal measured at the secondary, a pickup coil. Since no leads are attached to the specimen, its shape is usually unimportant. In addition, the sample can be placed inside a pressure bomb with all electrical leads for the pickup and modulation coils outside.

In order to describe experimental techniques, it is necessary to consider the pickup voltage obtained in a field modulation experiment. We shall present here the case in which the frequency of modulation ω is sufficiently low and the sample is sufficiently small that the phase and amplitude of the ac modulation field do not vary significantly throughout the sample volume. We shall also assume here that the sample is nonferromagnetic and that magnetic interaction effects can be ignored. In this case the magnetic induction is given by

$$\mathbf{B} = \mathbf{H}_{\text{ap}},$$

where

$$\mathbf{H}_{\text{ap}} = \mathbf{H}_0 + \mathbf{h}_{\text{m}}(t) \quad \text{and} \quad \mathbf{h}_{\text{m}}(t) = \mathbf{h}_0 \cos \omega t. \quad (2.3.1)$$

Here $\mathbf{h}_{\text{m}}(t)$ is the modulation signal and $\mathbf{H}_0$ is the large magnetic field, which can be varied slowly to obtain the dHvA oscillations.

The explicit form of the time-dependent oscillatory magnetization can be obtained by substituting Eq. (2.3.1) into Eq. (2.2.1) to obtain

$$M_{\text{osc}} = D_{jl} \sin[2\pi F_j l/|\mathbf{H}_0 + \mathbf{h}_{\text{m}}(t)| + \gamma_{jl}]. \quad (2.3.2)$$

Now if we assume $|\mathbf{h}_{\text{m}}|/|\mathbf{H}_0| \ll 1$ and consider only one term M_{jl} in Eq. (2.3.2), we find to lowest order in $|\mathbf{h}_0|/|\mathbf{H}_0|$ that

$$\mathbf{M}_{jl} = \mathbf{D}_{jl}(\mathbf{H}_0, T, T_{\text{D}}) \sin[(2\pi l F_j/|\mathbf{H}_0|) - \Lambda_{jl} \sin \omega t], \quad (2.3.3)$$

where

$$\Lambda_{jl} = (2\pi l F_j |\mathbf{h}_0|/|\mathbf{H}_0|^2) [\cos \alpha - \sin \alpha (\partial[\ln F_j]/\partial \alpha)]. \quad (2.3.4)$$

We have assumed that $\mathbf{h}_0$ makes an angle α with respect to $\mathbf{H}_0$.

The derivative $[\partial(\ln F_j)/\partial\alpha]$ is the relative change in F_j with angle in a plane containing both $\mathbf{h}_{\text{m}}$ and $\mathbf{H}_0$ and evaluated at $\mathbf{H}_0$. The second term in brackets in Eq. (2.3.4) is zero if the modulation field and quasistatic field $\mathbf{H}_0$ are parallel or if the frequency does not change with angle at $\mathbf{H}_0$.

The pickup signal produced by this component of the oscillatory magnetization is found by carrying out the standard Bessel function expansion[28] of Eq. (2.3.2) and taking the time derivative. The resulting voltage is given by

$$v_{jl} = 10^{-8}(8\pi N_L \Omega \eta)\hat{\mathbf{p}} \cdot \mathbf{D}_{jl}(\mathbf{H}_0, T, T_D)$$
$$\times \sum_{n=1}^{\infty} n\omega \cos[2\pi l F_j(\mathbf{H}_0)/|\mathbf{H}_0| - n\pi/2 + \gamma_{jl}] J_n(\Lambda_{jl}) \sin n\omega t, \quad (2.3.5)$$

where $J_n(\Lambda_{jl})$ is the nth-order Bessel function of the first kind with argument Λ_{jl}. Here v_{jl} is given in volts, Ω is the sample volume in cubic centimeters, $\hat{p}$ is a unit vector parallel to the pickup coil axis, N_L is the effective number of turns per unit length in the pickup coil (see Section 2.3.2), and η is a constant depending upon the geometry of the pickup coil and sample (η has a maximum value of unity for perfect coupling between the pickup coil and sample). The above expression is correct for a cylinder in which the dHvA magnetization is uniform and eddy-current effects can be ignored.

2.3.1. dHvA Frequency Selectivity

The induced voltage arising from the lth harmonic of the jth dHvA component F_j is seen from Eq. (2.3.5) to be a series of terms containing all harmonics of the modulation frequency ω. If a narrow-band detection system is used to select only that voltage component whose frequency is $n\omega$, the detected amplitude will have a field dependence which is proportional to

$$\hat{p} \cdot \mathbf{D}_{jl}(\mathbf{H}_0, T, T_D) J_n(\Lambda_{jl}) \cos[2\pi l F_j(\mathbf{H}_0)/|\mathbf{H}_0| - n\pi/2 + \gamma_{jl}]. \quad (2.3.6)$$

This expression is the appropriate starting point for our discussion of the techniques used to select individual dHvA frequency components from the complex superposition of terms which is found in most metals. For simplicity we will consider only the first dHvA harmonic case, $l = 1$, and omit l in the following discussion.

First we note that a Bessel function with argument Λ_j occurs in each term. If the modulation amplitude $|\mathbf{h}_0|$ can be varied so that $|\mathbf{h}_0|/|\mathbf{H}_0|^2$ is constant, then the Bessel function is a constant, independent of the value of $|\mathbf{H}_0|$. The maximum response to a particular dHvA

---

[28] M. Abramowitz and I. A. Segun, eds., "Handbook of Mathematical Functions." NBS Applied Mathematical Series 55, Superintendent of Documents, U.S. Govt. Printing Office, Washington, D.C., 1964.

frequency F_j will be achieved by adjusting $|\mathbf{h}_0|$ so that Λ_j corresponds to the first maximum of $J_n(\Lambda_j)$. It is often the case, however, that some strong dHvA term F_j completely swamps all the other dHvA frequency components. When this happens, the dominant term can be eliminated quite effectively by adjusting Λ_j to correspond to one of the zeros of $J_n(\Lambda_j)$. By a judicious choice of both Bessel function order n, i.e., harmonic for detection, and modulation amplitude it is often possible to null out effectively more than one dHvA frequency.

Moreover, if the modulation field is small enough that Λ_j is much less than the value of the first maximum of the Bessel function, it is possible to expand and keep only the first term, that is,

$$J_n(\Lambda_j) \approx (1/n!)(\Lambda_j/2)^n. \qquad (2.3.7)$$

Under these conditions, the relative contribution from the Bessel functions to two different dHvA components with frequencies F_1 and F_2 is approximately $(F_1/F_2)^n$ and, consequently, higher frequencies will be enhanced relative to lower frequencies.

One may also note here that Λ_j is a function of α and $\partial F_j/\partial \alpha$. Therefore, by varying α and $|\mathbf{h}_0|$ it may be possible to separate two frequency components for which $F_1 \approx F_2$ but $\partial F_1/\partial \alpha \neq \partial F_2/\partial \alpha$. Windmiller and Priestley[29] have successfully exploited this oblique modulation scheme to study the complicated dHvA wave forms observed in antimony. Unfortunately this technique complicates dHvA measurements and therefore has found limited application. Furthermore, in ferromagnetic samples a spurious background signal is produced whenever $\alpha \neq 0$ due to perturbation of the ferromagnetic magnetization, and this appears to obscure small dHvA signals.[7]

Finally, it should be pointed out that since the $\mathbf{D}_j$ are not usually all parallel, further selectivity can be obtained from the $\hat{p} \cdot \mathbf{D}_j$ factor in Eq. (2.3.6) by appropriate choice of pickup coil direction. Very little use has been made of pickup coil adjustment in dHvA experiments.

2.3.2. Block Diagram

In Fig. 5 is shown a block diagram of the electronics for a typical field modulation dHvA system in which the narrow-band detection is accomplished by means of a lock-in amplifier with a reference frequency

---

[29] L. R. Windmiller and M. G. Priestley, *Solid State Commun.* **3**, 199 (1965).

$n\omega$, where ω is the modulation frequency. In order to permit the dHvA frequency selection discussed in the previous section, the modulation signal amplitude is constrained to remain proportional to $|\mathbf{H}_0|^2$ during a field sweep. Schemes for maintaining this proportionality have been described by Windmiller and Ketterson[12] and Stark and Windmiller.[13]

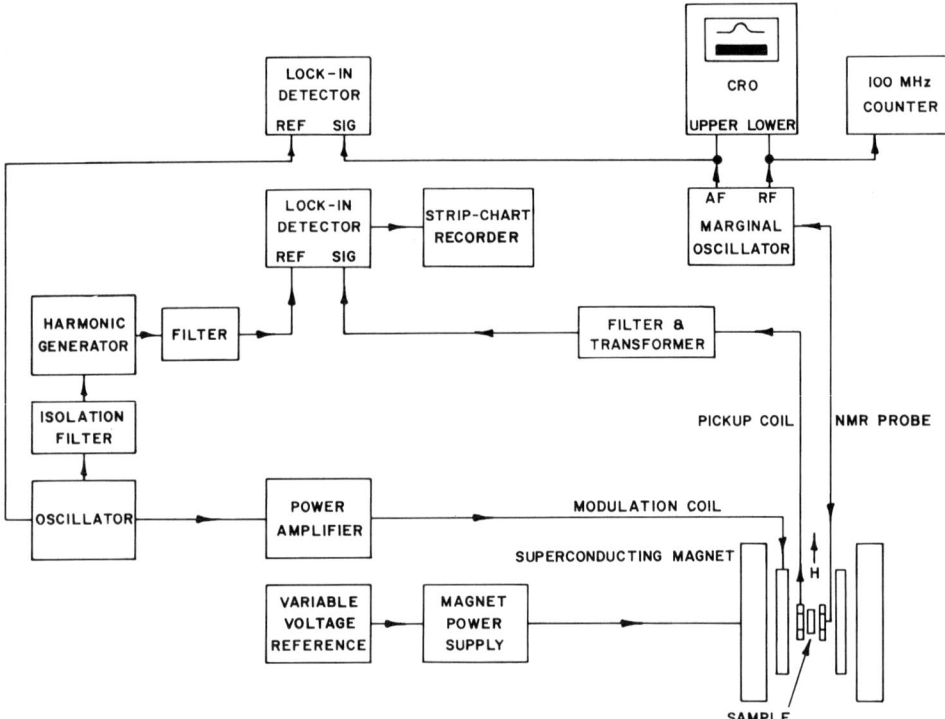

FIG. 5. Block diagram for the field modulation experiment.

The dHvA signal is induced in a pickup coil, filtered, and fed to a lock-in amplifier. The output from the lock-in can be displayed or recorded digitally. Details of the components in Fig. 5 are described below.

2.3.2.1. *Pickup Coil.* A properly designed pickup coil is very important for optimization of the dHvA signal. Usually two coils, wound in series opposition, are desirable in order to cancel signals from any external time-varying field that is approximately uniform over the volume of the pickup coil. Such "compensated" pickup coils have two advantages: (1) the noise contribution due to mechanical vibrations is reduced and (2) the low-level impedance matching transformer and preamplifier are less likely to be overloaded.†

Two coil compensation arrangements are in common use: (1) concentric—a compensating coil wound over the pickup coil, and (2) coaxial—pickup and compensating coils placed end-to-end. Although the second arrangement may be somewhat easier to construct, the first has been found to exhibit superior signal-to-noise characteristics,[13] particularly with respect to noise produced by mechanical vibrations. It can be shown that the concentric arrangement allows a signal-to-noise ratio better by at least 10^4 than the coaxial configuration for typical laboratory vibration environments and magnets. In addition, with concentric compensation a much lower inductive component of source impedance is obtained, which is important to the voltage gain that can be obtained with the use of impedance-matching transformers (see below).

For concentric coils the optimum compensation condition is given by $r^3 = \frac{1}{2}(r_0^3 + a^3)$, where r_0 is the inside radius of the inner coil, r is the outer radius of the inner coil, and the outer coil is wound directly on top of this inner coil. The radius of the outer coil a is assumed to limit the size of the pickup coil. It is also true that very little is gained by making the pickup coil longer than the sample. The signal picked up is proportional to the turns difference between the two coils, typical values of which are 1000–5000 with an overall coil volume of about 0.1 cm³.

2.3.2.2. Impedance Matching and Filtering. In order to match properly the preamplifier to the compensated pickup coil impedance (usually less than 100 Ω for typical low-frequency modulation experiments), a well-shielded transformer is required. While commercially available low-level input transformers‡ can be used for this purpose, Johnson noise and external pickup can be reduced by enclosing the transformer in a superconducting shield at liquid helium temperatures. This scheme also minimizes the length of low-voltage leads from the pickup coil.

Since complete compensation of the pickup coil is never achieved, filtering of the pickup signal to remove the fundamental component is usually necessary. Although this filtering can be done with a twin or

---

† If harmonics of the fundamental frequency are detected, as is the usual case, an uncompensated pickup coil can indeed be used in conjunction with appropriate filtering. However, such a system will exhibit signal to noise which is inferior to that obtained from compensated coils.

‡ Triad Geoformers with turn ratios up to 500 to 1 are convenient for this purpose. The μ metal shields can be removed for low-temperature operation.

bridge tee[13] between the transformer and preamplifier, filter noise and loading can be eliminated with a superconducting LC filter.† If this filter is introduced between the pickup coil and transformer primary, transformer overloading will be prevented.

2.3.2.3. **Preamplifier.** In order to minimize eddy current mixing effects,[15] it is convenient to choose a low modulation frequency so that the classical skin depth is larger than the specimen diameter. For materials for which the magnetoresistance does not saturate, modulation frequencies between 20 and 100 Hz are usually sufficient. At these frequencies commercially available preamplifiers‡ generate on the order of 5 $nV/Hz^{1/2}$ of noise for megohm source impedances. At liquid helium temperatures a typical pickup coil will have a resistance of the order of 25 Ω, implying Johnson noise of about 10^{-10} $V/Hz^{1/2}$. Therefore a transformer with a turn-ratio in excess of 100 is required to achieve a Johnson noise-limited experimental situation. Unfortunately it is usually found that noise produced by mechanical vibrations and external sources exceeds the Johnson noise by at least an order of magnitude, unless special precautions are taken. Stark[30] has reported detection of signals as low as 3×10^{-15} V at 0.02 K in a 40-kG magnetic field by using a magnetically shielded, acoustically decoupled "quiet" room.

2.3.2.4. **Modulation Field.** Conventional techniques with solenoids or Helmholtz coils have been used for producing modulation fields. (For oblique modulation two sets of coils are used.) One minor modification has been the use of superconducting coils for the large modulation fields required to observe some of the low dHvA frequencies at high fields. We have successfully obtained modulation amplitudes in excess of 1500 Oe at 44 Hz in dc fields greater than 50 kOe by using Formvar-insulated, 0.013–cm diameter, Nb–Ti wire, which had no copper cladding. This technique has proved quite successful in our studies of the dHvA effect in cobalt.[19] In these experiments it was found that excessive

---

[30] R. W. Stark, Final Rep., U.S. Army Res. Office, Durham, DA-ARO-D-31-124-G926 (1971).

---

† We have used 0.5-H Nb–Ti superconducting coils and metallized polycarbonate film capacitors to obtain LC networks with Q's in excess of 1000 at 40 Hz. These film capacitors are ideally suited for cryogenic filters since both the capacitance and dissipation factor change by less than 20% in going from room to liquid helium temperatures. Moreover, these capacitors are of relatively small size, 2 cm^3 for a 10-μF–50-V capacitor.

‡ For example, Princeton Applied Research Model 185.

mechanical vibrations rather than quenching of the modulation coil limited the maximum modulation amplitude that could be used.

2.3.3. Frequency Measurements

2.3.3.1. *Absolute Frequency Measurements.* In order to determine accurate dHvA frequencies, simultaneous measurements of dHvA oscillations and magnetic fields are required. Because of flux trapping, the field of a superconducting magnet probably cannot be obtained from a measurement of current through the magnet to better than about 0.05%. Hall probes and magnetoresistance sensors also have limited accuracy and are sensitive to temperature changes as well. Therefore, we have been routinely using *in situ* NMR for accurate field measurements[31] with tiny (0.001 cm³) NMR specimens located in close proximity to the dHvA sample. The accuracy with which the field can be determined is limited by the accuracy with which the NMR field–frequency relation is known. For these measurements we have used a marginal oscillator[31] which is continuously tunable over a range from 8 to 45 MHz under the conditions of our experiment (NMR coil at the end of about 120 cm of coaxial cable). The specimens used have been Al powder and the alkali halides, CsCl and NaCl doped with about 1% $NiCl_2$.

The Na^{23}, Cl^{35}, and Cs^{133} frequency–field relations have been obtained from calibration with Al^{27} (1.1112 MHz/kOe). The aluminum resonance is very strong, but has a rather wide line ($\sim$8 G) and is limited to relatively low fields, less than 40 kG with our marginal oscillator system. We have, however, used Na and Cs quite successfully between 4.2 and 1 K; both give line widths less than 1 G and both resonances can be seen at least weakly on the oscilloscope. In addition, with the Cs resonance, field measurements to 80 kG can be made. Figure 6 shows an example of the α oscillations in a [110] Pb sample along with the Cs NMR resonances. It is clear that accurate frequencies can be obtained from such measurements, especially if a single frequency can be isolated. Since magnetic fields can be measured to better than 1 part in 10^4 in this manner, the main limitation in measurement of an isolated dHvA component is the accurate determination of the number of cycles between two NMR determined field points. If the oscillations have a simple shape, the counting of, for example, 100 cycles can be done to about 1/10 of a cycle and thus 0.1% accuracy is almost routine. In fact, Coleridge *et*

---

[31] J. R. Anderson and D. C. Hines, *Phys. Rev.* **B2**, 4752 (1970). See also J. R. Anderson, R. L. Sandfort, and D. R. Stone, *Rev. Sci. Instrum.* **43**, 1129 (1972).

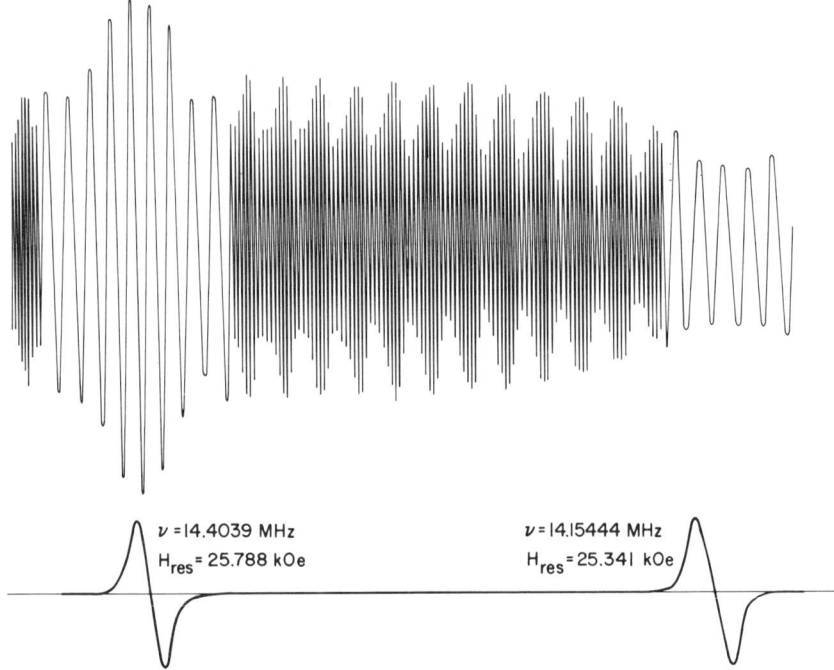

FIG. 6. Upper trace—α oscillations ($F = 159$ MG) in lead for $H_{ap} \parallel [110]$. Lower trace—^{133}Cs NMR signal used to determine the field. For the duration of each NMR measurement the sweep rate was decreased; the increase in dHvA amplitude was a result of an excessively long time constant setting on the lock-in amplifier. In this case the NMR line width actually appears to be about 6 G, the result of overmodulating the narrow ^{133}Cs line.

al.[32] have measured some selected frequencies in copper to 1 part in 50,000 and similar measurements have been made in lead to 2 parts in 50,000.[33] For ferromagnetic specimens, on the other hand, inhomogeneities are produced because of the sample magnetization, and *in situ* NMR measurements are more difficult to perform. Small Al NMR specimens have been used in conjunction with our experiments on iron spheres, but it has not been possible to measure ferromagnetic dHvA frequencies to much better than 0.5% because of the difficulties in determining B within the specimen.

[32] P. T. Coleridge, A. A. M. Croxon, G. B. Scott, and I. M. Templeton, *J. Phys. E.* **4**, 414 (1971).

[33] J. R. Anderson, W. J. O'Sullivan, and J. E. Schirber, *Phys. Rev.* **B12**, 4683 (1972).

2.3. FIELD MODULATION TECHNIQUE

2.3.3.2. Relative Frequency Measurements. Another useful approach to dHvA measurements is a study of "rotation oscillations," i.e., rotation of the sample at constant field, in order to measure the frequency variation with crystallographic orientation. If a particular dHvA frequency can be isolated by proper choice of pickup harmonic and modulation amplitude, the angular variation of dHvA frequency can be measured quite precisely; that is, because each time the frequency changes by the magnitude of the magnetic induction, the "rotation oscillations" undergo one complete cycle. Thus the relative change in frequency for one cycle $\Delta F/F$ equals B/F, which is usually less than 10^{-3}.

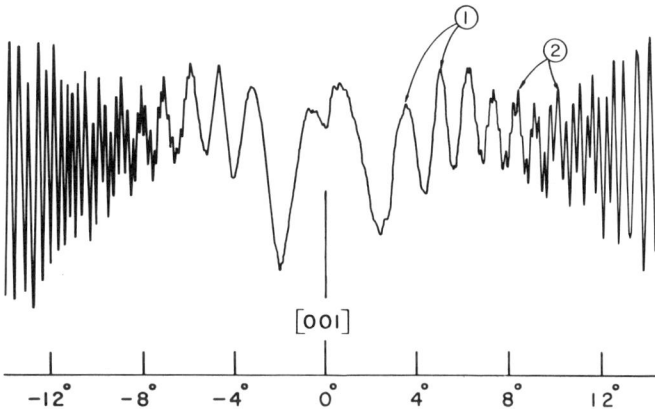

FIG. 7. Rotation diagram for a nickel sphere in a (110) plane for $B = 51.7$ kG. Two frequencies are present as indicated by (1) and (2) ($F_1 \approx 10$ MG, $F_2 \approx 5$ MG).

In Fig. 7 an example of nickel oscillations resulting from rotation in a (100) plane is shown. In this case the modulation has been chosen so that two dHvA frequencies differing by a factor of 2 are evident in the diagram; the "rotation oscillation" cycles correspond to relative frequency changes of 1% and 0.5% for the lower and higher frequencies, respectively. It is interesting to note that the more rapidly oscillating rotation component is actually the lower dHvA frequency (due to [111] necks on a Cu-like sheet of FS). Although these lower frequency oscillations are seen to vanish at ~7° from [100], the more slowly varying, higher-frequency oscillations (which arise from hole ellipsoids centered at X) remain in evidence all the way to [100]. It is clear from Fig. 7 that rotation diagrams are also very useful in locating symmetry orientations.

2.3.4. Amplitude Measurements

Up to now we have not discussed measurements of dHvA amplitudes in any detail. These measurements are very important, even though they are much less accurate than the frequency measurements, and fortunately absolute amplitudes are not usually required.

2.3.4.1. Temperature Dependence.
Cyclotron masses can be obtained from the variation of amplitude with temperature as was mentioned in Section 2.2. For sufficiently large argument, the $\sinh(l\lambda\mu_j T\,|\,\mathbf{B}\,|^{-1})$ term in Eq. (2.2.2) can be replaced by an exponential, and in this case a plot of $\ln(D_{jl}/T)$ versus T will have a slope of $-\lambda\mu\,|\,\mathbf{B}\,|^{-1}$. More generally, μ is found from the slope of

$$\ln(D_{jl}/T)\,[1 - \exp(2\lambda\mu T\,|\,\mathbf{B}\,|^{-1}]\text{ versus }T \qquad (2.3.8)$$

obtained by an iterative least-squares analysis. If there is an error in temperature ΔT and if $\Delta T/T$ is independent of temperature, then the error in the mass ratio μ becomes

$$\frac{\Delta\mu}{\mu} = -\left(1 - \frac{\tanh(\lambda\mu T/|\,\mathbf{B}\,|)}{\lambda\mu T/|\,\mathbf{B}\,|}\right)\frac{\Delta T}{T} = -K(T)\frac{\Delta T}{T}. \qquad (2.3.9)$$

We note that $K(T) > 0$ and thus if the measured temperature is lower than the actual temperature, the measured mass will be too large. This is not the whole story, since the percentage error in temperature may be greater at lower temperatures. In addition, temperature measurements below the λ point are usually more accurate because superfluid helium has such a high thermal conductivity that sample and bath temperatures are more nearly equal. With He^3 systems thermometry problems are even more severe, since thermal equilibrium is difficult to establish. In Section 2.4 an example of mass measurements in indium is given to indicate pitfalls that may be encountered.

2.3.4.2. Field Dependence.
At first sight, Dingle temperatures, which result from measurements of the variation of dHvA amplitude with magnetic field, can be obtained in a very straightforward manner. The magnetic field can be accurately determined and, if the sample temperature is known and remains constant, the main problem is isolation of a single dHvA component. When only two dHvA frequencies are present in the data, it is possible to obtain both Dingle temperatures by measuring amplitudes at the waists and maxima of each beat cycle; for more com-

plex wave forms such an analysis has not yet been carried out successfully. In addition, special care must be taken to avoid contributions from harmonics of the fundamental at the high end of any field sweep. As an example, in Section 2.4 we describe some work on lead alloys in which neglect of the presence of 600 cycle beats gave Dingle temperatures that were too small by a factor of about 2.5.

Since both Dingle temperatures and effective masses are essentially averages around cyclotron orbits, it is necessary to invert the data in order to obtain the more fundamental quantities, local scattering times and local Fermi velocities. Ketterson et al.[34] have used a cubic harmonic expansion to obtain local velocities from dHvA areas and cyclotron masses in platinum. Bosacchi et al.[35] and Miller et al.[36] have used a symmetrical Fourier series inversion scheme to invert the Dingle temperature data in Au:Cu alloys in order to obtain local values of scattering times. Both the cubic harmonic and the Fourier series approaches appear to require a large number of terms, more than 12 in the case of cubic harmonics,[37] in order to fit the data. However, recently Shaw et al.[38] successfully used a partial wave analysis with only a small number of phase-shift parameters to invert their cyclotron mass and Dingle temperature results in platinum and palladium.

2.3.5. Computer Analysis

In addition to analog measurements taken on X–Y or strip chart recorders, it is frequently useful to record amplitude and field information in digital form for computer analysis of dHvA wave forms and frequencies. In those cases for which dHvA frequencies cannot be studied conveniently by means of the Bessel function technique, a Fourier spectrum of the complex dHvA pattern can be obtained with the aid of a computer. We have successfully applied fast Fourier transform techniques for such analyses. In addition, we have used least-squares routines to fit the dHvA amplitudes in order to determine Dingle temperatures. For ferromagnetic specimens, computer analyses are almost indispensable

---

[34] J. B. Ketterson, L. R. Windmiller, J. B. Ketterson, and S. Hörnfield, *Phys. Rev.* **B3**, 4213 (1970).
[35] B. Bosacchi, J. B. Ketterson, and L. R. Windmiller, *Phys. Rev.* **B5**, 3816 (1972).
[36] K. M. Miller, R. G. Poulsen, and M. Springford *J. Low. Temp. Phys.* **6** (1971).
[37] M. Springford, *Advan. Phys.* **20**, 493 (1971).
[38] J. C. Shaw, J. B. Ketterson, and L. R. Windmiller, *Bull. Amer. Phys. Soc. II* **17**, 302 (1972).

in obtaining the contributions of the saturation magnetization, and shape and crystalline anisotropy to dHvA data.

2.4. Examples of dHvA Experiments

In this section some examples of dHvA studies are given in order to demonstrate applications of the experimental techniques described in the preceding sections.

2.4.1. dHvA Studies in Lead

dHvA frequencies have been accurately determined in lead for $\mathbf{H}_{ap}$ along symmetry directions by making use of *in situ* NMR. The primary purpose of such measurements was to provide a stringent test for models of one-electron potentials for lead. In Table II examples of these frequencies and estimated errors are given.[33] The errors are primarily due to uncertainty in crystallographic orientation with respect to the direction of magnetic field and to imperfect isolation of a single dHvA frequency component. Such accurate measurements are the basis of two other types of experiments in lead, alloy studies and pressure experiments.

TABLE II. dHvA Frequencies in Lead

Oscillation orientation	Frequency (MG)
α [100]	204.36 $\pm$ 0.02
β [100]	51.245 $\pm$ 0.002
α [110]	159.11 $\pm$ 0.05
γ [110]	18.06 $\pm$ 0.02

A series of experiments has been carried out in dilute alloys of lead containing either bismuth or thallium or equal amounts of bismuth and thallium. With accurate field measuring techniques it has been possible to measure changes in frequency and Dingle temperature in lead alloyed with as little as 0.05% of the second constituent. For example, a plot of frequency $\Delta F/F$ versus T_D is shown in Fig. 8 for the [100]β oscillations in $Pb_{1-x}Bi_x$ alloys.[31] The dominant uncertainties in such experiments result from inhomogeneous samples, inexact sample orien-

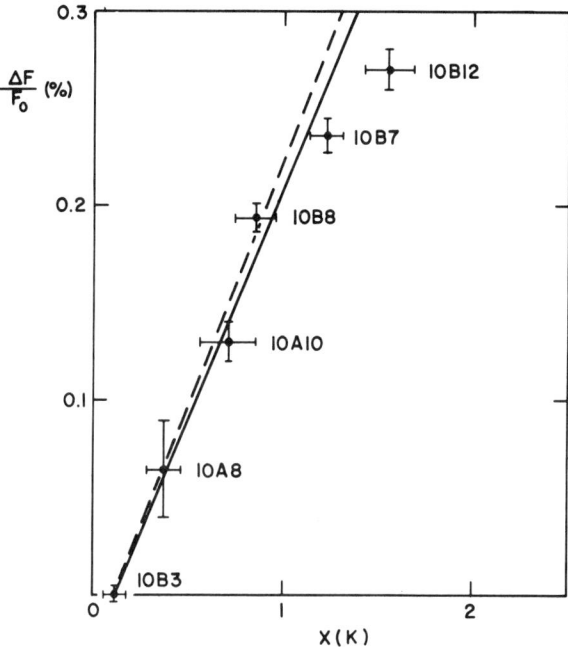

FIG. 8. Relative frequency change versus Dingle temperature for $Pb_{1-x}Bi_x$ alloys. The solid and dashed lines are rigid band predictions [from J. R. Anderson and D. C. Hines, Phys. Rev. **B2**, 4752 (1970)].

tation, and contributions from more than one dHvA component. Our lead alloys with x values between 0.0005 and 0.01 are prepared by the Czochralski technique; the impurity concentration has been found to vary by as much as 20% over the length of some of our specimens. However, this uncertainty should not be very significant in a plot such as Fig. 8 because F and T_D are measured from the same dHvA oscillations, which probably come from the purest portion of the sample.

As Coleridge et al. pointed out,[32] it is also important for alloy studies at symmetry orientations to be able to make changes in sample orientation of 1 to 2° in any direction during the course of an experiment. This in situ adjustment was not possible in the bismuth–alloy experiments shown in Fig. 8 but was added later and contributed greatly to the reproducibility of our recent alloy studies in Pb_xTl_{1-x}.

The effect of a second frequency is dramatically illustrated in Fig. 9, which shows a plot of $\ln(DH_{ap}^{1/2})$ versus $1/H_{ap}$ for the β [100] oscillations in Pb_xTl_{1-x} for $x = 0.0, 0.001, 0.002$, and 0.003. There are beats of about 600 cycles in these oscillations, which show that there are actually two frequencies present differing by about 0.17%. In the pure sample, dHvA oscillations can be observed over a field range that is large enough to show the beats unambiguously, but in the alloys, as shown in Fig. 9, a complete beat cycle is not observed. If one found the slope of a straight line through the points ignoring the presence of the second frequency, a large error in Dingle temperature could result.

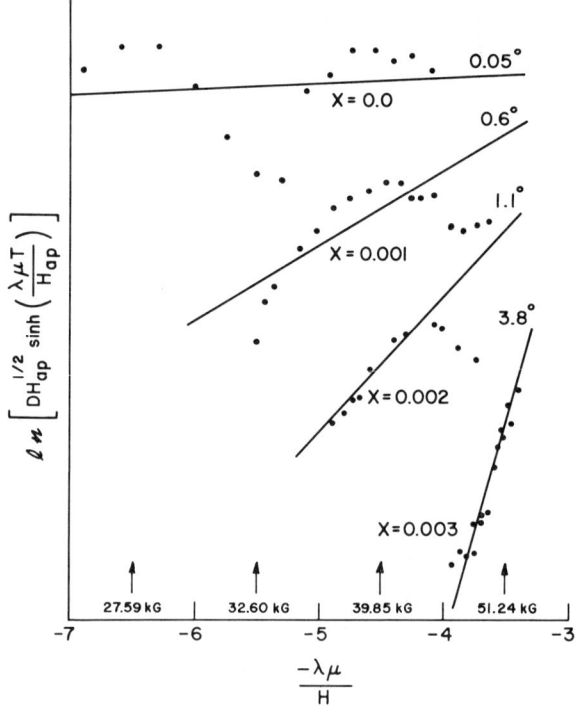

FIG. 9. Dingle temperature plots for $Pb_{1-x}Tl_x$ alloys. The upper curve for pure lead ($x = 0.0$) clearly shows the beats of about 600 cycles, but the beats are not as apparent in the alloys because the measurements were restricted to higher fields. The straight lines have slopes determined by the actual Dingle temperatures and would represent the data in the absence of beats.

Changes in the Fermi surface with lattice spacing can be determined from studies of the dHvA effect at high pressures. Schirber[39] has written a comprehensive review of high-pressure Fermi surface studies which

[39] J. E. Schirber, in "Materials Under Pressure." Honda Memorial Ser. on Mater. Sci. (to be published).

summarizes the work that has been carried out on most metals. Three methods have been used to study the effect of hydrostatic pressure on dHvA oscillations:

(1) fluid helium phase-shift technique,[40,41]
(2) solid helium technique,[42–44]
(3) frozen kerosene oil technique.[45]

In the fluid helium studies, the pressure is varied continuously and therefore must be less than that at which the helium solidifies, about 25 bars at 1 K. The effect of pressure on frequency is very small since typical metal compressibilities are of the order of 0.1% kbar$^{-1\dagger}$ and typical frequency shifts are also of this order. That is, if δH is the shift in the phase of a dHvA oscillation with pressure and ΔH is the field separation between dHvA cycles, then

$$\delta H / \Delta H = (F/B) \, \Delta F / F.$$

Here ΔF is the frequency shift with pressure and B is the magnetic induction. For example, for the [100] β oscillations in lead (orbit γ) at $\beta = 30$ kOe, one finds a phase shift $\delta H / \Delta H \approx 0.1$, which is relatively easy to measure.

Because the dHvA amplitude also can change with pressure, it is usually necessary to sweep reproducibility over at least one half an oscillation cycle in order to determine the actual phase shift. This has been accomplished with superconducting magnets by putting the magnet in the persistent mode and sweeping over a few Gauss with external coils. Note, also, that it is not necessary to determine the actual phase shift in Gauss, but only the shift relative to a dHvA cycle.

The second technique for studying Fermi surface changes with hydrostatic pressure makes use of solid helium as the pressure transmitting medium; pressures to about 10 kbars have been attained. In these experiments frequencies must be accurately measured at several pressures

---

[40] I. M. Templeton, *Proc. Roy. Soc. (London)* **A292**, 413 (1966).
[41] J. E. Schirber and W. J. O'Sullivan, *Phys. Rev.* **184**, 628 (1969).
[42] C. A. Swenson, *Solid State Phys.* **11** (1960).
[43] J. S. Dugdale and J. A. Hulbert, *Can. J. Phys.* **35**, 720 (1957).
[44] J. E. Schirber, *Cryogenics* (to be published).
[45] E. S. Itskevich, *Cryogenics* **4**, 365 (1964).
[46] D. L. Waldorf and G. A. Ahlers, *J. Appl. Phys.* **33**, 3206 (1962).

---

† In lead the compressibility is 0.204% kbar^{-1} (Ref. 46).

and errors are similar to those in alloy studies, the only additional complication being the possibility of a shift in sample orientation with pressure. The possibility of this occurrence has usually restricted such pressure measurements to symmetry orientations. The third approach, frozen kerosene oil, has permitted somewhat higher pressures (up to 15 kbars have been reported); the measurement problems are similar to those solid helium technique.

2.4.2. Cyclotron Mass Measurements in Indium

As an example of the difficulties which are encountered in making mass measurements to better than 1%, we will summarize our unsuccessful attempts to use the field modulation techniques to detect a variation of cyclotron mass with magnetic field.[†] The purpose of the experiment was to investigate the electron–phonon mass-enhancement factor in indium.

In indium there are dHvA oscillations from a third zone arm (110) cross section with frequency 4.59×10^6 G and mass ratio $\mu = 0.204$ measured at fields below 20 kOe. We measured the variation of amplitude with temperature at about 12 and 55 kOe expecting any difference in μ value to be less than 0.5%. Our first results yielded a value of μ at 55 kOe of about 0.196 and we were delighted to see such a large effect until we realized that our temperatures, determined from the helium vapor pressure at the top of the cryostat, were inaccurate. In order to improve these measurements the manometer was attached to a tube, which had been inserted into the cryostat and reached down to a point just above the liquid helium level. This corrected for the pressure difference between the top of the cryostat and the helium level since the tube was not part of the pumping system.[‡] All the measurements were made at temperatures below the λ point and the new temperatures were 2–10% larger than those measured previously. Because the largest corrections were found at the lowest temperatures, our value for the mass ratio became larger. In fact, the number seemed to be too large, $\mu \approx 0.23$, and it became necessary to look for sources of error in the measurement of amplitude.

[†] These experiments were carried out in collaboration with Dr. J. E. Schirber of Sandia Laboratories, Albuquerque, New Mexico.

[‡] Later, slightly better results were obtained with a carbon resistance, calibrated against the helium vapor pressure.

At fields of 50 kG, the dHvA oscillations are over 500 G apart and modulation amplitudes of about one half this magnitude are required in order to obtain the largest dHvA amplitudes at such high fields. However, since eddy-current heating could have been a problem with such large modulation, only about 50 G of modulation were actually used. It was concluded that the heating due to eddy currents was unimportant because no unusual changes in dHvA amplitudes were observed at the λ point. However, a slight distortion of the dHvA wave form was apparent, suggesting that higher dHvA harmonics were present in the signal. By filtering the output of the phase-sensitive detector it was possible to produce a signal with no apparent dsitortion and as a result the value of μ was reduced to approximately the 10-kOe value of 0.204. Thus, after the measurement errors had been corrected, the experiment could be performed with an accuracy of slightly better than 1%, but this was not adequate for observation of the predicted mass decrease at high fields.

2.4.3. Ferromagnetic Metals

The ferromagnetic metals present special measurement problems. In order to describe these difficulties we will now discuss some aspects of the experiments on cobalt and nickel. Iron[7] has been studied more completely, especially with respect to large portions of the Fermi surface.

2.4.3.1. Cobalt. In hcp cobalt the anisotropy constants are large,[†] and the easy direction of magnetization is along the c axis. Thus, even for spherically shaped samples above saturation, the magnetization does not line up with $\mathbf{H}_{ap}$ unless $\mathbf{H}_{ap}$ is parallel to the c axis or lies in the basal plane. In Fig. 10 the direction of $\mathbf{B}$ and the deviation of $|\mathbf{B}|$ from the value for a perfect sphere with no anisotropy, Eq. (2.2.8) are shown as a function of the magnitude and direction of the applied field. For an applied field greater than 30 kOe, the error in $|\mathbf{B}|$ is less than 0.5% at all orientations of $\mathbf{H}_{ap}$ and can probably be neglected. However, we have been able to follow certain dHvA low frequencies (e.g., α oscil-

---

[47] R. Pauthenet, Y. Barnier, and G. Rinef, *J. Phys. Soc. Japan Suppl. B-I* **17**, 309 (1962).

[†] The anisotropy constants at 4.2 K are $K_1 = 6.8 \times 10^6$ erg/cm^3 and $K_2 = 1.7 \times 10^6$ erg/cm^3 (Ref. 47).

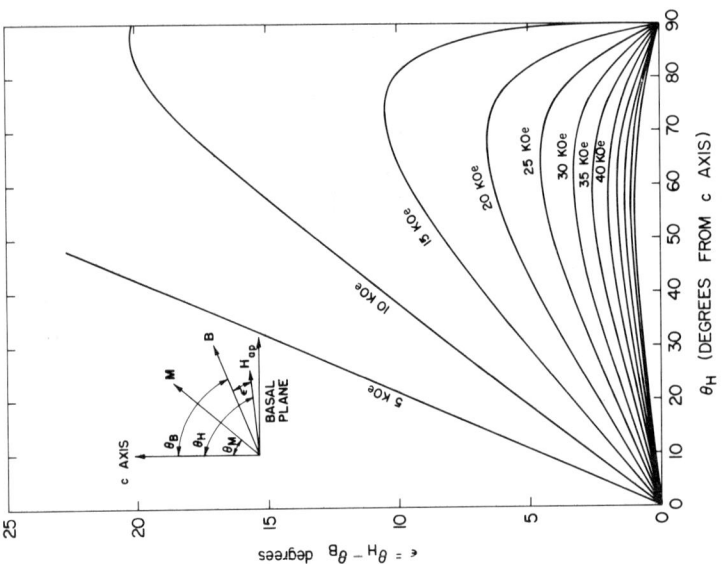

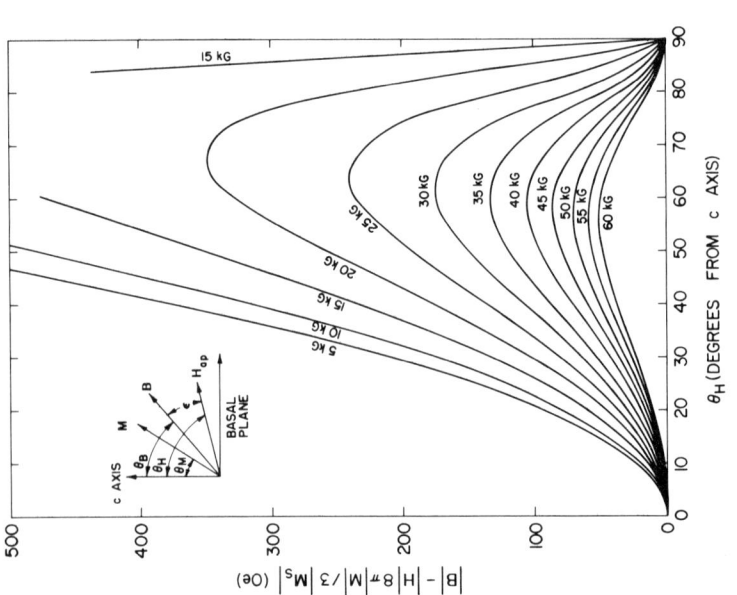

FIG. 10. (a) Correction to **B** in a cobalt sphere for magnetic anisotropy. θ_H is the direction of the applied magnetic field ($M_s = 1430$ esu[19]). (b) Angle between **B** and $\mathbf{H}_{ap}$ for a cobalt sphere.

lations with $F \approx 1$ MG [19]) down to applied fields below 10 kOe for $\mathbf{H}_{\mathrm{ap}}$ within $50°$ of the c axis, and under these circumstances the corrections to $|\mathbf{B}|$ are larger and must be included.

The angle $\theta_H - \theta_B$ in Fig. 10b, is less easily taken into account. Since the direction of $\mathbf{B}$ changes during a field sweep, the dHvA oscillations will not appear to be periodic in $|\mathbf{B}|^{-1}$ and unfortunately the correction depends upon the shape of the piece of Fermi surface under investigation. For cobalt the effect should be most apparent in the α oscillations. However, over the orientations at which these oscillations could be unambiguously studied (within about $30°$ of the c axis), the variation of frequency with angle was so slight that the lack of periodicity was barely in evidence. The change in angle with $\mathbf{H}_{\mathrm{ap}}$ does appear to complicate our attempts to disentangle the low-frequency oscillations for field orientations at larger angles from the c axis, but experiments in this regime are not yet complete.

A final point about the large magnetic anisotropy in cobalt is that, even for a spherical sample, the torque is large; at arbitrary orientations the maximum torque can be greater than 10^5 dyn cm for an applied field of more than 7 kOe and a sample diameter of about 2 mm. As a consequence, an optimum mounting technique has not been devised. In our work we have held the cobalt spheres with Hysol epoxy, but, since the cobalt crystals are of inferior quality, it is not known whether this method of mounting produces excessive strain.

2.4.3.2. *Nickel.* Since good single crystals of Ni are available, this material has been studied by a number of investigators, and there is a rather complete model for the Fermi surface.[17,18] In spite of these facts, many predictions of the model have not been verified directly by experiment. For example, dHvA data are available for the necks on a copper-like surface and for small hole pockets centered at the symmetry point X in the Brillouin zone for fcc structures, but no results have been reported for the "belly" orbits expected from the copper-like surface or for the orbits expected from the two larger electron surfaces centered at Γ. It is believed that the difficulty is caused by the large masses of these pieces of the Fermi surface. The lower temperatures provided by a He^3 dilution refrigerator might permit a more complete study of nickel.

A large ferromagnetic splitting of the energy levels can be expected to complicate Fermi surfaces and, in addition, the accidental degeneracies between "spin-up" and "spin-down" energy states will be lifted when spin–orbit interactions are taken into account. However, since the spin–orbit splitting is fairly small, orientation-dependent magnetic breakdown effects can be expected to occur. For example, Tsui[18] has observed that the amplitude of one dHvA signal, resulting from small hole pockets of the Fermi surface in nickel, decreases abruptly to zero and then increases to its original value for a variation of sample orientation of less than 2°. This is attributed to an accidental band degeneracy at the Fermi surface at a (110) symmetry plane that is removed as the magnetic field is tilted away from the [110] symmetry direction. Observation of this effect required precisely controlled rotations of the magnetic field relative to the sample axes.

Finally, the experimental studies in nickel point out the necessity for simple theoretical models of the electronic structure in order to interpret experimental results and to suggest new approaches. Hodges *et al.*[17] showed theoretically that the unexpected large anisotropy of the hole-pocket dHvA frequencies resulted from the change in Fermi surface with magnetic field direction. This dependence of Fermi surface upon magnetic field arises because of a combination of spin–orbit and ferromagnetic exchange effects.

2.5. Summary

The field modulation technique with the use of superconducting magnets has become a quite sophisticated method for studying dHvA oscillations. As a result, there are very few metals in which dHvA signals have not been observed. (For example, dHvA oscillations have just recently been observed in titanium,[48] which appears to be similar to zirconium.[49]) In this review we have attempted to describe some of the more recent successful approaches that have been used with field modulation experiments. For background and many of the basic ideas we have relied heavily upon the work of Stark and Windmiller.[13]

In addition to the analogue techniques, the use of digital data acquisition systems has become almost routine in dHvA experiments. This allows one to use the computer to do comprehensive analyses of data in order to enhance signal-to-noise and to separate particular dHvA components for study.

---

[48] G. N. Kamm and J. R. Anderson, *Low Temp. Conf.* **LT-13** (to be published).
[49] A. C. Thorsen and A. S. Joseph, *Phys. Rev.* **131**, 2078 (1963).

Acknowledgments

In our experimental research program at the University of Maryland, we have benefitted from discussions on enhancement of signal to noise with Professor J. Weber and members of his group. In addition we refer here to several research projects in progress at the University of Maryland and make use of some as yet unpublished work. We wish to acknowledge helpful discussions with Mr. J. Ya-Min Lee, Mr. John Hudak, and Dr. J. H. P. van Weeren. Specifically we wish to thank Mr. Lee for the use of Figs. 6 and 9, Mr. Hudak for the use of Figs. 1 and 10, and Dr. van Weeren for Fig. 3.

3. EXPERIMENTAL TECHNIQUES FOR VISIBLE AND ULTRAVIOLET PHOTOEMISSION*

3.1. Introduction

3.1.1. Photoemission Measurements Past and Present

The early photoemission measurements on metals by Richardson and Compton[1] and independently by Hughes[2] provided the first adequate test of Einstein's[3] theory of the photoelectric effect. Einstein explained the photoelectric effect by assuming light to consist of quantized particles, photons, each with a certain energy $h\nu$ which could be imparted to an electron. Figure 1 illustrates the photoemission event for a metal. The work function ϕ gives the minimum energy which an electron at the Fermi surface must receive from an incident photon in order to escape

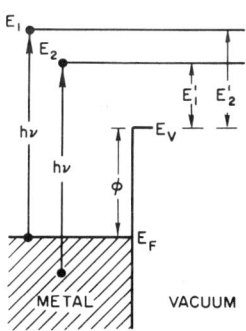

FIG. 1. Photoemission from a metal. The vacuum level E_v equals $\phi + E_F$, where ϕ is the work function and E_F is the Fermi level. The final energy to which the electrons are excited is given by the initial energy plus $h\nu$. E_1' and E_2' are the kinetic energies of the electrons in vacuum.

[1] O. W. Richardson and K. T. Compton, *Phil. Mag.* **24**, 575 (1912).
[2] A. L. Hughes, *Phil. Trans. Roy. Soc.* **A212**, 205 (1912).
[3] A. Einstein, *Ann. Phys.* **17**, 145 (1905).

* Part 3 is by G. F. Derbenwick, D. T. Pierce, and W. E. Spicer.

from the solid. At $T = 0$ K, the quantum condition on the minimum frequency of light for which photoemission can be obtained is

$$h\nu \geq \phi. \qquad (3.1.1)$$

The maximum kinetic energy of an electron in vacuum is then

$$E_{\max} = h\nu - \phi. \qquad (3.1.2)$$

In contrast to optical absorption and reflection, which had been quite successfully described by the classical Drude and Lorentz theories, photoemission demonstrated the quantum nature of the optical excitation process.

The 1920's and early 1930's saw considerable activity in the field of photoemission. For example, Ives and co-workers[4] calculated electron orbits for a retarding field spherical analyzer and verified their results experimentally by measuring the angular dependence of photoemission on the best samples obtainable in those days. Fowler[5] used Fermi–Dirac statistics to account for the temperature dependence of yield near the photoelectric threshold. These calculations were extended by DuBridge[6] to account for the tails observed in experimental energy distributions of photoelectrons. The book of Hughes and DuBridge[7] contains a fairly complete account of early photoemission experiments and theories.

In the last two decades the photoemission measurement has become increasingly important as a powerful tool to probe the electronic structure of atoms and condensed matter. In this chapter we will limit ourselves to photoemission experimental methods as applied to solids and liquids.[8] Moreover, our discussion will be concerned with photoemission experiments in which electrons are excited by visible or ultraviolet light. X-ray photoelectron spectroscopy has been extensively treated elsewhere.[9]

---

[4] H. E. Ives and T. C. Fry, *Astrophys. J.* **56**, 1 (1922); T. C. Fry and H. E. Ives, *Phys. Rev.* **32**, 44 (1928); H. E. Ives, A. R. Olpin, and A. L. Johnsrud, *ibid.* **32**, 57 (1928).

[5] R. H. Fowler, *Phys. Rev.* **38**, 45 (1931).

[6] L. A. DuBridge, *Phys. Rev.* **43**, 727 (1933).

[7] A. L. Hughes and L. A. DuBridge, "Photoelectric Phenomena." McGraw-Hill, New York, 1932.

[8] Photoelectron spectroscopy of gases has been extensively discussed by Turner; see, for example, D. W. Turner, C. Baker, A. D. Baker, and C. R. Brundle, "Molecular Photoelectron Spectroscopy." Wiley, New York, 1970.

[9] K. Siegbahn, C. Nordling, A. Fahlman, R. Nordberg, K. Hamrin, J. Hedman, G. Johansson, T. Bergmark, S. Karlsson, I. Lindgren, and B. Lindberg, "ESCA:

3.1. INTRODUCTION

Note that throughout this chapter we are dealing with the external photoeffect, in which an incident photon excites an electron which escapes from the solid into vacuum.

The two most important photoemission measurements are (1) the energy distribution of photoelectrons (frequently referred to as an *energy distribution curve* or EDC), and (2) the quantum yield, i.e., number of photoelectrons emitted per photon. The richness of structure in EDC's was not observed in the early days of photoemission primarily because of poor vacuum conditions. Through studying the energy distribution of the electrons $N(E, h\nu)$ over a wide range of photon energy, one can obtain considerable insight into the character of the electronic states of the solid as well as the nature of the optical excitation process. The photoelectron energy distribution $N(E, h\nu)$ is related to $P(E, h\nu)$, the probability of a photon with energy $h\nu$ exciting an electron to final energy E. $P(E, h\nu)$ is modified by an escape function $T(E)$ (see Section 1.2), which equals zero for $E < E_\mathrm{f} + \phi$ and normally varies smoothly and monotonically for $E > E_\mathrm{f} + \phi$. Thus if one measures $N(E, h\nu)$, one has information on $P(E, h\nu)$, which in turn reflects features of the electronic structure.

In contrast to the energy distribution curve $N(E, h\nu)$, the optical constants are determined by the integral of the transition probability over all final states, i.e.,

$$\varepsilon_2 \propto \int_{E_\mathrm{F}}^{E_\mathrm{F}+h\nu} P(E, h\nu)\, dE.$$

The integral nature of $\varepsilon_2(h\nu)$ is illustrated in Fig. 2, where $\varepsilon_2(h\nu)$ for GaAs is shown.[10] The insert shows an EDC at 10.4 eV, which is in a photon energy region where $\varepsilon_2(h\nu)$ varies smoothly with little structure. The EDC, on the other hand, is rich in structure. Thus $P(E, h\nu)$ may have strong structure even though $\varepsilon_2(h\nu)$ does not.

In a typical study the EDC's are measured at many photon energies over as wide a range as possible. The strength and the photon energy dependence of the structure in the EDC of Fig. 2 is governed by the band structure of GaAs through the conditions on conservation of wave vector **k** and energy E. Thus, by following the behavior of the structure in the EDC's with increasing photon energy, we can learn much about the band structure of the material. This approach is particularly fruitful

Atomic, Molecular and Solid State Structure Studied by Means of Electron Spectroscopy." Almqvist and Wiksells Boktryckeri AB, Upsula, 1967.

[10] W. E. Spicer and R. C. Eden, *Proc. Int. Conf. Phys. Semiconductors, 9th, Moscow* **1**, 65, Nauka, Leningrad, 1968; R. C. Eden, Ph.D. Dissertation, Stanford Univ. (1967).

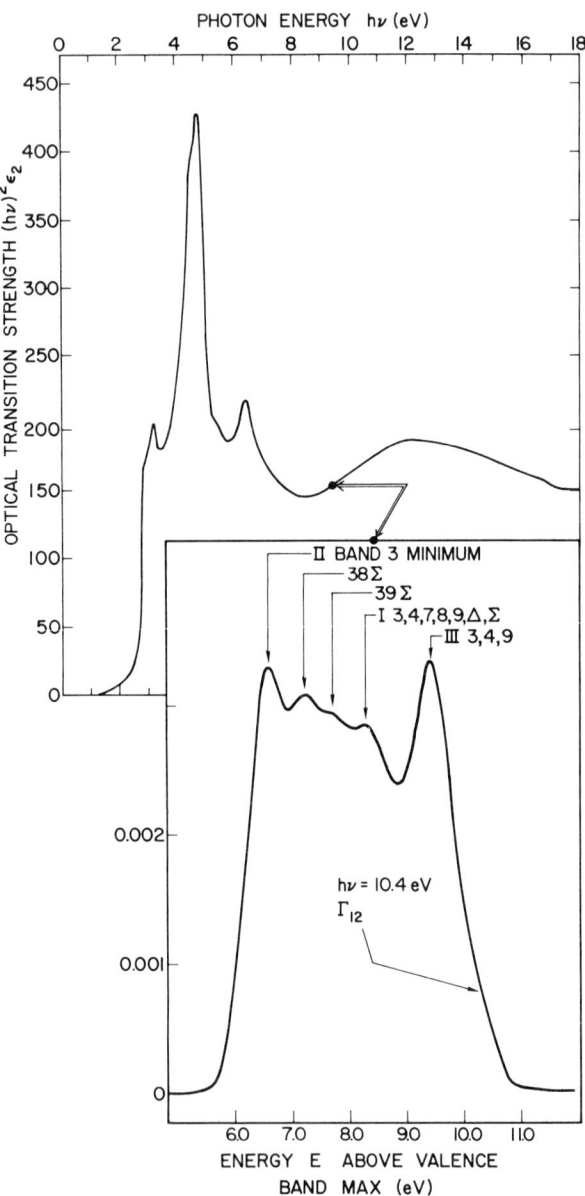

FIG. 2. Information contained in EDC's contrasted to information contained in ε_2. The upper curve shows ε_2 versus $h\nu$ for GaAs while the lower curve is a photoemission EDC (without cesium) for one value of $h\nu$. Note the large amount of structure contained in the EDC. Similar EDC's are obtained over a wide range of photon energy. [W. E. Spicer and R. C. Eden, *Proc. Int. Conf. Phys. Semiconductors*, 9th, Moscow **1**, 65. Nauka, Leningrad, 1968; R. C. Eden, Ph. D. Dissertation, Stanford Univ. (1971).]

3.1. INTRODUCTION

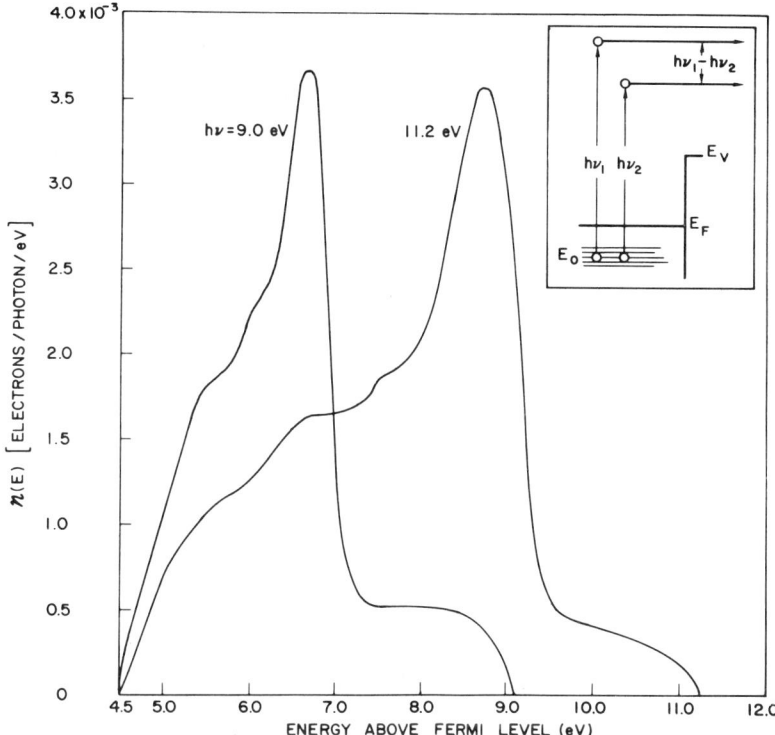

FIG. 3. Energy distribution curves from Cu for $h\nu_1 = 9.0$ eV and $h\nu_2 = 11.2$ eV. Note that the difference in the positions of the dominant peak in each curve is just $h\nu_2 - h\nu_1 = 2.2$ eV. The insert shows schematically the high density of states at E_0 which gives rise to the strong peak in the EDC's.

when fairly accurate band structure calculations are available. The calculated E versus **k** curves can be used to interpret the photoemission results and the photoemission results can be used to check the band calculations and indicate where adjustments can be made in the band calculations. This has been extensively demonstrated for GaAs,[10] Ge,[11] and many other materials.

The power of the photoemission technique is further illustrated in Fig. 3. Two EDC's from copper[12] at photon energies 9 eV and 11.2 eV are shown. The final state energy is taken relative to the Fermi level (measured electron energy plus work function). Note that the increase

[11] T. M. Donovan, J. E. Fischer, J. Matsuzaki, and W. E. Spicer, *Phys. Rev.* **3B**, 4292 (1971).

[12] W. F. Krolikowski and W. E. Spicer, *Phys. Rev.* **185**, 882 (1969).

in the high-energy cutoff is just equal to the increase in the photon energy as expected from the Einstein relation [Eq. (3.1.2)]. The striking feature of these EDC's is the large peak, the position of which increases by an amount equal to the increase in photon energy. The peak remains at a constant position below the high-energy cutoff. The situation is presented schematically in the insert of Fig. 3, where the lines at E_0 correspond to the initial states which gives rise to the peak in the EDC's.

E_0 is the energy of a feature in the electron structure of copper from which the electrons are excited. Note that since the electron energy as well as the photon energy is measured, the absolute energies of the initial and final states of the optical transition are obtained. In fact, this EDC peak corresponds to a peak in the density of states of Cu due to the d bands, which lie several electron volts below the Fermi level in copper. The large rise in the EDC about 2 eV below the high-energy cutoff of the EDC marks the upper edge of the d bands. This is confirmed by one-electron energy band calculations.[13]

The ordinate on Fig. 3 is given in terms of (electrons/absorbed photon)/electron volt. To obtain such a scale it is necessary to know the efficiency (quantum yield) with which the incident photons produce photoelectrons. The quantum yield can be defined either in terms of incident or absorbed photons. The former definition is most useful for practical light detectors, including the calibration standards discussed in Section 3.2, where one wants to relate the emitted current to an incident light flux. Quantum yield in terms of absorbed photons is most useful for fundamental studies of the electron structure of a material. Absolute or normalized EDC's may be given in terms of the electrons emitted per *incident* photon or, if correction is made for the partial reflection of the incident radiation, in terms of electrons emitted per *absorbed* photon. Methods for determining the quantum yield will be discussed in Section 3.2.

A quantum yield curve for Cu is given in Fig. 4. One can obtain considerable information about the potential barrier at the surface of solids by studying the low-energy threshold of such curves. For a metal, the low-energy cutoff of the spectral yield curves is determined by the work function. It is customary to use Fowler theory[5] and a Fowler plot of the experimental data to determine the work function from the yield data.

In contrast to a metal the work function and photoelectric threshold generally do not coincide for a nonmetal. In semiconductors and in-

[13] B. Segall, *Phys. Rev.* **125**, 109 (1962).

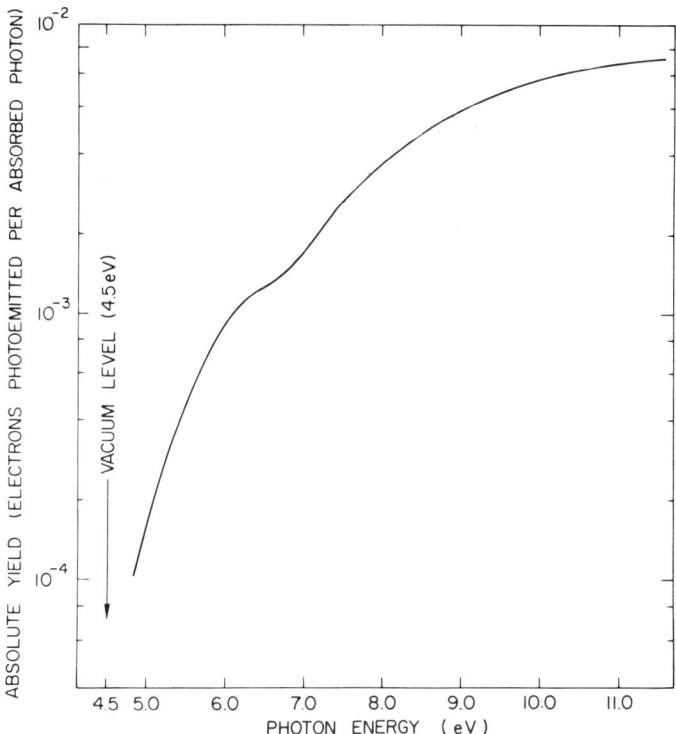

FIG. 4. Quantum yield from a clean Cu sample. [W. F. Krolikowski and W. E. Spicer, *Phys. Rev.* **185**, 882 (1969); W. F. Krolikowski, Ph. D. Dissertation, Stanford Univ. (1967).]

sulators, the photoelectric threshold is the energy from the valence band maximum to the vacuum level or equivalently the energy gap plus the electron affinity. A careful comparison of the energy relationship in metals and nonmetals is given in Section 3.3.2. While the Fowler theory explains the photoelectric emission near threshold for metals, there is, for semiconductors and insulators, no simple, empirically well-established theory. Several factors make the determination of the photoemission threshold from yield data difficult in semiconductors. First of all, band structure effects and optical selection rules often appear to play a more important part in photoemission from semiconductors than in that from metals. Kane[14] has considered these effects theoretically and obtained a number of power law relationships for the yield in various cases. Another

[14] E. O. Kane, *Phys. Rev.* **127**, 131 (1962).

complexity is that produced by band bending, which will complicate the yield curve near threshold. A third complicating factor is the possibility of photoemission from impurities in the forbidden gap as well as from surface states.[15] It is sometimes difficult to distinguish between such emission and valence band emission near threshold. When one considers all of these complicating factors, it is not surprising that it is difficult to determine the photoelectric threshold precisely unless all the parameters of the nonmetal are well known. The determination of the threshold, electron affinity, and work function from EDC's is discussed in Section 3.3.2.

Whereas most of the experimental work in the last decade has concentrated on studies of energy distributions and quantum yield, other techniques are of interest and may prove very fruitful in the coming years. One of these is the study of the angular dependence of photoemitted electrons. Theoretical predictions are available,[16] and it appears that this may provide an important area of work. Polarized light measurements are also of interest. Recent experiments[17] have shown that EDC's from noncubic crystals can be a function of light polarization. One would, of course, like to measure EDC's as a function of both angle of emission and direction of polarization of light. Modulation techniques, where the pressure, temperature, or wavelength is modulated, may also be applied to photoemission measurements. Recent measurements of the spin of photoelectrons from ferromagnetic Ni have been reported.[18]

3.1.2. Physics of the Photoemission Process

In the first few decades after the development of quantum mechanics, photoemission was looked upon as a surface rather than a bulk effect, i.e., it was assumed that conservation of wave vector **k** in the excitation process was provided by interaction with the surface. Perhaps the first theorist to emphasize the point that **k** conservation can be provided by volume as well as surface processes was Fan[19] in the 1940's. However, it was not until the 1950's that it became well established experimentally that photoemission from a wide variety of materials was predominantly

---

[15] T. E. Fischer, *Surface Sci.* **13**, 31 (1969).

[16] G. D. Mahan, *Phys. Rev. Lett.* **24**, 1068 (1970).

[17] L. D. Laude, B. Fitton, M. Anderegg, *Phys. Rev. Lett.* **26**, 637 (1971).

[18] U. Banninger, G. Busch, M. Campagna, and H. C. Siegmann, *Phys. Rev. Lett.* **25**, 585 (1970); K. Sattler and H. C. Siegmann, *ibid.* **29**, 1565 (1972).

[19] H. Y. Fan, *Phys. Rev.* **68**, 44 (1945).

3.1. INTRODUCTION

a bulk or volume process rather than a surface effect.[†,20] It was in this period that a simple model for photoemission was proposed by Spicer[21] and has provided the basis for understanding much of the subsequent work. In this model, the photoemission is usually viewed as a semi-classical three-step process in which each step can be treated independently. The steps are: (1) optical excitation, (2) transport through the solid to the surface, and (3) escape over the potential barrier at the surface. In this way the complex quantum-mechanical description of photoemission is reduced into simpler parts which can be handled separately. Despite its successful use over the past decade, it must be recognized that this model is at best just a good first approximation and that there may be important cases where the complete decoupling of the various steps is not a good approximation, such as when the electron escape depth becomes comparable to the lattice constant.

Detailed treatment of the three-step model are available in the literature,[22] and we limit ourselves here to a brief discussion of the most significant aspects of this model.

3.1.2.1. Optical Excitation. An important result of quantum-mechanical time-dependent perturbation theory is that the probability per unit time P for optical transitions from initial states ψ_i to final states ψ_f is

$$P(E, h\nu) = (2\pi/\hbar) |\langle \psi_i | H' | \psi_f \rangle|^2 N_f(E) \delta(E - E_i - h\nu), \quad (3.1.3)$$

where E is the energy of the final state, $N_f(E)$ is the density of the final states, and H' is the operator $(1/m)(\mathbf{A} \cdot \mathbf{p})$. If the initial and final state wave functions are well described by Bloch functions ($\psi = \alpha e^{i\mathbf{k} \cdot \mathbf{r}}$, where α is a function with the periodicity of the lattice), then a standard result of one-electron band theory is that the matrix element vanishes for $\mathbf{k}_i \neq \mathbf{k}_f$, where $\mathbf{k}_i$ and $\mathbf{k}_f$ are the wave vectors of the initial and final

---

[20] J. G. Endriz and W. E. Spicer, *Phys. Rev. Lett.* **24**, 64 (1970); J. G. Endriz, Ph.D. Dissertation, Stanford Univ. (1970).

[21] W. E. Spicer, *Phys. Rev.* **112**, 114 (1958).

[22] C. N. Berglund and W. E. Spicer, *Phys. Rev.* **136**, A1030, A1044 (1964).

---

† In retrospect, this is not at all surprising since most of the photoemission experiments were done with light at near normal incidence, and examination of the surface effect theories shows that they predict no photoemission for light at normal incidence. In certain cases (Ref. 20), however, evidence has been obtained recently of surface photoemission effects.

one-electron states, respectively. It is assumed that the wave vector associated with the photon is much smaller than $|\mathbf{k}_i|$ and $|\mathbf{k}_f|$. Such optical transitions are called *direct transitions* since they represent vertical transitions on an E versus $\mathbf{k}$ diagram. We see from Eq. (3.1.3) that, for direct transitions, the important optical selection rules are conservation of energy and conservation of crystal momentum. Since there are many theoretical treatises on direct transitions available in the literature,[23] there seems to be little point in going into detail here. Detailed applications to photoemission can be found, for example, by Kane,[24] Brust,[25] Spicer and Eden,[10] Smith,[26] Janak *et al.*,[27] and Christensen.[28]

For materials such as the covalent semiconductors and free electron metals, where the wave functions of the valence electrons are well extended, the use of the direct transitions seems well established. The situation may not be so clear for valence states, such as the d band of Cu, where the wave functions retain more of their atomic nature and are not so well extended. Early work of Spicer and his co-workers,[22,29] and Eastman[30] showed that surprisingly good agreement could be obtained for certain classes of states in crystalline solids[31] if it was assumed that the number of electrons n excited to an energy E was given by

$$n(E) \propto N_c(E) N_v(E - h\nu), \qquad (3.1.4)$$

where $N_c(E)$ and $N_v(E - h\nu)$ are the optical density of states at the final state energy E and initial state energy $E - h\nu$, respectively. The term "optical density of states" is used since the relative optical matrix elements may be a function of energy. However, (1) if data are taken over a wide enough photon energy range so that the sum rule for initial states is almost completely satisfied, and (2) if the data fits Eq. (3.1.4),

---

[23] J. C. Phillips, *Solid State Phys.* **18**, 56 (1966).

[24] E. O. Kane, *Phys. Rev.* **175**, 1039 (1968).

[25] D. Brust, *Phys. Rev.* **139**, A489 (1965).

[26] N. V. Smith, *Critical Rev. Solid State Sci.* **2**, 45 (1971).

[27] J. F. Janak, D. E. Eastman, and A. R. Williams, *Solid State Commun.* **8**, 271 (1970).

[28] N. E. Christensen, Phys. Lab. I, Tech. Univ. of Denmark, Rep. #85 (May 1971) (to be published).

[29] W. E. Spicer, *Phys. Rev. Lett.* **11**, 243 (1963); *J. Res. Nat. Bur. Std.* **74A**, 397 (1970).

[30] D. E. Eastman, *Solid State Commun.* **7**, 1697 (1969); *J. Appl. Phys.* **40**, 1387 (1969).

[31] W. E. Spicer and T. M. Donovan, *J. Non-Crystalline Solids* **2**, 66 (1970).

3.1. INTRODUCTION

the optical density of states is closely related to the actual density of states.[22,29,30]

Equation (3.1.4) gave best results when applied to disordered materials[31] and materials where the valence states were characterized by relatively flat bands and high effective masses, such as NiO.[82] The major departure of Eq. (3.1.4) from classical one-electron band theory is that the selection rule on **k** arising from the matrix element and Bloch wave functions has been relaxed—only conservation of energy remains as a strong selection rule. Hence the model associated with Eq. (3.1.4) was called the *nondirect model*. Spicer[32] suggested that in cases where the nondirect model might apply to crystalline solids, **k** could be conserved through many-body effects. Recent many-body calculations give some support to this point of view.[33] On the other hand, other investigators[26,30] have found experimental and theoretical evidence against the nondirect model as applied to crystalline materials. Although these more refined experiments have uncovered additional evidence for some direct transitions, there is often a large background of photoelectrons that have not been discussed in terms of direct transitions. In agreement with the theory of Doniach,[33] it appears that direct transitions may be providing much of the sharp structure, whereas nondirect transitions may be providing the rest of the unscattered electrons in the EDC's. However, more work is needed before this question is adequately resolved.

3.1.2.2. *Transport and Scattering Events*. The second process in the three-step model of photoemission is that in which the photoexcited electron moves through the solid to the surface. During this movement, the electron may be scattered. Two types of scattering processes must be considered: (1) strong inelastic scattering, in which the primary electron can lose a large fraction of the energy which it received from the photon, and (2) weak inelastic or elastic scattering, in which the electron energy change is small. An example of the former is electron–electron scattering, and an example of the latter is scattering from the lattice with phonon creation or destruction.

Let us first examine the electron–electron scattering event. This event has two important characteristics. First, the scattered electron loses a large fraction of its energy, and second, the probability of scattering is a strong function of the energy of the primary electron. In a nonmetal, the minimum photon energy threshold for electron–electron scattering is

---

[32] W. E. Spicer, *Phys. Rev.* **154**, 385 (1967).
[33] S. Doniach, *Phys. Rev.* B **2**, 3898 (1970).

twice the band gap; whereas, for a metal, any electron excited above the Fermi energy can suffer electron–electron scattering. However, in either case, the probability for electron–electron scattering is small near threshold and increases strongly with the energy of the primary electron. The electron–electron scattering length can usually be expressed in terms of a characteristic length $L(E)$[22]

$$P_{\rm s}(E, x) = e^{-x/L(E)\cos\theta}, \qquad (3.1.5)$$

where $P_{\rm s}(E, x)$ is the probability of an electron with energy E traveling a distance x at an angle θ to the surface normal without scattering. A plot of $L(E)$ experimentally determined for gold is presented in Fig. 5. As

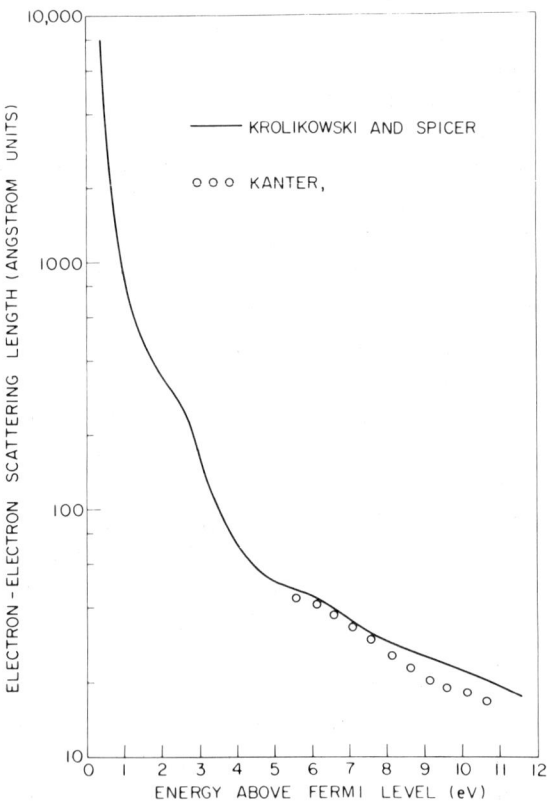

FIG. 5. Electron–electron scattering length in Au. The Krolikowski–Spicer curve is calculated on an approximate model. The points are from H. Kantor, *Phys. Rev. B* **1**, 522 (1970). [W. F. Krolikowski and W. E. Spicer, *Phys. Rev. B* **1**, 478 (1970); W. F. Krolikowski, Ph. D. Dissertation, Stanford Univ. (1967).]

can be seen, $L(E)$ decreases from $\sim 10^3$ Å at 1.0 eV to ~ 20 Å at 10 eV. Values of less than 10 Å have been obtained from some alkali metals for electrons 10 eV above the Fermi level. Consequently, where electron–electron scattering is strong, unscattered electrons with high energies are optically excited much closer to the surface than unscattered low energy electrons. By varying the photon energy, this effect could perhaps be used to detect changes in the electron structure of a solid as the surface is approached.

Secondary electrons resulting from electron–electron scattering events produce a peak in the EDC's near the low-energy cutoff, which is modulated by the conduction band density of states. Usually this peak remains near the low-energy cutoff and increases in strength as the photon energy is increased.[22] This characteristic helps greatly in identifying peaks in EDC's that are due to such scattered electrons.

In addition to the electron–electron scattering event, scattering with the resulting excitation of a plasma oscillation (plasmon) is strongly inelastic. Either bulk or surface plasmons may be excited. In either case, the energy of the primary electron is decreased by the energy of the plasma oscillation that is excited in the scattering event. Thus ordinarily electron–plasmon scattering leaves a well-defined signature in the EDC's in that the scattered photoelectrons are displaced to lower energy with respect to the primary electrons by an amount equal to the energy of the plasma oscillation. There are surprisingly few examples of well-established scattering due to plasma excitation.

If the mean free paths for the inelastic processes discussed above are sufficiently long, elastic or mostly elastic processes become important.[10] The process usually considered is optical phonon scattering. In this event a phonon is absorbed or emitted so that the electron's energy is changed by $\pm E_p$, where E_p is the phonon's energy. Normally, E_p is a small fraction of the primary electron's energy. Whereas the electron–electron and electron–plasmon scattering events are temperature-independent, electron–phonon scattering will be temperature-dependent. The probability of scattering where the electron loses energy and a phonon is created varies as

$$1 + 1/(e^{E_p/kT} - 1), \qquad (3.1.6)$$

whereas that for scattering with phonon absorption varies as

$$1/(e^{E_p/kT} - 1), \qquad (3.1.7)$$

which is the average number of phonons present with energy E_p at temperature T. Since E_p is small (in the range 0.01–0.05 eV), the principle effect of a single scattering event is to change the direction of motion of the excited electron. To examine this effect, consider the probability of emission from an arbitrarily thick sample where only phonon scattering is possible, and take the case where the probability of escape is large for any electron that reaches the surface. Were it not for phonon scattering, the maximum yield would be 50%, because on the average 50% of the electrons are excited with velocity away from the surface. However, phonon scattering events can change each electron's direction of motion so that those excited with velocity away from the surface can be "turned around" and may reach the surface. Thus, with phonon scattering, the maximum yield can approach 100%.[31] Such yields have been obtained for CsI.[35]

In the intermediate case where both phonon and electron–electron scattering must be considered, the electrons that are "turned around" by phonon scattering will travel a relatively long distance before escape and thus have an increased probability to suffer electron–electron scattering before escape. This case is particularly complex and is probably best treated by Monte Carlo computer calculations.[36]

3.1.2.3. Escape at the Surface. Perhaps less is known about the physics governing the escape of electrons at the surface of solids than about the optical excitation or transport and scattering events. Fortunately, this has a relatively small effect on much of the interpretation of the data. Normally one defines an escape function $T(E)$ that is a function only of the energy E of the electron. For most materials, it seems from experimental EDC's that $T(E)$ increases monotonically from zero at the vacuum level. The slope of $T(E)$ decreases with E until $T(E)$ approaches a constant value several electron volts above the threshold. A semiclassical equation based on the Sommerfeld free electron model of a metal can be derived:

$$T(E) = \begin{cases} 0, & E \leq W_0, \\ \tfrac{1}{2}[1 - (W_0/E)^{1/2}], & E \geq W_0. \end{cases} \qquad (3.1.8)$$

Here W_0 is the depth of the Sommerfeld potential well ($W_0 = E_f + \phi$,

---

[34] S. W. Duckett, *Phys. Rev.* **166**, 302 (1968).

[35] W. F. Krolikowski, Ph.D. Dissertation, Stanford Univ., Stanford Electron. Lab. Rep. #SU-SEL-67-039 (1967).

[36] R. N. Stuart and F. Wooten, *Phys. Rev.* **156**, 364 (1967).

where E_f is the Fermi energy and ϕ is the work function). Since it is difficult to derive the escape function for anything except the Sommerfeld model of a free electron metal, for materials with real band structures Eq. (3.1.8) is sometimes used with W_0 as an adjustable parameter.

Much research remains to be done concerning the escape of electrons at the surface of solids. Not only is it important to consider escape at clean surfaces of materials with real band structures, but escape at cesiated or even dirty surfaces is important when practical photoemitters are considered.

3.2. Photoelectric Quantum Yield

3.2.1. Definition and Measurement

The absolute quantum yield $Y(\hbar\omega)$ of a material at a given photon energy is defined as the ratio of the number of photoemitted electrons to the number of absorbed photons. Since measurements of the spectral distribution of quantum yield characterize the cathode efficiency, they are of great importance in the study of practical photoemitters. Yield measurements are also useful in normalizing photoelectron energy distribution curves. Fowler plots[5] of the spectral distribution of the yield can be used to determine the work function of metals.

The number of photoemitted electrons is easily measured by biasing the collecting electrode positive with respect to the emitter so that all electrons are collected. The positive voltage required depends on the geometry as well as the photon energy and contact potential between emitter and collector. To find the proper collecting voltage, one measures the anode current as a function of applied voltage. As the voltage is increased from zero, the current first increases sharply and then tends to saturate. The proper collector voltage is that just necessary and sufficient to reach the region of saturation. If more voltage is applied, the current will increase slowly due to the electric field lowering the potential barrier at the surface of the emitter.

The number of absorbed photons is determined by the incident photon flux, by the transmission of any window(s) and/or grids the light must go through $T_w(\hbar\omega)$, and by the reflectance of the sample $R(\hbar\omega)$. The window transmission and sample reflectance are determined from separate measurements. The incident photon flux can be measured with a known standard. The absolute quantum yield of the sample is then given by

$$Y(\text{sample}) = I(\text{sample})\, Y(\text{standard})/I(\text{standard})\, T_w(1-R), \quad (3.2.1)$$

where I(sample) and I(standard) are the measured photocurrents of the sample and the standard. Both the sample and standard must intercept the entire incident beam from the monochromator.

Thermopiles or thermocouples could be used as standards because they can be calibrated relatively easily and their sensitivity (in voltage per incident energy flux) is usually independent of wavelength. However, these devices are relatively insensitive and often clumsy to use. This is particularly true in the vacuum ultraviolet. Thus it is often advantageous to use a photoemission cell that has been properly calibrated.

A suitable reference phototube will have the following characteristics: (1) a compact geometry that allows insertion and removal from the monochromator beam, (2) adequate sensitivity in the spectral range of interest, and (3) a response that is constant in time. A reference phototube employing a Cs_3Sb emitter in a glass cell with a LiF window serves quite satisfactorily. A thin Pt disk is used as the emitter substrate and the silver expansion ring used in the LiF–glass seal acts as the collector. The reference phototube is mounted in a brass ring which fits between the monochromator exit flange and the high vacuum photoemission chamber. The fabrication of this particular type of reference tube has been described in some detail by Blodgett.[37]

3.2.2. Calibration of Reference Phototube

Any reference cell of suitable size, sensitivity, and stability can be used for yield measurements if it is properly calibrated. However, considerable care must be taken in determining the calibration. It is often convenient to determine the relative spectral response separately from the absolute response of the cell.

The relative spectral response of the reference tube above approximately 4 eV can be determined by comparing its response to the response of a freshly deposited sodium salicylate film. Sodium salicylate absorbs ultraviolet photons and reemits light at 2.9 eV in a band with full width at half maximum of approximately 0.5 eV.[38] Sodium salicylate has been reported to have a constant quantum efficiency independent of wavelength from 900 to 3600 Å.[38–40] However, decreases in quantum efficiency at

---

[37] A. J. Blodgett, Ph.D. Dissertation, Stanford Univ. (1965).
[38] R. Allison, J. Burns, and A. J. Tuzzolino, *J. Opt. Soc. Amer.* **54**, 747 (1964).
[39] K. Watanabe and E. C. Y. Inn, *J. Opt. Soc. Amer.* **43**, 32 (1953).
[40] J. F. Harmmann, *Z. Angew Phys.* **10**, 187 (1958).

6.2 eV and higher photon energies have also been reported.[41-44] Despite these variations in spectral response, sodium salicylate films are extremely useful for calibration in the vacuum ultraviolet if proper care is taken.

The response of sodium salicylate films is critically dependent on the environment in which they are stored, so fresh films must be used.[44,45] Uniform films can be prepared by using an atomizer to spray a solution of sodium salicylate (80 g in one liter absolute methyl alcohol) onto a clean glass substrate.[46] The optimum film thickness is 2–4 mg/cm².[38] Very thin films do not absorb all the ultraviolet radiation and in very thick films reabsorption of the fluorescent emission by the phosphor causes a decrease in efficiency.

In calibrating the response of the reference phototube, any detector can be used to measure the sodium salicylate response. Where possible it is advantageous to measure the sodium salicylate response with the reference tube itself by moving the sodium salicylate in and out of the light beam in front of the reference tube. This eliminates the need for an additional photomultiplier, the same photon flux is incident on the tube being calibrated and the calibrating phosphor, and the same ammeter is used for all current measurements.[47] The sodium salicylate is sprayed onto a narrow band-pass optical filter (such as Corning Glass Filter No. 5-60) to prevent any ultraviolet radiation that may pass through the phosphor from being detected by the reference cell.

In the visible and near uv, where intense light sources are available, the absolute spectral response of the reference tube can be measured with a calibrated thermopile. The relative response as determined from the sodium salicylate measurements can be fixed by comparing to the absolute measurements in the 4.0- to 5.0-eV range, where the two methods overlap. Absolute measurements with the thermopile in the vacuum uv are possible but much more difficult because the light sources are much weaker than in the visible region. The best method of measuring absolute intensities in the uv makes use of gas photoionization properties.[44]

---

[41] N. Kristianpoller and R. A. Knapp, *Appl. Opt.* **3**, 915 (1964).

[42] E. C. Bruner, Jr., *J. Opt. Soc. Amer.* **59**, 204 (1969).

[43] J. A. R. Samson, *J. Opt. Soc. Amer.* **54**, 6 (1964).

[44] J. A. R. Samson, "Techniques of Vacuum Ultraviolet Spectroscopy." Wiley, New York, 1967.

[45] R. Koyama, Ph.D. Dissertation, Stanford Univ. (1969).

[46] R. A. Knapp, *Appl. Opt.* **2**, 1334 (1963).

[47] R. Bauer, Ph.D. Dissertation, Stanford Univ. (1970).

A relatively easy calibration can be made on the intense Lyman α line at 10.2 eV with a nitric oxide ionization chamber.†

3.3. Measurement of Energy Distribution Curves

3.3.1. The Retarding Field Analyzer

3.3.1.1. Practical Analyzer. In addition to measurement of the quantum yield, the measurement of the energy distribution of photoemitted electrons is central to the photoemission experiment. Most previous photoemission measurements have utilized retarding field energy analyzers. Practical analyzers, of course, differ from the ideal geometry of a point emitter in a spherically symmetric retarding field. Examples of some photoemission analyzers are shown in Fig. 6.

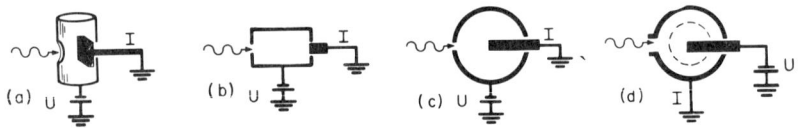

FIG. 6. Examples of practical photoemission analyzers: (a) cylindrical collector, (b) Faraday cage type, (c) spherical collector, and (d) screened emitter analyzer.

The easily fabricated cylindrical geometry of Fig. 6a has been widely used since the early 1950's.[48,49] Both the cylindrical analyzer and the Faraday cage analyzer[50,51] shown in Fig. 6b give surprisingly good results, in spite of their apparent departure from spherical symmetry.[52] The spherical collectors of Figs. 6c and d are also easily constructed from spun hemispheres. The screen electrode added to the analyzer in Fig. 6d improves the spherical symmetry as discussed later in this section. Both of these spherical analyzers are approximations to the spherical capacitor analyzer, the resolution of which is amenable to analysis.

[48] L. Apker, E. Taft, and J. Dickey, *J. Opt. Soc. Amer.* **43**, 78 (1952).
[49] W. E. Spicer, *Phys. Rev.* **125**, 1297 (1962).
[50] F. G. Allen and G. W. Gobeli, *Phys. Rev.* **144**, 558 (1966).
[51] T. A. Calcott, *Phys. Rev.* **161**, 746 (1967).
[52] T. H. Di Stefano and D. T. Pierce, *Rev. Sci. Instrum.* **41**, 180 (1970).

† Available commercially from Melpar, 7700 Arlington Blvd., Falls Church, Virginia.

3.3.1.2. Resolution.
The retarding field actually measures the electron momentum parallel to the field lines. The difference between the total kinetic energy of the electron and the component of kinetic energy along the retarding field is a measure of the imperfect resolution of the analyzer. A comprehensive discussion of the energy resolution of the photoemission analyzer has been given by DiStefano and Pierce,[52] and of retarding field analyzers in general by Simpson.[53]

Some major factors which limit the resolution of a practical retarding field analyzer are (1) departure from perfect spherical geometry, (2) nonuniform work function on the collector (patch effect), (3) nonuniform work function on the emitter (emitter fringing fields), (4) reverse photoemission, (5) residual magnetic fields, and (6) electronics used for measurement of EDC's. The latter factor will be discussed in Section 3.3.2. The others are discussed below.

The resolution of an ideal spherical capacitor analyzer with an emitter sphere of radius a and a collector of radius b has a theoretical maximum energy error of[54]

$$\Delta E = (a/b)^2 E_0, \qquad (3.3.1)$$

where E_0 is the initial kinetic energy. The energy error ΔE is the tangential kinetic energy of the electron at the point of collection. The retarding potential just sufficient to repel an electron at energy E_0 is not E_0/e but $(E_0 - \Delta E)/e$. Note that the energy error due to the finite emitter radius is greatest for high-energy electrons and zero for electrons with zero kinetic energy.

The chief deviations of a practical photoemission analyzer from the ideal spherical capacitor are the flat emitter surface, the emitter support structure, and the light aperture. Some of these departures from sphericity have been investigated using an electrolytic tank analog. Ganichev and Umkin[55] found that for a support rod with half the diameter of the emitter disk the energy resolution is $\Delta E/E = 3\%$. Without the emitter support the error $\Delta E/E$ was only 1%. The ratio of the disk diameter to the sphere diameter in each case was 1 to 10. Effects of the light aperture can be minimized by placing a fine-wire, coarse-mesh screen grid at collector potential over the light aperture, or by providing a cylindrical extension of the aperture (see Fig. 6d).

---

[53] J. A. Simpson, *Rev. Sci. Instrum.* **32**, 1283 (1961).
[54] P. Lukirski, *Z. Phys.* **22**, 351 (1924).
[55] D. A. Ganichev and K. C. Umkin, *Sov. Phys. Solid State* **1**, 590 (1958).

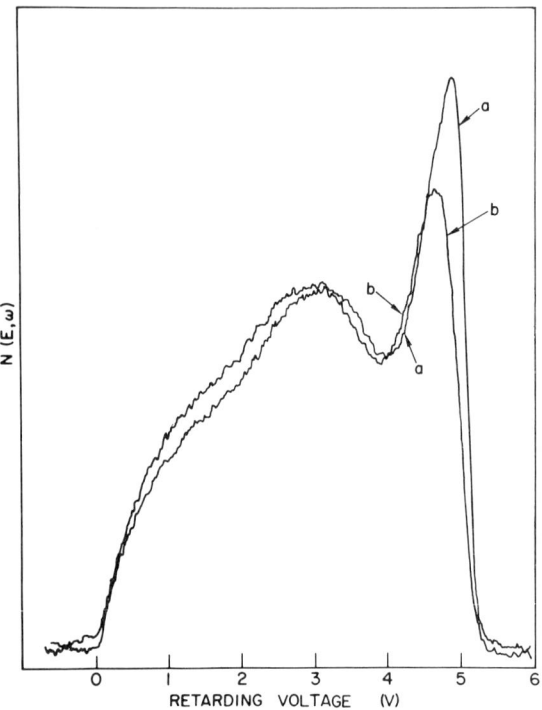

FIG. 7. The effect of analyzer resolution on EDC's. EDC's from Ni with $\hbar\omega = 10.2$ eV obtained with (a) high-resolution screened emitter analyzer and (b) simple spherical collector show how lower resolution typically causes structure to be broadened and shifted to lower energy. [D. T. Pierce, Ph. D. Dissertation, Stanford Univ. (1970).]

The resolution of the analyzer of Fig. 6c can be improved by surrounding the flat emitter with a fine spherical screen as shown in Fig. 6d. The screen at emitter potential produces a field-free region between the emitter and screen. The electrons are retarded between the screen and collector and the collected photocurrent is measured. Hence the screen restores spherical symmetry to the retarding field for any size or shape emitter, provided the work functions in the vicinity of the emitter are sufficiently uniform for the drift-free region to exist. The construction and operation of a prototype screened emitter analyzer has been described elsewhere.[56] An example of the improved resolution is seen in Fig. 7. EDC's with poorer resolution exhibit structure that is broadened and shifted to lower energy.

[56] D. T. Pierce and T. H. DiStefano, *Rev. Sci. Instrum.* **41**, 1740 (1970).

All the advantages of good design can be wiped out if the work function of the collector is not sufficiently uniform. This is the case since the collector work function provides the "zero of potential" to which the electron energy is referred. Contaminated surfaces usually have wide variations in work function. It has been found that evaporating a clean metal (usually, Au or Cu) on the collector *in situ* just before the atomically clean emitter surface is prepared gives considerable improvement in resolution.

If the work function of the sample holder or the sides of the sample is higher than the sample work function, fringing emitter fields produce a potential barrier in front of the emitter which distorts the *low-energy* part of an EDC.[52] Figure 8a illustrates the potential barrier at different retarding voltages. The collector must be made quite positive to collect all the low energy electrons. The low-energy portion of the EDC is

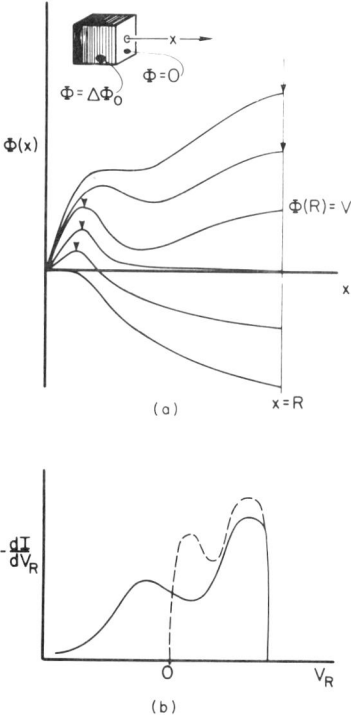

FIG. 8. (a) Retarding potentials along a line normal to the emitting surface for various collector voltages. The emitter is a cube with a lower work function on its face than on its sides. (b) Distortion of a true EDC (dashed) caused by fringing emitter fields. [T. H. Di Stefano and D. T. Pierce, *Rev. Sci. Instrum.* **41**, 180 (1970).]

shifted and broadened as in Fig. 8b. Nonuniformity of work function on the actual emitting surface also distorts the low-energy part of an EDC. Insulating materials should not be used inside the collector because of perturbing fields produced by charged insulator surfaces.

Reverse photoemission is photoemission from the collector caused by scattered light. A specular sample surface and careful alignment minimize the problem of scattered light onto the collector. Typically, reverse photoemission produces a shoulder around $V_R = 0$ in the measured EDC.[52] The tail of the reverse photoemission component extends through the EDC and above the normal high-energy cutoff, since the geometry for reverse photoemission is far from ideal. The amount of reverse photocurrent is measured by increasing V_R sufficiently so that all electrons photoemitted from the collector are collected by the emitter. Ratios of forward photocurrent to reverse photocurrent (front-to-back ratio) are typically at least 10:1 for metallic samples and over 100:1 for samples with yields much larger than the yield of the collector surface.

Equation (3.1.1) has been extended to include the effect of a residual magnetic field.[52] The residual magnetic field causes an error proportional to the square of the collector radius and the square of the magnetic field. The error is independent of the variation of retarding potential $\Phi(r)$ with r and independent of the electron kinetic energy to the first order. Depending on the resolution required, external magnetic coils or mu-metal shielding may be necessary to reduce fields in the collector region to an acceptable level.

The sum of these contributions to resolution error tends to broaden structure in an EDC and shift it to lower energy. An important concept in considering resolution error is that of the broadening (or resolution) function of a given energy analyzer. The broadening function is equivalent to the EDC that would be measured if the true distribution were a delta function of monoenergetic electrons as shown schematically in Fig. 9a. Only a small fraction of the electrons emitted at an energy V_0 are measured at that energy, while most are measured at lower energies. The shape of the broadening function can only be guessed. Its width can be estimated from the total of the contributions to the resolution errors. A measure of the width of the broadening function can also be obtained from experimental EDC's of a metal where, in the absence of broadening, the high-energy cutoff is determined by the Fermi distribution, as shown schematically in Fig. 9b. As indicated in this figure, the width of the broadening function is expected [from Eq. (3.3.1)] to vary throughout

3.3. MEASUREMENT OF ENERGY DISTRIBUTION CURVES

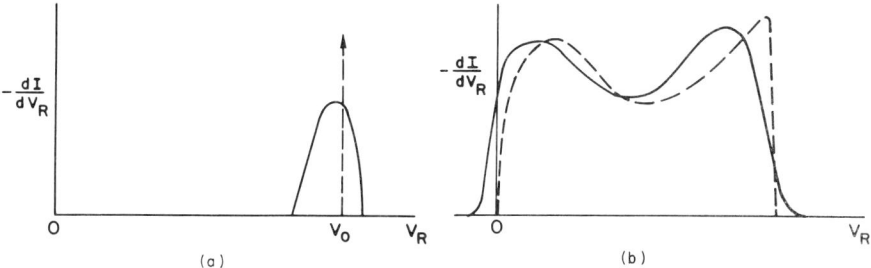

FIG. 9. (a) Schematic representation of the resolution function. A delta function distribution of monoenergetic electrons (arrow) would give a measured distribution equivalent to the resolution function. (b) The effect of the resolution function of (a) on the true EDC (dashed) is shown schematically by the full curve.

the EDC. In analyzing experimental EDC's, corrections should be made for the effect of the broadening function.

3.3.1.3. Design Considerations. In choosing the size and type of retarding field analyzer and integrating it with the vacuum chamber–monochromator system, some important points should be remembered. A spherical collector with a small emitter is desirable. The minimum emitter size is determined by the size of the light spot from the monochromator. Since, in general, the monochromator exit beam diverges, the spot size is reduced by constructing the experimental chamber so the analyzer is as near to the monochromator exit slits as possible. Alternatively, a mirror can be used to image the exit slits on the sample. The reflected light from the sample should return through the collector light aperture in order to avoid reverse photoemission from the collector. As the collector diameter is increased, ambient magnetic fields must be more carefully compensated using Helmholz coils. In situations where one is constrained to use a large emitter, it would be worthwhile to consider using a screened emitter analyzer.

3.3.2. Electronics for the Retarding Field Analyzer

3.3.2.1. Principle. The kinetic energy distribution of electrons photoemitted from the sample is obtained by varying the retarding voltage V_R applied to the collector and taking the derivative of the current voltage characteristic which is generated. The negative derivative $-dI/dV_R$ gives the number of electrons per unit energy range which is the energy distribution desired.

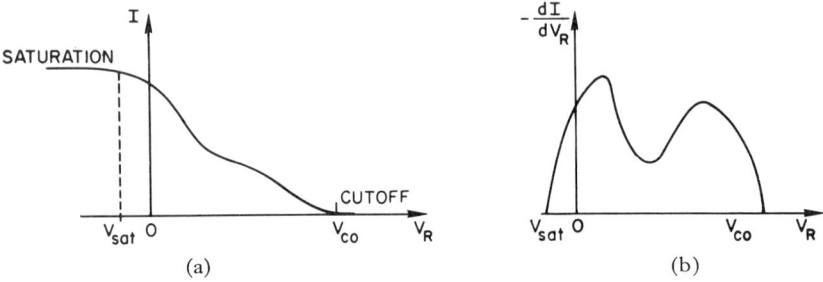

FIG. 10. The current–voltage curve (a) is differentiated (b) to give the energy distribution curve. Note the corresponding saturation V_{sat} and cutoff V_{co} voltages in (a) and (b).

As shown in Fig. 10, when a sufficiently positive voltage is applied to the collector (negative V_R, $V_R \leq V_{sat}$), all the photoemitted electrons are collected and the photodiode is in saturation. As the retarding potential is increased, fewer electrons are collected until, at cutoff ($V_R = V_{co}$), all photoelectrons are retarded.

Various parameters of the emitter and collector can be determined from the photoemission EDC's. Referring to Fig. 11, when the photodiode is just in saturation, the retarding voltage is equal to the negative of the contact potential between the collector and emitter V_{cpo}:

$$V_{sat} = -V_{cpo} = -(\phi_C - \phi_E), \quad (3.3.2)$$

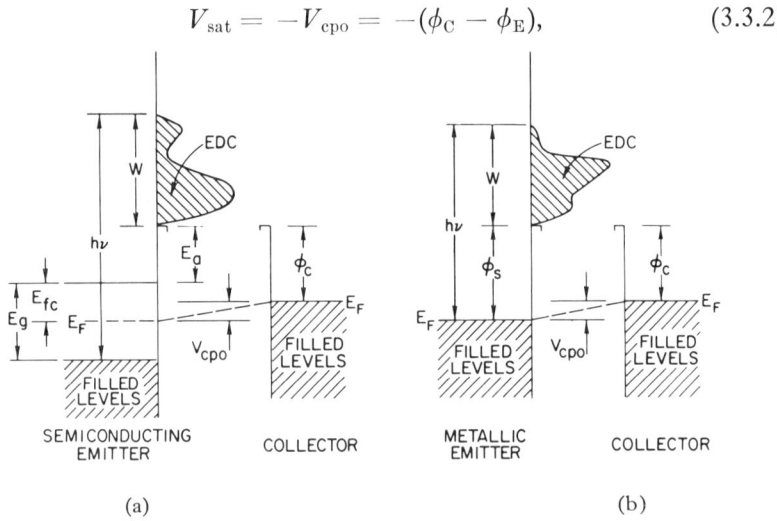

FIG. 11. (a) Energy level diagram for a semiconducting emitter (assuming no band bending) and (b) for a metallic emitter. The emitter and collector are shown just in saturation bias $V_R = V_{sat}$.

3.3. MEASUREMENT OF ENERGY DISTRIBUTION CURVES

where ϕ_C and ϕ_E are the work functions of the collector and emitter, respectively.[†] The cutoff voltage V_{co} is related to the photon energy $\hbar\omega$, the work function of the collector, and the energy separation between the Fermi level and filled valence levels of the emitter. For the case of a metallic emitter,

$$V_{co} = \hbar\omega - \phi_C, \quad (3.3.3)$$

and hence the work function of the collector can be inferred from the high-energy cutoff of the EDC. For the case of a semiconducting emitter, assuming no band bending near the surface of the sample,

$$V_{co} = W - V_{cpo} = \hbar\omega - E_a - E_g - V_{cpo}, \quad (3.3.4)$$

where E_a is the electron affinity of the emitter (the energy difference between the vacuum level and bottom of the conduction band), E_g is the energy gap, and W is the total width of the EDC. Since E_a can be determined from the photon energy, width of the EDC, and energy gap ($E_a = \hbar\omega - W - E_g$), the position of the Fermi level below the conduction band E_{fc} can be determined if the collector work function is measured independently:

$$E_{fc} = \phi_E - E_a = \phi_C - V_{cpo} - E_a. \quad (3.3.5)$$

Consequently, if semiconducting emitters are to be studied, it is worthwhile to include some provision for measurement of the work function of the collector.

Band bending near the surface of a semiconducting emitter can cause some difficulty in the determination of the parameters associated with a semiconducting emitter. If the band-bending region is large compared to the escape depth of the electrons, the value of E_{fc} obtained will apply to the bands at the surface. If the band-bending region is small compared to the escape depth of the electrons, the value of E_{fc} obtained will apply to the bands in the bulk. If the band-bending region is comparable to the escape depth, E_{fc} might appear to be a function of the photon energy and will not be well defined.

The x and y axes of the EDC's are given in terms of volts (retarding voltage or electron energy) and electrons per photon per electron volt number of electrons per unit energy range), respectively. The area under

---

[†] In the remainder of this chapter, voltage will be considered a measure of energy in electron volts.

the EDC has units of electrons per photon and is usually normalized to the absolute photoelectric quantum yield (see Section 3.2). If the yield has been corrected for the reflectivity of the sample [see Eq. (3.2.1)], the units are in terms of absorbed photons; if the yield has not been corrected for reflectivity, the units are in terms of incident photons. The quantity directly measured on the x axis in the experiment is retarding voltage. This can be converted to electron kinetic energy E_e using the relation

$$E_e = V_R - \phi_C + \phi_E = V_R - V_{cpo}. \tag{3.3.6}$$

For display and interpretation, several EDC's may be plotted on the same graph with photon energy as a parameter. One can plot the EDC's using final state energy with respect to the valence band maximum E as the abscissa. On this plot, structure in the EDC's associated with structure in the final density of states remains stationary as $\hbar\omega$ is varied. The following relations are useful:

$$E = V_R + \phi_C \qquad \text{(metallic emitter)}, \tag{3.3.7}$$

$$E = V_R + \phi_C + (E_g - E_{fc}) \qquad \text{(semiconducting emitter)}. \tag{3.3.8}$$

It is often useful to plot the EDC's with "initial" state energy $E - \hbar\omega$ as the abscissa so that structure in the EDC's associated with the initial density of states remains stationary as $\hbar\omega$ is varied.

For many years, EDC's were obtained by measuring an I–V curve and differentiating it graphically, a rather tedious practice which is fraught with experimental difficulties.[1] Discussed here are two direct methods of obtaining the EDC electronically. The first is an ac method in which the ac conductance of the photodiode (energy analyzer) is measured as a function of retarding voltage. The second is a time differentiation method in which the output of an electrometer that measures the photocurrent is differentiated using an analog differentiator as the retarding voltage is swept linearly in time.

3.3.2.2. ac Method for Obtaining EDC. In the ac method used for taking the EDC,[57] a small ac voltage v_{ac} (typically less than 0.2-V peak-to-peak) is superimposed on the retarding voltage. The resulting in-phase ac component of the photocurrent i_{ac} is proportional to $-dI/dV_R$ as illustrated in Fig. 12. The ac component of the photocurrent is meas-

---

[57] W. E. Spicer and C. N. Berglund, *Rev. Sci. Instrum.* **35**, 1685 (1964).

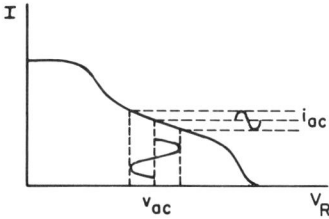

FIG. 12. ac differentiation of the $I - V_R$ curves is obtained by superimposing a small ac voltage V_{ac} on V_R. The derivative is given by the resulting in phase component of the ac current and is proportional to the EDC.

ured using phase-sensitive detection techniques. The ac voltage causes a capacitive current through the energy analyzer, which can be much larger than the desired in-phase component of current. It is desirable to cancel the capacitive current in order to reduce noise and prevent overloading of the sensitive meter ranges. This can be accomplished with the current bridge circuit of Fig. 13.[58]

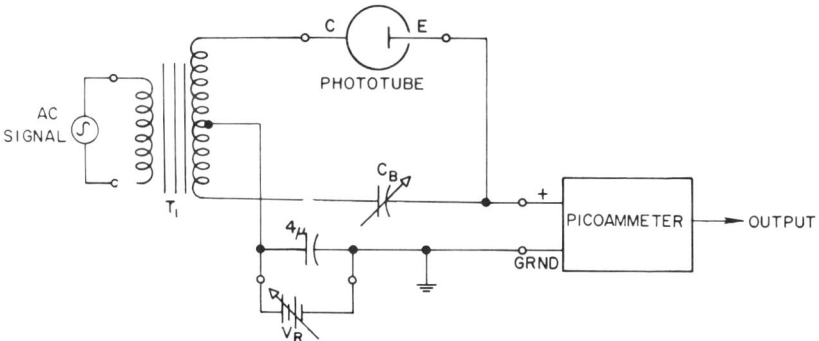

FIG. 13. Capacitive balance circuit for EDC measurement featuring the use of a single-ended input detector. The use of transformer phase inversion allows capacitive current rejections of the order of 70 dB. [R. C. Eden, *Rev. Sci. Instrum.* **41**, 252 (1970); R. C. Eden, Ph. D. Dissertation, Stanford Univ. (1967).]

A center-tapped secondary transformer is used for superimposing v_{ac} on V_R. With C_B adjusted equal to the photodiode capacitance, the current arriving at the picoammeter through C_B will be out of phase with and just cancel the capacitive current through the photodiode.

[58] R. C. Eden, *Rev. Sci. Instrum.* **41**, 252 (1970).

A block diagram of the electronics for the ac method of EDC measurement is shown in Fig. 14. The capacitive balance circuit and energy analyzer are in a shielded enclosure to minimize noise. The total photocurrent is amplified by the picoammeter and the ac component is detected by the lock-in amplifier. The subharmonic generator[59] provides an even subharmonic of the line frequency (usually the fourth or eighth), which is applied to both the photodiode and the reference frequency input to

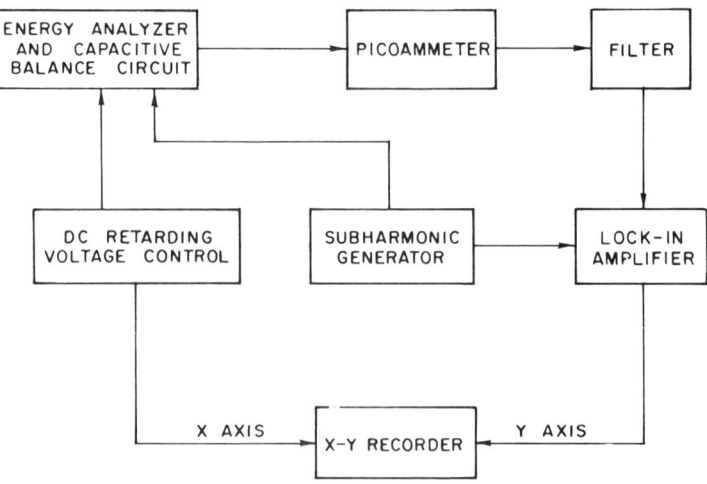

FIG. 14. Block diagram of EDC measurement electronics for ac method.

the lock-in amplifier. The use of an exact subharmonic minimizes noise at the line frequency. In the absence of the subharmonic generator, the ac voltage is generated internally by the lock-in amplifier, and a filter that removes 60- and 120-Hz noise is inserted between the picoammeter and lock-in amplifier. The output of the lock-in amplifier drives the y axis of an X–Y recorder, and the retarding voltage drives the x axis. Great care must be taken to eliminate noise due to pickup and microphonics if the system performance is to approach the theoretical noise limits of the Johnson noise in the picoammeter feedback resistor and the shot noise of the energy analyzer.

The electronics used with the screened emitter analyzer of Fig. 6d differ only slightly from Figs. 13 and 14. The collector current is measured instead of emitter current. The screen is connected to the emitter

[59] T. H. DiStefano, Ph.D. Dissertation, Stanford Univ. (1970).

through a small bias battery to compensate contact potential differences. The ac voltage is applied to the emitter but not the screen. The details of the circuits used with the screened emitter analyzer are discussed elsewhere.[56]

It is important to ensure that the resolution of the EDC as measured by the energy analyzer is not unnecessarily decreased by the measurement electronics. In particular, the scan rate of the retarding potential must be appropriate for the system response time. Also the ac peak-to-peak voltage must be as small as possible so as not to broaden structure in the EDC, and thus degrade the resolution. The decrease in resolution caused by the ac voltage can be deduced from Fig. 15, in which an artifically generated I–V curve has been differentiated by the electronics of Fig. 14. Ultimately the tradeoff is between signal-to-noise ratio and resolution.

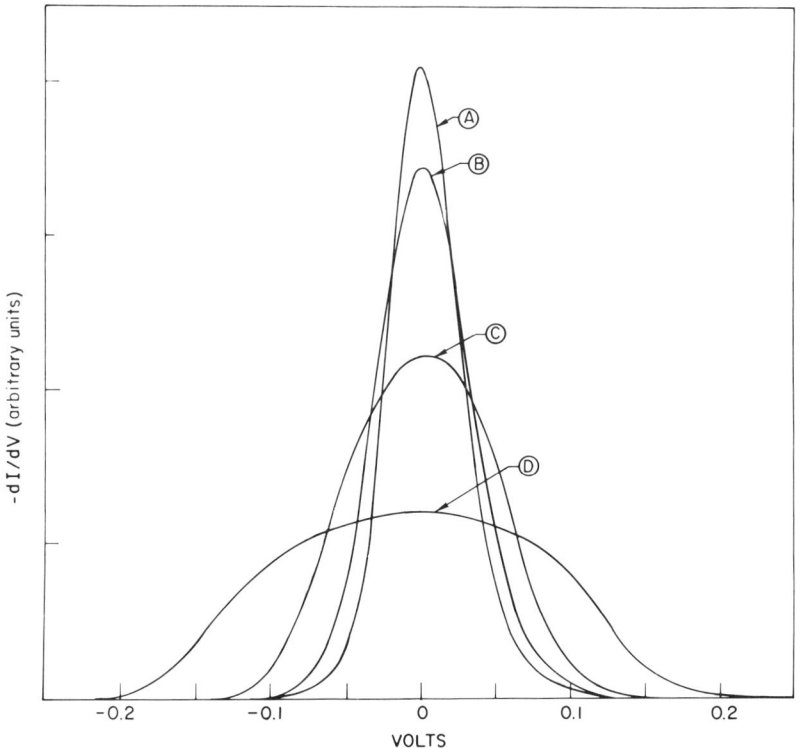

FIG. 15. The effect of ac peak-to-peak voltage on the resolution of a peak. (A): 0.005, 0.010, 0.020; (B) 0.050; (C) 0.100; (D) 0.200. [D. T. Pierce, Ph. D. Dissertation, Stanford Univ. (1970).]

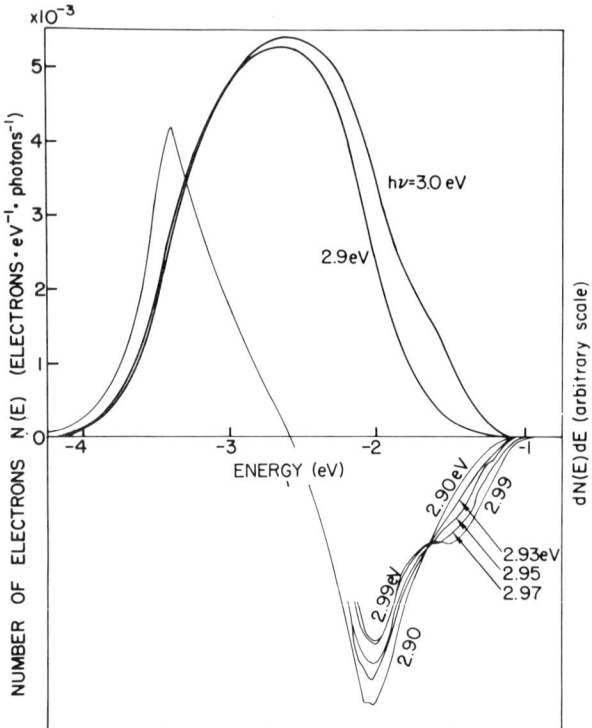

FIG. 16. Location of weak structure using the derivative of the EDC. In this example, the change in curvature of the leading edge associated with the Γ'_{15} threshold is first observed at $hz = 2.93$ eV. The sample is a p-type Ge crystal. [T. M. Donovan, J. E. Fischer, J. Matsuzaki, W. E. Spicer, *Phys. Rev. B* **3**, 4292 (1971).]

When two peaks in an EDC are close together in energy, the weaker peak often appears only as a shoulder. The difficulty in locating such weak structure in an EDC is overcome if the derivative of the EDC is measured.[60] The second harmonic of the ac photocurrent is proportional to the second derivative of the *I–V* curve or the derivative of the EDC. The second derivative is measured by synchronously detecting the second harmonic of the photocurrent with the lock-in amplifier. The advantages of the derivative for locating structure are illustrated in Fig. 16, which shows the determination of the Γ'_{15} threshold of Ge.

Similarly, second derivatives of EDC's have been measured by synchronously detecting the third harmonic of the ac photocurrent.[61] In

[60] L. W. James, R. C. Eden, J. L. Moll, and W. E. Spicer, *Phys. Rev.* **174**, 909 (1968).
[61] N. V. Smith and M. M. Traum, *Phys. Rev. Lett.* **25**, 1017 (1970).

this case, regions of negative curvature (i.e., peaks and shoulders) in an EDC appear as peaks in the second derivative curve.

3.3.2.3. Time Differentiation Method of Obtaining EDC.

An alternate method of obtaining an EDC is by direct time differentiation of the photocurrent as the retarding voltage is swept linearly in time.[62] A block diagram of the electronics for the time differentiation method of EDC measurement is shown in Fig. 17. The linear sweep (ramp generator)

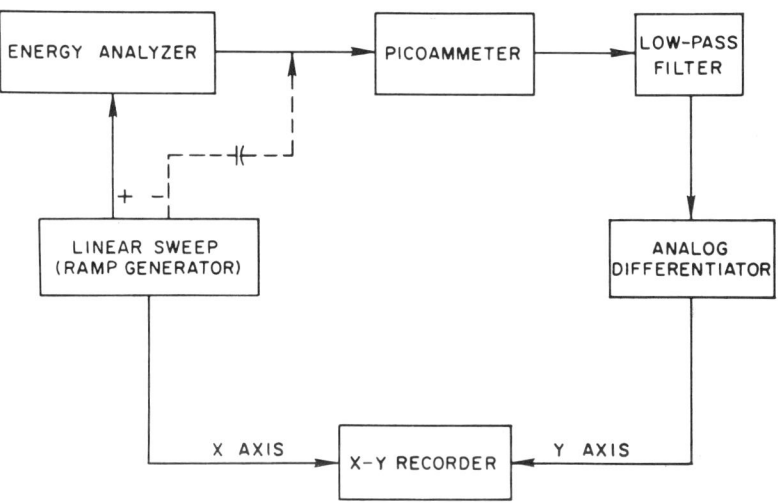

FIG. 17. Block diagram of EDC measurement electronics for the time differentiation method.

provides a retarding voltage V_R that is directly proportional to time, i.e., $V_R = at +$ constant, where a is some constant related to the sweep rate. The EDC is then given by

$$-\frac{dI}{dV_R} = -\frac{dI}{dt}\frac{dt}{dV_R} = -\frac{1}{a}\frac{dI}{dt}. \quad (3.3.9)$$

The photocurrent is amplified by the picoammeter and high-frequency noise is reduced by a sharp cutoff filter with a nearly linear phase characteristic. A good filter is a necessity, since the time differentiation method is very sensitive to high-frequency noise. An analog differentiator constructed from an operational amplifier takes the required time derivative and provides an output proportional to dI/dV_R.

[62] F. G. Allen and G. W. Gobeli, *Phys. Rev.* **144**, 558 (1966).

When using the time differentiation method, resolution is increased by decreasing the retarding voltage scan rate. Signal-to-noise ratio can then be increased by decreasing the bandpass of the low-pass filter. The limitation here is on the noise at the output of the differentiator and the drift associated with the electrometer and light source. For fast scan rates, the displacement current of the photodiode must be canceled. This can be accomplished by applying a voltage $-dV/dt$ to a capacitor in parallel with and with the same capacitance as the photodiode as shown in Fig. 17. Note that the analog differentiator eliminates the effects of small constant displacement currents.

Provided the low-pass filter has sufficiently linear phase and constant amplitude characteristics, the output of the filter will be a time-delayed replica of the input. This causes the recorded EDC to be shifted with respect to the retarding voltage, i.e., the true EDC is $N(V_R - \Delta V_R)$, where ΔV_R is related to the time delay through the filter and the sweep rate. This shift is negligible for short time constants, but must be corrected for when long time constants (long compared to the desired resolution divided by the scan rate) are used.

If first or second derivatives of EDC's are desired, additional analog differentiators are required on the output. Noise problems increase with each additional differentiation and extreme care must be taken to maximize the signal-to-noise ratio.

The major disadvantage of the time differentiation approach as compared to the ac method of EDC measurement is that the desired gain of the system cannot be set without sweeping the retarding voltage. In addition, the recorded height of the structure in the EDC is a function of the scan rate and it is often necessary to take a preliminary scan of the EDC in order to adjust the gain at the output of the analog differentiator. In contrast, in the ac method, the retarding voltage can be adjusted to a peak in the EDC, and the height of the curve set by varying the ac gain in the system.

Error analyses have been made of both the ac[58] and time differentiation methods[63] of EDC measurement. An advantage of the time differentiation method is that some of these errors can be corrected using additional simple electronic circuitry. Further advantages are that a lock-in amplifier is not required for the time differentiation circuitry, and the time differentiation method does not suffer from a decrease in resolution due to a finite ac signal. However, the linearity of the retarding voltage sweep is critical in the time differentiation method.

---

[63] K. A. Kress and G. J. LaPeyre, *Rev. Sci. Instrum.* **40**, 74 (1969).

3.3.3. Other Types of Energy Analyzers

A simple retarding field energy analyzer has served satisfactorily for most photoemission measurements at photon energies below the LiF window cutoff at 12 eV. For measurements of photoelectrons excited by higher photon energies and for angular distribution measurements, electrostatic and magnetic electron spectrometers, which have been widely studied,[64,65] offer certain advantages. Such analyzers generally can be built for higher resolution and accept a smaller, well-defined, solid angle of electrons than simple retarding field analyzers.

Electron multipliers and counting techniques can be used in detecting the electrons in the narrow energy range selected by such analyzers, making these analyzers particularly suited for electron energy distribution measurement when the total photocurrent is very small (less than 10^{-12} A). Eastman,[66] for example, has used a 90° cylindrical electrostatic analyzer in photoemission measurements at high photon energies (16–41 eV) with satisfactory results, although secondary emission from the analyzer prevented meaningful measurements of electron energies less than 4 eV.

Several types of magnetic and electrostatic spectrometers have been used in X-ray photoelectron spectroscopy[9] and could be applied to photoemission measurements in the ultraviolet and extreme ultraviolet range. Of the magnetic spectrometers, the semicircular uniform field type[67] is one of the simplest, but has the disadvantage of only focusing in a plane. A double focusing magnetic spectrometer introduced by Siegbahn and Svartholm[68] focuses in three dimensions and has been used in many variations. Of the electrostatic spectrometers, focusing in a plane, as in the uniform field magnetic spectrometer, can be obtained by a cylindrical condenser that subtends an angle of $\pi/2^{1/2}$.[69] The spherical condenser spectrometer described by Purcell[70] is an extremely versatile

---

[64] T. R. Gerholm, in "Methods of Experimental Physics," Vol. 5A, Sec. 2.2.1, p. 341. Academic Press, New York, 1961.

[65] K. Siegbahn, "Alpha-, Beta-, and Gamma-Ray Spectroscopy" (K. Siegbahn, ed.), Chapter 3. North-Holland Publ., Amsterdam, 1965.

[66] D. E. Eastman and J. K. Cashion, *Phys. Rev. Lett.* **24**, 310 (1970).

[67] J. Danysz, *Radium* **9**, 1 (1912); **10**, 4 (1913).

[68] K. Siegbahn and N. Svartholm, *Nature (London)* **157**, 872 (1946).

[69] A. L. Hughes and V. Rojansky, *Phys. Rev.* **34**, 284 (1929).

[70] E. M. Purcell, *Phys. Rev.* **54**, 818 (1938). A very good discussion of the practical application of this analyzer to photoemission has been given by I. Lindau and S. B. M. Hagstrom, *J. Phys.* **E4**, 936 (1971).

electrostatic spectrometer of the double focusing type. The cylindrical mirror analyzer[71] is well suited to photoemission experiments because it offers high resolution at a larger acceptance angle than the spherical condenser spectrometer.[72]

3.3.4. Angular EDC Measurements

Kane,[73] Mahan,[74] and Smith[26] have pointed out the desirability of measuring EDC's as a function of the direction of electron emission (angular EDC) in order to determine characteristics of band structures. When measuring angular EDC's, the collected current must be measured. There are three general techniques which can be used to obtain angular EDC's: (1) a movable probe at collector potential can be used, (2) a retarding field analyzer can be segmented and the collected current intercepted by each segment can be measured, and (3) an electrostatic or magnetic deflection analyzer can be used.

For the movable probe technique, a mesh can be used for the collector and the probe can be located just outside of the mesh and collect those photoelectrons which have enough energy to overcome the retarding potential and pass through the mesh. Or a solid collector can be used with the movable probe just skimming inside the collector. While the movable probe technique provides large flexibility, the segmented collector technique eliminates the geometrical difficulties associated with constructing a probe that can be moved over a spherical surface. Because of the equivalence of several directions of electron emission when single crystals are studied, a few of the segments of the segmented collector can be electronically connected, thus providing substantial increases in the signal-to-noise ratio. More than one movable probe would have to be used to obtain the same result when using the movable probe technique.

Electrostatic or magnetic deflection analyzers used in conjunction with electron counting techniques might perhaps provide the best signal-to-noise ratio when measuring angular EDC's. However, the analyzer must have the capability of being rotated with respect to the sample, and this could be a serious problem in designing the high-vacuum chamber.

Angular photoemission studies have been reported as early as 1928[75]

---

[71] E. Blauth, *Z. Phys.* **147**, 228 (1951).
[72] H. Hafner, J. A. Simpson, and C. E. Kugatt, *Rev. Sci. Instrum.* **39**, 33 (1968).
[73] E. O. Kane, *Phys. Rev. Lett.* **12**, 17 (1964).
[74] G. D. Mahan, *Phys. Rev. Lett.* **24**, 1068 (1970).
[75] H. E. Ives, A. R. Olpin, and A. L. Johnsrud, *Phys. Rev.* **32**, 57 (1928).

using the movable probe and segmented collector techniques. More recently, Gobeli et al.[76] have measured photoelectric yield as a function of electron emission using a variation of the segmented collector. A cylindrical screen mesh served as the collector and six symmetrically spaced rigid collecting probes enabled the angular photoelectric yield to be obtained. Seah and Forty[77] have reported using a movable probe in conjunction with a sample which can be rotated in order to obtain angular energy distribution measurements of electrons. Recent experimental results have also been reported on Cu,[78] GaAs,[79] and Ag.[80]

3.3.5. EDC Measurements As a Function of Temperature.

Photoemission measurements have been made at temperatures ranging from liquid nitrogen temperature (77 K) to about 400°C. As discussed in Section 3.1, many interesting properties of materials can be investigated through photoemission measurements as a function of temperature. A common problem in high- and low-temperature measurements is maintaining a temperature different from room temperature without introducing excessive electrical or microphonic noise.

For low-temperature measurements, liquid nitrogen is typically carried by a tube to a heat sink while another tube carries away the nitrogen as it begins to boil. A slightly different approach is to use a single "cold finger" attached to the heat sink. The heat sink is in thermal contact with, but electrically isolated from, the sample by a piece of sapphire or other suitable electrical insulator. Care should be taken to thermally insulate the nitrogen tubes or cold finger to minimize boiling, and the cooling system and emitter support should be constructed to minimize vibrations which cause $V_R \, dC/dt$ microphonic noise. Temperatures between liquid nitrogen temperature and room temperature have been attained by using a mixture of He and forming gas cooled by a liquid nitrogen heat exchanger.[47]

Measurements above room temperature can be made by using heated air or fluids, such as oil or water, in a system similar to that used for attaining low temperatures. A resistance heater can be used to heat the sample if the circuit of Fig. 13 is modified so that collector current rather

---

[76] G. W. Gobeli, F. G. Allen, and E. O. Kane, *Phys. Rev. Lett.* **12**, 94 (1964).
[77] M. P. Seah and A. J. Forty, *J. Phys.* **E3**, 833 (1970).
[78] U. Gerhardt and E. Dietz, *Phys. Rev. Lett.* **26**, 1477 (1971).
[79] F. Wooten, T. Huen, H. V. Winsor, *Phys. Lett. A* **36A**, 351 (1971).
[80] T. Gustafsson, P. O. Nilson, and L. Wallden, *Phys. Lett.* **37A**, 121 (1971).

than emitter current is measured. The heater and emitter are connected through a large capacitor to ground to minimize ac disturbances. A regulated, low-ripple, dc current is used to heat a noninductively wound heater.

In making photoemission measurements as a function of temperature, one must be careful to avoid errors due to electrical junctions at different temperatures. The contact potential difference between an emitter and collector at different temperatures also contains a thermoelectric emf contribution.[81]

3.4. Sample Preparation

For photoemission experiments to be successful, it is normally necessary to prepare a sample with a surface that approaches atomic cleanliness. Since electron transport occurs across the surface, the photoelectric yield and EDC's are often very sensitive to surface contamination and oil-free vacuums below about 10^{-9} Torr are usually required to obtain good results. The degree of sensitivity of the photoemission with respect to surface contamination is somewhat dependent upon the material being studied. EDC's obtained from gettering materials such as Ba and Sr and metals such as Ag and Cu are quite sensitive to surface contamination,[12] whereas EDC's from some oxides[82] and sulfides[83,84] are not nearly as sensitive. Au EDC's (and hence also its work function) are quite insensitive to surface contamination, so gold is often evaporated on the collector to provide a uniform collector work function which is insensitive to contamination that might arise from cleaning the samples to be studied.

Other more subtle aspects of preparing an atomically clean surface are important. It has been found that EDC's and yield from Al are affected by surface roughness.[20] The rough surface provides a mechanism by which incoming normally incident photons can create surface plasmons. The exact nature of the surface could give rise to surface states that could provide an additional contribution to the EDC's. Also, the crystal structure near a free surface may differ from that of the bulk material, causing the EDC's to reflect surface properties rather than bulk properties whenever the electrons escape depth is sufficiently small.

---

[81] C. Herring and M. H. Nichols, *Rev. Mod. Phys.* **21**, 185 (1949).
[82] R. J. Powell, Ph.D. Dissertation, Stanford Univ. (1967).
[83] J. L. Shay, W. E. Spicer and F. Herman, *Phys. Rev. Lett.* **18**, 649 (1967).
[84] J. L. Shay and W. E. Spicer, *Phys. Rev.* **161**, 799 (1967); **138**, 561 (1965).

In the remaining part of this section several different methods of sample preparation will be discussed.

3.4.1. Cleavage

Perhaps one of the simplest and quickest methods of preparing a surface approaching atomic cleanliness is by cleavage in ultrahigh vacuum. The sample must be a single crystal and have one or more cleavage planes; cleavage planes of many minerals may be found in "Dana's Manual of Mineralogy."[85] In general, samples with the same crystal structures will have the same cleavage planes, although the quality of the cleaves may vary. At Stanford University and elsewhere many samples

TABLE I. Examples of Cleavage

Sample	Cleavage plane	Crystal structure	Space group
Silicon	(111)	Diamond	O_h^7 ($Fd3m$)
Germanium	(111)		
GaAs	(110)	Zincblende	T_d^2 ($F\bar{4}3m$)
CdTe	(110)		
SrTiO$_3$	(100)	Perovskite	O_h^1 ($Pm3m$)
TiO$_2$	(110)	Rutile	D_{4h}^{14} ($P4_2/mnm$)
NiO	(100)	Sodium chloride	O_h^5 ($Fm3m$)
CoO	(100)		
CdS	(10$\bar{1}$0) and (1$\bar{2}$10)	Wurtzite	C_{6v}^4 ($P6_3mc$)
CdSe	(1$\bar{2}$10)		
ZnO	(10$\bar{1}$0)		

have been successfully prepared by this technique, some of which are listed in Table I. The sample to be cleaved is placed between a tungsten carbide blade and an annealed copper anvil or gold anvil. The sample is squeezed between those along a cleavage plane, causing the sample to cleave. The details of the mechanism will be described more fully in the section on high-vacuum chambers.

[85] C. S. Hurlbut, "Dana's Manual of Mineralogy." Wiley, New York, 1971.

The nature of a cleaved surface is of interest. Lander et al.[86] have studied cleaved Ge and Si (111) surfaces using low-energy electron diffraction techniques. It was found that, after cleavage at room temperature, a rectangular mesh existed at the surface with the first and second surface layer atoms displaced from their equilibrium positions. Consequently, the surface atoms had a different symmetry from the threefold symmetry of the (111) plane. Where the escape depth of the photoelectrons becomes comparable to the number of atomic layers involved in the rectangular surface structure, the EDC's might not reflect bulk properties of the silicon or germanium. An interesting result of this work was that when the cleaved sample was heated, the surface had irreversible transitions to structures observed on argon bombarded and annealed Ge and Si samples. These new surface structures had the threefold symmetry of the (111) plane.

3.4.2. Evaporation

Samples evaporated *in situ* for photoemission experiments are widely used where possible, since they are often easily and reliably prepared. The quality of the evaporated film depends on the speed of evaporation, angle of evaporation, thickness of the film, preparation of the substrate, temperature of the substrate, and residual gas pressure. The residual gas pressure allowable depends on the rate of evaporation and reaction rate with the residual gas of the material being evaporated. Base pressures below 10^{-9} Torr are usually required, but pressures above 10^{-9} Torr are sometimes allowable if the evaporation rates are sufficiently high. The optimum temperature of the substrate during evaporation varies with the type of sample.

A clean substrate which has been well degreased and sometimes chemically etched is necessary. The substrate is almost always further cleaned *in situ* by heating or ion bombardment. Polished metal substrates or glass substrates with preevaporated electrical contacts are most common. Specific substrates may be required for special purposes such as for epitaxial growth or to minimize diffusion from the substrate if the evaporated film has to be heated.

Resistance heating, electron bombardment, and high-frequency heating can all be used to achieve the temperatures required for evaporation. The material (usually Mo, W, Ta) and shape of the filament or boat

---

[86] J. J. Lander, G. W. Gobeli, and J. Morrison, *J. Appl. Phys.* **34**, 2298 (1963).

used in resistance heating depends on the wetting and alloying properties of the material to be evaporated. Filament-heated conical quartz evaporators have been successfully used to evaporate germanium, alkali halides,[59] and silver halides.[47] Alkali metals can be evaporated from glass ampules.[87] Gold and copper can be evaporated from a metallic bead that has been preheated and wetted to a hairpin molybdenum filament, and silver can be evaporated from a helical tungsten filament. The transition metals, for which high melting points are encountered, have been evaporated most successfully using electron beam bombardment.[30] Alloying and wetting problems are avoided because the sample material is locally heated, the rest of the sample forming the "boat" as it sets in a water-cooled copper crucible. Commercial intruments, such as the Varian e-gun,[†] have worked well in photoemission experiments. When photoemission measurements are taken, care must be taken to see that the e-gun focusing magnets do not produce undesirable residual magnetic fields in the collector region.

In many experiments film thickness is quite critical. This is especially true in photoemission studies of insulators, where charging of thick films presents serious problems. Charging problems have been overcome in photoemission studies of the alkali halides, for example, by evaporating films of about 100 Å and using small photon fluxes.[59] Controlled evaporation of thin films can be achieved using a quartz crystal thickness monitor. The change in frequency of the quartz crystal as a film is deposited on it measures the thickness. Such devices are available commercially[‡] or can be constructed to meet the particular constraints of a given photoemission experiment.[59]

3.4.3. Heat Cleaning

Simply heating a sample in high vacuum may be sufficient to obtain a clean surface. An EDC from a bulk copper sample heated for 12 hours at 650°C in a vacuum of $\sim 10^{-9}$ Torr is seen in Fig. 18 to be in good agreement with an EDC from an evaporated Cu film.[88] At what temperature and for how long a sample needs to be heated varies greatly. For

---

[87] N. V. Smith, *Phys. Rev.* **183**, 634 (1969).
[88] D. H. Seib, Ph.D. Dissertation, Stanford Univ. (1969).

---

† Varian Associates, Palo Alto, California.
‡ Sloan Instruments Corp., Santa Barbara, California.

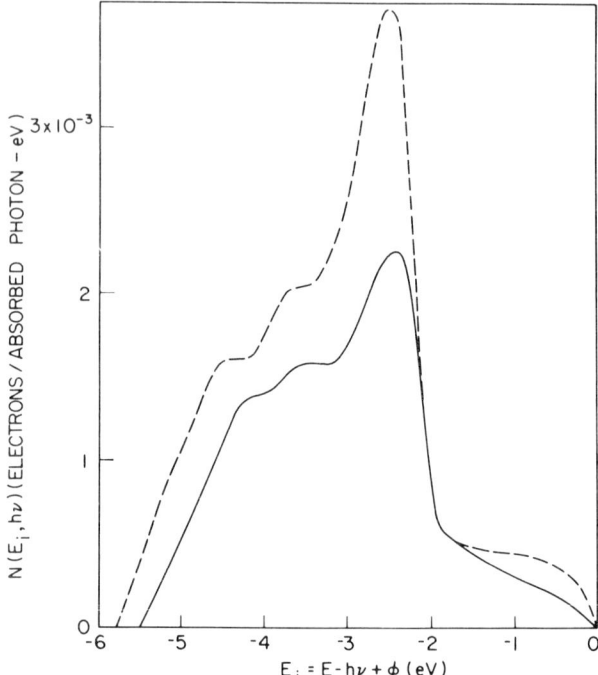

FIG. 18. Comparison of EDC's from a heat-cleaned Cu bulk sample (—) and an evaporated Cu film (- - -) with $h\nu = 10.2$ eV. [D. H. Seib, Ph. D. Dissertation, Stanford Univ. (1969).]

example, a comparison of an EDC from a rhodium sample heat cleaned for 12 hours at 710°C and an EDC from an evaporated film shows that such heat cleaning is clearly insufficient.[89] Ideally, in the heat cleaning process, surface contaminants desorb into the vacuum and bulk impurities diffuse to the surface to be likewise desorbed. Different behavior of impurities is also possible, as in the case of sulfur and carbon impurities in nickel. It appears that carbon diffuses into the bulk and that sulfur diffuses to and collects on the surface.[90] This prevents heat cleaning from being an effective means of preparing Ni samples for photoemission experiments.

Different surface contaminants are removed by heat cleaning at different temperatures. For example, chlorine, sulfur, oxygen, and carbon have all been observed as contaminants on GaAs surfaces. Measurement

[89] D. T. Pierce and W. E. Spicer, *Phys. Rev.* **85**, 2125 (1972).
[90] L. A. Harris, *J. Appl. Phys.* **39**, 1428 (1968).

of the contaminants by Auger electron spectroscopy has shown that the chlorine, sulfur, and oxygen can be removed by heating to temperatures below the congruent evaporation temperature of GaAs (627°C) but that carbon remains.[91] Thus heat cleaning alone is not sufficient in preparing an atomically clean surface on GaAs.

When oxides are heated, it is often necessary to maintain a specific partial pressure of oxygen to prevent the sample from dissociating. If the oxide starts to dissociate at a low temperature, as in the case of VO_2, it can indeed be difficult to control the heat-cleaning temperature and oxygen partial pressures accurately to maintain the desired stoichiometry of the sample.[82,92]

3.4.4. Ion Bombardment Cleaning

Atomically clean surfaces of many materials can be obtained by inert gas ion bombardment followed by heating or annealing.[93,94] The ion bombardment physically sputters off surface layers. The subsequent heating is necessary to anneal the surface damage and remove the embedded gas. Ion bombardment cleaning is usually used in conjunction with a prior heat cleaning. If the sample has not been thoroughly outgassed, the annealing process may cause bulk impurities to diffuse to the surface to form a new layer of contamination.[93] The influence of the gas pressure, ion energy, current density, and target temperature on the sputtering process is discussed in a review by Wehner.[95]

The simplest method of obtaining ion bombardment is by establishing a glow discharge. Typical parameters for such a glow discharge ion bombardment are: sample at a negative potential of approximately 1 kV, argon pressure 30 μ, and argon ion current density of the order of 100 μA/cm². More control over the discharge and lower ion energies can be achieved by using a hot filament. The sample is shielded from the filament to prevent any material evaporated from the filament from reaching the sample. The arc discharge is typically maintained at 60 V and 120 mA at a pressure of 10 μ. When the sample is maintained at −100 V, it is bombarded with an ion current of 3 to 5 mA/cm², which removes 6–60

---

[91] J. J. Uebbing, *J. Appl. Phys.* **41**, 802 (1970).
[92] G. F. Derbenwick, Ph.D. Dissertation, Stanford Univ. (1970).
[93] H. F. Farnsworth, R. E. Schlier, T. H. George, and R. M. Burger, *J. Appl. Phys.* **29**, 1150 (1958).
[94] H. D. Hagstrum and C. D'Amico, *J. Appl. Phys.* **31**, 715 (1960).
[95] G. K. Wehner, *Advan. Electron. Electron Phys.* **7**, 239 (1955).

atomic layers per minute.[94] With an unconfined glow discharge, care must be taken to minimize sputtering of the sample holder and high-voltage leads to the sample by appropriate shields. High-purity argon must be used. The results obtained may be improved by precleaning high-purity bottled argon by passing it through a chamber (titanium sublimation pump) with a freshly deposited titanium layer. The sample chamber can also be left open to a titanium sublimation pump during the glow discharge to help remove impurities produced by the sputtering.

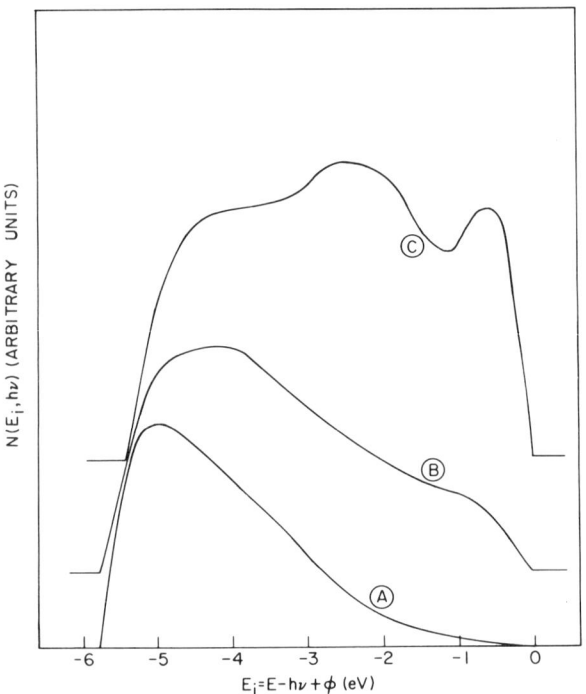

FIG. 19. EDC's from a Ni sample with $h\nu = 10.2$ eV for (A) no surface treatment, (B) after argon bombardment for 25 min with no anneal, and (C) after being heated to 650°C for 13 hr to anneal the surface and drive off imbedded argon. [D. H. Seib, Ph. D. Dissertation, Stanford Univ. (1969).]

Figure 19 shows three energy distribution curves obtained from a (100) Ni single crystal: (a) before surface treatment, (b) after 25-min argon bombardment, and (c) after it was then heated 13 hours at 650°C.[87]

Instead of a dc glow discharge, an rf discharge may be used. The rf discharge makes it possible to maintain more control over the ion energy and also makes possible the ion bombardment of dielectrics.

3.4. SAMPLE PREPARATION

An ion bombardment gun may have advantages in many applications. Lower pressures are possible because electrons from a filament generate ions in a confined region. The increased control of ion energies and current densities on the ion gun may be desirable in the cleaning of sensitive materials. The ion beam can be directed on the sample, avoiding undesirable sputtering of the sample holder, etc. On the other hand, very high ion current densities, when required, are more easily obtained in a glow discharge.

The problem of determining when a clean surface has been obtained is a most difficult one. At worst, the sample can be sputtered and annealed and recycled until no further changes in the EDC's are observed. A LEED or Auger spectroscopy apparatus used in conjunction with the sputtering could prove very useful in determining when an atomically clean surface has been approached. Sometimes another standard of comparison may be available, as in the case of Ni shown in Fig. 20. The solid curve is an EDC from a Ni film e-gun evaporated at a pressure

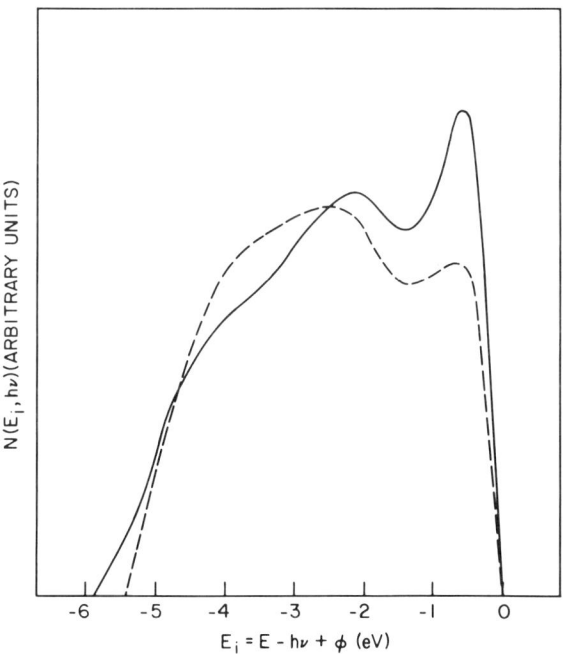

FIG. 20. Comparison of Ni EDC's from an electron gun evaporated sample [—: D. E. Eastman and W. F. Krolikowski, *Phys. Rev. Lett.* **21**, 623 (1968)] and an argon bombarded sample which gave clean Ni LEED patterns [- - -: T. A. Calcott and A. U. MacRae, *Phys. Rev.* **178**, 966 (1969)].

less than 5×10^{-9} Torr.[96] The dashed curve is an EDC from a sample which was argon bombarded and heat cleaned until a LEED pattern characteristic of the clean surface was obtained.[97] So sensitive is photoemission to surface preparation that even a surface which appears atomically clean by LEED does not necessarily give good photoemission results.

With some materials, notably anisotropic oxides, ion bombardment results in a quasi-amorphous layer at the surface.[98,99] It is necessary that the annealing temperature be high enough to recrystallize this damage. However, if the sample dissociates at a lower temperature, it will not be possible to use the ion bombardment cleaning technique unless a carefully controlled oxygen partial pressure can be maintained during the annealing phase.

3.4.5. Lowering Electron Affinity by Applying Surface Layers

The electron affinity of the photoemission sample determines the threshold for photoemission and places a lower limit on the useful range of photon energies. If the electron affinity can be lowered by applying surface layers of an appropriate material, the photon energy range can be extended to lower energies and deeper valence bands and lower energy conduction bands can be probed using photoemission techniques. In Fig. 21, EDC's from clean and cesiated GaAs are shown and in Fig. 22, EDC's from clean and cesiated copper are shown. Note how transitions from deep valence bands to low-energy conduction bands are observed. Electron–electron scattered electrons often contribute significantly to the magnitude of the EDC's at low electron energies—the peaks in the EDC's due to these scattered electrons remain stationary with photon energy and are generally much more apparent in EDC's from cesiated samples than in EDC's from clean samples.

For studying bulk properties of the sample, it is necessary that the application of the surface layers does not greatly affect the EDC's but primarily lowers the threshold for photoemission. For practical photocathodes, the EDC's are not as important—the major consideration here is to extend the operating range of the photocathode to lower photon

---

[96] D. E. Eastman and W. F. Krolikowski, *Phys. Rev. Lett.* **21**, 623 (1968).

[97] T. A. Calcott and A. U. MacRae, *Phys. Rev.* **178**, 966 (1969).

[98] Hj. Matzke and J. L. Whitton, *Can. J. Phys.* **44**, 995 (1966); Hj. Matzke, *ibid.* **46**, 621 (1968).

[99] C. Jech and R. Kelly, *J. Nucl. Mater.* **20**, 269 (1967).

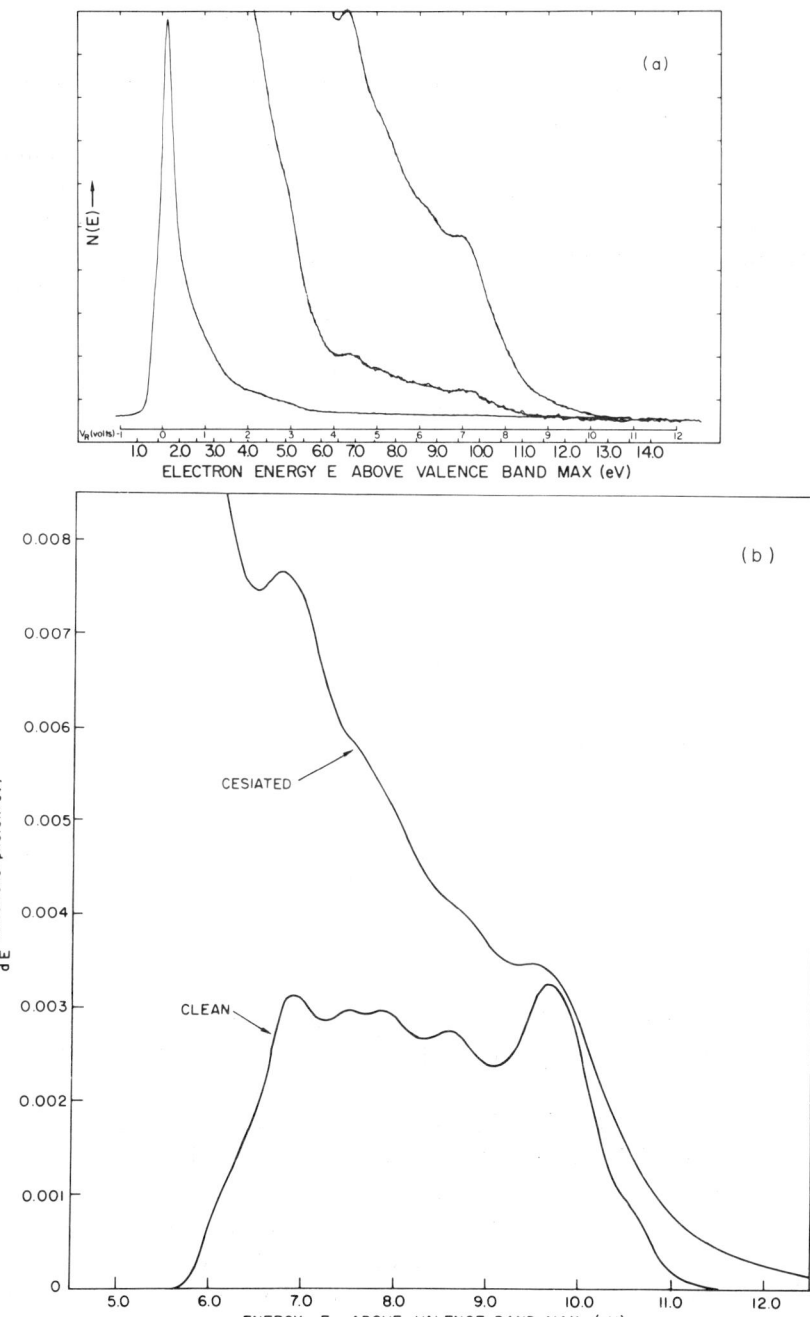

FIG. 21. (a) EDC from cesiated GaAs and (b) comparison of EDC's from clean and cesiated GaAs. [W. E. Spicer and R. C. Eden, *Proc. Int. Conf. Phys. Semiconductors, 9th, Moscow* **1**, 65. Nauka, Leningrad, 1968; R. C. Eden, Ph.D. Dissertation, Stanford Univ. (1971)]. Note the greatly reduced threshold for cesiated GaAs and the agreement between peak positions in the cesiated and clean GaAs EDC's. $h\nu = 10.6$ eV.

energies and to take advantage of the higher quantum yield which usually results at higher photon energies.

Normally surface layers of cesium are used to lower the electron affinity of photoemission samples. This is because it is fairly easy to apply carefully controlled surface layers of cesium in high-vacuum chambers and the decrease in electron affinity is generally larger with cesium than with other materials (probably because of the small electronegativity of cesium). Often, by applying several alternate layers of cesium and oxygen

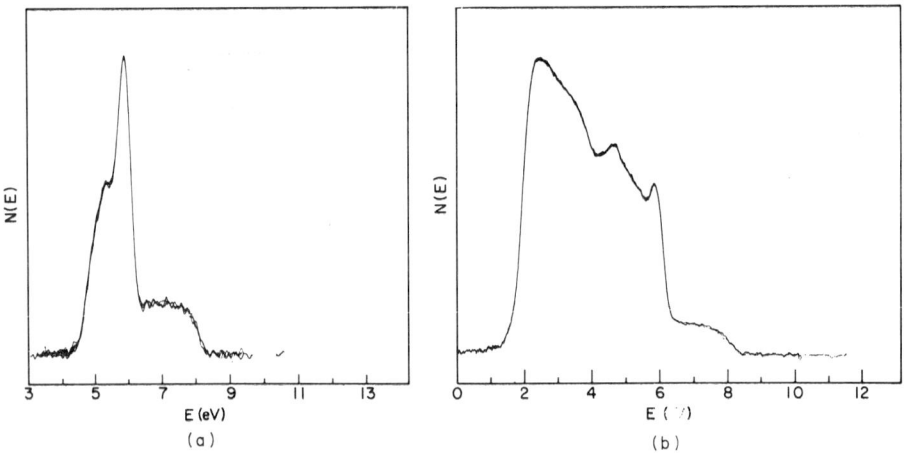

FIG. 22. Comparison of EDC's from (a) clean and (b) cesiated Cu [G. F. Derbenwick, Ph.D. Dissertation, Stanford Univ. (1970)]. The photon energy in each case was 8.2 eV.

(ending with a cesium layer), lower electron affinities can be obtained than by applying a single cesium layer. The techniques of cesiation with or without alternate oxygen layers have been used successfully in studying GaAs,[100] Si,[50,101] copper,[102] and other metals and III–V compounds. Cesiation has not been found to be effective in studying II–VI compounds or the silver halides, due to strong reactions between the samples and the cesium, which prevented a surface monolayer of cesium from forming.

When p-type semiconductors are cesiated, advantageous band bending near the surface often results in a much reduced vacuum level. Negative effective electron affinities have been obtained when cesium and oxygen

[100] R. L. Bell and W. E. Spicer, *Proc. IEEE* **58**, 1788 (1970).
[101] J. van Laar and J. J. Scheer, *Phillips Res. Rep.* **17**, 101 (1962).
[102] I. Lindau and L. Wallden, *Phys. Scripta* **3**, 77 (1971).

3.4. SAMPLE PREPARATION

surface layers have been applied to GaAs,[100] permitting an efficient infrared detector to be constructed.

Cesium channels (obtained from RCA or Varian Associates), cesium ampoules, or a cesium gun can be used to apply monolayers of cesium to a clean photoemission sample. Cesium channels consist of a tantalum tube containing cesium chromate and silicon. When the channel is heated by passing a current of a few amperes through it, a reaction occurs which liberates atomic cesium while silicon chromate is formed. Before using a cesium channel, it must be thoroughly outgassed in the high-vacuum chamber prior to cleaning the photoemission sample if best results are to be obtained. Cesium ampoules are not very often used for cesiation, since once the ampoule is broken it must be kept chilled or valved off to prevent unwanted deposition of cesium. Construction of a cesium gun has been described elsewhere.[103] Its advantages over the cesium channels are that the exact amount of cesium deposited can be determined from the operation of the gun, and very little contamination results from its operation. In addition the cesium can be directed at the sample, thereby avoiding cesiation of the collector. A cesiated collector sometimes raises havoc with reverse photoemission because of its lowered work function and increased yield. When cesiating, large differences in electron affinities or work functions of the emitter must be avoided, such as between a cesiated sample surface and uncesiated sides of the sample or sample holder. Such differences in work functions can create bothersome fringing fields in the region of the sample, which can seriously degrade the resolution of the photoemission analyzer.

When cesium channels are used, line-of-sight between the sample surface and cesium channel should be avoided to provide for a "cleaner" cesiation. The vapor pressure of the cesium is sufficiently high that the cesium readily moves about the chamber, whereas emitted contaminants are probably adsorbed on the walls of the chamber immediately. A monolayer of cesium is readily formed, since the first monolayer generally has a high sticking coefficient and successive layers have low sticking coefficients. With cesium channels, virtually everything in the high-vacuum chamber is coated with cesium. The optimum quantity of cesium to be deposited can be determined by monitoring the photocurrent near threshold while cesiating. Generally the photocurrent increases rapidly, reaches a peak, and then slowly decreases as cesiation progresses. The quantity of cesium deposited is roughly one monolayer or slightly

---

[103] F. G. Allen and G. W. Gobeli, *Rev. Sci. Instrum.* **34**, 184 (1963).

more when the peak in photocurrent occurs. This is generally the instant when cesiation is terminated. If too little cesium is deposited, the vacuum level is not optimally lowered, and if too much cesium is deposited, often excessive electron scattering occurs at high photon energies, broadening much of the structure in the EDC's.

For studies of the electronic structure of solids, it is important that the EDC's are not greatly distorted by diffusion of cesium into the sample or by chemical reaction with the cesium. An apparently successful cesiation results if some or several of the following are observed:

(1) Comparison of EDC's obtained from clean and cesiated samples in the overlapping regions of the EDC's show that the EDC structure changes with photon energy in the same manner and the peaks and shoulders in the EDC's are located at approximately the same energies (see Figs. 21 and 22).

(2) Low-energy structure in the EDC's from the cesiated sample agrees reasonably well with a reliable band or density of states calculation.

(3) Application of surface layers of alkali metals other than cesium may affect the vacuum level of the cesiated sample but does not greatly affect the structure observed in the EDC's.

(4) Inspection of the cesiated sample suggests that chemical reactions have not occurred.

Materials other than cesium, such as other alkali metals, can be used to lower the vacuum level of photoemission samples. However, lower vapor pressures makes deposition of the surface layers more difficult and very few experiments of this nature have been reported in the literature.

3.5. High Vacuum Photoemission Chambers

A comprehensive treatment of high-vacuum techniques can be found in an article by G. E. Becker in Volume 4, Chapter 5 of "Methods of Experimental Physics." Here we will be concerned only with a brief survey of special techniques that are applicable to the design of photoemission chambers.

3.5.1. Vacuum Pumps

For photoemission studies requiring only photon energies below 11.8 eV, a LiF window can be used to separate the photoemission chamber from the monochromator. These windows consist of a polished or cleaved disk of LiF which is attached to a high-vacuum flange using silver chloride and a silver ring, and are available commercially.[†] Normally the polished LiF disks have somewhat lower transmission in the vacuum ultraviolet than the cleaved LiF disks. The AgCl in the window assembly limits the bakeout temperature of the photoemission chamber to about 175°C. For photoemission studies requiring maximum photon energies substantially below 11.8 eV, it may be possible to use quartz or sapphire windows instead of an LiF window.

When the photoemission chamber is separated from the monochromator by a window, "dry" high-vacuum pumps are almost always used for pumping the photoemission chamber. Often these consist of one or more cryogenic molecular sieve pumps for rough pumping and sputter ion and titanium sublimation pumps for ultrahigh-vacuum pumping. Sometimes auxiliary sputter ion pumps and titanium sublimation pumps are also used for pumping during the bakeout cycle. During actual measurements, only the high-vacuum sputter ion pump is used. Magnetic shielding of the sputter ion pumps is often required to reduce the magnetic field in the region of the energy analyzer to a sufficiently low value. In place of the sputter ion and titanium sublimation pumps an NRC orb-ion pump can be used for ultrahigh-vacuum pumping. Although the orb-ion pump produces no undesirable magnetic field, stray low-energy light and electrons are created by filaments which are run continuously. Both of these effects can be minimized by proper placement of water-cooled baffles and electrostatic shields.

The sizes of the pumps selected depend primarily upon the sensitivity of the samples to residual gas contaminants. Gettering materials require high pumping speeds to maintain sufficiently low pressure during evaporation and measurements, and large vacuum pumps are required. On the other hand, sputter ion pumps as small as 7 or 15 liters/s have served well when materials less sensitive to surface contamination are studied.

For photoemission experiments where photon energies above 11.8 eV are to be used, the photoemission chamber cannot be separated from the monochromator by a window. Differential oil or mercury diffusion pumping systems can often be used to attain sufficiently low pressures, although the vacuums obtained are not as acceptable as dry vacuums. Where differential pumping is not available and the samples to be studied are not especially sensitive to surface contamination, an alternative technique

---

[†] Harshaw Chemical Co., Cleveland, Ohio.

116   3. VISIBLE AND ULTRAVIOLET PHOTOEMISSION

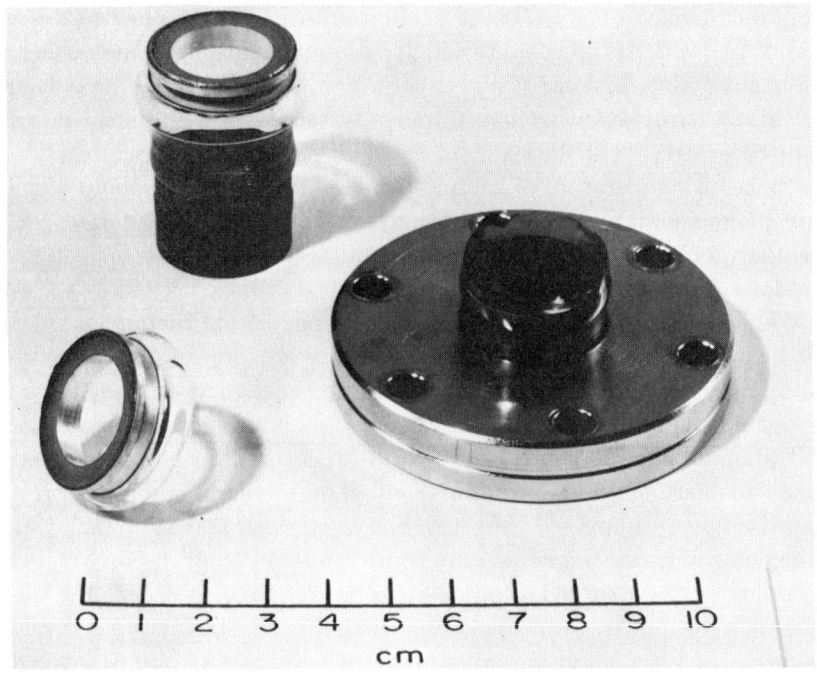

FIG. 23. LiF knock-off window. Shown is a new window prior to welding to a flange, and a window which has been knocked off. [G. F. Derbenwick, Ph. D. Dissertation, Stanford Univ. (1970).]

has been found useful. The sample chamber is separated by a LiF window mounted on a thin piece of glass tubing. Such a window is shown in Fig. 23. After the sample is prepared under high-vacuum conditions and EDC's below 11.8 eV are obtained, the window is sheared off using a cleaving blade, exposing the sample to the monochromator vacuum. High-energy EDC's are immediately obtained before sample contamination becomes too serious. EDC's below 11.8 eV should then be taken to compare with those previously taken under high-vacuum conditions. This comparison should give some measure of how badly the photoemission results were affected by the poor vacuum conditions.

3.5.2. Photoemission Chambers

The design of photoemission chambers varies widely depending upon the requirements of specific experiments. An important factor to consider is the method of sample preparation. In Figs. 24–26, representative

3.5. HIGH VACUUM PHOTOEMISSION CHAMBERS

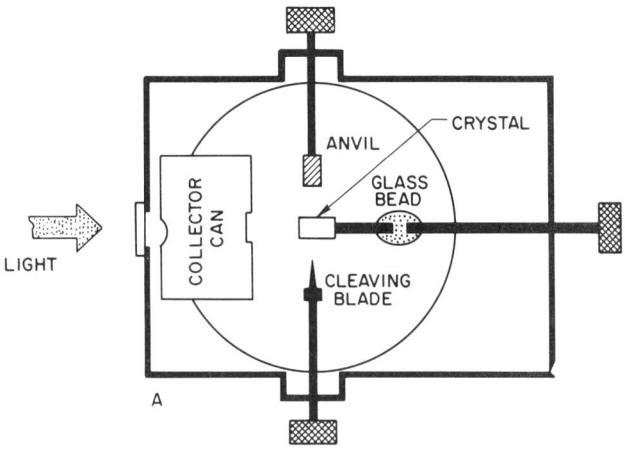

FIG. 24. Typical cleavage photoemission chamber [R. J. Powell, Ph.D. Dissertation, Stanford Univ. (1967); G. F. Derbenwick, Ph.D. Dissertation, Stanford Univ. (1970)]. The linear motion sample and cleavage arms are mounted on stainless steel bellows. (A) Schematic top view. (B) Actual chamber.

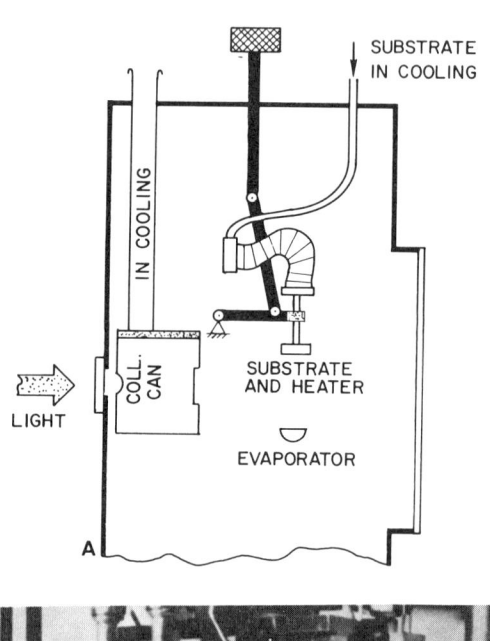

Fig. 25. Typical evaporation photoemission chamber [W. F. Krolikowski, Ph.D. Dissertation, Stanford Univ., Stanford Electron. Lab. Rep. # SU-SEL-67-039 (1967)]. (A) Schematic side view. (B) Actual chamber.

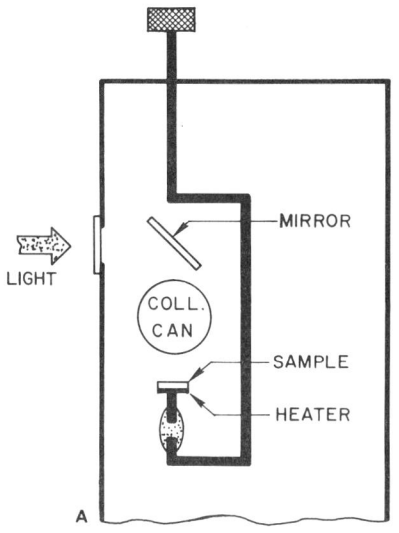

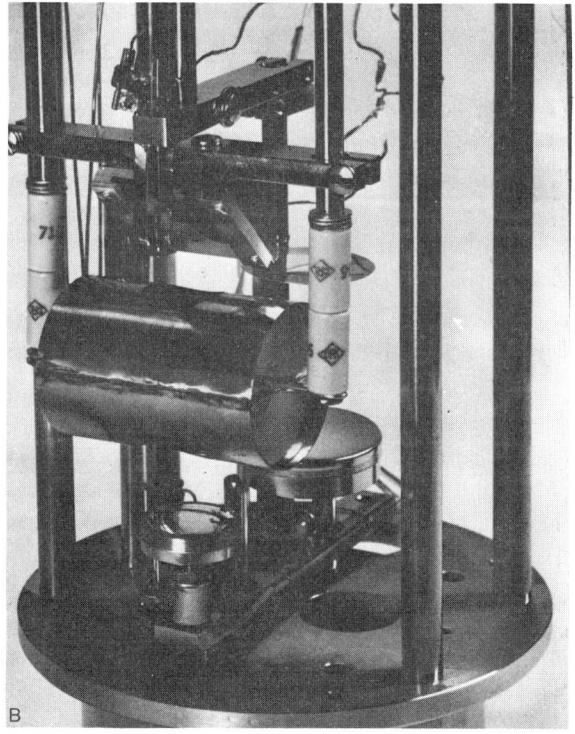

Fig. 26. Photoemission chamber for studying liquids [R. Koyama, Ph.D. Dissertation, Stanford Univ. (1970)], such as Ga and In. (A) Schematic side view. (B) Actual chamber.

designs for studying samples prepared by cleavage and by evaporation, and for studying liquids are shown. Heat-cleaning and/or sputtering can be utilized for sample preparation in any of these chambers by mounting a filament on the sample holder and mounting appropriately located electrodes on the flange. Rotary and linear motions can be obtained using magnetic feedthroughs and Heli-arc welded stainless steel bellows, respectively. It has been found that excellent cleavages (see Table I) are obtained using a tungsten carbide blade carefully aligned with the desired cleavage plane (Fig. 24). Different types of evaporators, such as the electron beam evaporator, can be used with the chamber section in Fig. 25.

For measurements at temperatures other than room temperature, heaters or liquid reservoirs in thermal contact with the sample holder can be used. These techniques are described in some detail in Section 3.3.5.

The photoemission chambers described above utilize a retarding field analyzer as described in Section 3.3.1. When other analyzers are used, the detailed design of the photoemission chamber can differ somewhat from that described here. For these details it is best to refer directly to the references given in Section 3.3.3.

3.6. Monochromator and Light Sources

A comprehensive survey of monochromators and light sources can be found in the book "Techniques of Vacuum Ultraviolet Spectroscopy" by Samson.[44] A typical set up for a photoemission experiment might include a 1-m normal incidence monochromator such as the McPherson 225 with a 600-line/mm grating. In the low-intensity region of the hydrogen lamp spectrum where maximum slit width is used, the extra intensity obtained from the 600 line/mm grating as compared to the 1200 line/mm grating is welcome. An increase in intensity is obtained from the tripartite grating blazed at 1500 Å compared to a bipar grating, which is only available blazed at 2000 Å. The very slight decrease in the focus and hence resolving power of the tripar grating is of no consequence at the slit widths required for photoemission experiments.

The Hinteregger hydrogen discharge lamp is used up to the LiF window cutoff at 11.8 eV. Quartz or Pyrex filters remove higher-order light at longer wavelengths. A resolution of 0.1 to 0.15 eV full width at half maximum can be maintained with sufficient light for most experiments. The intensity of the lamp can be increased by about three to ten times by using a hot filament to maintain the arc discharge. A lower voltage, more easily regulated, and less expensive power supply is then adequate. A lamp designed especially for hot filament operation has been described by Hartman,[104] but the Hinteregger discharge lamp is easily converted.[105]

Often good photoemission results at high photon energies can be obtained from windowless experiments in adequately pumped systems, as described in the previous section. The hydrogen lamp can be used up to about 14 eV. Other gases can be used in the Hinteregger discharge lamp to obtain higher photon energies. Neon has lines at 14.7, 16.8, and 26.9 eV, He at 21.2, 23.1, and 40.8 eV, and N_2 at 18.5 eV. The 16.8-eV line of Ne and 21.2-eV line of He are by far the strongest.

Photoemission measurements are made in the visible and near ultraviolet regions of the spectrum on samples with low work functions such as where a Cs layer has been applied. Measurements in this region are facilitated by intense light sources, such as a Hg arc or tungsten lamp, and by the absence of a need for the vacuum monochromator.

3.7. Directions for Future Research

It has been seen that visible and ultraviolet photoemission is a valuable tool in probing the electronic structure of solids and the associated optical excitation processes. The technology is sufficiently advanced that sample surfaces approaching atomic cleanliness can routinely be obtained under ultrahigh-vacuum conditions using a variety of methods of sample preparation. New experimental innovations can substantially increase the information obtained in photoemission experiments. However, as the experiments become more elaborate and complex, one should exercise care and judgment in applying the technology and interpreting the data.

Since photoemission experiments are restricted by the limited photon energies available from conventional light sources, immediate gains result from using synchrotron radiation as a high-energy photon source.[106,107] With synchrotron radiation EDC's closely spaced in photon energy up to

---

[104] P. L. Hartman, *J. Opt. Soc. Amer.* **51**, 113 (1961).
[105] D. E. Eastman, "Techniques of Metals Research" (E. Passaglia, ed.), Vol. 4. Wiley (Interscience), New York, 1971.
[106] D. E. Eastman and W. D. Grobman, *Phys. Rev. Lett.* **28**, 1327 (1972).
[107] R. Y. Koyama and L. R. Hughey, *Phys. Rev. Lett.* **29**, 1518 (1972).

extremely high energies can be obtained. Over such a wide photon energy range core levels, as well as deeper lying valence bands and more energetic conduction bands, can be probed. By appropriate choice of photon energy it may be possible to look preferentially at surface or bulk excitation by studying changes in the EDC's as the electron escape depth varies. Also, the inherent polarization of the synchrotron radiation source can provide a powerful tool when studying oriented noncubic crystals.

Surface chemistry is another area for exciting new developments. The normal problem that photoemission can be extremely sensitive to surface conditions and contamination can be used to advantage in studying surface properties. This might include chemical reactions of the surface with ambient gases or adsorption of atoms or molecules on the surface. With increases in resolution and signal-to-noise ratio intrinsic surface states can be studied.

The design of analyzers which permit measurement of EDC's as a function of the direction of electron emission (discussed in Section 3.3.4) and the design of vacuum ultraviolet light polarizers which permit the measurement of EDC's as a function of light polarization[92] perhaps will permit more detailed interpretation of EDC's with regard to energy band structures. This would include studies of the optical selection rules as well as energy bands at different points in **k** space.

Another promising new area is photoinjection of electrons through wide gap (insulating) layers.[59,108–111] Here electrons are photoinjected from a substrate through an insulating film and either collected by another electrode or injected into vacuum to be analyzed in a conventional manner. This form of photoemission experiment can yield information about the interfacial barrier heights, properties of the conduction bands of the insulating layer, and scattering lengths of the electrons in the insulator conduction bands.

---

[108] R. Williams, *Phys. Rev.* **140**, 269 (1965).
[109] A. V. Goodman, "Optical Properties of Insulating Films" (Norman N. Axelrod, ed.), Chapter 5, The Electrochemical Society, Inc., 1968.
[110] R. J. Powell, *J. Appl. Phys.* **40**, 5093 (1969).
[111] R. J. Powell, *J. Appl. Phys.* **41**, 2424 (1970).

4. EXPERIMENTS ON ELECTRON TUNNELING IN SOLIDS[†][*]

4.1. Introduction

The concept of particle tunneling has been known for a long time and has formed an essential element in the appropriate quantum-mechanical explanation of many phenomena observed in solids. An electron represented by a wave function can, according to quantum mechanics, penetrate into a classically forbidden region and thereby pass through a barrier rather than over it, in other words "tunnel." This concept can be applied on the atomic scale to explain fundamental properties of solids, or on a more macroscopic scale to explain tunneling that occurs between electrodes separated by a barrier. In the latter case the barrier is usually formed from the forbidden energy gap in a semiconductor or an insulator, and the top or the bottom of the barrier is to be located at the bottom of the conduction band or at the top of the valence band. The macroscopic type of tunneling will be the main subject of this chapter and we will explore the various techniques used to fabricate and measure such tunnel junctions and review the essential information on physical properties of solids that one can expect to learn from such experiments.

A basic tunneling experiment is very easy to perform and can be carried out on a circuit such as the one shown in Fig. 1, which is sufficient to determine the current versus voltage characteristic of the junction.

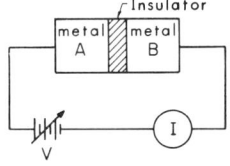

Fig. 1. Schematic model of a tunnel junction and measuring circuit.

[†] Work supported by the United States Atomic Energy Commission.
[*] Part 4 is by R. V. Coleman, R. C. Morris, and J. E. Christopher.

This characteristic alone can tell us quite a bit about both the barrier and the electrodes forming the junction, particularly if a reasonably accurate theoretical model is available. An adequate model will involve parameters such as the barrier height, the barrier thickness, the barrier shape, and the density of electronic states in the metals or semiconductors forming the electrodes. The appropriate models have been discussed at length by many authors and detailed information can be found in two recent books on tunneling,[1,2] which will be liberally referred to in the course of this chapter.

The I–V characteristic of a tunnel junction can be examined essentially on two different scales of energy. One corresponds to the general shape observed over a range of hundreds of millivolts and is generally determined by the elastic tunneling process where a single electron tunnels at a constant energy. This is described by an effective mass approximation for the electronic wave functions and some potential-barrier model based on a local average potential in the insulator that varies slowly in the junction due to macroscopic electric fields.

The second scale of energy occurs over regions of a few millivolts and consists of fine structure due to rapid variations in the I–V characteristic which can occur due either to many-body effects that occur, for example, when one or both of the electrodes are superconducting, or to inelastic tunneling processes. This fine structure can occur from basic excitations of the solid such as phonons, spin waves, plasmons, etc., which reflect in the density of states and thus modify the tunneling. Direct interaction of the tunneling electrons with impurities in the barrier or at the interface can activate various inelastic tunneling channels at specific energies. Both ranges of energy have proven to be rich fields for experimental study and we will discuss the significant experimental work in both cases.

The first and most important step in experiments of this type is the fabrication of the junction. This consists of fabricating two electrodes in intimate contact with the barrier, which should be thin enough to allow appreciable tunnel current to flow for reasonable applied voltages. The barrier should be of uniform thickness, homogeneous, and free of any structural defects such as pinholes. In practice the typical barrier thick-

---

[1] E. Burstein and S. Lundquist, eds., "Tunneling Phenomena in Solids." Plenum Press, New York, 1969.

[2] C. B. Duke, "Tunneling in Solids" (*Solid State Phys. Suppl. 10*) (F. Seitz, D. Turnbull, and H. Ehrenreich, eds.). Academic Press, New York, 1969.

4.1. INTRODUCTION

ness is in the range 10–100 Å with a typical dc resistance in the range 1–10,000 Ω. The requirements on the barrier have proved to be the most difficult part of the junction fabrication with the result that most experiments are done with a barrier formed from the natural oxide of one of the electrodes. This is either formed by thermal oxidation in various atmospheres or by gas discharge in the vacuum chamber with a reduced pressure of oxygen or air. The second electrode is then evaporated over this natural oxide barrier. The most common junction configuration is produced by evaporating both electrodes onto a glass substrate in the form of an X and carrying out the oxidation between evaporations. Figure 2 shows the steps for producing a typical junction configuration in which the junction occurs where the two strips cross. Numerous

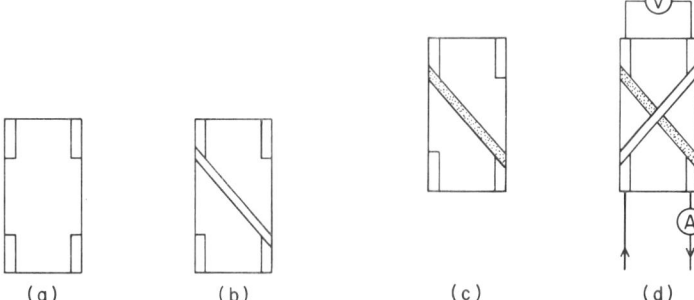

FIG. 2. Steps in the production of a tunnel junction. Electrodes and contacts are evaporated onto a clean glass substrate. Oxidation to form the barrier is carried out between steps (b) and (c). The tunnel junction is formed where the metal electrodes cross in (d).

variations on these techniques have been reported as well as procedures for producing a satisfactory oxide layer. In Section 4.5.1 we have listed specific details for fabricating junctions of various metals and semiconductors that have been successfully used for tunnel junctions. This list is not exhaustive, but should give a representative coverage of the usual junctions discussed in the literature.

After fabrication of a tunnel junction it is still necessary to test the characteristic behavior thoroughly to be sure that the majority of the electrons contributing are indeed flowing via a tunneling process. Even the most successful fabrication techniques do not produce a very high percentage of junctions that meet acceptable standards for proving that tunneling is occurring. Therefore any experimenter in this field must be prepared to include these tests along with data obtained from a given tunneling experiment. The most useful and definitive tests involve at least one electrode in the superconducting state, since the appearance of the superconducting energy gap is a very direct indication of tunneling. A comparison of the current flowing in the normal state versus the superconducting state can also be used to obtain a quantitative measure of the amount of tunneling taking place. The detailed tests for application to specific junctions are listed in and discussed in Section 4.5.2, along with the appropriate quantitative criterion.

Once a satisfactory junction is obtained and has passed the tests for tunneling the measurements can proceed. In addition to the I–V curve, a complete set of data will usually include dI/dV versus V and dI^2/dV^2 versus V. The second quantity is usually called the *differential tunnel conductance* $G \equiv dI/dV$, and the derivative of this quantity is particularly useful for looking at the fine structure introduced by the inelastic tunneling events. Many circuits have been devised for measuring these quantities and the basic operation of these circuits along with the useful experimental parameters is included in Section 4.5.3.

This chapter is divided into three main sections covering tunneling in superconductors, in normal metals, and in semiconductors. The technique of tunneling into superconductors has been developed with astounding success in recent years and will be discussed first, although the two other areas are also contributing equally important information to our knowledge of solids.

4.2. Tunneling in Superconductors

4.2.1. Summary of Superconducting Tunnel Characteristics

The method of electron tunneling has been applied to the study of superconductors with enormous success. Giaever[3,4] was the first to introduce the method as a means of studying the energy gap in the superconducting density of states and his original papers should be consulted for an account of these pioneering experiments.

When tunnel junctions are used to study superconductors, one or both of the metal electrodes can be in the superconducting phase, and in this section we outline the techniques that have been applied specifically to

---

[3] I. Giaever, *Phys. Rev. Lett.* **5**, 464 (1960).
[4] I. Giaever, *Phys. Rev. Lett.* **5**, 147 (1960).

4.2. TUNNELING IN SUPERCONDUCTORS

obtaining information about superconductors. There is, of course, some overlap with the tunneling techniques used for studying semiconductors and normal metals and later sections will include some repetition, although the emphasis will be on methods for obtaining fundamental parameters of metals in their superconducting phase. The general field of tunneling into superconducting metals has been covered in a number of excellent books and reviews, and the reader can find more detailed discussion of the physics in Ref. 1 and the many references listed therein.

Junctions for superconducting tunneling measurements are prepared in the form of a capacitor-like arrangement in the same fashion as was described for all general types of junctions in Section 4.1. The junctions should also be tested according to the standard tests that have been reviewed in Section 4.5.2.

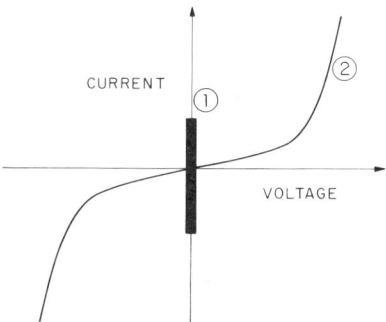

FIG. 3. Current–voltage characteristic for a typical superconducting tunnel junction with both electrodes in the superconducting phase. Characteristic (1) is due to the Josephson current, while characteristic (2) is the finite voltage characteristic due to quasiparticle tunneling.

When both electrodes are in the superconducting phase, the basic I–V characteristic is as shown in Fig. 3. The characteristic curve has two parts, one for which the voltage across the barrier is zero and a supercurrent flows through the barrier, and one for which a finite voltage appears across the barrier and the junction shows a resistive behavior.

The zero-voltage characteristic is due to the Josephson effect,[5] which was first experimentally demonstrated by Rowell and Anderson.[6] This corresponds to the flow of paired superconducting electrons only through the barrier and no excited quasi-particles contribute to this current. The

---

[5] B. D. Josephson, *Phys. Rev. Lett.* **1**, 251 (1962).
[6] P. W. Anderson and J. M. Rowell, *Phys. Rev. Lett.* **10**, 230 (1963).

finite voltage characteristic on the other hand involves the flow of excited quasi-particles and gives rise to a normal current flow through the barrier. The finite voltage characteristic is the one used to determine the energy gap, and we discuss this type of experiment in some detail along with the significant experimental techniques. The Josephson effect will be treated in Chapter 13 along with superconducting device applications.

4.2.2. Tunneling between Normal Metal and a Superconductor

One can understand a good deal about the finite voltage characteristic by considering only the single-particle tunneling. The results are simply related to an effective tunneling density of states $N(E)$, which plays a similar role to that of the conventional density of states in normal metals and semiconductors. A semiconductor-type of energy level diagram can be used to describe the major aspects of the situation, and Fig. 4 shows the case where the left metal electrode A is in the normal phase and the right metal electrode B is in the superconducting phase. With the applied voltage $V = 0$, the Fermi levels E_F^A and E_F^B are equal and a gap of width 2Δ is present at the Fermi level in the superconducting density of states shown on the right. At $T = 0$, quasi-particles can flow between A and B only if a voltage eV greater than Δ or less than $-\Delta$ is applied to the junction. As shown in Figs. 4b and c, the first case corresponds to the flow

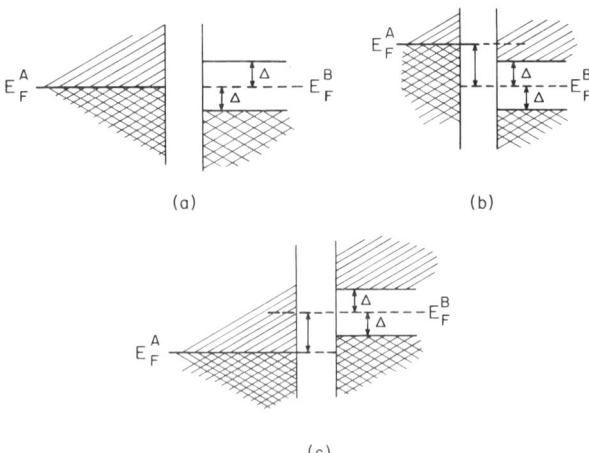

FIG. 4. Semiconductor-type of energy level diagrams describing tunneling between normal and superconducting electrodes. Metal A is normal and metal B is superconducting. (b) and (c) indicate positive and negative voltage bias with the flow of quasi-electrons or quasi-holes, respectively.

of quasi-electrons into the superconductor and the second case to the flow of quasi-holes into the superconductor. The existence of the gap therefore explains the qualitative features of the tunneling I–V curve when a finite voltage is applied.

The use of a semiconductor analogy as in the diagrams of Fig. 4 gives only a general correspondence to the electron states of a superconductor. A more complete picture can be obtained by considering the energy–momentum distribution of the electrons. At absolute zero the mean occupation number $\langle n_k \rangle$ for a superconductor does not change abruptly from 1 to 0 at the Fermi level, but changes smoothly over a range of k as shown in Fig. 5a. This smearing of the Fermi surface is due to the pairing interaction between the electrons and means that in the vicinity

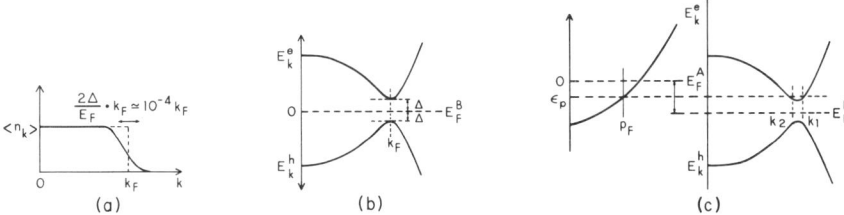

FIG. 5. (a) Mean occupation number $\langle n_k \rangle$ as a function of wave vector k for a superconductor at $T = 0$. (b) Energy versus k for the addition of a single quasi-particle to a superconductor. Upper curve is for quasi-electrons and lower curve is for quasi-holes. Both require positive energies. (c) Energy momentum diagram representing tunneling between a normal and superconducting metal. Applied voltage bias $eV > \Delta$.

of the Fermi level one must think of a superposition of holes and electrons such that the occupation number does not go abruptly from 1 to 0 at k_F. One can then consider the excitation curves directly in terms of the energy required to add an electron to the system or remove an electron from the system with a consequent increase or decrease of the total momentum of the system by k. The energy versus k curves for the addition of a quasi-electron or quasi-hole to the system are shown in Fig. 5b where the excitation energies are positive for both types of excitations. Positive energy plotted upward for electrons and downward for holes. The minimum energy to insert a quasi-particle into the superconducting system is therefore $\Delta = E^e_{k_F} = E^h_{k_F}$ and occurs for $k = k_F$. This minimum energy must be supplied by the voltage in order to tunnel a single quasi-particle at $T = 0$. For a given electron energy there are, therefore, two k states having the same energy in the superconductor and also for a given hole energy there are two k states with the same energy.

The states above the energy gap in the semiconductor description of Fig. 4 correspond to the electron excitation curve of Fig. 5b and the states below the energy gap correspond to the hole excitation curve of Fig. 5b. In either case all k states enter including both $|k| < k_\mathrm{F}$ and $|k| > k_\mathrm{F}$ as contrasted to the true semiconductor case where the holes and electrons are distinctly different and arise from separate single particle energy bands. If the energy–momentum picture for the superconductor as indicated in Fig. 5b is used along with that appropriate to the normal metal, then a junction between a normal and superconducting metal can be represented as in Fig. 5c, where a voltage bias $eV > \Delta$ has been applied so that electron flow will occur from A to B. The detailed energy–momentum curve for the superconducting quasi-particle excitations as shown in Fig. 5b plays an essential role in determining the correct tunneling characteristic and must be taken into account in following the procedure to compute the current or conductance of the junction. This twofold degeneracy of the excited quasi-particle states plays a fundamental role in geometrical resonances which will be discussed in Section 4.2.9.

In the case of a normal metal–insulator–superconductor junction, the resulting tunnel conductance is given by

$$dI_{\mathrm{NS}}/dV = \alpha N(0)[|eV|/(|eV|^2 - \Delta^2)^{1/2}], \quad (4.2.1)$$

where α is a constant involving the average of the tunneling matrix element $|T|^2$ and the momentum perpendicular to the junction interface. $N(0)$ is the density of states at the Fermi surface in the normal metal. The detailed calculation leading to the expression in Eq. (4.2.1) can be found, for example, in the work by Schrieffer.[7] The important thing to realize is that the normalized conductance given by

$$g(V) \equiv \frac{dI_{\mathrm{NS}}/dV}{dI_{\mathrm{NN}}/dV} = \frac{|eV|}{[(eV)^2 - \Delta^2]^{1/2}} = \frac{N(E)}{N(0)} \quad (4.2.2)$$

is the result that would be derived from the semiconductor model of Fig. 4 using a density of states in the superconductor of the form

$$N(E)_\mathrm{S} = N(0)|E|/(E^2 - \Delta^2)^{1/2}. \quad (4.2.3)$$

From an experimental point of view one can therefore measure the density-of-states in the superconductor by making a detailed measurement

---

[7] J. R. Schrieffer, "Tunneling Phenomena in Solids" (E. Burstein and S. Lundquist, eds.), p. 287. Plenum Press, New York, 1969.

of the tunnel conductance, and this technique will be discussed as well as the experiments on direct measurements of the energy gap.

The I–V curve for a Mg–MgO–Sn junction at 0.30 K is shown in Fig. 6a, and the rapid rise of current in the vicinity of $V = \Delta/e = 1.34 \times 10^{-3}/e$ V ($\Delta = 1.34 \times 10^{-3}$ eV) determines the energy gap 2Δ with fair accuracy. The quantity dI_{NS}/dV can also be determined by modulating I_{NS} with a small oscillating current and measuring the resulting ΔV produced by the modulation as described in Section 4.5.3. The resulting curve for an Mg–MgO–Pb junction at 0.33 K is shown in Fig. 6b and has been normalized to $(dI/dV)_{\mathrm{NN}}$. The large peak at $eV = \Delta$ is clear confirmation of the general behavior expected from Eq. (4.2.2), which derives from the BCS theory with a density of states as given by Eq. (4.2.3). The extra bumps observed at higher energies, however, represent deviations from the smooth curve predicted by Eq. (4.2.2) and it turns out that the phonon spectrum of the metal must also be taken into account since it is reflected in the electron density of states and produces deviations from Eq. (4.2.3). This observation immediately suggests that the superconducting tunneling experiment can also be used to obtain important information on the phonon spectrum in metals and this will be discussed in more detail in Section 4.2.7 and 4.2.8.

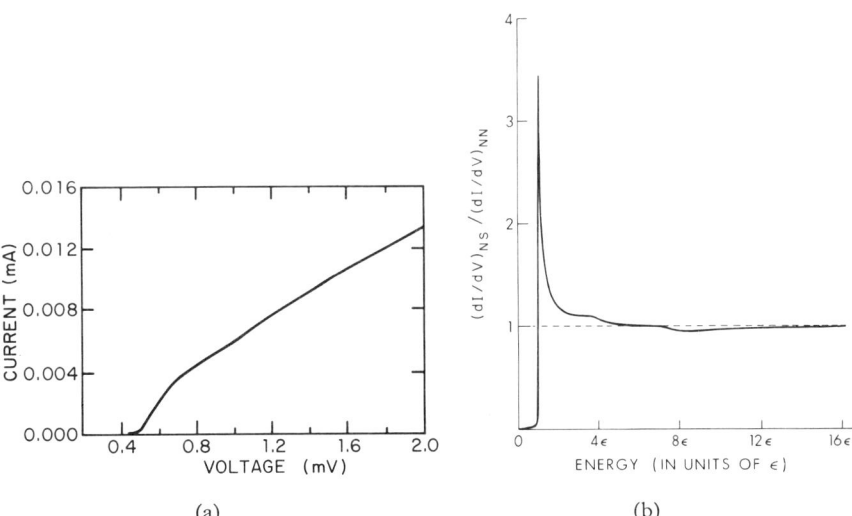

FIG. 6. (a) I–V curve for a Mg–MgO–Sn junction at $T = 0.30$ K. Rapid rise of current is observed when $eV = \Delta = 0.52$ meV. (b) Conductance versus voltage for a Pb–MgO–Mg junction at 0.33 K. Peak in conductance occurs at $eV = \Delta = 1.34 \times 10^{-3}$ eV [from I. Giaever, H. R. Hart, and K. Megerle, *Phys. Rev.* **126**, 941 (1962)].

4.2.3. Tunneling between Two Superconductors

So far we have considered the case where only one of the metal electrodes is in the superconducting phase. Tunneling experiments are often done with the same superconductor forming both electrodes or with two different superconductors separated by an oxide barrier. In either case the density of states on both sides of the barrier will now have a gap at the Fermi level and the tunneling characteristic will be modified accordingly. Figure 7a shows the semiconductor model for a junction made with the same superconductor forming both electrodes. At sufficiently low temperatures, where the thermal excitation of quasi-particles can be neglected, it is clear from Fig. 7a that a tunnel current will now flow only when a voltage is applied such that $eV = 2\varDelta$, as shown in Fig. 7b, and the I–V characteristic appears as in Fig. 7c, which shows data for a Sn–SnO$_x$–Sn junction at 0.30 K.

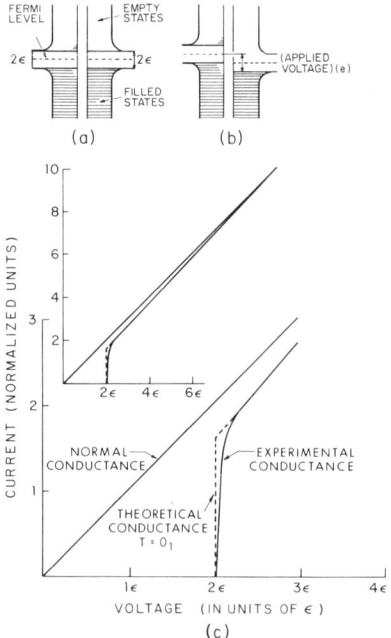

FIG. 7. (a) Semiconductor model of tunnel junction with both electrodes in the superconducting state. (b) Semiconductor model with voltage applied such that $eV = 2\varDelta$. (c) Experimental I–V curve for a Sn–SnO$_x$–Sn junction at 0.30 K with $\epsilon = 0.60 \times 10^{-3}$ eV. Lower curves show comparison to theoretical conductance at $T = 0$ and to normal state conductance.

If the two electrodes are formed from different superconductors with energy gaps $2\varDelta_1$ and $2\varDelta_2$, respectively, then the semiconductor model appears as shown in Fig. 8a, where in this case we have allowed for the presence of thermally excited quasi-particles which play a special role in determining the I–V characteristic in this case. As voltage is applied in the range up to $eV = \varDelta_2 - \varDelta_1$ as shown in Fig. 8b, thermally excited electrons on the left tunnel with an increasing rate as the density of states

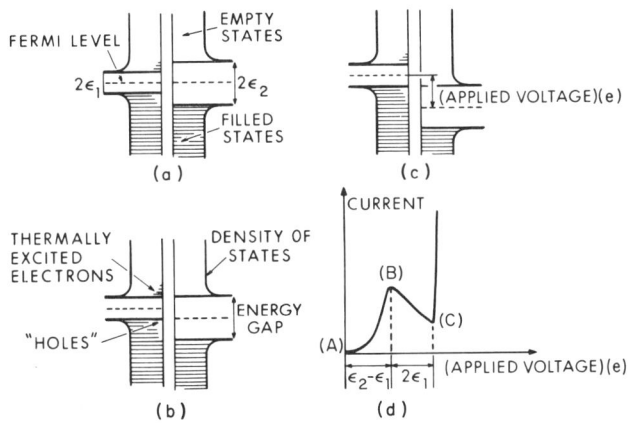

FIG. 8. Semiconductor models and current voltage characteristic for tunnel junction with dissimilar superconducting electrodes. (a) Zero applied voltage; (b) Voltage applied such that $eV = \varDelta_2 - \varDelta_1$; (c) Voltage applied such that $eV = \varDelta_1 + \varDelta_2$; (d) I–V characteristic for an Al–Al$_2$O$_3$–Pb junction at 1 K.

on the right above the energy gap become available. For further increase of voltage the number of excited electrons remains constant, but the density of states on the right decreases, causing a decrease in the tunneling current until a voltage such that $eV = \varDelta_1 + \varDelta_2$ is reached. At this point the electrons below the gap on the left can now tunnel into the empty states above the gap on the right (holes tunnel into the superconductor on the left) as shown in Fig. 8c, and the tunneling current increases again. The corresponding I–V characteristic will have a region of negative resistance and appears as shown in Fig. 8d, which shows the I–V curve measured at 1 K for an Al–Al$_2$O$_3$–Pb junction.

4.2.4. Deviations from Ideal Tunneling Behavior

The discussion above refers to the expected ideal behavior of superconducting tunnel junctions. In practice there are often factors that produce deviations from ideal behavior, and anyone attempting serious experiments on tunnel junctions should be aware of these. In the case where the junction has both electrodes made from the same superconductor, the voltage should theoretically make a discontinuous jump at $eV = 2\varDelta$ even for finite temperatures. As shown, for example, in Fig. 7c this is never observed (the current rise has a finite slope at $eV = 2\varDelta$). Explanations advanced for this observation include the finite lifetime of excited electrons in the superconductor, which may smear the density of states and the averaging of the energy gap anisotropy due to the use of polycrystalline films.

When two different superconductors are used as electrodes, the expected negative resistance region may be present at intermediate temperatures and then disappear at lower temperatures. This observation indicates that not all of the current is flowing by tunneling and that there is a measurable leakage current through the junction. The negative resistance depends on the thermally excited electrons whose density dies out exponentially as the temperature is lowered, while the leakage current is independent of temperature and may mask out the negative resistance region at low temperature. This behavior, in fact, suggests the following criterion for a good junction between two dissimilar superconductors. The cusp in the I–V characteristic at $\varDelta_2 - \varDelta_1$ should be quite sharp, and the negative resistance region from $\varDelta_2 - \varDelta_1$ to $\varDelta_1 + \varDelta_2$ should be well defined.

In the case where a normal metal–insulator–superconductor junction is measured well below the transition temperature of the superconductor, the junction can only be considered to be good if the conductance at zero bias in the superconducting state $(dI/dV)_S$ is very small compared to the zero bias conductance in the normal state $(dI/dV)_N$. This ratio can be checked at low temperature by applying a magnetic field to quench the superconductivity. Rowell[8] has suggested that in good junctions of Al–I–Pb, Al–I–Sn, or Al–I–In at 1 K this ratio should be less than 10^{-3}. If this ratio is large, it is questionable whether a valid gap measurement can be made and a large ratio is usually indicative of gaplessness or the presence of a metallic short. Further discussion of tests for good junction behavior is given in Section 4.5.2.

When a magnetic field is applied to the junction, one must also be on the lookout for leakage current due to trapped magnetic flux. If hysteresis is observed in the tunnel current with application and removal of the magnetic field, then small regions of trapped flux may be producing small normal areas in parallel with the superconducting areas and this will

4.2.5. Experimental Determination of the Energy Gap

The measurement of the gap from the *I–V* characteristic must still be done with some care, even though the junction has been established as a good junction. A method used by Rowell[8] for Pb–I–Pb junctions is illustrated in Fig. 9. The voltage that corresponds to a current equal to one half of the current value extrapolated from voltages just above 2Δ is picked as the appropriate value of 2Δ and is indicated by the vertical dotted line in Fig. 9. The size of the current jump at 2Δ should be $\pi/4$ times the normal current at that voltage for weak coupling superconduc-

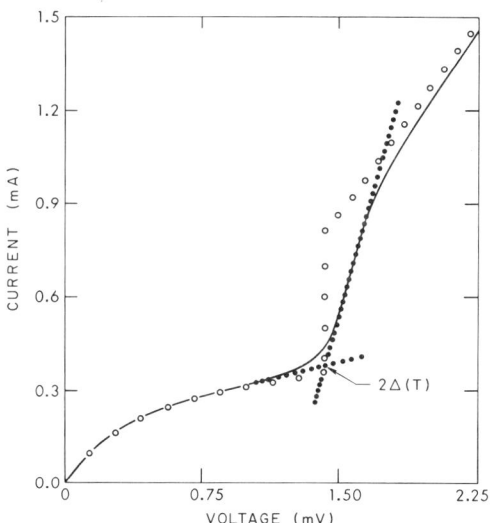

FIG. 9. Construction used for determining the correct value of 2Δ from the *I–V* curve for Pb–I–Pb at $T = 6.5$ K. Voltage value is picked where current is one half the value of the current extrapolated from values above 2Δ. This is then compared to the current voltage characteristic of the normal state. The values are for experimental (—), constructed (···), and calculated (○) results [from J. M. Rowell, "Tunneling Phenomena in Solids" (E. Burstein and S. Lundquist, eds.), p. 273. Plenum Press, New York, 1969].

[8] J. M. Rowell, "Tunneling Phenomena in Solids" (E. Burstein and S. Lundquist, eds.), p. 273. Plenum Press, New York, 1969.

tors and $1.05 \times \pi/4$ for Pb, and these points are indicated in Fig. 9. After picking the voltage for $2\varDelta$ by the one-half current method, the value of the corresponding normal current point can then be compared to the current jump at $2\varDelta$ to check if a good pick for the gap value has been made. The normal I–V characteristic is recorded by applying a magnetic field to the junction.

The method just outlined appears to be about the best for accurate determination of the gap although other acceptable methods have appeared in the literature. For example, Gasparovic et al.[9] used the construction as shown in Fig. 10 where the gap is identified by the point at which the current in the I–V curve first begins to increase rapidly.

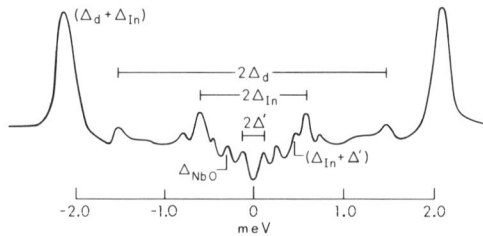

Fig. 10. Alternate construction used to determine $2\varDelta$ from I–V curve. Dotted line shows construction [from R. F. Gasparovic, B. N. Taylor, and R. E. Eck, *Solid State Commun.* **4**, 59 (1966)].

A number of additional methods for determining the energy gap from the experimental data have been discussed by McMillan and Rowell[10] and some of these are summarized briefly: (1) For junctions in which only one electrode is superconducting, an estimate of the gap can be made by determining the voltage at which $(dI/dV)_N = (dI/dV)_S$. This is done by first measuring the junction I–V curve at a temperature above T_c where the behavior is ohmic, and then measuring the superconducting characteristic. The point at which the two slopes match then determines $\varDelta$. This method is reasonably good if the second measurement is made at $T \ll T_c$, but can be seriously in error if the measurement is made close to T_c. (2) In the case of normal metal–insulator–superconductor junctions near T_c, an accurate measurement of the gap can be made by measuring the conductance at zero bias in the superconducting state

---

[9] R. F. Gasparovic, B. N. Taylor, and R. E. Eck, *Solid State Commun.* **4**, 59 (1966).
[10] W. L. McMillan and J. M. Rowell, "Superconductivity" (R. D. Parks, ed.), p. 561. Dekker, New York, 1969.

versus the conductance in the normal state. Then the resistance R is given by

$$R = (dI/dV)_S/(dI/dV)_N. \qquad (4.2.4)$$

At zero bias, R depends on Δ/kT as given by the following equation for dI_{SN}/dV:

$$\frac{dI_{SN}}{dV} = C_n \int_\Delta^\infty \frac{|\varepsilon|}{(\varepsilon^2 + \Delta^2)^{1/2}} \frac{\beta \exp[\beta(\varepsilon + V)]}{\{1 + \exp[\beta(\varepsilon + V)]\}^2} d\varepsilon$$
$$+ \int_{-\infty}^\Delta \frac{|\varepsilon|}{(\varepsilon^2 + \Delta^2)^{1/2}} \frac{\beta \exp[\beta(\varepsilon + V)]}{\{1 + \exp[\beta(\varepsilon + V)]\}^2} d\varepsilon, \qquad (4.2.5)$$

where $\beta = 1/KT$ and $\varepsilon =$ Bloch energy. This equation has been derived assuming a BCS density of states, and this method is therefore best for soft superconductors. More details can be found in Ref. 10.

In the case of dissimilar superconductors, the identification of critical voltage points is often considerably more difficult, particularly for the $\Delta_2 - \Delta_1$ point. According to Rowell,[8] the best way seems to be to find the origin of the logarithmic singularity at $\Delta_2 - \Delta_1$ by trial and error methods. The $\Delta_1 + \Delta_2$ point can then be found by the one-half current method used in the 2Δ case above. An alternative method consists of measuring the current–voltage characteristic of the junction as the helium is pumped down to low temperature. If the first electrode is superconducting already and the second electrode goes superconducting in this range, a sharp break will appear in the curve at exactly Δ_2. By adjusting the bath temperature, the location of this break can be determined with good accuracy and any thermal emfs can be eliminated by measuring both directions of bias. The gap Δ_2 is therefore known at the transition temperature of the second electrode and the correction to zero temperature can be made by using the BCS dependence of Δ on T. For Pb, Sn, and In, this temperature correction is quite small.

During the last ten years, a large number of papers have reported measurements of the energy gap in superconductors using the above techniques. Table I summarizes a number of these measurements and lists the type of junction used, the value of 2Δ measured, and the reference to the original paper which should be consulted for more details.

As a final point, it is often useful to know the exact transition temperature of the superconducting films, particularly since for a number of metals it may be quite different from the bulk value of T_c. One can do this using a straightforward dc resistance measurement by connecting

TABLE I. Measurements of the Energy Gap in Superconductors

Metal	T_c (K)	Junction	Energy gap 2Δ (meV)[b]	Reference
Pb	7.2	Pb–PbO$_x$–Pb	2.68	c,d
		Ta–TaO$_x$–Pb	2.67	e
		Ta–TaO$_x$–Pb	2.90	e
		Pb–PbO$_x$–Pb	2.80 ($2\Delta_0$)	f
Nb	9.2	Nb–NbO$_x$–Pb[a]	3.02	e
		Nb–NbO$_x$–Sn[a]	3.05	g
		Nb–Nb$_2$O$_5$–Sn[a]	2.90	h
Ta	4.5	Ta–TaO$_x$–Ag	1.42 ($2\Delta_0$)	i
		Ta–TaO$_x$–Pb[a]	1.40	j
		Ta–TaO$_x$–Pb[a]	1.40	g
		Ta–TaO$_x$–Pb[a]	1.34	k
		Ta–Ta$_2$O$_5$–Pb	1.35	h
Sn	3.8	Nb–NbO$_x$–Sn[a]	1.15	g
			1.11	l
			1.02	m
		Sn–SnO$_x$–Sn	1.21 ($2\Delta_0$)	f
In		Al–AlO$_x$–In	1.08 ($2\Delta_0$)	f,m
La	5.0	Al–Al$_2$O$_3$–La	1.4	o
		Mg–MgO$_x$–La	0.66 − 0.76	p
Hg	4.15	Al–Al$_2$O$_3$–Hg	1.65 ($2\Delta_0$)	q
		Al–Al$_2$O$_3$–Hg	1.66 ($2\Delta_0$)	r
Al		(50 Å) Al–Al$_2$O$_3$–Ag	0.86	s
		(420 Å) Al–Al$_2$O$_3$–Pb	0.80	t
		(9800 Å) Al–Al$_2$O$_3$–Pb	0.63	t
Re	1.7	Re–C–In	3.59 ($2\Delta_0/kT_c$)$_{AV}$	u
GeTe		Al–Al$_2$O$_3$–GeTe	0.15	v

TABLE I (*continued*)

Metal	T_c (K)	Junction	Energy gap 2Δ (meV)[b]	Reference
$Pb_{40}Tl_{60}$		$Al-AlO_x-Pb_{40}Tl_{60}$	1.61 $(2\Delta_0)$	f,w
$Pb_{60}Tl_{40}$		$Al-AlO_x-Pb_{60}Tl_{40}$	2.16 $(2\Delta_0)$	f
$Pb_{90}Bi_{10}$		$Al-AlO_x-Pb_{90}Bi_{10}$	3.08 $(2\Delta_0)$	f
$Pb_{60}Bi_{20}Tl_{20}$		$Al-AlO_x-Pb_{60}Bi_{20}Tl_{20}$	3.0 $(2\Delta_0)$	f
Ga	1.08	$\langle 001 \rangle$ $Ga-GaO_x-Pb$	3.63 $(2\Delta_0/KT_c)$	x
Tl	2.41	Mg–MgO–Tl	0.74	y
		$Al-Al_2O_3-Tl$	0.74	y

[a] First electrode formed from a bulk piece of material.
[b] Δ_0 is Δ extrapolated to $T = 0$ K.
[c] I. Giaever, *Phys. Rev. Lett.* **5**, 464 (1960).
[d] J. Nicol, S. Shapiro, and P. Smith, *Phys. Rev. Lett.* **5**, 461 (1960).
[e] M. D. Sherrill and H. H. Edwards, *Phys. Rev. Lett.* **6**, 460 (1961).
[f] J. M. Rowell, W. L. McMillan, and R. C. Dynes, Bell Lab. Rep.
[g] P. Townsend and J. Sutton, *Phys. Rev.* **128**, 591 (1962).
[h] I. Giaever, *Int. Conf. Low Temp. Phys., 8th*, p. 171. Butterworths, London and Washington, D.C., 1963.
[i] L. Y. L. Shen, *Phys. Rev. Lett.* **24**, 1104 (1970).
[j] A. F. G. Wyatt, *Phys. Rev. Lett.* **13**, 160 (1964).
[k] I. Dietrich, *Int. Conf. Low Temp. Phys., 8th*, p. 173. Butterworths, London and Washington, D.C., 1963.
[l] S. S. Shapiro, P. H. Smith, J. Nicol, J. L. Miles, and P. F. Strong, *IBM J. Res. Develop.* **6**, 34 (1962).
[m] I. Giaever and K. Megerle, *Phys. Rev.* **122**, 1101 (1961).
[n] R. C. Dynes, *Phys. Rev. B* **2**, 644 (1970).
[o] J. J. Hauser, *Phys. Rev. Lett.* **17**, 921 (1966).
[p] A. S. Edelstein and A. M. Toxen, *Phys. Rev. Lett.* **17**, 197 (1966).
[q] S. Bermon and D. M. Ginsberg, *Phys. Rev.* **135**, A306 (1964).
[r] W. N. Hubin and D. M. Ginsberg, *Phys. Rev.* **188**, 716 (1969).
[s] R. Meservey, P. M. Tedrow, and P. Fulde, *Phys. Rev. Lett.* **25**, 1271 (1970).
[t] D. H. Douglass, Jr. and R. Meservey, *Phys. Rev.* **135**, A19 (1964).
[u] S. I. Ochici, M. L. A. MacVicar, and R. M. Rose, *Phys. Rev. B* **4**, 2988 (1971).
[v] P. J. Stiles, L. Esaki, and J. F. Schooley, *Phys. Lett.* **23**, 206 (1966).
[w] J. M. Rowell and W. M. McMillan, *Phys. Rev.* **178**, 897 (1969).
[x] K. Yoshihiro and W. Sasaki, *J. Phys. Soc. Japan* **24**, 426 (1968).
[y] T. D. Clark, *J. Phys. C (Proc. Phys. Soc.)* **1**, 732 (1968).

leads to both ends of the electrode strip and measuring the resistance transition as the temperature is raised. Alternatively, one can look at the zero bias tunnel conductance as the temperature is raised and observe the point where it first starts to increase. The second method is often useful in that it can indicate a nonuniform film in which the edges have a higher transition temperature than the main part of the film. This will produce two distinct increases in conductance at different temperatures. Aluminum films, particularly when rather thin, can show a very great deviation in both T_c and H_c from bulk values. Values of T_c up to 2.8 K and critical fields up to 60 kOe have been obtained (bulk values T_c = 1.1 K, H_c = 1000 Oe). The high critical fields obtainable in thin aluminum films have in fact made it possible to do a number of superconducting tunneling experiments in high magnetic fields, and these interesting techniques will be described in Section 4.2.11.

4.2.6. Tunneling into Single Crystals

In a single-crystal superconductor the energy gap will in general be a function of direction in the crystal, and the tunneling measurements of the gap 2Δ in polycrystalline films will be some type of average value. A number of experiments have been carried out using single crystals as one electrode of a tunnel junction, and the energy gap anisotropy has been investigated successfully using this technique. Blackford and March[11] have carried out a series of tunneling experiments on Pb single crystals, which form in the shape of oblate spheriods when Pb is cooled in a vacuum melting process at 10^{-8} to 10^{-7} Torr. Flat facets 1–2 mm in diameter develop with (001) and (111) orientations, and these can be conveniently used to form a junction with an evaporated lead film. Their measurements showed the existence of two gaps labeled Δ_1 and Δ_2, which they interpreted as being due to two groups of tunneling electrons coming from different parts of the Fermi surface of the Pb crystal. Bennett[12] has worked out a detailed model based on tunneling contributions from the second-zone hole surface and the third-zone electron surface. If Δ_2 is identified with the hole surface and Δ_1 with the electron surface, then the tunneling data indicate a maximum for I_2/I_1 of 1.7 in the $\langle 111 \rangle$ direction and a minimum of 0.88 in the $\langle 001 \rangle$ direction. On the other hand, the gaps themselves did not show a very strong anisotropy. $2\Delta_1$ changes by a

---

[11] B. L. Blackford and R. H. March, *Phys. Rev.* **186**, 397 (1969).
[12] A. J. Bennett, *Phys. Rev.* **140**, A1902 (1965).

TABLE II. Energy Gap Anisotropy Measurements with Single Crystals

Metal	Crystal orientation	Junction	Energy gap 2Δ (meV)	Reference
Ta	⟨100⟩	Ta–TaO$_x$–Pb	1.40	a
	⟨110⟩	Ta–TaO$_x$–Pb	1.34	a
	⟨210⟩	Ta–TaO$_x$–Pb	1.36	a
Pb	⟨111⟩	Pb–PbO$_x$–Pb	2.36 ($2\Delta_1$)	b
	⟨101⟩	Pb–PbO$_x$–Pb	2.48 ($2\Delta_1$)	b
	⟨001⟩	Pb–PbO$_x$–Pb	2.44 ($2\Delta_1$)	b
	⟨111⟩	Pb–PbO$_x$–Pb	2.78 ($2\Delta_2$)	b
Sn		Sn–SnO$_x$–Sn	4.3 ($2\Delta/KT_c)_{max}$	c
			2.8 ($2\Delta/KT_c)_{min}$	
Nb		Nb–NbO$_x$–In	3.10 ($2\Delta_d$)	d
			0.30 ($2\Delta_s$)	
	⟨111⟩	Nb–NbO$_x$–In	3.20	e
	⟨112⟩	Nb–NbO$_x$–In	3.06	e
	⟨110⟩	Nb–NbO$_x$–In	3.08	e
Ga	⟨010⟩	Ga–GaO$_x$–Pb	3.78 ($2\Delta_0/KT_c$)	f
	⟨110⟩	Ga–GaO$_x$–Pb	3.78–3.63 ($2\Delta_0/KT_c$)	f
	⟨001⟩	Ga–GaO$_x$–Pb	3.63 ($2\Delta_0/KT_c$)	f
Re	[0001]	Re–C–In	3.35 ($2\Delta_0/KT_c$)	g
	[1$\bar{2}$10]	Re–C–In	3.91 ($2\Delta_0/KT_c$)	g
NbSe$_2$	[c axis]	NbSe$_2$–C–Pb	1.24	h
	[c axis]	NbSe$_2$–C–In	1.24	h

[a] I. Dietrich, *Int. Conf. Low Temp. Phys. 8th*, p. 173. Butterworths, London and Washington, D.C., 1963.
[b] B. L. Blackford and R. H. March, *Phys. Rev.* **186**, 397 (1969).
[c] N. V. Zavaritskii, *Sov. Phys. JETP* **21**, 557 (1965).
[d] J. W. Hafstrom and M. L. A. MacVicar, *Phys. Rev. B* **2**, 4511 (1970).
[e] M. L. A. MacVicar and R. M. Rose, *J. Appl. Phys.* **39**, 1721 (1968).
[f] K. Yoshihiro and W. Sasaki, *J. Phys. Soc. Japan* **24**, 426 (1968).
[g] S. I. Ochici, M. L. A. MacVicar, and R. M. Rose, *Phys. Rev. B* **4**, 2988 (1971).
[h] R. C. Morris and R. V. Coleman, *Phys. Lett.* **43A**, 11 (1973).

maximum of about 5% while $2\Delta_2$ shows a change considerably less than 5% as a function of crystal orientation. The values are listed in Table II, which also summarizes other energy gap measurements on single crystals along with appropriate references.

Dietrich[13] has made tunneling measurements on single crystals of tantalum obtained by cutting large grains from a slowly cooled tantalum block. The junctions were made by evaporating a lead film as the second electrode after the tantalum was oxidized using procedures similar to those listed in Section 4.5.1. The gap $2\varDelta$ in this case also showed a relatively small variation with crystal orientation being equal to about 4% as indicated by the values in Table II.

In the case of Sn, Zavaritskii[14] has observed a rather large anisotropy in the energy gap. The tin single crystals were prepared by casting against a glass plate in vacuum and the second electrode was an evaporated tin film. The anisotropy in $\varDelta$ was a rather complicated function of the crystal orientation and the data was not sufficient to make a complete identification with the topology of the Fermi surface. However, some correlation is suggested and his paper should be consulted for the detailed analysis. The tunneling measurements were made at 1.36 K and he took $eV_{\text{meas}} = \varDelta_k + \varDelta^*$, where $\varDelta^* = 0.56$ meV is the width of the gap for the polycrystalline tin film serving as the second electrode. His paper gives the data in terms of relative units defined as $(\varDelta_K) = 2\varDelta_K/kT_c$, for which he observed a variation from a maximum of $(\varDelta_K) = 4.3$ to a minimum of $(\varDelta_K) = 2.8$. There were fairly large regions of orientation where $(\varDelta_K)$ was observed to be fairly constant, and these were separated by rather sharp boundaries where an abrupt change of $(\varDelta_K)$ occurs. The general analysis indicates an identification of the regions of constant $(\varDelta_K)$ with specific bands of tin. The value of $(\varDelta_K)$ can also be used to indicate the amount of deviation from free electron behavior within each band. The boundaries of rapid change in $(\varDelta_K)$ would then correspond to singularities in the Fermi surface topology.

Hafstrom and MacVicar[15] have recently made a careful study of tunneling into high-purity single crystals of niobium having residual resistance ratios of 3000. Their experimental techniques for preparing the junctions are quite sophisticated and are described in Section 4.5.1. Their original article should also be consulted for an excellent description of the relevant experimental behavior and testing of niobium tunnel junctions. Their experimental results definitely support the existence of two energy gaps in niobium, one of which is due to electrons of s

---

[13] I. Dietrich, *Int. Conf. Low Temp. Phys. 8th*, p. 173. Butterworths, London and Washington D.C., 1963.

[14] N. V. Zavaritskii, *Sov. Phys. JETP* **21**, 557 (1965).

[15] J. W. Hafstrom and M. L. A. MacVicar, *Phys. Rev. B* **2**, 4511 (1970).

character and the other to electrons of d character. A conductance versus voltage plot for a niobium single crystal with an indium electrode is shown in Fig. 11. The large energy gap $2\varDelta_d$ is the usual one measured, while the one labeled $\varDelta'$ is the one to be associated with electrons of s-like character. The relative magnitudes are given by $\varDelta_s \approx (1/10)\,\varDelta_d$. The authors report that $\varDelta_s$ exhibits very little change with crystal orientation, while $\varDelta_d$ shows a minimum along $\langle 100 \rangle$ and is larger along $\langle 110 \rangle$ and $\langle 111 \rangle$.

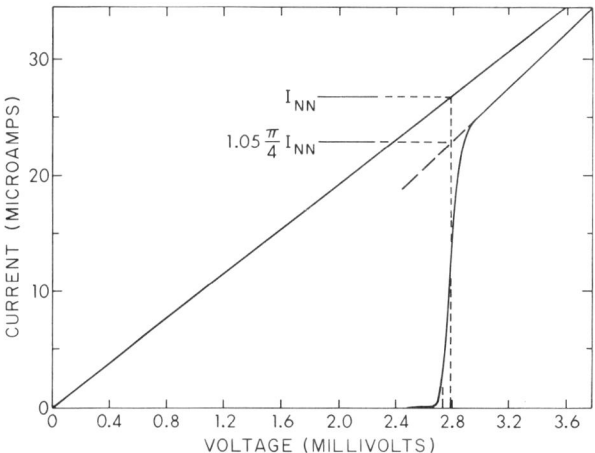

FIG. 11. Conductance versus voltage plot for a Nb–NbO$_x$–In junction using a niobium single crystal as one electrode with $T = 1$ K. The large gap $2\varDelta_d$ is the usual gap measured for niobium. The small gap $2\varDelta'$ is associated electrons of s-like character in niobium (from J. W. Hafstrom and M. L. A. MacVicar, *Phys. Rev. B* **2**, 4511 (1970)].

Ochiai et al.[16] have recently completed tunneling measurements on the energy gap anisotropy in single crystals of rhenium. They observe a variation of the reduced energy gap $2\varDelta_0/kT_c$ from 3.35 to 3.91 kT_c for the gap determined from single-particle tunneling. The authors also determined the energy gap from the double-particle tunneling peaks and find that they are consistently smaller than those determined from the single-particle peaks. This observation suggests that the double-particle tunneling selection rules may be different from those for single-particle tunneling. The anisotropies in the energy gap determined from the single-particle and double-particle tunneling peaks are 15.7 and 17.7%, respectively. These experiments show no indication that different

[16] S. I. Ochici, M. L. A. MacVicar, and R. M. Rose, *Phys. Rev. B* **4**, 2988 (1971).

sheets of the Fermi surface are contributing any variations or structure to the gap. The authors conclude that the electronic band structure does not play a dominant role in the gap anisotropy of hcp superconductors, but that phonon dispersion is the major factor responsible for the gap anisotropy. The tunneling experiments on single crystals of rhenium were made using a barrier made by evaporating high-purity carbon 20–50 Å thick. Further experimental details of this technique are given in Section 4.5.1.

4.2.7. Tunneling Measurements of Density of States and Phonon Spectra

As pointed out in Section 4.2.2, for a normal metal–insulator–superconductor junction the derivative of the tunneling I–V curve $(dI/dV)_S$ in the superconducting phase, divided by $(dI/dV)_N$ for the normal phase is equal to $N(E)/N(0)$, where $N(E)$ is the density of excited allowed states in the superconductor and $N(0)$ is the density of states in the normal metal. Tunneling experiments have definitely confirmed that the BCS expression given by Eq. (4.2.2) is a reasonably accurate description of the behavior. However, careful experiments at low temperatures, where thermal smearing is reduced, have shown that deviations in the smooth behavior expected from Eq. (4.2.2) occur at voltages well above the gap voltage. In strong-coupling superconductors such as lead, these are particularly easy to see and can produce up to a 10% charge in conductance over a small energy range. From the theoretical point of view these deviations from the BCS prediction imply that the energy gap Δ is energy dependent and that tunneling is a direct probe of this energy dependence, which in turn is a consequence of the detailed phonon spectrum in the particular metal. All of this suggested that tunneling experiments could therefore give important information on the normal metal properties as well as check the accuracy of various superconducting theories, particularly the modifications of BCS theory necessary to explain the behavior of the strong-coupling superconductors such as Pb and Hg.

Rowell and Kopf[17] have studied lead junctions from this point of view in great detail and have refined the second-derivative technique to such an extent that extensive information can be obtained on the location of both the phonon peaks and the Van Hove[18] critical points. They have used the harmonic detection techniques developed by Thomas and de-

---

[17] J. M. Rowell and L. Kopf, *Phys. Rev.* **137**, 907 (1965).
[18] L. Van Hove, *Phys. Rev.* **89**, 1189 (1953).

scribed in Section 4.5.3. The data have been generally taken on Pb–I–Pb junctions, since the thermal excitation across the energy gap ($\sim$20 kT) is small on both sides and this contributes to a very sharp tunneling characteristic, which is essential in order to obtain high resolution in the second derivative curves. For the junctions used, the width of the current rise at 2Δ was typically 0.15 mV for the region comprising 10–90% of the current jump. The authors believe that this width was probably limited by strains in the lead films due to differential contraction between the film and the substrate and that even sharper second derivative data could be obtained if the current rise at 2Δ were made sharper by eliminating strains.

The density of states variation with energy for lead obtained from the tunnel conductance is shown in Fig. 12, along with the smooth variation predicted by the BCS theory. The two relatively sharp drops to values below the BCS density of states suggest the existence of relatively sharp

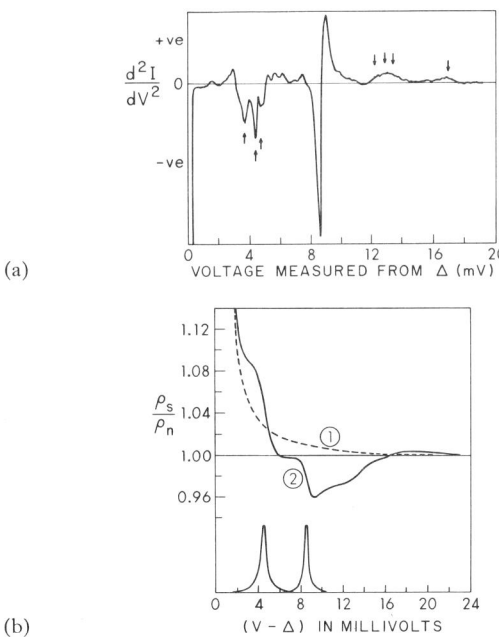

FIG. 12. (a) Second derivative d^2I/dV^2 of current with respect to voltage versus voltage for a Pb–I–Pb junction at 1.3 K and 50 μV. The sharp structure indicated by arrows is due to excitation of phonons. Sums and harmonics of the lower voltage peaks account for the higher voltage peaks. (b) Density of states for lead: (1) theoretical BCS density of states; (2) density of states measured from tunneling experiment. Lower peaks show phonon location [from J. M. Rowell and L. Kopf, *Phys. Rev.* **137**, 907 (1965)].

peaks in the phonon spectrum of lead at 4.4 and 8.5 mV, as shown in the estimated spectrum at the bottom of Fig. 12. These peaks show up clearly as negative minimum in the second derivative curve, which is also plotted in Fig. 12.

In order to explain the conductance result in further detail it is necessary to consider a detailed expression for the energy gap as a function of energy. The most successful theoretical expression has been developed by Eliashberg[19] and is given by

$$\Delta(E) = dn/d \int_{\Delta_0}^{\infty} dE' \, V(E - E') \, \text{Re}\{\Delta(E')/[E^2 - \Delta^2(E')]^{1/2}\}, \qquad (4.2.6)$$

where E is the quasi-particle energy and Δ_0 is the gap parameter at the energy gap. The interaction between the electrons including both the Coulomb repulsion and the phonon attraction is described by $V(E - E')$ and contains the appropriate phonon spectrum. For further reference, Culler et al.[20] have published theoretical curves of Δ/Δ_0 versus energy for both weak- and strong-coupling superconductors. In a tunneling experiment, the appropriate expression for the conductance ratio or the density of states, taking into account the energy dependence of Δ, was obtained by Schrieffer et al.[21] and is given by

$$N_S(E)/N_N(0) = \text{Re}\{E[E^2 - \Delta^2(E)]^{-1/2}\}. \qquad (4.2.7)$$

Schrieffer et al.[21] used a phonon spectrum for lead having two major peaks as shown in Fig. 12 and solved the gap equation using a coupling constant adjusted to give the correct value of the gap parameter Δ_0 at the gap edge. The resulting density of states is plotted in Fig. 13 along with the experimental curve, and generally good agreement is found. The sharp drops in the neighborhood of the phonon peaks are reproduced, as well as higher energy structure that is due to sum and harmonic effects that are generally predicted by solutions to the gap equation and appear to be observed as rather broad structure in the experiment at higher energies.

The Van Hove singularities in the phonon spectrum of the metal also produce structure in the plot of d^2I/dV^2 versus V. Scalapino and Ander-

---

[19] G. M. Eliashberg, *Zh. Eksperim. Teor. Fiz.* **38**, 966 (1960).
[20] G. J. Culler, B. D. Fried, R. W. Huff, and J. R. Schrieffer, *Phys. Rev. Lett.* **8**, 399 (1962).
[21] J. R. Schrieffer, D. J. Scalapino, and J. W. Wilkins, *Phys. Rev. Lett.* **10**, 336 (1963).

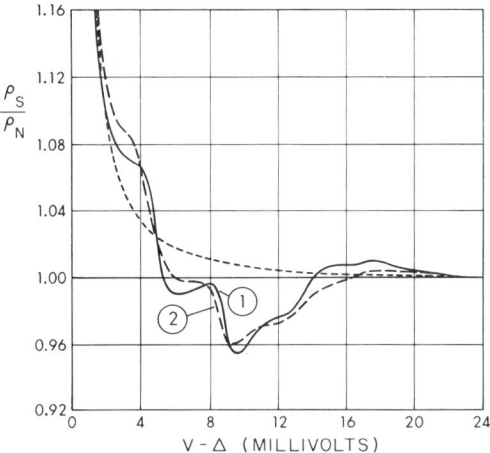

FIG. 13. Comparison between (1) the calculated density of states and (2) the density of states measured by tunneling [from J. M. Rowell and L. Kopf, *Phys. Rev.* **137**, 907 (1965)].

son[22] have analyzed these effects in some detail and concluded that there should be singularities superimposed on the density of states variations that occur due to the general shape of the phonon spectrum. For a critical point where an infinite discontinuity occurs in the derivative of the phonon spectrum with respect to the frequency dg/dV, the following singularities should occur in metal–insulator–superconductor and superconductor–insulator–superconductor junctions. In the case of M–I–S junctions, the d^2I/dV^2 versus voltage should exhibit a logarithmic singularity and jump discontinuity at a voltage equal to the gap Δ plus the energy E_c of the critical point. In the case of an S–I–S junction, an inverse square root singularity should be observed at $2\Delta + E_c$ if the two superconducting films are identical. For finite discontinuities in the derivative of the phonon spectrum, one should observe logarithmic singularities in the d^2I/dV^2 versus V for the S–I–S junction. Rowell and Kopf[17] have compared detailed second derivative plots for their Pb–I–Pb junctions with the critical points predicted by neutron diffraction results. They find a correspondence between the detailed structure and expected critical points although the resolution is not sufficient to make a complete identification. The reader should refer to the excellent paper by Rowell and Kopf[17] for more complete details.

[22] D. J. Scalapino and P. W. Anderson, *Phys. Rev.* **133**, A921 (1964).

This section has given a brief summary of the main points involved in using superconducting tunneling to obtain information on the phonon spectrum of metals. The main experimental consideration is the sharpness of the tunnel characteristic at the gap which in turn determines the resolution of the phonon structure above the gap. Homogeneous strain free films and very low temperatures help in this respect. In practice S–I–S junctions are the best, since thermal smearing of the characteristic

TABLE III. Density of States Measurements and Determination of Phonon Spectra

Metal	Junction	$N(E)$ information	Phonon spectrum information	Energy ranges phonon structure (meV)	Information on critical points	Ref.
Pb	Pb–MgO–Mg	Yes				a
	Pb–PbO$_x$–Pb	Yes	Yes	1.75, 4.4, 8.5	Yes	b
				0.3–5.5 meV		
				7.5–9.5 meV		
	Al–AlO$_x$–Pb	Yes	Yes			b
Sn	Sn–SnO$_x$–Sn	Yes	Yes	1.5	Yes	b
				3.0–19.0		
In	Al–AlO$_x$–In	Yes	Yes	3–7, 11–17		b
		Yes	Yes	6.0, 14.0		c
Tl	Al–AlO$_x$–Tl	Yes	Yes	4.0		b
				9.0–10.5		
	Mg–MgO$_x$–Tl	Yes	Yes	4.0, 9.45	Yes	d
	Al–AlO$_x$–Tl					
Ta	(Bulk) Ta–TaO$_x$–Ag	Yes	Yes	11.4, 18.0	Yes	e
Hg	Al–Al$_2$O$_3$–Hg	Yes	Yes	0.84–12.2	Yes	f,g
La	Al–Al$_2$O$_3$–La	Yes	Yes	4.5, 10.0		h

[a] I. Giaever, H. R. Hart, and K. Megerle, *Phys. Rev.* **126**, 941 (1962).
[b] J. M. Rowell and L. Kopf, *Phys. Rev.* **137**, 907 (1965).
[c] J. G. Adler and J. Rogers, *Phys. Rev. Lett.* **10**, 217 (1963).
[d] T. D. Clark, *J. Phys. C (Proc. Phys. Soc.)* **1**, 732 (1968).
[e] A. F. G. Wyatt, *Phys. Rev. Lett.* **13**, 160 (1964).
[f] S. Bermon and D. M. Ginsberg, *Phys. Rev.* **135**, A306 (1964).
[g] W. N. Hubin and D. M. Ginsberg, *Phys. Rev.* **188**, 716 (1969).
[h] J. S. Rogers and S. M. Khana, *Phys. Rev. Lett.* **20**, 1284 (1968).

is reduced to a minimum. At very low temperatures, 0.8 K or below, it is also possible to measure other superconducting metals against aluminum, since it is generally concluded that the density of states variation in aluminum contains no resolvable structure and follows the smooth curve predicted by BCS. Table III summarizes briefly information and references on phonon spectra obtained by tunneling.

4.2.8. Tunneling Used as a Probe of the Electron–Phonon Interaction

As described in the previous section, detailed tunneling experiments can potentially give very extensive information on the phonon spectrum of a metal as well as determine the basic superconducting gap parameter $\varDelta_0$. Such experimental information, when combined with the microscopic theory of superconductivity, can provide a wealth of information about the metal particularly in the case of the strong coupling superconductors, where phonon effects are easy to see. McMillan and Rowell[10] have developed such a technique by utilizing the Eliashberg[19] equations and appropriate computer programs for fitting the experimental data. The technique starts with the experimental tunneling determination of the gap $\varDelta_0$ and the density of states $N(E)$ for a particular metal. Then an initial guess is made for the values of the parameters needed to solve the Eliashberg equations for the energy-dependent gap parameter $\varDelta(\omega)$. These parameters include μ^*, the Coulomb pseudopotential, and $\alpha^2(\omega)F(\omega)$, where $\alpha^2(\omega)$ is the electron–phonon coupling parameter and $F(\omega)$ is the phonon density. The density of states $N(\omega)$ calculated from the Eliashberg equations is then compared to the experimental density of states and the two parameters μ^* and $\alpha^2(\omega)$ are adjusted until the calculated value of $N(\omega)$ agrees with the experimental value to within one part in 10^3.

Once having accurately determined the parameters $\alpha^2(\omega)F(\omega)$ and μ^*, these can then be used to make further calculations concerning the equilibrium properties of the metal. For example, the magnitude of the electron–phonon matrix elements and the electron–phonon coupling constant can be determined. This in turn determines the enhancement of the specific heat and the cyclotron mass. Properties of the superconductor such as T_c and H_c versus T_c can also be calculated.

Rowell *et al.*[23] have summarized in detail the results of this technique applied to the following metals: Pb, In, Sn, Hg, Tl, Ta, $Pb_{60}Tl_{40}$,

---

[23] J. M. Rowell, W. L. McMillan, and R. C. Dynes, Bell Lab. Rep.

$Pb_{40}Tl_{60}$, $Pb_{60}Tl_{20}Bi_{20}$, and $Pb_{90}Bi_{10}$, and the reader should refer to this work or to the one cited in Ref. 10 for a more detailed account of these impressive applications of tunneling techniques.

4.2.9. Phonon Generation and Detection

The single-particle tunneling experiments that we have been discussing characterize the tunneling of an excited quasi-particle in the superconducting electrode whose initial energy was supplied by the battery. This excited quasi-particle can subsequently decay to lower energies by emitting phonons with an energy spectrum in the range $eV-2\Delta$. Once having reached the top of the gap edge, the quasi-particle can undergo recombination with a quasi-particle of opposite spin to form a pair accompanied by phonon emission of energy 2Δ. The phonons generated by these relaxation and recombination processes will in turn undergo interaction and reach thermal equilibrium with the lattice. The existence of these primary decay phonons suggests that tunneling experiments might be used as a technique to study phonons in solids that are not in thermal equilibrium with the lattice, particularly if the lifetimes are sufficiently long.

Rothwarf and Taylor[24] have carried out experiments to measure the recombination lifetimes in superconductors using aluminum tunnel junctions. The experimental technique utilizes a double tunnel junction, as schematically indicated in the energy versus k diagram of Fig. 14. One

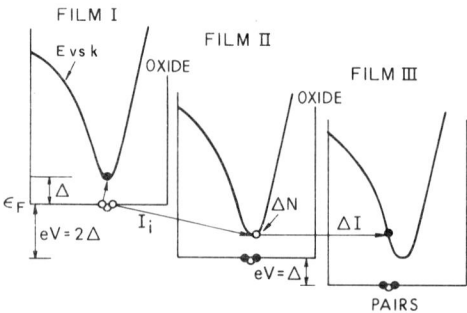

FIG. 14. Schematic energy versus k diagram representing a double tunnel junction. First junction injects quasi-particles into film II and second junction detects the increase in density of quasi-particles in film II [from A. Rothwarf and B. N. Taylor, *Phys. Rev. Lett.* **19**, 27 (1967)].

---

[24] Allen Rothwarf and B. N. Taylor, *Phys. Rev. Lett.* **19**, 27 (1967).

junction is used to inject quasi-particles into the superconducting film and the second junction is used to detect the resulting increase in the density of excited quasi-particles in the film. Injection currents used in the experiment were typically on the order of 5×10^{-2} A/cm², corresponding to a quasi-particle injection rate of 3.1×10^{17}/sec cm². The middle-film thickness was approximately 3×10^{-6} cm, resulting in an injected carrier density of $\sim 10^{23}$/cm³ sec. The theoretical estimate for the lifetime of excited quasi-particles in aluminum is approximately 10^{-7} sec, so that a steady-state density of injected quasi-particles was estimated to be of the order of 10^{16}/cm³. Analysis of the experiments, however, indicates that the recombination phonons also play an important role in that they will create quasi-particles at nearly the same rate as they are being injected. This is true at nearly all temperatures and the experimental technique must be refined to differentiate between these two sources. The authors suggest that pulse experiments may be able to discriminate against the recombination phonon production and thus measure the injected quasi-particle lifetime only.

Experiments to detect the recombination phonons have been carried out by Eisenmenger and Dayem.[25] For the phonon generator they have used a Sn–SnO₂–Sn tunnel diode evaporated onto an optically flat surface of a 1-cm-long sapphire single crystal. The detector was a similar tunnel diode evaporated onto the opposite face of the crystal. The generating diode was biased at a voltage $V_1 > 2\varDelta$, while the receiver is biased at $V_2 < 2\varDelta$. The tunneling process provides injected quasi-particles in both films of the generator and some of the relaxation phonons generated propagate through the sapphire in rectilinear trajectories with little attenuation. These phonons incident on the receiving diode produce a change in tunnel current, which is sensitive to both the frequency and number of phonons. A time-of-flight pulse technique was used to differentiate the generation and detection phenomena. The experiments definitely detect phonons at energies greater than $2\varDelta$, which produce excited quasi-particles by dissociating pairs in the receiver, which in turn produce an increase in conductance at bias $V_2 < 2\varDelta$. As the generator voltage bias V_1 is varied, sharp structure is observed in the receiver current when $V_1 = 4\varDelta$ and $6\varDelta$, in accordance with detection of phonons greater than $2\varDelta$. These experiments provide rather elegant tests of some of the ideas of superconducting tunneling and more complete details can be found in Chapter 26 of Ref. 1.

---

[25] W. Eisenmenger and A. H. Dayem, *Phys. Rev. Lett.* **18**, 125 (1967).

4.2.10. Geometrical Resonances in Tunneling

When a thick (several microns) superconducting film is used as one of the electrodes in a tunnel junction, it is possible to set up a type of standing wave resonance. The excitation of these resonances introduces structure into the d^2V/dI^2 curve, and such data can be used to yield experimental information on physical quantities such as $E(\bar{k})$, the Fermi velocity v_F, and various boundary phenomena.

The origin of these effects is connected with the twofold degeneracy of the energy–momentum distribution of excited quasi-particles in the superconductor. As indicated in Fig. 3b of Section 4.2.2, a single excitation with $E > \Delta$ can be associated with two wave vectors $k_1(E) < k_F < k_2(E)$. A composite excitation can result from a coherent composition of these two degenerate states and such a composite excitation may transport only a small momentum compared to $\hbar k_F$ and therefore have a deBroglie wavelength much greater than $\lambda = 2\pi/k_F$. These standing-wave resonances were first observed by Tomasch,[26] and the resulting structure in the d^2V/dI^2 curve is shown in Fig. 15. These effects have been observed in films up to 30 μ thick. They basically result from a perturbation of the density of states due to an electronic standing wave which occurs when the electronic mean free path in the superconductor is in the micron range. The experimental data confirm the electronic standing wave idea by exhibiting structure such that $h\nu \propto 1/d$, where d is the thickness of the film. The theory of the geometrical resonance was first presented by McMillan and Anderson[27] and a one-dimensional model of the basic idea can be found in Tomasch.[28]

The basic techniques involved in such an experiment proceed similarly to those for standard tunnel junctions except that one electrode is made several microns or more thick and care is taken to see that the electronic mean free path is equal to or exceeds the film thickness. This can usually be determined by making a resistivity measurement on the thick electrode at the temperature of the tunneling experiment. The experimental behavior can also be influenced by evaporating an additional metal film on top of the thick electrode, thus changing the boundary condition at the resonance termination surface far from the junction interface.

---

[26] W. J. Tomasch, *Phys. Rev. Lett.* **15**, 672 (1965).

[27] W. L. McMillan and P. W. Anderson, *Phys. Rev. Lett.* **16**, 85 (1966).

[28] W. J. Tomasch, "Tunneling Phenomena in Solids" (E. Burstein and S. Lundquist, eds.), p. 315. Plenum Press, New York, 1969.

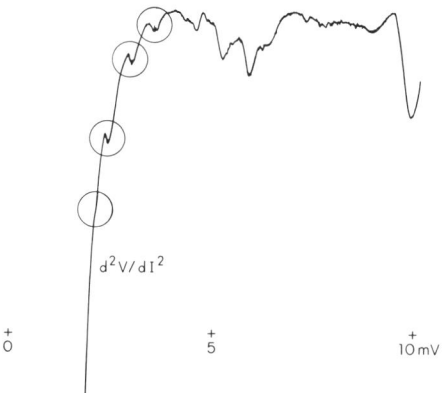

FIG. 15. Voltage dependence of the second derivative for a Pb^{210} film 2.79 μ in thickness with $T = 1.1$ K and rms voltage = 70 μV. The structure due to geometrical resonances is indicated by circles [from W. J. Tomasch, *Phys. Rev. Lett.* **15**, 672 (1965)].

4.2.11. Superconducting Tunneling in High Magnetic Fields

Meservey and Tedrow[29] have used tunneling in thin aluminum films to observe the magnetic field splitting of the quasi-particle states in the superconducting phase. This technique has worked very well for aluminum, since films about 50 Å thick evaporated onto a liquid-nitrogen-cooled substrate have critical fields extrapolated to 0 K of about 48 kOe. If the films are thin enough and provided the spin–orbit scattering rate is small, the magnetic field will effect the electron spins to a much greater extent than the electron orbits. The BCS quasi-particle spectrum will then be given by $E_k\uparrow, \downarrow = (\varepsilon_k{}^2 + \varDelta^2)^{1/2} \pm \mu H$, where μ is the magnetic moment of the electron and the spectrum has simply been split and shifted in energy by a Zeeman term. In the presence of a magnetic field, the tunneling density of states would therefore consist of the addition of two BCS-type density of states curves shifted in energy by $eV = \pm \mu H$ from the position of the single density of states curve in zero field. At temperatures of about 0.4 K, the thermal broadening is sufficiently reduced that this shift can be resolved and experimental data are shown in Fig. 16, which shows a plot of the normalized conductance versus voltage for an Al–Al$_2$O$_3$–Ag junction. As the field is increased two peaks in the conductance are clearly resolved, corresponding to the spin-up and spin-down density of states, respectively. Figure 17 shows the voltage

---

[29] R. Meservey and P. M. Tedrow, *Phys. Rev. Lett.* **25**, 1270 (1970).

154   4. ELECTRON TUNNELING IN SOLIDS

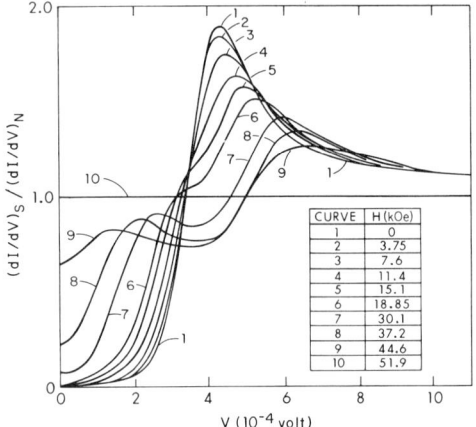

FIG. 16. Normalized conductance versus voltage for an Al–Al$_2$O$_3$–Ag junction at 0.4 K and successively increasing magnetic fields [from R. Meservey and P. M. Tedrow, *Phys. Rev. Lett.* **25**, 1270 (1970)].

value measured for the two peaks as a function of field and they clearly exhibit the Zeeman splitting expected from the straightforward analysis.

The above authors have also successfully applied this technique of field splitting to measure the spin-dependent tunneling into ferromagnetic nickel.[30] Since the density of states is rather sharp at the edge of the gap, the field splitting observed in the tunneling from thin aluminum films

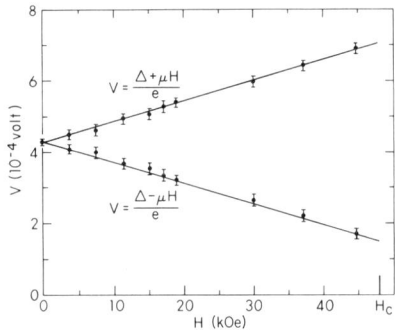

FIG. 17. Voltage corresponding to the conductance peaks observed in Fig. 16 plotted as a function of applied magnetic field. Zeeman splitting is clearly demonstrated [from R. Meservey and P. M. Tedrow, *Phys. Rev. Lett.* **25**, 1270 (1970)].

[30] P. M. Tedrow and R. Meservey, *Phys. Rev. Lett.* **26**, 192 (1971).

allows one to select tunneling electrons with a given spin direction. Such a polarized tunnel current can then be used to investigate the polarization in the density of states of a second metal used as the other junction electrode. In the case of nickel, the experiment was carried out by measuring the tunneling conductance $(dI/dV)_\text{S}/(dI/dV)_\text{N}$ as a function of voltage for an Al–Al$_2$O$_3$–Ni junction at 0.4 K and in fields up to 50 kOe, and the results are shown in Fig. 18. The nickel film was approximately 500 Å thick and is, of course, saturated as a single domain in the direction of the magnetic field which was parallel to the film surface. The two peaks in the tunnel conductance due to the spin split density of states in the

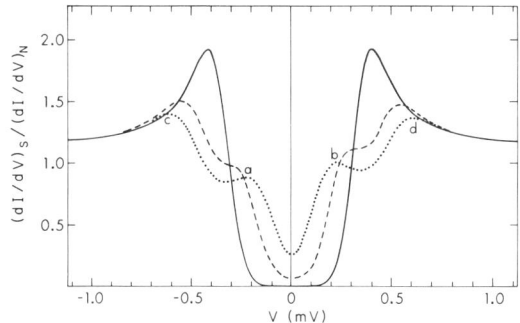

FIG. 18. Conductance versus voltage measured for an Al–Al$_2$O$_3$–Ni junction at 0.4 K in fields up to 50 kOe. The values for H are: 0 (—), 22.6 (– – –), and 33.7 (· · ·). The left–right asymmetry is assigned to the difference in the spin up and spin down density of states in Ni [from P. M. Tedrow and R. Meservey, *Phys. Rev. Lett.* **25**, 1270 (1970)].

superconducting aluminum are again observed. However, the peaks for positive and negative values of the applied voltage are not the same height. This can be explained in terms of the difference in the density of states for spin-up and spin-down electrons in the nickel film and can be qualitatively understood from the semiconductor model of Fig. 19. The top shows the field splitting of the zero-field aluminum density of states curve into two BCS-type curves shifted by $eV = \pm\mu H$. The dashed curve represents spin up (spin parallel to H, magnetic moment antiparallel to H, energy increased) and the dotted curve represents spin down (spin antiparallel to H, magnetic moment parallel to H, energy decreased). On the bottom is shown the density of states of ferromagnetic nickel, where it is assumed that the density of states for spin-down electrons is greater than for spin-up electrons. As a positive bias is applied the Fermi level of the nickel shifts up, and electrons tunnel into the aluminum from the spin-down bands of nickel when $eV = \Delta - \mu H$ and the peak at point b of Fig. 18 is observed. When a negative bias is applied, electrons tunnel into the aluminum from the spin-up bands of nickel when $eV = -(\Delta - \mu H)$, and the peak at point a of Fig. 18 is observed. In a similar way one explains the peaks at $eV = \pm(\Delta + \mu H)$, labeled c and d in Fig. 18, although at the higher voltages some spin mixing may be occurring. One can use the straightforward tunnel conductance formulas of Giaever and Megerle[31] for the normalized con-

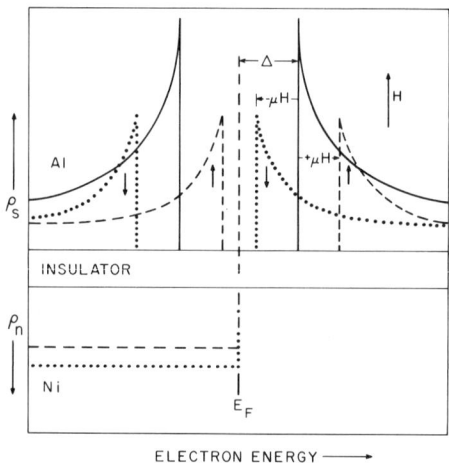

FIG. 19. Schematic density of states versus energy for an Al–Al$_2$O$_3$–Ni junction in a magnetic field. The superconducting densities are spin split as well as the densities in the ferromagnetic nickel [(· · ·) spin down, (– – –) spin up].

ductance in order to obtain an expression for the ratio of the heights of the inner peaks $G(b)/G(a)$ in terms of the relative fraction of spin-up and spin-down electrons in the nickel. This can then be used to obtain the net polarization $P \equiv (\varrho_{n\downarrow} - \varrho_{n\uparrow})/(\varrho_{n\downarrow} + \varrho_{n\uparrow})$ of the tunneling electrons in nickel. The number obtained in the above experiment was $p = 0.075$, which implies a majority of spin-down electrons in nickel (magnetic moments parallel to the magnetization and the external field). This appears to be a potentially very useful technique and hopefully can be applied to other magnetic metals in the near future.

[31] I. Giaever and K. Megerle, *Phys. Rev.* **122**, 1101 (1961).

4.3. Normal-Metal Tunneling

4.3.1. Introduction to Normal-Metal Tunneling

In the previous section we discussed tunneling behavior that was characteristic of junctions with electrodes in the superconducting phase. We shall now examine the behavior for physically similar types of junctions in which both electrodes are in the normal phase, or in which the tunneling processes are not related to the superconducting state of the electrodes.

In contrast to the rapid and detailed investigation of superconducting junctions, extensive investigation of normal-metal tunneling did not occur until much later. The new interest was spurred by the more critical application of derivative techniques (measurement of dI/dV and d^2I/dV^2 versus V) and the subsequent observation of new phenomena.

When compared with superconducting tunneling, normal-metal tunneling offers a number of new areas of investigation. The major characteristics associated with superconducting tunneling are confined to a voltage region close to zero bias. The effects of normal-metal tunneling, however, extend over the entire voltage range from zero to several volts or more. Observation of normal-metal tunneling at low voltages has revealed new phenomena that were not observed in superconducting junctions because of the dominance of the superconducting characteristics.

In the case of normal-metal tunneling, both elastic and inelastic tunneling processes are important. In particular, studies of inelastic tunneling have provided a number of new tools for examining the three factors that strongly affect tunneling characteristics: the metal electrodes, the insulating barrier, and impurities and imperfections in the area of the barrier.

4.3.2. Elastic Normal-Metal Tunneling

The preparation of normal-metal tunnel junctions is, in general, the same as described for superconducting junctions, and details can be found in Section 4.5.1. Observations of the tunneling characteristics are made in the form of either current, dynamic-resistance dV/dI, dynamic conductance dI/dV, or d^2I/dV^2 versus voltage using experimental procedures as described in Section 4.5.3. A conductance-versus-voltage curve for a normal-metal junction of aluminum–aluminum oxide–tin

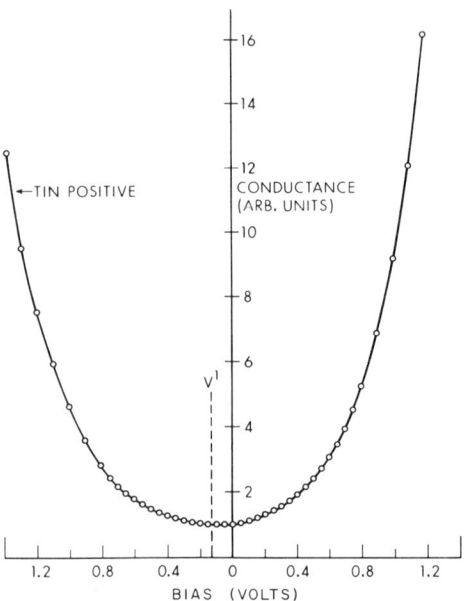

FIG. 20. Conductance versus voltage for an Al–AlO$_x$–Sn junction at 4.2 K. General parabolic shape as well as an offset from zero bias of the minimum in conductance can be observed [from J. M. Rowell, "Tunneling Phenomena in Solids" (E. Burstein and S. Lundquist, eds.), p. 385. Plenum Press, New York, 1969].

is shown in Fig. 20. The type of conductance behavior shown here is typical of normal-metal junctions. There are two main features that characterize the behavior: a general parabolic shape of the conductance versus voltage curve, and the offset from zero bias of the minimum conductance. Asymmetries in the conductance as in Fig. 20 result in offsets ranging from tens of millivolts to values approaching 300 mV.[8] For Al–Al oxide–metal junctions the offset appears to always occur in the base or oxidized aluminum negative bias direction.[32]

The tunneling conductance of Fig. 20 is obtained at liquid helium temperature; however, the general features of the curve are relatively independent of temperature. Hartman and Chivian[33] discuss the temperature dependence of the tunneling current and show how to use the measurement of the I–V characteristics as a function of temperature as a test of the quality of a tunnel junction. The validity of this test can be

[32] W. F. Brinkman, R. C. Dynes, and J. M. Rowell, *J. Appl. Phys.* **41**, 1915 (1970).
[33] T. E. Hartman and J. S. Chivian, *Phys. Rev.* **134**, A1094 (1964).

affected for low-resistance junctions, however, by the temperature dependence of the resistance of the metal films.[32,34]

Since the behavior shown in Fig. 20 can be considered as standard to normal-metal tunneling, it is worthwhile to examine the theoretical explanation of such behavior in order to better understand the deviations from this behavior, which we shall examine later. A detailed discussion of the tunneling problem is given by Duke[2]; therefore, only an outline will be given here. The general tunneling expressions of Simmons[35-42] and others can be applied in this voltage region to obtain an adequate description. However, a simpler and possibly clearer approach is to assume that the current can be expanded in powers of V and that progressively higher orders in V become important as the voltage increases.[8,43]

One can write

$$I = \alpha V + \beta V^2 + \gamma V^3 + \delta V^4 + \cdots$$

or, differentiate to obtain the conductance

$$G = \alpha + 2\beta V + 3\gamma V^2 + 4\delta V^3 + \cdots . \qquad (4.3.1)$$

In the low-voltage region we keep only the first three terms such that the conductance shows a parabolic dependence on voltage $G = \alpha + 3\gamma V^2$ with an offset from zero bias of $\beta/3\gamma$.

A more generalized approach to the problem of predicting the current–voltage relationship for metal–insulator–metal junctions contains descriptions of one or more of the following phenomena: (1) the uniform-field voltage drop across the insulator, (2) the image potential in the insulator, (3) the penetration of the electric field into the metal electrodes, and (4) charge accumulation at the metal–insulator interface.[2] The equation for the tunneling current density can be fairly easily derived from the above considerations, but can be solved only approximately. A

---

[34] I. Giaever, "Tunneling Phenomena in Solids" (E. Burstein and S. Lundquist, eds.), p. 255. Plenum Press, New York, 1969.

[35] J. G. Simmons, *J. Appl. Phys.* **34**, 1793 (1963).

[36] J. G. Simmons, *J. Appl. Phys.* **34**, 2581 (1963).

[37] J. G. Simmons, *Phys. Rev. Lett.* **10**, 10 (1963).

[38] J. G. Simmons, *J. Appl. Phys.* **35**, 2472 (1964).

[39] J. G. Simmons, *J. Appl. Phys.* **35**, 2655 (1964).

[40] J. G. Simmons, *Met. Soc. AIME* **233**, 485 (1965).

[41] J. G. Simmons, *Phys. Lett.* **16**, 233 (1965); **17**, 104 (1965).

[42] J. G. Simmons, *Brit. J. Appl. Phys.* **18**, 269 (1967).

[43] J. M. Rowell, W. L. McMillan, and W. L. Fedlmann, *Phys. Rev.* **180**, 658 (1969).

trapezoidal barrier is commonly used as the potential model with either an average barrier[35–42,44–48] or WKB[45,49–53] approximation at zero temperature applied in order to solve the tunneling current density equation. Modifications to account for image potentials,[35,36,39,40,47] field penetrations,[41–44] and finite temperature [38–42,44,45] have been introduced. Simmons[35–42] has used the average barrier approximation with a trapezoidal barrier and derives a solution as an explicit function of all quantities describing the tunnel junction in the form

$$J = J(s, \phi_1, \phi_2, m, \varepsilon, T, V, \lambda_1, \lambda_2), \qquad (4.3.2)$$

where s is barrier thickness, ϕ_1 and ϕ_2 are the respective barrier heights (or metal work functions) on either side of the trapezoidal barrier, m is the effective electron mass, ε is the dielectric constant of the barrier, T is the absolute temperature, V is the voltage across the junction, and λ_1 and λ_2 are the fractional drops in the respective electrodes of the applied voltage because of field penetration into the electrodes. The zero-temperature, zero-penetration limit of his equations is given by the expression[35]

$$J = J_0\{\bar{\phi}\exp(-A\bar{\phi}^{-1/2}) - (\bar{\phi} + eV)\exp[-A(\bar{\phi} + eV)^{1/2}]\}, \quad (4.3.3)$$

where

$$J_0 = (e/2\pi h)[1/(\beta\,\Delta s)^2], \qquad A = (4\pi\beta\,\Delta s/h)(2m)^{1/2},$$

β is a function related to the shape of the barrier and approximately equal to 1, Δs is the effective barrier thickness accounting for image charges, and $\bar{\phi}$ is the mean barrier height,

$$\bar{\phi} = (1/\Delta s)\int_{s_1}^{s_2} \phi(x)\,dx.$$

[44] R. Stratton, G. Lewicki, and C. A. Mead, *J. Phys. Chem. Solids* **27**, 1599 (1966).
[45] T. E. Hartman, *J. Appl. Phys.* **35**, 3283 (1964).
[46] R. Holm, *J. Appl. Phys.* **22**, 569 (1951).
[47] C. K. Chow, *J. Appl. Phys.* **34**, 2490 (1963).
[48] C. K. Chow, *J. Appl. Phys.* **36**, 559 (1965).
[49] K. H. Grundlach, *Solid-State Electron.* **9**, 949 (1966).
[50] K. H. Grundlach, *Phys. Lett.* **24A**, 731 (1967).
[51] K. H. Grundlach and G. Heldmann, *Phys. Status Solidi* **21**, 575 (1967).
[52] F. Forlani and N. Ninnaja, *Nuovo Cimento* **31**, 1246 (1964).
[53] C. A. Mead, "Tunneling Phenomena in Solids" (E. Burstein and S. Lundquist, eds.), p. 127. Plenum Press, New York, 1969.

Calculations of conductance versus voltage have been made[32] using the WKB approximation which agree with the simple description expressed in Eq. (4.3.1) above. The linear term in V, or offset term, is derived from the assumption that there exists a large asymmetry in the trapezoidal barrier.

The theory appears to explain adequately the data for junctions with small offsets ($\sim$50 mV). At large offsets, however, the theory does not give a good description of the observed behavior, and other mechanisms have been proposed to explain the offsets.[32]

It has been suggested by Rowell[32] that if the conductance of a normal-metal tunnel junction follows a parabolic dependence on voltage, this is a good test in itself of the junction quality in a high-voltage region, where most of the tests mentioned in Section 4.5.2 do not apply.

4.3.3. Dispersion Relations in the Barrier

The current equations governing elastic tunneling have been used as a basis to deduce the electron energy–momentum dispersion relation[44,53] in the forbidden energy region of insulating and semiconducting barriers from elastic tunneling data.[54,55] These attempts include studies of experimental $E(k)$ relationships for amorphous AlN thin films[54] and single crystal GaSe.[55] The process for obtaining $E(k)$ involves measuring the tunnel current as a function of voltage and then solving the inverse transform of the tunnel current equation to obtain a value for $E(k)$ that is explicitly present in the transmission probability. An example of such a dispersion relationship for a GaSe barrier is shown in Fig. 21.[55] The $E(k)$ relationships obtained agree with those expected.[55] However, the reliability of such calculations is questioned[56] because of the large uncertainty in the parameters associated with the barrier and with the uncertainty in the exact charge transfer mechanism, all of which are extremely dependent on the quality of the tunnel junction.

4.3.4. High-Voltage Tunneling

The behavior of tunnel junctions when the bias voltage is on the order of the barrier height (typically 2 V) is another area of interest. In this

---

[54] G. Lewicki and C. A. Mead, *Phys. Rev. Lett.* **16**, 939 (1966).
[55] S. Kurtin, T. C. McGill and C. A. Mead, *Phys. Rev. Lett.* **25**, 756 (1970).
[56] S. L. Sarnot and P. K. Dubey, *Phys. Rev. Lett.* **27**, 259 (1971).

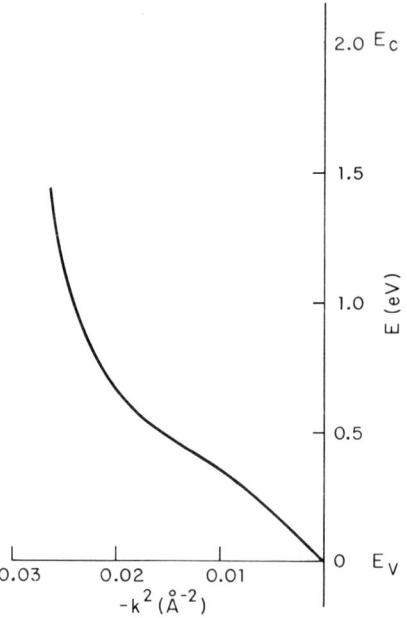

FIG. 21. The measured E-k^2 dispersion relation in the forbidden gap of GaSe. This curve is calculated using the known potential barrier shape of Al–GaSe–Au structures and the experimental $J(V)$ characteristics [from S. Kurtin, T. C. McGill, and C. A. Mead, *Phys. Rev. Lett.* **25**, 756 (1970)].

region, experimental work has been directed mainly at the determination of barrier heights and at checking the validity of theoretically derived tunnel current equations.

As one continues to raise the bias voltage across a tunnel junction, the condition is reached where $V = e\phi$. For voltages greater than this, an increase in the voltage corresponds to a decrease in the effective barrier thickness, since the electron can now tunnel directly into the conduction band of the insulating barrier rather than into the second electrode. The result of this effect on the tunneling current is expressed in the familiar Fowler–Nordheim tunneling equation[57]

$$J = (eV^2/8\pi h\phi s^2)\exp[-(8\phi s/3heV)(2m)^{1/2}\phi^{3/2}]. \qquad (4.3.4)$$

In order to determine accurate barrier heights, several methods of amplifying the changes that occur in the I–V characteristic at voltages

[57] R. H. Fowler and L. Nordheim, *Proc. Roy. Soc. (London)* **A119**, 173 (1928).

4.3. NORMAL-METAL TUNNELING

equal to the barrier height have been developed. Simmons[39] showed that the ratio of the tunneling currents at two temperatures $J = J(V, T_2)/J(V, T_1)$ plotted versus voltage peaks sharply at a voltage equal to the interfacial barrier height. However, such measured barrier heights proved unreliable because of thermal effects which change the electric field penetration in the electrodes,[58-60] and because of changing barrier height with temperature.[61] The method of Gundlach[62] for determining barrier heights involves a plot of the logarithmic derivative of the tunnel current against the applied voltage. The curve results in a pronounced maximum at the voltage equal to ϕ_+ (the barrier height at the positively biased electrode) and is independent of other parameters.

Measurement of barrier heights by the methods above have not been as helpful as first hoped. The accuracy is very dependent on junction preparation, which has yet to be developed to the point of giving completely reproducible results. Measured barrier heights in Al–Al oxide–metal junctions, for example, vary from 0.78 to 2.00 V for ϕ_1 and from 0.89 to 2.60 V for ϕ_2.[63] There is also great difficulty in obtaining suitable tunneling barriers other than Al oxide. These difficulties arise from the fact that most oxide barriers break down before voltages equal to the barrier height are reached.

It is interesting to note that oxide barriers with similar electrodes have different measured values of ϕ_1 and ϕ_2, with the lower value of ϕ corresponding to the film on which the oxide was grown. This result is in agreement with results obtained by analyzing the offset of the parabolic conductance behavior at low voltages.

4.3.5. Inelastic Tunneling

The discussion above has described the elastic tunneling processes in which the electron tunnels through the barrier without giving up energy, and this corresponds to process (a) in Fig. 22. The electron returns to the Fermi level in the second electrode by losing energy via phonon emission in some manner which is not observed to affect the normal tunneling characteristic. In addition, there are inelastic tunneling pro-

---

[58] Y. Kumagai, K. Inukai, and Y. Suzuki, *J. Appl. Phys.* **42**, 2981 (1971).
[59] H. Y. Du and F. G. Ullman, *J. Appl. Phys.* **35**, 265 (1964).
[60] J. G. Simmons, *Appl. Phys. Lett.* **6**, 54 (1965).
[61] K. H. Gundlach and A. Wilkinson, *Phys. Status Solidi* **2**, 295 (1970).
[62] A. D. Brailsford and L. C. Davis, *Phys. Rev. B* **2**, 1708 (1970).
[63] Z. Hurych, *Solid-State Electron.* **13**, 683 (1970).

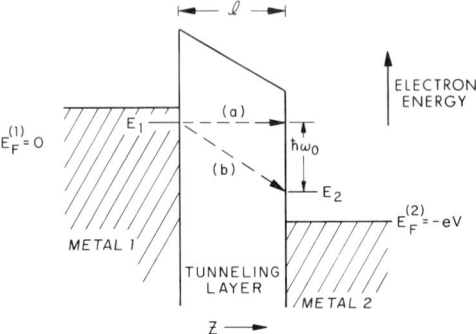

FIG. 22. Schematic energy-level diagram for tunneling between normal metals. Elastic processes (a) give rise to ordinary tunneling. In the inelastic case (b) the electron loses energy $\hbar\omega$ to an excitation in the barrier region. This can only occur if $eV \geq \hbar\omega$. The Fermi energies in the two metals are related by $E_F^{(1)} - E_F^{(2)} = eV$ [from J. Lambe and R. C. Jaklevic, *Phys. Rev.* **165**, 821 (1968)].

cesses[64,65] which occur when the tunneling electrons transfer energy to various excitations in the vicinity of the barrier, and these are indicated as process (b) in Fig. 22. Theoretical models for this process have been discussed by several authors.[2,62,66] In the inelastic process, a tunneling electron with energy $eV \geq \hbar\omega$, where ω is the frequency of the excitation, can create an excitation at the metal–insulator interface. Because such excitations provide an additional tunneling channel, there is an increase in the junction conductance $\sigma = dI/dV$ in the vicinity of $V = \hbar\omega/e$, and this can be observed as a peak in d^2I/dV^2. Several authors have shown that the size of this peak is roughly proportional to the spectral density of excitation in the barrier region.[10,67]

There are three different inelastic interactions in the barrier region that affect the tunneling conductance. These are interactions with impurities in the vicinity of the barrier, interaction with phonons or vibrational modes of the molecules in the oxide barrier itself, and interaction with excitations in the neighboring electrodes. All of these interactions cause rather small increases in conductance in contrast with the superconducting case. Such increases are on the order of 1% or less, but can be observed fairly easily using standard techniques for obtaining d^2I/dV^2 versus V as described in Section 4.5.3.

[64] R. C. Jaklevic and J. Lambe, *Phys. Rev. Lett.* **17**, 1139 (1966).
[65] J. Lambe and R. C. Jaklevic, *Phys. Rev.* **165**, 821 (1968).
[66] J. A. Appelbaum and W. F. Brinkman, *Phys. Rev.* **186**, 464 (1969).
[67] D. J. Scalapino and S. M. Marcus, *Phys. Rev. Lett.* **18**, 459 (1967).

4.3.6. Molecular Excitations in Barriers

Jaklevic and Lambe[64,65,68] have studied excitations due to organic contaminants introduced into the oxide barrier of tunnel junctions. Their standard junction preparation procedure consists of oxidizing an aluminum film several thousand angstroms thick by glow-discharge oxidation to give an oxide barrier tens of angstroms thick with an impedance on the order of 100 Ω. The glow-discharge method of oxidation rather than thermal oxidation was necessary in order to ensure that the entire procedure of growing the film, oxidizing it, contaminating it with a known organic compound, and finally depositing the second electrode could be accomplished without exposing the junction to air. It was found that exposure to air for times as short as one minute resulted in the deposition of at least a monolayer of organic contaminate from the air, an amount which was sufficient to substantially modify the tunneling characteristic. A number of organic materials were introduced onto the oxidized film, by exposing the films directly to the organic vapor. Strong bonding materials (acids, alcohols) and materials with high molecular weights have shown the best spectra, most probably because they form a stable absorbed layer. Some materials (e.g., acetone, benzene, chloroform) did not absorb in sufficient quantity to produce detectable spectra.[68] Figure 23 shows the tunneling spectrum d^2I/dV^2 versus V for a large molecular weight hydrocarbon with a comparison to the IR absorption spectrum. From spectral data as in Fig. 23 on a number of organic compounds, Jaklevic and Lambe were able to identify a number of specific vibrational modes in tunneling spectra. Figure 23 shows peaks identified with the C—H stretching and bending modes, while in other compounds they were able to distinguish OH stretch and bend modes, OD stretch and bend modes, and $C \equiv N$ stretch modes. Such vibrational modes in organic molecules generally occur in the voltage range 75–500 mV, although harmonics as high as 950 mV have been observed.[65]

Organic dopants have most often been added to aluminum–aluminum oxide–lead junctions; however, spectra have been observed with Au, Sn, Al, Mg, and Cr in place of the Pb and in Ta–Ta oxide–Pb,[91] and Mg–Mg oxide–Pb[69] tunnel junctions.

In order to interpret correctly the tunneling spectrum of a junction doped with organic molecules, it is essential to establish the true charac-

---

[68] R. C. Jaklevic and J. Lambe, "Tunneling Phenomena in Solids" (E. Burstein and S. Lundquist, eds.), pp. 233–253. Plenum Press, New York, 1969.

[69] A. L. Geiger, B. S. Chandrasekhar, and J. G. Adler, *Phys. Rev.* **188**, 1130 (1969).

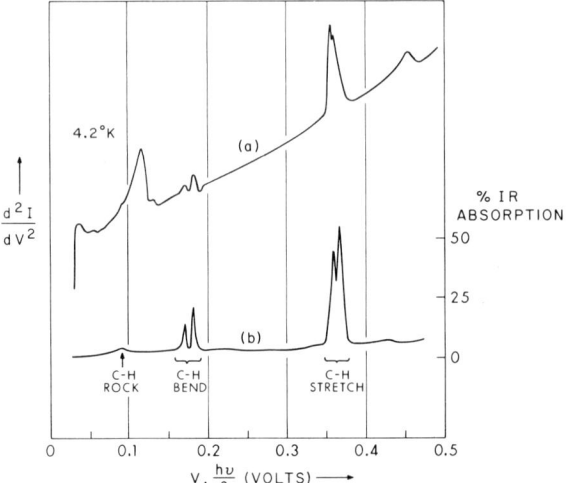

FIG. 23. (a) Tunneling spectrum d^2I/dV^2 versus V with a monolayer of a large molecular weight hydrocarbon (pump oil) at 4.2 K. For comparison (b) an IR absorption spectra of the same material is also shown. In addition to OH lines, CH stretching and bending modes at 360 mV and 175 mV, respectively, are clearly visible [from J. Lambe and R. C. Jaklevic, *Phys. Rev.* **165**, 821 (1968)].

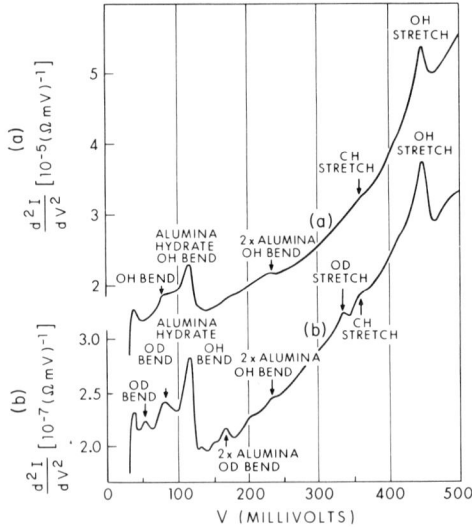

FIG. 24. Tunneling spectra d^2I/dV^2 versus V of Al–Al oxide–Pb junctions at 4.2 K in which (a) H_2O and (b) D_2O were present in O_2 discharge used to grow the oxide films. The different modes are identified [from A. L. Geiger, B. S. Chandrasekhar, and J. G. Adler, *Phys. Rev.* **188**, 1130 (1969)].

teristic for a "clean" Al–Al oxide–metal junction. Figure 24 shows the tunneling spectra for an Al–Al oxide–Pb junction in which H_2O and D_2O, respectively, were present in the O_2 discharge used to oxidize the films.[69] From this data, Geiger et al.[69] have been able to distinguish that part of the spectrum due to H_2O contamination from that part due to the oxide itself. The observation of the two modes due to the OH bond led them to speculate that the barrier in a junction formed by the oxidation of an aluminum film may not consist entirely of aluminum oxide but may include aluminum hydrates.

For tunneling experiments, the spectral line widths of the observed inelastic excitations are determined by three factors: (1) the natural line width of the molecular excitation, (2) the temperature broadening due to smearing of the electron distribution in the metals, and (3) density of states effects due to the electrodes when they are in the superconducting phase.[69] Line width broadening from both theory and experiment increases from typically 15 mV at 4.2 K to 40 mV at 77 K and to greater than 100 mV at room temperature.[65] Thus it would appear that in order to obtain significantly detailed organic spectra, liquid helium temperatures are required for the experiments, although some information is attainable at nitrogen temperature.

4.3.7. Barrier Excitations in Normal-Metal Tunneling

At energies less than 200 mV vibrational spectra are observed in the tunneling conductance, which can be associated with excitations of the oxide barrier itself.[43,69–75] Such excitations, considered either as phonons in the oxide lattice or as molecular vibrations of the oxide molecules themselves, have been observed for a number of standard junctions with barriers of Pb oxide,[43] Al oxide,[43,69,72,73] Mg oxide,[71,72] Ni oxide,[75] Yt oxide,[74] and Cr oxide.[74,76]

Giaever and Zeller[70] have identified such barrier emission processes in thermally grown zinc, magnesium, and cadmium oxides, in evaporated insulators of zinc and cadmium sulfides, and in amorphous germanium.

---

[70] I. Giaever and H. R. Zeller, Phys. Rev. Lett. **21**, 1385 (1968).
[71] J. G. Adler, Solid State Comm. **7**, 1635 (1969).
[72] J. G. Adler, Phys. Lett. **29A**, 675 (1969).
[73] T. T. Chen and J. G. Adler, Solid State Commun. **8**, 1965 (1970).
[74] R. C. Jaklevic and J. Lambe, Phys. Rev. B **2**, 808 (1970).
[75] J. G. Adler and T. T. Chen, Solid State Commun. **9**, 501 (1971).
[76] G. I. Rochlin and P. K. Hansma, Phys. Rev. B **2**, 1460 (1970).

Figure 25 shows d^2V/dI^2 data for Cd and Zn oxides and sulfides as obtained by Giaever and Zeller.[70] The oxide barriers were obtained by simple oxidizing of Cd and Zn films. The sulfide barriers were obtained by evaporating a thin layer of the appropriate substance onto a freshly evaporated metal film. These CdS and ZnS films contained pinholes (places where the film was noncontinuous), and thus it was necessary to oxidize the substrate after barrier deposition as described by Giaever.[77]

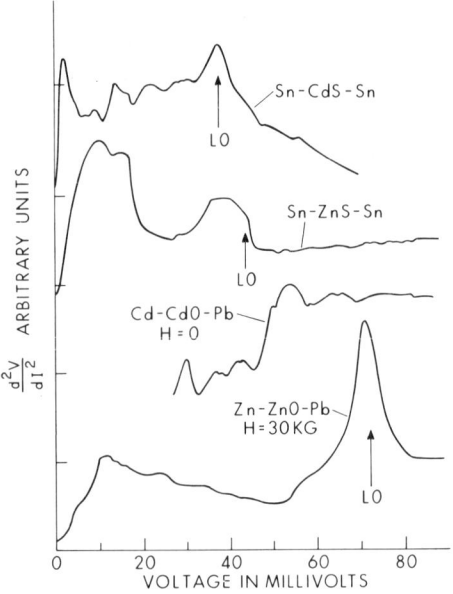

FIG. 25. Tunneling spectra d^2I/dV^2 versus V for junctions with cadmium and zinc oxides and sulfides as barriers at $T = 4.2$ K. Structure is observed which correlates well with the long-wavelength LO phonon energies obtained from Raman scattering data as indicated by the arrows [from I. Giaever and H. R. Zeller, *Phys. Rev. Lett.* **21**, 1385 (1968)].

While the Raman scattering and infrared spectroscopy select out phonons of long wavelengths, no such selection rule is known to exist for tunneling, and phonons of all wavelengths should contribute to the observed peaks. The agreement between the tunneling data and existing data for the LO phonons in Fig. 25 emphasizes that the optical-phonon spectrum must depend upon short-range order rather than long-range

[77] I. Giaever, *Phys. Rev. Lett.* **20**, 1286 (1968).

order in the crystals, since the tunneling barriers formed by the oxides and sulfides are very disordered.

Because of the asymmetry of the background conductance about $V = 0$ for a normal tunnel junction, a convenient analysis of the data can be carried out by calculating the even $G^{\mathrm{E}} = \frac{1}{2}[G(+V) + G(-V)]$, and odd $G^{\mathrm{O}} = \frac{1}{2}[G(+V) - G(-V)]$ tunneling conductances[43] as described in Section 4.5.3. All symmetrical processes, such as vibrational spectra, would only appear in G^{E}, whereas any asymmetrical processes will also appear in G^{E} but more obviously in G^{O}. One possible asymmetrical process that shows up in the odd conductance is the self-energy effects of the electron in the normal metal electrode.[43]

4.3.8. Electrode Excitations in Normal-Metal Tunneling

The interaction of tunneling electrons with excitations in the metal electrodes is a well-known phenomena in superconducting junctions, a process which has been discussed earlier in this chapter, but has only recently been observed to any great extent in normal metal electrodes.[13,72,73,75]

Figure 26 shows the experimentally observed tunneling spectra for standard Al–Al oxide junctions in which different metals have been used as the counter electrode. As seen in the figure, the interactions appear as peaks in a plot of d^2I/dV^2 versus V in the same manner as the previously discussed inelastic processes. The arrows indicating the position of phonon peaks in the metals as determined by other means show good

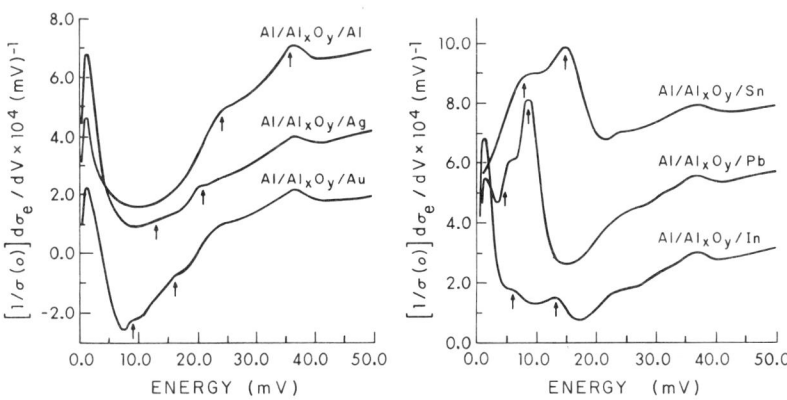

FIG. 26. d^2V/dI^2 curves for a number of Al–Al oxide–metal junctions at 4.2 K. Arrows indicate energy positions of the phonons of the respective counter metal electrodes as measured by other means [from T. T. Chen and J. G. Adler, *Solid State Commun.* **8**, 1965 (1970)].

agreement with the tunneling data. The peaks corresponding to phonons in metals with a strong electron phonon coupling (Pb, Sn) are much larger than those in weak coupling materials (see Fig. 26). Present theory, however, is not good enough to allow for substracting out the background (elastic) conductance with sufficient accuracy to make quantitative determinations of the electron–phonon coupling strength.

The phonon interaction problem in normal-metal junctions differs slightly from that in superconducting junctions. In the superconducting case, the phonon interaction is essentially a determination of the bulk properties of the metal film, while in the normal-metal junction the electron–phonon interaction must occur near the barrier for the effect to be observed in the tunneling conductance. Normal-metal tunneling thus reflects the surface or interface properties of the film. This difference in interaction has been used to account for a slight shifting to higher energy of the phonon peak in normal-metal tunneling as compared with superconducting tunneling.[43]

Adler et al.[78] have developed a theory for a one-dimensional, two-channel (elastic and inelastic) tunneling process to explain the phonon interaction in normal-metal tunneling. They have applied the description to Al–Al oxide–Pb and Pb–Pb oxide–Pb junctions, and have obtained good agreement between experiments and their theory. They attribute the shift to higher energy of the phonon peaks in normal-metal tunneling to a factor which arises from the difference in the k vector of the incident plane wave and the transmitted and reflected inelastic wave.

4.3.9. Zero-Bias Anomalies

The previous discussion of deviations from perfect tunneling behavior has dealt with excitation processes that occur at some finite bias. There is another type of deviation that is symmetric and occurs around zero bias and is usually given the name "zero-bias anomaly." Two types of zero-bias anomalies have been observed. The first type is characterized by a local increase in the conductance at zero bias by ~10% and is only a few millivolts wide. The second type or "giant zero-bias anomaly," on the other hand, is observed as a strong resistance peak in the tunneling characteristic and may have a width much greater than that of the conductance peak anomaly.

[78] J. G. Adler, H. J. Kreuzer, and W. J. Wattamaniuk, *Phys. Rev. Lett.* **27**, 185 (1971).

TABLE IV. Systems with Zero-Bias Anomalies

Junction		Dopant	Type of peak[r]	Reference
Ta–Ta	Oxide–Pb, Al, Au, Sn, Ag	None	G	a–c
Nb–Nb	Oxide–Al, Ag, Au	None	G	a,b
Mg–Mg	Oxide–Pb, Mg	None	G	a
Cr–Cr	Oxide–Ag, Pb	None	R	a,c
Fe–Fe	Oxide–Al	None	R	d
Fe–Fe	Oxide–Fe	None	R	d
Co–Co	Oxide–Al	None	R	d
Ni–Ni	Oxide–Al	None	R	d
Ni–Ni	Oxide–Pb, In, Ag, Au	None	R	e
Sn–Sn	Oxide–An	None	G	a
Sn–Sn	Oxide–Sn	Pt, Ni, Al	G, R	f
Sn–Sn	Oxide–Sn	Sn	R	e,g
Al–Al	Oxide–Fe	None	R	d,f
Al–Al	Oxide–Ni	None	R	h
Al–Al	Oxide–Al	TiO_2	G, R	i
Al–Al	Oxide–Ag	TiO_2	R	i
Al–Al	Oxide–Al, Ag	Cr_2O_3	R	i
Al–Al	Oxide–Al, Ag	CuO	R	i
Al–Al	Oxide–Al, Ag	Ge	R	i
Al–Al	Oxide–Al	Cr	G, R	j–k
Al–Al	Oxide–Ag	Ti	G, R	i,l,m
Al–Al	Oxide–Al	Co	G, R	i
Al–Al	Oxide–Ag	Cu	R	m
Al–Al	Oxide–Al	Fe	G, R	d,n
Al–Al	Oxide–Al	Mn	R	o
Al–Al	Oxide–Al	Acenapthene	G	p
Al–Al	Oxide–Al	Al	G, R	q

[a] J. M. Rowell and L. Y. L. Shen, Phys. Rev. Lett. **17**, 15 (1966).
[b] A. F. G. Wyatt, Phys. Rev. Lett. **13**, 160 (1964).
[c] L. Y. L. Shen and J. M. Rowell, Phys. Rev. **165**, 566 (1968).
[d] J. E. Christopher, R. V. Coleman, A. Isin, and R. C. Morris, Phys. Rev. **172**, 485 (1968).
[e] J. G. Adler and T. T. Chen, Solid State Commun. **9**, 501 (1971).
[f] L. Y. L. Shen, Phys. Rev. Lett. **21**, 361 (1968).
[g] H. R. Zeller and I. Giaever, Phys. Rev. **181**, 789 (1969).
[h] J. M. Rosell, J. Appl. Phys. **40**, 1211 (1969).
[i] P. Nielsen, Phys. Rev. B **2**, 3819 (1970).
[j] F. Mezei, Phys. Lett. **25A**, 534 (1967).
[k] P. Nielsen, Solid State Commun. **7**, 1429 (1969).
[l] D. J. Lythall and A. F. G. Wyatt, Phys. Rev. Lett. **20**, 1361 (1968).
[m] A. F. G. Wyatt and D. J. Lythall, Phys. Lett. **25A**, 541 (1967).
[n] R. C. Morris, J. E. Christopher, and R. V. Coleman, Phys. Lett. **30A**, 396 (1969).
[o] N. Kroo and Zs. Szentirmay, Phys. Lett. **32A**, 543 (1970).
[p] A. A. Galkin and O. M. Ignatiev, Phys. Status Solidi **A4**, K75 (1971).
[q] F. Mazei, Solid State Commun. **7**, 771 (1969).
[r] G is conductance, R is resistance.

Zero-bias anomalies have been observed in a number of different systems: semiconductor–metal Schottky-barrier junctions, metal–insulator–metal junctions, and metal–insulator–metal junctions in which the insulating barrier was doped with some material before the deposition of the second electrode. Table IV lists systems in which zero-bias anomalies have been observed and the type of anomaly.

The zero-bias conductance peaks listed in Table IV have a number of common characteristics. These peaks, which are only observable in the liquid helium temperature range, are typically a few millivolts wide and may represent changes in the total conductance of from less than 1% to maximums of 10 to 20%. The conductance peaks have a logarithmic voltage dependence and a magnitude that increases logarithmically with decreasing temperature for temperatures below 4.2 K.

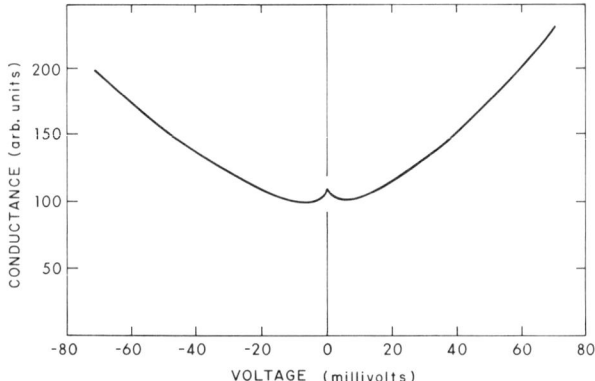

FIG. 27. Conductance versus voltage for a Ta–Ta oxide–Al junction at 1 K showing zero-bias conductance anomaly. A magnetic field sufficient to destroy superconductivity was applied [from L. Y. L. Shen and J. M. Rowell, *Phys. Rev.* **165**, 566 (1968)].

Rowell and Shen[79] have done extensive work on zero-bias conductance anomalies in metal–insulator–metal junctions. Figure 27 shows their data on a typical zero-bias conductance anomaly in a Ta–Ta oxide–Al junction. They found that sample preparation technique, particularly when using bulk material, was very important in determining the size of the observed anomaly.

Figure 28 shows their results on several different types of samples with conductance peak anomalies when magnetic fields are applied to

[79] L. Y. L. Shen and J. M. Rowell, *Phys. Rev.* **165**, 566 (1968).

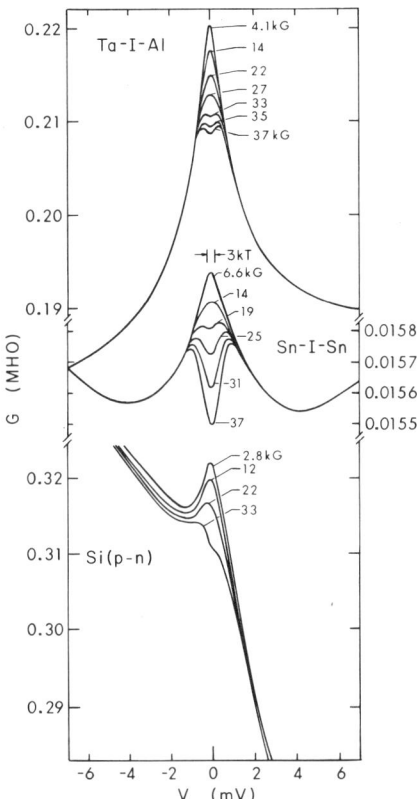

FIG. 28. Experimental results for the dependence of the zero-bias conductance anomaly on magnetic field at 1.4 K. Junctions are Ta–Ta oxide–Al, Sn–Sn oxide–Sn, and a silicon p–n diode [from L. Y. L. Shen and J. M. Rowell, *Phys. Rev.* **165**, 566 (1968)].

the junction.[79] As can be seen, the magnetic field first reduces the conductance peak, and then, for higher fields, a dip in the conductance at zero-bias appears.

Nielsen[80,81] has studied a number of Al–Al oxide–metal junctions in which controlled amounts of Ca, Ti, Cr, Co, Cu, their oxides, and Ge, have been deposited at the interface between the Al oxide barrier and the top metal electrode. He used a constant-speed rotating shutter with different length slots in order to control the amount of dopant on several

---

[80]*P. Nielsen, *Phys. Rev.* **B2**, 3819 (1970).
[81] P. Nielsen, *Solid State Commun.* **7**, 1429 (1969).

junctions at one time. A quartz microbalance was used to monitor the thickness.

Using this method of preparation he was able to show that conductance peaks were observable for a certain range of doping, depending on the material, but usually in the equivalent thickness range of 0.3 to 10 Å. Figure 29 shows the percentage change in conductance at zero bias for different chromium dopings. As this figure suggests, Nielsen found that with doping above a certain concentration or with oxidation of the

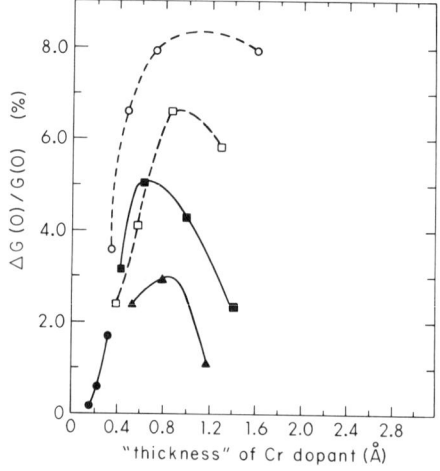

FIG. 29. Magnitude of the zero-bias conductance peak at 2 K for chromium doped Al–Al oxide–Al junctions as a function of the amount of chromium and the extent of its oxidation with a chamber pressure of $5 \cdot 10^{-5}$ Torr O_2 during deposit. The data points are: ■ Cr^{185}, ▲ Cr^{186}; ● Cr^{191}; ○ Cr^{198}, □ Cr^{199}; with Cr^{198} and Cr^{199} being deposited at five times the rate of the preceding ones [from P. Nielson, Solid State Commun. 7, 1429 (1969)].

dopant, the conductance peak at zero bias disappeared and was replaced by a resistance peak. The magnetic field dependence and temperature dependence of the conduction peak for doped junctions appear to be much the same as in metal–insulator–metal junctions.

Wolf and Losee[82] have made an extensive study of conductance peaks in semiconductor–metal Schottky-barrier junctions. Their experimental results are in basic agreement with the observed data on normal-metal junctions.

[82] E. L. Wolf and D. L. Losee, Phys. Rev. B 2, 3660 (1970).

4.3. NORMAL-METAL TUNNELING

In order to explain the experimental results on zero-bias conductance peaks in tunneling experiments, Anderson[83] and Appelbaum[84] applied the Kondo method used for scattering by magnetic moments in dilute alloys to the tunneling problem. There are a number of discussions of this theory[8,79] in the literature so we shall only briefly outline it here. Appelbaum[84] assumes that a number of noninteracting localized magnetic impurities exist at the metal–oxide interface. There are two primary interactions that are relevant to the conductance peak anomaly. The first contribution to the conductance is due to electrons which scatter in going from one electrode to the other and in doing so flip the spin of the impurity, and second, a contribution which expresses the interference between these transmitted spin-flip electrons and the electrons which are scattered back into the first electrode with spin-flip. This interference term of the theory predicts a conductance peak at zero bias given by

$$\Delta G/G = -G \ln\{[(eV)^2 + (nkT)^2]^{1/2}/E_0\}, \qquad (4.3.5)$$

where n is a constant approximately equal to 2 and E_0 is a cutoff energy. This expression is in good agreement with the experimentally observed voltage and temperature dependence.

The real value of the theory appears to be in its ability to explain the magnetic field dependence of the conductance peak. The conductance term due to spin-flip scattering will develop a hole about zero bias of width $2g\mu H$ because the magnetic field splits the Zeeman levels of the impurity by $g\mu H$, so that the tunneling electron which flips the spin will also have to lose energy $g\mu H$ in a manner analogous to the inelastic tunneling processes described previously. The strength of this depression depends on the spin of the impurity as $S/S(S+1)$.[82] The interference conductance term would also be expected to split into two peaks, but this effect appears to be very weak and not observable in the splitting of the conductance peak.[79,82]

$\Delta G(V, H, T)$ can be expressed as[79]

$$h[(\Delta + eV)/kT] + h[(\Delta - eV)/kT] \qquad (4.3.6)$$

where $\Delta = g\mu H$ and

$$h(x) = (-1 + e^{2x} - 2xe^x)/(1 - 2e^x + e^{2x}).$$

---

[83] P. W. Anderson, *Phys. Rev. Lett.* **17**, 95 (1966).
[84] J. A. Appelbaum, *Phys. Rev. Lett.* **17**, 91 (1966); *Phys. Rev.* **154**, 633 (1967).

The only unknown parameter in this expression is the g value for the impurity appearing in $\varDelta = g\mu H$. Thus we have a method of determining the g value of the impurity. Wolf and Losee[82] have shown that in order to obtain the correct agreement, the effective g value must be written as $g = g_0 + \varDelta g$, where $\varDelta g$ is a g shift associated with the exchange coupling. In general, all the conductance peak observations listed in Table IV appear to show good agreement with this theory. There are, of course, a few disagreements,[80,82] but they appear to be minor.

The characteristic behavior of the "giant" zero-bias anomaly is demonstrated in Fig. 30.[79] It is characterized at 4.2 K by a large resistance peak

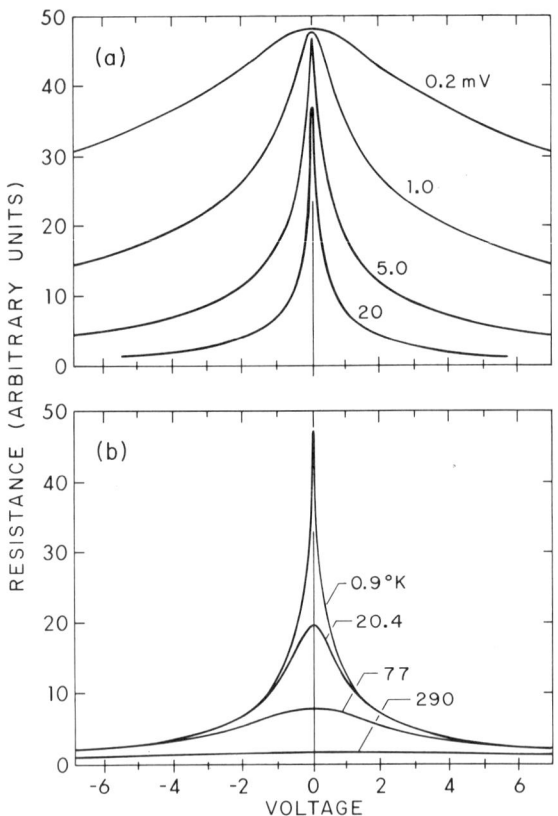

FIG. 30. Dynamic resistance versus voltage for Cr–Cr oxide–Ag junction showing "giant" zero-bias resistance anomaly: (a) for a temperature of 0.9 K; the numbers on each curve give the voltage in millivolts per division of the lower scale; (b) the voltage dependence at the four temperatures indicated, the voltage being 10 mV/division [from J. M. Rowell and L. Y. L. Shen, *Phys. Rev. Lett.* **17**, 15 (1966)].

at zero bias. The height of this peak decreases as the temperature is raised, but appears in most cases to be independent of temperature in the helium temperature range. The height of the resistance peak varies from system to system, but in some cases may correspond to zero-bias resistances tens to hundreds times greater than resistances at tens of millivolts. Behavior such as that of Fig. 30 has been observed in a large number of junction types. Some of these are listed in Table IV. The types of junction systems include metal–metal oxide–metal junctions and metal–metal oxide–metal junctions in which the oxide is doped with particles of material.[80]

Examination of Table IV shows that there is a great deal of data on these resistance peak anomalies. In spite of this there appears to be no definite conclusion as to the cause of the effect. A number of different theories have been proposed but none appears to have universal applicability.

Explanations based on: (1) strong coupling between the tunneling electron and magnetic moments of impurities in the oxide,[85] (2) low barrier heights,[86] (3) small metal particles imbedded in the tunneling barrier,[87] and (4) mobility gaps and localized states at the oxide–metal interface[80] all appear to fall short of universal applicability. In general it can be said that "giant" zero-bias anomalies are not well understood and the search for the correct mechanism of conduction continues.

4.4. Semiconductors in Tunnel Junctions

4.4.1. Summary

Because of the special electronic properties of semiconductors, they have been used in tunnel junctions not only as electrodes in electrode–barrier–electrode junctions but also as barrier materials. This has been realized in metal–semiconductor–metal configurations, in metal–semiconductor configurations where the inversion layer of the semiconductor serves as the barrier, and also in p-n tunnel junctions where the transition region between the p- and n-type material serves as the barrier. The preparation and general description of the behavior of junctions involving

---

[85] J. Solyom and A. Zawadowski, *Phys. Konen. Mater.* **7**, 325, 342 (1968).
[86] J. E. Christopher, R. V. Coleman, A. Isin, and R. C. Morris, *Phys. Rev.* **172**, 485 (1968).
[87] H. R. Zeller and I. Giaever, *Phys. Rev.* **181**, 789 (1969).

semiconductors will be given. Discussion of the behavior of the junctions will mostly be in terms of simple models—noninteracting electrons tunneling at constant energy[88] emphasizing the relation of the behavior to properties of the semiconductor. The experimental techniques used in measuring the characteristics of junctions are similar to those described above. The most useful information is the I–V characteristic, the dI/dV (G) versus V characteristic, and the d^2I/dV^2 versus V characteristic. As in the case of all tunnel junctions, care must be taken to ensure that essentially all the current is flowing via tunneling and that conduction by the way of pinholes, heating effects, and so forth are absent.

4.4.2. Metal–Insulator–Semiconductor Tunnel Junctions

The formation of the insulating layer in metal–insulator–semiconductor junctions can be carried out either by oxidizing the metal or through oxidizing the semiconductor. Many authors use the terminology semiconductor–insulator–metal for the latter configuration, the insulator being derived from the first named material. This convention will be followed here. The preparation of metal–insulator–semiconductor junctions is, of course, identical except for the last electrode to many of the junctions described above, and therefore will not be described in detail here. Techniques for the evaporation of semiconductors have been described in many books and articles.[89]

The conditions for oxide growth on several semiconductors for preparation of semiconductor–insulator–metal junctions are given in Section 4.5.1.4 with references. Most, but not all, of the semiconductors measured were degenerate. The tests for tunneling in the junctions were not investigated in all cases.

A band diagram for a degenerate p-type semiconductor and theoretical V–I curves[90] for a simple model are shown in Fig. 31, along with a comparison of experiment and theory. The dip in conductance is explained with the simple model in terms of the lack of unoccupied states across the barrier from occupied states, thereby, because of the requirement of the elastic tunneling, reducing the current. As can be

---

[88] An excellent review of the theoretical and experimental work on tunnel junctions is contained in C. B. Duke, "Tunneling in Solids." Academic Press, New York, 1969. In addition, for semiconductors see C. B. Duke, *J. Vacuum Sci. Technol.* **7**, 22 (1970).

[89] See Chapter 17 of this book or Leon I. Maissel and B. Glang, "Handbook of Thin Film Technology." McGraw-Hill, New York, 1970.

[90] L. L. Chang, P. J. Stiles, and L. Esaki, *J. Appl. Phys.* **38**, 4440 (1967).

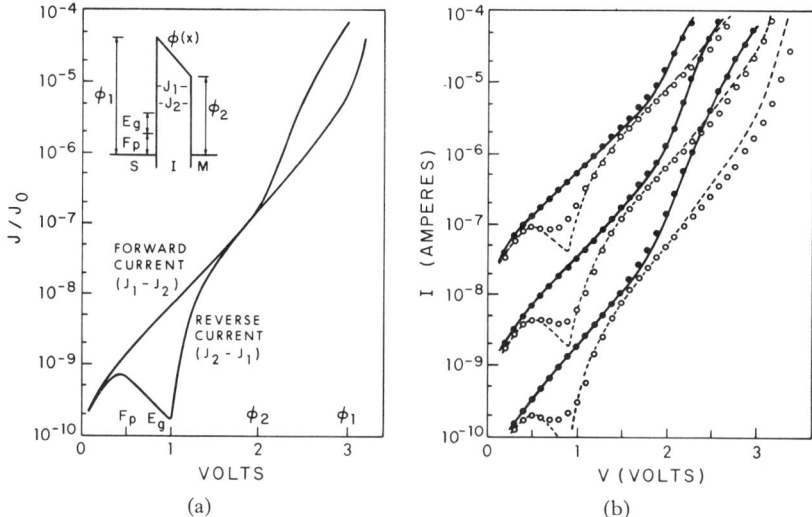

FIG. 31. (a) Theoretical current–voltage characteristic for a metal–insulator–semiconductor junction. The semiconductor is degenerate p-type and the conduction band of the insulator provides the tunneling barrier. (b) Current–voltage characteristics for Al–Al oxide–SnTe junctions at 4.2 K for three junctions with different oxide thicknesses. The values shown are: ● forward experimental, ○ reverse experimental, — forward theoretical, and – – – reverse theoretical. Curve at highest current corresponds to thinnest oxide [from L. L. Chang, P. J. Stiles, L. Esaki, *J. Appl. Phys.* **38**, 4440 (1967)].

seen, experiment and theory agree qualitatively. Extraction of semiconductor properties such as the Fermi energy, the gap energy, and the work function from the data would also be qualitatively correct, but only qualitatively. The relation of the conductance extrema to the semiconductor properties has been discussed by Chang[91] and Chang and Moore.[92] The discussion above has centered on the degenerate p-type semiconductors. Similar but not identical results should be obtained for degenerate n-type semiconductors.[93]

In the case of nondegenerate semiconductors, a significant amount of the bias voltage appears at the semiconductor surface with associated band bending. This is illustrated in Fig. 32, along with experimental data. Because of the gap in the semiconductor density of states, the low conductance for low bias is expected. The electrons would tunnel elas-

[91] L. L. Chang, *J. Appl. Phys.* **39**, 1415 (1968).
[92] L. L. Chang and S. Moore, *J. Appl. Phys.* **40**, 5315 (1969).
[93] C. B. Duke, "Tunneling in Solids" (*Solid State Phys. Suppl. 10*) (F. Seitz, D. Turnbull, and H. Ehrenreich, eds.), p. 68. Academic Press, New York, 1969.

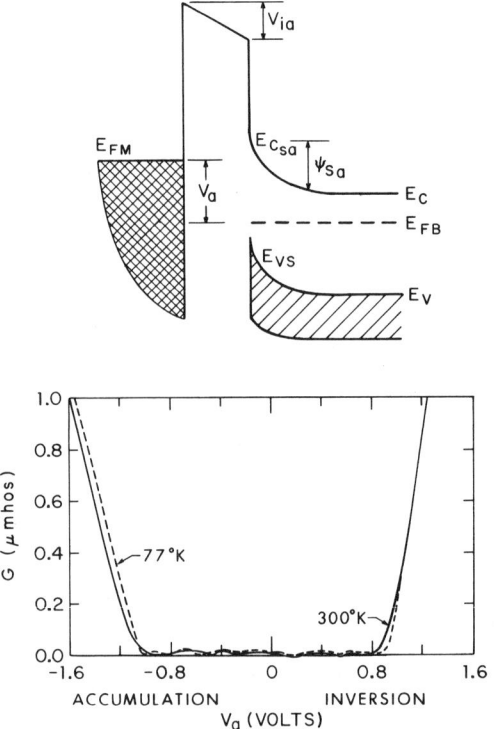

FIG. 32. Energy-band diagram at negative bias and experimental low-frequency conductance characteristic for a metal–insulator–semiconductor tunnel junction for p-type Si at 10 Ω cm with $\omega = 46$ Å and $f = 17$ Hz [from J. Shewchun, A. Waxman, G. Warfield, *Solid State Electron.* **10**, 1165 (1967) and A. Waxman, J. Shewchun, and G. Warfield, *Solid State Electron.* **10**, 1187 (1967)].

tically both from the semiconductor and the metal, but there are no available states until the bias has been increased to the point where either electrons tunnel from the valence band of the semiconductor to approximately the Fermi level of the metal, or from the Fermi level of the metal to the conduction band of the semiconductor. The width of the low-conductance region is due to the band bending and is not simply related to the gap.[94]

In both degenerate and nondegenerate semiconductors, surface states arising from a variety of sources can play large roles. For example, elastic tunneling to a surface state with later recombination can take place opening a tunneling channel not otherwise allowed. (See Shewchun

---

[94] A. Waxman, J. Shewchun, and G. Warfield, *Solid-State Electron.* **10**, 1187 (1967).

4.4. SEMICONDUCTORS IN TUNNEL JUNCTIONS

et al.[95] for a discussion of the effect of surface states in the case of nondegenerate semiconductors and Chang *et al.*[90] in the degenerate case.) The bound state and two-dimensional energy bands at an InAs–oxide interface have been studied using tunneling.[96]

It should be pointed out that although structure in the characteristic curves for tunnel junctions involving semiconductors have been related to band edges, this seems to be the limit, that is, tunneling does not measure the density of states.[97] The conductance structure in metal–insulator–semimetal junctions has been interpreted in terms of band edges.[98] Some information about bands in amorphous germanium has been obtained using electron tunneling.[99]

Additional structure in the characteristic curves of tunnel junctions can also be expected from electron interactions through either inelastic tunneling or a change in electronic dispersion relation. These effects will be discussed below in the description of tunneling in metal–semiconductor tunnel junctions.

4.4.3. Metal–Semiconductor Tunnel Junctions

Metal–semiconductor tunnel junctions can be prepared in many ways. The best technique, for obvious reasons of low density of impurities, is to cleave the semiconductor in a vacuum system in a stream of evaporating material. This technique has been used for a variety of substances. Reproducible tunnel junctions which have V–I characteristics as predicted by theory using parameters determined by independent measurements[100] have been constructed in this manner. An apparatus used successfully in vacuum systems for cleavage has been described by Wolf and Compton.[101] Vacuum cleavage of the semiconductor followed by pressing against it a pointed piece of a soft metal has also been used.[102] Of course, not all materials and crystal faces lend themselves to cleavage.

---

[95] J. Shewchun, A. Waxman, and G. Warfield, *Solid-State Electron.* **10**, 1165 (1967).
[96] D. C. Tsui, *Phys. Rev. Lett.* **24**, 303 (1970).
[97] C. B. Duke, "Tunneling in Solids" (*Solid State Phys. Suppl. 10*) (F. Seitz, D. Turnbull, and H. Ehrenreich, eds.), p. 55. Academic Press, New York, 1969.
[98] L. Esaki and J. Stiles, *Phys. Rev. Lett.* **16**, 574 (1966); J. L. Hauser and R. L. Tesfandi, *ibid.* **20**, 12 (1968); J. R. Vaisnys, D. B. McWhan, and J. M. Rowell, *J. Appl. Phys.* **40**, 2623 (1969); Y. Sawatori and U. Arai, *J. Phys. Soc. Japan* **28**, 360 (1970).
[99] J. W. Osmun and H. Fritzshe, *Appl. Phys. Lett.* **16**, 87 (1970).
[100] L. C. Davis and F. Steinrisser, *Phys. Rev. B* **1**, 614 (1970).
[101] E. L. Wolf and W. D. Compton, *Rev. Sci. Instrum.* **40**, 1497 (1969).
[102] Z. Stroubek, *Solid State Commun.* **7**, 1561 (1970).

Mechanical polishing and chemical cleaning followed by heating in a vacuum with subsequent cooling and evaporation of the metal contact has led to reported good results for GaAs, CdS, and GaSb.[103] A discussion of the merits of vacuum heating is described in this reference as well as techniques for preparing SnO_2 junctions. Chemical cleaning alone has been used successfully on GaAs[104] for preparation of GaAs–Pb (evaporated) junctions. Sharp etched points of GaAs have been used to prepare tunnel junctions by pushing them into a clean metal (Pb) at 4.2 K.[105] This list in no way exhausts the semiconductors or the techniques, nor is the list of references inclusive. The best and most commonly used techniques have been illustrated.

Where possible for the above junctions, the checks for tunneling were carried out. In all cases the semiconductor was degenerate so that the barrier arising from the space charge region would be thin enough for tunneling to dominate.

A band diagram for a metal–semiconductor (n-type) tunnel junction and a theoretical V–I curve are shown in Fig. 33. As indicated, the break in the V–I curve appears approximately at the Fermi energy (measured relative to the conduction band). The lack of significant increase in current for voltages greater than this is expected, due to the lack of an increased number of occupied states opposite (at constant energy) unoccupied states. A discussion of the limitations for use of this knee to determine the Fermi energy is given by Stratton and Padovani.[106]

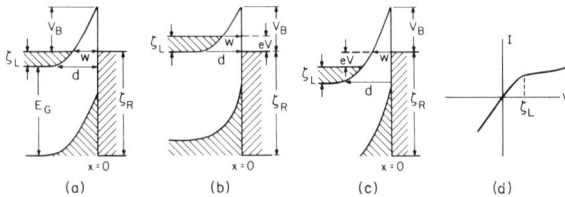

FIG. 33. Schematic diagram of the potential energy as a function of position and (d) a theoretical current–voltage characteristic at zero temperature for a metal–semiconductor contact for which V_B is independent of the bias V. Diagrams are for (a) zero bias (b) forward bias, (c) reverse bias [from C. B. Duke, *in* "Tunneling in Solids," *Solid State Phys. Suppl. 10* (F. Seitz, D. Turnbull, and H. Ehrenreich, eds.). Academic Press, New York, 1969)].

[103] M. Mikkor and W. C. Vassell, *Phys. Rev. B* **2**, 1875 (1970).
[104] D. C. Tsui, *Phys. Rev. Lett.* **21**, 994 (1968).
[105] D. C. Tsui, *J. Appl. Phys.* **41**, 2651 (1971).
[106] R. Stratton and F. Padovani, *Solid-State Electron.* **10**, 813 (1967).

The tunnel current depends on the barrier, and in fact depends on the k dependence of energy in the barrier. This has been exploited to determine the dispersion relation $[E(k)]$ in the gap.[107] Duke[108] estimates that the accuracy in the determination of the dispersion relation is approximately 30%.

The discussion above has centered on n-type semiconductors; similar results are expected for p-type.

Impurities in the junction area can affect the tunneling current through interaction of the tunneling electrons with the impurities. An example of this is the so-called zero-bias anomalies in In, Au, or Ni: Si(P) metal semiconductor contact,[109] where conductance changes on the order of 10% are observed at zero bias. The results follow rather closely a theory based on magnetic interactions[84] and the g value for the impurities has been extracted. This effect has been seen in many junctions. This is not the only impurity effect that can be studied using tunneling. For example, it has been suggested that tunneling can be used to study impurity bands[110] and traps.[111]

Other variations from the characteristic curves for the simple model have also been observed. These variations, which usually are rather small and have been observed only at low (helium) temperatures, have been described in terms of many-body effects. The effects of electron–phonon interactions have been studied the most. As illustrated in Fig. 34, the symmetry of the dI/dV and d^2I/dV^2 versus V curves often makes it possible to separate interaction effects on the electron dispersion relation in the electrode (self-energy effects) and inelastic tunneling. Inelastic tunneling describes the process where the tunneling electrons absorb or emit energy, for example, by emitting or absorbing a phonon, thereby opening up a new channel for tunneling and giving a step in dI/dV. Phonons have been studied using either or both effects in Ge,[112] GaAs,[103,113]

---

[107] F. A. Padovani and R. Stratton, *Phys. Rev. Lett.* **16**, 1202 (1966).

[108] C. B. Duke, "Tunneling in Solids" (*Solid State Phys. Suppl. 10*) (F. Seitz, D. Turnbull, and H. Ehrenreich, eds.), p. 76. Academic Press, New York, 1969.

[109] E. L. Wolf and P. L. Losee, *Solid State Commun.* **7**, 665 (1966).

[110] G. D. Mahan and J. W. Conley, *Appl. Phys. Lett.* **11**, 29 (1967).

[111] G. H. Parker and C. A. Mead, *Phys. Rev.* **184**, 780 (1969); *Appl. Phys. Lett.* **14**, 21 (1969).

[112] F. Steinrisser, L. C. Davis, and C. B. Duke, *Phys. Rev.* **176**, 912 (1968); L. C. Davis and F. Steinrisser, *Phys. Rev. B* **1**, 614 (1970).

[113] P. Thomas and H. J. Queisser, *Phys. Rev.* **175**, 983 (1968); J. W. Conley and G. D. Mahan, *ibid.* **161**, 681 (1967).

184   4. ELECTRON TUNNELING IN SOLIDS

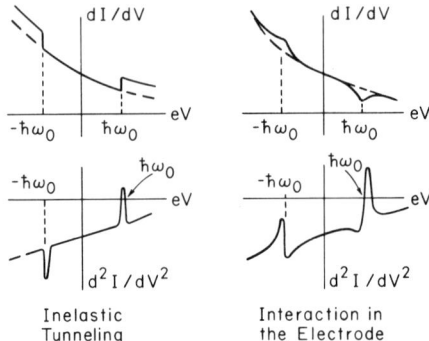

FIG. 34. Schematic illustration of the distinction in the dI/dV and d^2I/dV^2 versus V curves between the tunneling characteristics associated with optical phonon interactions in the barrier and those associated with interactions in the electrodes [from C. B. Duke, "Tunneling in Solids." Academic Press, New York, 1969].

GaSb,[103] CdS,[103,114] SnO_2.[103] The observation of surface plasmons[103,115] has also been reported.

By looking at the odd conductance $\frac{1}{2}[G(V) - G(-V)]$ and the even conductance, $\frac{1}{2}[G(V) + G(-V)]$, the differences in self-energy effects and inelastic tunneling are accentuated.[43]

This discussion has illustrated, but in no way exhausted, the properties and effects that can be studied using tunneling in metal–semiconductor contacts. It is interesting to note that ohmic contacts prepared by doping at the contact or on heavily doped materials have been explained in terms of low-resistance tunneling through the thin space–charge region.[116]

4.4.4. p-n Tunnel Junctions

A p-n tunnel junction consists of degenerate p-type material separated from degenerate n-type material by a thin transition region—a band gap—through which electrons tunnel. Because of the commercial interest in p-n junctions they have received a lot of attention, and many articles

[114] D. L. Losee and E. L. Wolf, *Phys. Rev.* **187**, 925 (1969).
[115] D. C. Tsui, *Phys. Rev. Lett.* **22**, 293 (1969).
[116] C. Y. Chang and S. M. Sze, *Solid-State Electron.* **13**, 727 (1970). A. Y. C. Yu, *ibid.* **13**, 239 (1970).

4.4. SEMICONDUCTORS IN TUNNEL JUNCTIONS

have been published describing their preparation.[117] For this reason these techniques will not be reviewed again here. It should be pointed out that a thin transition region between the p- and n-type material is required for tunneling, and the alloying method seems to be the best way to obtain this.[118] This method will be illustrated by giving the details of one of the techniques for silicon. Beginning[119] with a degenerately doped n-type silicon crystal wafer to which an ohmic contact is attached to the back, the junction is formed by heating to 725°C and alloying an aluminum wire or dot to the wafer, the solution of aluminum and silicon forming p-type silicon. Using aluminum plus 1% boron gives higher p-type doping. Usually the heating times are on the order of 1 min to reduce diffusion. Usually the outer edge of the alloy–semiconductor interface is etched away. This description has been for so-called *homodiodes*, where the p and n regions differ only in doping. Heterojunction diodes have also been prepared where the p- and n-type regions are different not only in doping but also in other material properties, most notably the band gap. Again not all the p-n junction formation techniques[120] are applicable, because of the requirement of a thin transition region.

A band diagram and a theoretical V–I curve for a p-n homodiode junction are shown in Fig. 35. Again the variation of current with voltage is seen to be sensitive to band edges. The occurance of the negative resistance region is a test for tunneling. As indicated in Fig. 35, the lowest current is greater than expected. This has been explained in terms of two-step tunneling through impurity states as indicated in (d).[119]

The absorption and emission of phonons also has been observed in p-n tunnel junctions through steps in the conductance–voltage characteristics or peaks in the d^2I/dV^2–voltage characteristics. Again it should

---

[117] See for example A. Platt, PN-junction formation techniques, *in* "Handbook of Semiconductor Electronics" (L. P. Hunter, ed.), 3rd ed., Sect. 7. McGraw-Hill, New York, 1970.

[118] L. Esaki, "Tunneling Phenomena in Solids" (E. Burstein and S. Lundquist, eds.), p. 51. Plenum Press, New York, 1969.

[119] R. A. Logan, "Tunneling Phenomena in Solids" (E. Burstein and S. Lundquist, eds.), p. 150. Plenum Press, New York, 1969.

[120] R. L. Longini and D. L. Feucht, *Trans. Met. Soc. AIME* **233**, 433 (1965). See C. B. Duke, "Tunneling in Solids" (*Solid State Phys. Suppl. 10*) (F. Seitz, D. Turnbull, and H. Ehrenreich, eds.), p. 133. Academic Press, New York, 1969, for reference to original papers on preparation of heterojunctions. See also A. Laugier and G. Mesnard, *Solid State Commun.* **8**, 83 (1970), and P. Durupt, J. P. Raymond, and G. Mesnard, *Solid State Electron.* **12**, 469 (1969).

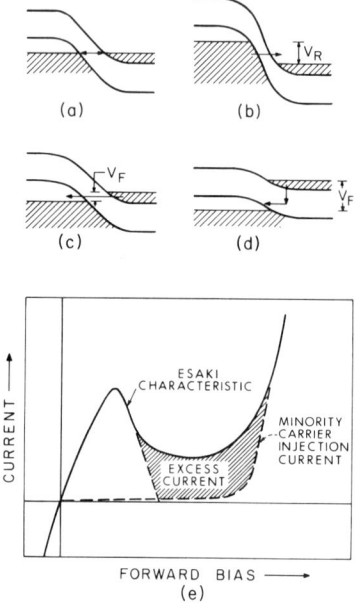

Fig. 35. The band diagram and (e) a theoretical V–I curve for a p-n homodiode junction. The potential energy diagrams correspond to (a) zero bias, (b) reverse bias V_R, (c) small forward bias V_F, and (d) large forward bias V_F. In each case the arrows indicate the tunneling path [from R. A. Logan, "Tunneling Phenomena in Solids" (E. Burstein and S. Lundquist, eds.), p. 150. Plenum Press, New York, 1950].

be pointed out that these effects give sharp structure only at low temperatures. Phonons in many semiconductors have been studied through this effect. Because the conduction and valence bands are centered at different points in the Brillouin zone in many semiconductors, most notably Ge and Si, the effect of phonons is different there than in materials like GaAs where both bands are centered at the same k value.[121] Photon-assisted tunneling has also been observed in the latter materials.[122]

Small resistance and conductance peaks at zero bias (zero-bias anomalies) have also been observed in p-n tunnel junctions.[123] In addition to the

---

[121] C. B. Duke, "Tunneling in Solids" (*Solid State Phys. Suppl. 10*) (F. Seitz, D. Turnbull, and H. Ehrenreich, eds.), Sect. 8. Academic Press, New York, 1969.

[122] C. B. Duke, "Tunneling in Solids" (*Solid State Phys. Suppl. 10*) (F. Seitz, D. Turnbull, and H. Ehrenreich, eds.), Sect. 14. Academic Press, New York, 1969.

[123] See for example R. A. Logan and J. M. Rowell, *Phys. Rev. Lett.* **13**, 404 (1964).

impurity effects as an explanation of this, for materials like GaAs phonon assisted tunneling has been suggested.[124]

4.4.5. Electrode–Semiconductor–Electrode Tunnel Junctions

The efforts to prepare good tunnel junctions with a semiconductor barrier, for example, metal–semiconductor–metal junctions, have largely met with failure[125] because of the inability to obtain a thin continuous semiconductor layer. This problem has been circumvented[126] in special cases by oxidizing the base metal through the holes in the semiconductor layer, resulting in dominantly tunneling through the semiconductor. Some success has been observed[16,127] with deposited carbon films. In these metal/carbon/metal sandwiches, tunneling is the major conduction process only at low temperatures and other conduction processes are clearly active even there. From a research application standpoint, however, this is not a serious drawback because most of the applications of tunneling require low temperatures.

4.5. Special Topics on Experimental Techniques

4.5.1. Junction Fabrication

4.5.1.1. General Procedures for Junction Preparation.
The majority of junctions used for tunnel experiments are prepared by evaporating both metal electrodes in a standard vacuum system. An intermediate process is carried out between evaporations in order to form the oxide barrier. When single crystal or polycrystalline bulk electrodes are used, the barrier is prepared on the bulk material and the second electrode is then evaporated. The apparatus and techniques applied to evaporation of the metal electrodes are the same as those used in the preparation of thin films and details of these techniques can be found in Chapter 12.

The general preparation should proceed in an oil-free high-vacuum system, since any impurities at the metal–barrier interface can strongly influence the tunnel characteristic. To reduce the impurity effects it

---

[124] C. B. Duke, "Tunneling in Solids" (*Solid State Phys. Suppl. 10*) (F. Seitz, D. Turnbull, and H. Ehrenreich, eds.), p. 126. Academic Press, New York, 1969.
[125] I. Giaever, "Tunneling Phenomena in Solids" (E. Burstein and S. Lundquist, eds.), p. 19. Plenum Press, New York, 1969.
[126] I. Giaever and H. R. Zeller, *J. Vacuum Sci. Technol.* **6**, 502 (1969).
[127] M. L. A. MacVicar, *J. Appl. Phys.* **41**, 1965 (1970).

would therefore be desirable to complete the entire fabrication without exposure to the laboratory atmosphere. However, a major portion of the successful junction preparation techniques simply involve thermal oxidation in room air or flowing oxygen after removal of the first electrode from the vacuum system. After oxidation the substrate is then replaced in the vacuum system for evaporation of the second electrode. Experiments using this type of junction preparation have been used for the bulk of the experimental data collected to date. The most highly sensitive tunnel experiments are those involving molecular spectroscopy and, if possible, these junctions should be fabricated without removal from the vacuum system. This is most convenient when the glow-discharge type of oxidation is used since this can be carried out by admitting a small oxygen pressure to the vacuum system used for electrode evaporation.

The evaporation of thin film strips through masks generally produces a strip in which the edges are thinner than the central part of the strip. Therefore the edges may have quite different properties from those of the central portion and it is desirable to remove them from the effective junction area. This can be accomplished by painting an insulating layer such as collodion, G. E. varnish, or Formvar along the edge of the first electrode after oxidation and before evaporation of the second electrode. This technique can also be employed in the case of bulk electrodes in order to insulate the larger area of the bulk electrode and define the small area which will be used for the junction. Evaporation of a second electrode somewhat wider than the defined area then removes the thin edges of the second electrode from participation in the junction area. In addition to painting on the insulating material various insulating materials such as ZnS or SiO can also be evaporated through an appropriate mask system onto the prepared electrode.

4.5.1.2. Barrier Preparation. The most critical experimental technique involved in tunnel junction fabrication is the preparation of the insulating barrier between the electrodes. The quality of the tunnel junction is so sensitive to the uniformity and perfection of this layer that the majority of successful barriers have been formed by oxidizing one of the electrodes to form the barrier. The critical part of the technique in this case involves the selection of the right time, temperature, pressure, etc. in order to form a thin continuous oxide of correct thickness on the particular metal involved. The two major methods are thermal oxidation and glow-discharge oxidation. We will summarize below the detailed information on oxidation techniques that have proved successful for a number of

4.5. SPECIAL TOPICS ON EXPERIMENTAL TECHNIQUES

specific metals. For some metals we list several methods that have proven to be satisfactory. In a few cases the use of a barrier other than an oxide of the metal electrodes have produced successful junctions and we will also summarize these cases, although the reproducibility of this type of junction is usually much less than for junctions with the natural oxide as the barrier. It should be emphasized that none of the procedures for junction barrier preparation are sufficiently reliable that one can expect the same conditions to produce identical and reliable junctions each time. For this reason it is advisable to produce as many junctions as possible in a given experiment and to measure a number of them so that the data can be compared and a basis for junction quality and reproducibility can be developed in each case.

The oxidation process must be controlled in such a way that the completed junction has a resistance in the convenient range of one ohm to a few thousand ohms. Junctions with resistance greater than $10 \text{ k}\Omega$ show large capacitance effects when measured on the standard electronic circuits and the data is then influenced by the electronic response of the measuring circuit.

The major defect encountered in barrier preparation is shorting of the junction by metallic filaments connecting the electrodes. This can be reduced to a minimum by reducing the junction area to the smallest possible useable size. It is often useful to make a number of junctions in the same evaporation with successively smaller areas. The measured resistances should then be proportional to the area within approximately 20% if the barrier fabrication is free of shorts.

Once the junction fabrication is complete the tunneling experiment should be carried out as soon as possible in order to avoid deterioration of the junction particularly after it has been removed from the vacuum system. The junction can usually be preserved indefinitely at liquid nitrogen temperature, but frequent thermal cycling may destroy it. It is generally agreed that Al–AlO$_x$–M junctions survive thermal cycling better than junctions with other types of barriers.

4.5.1.3. Oxidation Procedures for Barrier Formation on Metal Electrodes.

Pb–PbO$_x$

(a) Oxidation can be carried out for $1\frac{1}{2}$ to 2 hr in air in an oven at 55 °C. Relative humidity at 25 °C should be between 30 and 35%.

(b) Oxidation can be carried out in a stream of dry oxygen for 5 to 30 min.

(c) Oxidation can be carried out in O_2 at atmospheric pressure for 16 hr at 60°C (applied to Pb single crystals).

(d) Pure dry O_2 is bled into the vacuum system to a pressure of ~30 mTorr. A glow discharge is fired for ~10 sec.

Al–AlO$_x$

(a) Oxidation in room air at 20°C for 1 min is sufficient. Times of 5 to 10 min are also commonly used and times up to many hours produce useable junctions.

(b) Glow discharge oxidation can be done in O_2 at 100-μ pressure and discharge is fired for a few minutes.

Sn–SnO$_x$

(a) Oxidation can be carried out in air at 110°C for 12 hr to 2 days. Flowing air has been used.

Mg–MgO$_x$

(a) Oxidation can be carried out in air at 20°C for 1 min.

(b) Oxidation can be carried out by flow discharge in O_2 at 100-μ pressure for 20 min to 1 hr.

Nb–NbO$_x$ and Ta–TaO$_x$

(a) Oxidation can be carried out at 40 to 60°C in an atmosphere of pure oxygen for 2 hr.

(b) Oxidation has also been carried out by exposure to air at room temperature for $\frac{1}{2}$ hr.

Cr–CrO$_x$

(a) Oxidation in air at 100°C for 3 hr or at 200°C for 2 hr has produced satisfactory junctions.

Be–BeO$_x$

(a) Oxidation can be carried out at 20°C in pure O_2 atmosphere.

Ni–NiO$_x$

(a) Oxidation can be carried out in air at 375 K for 3 hr.

(b) Glow discharge oxidation can be used at an O_2 pressure of ~200 μ for 10 hr.

Y–YO$_x$

(a) Oxidation can be carried out in $\frac{1}{2}$ atmosphere of O_2 for 2 hr.

4.5.1.4. Oxidation Procedure for Semiconductor–Insulator–Metal Tunnel Junctions.
A summary of oxidation procedures for S–I–M tunnel junctions is given in Table V.

TABLE V. Oxidation Procedure

Material	Atmosphere	Time	Temperature (°C)	Reference
InAs	Dry O_2		120	a
Si	Steamflow	0.5–1 min	1080	b
Si	Dry O_2 or steam (with 1500-V bias)	10–30 sec	700–800	c
Si	O_2 at 150 mm Hg pressure	10 hr	100	d
InSb	O_2 at 200 μ Hg pressure			e

[a] D. C. Tsui, *Solid State Commun.* **8**, 113 (1970).
[b] A. Waxman, J. Shewchun, and G. Warfield, *Solid State Electron.* **10**, 1187 (1967).
[c] W. E. Dahlke and S. M. Sze, *Solid State Electron.* **10**, 865 (1967).
[d] P. V. Gray, *Phys. Rev.* **140**, A179 (1965).
[e] L. L. Chang, L. Esaki, and F. Jona, *Appl. Phys. Lett.* **9**, 21 (1966).

4.5.1.5. Preparation of Artificial Barriers

4.5.1.5.1. CdS AND ZnS BARRIERS. Cadmium sulfide or zinc sulfide barriers have been successfully prepared by Giaever and Zeller[70] for use in measuring the superconducting tunnel characteristics of a number of M–I–M junctions. After evaporating the first metal electrode, a layer of CdS (ZnS) is evaporated onto the electrode to serve as the barrier. The CdS (ZnS) layer may have pinholes, which expose the first electrode and represent potential shorts, but these can be effectively blocked by exposing the partially completed junction to an oxygen atmosphere at this point. Any exposed parts of the first electrode will then oxidize and prevent the second electrode evaporation from forming a short. The CdS (ZnS) barrier is also photoconductive, so that the electronic conductivity of the barrier can be changed during the experiment by exposure of the junction to light. The CdS and ZnS barriers prepared in this fashion were up to 400 Å thick.

4.5.1.5.2. CARBON BARRIERS. After evaporation of the first electrode, a carbon film is flash evaporated onto the electrode using a carbon arc similar to those employed for the preparation of carbon replicas for use in an electron microscope. Barriers in the range 90–300 Å have been demonstrated to give observable tunneling in a number of cases.[16,128] The main problem is again the occurrence of microshorts through the carbon film so that a certain fraction of the current flows by other processes than tunneling. The carbon films become more discontinuous as the thickness is reduced below 100 Å, but are fairly continuous above 100 Å and show advantageous metallurgical behavior with most common materials.

4.5.1.6. Attaching Contacts.

(a) Pure indium can be used to solder contacts directly to the thin film electrodes. A thick electrode is desirable in this case in order to avoid damage during soldering.

(b) Silver contact areas can evaporate onto the substrate spaced so that the subsequently evaporated electrodes will overlap these areas. Fine gold wires can then be connected to these silver contacts with indium solder. This method is best for avoiding damage to the electrodes.

(c) Silver paste or pressure contacts have also been used with success, but for low-noise measurements the solder methods outlined above are best.

4.5.2. Junction Testing

Even after following the best possible junction fabrication techniques, it is still necessary to test the individual junctions in order to establish that the current is indeed flowing via a tunneling mechanism. As previously mentioned, the easiest and most reliable test involves having one or both of the metal electrodes in the superconducting phase, and we summarize these tests below although they have already been mentioned previously. In addition, we list a few tests that also apply when neither of the electrodes are in the superconducting phase.

(a) Junctions of different areas can be evaporated simultaneously. Upon cooling to liquid nitrogen temperature, the junction resistance should be proportional to area within $\pm 20\%$.

---

[128] R. C. Morris and R. V. Coleman, *Phys. Lett.* **43A**, 11 (1973).

(b) When the junction is cooled to liquid helium temperature, its resistance should increase only slightly.

(c) At 1 K, when one electrode is in the superconducting phase, the conductance at zero bias should be 10^{-3} of the conductance observed when the superconductivity is quenched by application of a magnetic field.

(d) If both electrodes are in the superconducting phase, then the cusp in the I–V characteristic at $\Delta_2 - \Delta_1$ should be quite sharp and the negative resistance region from $(\Delta_2 - \Delta_1)$ to $(\Delta_1 + \Delta_2)$ should be well defined.

(e) For junctions with electrodes in the normal state, the conductance should follow a parabolic dependence on voltage. This is a reasonably good test of tunneling and can also test a junction to fairly high voltage as compared to the superconducting characteristics, which are observed only at low voltages.

(f) Structure in the derivative curves due to phonon excitation in the electrodes and the barrier should also be observable in a good tunnel junction. For example, junctions of the type Al–AlO$_x$–M should show a strong peak at 120 mV, which has been identified with the OH bending mode in aluminum hydrate.

4.5.3. Measurement Circuits

In the study of tunneling behavior there are three characteristics that are measured in order to obtain information on the basic tunneling process. There are: (1) I versus V, (2) the dynamic conductance dI/dV versus V, and (3) the derivative of the conductance d^2I/dV^2 versus V.

The simple circuit of Fig. 36 can be used to obtain an I–V characteristic. Amplifiers with high input impedance should be used if the junction resistance approaches the input impedance of the recording instrument.[10]

When it is desired to measure dI/dV or d^2I/dV^2, the commonly used procedure is not to measure these quantities directly but rather to measure the dynamic resistance dV/dI and its derivative d^2V/dI^2, and then use these values to calculate dI/dV and d^2I/dV^2. The method employed to determine dV/dI and d^2V/dI^2 is to apply a small ac signal on top of the dc bias voltage across the tunnel junction and then to detect the voltage at the fundamental and second harmonic frequency of this ac signal. If the current modulation amplitude δ is kept constant, then the voltage

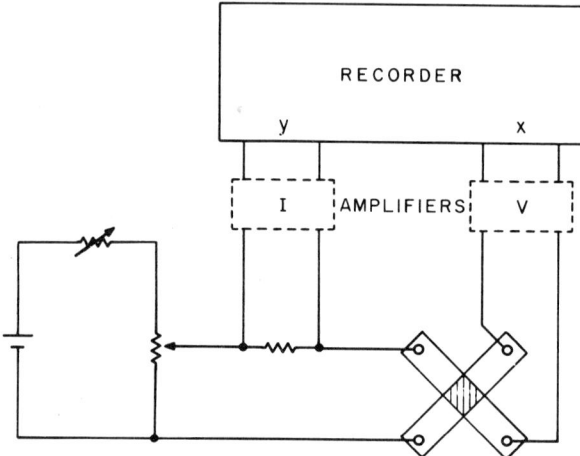

FIG. 36. Simple circuit to display I–V characteristic of a junction on an X–Y recorder [from W. L. McMillan and J. M. Rowell, "Superconductivity" (R. D. Parks, ed.), p. 561. Dekker, New York, 1969].

across the junction may be written[129] in a Taylor series as

$$V(I) = V(I_0) + (dV/dI)_{I_0}\delta \cos \omega t + \tfrac{1}{2}(d^2V/dI^2)_{I_0}\delta^2 \cos^2 \omega t + \cdots$$
$$= V(I_0) + (dV/dI)_{I_0}\delta \cos \omega t + \tfrac{1}{4}(d^2V/dI^2)_{I_0}\delta^2(1 + \cos 2\omega t) + \cdots,$$

where $V(I_0)$ is the dc bias of the tunnel junction and ω is 2π times the modulating frequency. Thus if δ is constant, the component of voltage at the primary frequency ω is proportional to $(dV/dI)_{I_0}$ and the component at the second harmonic 2ω is proportional to $(d^2V/dI^2)_{I_0}$.

With a stable amplitude oscillator and a good lock-in amplifier, this method of harmonic detection can be applied directly to give measurement of dV/dI with an accuracy one in 10^3. A typical circuit is shown in Fig. 37.[10] However, in order to obtain better accuracy and more stability it becomes necessary to use a bridge circuit as shown in Fig. 38.[10,129,130] A bridge system such as that in Fig. 38 is capable of resolving changes in the tunneling conductance of about one part in 10^5. Typical modulating frequencies used in such bridge systems run from 500 Hz to 10 kHz. It is essential that the amplitude of the modulating signal be kept low in order that the only smearing in the characteristic be that due to thermal

[129] J. G. Adler and J. E. Jackson, *Rev. Sci. Instrum.* **37**, 1049 (1966).
[130] J. G. Adler, T. T. Chen, and J. Straus, *Rev. Sci. Instrum.* **42**, 362 (1971).

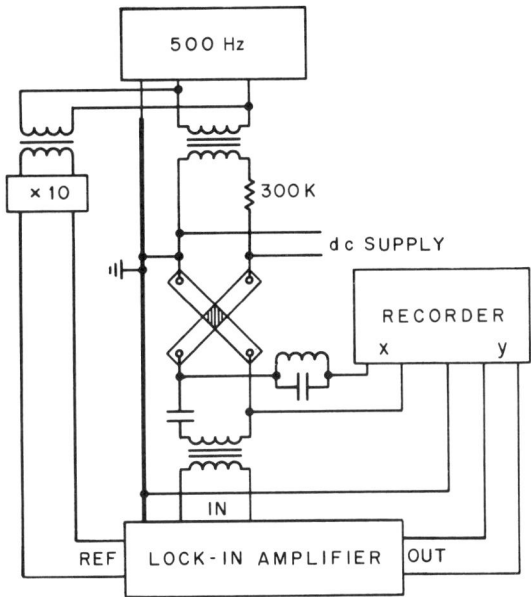

Fig. 37. Circuit to measure directly dV/dI of a junction by the ac modulation technique [from W. L. McMillan and J. M. Rowell, "Superconductivity" (R. D. Parks, ed.), p. 561. Dekker, New York, 1969].

smearing. The thermal energy kT at 1 K is 86 μV rms. For superconducting junctions one is particularly interested in the normalized conductance $\sigma = (dV/dI)_N/(dV/dI)_S$. Since it is necessary to measure the dynamic resistance in both the superconducting and normal state, it is necessary that the measuring circuit be very stable between measurements in order to get an accurate value of σ. The bridge circuit can give this stability.

If one is interested in obtaining a value of the conductance dI/dV from the measured resistance dV/dI, it is necessary to mathematically invert dV/dI. This can be done point by point or by the use of a computer.

It is also possible to measure $(dI/dV)_{V_0}$ directly.[129,131] Figure 39 shows a typical circuit for this measurement. A constant voltage ac signal (obtained by using a very low-input impedance power supply) is now applied to the junction and the methods of harmonic detection are employed to measure the current changes by measuring the voltage across a fixed value resister. Very low modulation levels (5–10 μV rms) are

[131] J. S. Rogers, *Rev. Sci. Instrum.* **41**, 1184 (1970).

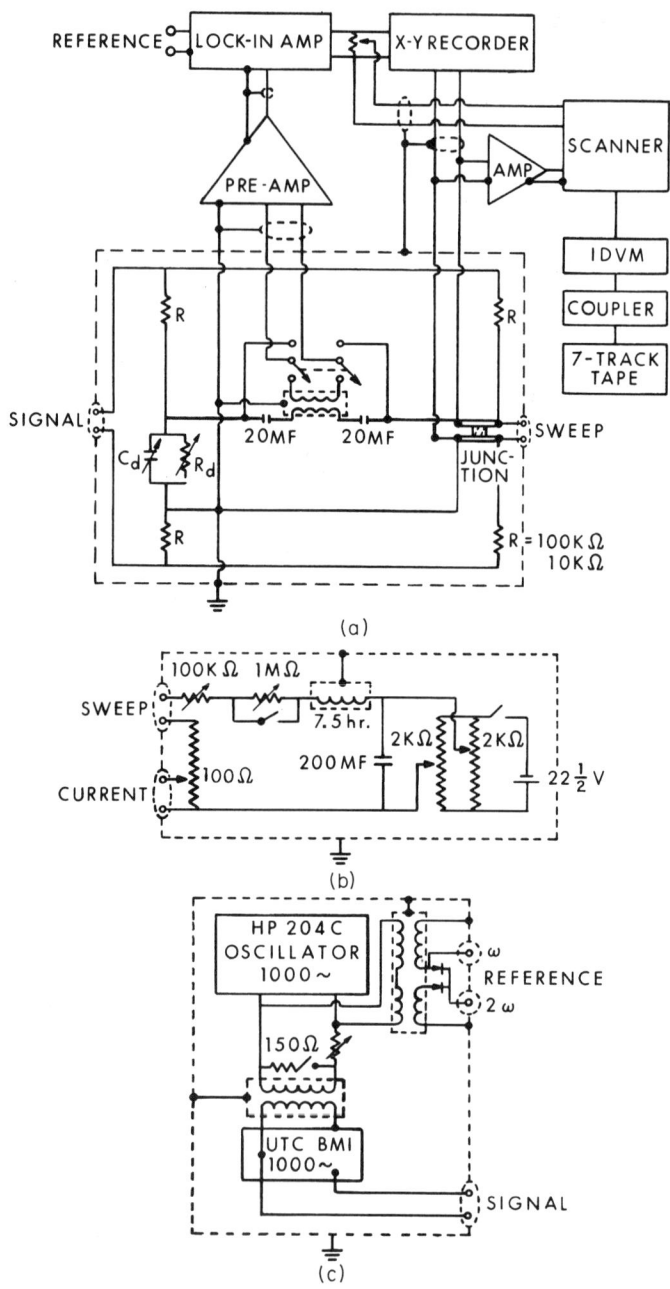

FIG. 38. (a) Bridge circuit, (b) sweep circuit, and (c) modulation supply for high-resolution resistance bridge capable of measuring charges in σ of one part in 10^5 [from J. G. Adler, T. T. Chen, and J. Straus, *Rev. Sci. Instrum.* **42**, 362 (1971)].

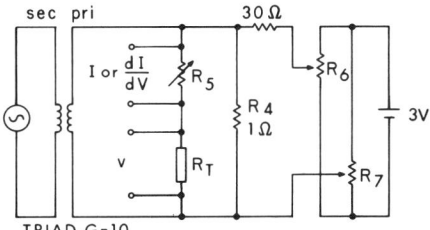

FIG. 39. Circuit for determining I versus V and dI/dV versus V curves at low biases [from J. G. Adler and J. E. Jackson, Rev. Sci. Instrum. **37**, 1049 (1966)].

employed in the circuit of Fig. 39. This method of measuring dI/dV suffers from the fact that it is less sensitive than a measurement of dV/dI.

Because of the asymmetry of the background conductance about V for most junctions, a convenient analysis of the data can be carried out by calculating the even and odd conductances.[43] These are defined as $G^E = \frac{1}{2}[G(+V) + G(-V)]$ and $G^O = \frac{1}{2}[G(+V) - G(-V)]$, where $G(+V)$ is the conductance at positive voltage V and $G(-V)$ the conductance at the same negative voltage. This type of analysis would require the use of a computer.

When the function of interest in a tunneling measurement is d^2I/dV^2, there are several different methods by which this value can be obtained. A measurement of d^2V/dI^2 can be obtained by using a circuit as in Fig. 38 and monitoring the second harmonic (signal 2ω) rather than the primary frequency. Having obtained a value for d^2V/dI^2, one can then use the identity $d^2I/dV^2 = (d^2V/dI^2)(dI/dV)^3$ to obtain a value for the derivative of the conductance.

Rowell et al.[43] describe another method for obtaining d^2I/dV^2. As the dc voltage is slowly swept in a bridge circuit similar to Fig. 38, the fundamental ac signal proportional to dV/dI is monitored across the junction. This signal as the dc output from the lock-in amplifier is integrated using a digital voltmeter with a 10-sec time constant and this value, proportional to dV/dI, is recorded every 10 sec. Using these points, a computer is used to calculate d^2I/dV^2.

Adler et al.[130] describe in detail the method for using the circuit of Fig. 38 in a system with analog to digital conversion and the subsequent use of a computer to give calibrated values of d^2I/dV^2.

5. EXPERIMENTS USING WEAKLY LINKED SUPERCONDUCTORS[†*]

5.1. Introduction

The extraordinary properties of weakly coupled superconductors were discovered by Brian Josephson, who published the results of his calculations in a sequence of remarkable papers beginning in 1962.[1-4] He found that when two superconductors are separated by a barrier (thin insulator, normal metal film, or narrow neck of superconductor) that permits the exchange of electrons between the superconductors, it is possible to have a steady current of correlated pairs of electrons, i.e., a dc supercurrent, with no voltage across the barrier. Further, when there is voltage across the barrier, there is an alternating supercurrent whose frequency is proportional to the voltage. These two phenomena are commonly called the *Josephson effects*, and the barrier region is called a *Josephson junction* or a *weak link*.

Immediately after Josephson's initial paper, there followed a number of experiments verifying his predictions. Closely related phenomena are continuing to be discovered a decade later. The Josephson effects and the experiments by which they have been studied are discussed in the first section of this chapter. In the second section, the specific characteristics of various types of weak links are examined. The third section is devoted to experimental techniques using weakly linked superconductors.

[1] B. D. Josephson, *Phys. Lett.* **1**, 251 (1962).
[2] B. D. Josephson, *Rev. Mod. Phys.* **36**, 216 (1964).
[3] B. D. Josephson, *Advan. Phys.* **14**, 419 (1965).
[4] B. D. Josephson, *in* "Quantum Fluids" (D. F. Brewer, ed.), p. 174. North-Holland Publ., Amsterdam, 1966.

[†] Work supported by National Science Foundation, Office of Naval Research, and the Alfred P. Sloan Foundation.

[*] Part 5 is by **B. S. Deaver, Jr.** and **D. A. Vincent**.

An expansive literature exists on both the phenomena themselves and the rapidly developing technology. Anderson[5,6] has presented an extensive treatment of the Josephson effects and macroscopic coherence in superconductors. Bloch[7] has examined the fundamental basis for the Josephson effects from a very general point of view. A number of useful summaries and reviews are available.[8-13] Devices based on the Josephson effects are emerging at an increasing rate. The various applications have been summarized by several authors[14-21] and details are available in the proceedings of several conferences and symposia.[22-26] A comprehensive

[5] P. W. Anderson, in "Lectures on the Many-Body Problem" (E. R. Caianiello, ed.), Vol. 2, p. 113. Academic Press, New York, 1964.

[6] P. W. Anderson, in "Progress in Low Temperature Physics" (C. J. Gorter, ed.), Vol. 5, p. 1. North-Holland Publ., Amsterdam, 1967.

[7] F. Bloch, Phys. Rev. B **2**, 109 (1970).

[8] B. D. Josephson, in "Superconductivity" (R. D. Parks, ed.), Vol. 1, p. 423. Dekker, New York, 1969.

[9] D. N. Langenberg, D. J. Scalapino, and B. N. Taylor, Proc. IEEE **54**, 560 (1966).

[10] G. F. Zharkov, Sov. Phys. Usp. **9**, 198 (1966) [English transl. of: Usp. Fiz. Nauk **88**, 419 (1966)].

[11] B. W. Petley, Contemp. Phys. **10**, 139 (1969).

[12] J. Clarke, Amer. J. Phys. **38**, 1071 (1970).

[13] I. M. Dmitrenko, V. M. Dmitriev, I. O. Kulik, and I. K. Yanson, Vestnik Acad. Nauk Ukr. SSR **35**, 36 (1971). Translation available as FSTCH-HT-23-1901-72 from Dept. A, NTIS, Springfield, Virginia 22151.

[14] B. N. Taylor, J. Appl. Phys. **39**, 2490 (1968).

[15] A. H. Silver, IEEE J. Quantum Electron. **QE4**, 738 (1968).

[16] J. Matisoo, Anal. Chem. **41**, 83A, 85A (1969).

[17] J. E. Zimmerman, J. Appl. Phys. **42**, 30 (1971).

[18] R. A. Kamper, L. O. Muller, and D. B. Sullivan, NBS Tech. Note 381 (October 1969).

[19] J. Clarke, Phys. Today **24**, 30 (1971).

[20] J. Clarke, G. Dick, D. Langenberg, W. Little, J. Mercereau, P. Richards, and D. Scalapino, Device Applications of Cryogenics Part I. Superconducting Electronics. Research Advisory Institute (1971). AD 729697 available from Nat. Tech. Information Service, Springfield, Virginia 22151.

[21] V. L. Newhouse, ed., "Treatise on Applied Superconductivity." Academic Press, New York (to be published).

[22] Proc. Symp. Phys. Superconducting Devices. Univ. of Virginia, Charlottesville, 1967. AD 661848 available from Nat. Tech. Information Service, Springfield, Virginia 22151.

[23] Papers presented at Applied Superconductivity Conferences: Austin, Texas, Nov. 1967, J. Appl. Phys. **39**, 2489–2686 (1968); Gatlinburg, Tennessee, Oct. 1968, ibid. **40**, 1994–2163 (1969); Boulder Colorado, June 1970, ibid. **42**, 1–189 (1971); Annapolis, Maryland, May 1972, Applied Superconductivity Conference Record 1972, IEEE Pub. No. 72-CHO682-5-TABSC.

and continuing survey of the literature is available as a joint publication of the National Bureau of Standards and the Office of Naval Research.[27]

5.2. Experiments That Study the Josephson Effects

5.2.1. The Josephson Effects

Although the primary intent of this chapter is to discuss experiments, it is convenient to collect some theoretical results in outline at the beginning. In his first paper,[1] Josephson used a microscopic point of view to calculate the tunneling current between two superconductors separated by an insulating barrier, and he found, in addition to the normal tunnel current, a current due to the tunneling of electron pairs. The electron pair current depends on the macroscopic occupation of the quantum states and on the coherence between the states on the two sides of the barrier. One remarkable feature of Josephson's papers is the degree to which he discussed experimentally observable consequences of this pair current, and anticipated many of the experiments by which the phenomenon would be studied.

In a subsequent paper,[2] Josephson considered weakly coupled superconductors from the point of view of the Landau–Ginsburg theory. With this approach each superconductor is described by a macroscopic order parameter

$$\Psi = \psi^{i\chi}, \qquad (5.2.1)$$

where $|\Psi|^2 = n_s$, the electron pair density, and where the phase χ has the same value over the entire superconductor. This function is interpreted as the macroscopic wave function for the superconductor, and the

---

[24] *Proc. Symp. 1969 Spring Superconducting Symp. 4th*, Naval Res. Lab., Washington, D.C., NRL Rep. 7023 (1969).

[25] *Proc. Workshop Naval Appl. Superconductivity*, Panama City, Florida (J. E. Cox and E. A. Edelsack, eds.), NRL Rep. 7302 (Nov. 1970).

[26] W. D. Gregory, W. N. Mathews Jr., and E. A. Edelsack, eds., "Science and Technology of Superconductivity." Plenum Press, New York, 1972.

[27] A survey of the literature from 1959 to March 1967 was compiled by W. S. Goree and E. A. Edelsack and issued as a cooperative effort of Stanford Research Institute and the Office of Naval Research. This survey has been continued through a joint publication of the National Bureau of Standards and the Office of Naval Research issued quarterly. It is entitled "Superconducting Devices and Materials" and is currently being compiled by W. S. Goree, E. E. Takken, R. A. Kamper, and Neil A. Olien. Copies can be obtained from Cryogenic Data Center, Nat. Bur. of Std., Boulder, Colorado 80302.

requirement that it be single-valued leads to the quantized fluxoid property of superconductors.[28] This can be seen quickly by calculating the current in the state Ψ

$$\mathbf{j}_s = (n_s e/m)(\hbar \nabla \chi - 2e\mathbf{A}), \tag{5.2.2}$$

where $\mathbf{A}$ is the magnetic vector potential. (Throughout this chapter e and m designate the charge and mass of a single electron, and all equations are in mks units.) The gradient of the phase is then

$$\nabla \chi = \hbar^{-1}[(m/n_s e)\mathbf{j}_s + 2e\mathbf{A}]. \tag{5.2.3}$$

For Ψ to be single-valued, we require for every closed path within the superconductor

$$\oint \nabla \chi \cdot d\mathbf{s} = n2\pi, \tag{5.2.4}$$

where n is an integer. It follows from Eq. (5.2.3) that

$$\oint (m/2n_s e^2)\mathbf{j}_s \cdot d\mathbf{s} + \oint \mathbf{A} \cdot d\mathbf{s} = n\Phi_0. \tag{5.2.5}$$

This is the quantity London called the "fluxoid" and Φ_0 is the flux quantum,

$$\Phi_0 = h/2e = 2.07 \times 10^{-15} \quad \text{W}. \tag{5.2.6}$$

The requirement that the fluxoid be quantized is equivalent to

$$\oint \mathbf{p}_s \cdot d\mathbf{s} = \oint (2m\mathbf{v}_s + 2e\mathbf{A}) \cdot d\mathbf{s} = nh, \tag{5.2.7}$$

where $\mathbf{p}_s$ is the canonical momentum associated with the supercurrent, and $\mathbf{v}_s$ is the pair velocity and

$$\mathbf{j}_s = 2n_s e \mathbf{v}_s. \tag{5.2.8}$$

For two isolated superconductors, the phases of the wave functions are independent. If the two are placed in intimate contact, phase coherence is established and χ has the same value in both superconductors,

---

[28] F. London, "Superfluids," Vol. I, p. 152. Wiley, New York, 1950; B. S. Deaver, Jr., and W. M. Fairbank, *Phys. Rev. Lett.* **7**, 43 (1961); R. Doll and M. Näbauer, *Phys. Rev. Lett.* **7**, 51 (1961).

just as for a single superconductor. For the intermediate case of weakly coupled superconductors, i.e., ones permitted to exchange electrons through tunneling or through some barrier that prevents full phase coherence, Josephson showed that there is a coupling energy

$$f = -(\hbar/2e)i_c \cos\theta + \text{constant}, \tag{5.2.9}$$

where

$$\theta = \chi_2 - \chi_1 - (2e/\hbar)\int_1^2 \mathbf{A}\cdot d\mathbf{s} \tag{5.2.10}$$

is the gauge-invariant phase difference between the two superconductors. This coupling energy results in an electron pair current flowing between the two superconductors

$$i = i_c \sin\theta. \tag{5.2.11}$$

This is the very general result first obtained by Josephson from microscopic theory[1] and subsequently from Landau–Ginsburg theory[2] and derived in various ways by others.[5,7,29,30]

The maximum pair current i_c can be calculated using the microscopic theory of Bardeen, Cooper, and Shrieffer (BCS) and is proportional to the matrix element for the transfer of pairs. For the case of tunneling between identical superconductors, Ambegaokar and Baratoff[30] showed that

$$i_c = (\pi/2)[\Delta(T)/R_n]\tanh[\tfrac{1}{2}\Delta(T)/kT], \tag{5.2.12}$$

where $\Delta(T)$ is the BCS gap function[31] and R_n is the resistance of the junction when both superconductors are in the normal state. For the case of unequal gaps Δ_1 and Δ_2, Anderson[5] showed that at $T = 0$

$$i_c = \frac{\pi}{R_n} \frac{\Delta_1 \Delta_2}{\Delta_1 + \Delta_2} K\!\left(\left|\frac{\Delta_1 - \Delta_2}{\Delta_1 + \Delta_2}\right|\right), \tag{5.2.13}$$

where K is a complete elliptic integral.

---

[29] R. P. Feynman, "Lectures on Physics," Vol. 3. Addison-Wesley, Reading, Massachusetts, 1965.

[30] V. Ambegaokar and A. Baratoff, *Phys. Rev. Lett.* **10**, 486 (1963); **11**, 104 (1963).

[31] A convenient tabulation of $\Delta(T)$ is available in: B. Mühlschlegel, The thermodynamic functions of the superconductor, *Z. Phys.* **155**, 313 (1959). Translated in: "The Theory of Superconductivity" (N. N. Bogoliubov, ed.). Gordon and Breach, New York, 1962. Reprinted in: J. Bardeen and R. Schrieffer, "Progress in Low Temperature Physics," (C. J. Gorter, ed.), Vol. 3, p. 235. North-Holland Publ., Amsterdam, 1961.

A very general relationship exists between the phase χ of the order parameter and the chemical potential μ [8], namely,

$$\partial \chi / \partial t = -2\mu/\hbar. \qquad (5.2.14)$$

It follows that if there is a potential V between two superconductors, the phase difference θ is changing at a rate

$$\partial \theta / \partial t = 2eV/\hbar. \qquad (5.2.15)$$

This result together with Eq. (5.2.11) establishes the relationship between the supercurrent i and the potential V, and immediately leads to two features that are most often called the "Josephson effects": (1) A steady supercurrent can flow through the junction corresponding to constant phase difference between the superconductors and zero voltage across the junction. The maximum zero-voltage current is i_c, the critical current. (2) A steady voltage V_0 across the barrier produces a phase difference that increases linearly in time, and an oscillating supercurrent

$$i = i_c \sin((2eV_0/\hbar)t + \text{constant}). \qquad (5.2.16)$$

The frequency

$$\nu_J = 2eV_0/h \qquad (5.2.17)$$

is called the *Josephson frequency* and is 483.6 MHz/μV. (To be more precise we should write for the supercurrent amplitude $i_J(\nu_J)$, since it is a function of frequency and not simply a constant i_c. This dependence is discussed in Section 5.2.5.)

Another characteristic feature appears if, in addition to a steady potential V_0, an alternating potential $V_1 \cos \omega t$ is applied to the junction. Then from Eqs. (5.2.11) and (5.2.15)

$$i = i_c \sin[(2eV_0/\hbar)t + (2e/\hbar)(V_1/\omega) \sin \omega t + \text{constant}]. \qquad (5.2.18)$$

The oscillating supercurrent is then frequency modulated and can be Fourier analyzed and written as

$$i = i_c \sum_{n=-\infty}^{\infty} (-1)^n J_n(2eV_1/\hbar\omega) \sin[(2eV_0/\hbar - n\omega)t + \theta_n], \qquad (5.2.19)$$

where J_n is the nth-order Bessel function. This result indicates that there will be dc components of the supercurrent when the Josephson frequency is equal to ω or any of its harmonics, and the magnitudes depend on the amplitude V_1 of the ac voltage. Equation (5.2.19) is the basis for the first experimental demonstration of the existence of the alternating supercurrent.

To describe the supercurrent completely, it is necessary to know how it varies in space as well as time. For this purpose Eq. (5.2.11) is written as

$$\mathbf{j} = \mathbf{j}_c(z) \sin \theta(z) \tag{5.2.20}$$

and regarded as a local relation for the current density, which depends on the local values of the chemical potential, phase of the order parameter, and the resistivity. The spatial variation of the current density is affected by the magnetic field $\mathbf{H}$. For a uniform tunnel junction with barrier thickness l (Fig. 1), the variation of the phase difference along the barrier is

$$\nabla \theta = (2\mu_0 ed/\hbar)(\mathbf{H} \times \mathbf{n}), \tag{5.2.21}$$

where $\mathbf{n}$ is a unit vector perpendicular to the barrier and $d = \lambda_1 + \lambda_2 + l$, where λ_1 and λ_2 are the penetration depths in the two superconductors.

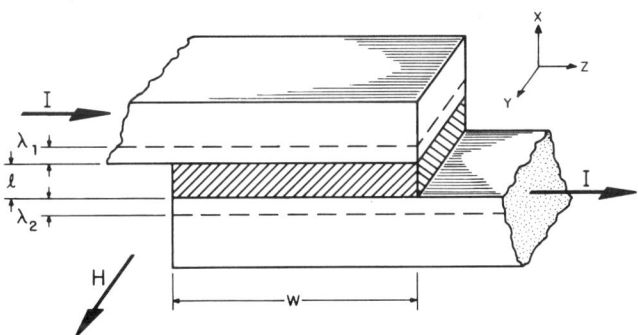

FIG. 1. Tunnel junction geometry.

Using Eqs. (5.2.20) and (5.2.21) together with a Maxwell equation to eliminate the field and current yields an equation for θ,

$$\nabla^2 \theta = (1/\lambda_J^2) \sin \theta. \tag{5.2.22}$$

The quantity

$$\lambda_J = (\hbar/2\mu_0 ed j_c)^{1/2} \tag{5.2.23}$$

is called the *Josephson penetration depth*. Since λ_J can be very large (about 1 mm for typical junctions), the barrier behaves in many respects like a type II superconductor. For barriers that are wide with respect to λ_J, the current is confined primarily to an effective width λ_J at each edge of the barrier, and there is diamagnetic screening of the magnetic field from the interior of the junction (Fig. 2a). This screening is effective up to a critical field

$$H_{c_1} = (8\hbar j_c/\mu_0 \pi^2 ed)^{1/2} \qquad (5.2.24)$$

at which quantized flux lines begin to enter the junction producing a line of current vortices along the barrier (Fig. 2b). The separation a between the vortices decreases as the field is increased.

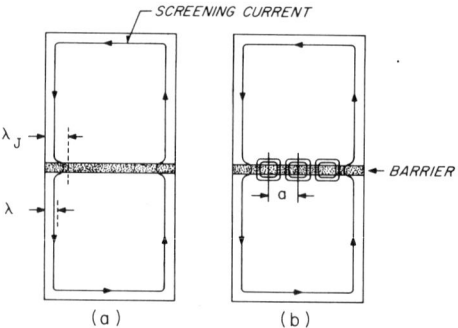

FIG. 2. Screening current distributions in a wide junction with a magnetic field H applied perpendicular to the plane of the page (a) $H < H_{c_1}$ (b) $H > H_{c_1}$. In the bulk superconductor the penetration depth is λ.

When there is a bias current through the junction, there is a force on the vortices directed along the barrier and at finite voltage there is a flow of vortices along the barrier. If ν vortices/sec cross the junction, the voltage is $V = \Phi_0 \nu$, corresponding to the Josephson frequency condition, Eq. (5.2.17). This flux flow description is often applicable to metallic barriers.

For junctions that are narrow with respect to λ_J, the magnetic field is almost uniform inside the barrier and the primary effect of the applied magnetic field is to cause a spatial variation of the phase θ and consequently of the current density j. For a field **H** in the y direction (Fig. 1), Eq. (5.2.21) gives

$$\partial \theta / \partial z = -(2\mu_0 ed/\hbar)H, \qquad (5.2.25)$$

from which

$$\mathbf{j} = \mathbf{j}_c \sin[\theta_0 - (2\mu_0 ed/\hbar)Hz]. \qquad (5.2.26)$$

5.2. EXPERIMENTS THAT STUDY THE JOSEPHSON EFFECTS

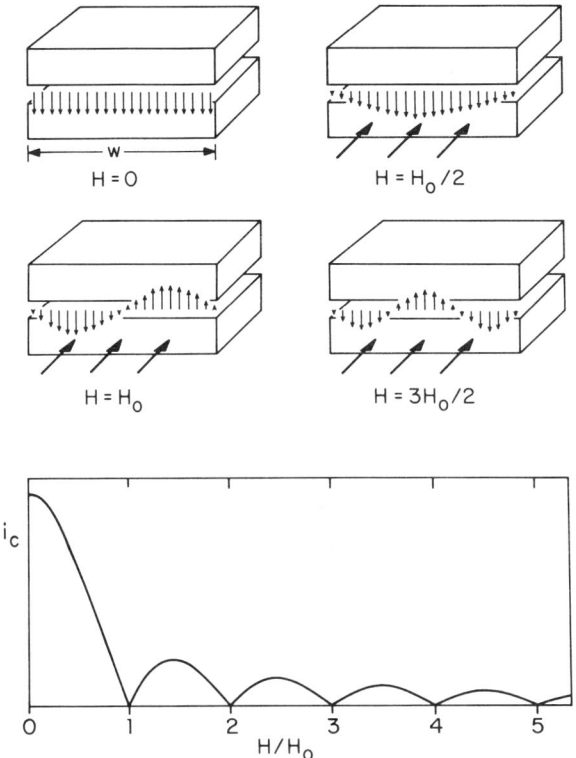

FIG. 3. Effect of a magnetic field on the current distribution and critical current of a tunnel junction.

Thus the direction of the current alternates and when the junction contains an integral number of cycles, the total current is zero. Integrating the current distribution across the junction to obtain the total current gives for the maximum zero-voltage current

$$i_c \propto \left| \frac{\sin(\pi H/H_0)}{\pi H/H_0} \right|, \qquad (5.2.27)$$

where $\mu_0 H_0 w d = \Phi_0$ and w is the width of the junction (Fig. 3).

Combining Eqs. (5.2.15), (5.2.20), and (5.2.21) and Maxwell's equations and eliminating electromagnetic quantities leads to a general equation for the phase difference θ

$$\nabla^2 \theta - v^{-2} \partial^2 \theta/\partial t^2 = \lambda_J^{-2} \sin \theta, \qquad (5.2.28)$$

where $v^2 = 1/\mu_0 \, dC$ and C is the capacitance per unit area of the junction. Linearizing this equation leads to the possibility of small amplitude oscillations with frequency

$$\omega_0 = v/\lambda_J = (2ej_c/\hbar C)^{1/2}. \tag{5.2.29}$$

This is a plasma oscillation,[4] a longitudinal oscillation with the current and electric field normal to the barrier, and corresponds to the usual plasma oscillations but with very much reduced carrier density.

To summarize these theoretical preliminaries, we can say that for Josephson junctions in which the dissipation is negligible, the behavior of the electron pair current is completely described by the set of equations

$$\mathbf{j} = \mathbf{j}_c(z) \sin \theta(z),$$
$$\partial \theta / \partial t = (2e/\hbar) V, \tag{5.2.30}$$
$$\nabla \theta = (2\mu_0 ed/\hbar)(\mathbf{H} \times \mathbf{n}),$$

together with Maxwell's equations. In many circumstances in which the dissipation is not negligible, an adequate description is obtained by assuming that the total current is the sum of the supercurrent and a normal current and to treat the two separately. This simple two-fluid picture makes possible very successful equivalent circuit models of weak links. To describe the dynamics of the Josephson oscillation it is necessary to use the time-dependent Landau–Ginsburg theory to follow the time evolution of the order parameter. To analyze experiments in which there are time variations at frequencies greater than Δ/h or spatial variations over distances less than the superconducting coherence length ξ, it is necessary, in general, to use the full microscopic theory of BCS.

5.2.2. General Comments on Experiments

The Josephson effects result from the macroscopic phase coherence of the superconducting state and are expected to be exhibited by *all* weakly linked superconductors. Properties of specific superconductors appear in the determination of the magnitude of the maximum supercurrent i_c and in measurements involving excited quasi-particles. The weak coupling required to produce the Josephson effects can be achieved in a variety of ways. The most common ones are illustrated in Fig. 4. The fabrication and specific characteristics of each are described in Section 5.3.1. Throughout this chapter the terms "weak link" and "Josephson junction" will be used in a generic sense referring to any of these types. "Tunnel junction" will be used to denote sandwich or capacitor-like structures with an insulating barrier through which electron tunneling is the dominant conduction mechanism. "SNS junction," "point contact," "Clarke slug," and "thin film bridge" can be considered to be defined by the features depicted in Fig. 4.

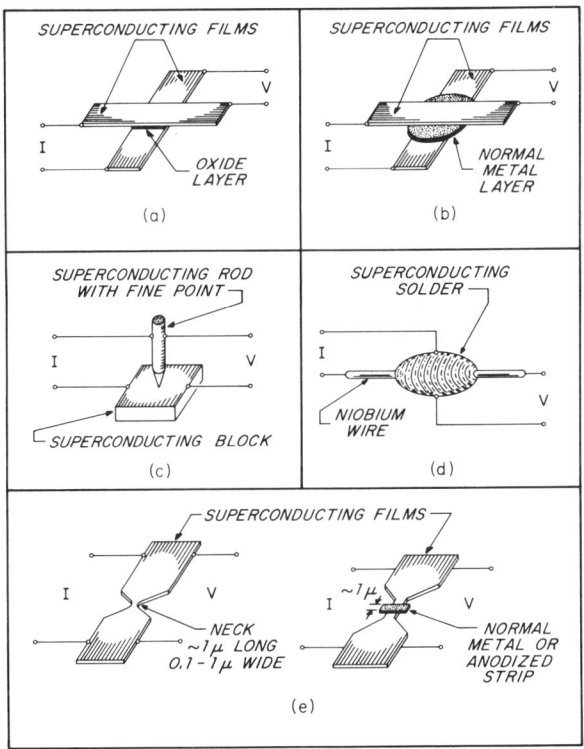

FIG. 4. Types of weakly linked superconductors: (a) tunnel junction, (b) SNS junction, (c) point contact, (d) Clarke slug, (e) thin film bridges.

The most universal measurement in the study of the Josephson effects is the determination of the current–voltage characteric or I–V curve for the weak link. In its simplest form the experiment consists of passing a direct current through the weak link and measuring the dc voltage across it, typically using a ramp generator and X–Y recorder. Alternatively, the current is supplied at low audio frequency and the I–V curve

displayed on an oscilloscope. Many specific schemes for obtaining the
I–V curve and its derivatives have been described.[32] Some of them are
discussed in Chapter 4.

Many of the features of the I–V curves can be exhibited with rudimentary equipment. However, because of the extreme sensitivity of the Josephson effects to magnetic fields and electromagnetic radiation, or, for that matter, any alternating potentials, it is often necessary to take somewhat elaborate precautions against these interferences for serious detailed measurements. Some typical precautions are: (1) careful attention to grounds, which may include using battery-powered electronics with isolated inputs; (2) operating the weak link in a metallic shield and bringing all leads in through rf interference filters (for some experiments it is desirable to use helium gas for heat exchange inside this closed shield rather than immersing the weak link directly in liquid helium); (3) multiple-layer magnetic shielding or Helmholtz coils to reduce magnetic fields to the milligauss level; (4) superconducting shields surrounding the weak link for even more effective magnetic shielding; (5) operating in a screen room to further shield from electromagnetic interference.

The experimentally observed I–V curves for weak links fall generally into two classes as shown in Fig. 5, one characteristic of nearly ideal tunnel junctions, the other typical of metallic links. Both types exhibit supercurrent at zero voltage with a maximum value i_c. If the bias current is supplied from a source capable of maintaining the voltage precisely constant across the junction, there will be no average supercurrent when the voltage is nonzero since the supercurrent oscillates sinusoidally. Thus at non zero voltage there will be only quasi-particle current due to the normal conductance of the junction. For a tunnel junction, the quasi-particle current has the distinctive form shown in Fig. 5a, with a sharp rise at $V = 2\Delta/e$ due to the gap 2Δ in the density of states for the electrons in the superconductors. For a metallic link, the I–V curve would be expected to be linear because of the essentially ohmic normal conductance.

In practice, it is impossible to maintain an instantaneously constant

---

[32] J. S. Rogers, J. G. Adler, and S. B. Wood, *Rev. Sci. Instrum.* **35**, 208 (1964); J. W. T. Dabbs, *Proc. Int. Conf. Low Temp. Phys.* **LT9**, 428 (1965); J. J. Tiemann, *Rev. Sci. Instrum.* **32**, 1093 (1961); D. E. Thomas and J. M. Klein, *ibid.* **34**, 920 (1963); R. Varteresian, *ibid.* **34**, 1265 (1963); D. E. Thomas and J. M. Rowell, *ibid.* **36**, 1301 (1965); A. Gaudefroy-Demonbvnes, E. Guyon, A. Martinet, and J. Sanchez, *Rev. Phys. Appl. (France)* **1**, 18 (1966); J. G. Adler and J. E. Jackson, *Rev. Sci. Instrum.* **37**, 1049 (1966); A. Longacre, Jr., *ibid.* **41**, 448 (1970).

voltage across the junction because of the finite impedance of the bias source. As a result, the supercurrent does not oscillate sinusoidally and can have a nonzero average value; thus there will be both supercurrent and normal current contributions to the total current at finite voltage. Because of the relatively low conductance and large shunt capacitance of tunnel junctions, the constant voltage condition can be more nearly achieved than with metallic links and the current at finite voltage can be almost entirely quasi-particle current. For metallic links typically having high normal conductance and low shunt capacitance, the usual bias source acts more like a constant current source. This results in a large

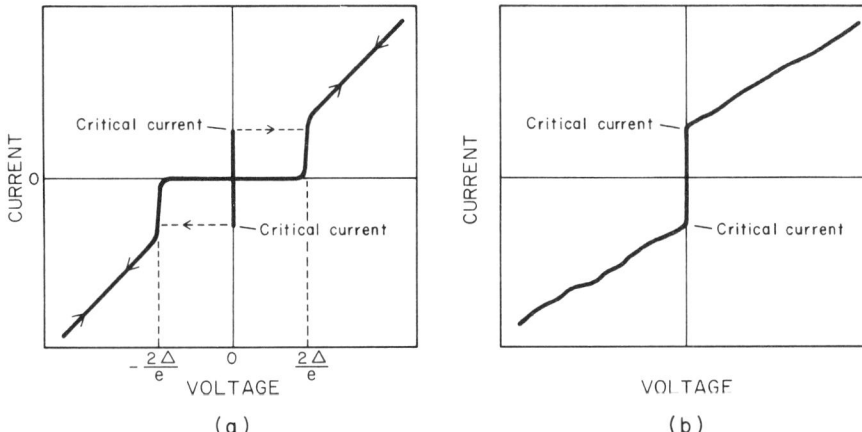

FIG. 5. Typical I–V curves: (a) tunnel junction, (b) weak links with a conducting barrier.

supercurrent contribution and a departure from ohmic behavior, particularly for $V < i_c R_n$, where R_n is the normal state resistance of the weak link. (For some thin film bridges, the nature of the supercurrent oscillation itself results in an average supercurrent even at high bias voltage, as will be discussed in Sections 5.3.1.3 and 5.3.2.)

If a tunnel junction is biased with a constant current source, there is a discontinuous switch at i_c to the finite voltage curve (as shown in Fig. 5a) as the current is increased. For decreasing current, the finite voltage characteristic is followed to low voltage and may be discontinuous again. As this behavior demonstrates vividly, the form of the I–V curve depends on the particular type of weak link and on the nature of the bias source. The effects of finite source impedance on the I–V curves of various types of weak links have been analyzed by McCumber,[33] Stewart,[34] and Warman and Blackburn.[35]

As implied in the preceding discussion, it is often convenient to think of a weak link as an ideal Josephson junction, described by

$$i = i_c \sin \theta \quad \text{and} \quad \partial \theta / \partial t = (2e/\hbar)V, \quad (5.2.31)$$

in parallel with a normal conductance and perhaps including other circuit elements to account for the physical characteristics of the weak link. These ideas are discussed more completely in Section 5.3.3.

A large part of the experimental study of the Josephson effects consists of examining the behavior of the I–V curve as a function of the type of weak link, geometry, temperature, magnetic field, and with applied ac voltage or incident electromagnetic radiation. The next section describes measurements of the dc supercurrent and how it is characterized by its dependence on magnetic field and temperature as well as geometry. Most of the subsequent sections are concerned with the alternating supercurrent, whose existance has been confirmed in many ways including the appearance of microwave-induced steps in the pair current, self-induced steps due to resonant electromagnetic modes in the junction structure itself, and direct observation of radiation from the link. Detailed studies of the microwave-induced steps have provided a method for determining the frequency dependence of the supercurrent amplitude first calculated by Riedel. The experimental demonstration of plasma oscillations verifies another of Josephson's predictions. Structure on the I–V curve at dc bias values $V = 2\Delta/ne$, $n = 2, 3, \ldots$, the so-called *subharmonic structure*, and phonons emitted from junctions biased at these same voltages give further information about the alternating supercurrent. In Section 5.2.11 the experimental consequences of thermodynamic fluctuations are discussed.

5.2.3. The dc Supercurrent

The first experimental evidence for the Josephson effects was obtained by Anderson and Rowell[36] and by Rowell,[37] who observed the dc super-

---

[33] D. E. McCumber, *J. Appl. Phys.* **39**, 3113 (1968).
[34] W. C. Stewart, *Appl. Phys. Lett.* **12**, 277 (1968).
[35] J. Warman and J. A. Blackburn, *Appl. Phys. Lett.* **19**, 60 (1971).
[36] P. W. Anderson and J. M. Rowell, *Phys. Rev. Lett.* **10**, 230 (1963).
[37] J. M. Rowell, *Phys. Rev. Lett.* **11**, 200 (1963).

current and its variation with magnetic field in Sn–Sn oxide–Sn and Pb–Pb oxide–Pb tunnel junctions. These were fabricated using evaporated metal films and thermally grown oxides in the usual capacitor-type structure. The dc current–voltage characteristic was measured and found to be of the general form shown in Fig. 5a, with a zero-voltage current typically about 10^{-4} A and a discontinuous switch to finite voltage. The existence of a zero-voltage current is not in itself proof of pair tunneling. There could be small superconducting shorts. However, the variation of the maximum zero-voltage current i_c with magnetic field is essentially conclusive.

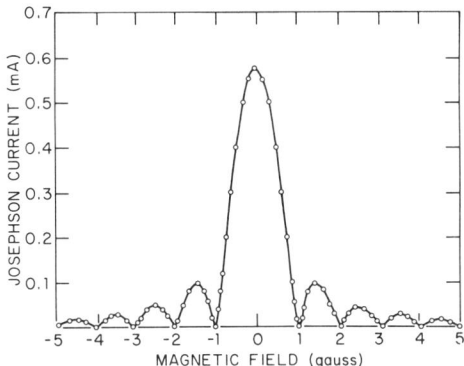

FIG. 6. The maximum current I_c as a function of magnetic field for a Sn–SnO–Sn tunnel junction with area 0.18×0.18 mm² [after P. K. Hansma, G. I. Rochlin, and J. N. Sweet, *Phys. Rev. B* **4**, 3003 (1971)].

An example of the experimentally observed variation of the critical current with magnetic field $i_c(H)$ for a narrow (i.e., $w < \lambda_J$) rectangular tunnel junction with a field applied perpendicular to one edge of the junction is shown in Fig. 6.[38] For carefully prepared tunnel junctions, the agreement with the calculated variation Eq. (5.2.27) is excellent. The critical current exhibits sharp minima at values of the field H for which there are an integral number of flux quanta in the junction, that is, for

$$H = nH_0, \quad n \text{ an integer}, \tag{5.2.32}$$

where $\mu_0 H_0 w (\lambda_1 + \lambda_2 + l) = \Phi_0$ and w is the junction width, l the barrier thickness (typically 10–20 Å), λ_1, λ_2 the penetration depths in the two

---

[38] P. K. Hansma, G. I. Rochlin, and J. N. Sweet, *Phys. Rev. B* **4**, 3003 (1971).

superconductors, and Φ_0 the flux quantum. Fiske[39] measured the field increment H_0 corresponding to the spacing of the minima as a function of temperature, and observed the variation expected for the temperature dependence of the effective area through $\lambda_1(T)$ and $\lambda_2(T)$.

The $i_c(H)$ curve for junctions containing nonuniformities or filamentary superconducting shorts can exhibit complicated interference phenomena rather than the Fraunhofer diffraction form. Thus the form of $i_c(H)$ has become a standard test for the quality of tunnel junctions. Dynes and Fulton[40] have used the relation between $i_c(H)$ and the supercurrent density distribution to calculate the supercurrent distribution from measurements of $i_c(H)$ for various types of tunnel junctions.

For wide tunnel junctions (i.e., $w > \lambda_J$) there are significant screening currents when a field is applied, resulting in a Meissner effect in the junction.[6,41] At higher fields, flux enters the junction and the current distribution corresponds to a linear array of quantized vortices.[6] Owen and Scalapino[42] have calculated and plotted the current and field distributions and the critical current for wide tunnel junctions. For a given magnetic field there are several allowed current distributions with different numbers of vortices in the junction, and each has its own critical current. Goldman and Kreisman[43] first reported observations of the Meissner effect and the mixed state by measurements of the critical current for wide tunnel junctions. Subsequently techniques for producing extremely high-quality tunnel junctions have been developed and extensive measurements of the field dependence of the critical current for wide junctions have been reported.[44–46] One example is shown in Fig. 7. In general, the agreement with the calculations of Owen and Scalapino is excellent.

Although the $i_c(H)$ curves exhibited in Figs. 6 and 7 are nicely symmetric, this will not, in general, be the case, because of the field produced by the tunnel current itself. For example, with the crossed film geometry the symmetric case obtains only if the tunnel current is shared equally between both arms on each side of the junction (see Fig. 31) rather than with current entering only one arm and exiting from one (see Fig. 4).

---

[39] M. D. Fiske, *Rev. Mod. Phys.* **36**, 221 (1964).
[40] R. C. Dynes and T. A. Fulton, *Phys. Rev. B* **3**, 3015 (1971).
[41] R. A. Ferrell and R. E. Prange, *Phys. Rev. Lett.* **10**, 479 (1963).
[42] C. S. Owen and D. J. Scalapino, *Phys. Rev.* **164**, 538 (1967).
[43] A. M. Goldman and P. J. Kreisman, *Phys. Rev.* **164**, 544 (1967).
[44] W. Schroen and J. P. Pritchard, Jr., *J. Appl. Phys.* **40**, 2118 (1969).
[45] J. Matisoo, *J. Appl. Phys.* **40**, 1813 (1969).
[46] K. Schwidtal, *Phys. Rev. B* **2**, 2526 (1970).

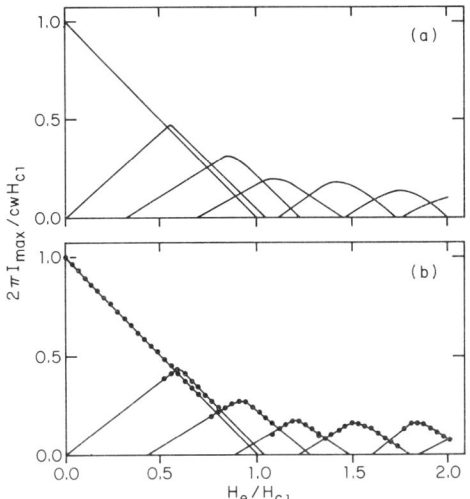

FIG. 7. Critical current versus applied magnetic field for a symmetrically biased wide tunnel junction. (a) Theoretical curve from Owen and Scalapino with $w/\lambda_J = 10$. (b) Experimental data showing fit to theory with $(w/\lambda_J)_{av} = 8.24$ [after K. Schwidtal, Phys. Rev. B **2**, 2526 (1970)].

The effects of the self field of the tunnel current on $i_c(H)$ have been investigated in detail by Yamashita et al.,[47] and data for asymmetric as well as symmetric cases are given by Matisoo[45] and Schwidtal.[46] Schwidtal and Finnegan[48] have studied self-field effects as well as the effects of varying the barrier thickness.

The maximum value of i_c (i.e., for $T = 0$ and $H = 0$) was calculated by Josephson[1] and by Ambegaokar and Baratoff[30] to be $(\pi/2)(\Delta/R_n)$, where R_n is the normal resistance of the junction formed between two identical superconductors with gap parameter Δ. The early experiments yielded values far smaller than this. However, Ferrell and Prange[41] pointed out that for wide junctions ($w > \lambda_J$) it is necessary to account for the effects of the self field of the tunnel current. For such junctions the current is confined essentially to within λ_J of the edge, thus reducing the effective width and correspondingly the critical current. (Noise currents and fluctuations also reduce the critical current. See Section 5.2.11.) Measurements on small junctions have yielded values of i_c within a few percent of the calculated value.

[47] T. Yamashita, M. Kunita, and Y. Onodera, *J. Appl. Phys.* **39**, 5396 (1968).
[48] K. Schwidtal and R. D. Finnegan, *J. Appl. Phys.* **40**, 2123 (1969).

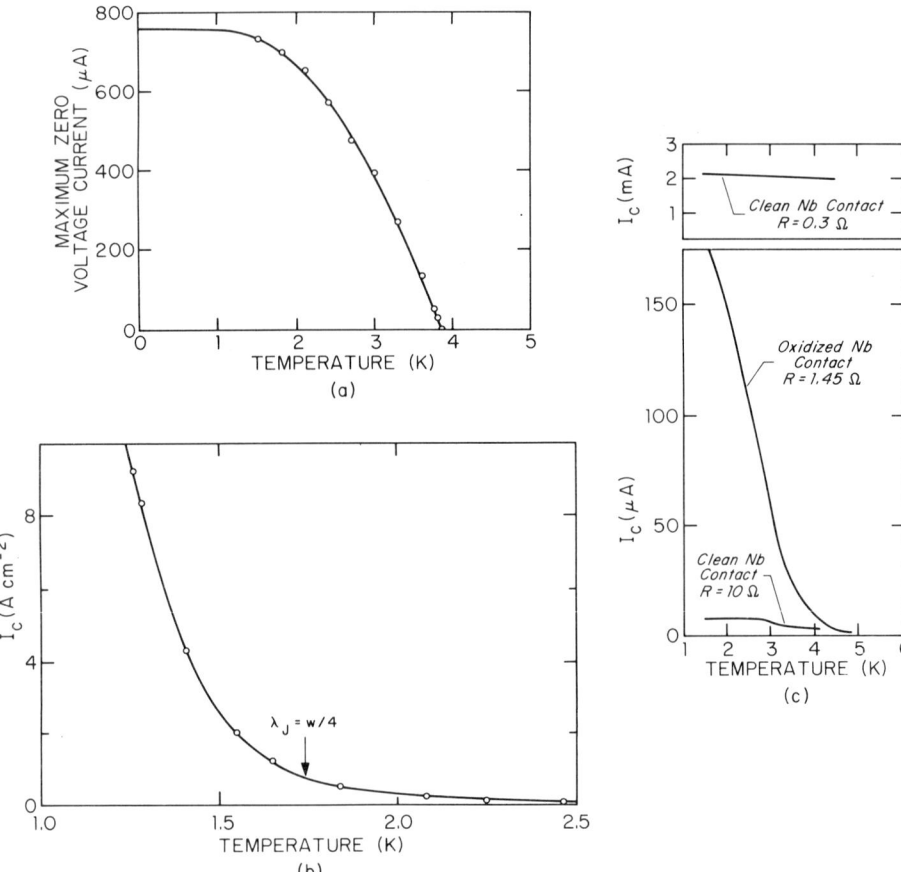

FIG. 8. Critical current versus temperature. (a) Nearly ideal tunnel junction (Sn–Sn oxide–Sn) showing fit to theory of Ambegaokar and Baratoff [after J. T. Anderson and A. M. Goldman, *Phys. Rev. Lett.* **23**, 128 (1969)]. (b) SNS junction (Pb–Cu/Al–Pb) with area 6.64×10^{-3} cm² and thickness of Cu/Al = 6500 Å [after J. Clarke, *Proc. Roy. Soc.* **A308**, 447 (1969). (c) Nb point contacts [after J. E. Zimmerman, Applied Superconductivity Conference Record, 1972. IEEE Publ. No. 72-CHO682-5-TABSC].

The temperature dependence of the critical current $i_c(T)$ of tunnel junctions was first measured by Fiske,[39] who found reasonable agreement with the variation calcuiated by Ambegaokar and Baratoff,[30] Eq. (5.2.12). More recent data[49] on nearly ideal Sn–Sn oxide–Sn junctions is shown in Fig. 8a. To obtain good agreement between measurements of $i_c(T)$

---

[49] J. T. Anderson and A. M. Goldman, *Phys. Rev. Lett.* **23**, 128 (1969).

5.2. EXPERIMENTS THAT STUDY THE JOSEPHSON EFFECTS

and theory for Pb–Pb oxide–Pb junctions it is necessary to take into account the energy dependence of the gap parameter using strong coupling theory.[50–52]

For weak links other than tunnel junctions, the variation of the critical current with temperature and magnetic field are not so readily characterized. For SNS junctions the critical current is typically a rapidly varying function of temperature[53] (Fig. 8b) and the precise form of the variation is strongly dependent on the mean free path in the normal barrier. The variation of i_c with magnetic field[53] is very similar to that for wide tunnel junctions (Fig. 7).

Some point contacts exhibit critical current versus temperature in good agreement with the Ambegaokar and Baratoff formula, probably indicating that they are essentially tunnel junctions. For others, $i_c(T)$ is nearly linear and may be related to pinning of the vortex structure.[54] For oxidized Nb point contacts there can be a very rapid variation, whereas for extremely clean ones the critical current can be almost independent of temperature at low temperatures (Fig. 8c).[55] For point contacts made by breaking a fine wire immersed in liquid helium and then remaking the contact, Zimmerman[55] finds a remarkable constancy of the ratio i_c/R_n at the value $\pi\Delta/2R_n$ which would be expected for good tunnel junctions rather than these "perfectly" clean contacts (see Fig. 37 and Section 5.3.1.4).

5.2.4. Microwave-Induced Steps on the I–V Curve

5.2.4.1. Supercurrent Steps. The first experimental evidence for the existence of the alternating supercurrent was obtained by Shapiro,[56] who observed step-like structure on the I–V curves of a tunnel junction when it was irradiated with microwaves, and the distinctive variation of the steps with microwave power was first studied by Shapiro, Janus, and Holly.[57] Subsequently supercurrent steps induced by an applied ac

---

[50] T. A. Fulton and D. E. McCumber, *Phys. Rev.* **175**, 585 (1968).
[51] K. Schwidtal and R. D. Finnegan, *Phys. Rev.* B **2**, 148 (1970).
[52] C. S. Lim, J. D. Leslie, H. J. T. Smith, P. Vashishta, and J. P. Carbotte, *Phys. Rev.* B **2**, 1651 (1970).
[53] J. Clarke, *Proc. Roy. Soc.* A **308**, 447 (1969).
[54] I. Taguchi and H. Yoshioka, *J. Phys. Soc. Japan* **29**, 371 (1970).
[55] J. E. Zimmerman, Applied Superconductivity Conference Record, 1972, IEEE Pub. No. 72-CHO682-5-TABSC.
[56] S. Shapiro, *Phys. Rev. Lett.* **11**, 80 (1963).
[57] S. Shapiro, A. R. Janus, and S. Holly, *Rev. Mod. Phys.* **36**, 223 (1964).

voltage have been observed with point contacts,[58] slugs,[59] thin film bridges,[60] and SNS junctions,[61] and detailed measurements on this structure constitute a major technique for studying the Josephson effects. The microwaves induce a voltage across the junction which frequency modulates the ac Josephson current; the current steps are the zero-frequency side bands. The supercurrent in a junction with an applied voltage $V = V_0 + V_1 \cos \omega t$ has dc components when the Josephson frequency is equal to the applied frequency or any of its harmonics, as shown in Section 5.2.1.

If the phase θ_n in Eq. (5.2.19) is chosen to be $\pm \pi/2$ to maximize these components, since it is these maximum values which are observed on the I–V curve, the dc components are found to be

$$I(V) = i_c \sum_{n=1}^{\infty} |J_n(2eV_1/\hbar\omega)| \, \delta(V_0 \pm n\hbar\omega/2e), \qquad (5.2.33)$$

where δ is the usual delta function and J_n is the nth-order Bessel function. If the junction is biased with a current source, these components appear

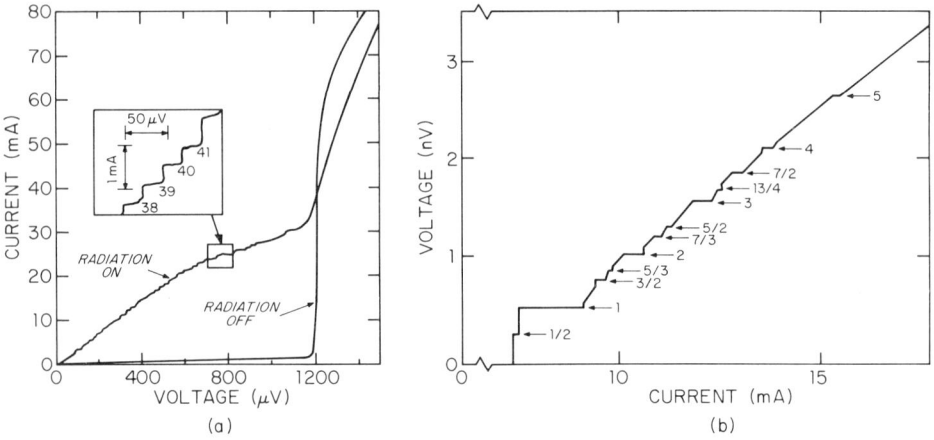

FIG. 9. Curves with rf-induced supercurrent steps. (a) Sn–Sn oxide–Sn tunnel junction at $T = 1.2$ K with $\nu = 10$ GHz [after W. H. Parker, D. N. Langenberg, A. Denenstein, and B. N. Taylor, Phys. Rev. 177, 639 (1969)]. (b) Pb–Cu–Pb junction at $T = 4.2$ K with $\nu = 250$ kHz [after J. Clarke, Phys. Rev. Lett. 21, 1566 (1968)].

[58] C. C. Grimes and S. Shapiro, Phys. Rev. 169, 397 (1968).
[59] S. Shapiro, Phys. Lett. 25A, 537 (1967).
[60] P. W. Anderson and A. H. Dayem, Phys. Rev. Lett. 13, 195 (1964).
[61] J. Clarke, Phys. Rev. Lett. 21, 1566 (1968).

5.2. EXPERIMENTS THAT STUDY THE JOSEPHSON EFFECTS

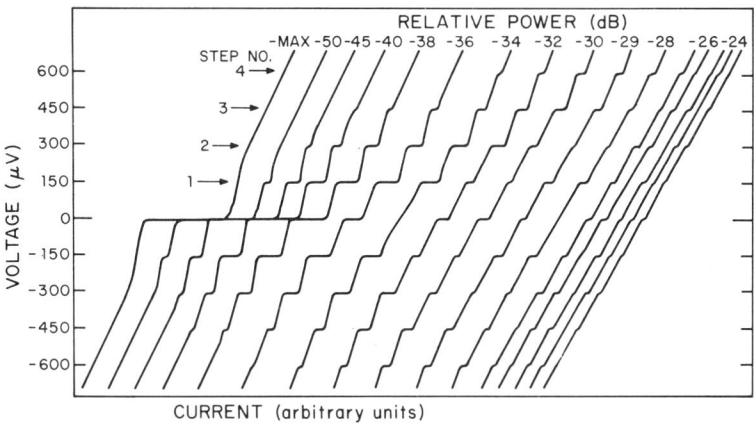

FIG. 10. Voltage–current curves for a Nb point contact showing rf-induced supercurrent steps at various rf power levels with $T = 4.2$ K and $\nu = 72$ GHz [after C. C. Grimes and S. Shapiro, *Phys. Rev.* **169**, 397 (1968)].

as steps on the I–V curve at discrete voltages

$$V = n\hbar\omega/2e \tag{5.2.34}$$

and the height of the nth step is predicted to vary as the nth-order Bessel function of V_1, the amplitude of the applied ac voltage.

Typically for these experiments the weak link is mounted in a waveguide or microwave cavity and the I–V curve is plotted for various levels of incident microwave power. Examples of I–V curves with supercurrent steps are shown for a tunnel junction (Fig. 9a),[62] an SNS junction (Fig. 9b),[61] and for a point contact (Fig. 10).[58] Steps have been observed for applied ac voltages at frequencies from a few kilohertz[61] up to 3.8 THz.[63] Most of the experiments have used microwave frequencies from 10 to 100 GHz; however, several have used lasers at 891 GHz, 964 GHz, 1.5 THz, and 3.8 THz.[63–66]

---

[62] W. H. Parker, D. N. Langenberg, A. Denenstein, and B. N. Taylor, *Phys. Rev.* **177**, 639 (1969).

[63] D. G. McDonald, A. S. Risley, J. D. Cupp, K. M. Evenson, and J. R. Ashley, *Appl. Phys. Lett.* **20**, 296 (1972).

[64] D. G. McDonald, V. E. Kose, K. M. Evenson, J. S. Wells, and J. C. Cupp, *Appl. Phys. Lett.* **15**, 121 (1969).

[65] T. G. Blaney and C. C. Bradley, *J. Phys. D: Appl. Phys.* **5**, 180 (1972).

[66] D. G. McDonald, K. M. Evenson, J. S. Wells, and J. D. Cupp, *J. Appl. Phys.* **42**, 179 (1971).

The frequency relation, Eq. (5.2.34), has been found to be exact to as great accuracy as the fundamental ratio e/h is known and to be the same for all types of weak links.[62] In fact, the experiment has been inverted and with precisely measured frequency and voltage used to determine e/h (see Section 5.4.1).

Clarke[61] has compared the Josephson voltage–frequency relation for lead, tin, and indium junctions and found it is the same to at least one part in 10^8. Using the extremely sharp steps obtained with SNS junctions (Fig. 9b) he compared the voltages across two junctions with the same incident frequency when they were biased to the same steps. Using a superconducting voltmeter he determined that the difference in potential was less than 1.7×10^{-17} V for steps at 2.1×10^{-9} V.

Another high-precision check of the Josephson frequency relation was accomplished by Finnegan et al.,[67] who compared the voltage at the 94th step of a tunnel junction irradiated with microwaves with the first step of a point contact irradiated with 891 GHz from an HCN laser. Using a potentiometer as a transfer standard to compare the two voltages, they established the precision of the frequency relation to one part per million.

Microwave-induced steps were studied using thin film bridges by Anderson and Dayem[60] and Dayem and Wiegand[68] who found, in addition to the steps satisfying Eq. (5.2.34), steps at

$$V = (n/m)(\hbar\omega/2e),$$

where n and m are integers corresponding to subharmonics of the incident frequency. They attribute these additional steps to synchronized flux flow across the bridge with multiple flux units traversing the bridge at each cycle of the rf. Subharmonic steps are also observed with SNS junctions as shown in Fig. 9b. This behavior can occur in general for weak links having current phase relationships that depart significantly from $i = i_c \sin \theta$. For very narrow bridges, Gregers–Hansen et al.[69] find only harmonic steps and behavior essentially like that for tunnel junctions. An interesting characteristic of thin film bridges is the enhancement of

---

[67] T. F. Finnegan, A. Denenstein, D. N. Langenberg, J. C. McMenamin, D. E. Novoseller, and L. Cheng, *Phys. Rev. Lett.* **23**, 229 (1969).

[68] A. H. Dayem and J. J. Wiegand, *Phys. Rev.* **155**, 419 (1967).

[69] P. E. Gregers-Hansen, M. T. Levinsen, L. Pedersen, and C. J. Sjöström, *Solid State Commun.* **9**, 661 (1971).

5.2. EXPERIMENTS THAT STUDY THE JOSEPHSON EFFECTS   221

the critical current in the presence of small amounts of rf power.[60,70,71] This has been attributed to an increase in phase coherence between the two bulk superconductors when the bridge is being irradiated.[72]

The variation of step height with rf power in qualitative agreement with the simple Bessel function form of Eq. (5.2.33) is clear in the data in Fig. 10. The dependence of step height on rf power was first studied by Shapiro et al.[57] using tunnel junctions and they obtained reasonable agreement with the Bessel function form they calculated assuming a voltage biased condition. Similar results have been obtained for thin film bridges.[69]

There are several reasons why the observed variations of step height fail to fit precisely the form of Eq. (5.2.33). Experimentally it is difficult to know the actual rf voltage across the weak link. In wide tunnel junctions it is necessary to account for the standing wave pattern of the electromagnetic field within the junction.[73] For point contacts and thin film bridges the finite normal conductance must be taken into account. This can be done using simple lumped parameter models and calculations based on such models give much better agreement with experiments.[74-78] As shown in Fig. 11, Taur et al.[79] obtain very good agreement between experimental data for a Nb point contact and calculations for a bias current $I = I_{dc} + i_{rf}$ applied to an ideal Josephson junction [an element completely described by Eq. (5.2.31)] shunted by a normal resistor. (For further discussion of equivalent circuit models see Section 5.3.3.) For small tunnel junctions good agreement is obtained with Eq. (5.2.33) particularly at high rf levels although there are small deviations due to the frequency dependence of the supercurrent amplitude[80] (see Section 5.2.5).

Another effect of irradiating a tunnel junction with microwaves has

---

[70] A. F. G. Wyatt, V. M. Dmitriev, W. S. Moore, and F. W. Sheard, *Phys. Rev. Lett.* **16**, 1166 (1966).
[71] P. E. Gregers-Hansen and M. T. Levinsen, *Solid State Commun.* **7**, 1215 (1969).
[72] P. V. Christiansen, E. B. Hansen and C. J. Sjöström, *J. Low Temp. Phys.* **4**, 349 (1971).
[73] C. A. Hamilton and S. Shapiro, *Phys. Rev. B* **2**, 4494 (1970).
[74] P. Russer, *Acta Physica Austriaca* **32**, 373 (1970).
[75] H. Fack, V. Kose, and H.-J. Schrader, *Messtechnik* **79**, 31 (1971) (in German).
[76] P. E. Gregers-Hansen and M. T. Levinsen, *Phys. Rev. Lett.* **27**, 847 (1971).
[77] H. Fack and V. Kose, *J. Appl. Phys.* **42**, 320 (1971).
[78] P. Russer, *J. Appl. Phys.* **43**, 2008 (1972).
[79] Y. Taur, P. L. Richards and F. Auracher, *in Proc. Int. Conf. Low Temperature Physics, 13th, Boulder, Colorado*, 1972, University of Colorado Associated Press, Boulder, Colorado (to be published).
[80] C. A. Hamilton, *Phys. Rev. B* **5**, 912 (1972).

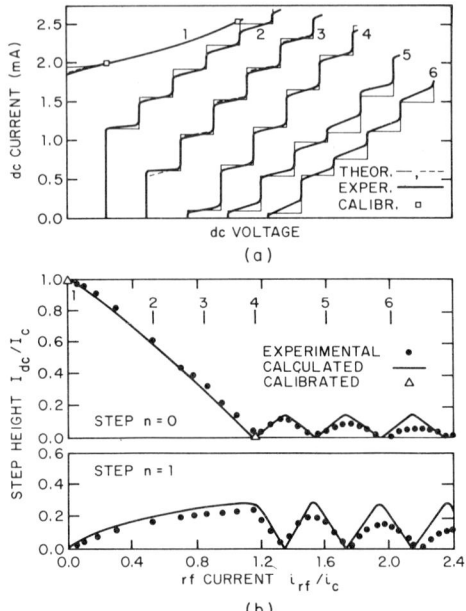

FIG. 11. (a) Experimental and theoretical I–V curves for a Nb point contact at $T = 4.2$ K irradiated with 36 GHz microwaves. The solid line was calculated for a resistively shunted ideal junction without noise and the dashed line was calculated using the same model with 100 K thermal noise. (b) Experimental and theoretical dependence of the zero voltage and the $n = 1$ step heights on rf power. Curve 1 is for zero rf power and curves 2–6 for progressively increasing rf power. The theory was fitted to the data at the two marked calibration points. This two point calibration was used to fit all the curves in (a) and (b). (After Y. Taur, P. L. Richards and F. Auracher, in *Proc. Int. Conf. Low Temperature Physics, 13th*, Boulder, Colorado. University of Colorado Associated Press, Boulder, Colorado, 1972.)

been studied by Chen et al.,[81] who observed discrete dc voltages across a junction with no external dc bias but with a large resistor across it. Using an analysis closely related to that used above to describe microwave-induced current steps, they calculate the set of allowed dc voltages as a function of the amplitude of the microwave voltage and obtain general agreement with the experiment.

5.2.4.2. Quasi-particle Steps. In addition to the supercurrent steps due to the Josephson effect, other step-like structure is found at

$$V = 2\Delta/e \pm nh\nu/e \qquad (5.2.35)$$

[81] J. T. Chen, R. J. Todd, and Y. W. Kim, *Phys. Rev.* B **5**, 1843 (1972).

5.2. EXPERIMENTS THAT STUDY THE JOSEPHSON EFFECTS

on the I–V curves of tunnel junctions irradiated with microwaves of frequency ν (Fig. 12). These steps, first reported by Dayem and Martin[82] and first explained quantitatively by Tien and Gordon[83] result from a process that has been called "photon-assisted tunneling." There have been additional experiments and extensions of the theory.[84-88] The I–V

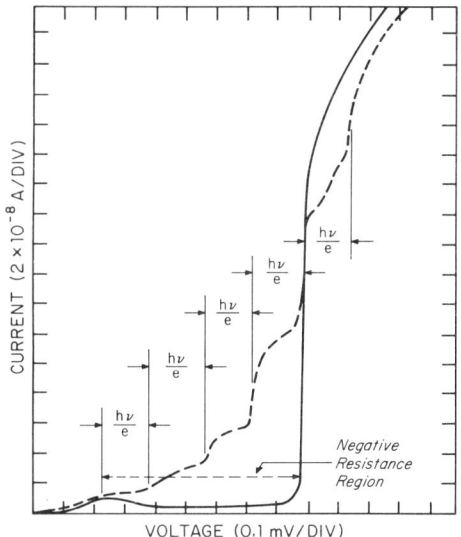

FIG. 12. I–V curve with quasi-particle steps for an Al–Al$_2$O$_3$–In tunnel junction with $h\nu/e = 0.16$ mV as measured by Dayem and Martin (– – –: with microwave field; —: without microwave field) [after P. K. Tien and J. P. Gordon, *Phys. Rev.* **129**, 647 (1963)].

curve of an unperturbed tunnel junction has the characteristic step due to the sharp increase in quasi-particle tunneling at $V = 2\Delta/e$. The steps that appear spaced at integral multiples of $h\nu/e$ on each side of the step at $2\Delta/e$ when microwaves of frequency ν are applied have the same shape as the step at $2\Delta/e$, and can be considered to result from quasi-particle tunneling with the absorption or emission of one or more photons.

[82] A. H. Dayem and R. J. Martin, *Phys. Rev. Lett.* **8**, 246 (1962).
[83] P. K. Tien and J. P. Gordon, *Phys. Rev.* **129**, 647 (1963).
[84] N. R. Werthamer, *Phys. Rev.* **147**, 255 (1966).
[85] C. F. Cook and G. E. Everett, *Phys. Rev.* **159**, 374 (1967).
[86] J. N. Sweet and G. I. Rochlin, *Solid State Commun.* **8**, 1341 (1970).
[87] N. Zeuthen Heidam, *Phys. Lett.* **35A**, 378 (1971).
[88] O. H. Soerensen and M. R. Samuelsen, *Phys. Lett.* **39A**, 137 (1972).

A detailed theoretical analysis by Werthamer[84] verified the Tien–Gordon results and showed an intimate theoretical relation between the alternating supercurrent and the quasi-particle current. The calculated contribution to the dc I–V curve from the quasi-particles in the presence of an rf field is

$$I(V) = \sum_n J_n^2(eV_1/h\nu) I_0(V \pm nh\nu/e), \qquad (5.2.36)$$

where V_1 is the amplitude of the rf voltage, J_n is the Bessel function of order n and $I_0(V)$ is the quasi-particle current with no applied rf.

In general, then, there are two rf-induced contributions to the dc I–V curves of tunnel junctions: (1) supercurrent steps at $V = nh\nu/2e$, with the size of the step varying like a Bessel function of the rf voltage

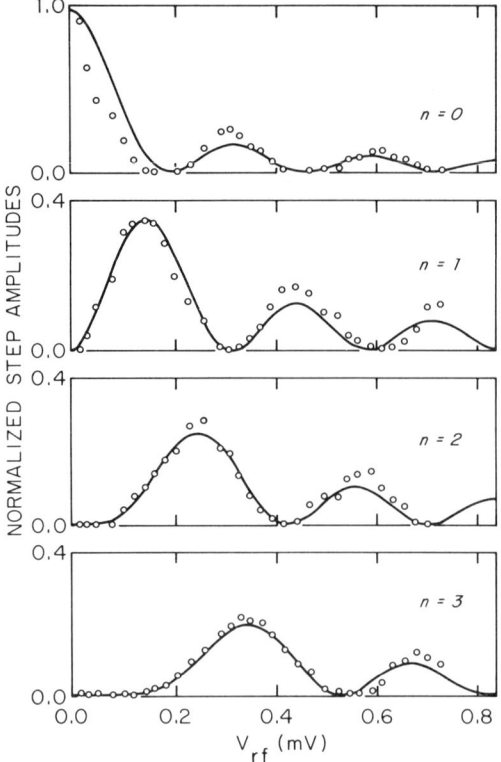

FIG. 13. Quasi-particle step heights as a function of rf voltage for a Sn microjunction of the form shown in Fig. 31 with $\nu = 20$ GHz. The solid curve is for $J_n^2(eV_{rf}/h\nu)$ and the points are experimental results [after C. A. Hamilton and S. Shapiro, *Phys. Rev. B* **2**, 4494 (1970)].

5.2. EXPERIMENTS THAT STUDY THE JOSEPHSON EFFECTS

[Eq. (5.2.33)]; and (2) quasi-particle steps at $V = 2\Delta/e \pm nh\nu/e$, with step size varying like the square of a Bessel function with rf voltage [Eq. (5.2.36)].

Hamilton and Shapiro[73] and Hamilton[80] have reported extensive experiments in which they measured both effects in the same junctions. Using very small tunnel junctions to ensure uniform rf fields in the junction, they find very good agreement with the calculated dependence on rf voltage as exemplified by Fig. 13 for quasi-particle steps. They also point out that at low frequencies the quasi-particle contribution to the dc I–V curve is equivalent to classical rf detection and they obtain good quantitative agreement with the calculated I–V curve for a low frequency at which the actual ac voltage across the junction can be measured.

By exposing a tunnel junction to microwave radiation at two frequencies simultaneously, Hamilton and Shapiro[89] have also observed quasi-particle steps corresponding to the sum and difference frequencies as expected from an extension of the theory.

5.2.5. Frequency Dependence of the Josephson Current and the Riedel Singularity

The frequency dependence of the amplitude $i_J(\nu_J)$ of the Josephson current was first investigated theoretically by Riedel,[90] who showed that the amplitude should be singular when $V = 2\Delta/e$, corresponding to the Josephson frequency $\nu_J = 4\Delta/h$. The predicted variation with frequency is shown in Fig. 14. Werthamer[84] derived analytic expressions for the amplitude of the Josephson current as a function of frequency for tunnel junctions at $T = 0$ K, and he showed that this frequency dependence modifies the behavior of the microwave-induced supercurrent steps. Implicit in the derivation of the usual expression, Eq. (5.2.33), is the assumption that i_c is independent of frequency. If the frequency dependence found by Riedel is included, the corresponding result for the supercurrent contribution to the dc I–V curve is

$$I(V) = \sum_N \left| \sum_n J_n(\alpha) J_{N-n}(\alpha) i_c[(2n - N)\nu] \right| \times \delta(V \pm Nh\nu/2e), \quad (5.2.37)$$

where $\alpha = eV_1/h\nu$.[80]

---

[89] C. A. Hamilton and S. Shapiro, *Phys. Lett.* **32A**, 223 (1970).
[90] E. Riedel, *Z. Naturforsch.* **19a**, 1634 (1964).

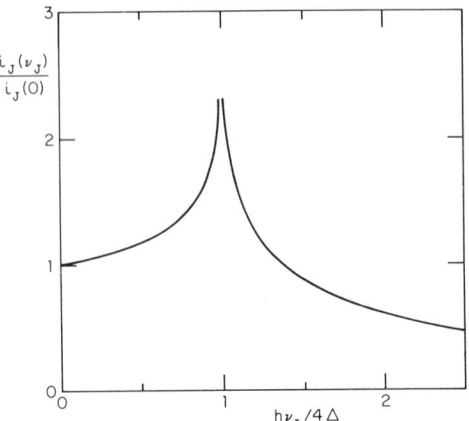

FIG. 14. Josephson current amplitude as a function of Josephson frequency as calculated by E. Riedel, *Z. Naturforsch.* **19a**, 1634 (1964).

For small values of α, this result is nearly identical to Eq. (5.2.33). However, for large V_1 there are large departures from simple Bessel function behavior if one of the induced steps lies at the gap voltage $V = 2\Delta/e$, as illustrated by the data in Fig. 15.[80] In general, if an even-numbered step falls at the gap, all the even-numbered steps will be singular, and if an odd one is at the gap, all the odd steps are singular.

By analyzing the behavior of the microwave-induced step heights as a function of frequency and rf power, Hamilton and Shapiro[91] obtained the first experimental demonstration of the Riedel singularity. They used very small (less than 10^{-6} cm^2) Sn–Sn oxide–Sn tunnel junctions to obtain data like that shown in Fig. 15 for frequencies from 20 to 26 GHz. To obtain the large amount of data required they took motion pictures of the I–V curves displayed on an oscilloscope as the rf power was varied continuously. Hamilton[80] has reported extensions of these measurements, from which he obtains a determination of the Riedel peak as shown in Fig. 16a. Rounding of the singularity is attributed to gap anisotropy and quasi-particle damping.

Another measurement of the Riedel singularity has been made by Buckner *et al.*[92] also using very small Sn–Sn oxide–Sn tunnel junctions but with an applied radiation frequency near 135 GHz. Since for tin at

---

[91] C. A. Hamilton and S. Shapiro, *Phys. Rev. Lett.* **26**, 426 (1971).
[92] S. A. Buckner, T. F. Finnegan, and D. N. Langenberg, *Phys. Rev. Lett.* **28**, 150 (1972).

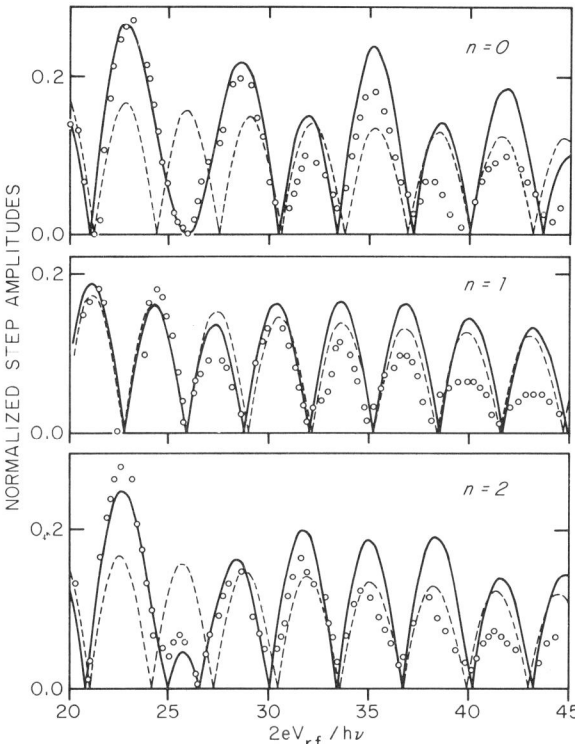

Fig. 15. Supercurrent step heights as a function of rf voltage showing the effects of the frequency-dependent supercurrent amplitude at high rf voltage. Dashed lines are calculated from Eq. (5.2.33); solid lines are calculated from Eq. (5.3.37); points are experimental data at 25.7 GHz for Sn tunnel junctions with area $\sim 10^{-6}$ cm^2 [after C. A. Hamilton, *Phys. Rev. B* **5**, 912 (1972)].

$T = 0$, $2\varDelta$ corresponds to about 290 GHz, the fourth step lay just below the gap. Their experimental procedure was to hold the rf power constant and increase the temperature starting at $T/T_c \approx 0.44$, thus sweeping the gap past the fourth microwave-induced step. The temperature dependence of the amplitudes of the steps were obtained from photographs of an oscilloscope display of the I–V curves. They observed the expected enhancement of the even-order steps. Data for the zero-voltage current are shown in Fig. 16b, where the resonance-like enhancement due to the Riedel singularity is apparent. They analyzed their data using the theory of Scalapino and Wu,[93] for which the curve shows a fit in

[93] D. J. Scalapino and T. M. Wu, *Phys. Rev. Lett.* **17**, 315 (1966).

Fig. 16b. This theory includes effects of quasi-particle damping on the shape of the singularity. For these particular experiments, they deduced that gap anisotropy dominates the rounding of the singularity. However, they proposed that for junctions in which the anisotropy has been reduced by impurities, it may be possible to probe the damping effects.

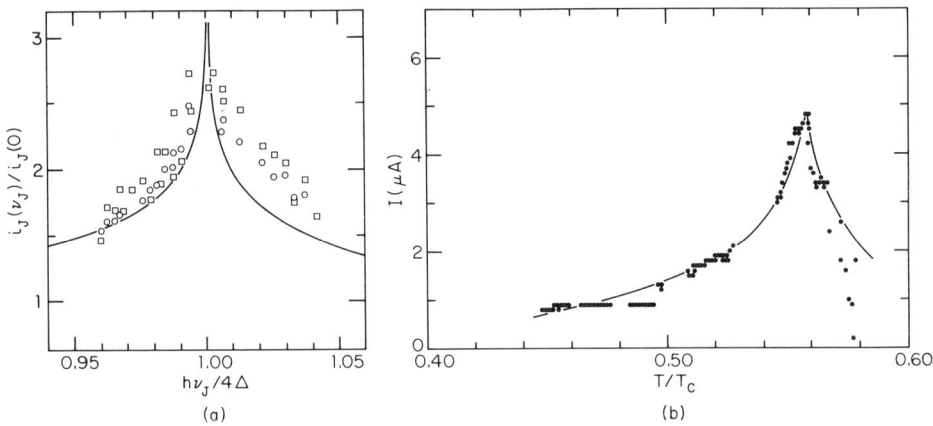

FIG. 16. Experimental evidence for the Riedel singularity. (a) Supercurrent amplitude as a function of frequency as determined from measurements of the supercurrent step heights at frequencies from 23 to 26 GHz from data like that shown in Fig. 15. Curve is $i_J(\nu_J)$ as shown in Fig. 14 [after C. A. Hamilton, *Phys. Rev. B* **5**, 912,(1972)]. (b) Zero-voltage current versus temperature with $\nu \approx 135$ GHz for which $2\Delta_0 \approx 2h\nu$ [after S. Z. Buckner, T. F. Finnegan, and D. N. Langenberg, *Phys. Rev. Lett.* **28**, 150 (1972)]. Both experiments used extremely small Sn tunnel junctions.

In addition to the Riedel singularity, another feature of great interest is the maximum frequency of oscillation of the Josephson current. For tunnel junctions, the frequency is probably limited by the large capacitance and thus experiments have concentrated on point contacts. However, there is no adequate theory for guidance for point contacts. Several methods have been used to investigate the high-frequency limit. One is based on observing the highest possible harmonic step generated in the junction by a relatively low-frequency applied signal. Using an applied signal at 70 GHz, McDonald et al.[64] have observed steps out to 8 THz. They suggest that the Josephson effect may be strongly attenuated at frequencies above the phonon modes, which also extend to approximately this frequency.

A second method is to determine the highest frequency for which microwave-induced steps can be produced. Steps have been observed at 3.8 THz, the 78-μ line of the water vapor laser.[63] In earlier experiments, steps at 2.5 THz were reported.[66]

A third method makes use of the current-dependent inductance of the weak link rather than the self oscillation. This technique consists of irradiating the junction with two high-frequency signals and through mixing in the nonlinear inductance to produce a low-frequency beat note which can be detected with ordinary rf amplifiers. Using this method McDonald *et al.*[63] have mixed the 401st harmonic of an X-band frequency with the 3.8-THz output of a water vapor laser to produce a beat signal at 9 GHz. All these results were obtained with niobium point contacts operated at about 2 K. As yet no well-defined upper limit has been determined for the Josephson oscillation.

5.2.6. Self-Induced Supercurrent Steps—Cavity Modes

Even with no external radiation the *I–V* curve can exhibit peaks or step-like features related to the presence of the alternating supercurrent. Most commonly these features result from coupling of the Josephson oscillation to electromagnetic resonances at frequencies ν_n, and produce singularities in the dc *I–V* curve at voltages $V_n = h\nu_n/2e$.

For thin film tunnel junctions, the junction structure itself constitutes a resonant cavity or a transmission line in which the oscillating supercurrent can excite large oscillating electromagnetic fields. These fields in turn act on the tunnel current in much the same way as an applied oscillating field, and produce dc contributions to the supercurrent. A good summary of these effects has been given by Langenberg *et al.*[9]

Although their explanations are closely related, two cases are distinguished experimentally for tunnel junctions. In one case sharp step-like features are found on the *I–V* curves (Fig. 17).[94] These steps first observed by Fiske[39] and studied by Coon and Fiske[95] are often called "Fiske steps." They occur for junctions for which the Josephson oscillation couples to discrete cavity modes of the structure. Since the spatial distribution of the supercurrent is determined by the magnetic field in the plane of the junction, the coupling to specific modes can be changed by varying the magnetic field.[9] Self-induced steps due to cavity modes

---

[94] D. N. Langenberg, D. J. Scalapino, B. N. Taylor, and R. E. Eck, *Phys. Rev. Lett.* **15**, 294 (1965).

[95] D. D. Coon and M. D. Fiske, *Phys. Rev.* **138**, A744 (1965).

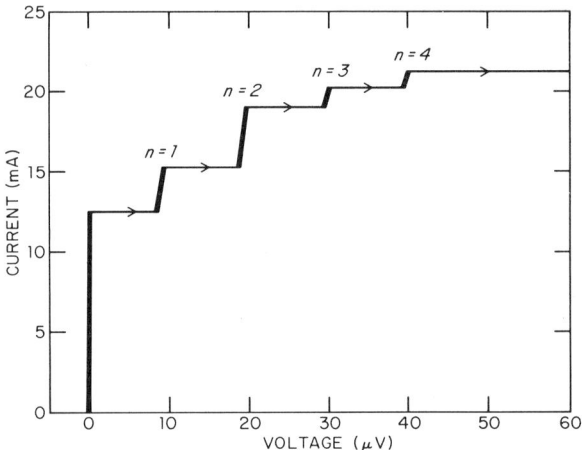

FIG. 17. I–V curve with self-induced supercurrent steps due to cavity modes. The data is for a Sn tunnel junction at $T = 1.2$ K with an applied magnetic field of 1.9 G. The modes are labeled by n and have a frequency separation of about 4.6 GHz [after D. N. Langenberg, D. J. Scalapino, B. N. Taylor, and R. E. Eck, *Phys. Rev. Lett.* **15**, 294 (1965)].

have been studied in detail in rectangular junctions[94,96,97] and in circular junctions.[98]

For large tunnel junctions, the structure is more like a semiinfinite transmission line than like a cavity with discrete modes. In this case the self-induced structure on the I–V curve appears as a resonance-like peak which can be tuned in voltage by changing the magnetic field. For a particular magnetic field, the peak occurs at that bias voltage for which the phase velocity of the Josephson current density wave matches that of the electromagnetic waves propagating along the transmission line. This effect was first observed and analyzed by Eck *et al.*[99]; an example of their data is shown in Fig. 18.

For perfectly uniform tunnel junctions in zero magnetic field, no resonant behavior of the junction should be possible. However, zero-field

---

[96] R. E. Eck, D. J. Scalapino, and B. N. Taylor, *Proc. Int. Conf. Low Temp. Phys.*, *9th* (J. G. Daunt, D. O. Edwards, F. J. Milford, and M. Yaqub, eds.), p. 415. Plenum Press, New York, 1965.

[97] I. M. Dmitrenko, I. K. Yanson and V. M. Svistunov, *JETP Lett.* **2**, 10 (1965) [*English transl. of: ZhETF Pis. Red.* **2**, 17 (1965)].

[98] S. Bermon and R. M. Mesak, *J. Appl. Phys.* **42**, 4488 (1971).

[99] R. E. Eck, D. J. Scalapino, and B. N. Taylor, *Phys. Rev. Lett.* **13**, 15 (1964).

5.2. EXPERIMENTS THAT STUDY THE JOSEPHSON EFFECTS

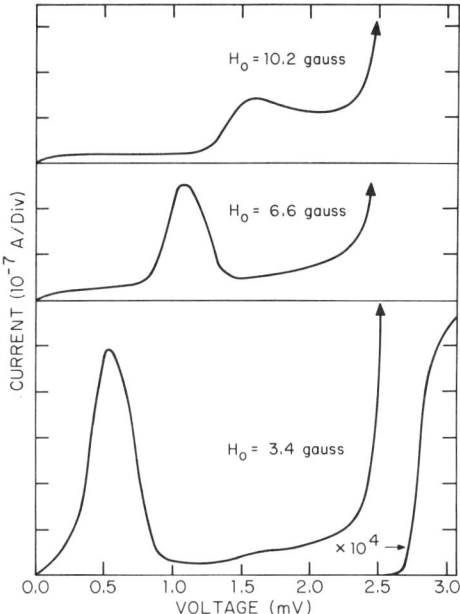

FIG. 18. I–V curves with peak due to electromagnetic modes in a tunnel junction showing essentially linear variation of the peak position with magnetic field. Data is for a Pb–Pb oxide–Pb junction at $T = 1.2$ K (after R. E. Eck, D. J. Scalapino, and B. N. Taylor, *Phys. Rev. Lett.* **13**, 15 (1964)].

resonances are observed and have been interpreted as cavity resonances excited because of inhomogeneities in the junction.[100,101]

In addition to the theoretical analysis included in papers reporting the experimental results described above, there have been a number of theoretical papers treating self-induced supercurrent steps. Werthamer carried out a detailed calculation of the total tunneling current, including the effects of self-coupling of the Josephson radiation in tunnel junctions.[84] Other theories based on the Josephson equations[102,103] and a numerical

---

[100] J. T. Chen, T. F. Finnegan, and D. N. Langenberg, *Proc. Int. Conf. Sci. Super-conductivity, Stanford*, 1969 (F. Chilton, ed.), p. 413. North-Holland Publ., Amsterdam, 1971.

[101] J. Matisoo, *Phys. Lett.* **29A**, 473 (1969).

[102] I. O. Kulik, *JETP Lett.* **2**, 84 (1965). [*English transl. of: ZhETF Pis. Red.* **2**, 134 (1965)].

[103] Yu. M. Ivanchenko, A. V. Svidzinskii, and V. A. Slyusarev, *Sov. Phys. JETP* **24**, 131 (1967) [*English transl. of: Zh. Eksp. Teor. Fiz.* **51**, 194 (1966)].

technique for calculating the self-resonant current peaks[104] have been reported.

Supercurrent steps due to cavity modes are an inherent part of the behavior of tunnel junctions because of their geometry, whereas the I–V curves of point contacts, for example, do not, in general, exhibit this type of structure. However, by placing a point contact in a closely coupled resonant cavity it is possible to produce steps on the I–V curve at those voltages at which the cavity is excited, namely when the Josephson frequency or one of its harmonics is equal to the cavity frequency.

Dayem and Grimes[105] first successfully coupled a point contact to an external cavity and observed the corresponding structure on the I–V curve. Werthamer and Shapiro[106] used an analog computer for a thorough study of the equations characterizing a Josephson junction coupled to a resonant cavity and also driven by an rf voltage. The effects of coupling a microwave cavity to a point contact have been studied extensively by Longacre and Shapiro,[107] and by Longacre,[108] who used an electronic

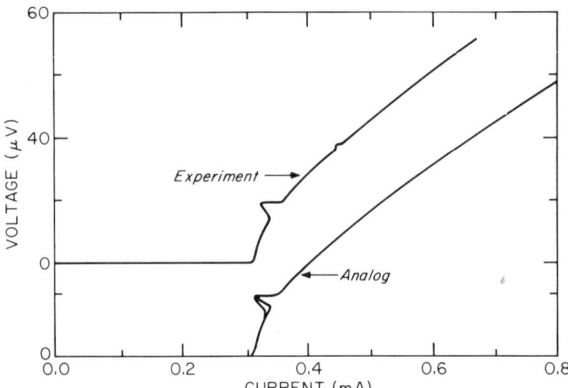

FIG. 19. Voltage–current curve for a point contact showing structure at 19 and 38 μV induced by coupling to a resonant coaxial cavity. The curve labeled "analog" was obtained with an electronic simulator (see Fig. 44) with a single resonance in the model circuit [after A. Longacre, Jr., Applied Superconductivity Conference Record, 1972. IEEE Publ. No. 72-CHO682-5-TABSC].

[104] J. D. Blackburn, J. D. Leslie, and H. J. Smith, *J. Appl. Phys.* **42**, 1047 (1971).
[105] A. H. Dayem and C. C. Grimes, *Appl. Phys. Lett.* **9**, 47 (1966).
[106] N. R. Werthamer and S. Shapiro, *Phys. Rev.* **164**, 523 (1967).
[107] A. Longacre, Jr. and S. Shapiro, *Proc. Symp. Submillimeter Waves* **20**. Microwave Res. Inst. Symp. Ser., Brooklyn Polytechnic Inst., 1970.
[108] A. Longacre, Jr., Applied Superconductivity Conference Record, 1972, IEEE Pub. No. 72-CHO682-5-TABSC.

analog developed by Hamilton[109] to simulate and analyze the results of their experiments. Figure 19 shows one of their experimental results, together with its analog simulation.

5.2.7. Radiation from the Oscillating Supercurrent

The most direct confirmation of the existence of the alternating Josephson current is generally considered to be by direct observation of radiation emitted from a biased junction at the Josephson frequency. Experimentally, this observation was hindered both by the small amount of power in the field of the oscillating Josephson current and by the severe impedance mismatch between typical tunnel junctions and free space, which severely reduced the amount of radiated power. Yanson et al.[110] were the first to observe radiation emitted from a Josephson junction. They mounted an oxide junction in a rectangular waveguide and observed radiation at 9.8 GHz. Subsequently Langenberg et al.[94] studied the radiation using a similar scheme diagrammed in Fig. 20, and they detected about 10^{-12} W at 9.2 GHz. Both of these experiments depended on biasing the junction on the Fiske steps so that large electromagnetic fields were present within the junction.

Another technique for detecting the alternating Josephson current was originated by Giaever.[111] He used two superimposed junctions with continuous oxide layers and detected the electromagnetic field generated by one of the junctions by means of the quasi-particle steps it produced on the I–V characteristic of the second junction.

Dayem and Grimes[105] first observed radiation emitted from a point contact by coupling it to an X-band cavity (Fig. 21). In their experiments they biased the contact to one of the steps produced on the I–V curve by interaction with the cavity and were able to obtain 10^{-10} W at 9.2 GHz. A number of similar experiments have been reported[112,113] and the harmonics generated by resistive feedback in the point contact have been observed.[114]

---

[109] C. A. Hamilton, *Rev. Sci. Instrum.* **43**, 445 (1972).

[110] I. K. Yanson, V. M. Svistunov, and I. M. Dmitrenko, *Sov. Phys. JETP* **21**, 650 (1965) [*English transl. of: Zh. Eksp. Teor. Fiz.* **48**, 976 (1965)].

[111] I. Giaever, *Phys. Rev. Lett.* **14**, 904 (1965).

[112] I. Ya. Krasnopolin and M. S. Khaikin, *JETP Lett.* **6**, 129 (1967) [*English transl. of: ZhETF Pis. Red.* **6**, 633 (1967)].

[113] J. E. Zimmerman, J. A. Cowen, and A. H. Silver, *Appl. Phys. Letters* **9**, 353 (1966)

[114] R. Adde and G. Vernet, *J. Appl. Phys.* **43**, 2405 (1972).

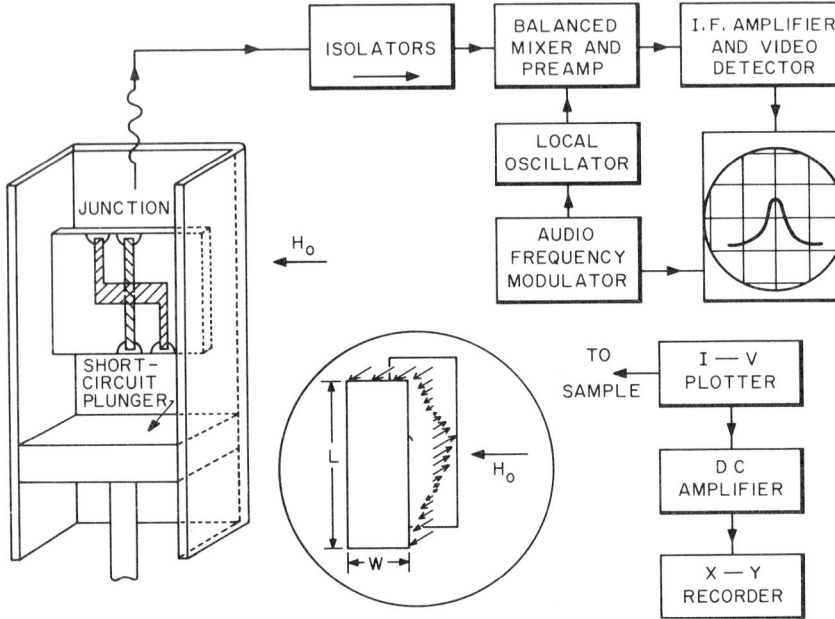

Fig. 20. Experimental scheme used for observing Josephson radiation from a tunnel junction. The inset shows the electric field distribution in the junction when it is biased on the second self-resonant mode ($n = 2$ in Fig. 17) [after D. N. Langenberg, D. J. Scalapino, B. N. Taylor, and R. E. Eck, Phys. Rev. Lett. **15**, 294 (1965)].

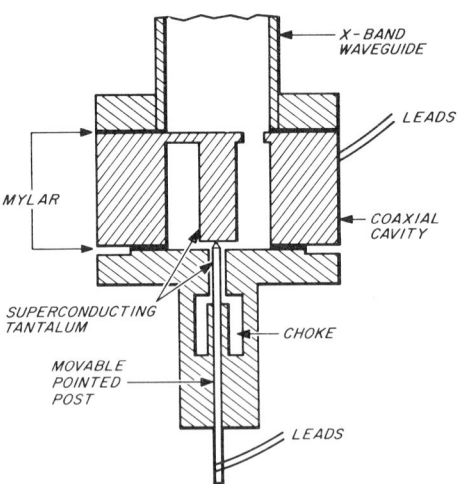

Fig. 21. Configuration used for coupling a point contact to a resonant cavity to observe Josephson radiation at 9.2 GHz [after A. H. Dayem and C. C. Grimes, Appl. Phys. Lett. **9**, 47 (1966)].

5.2. EXPERIMENTS THAT STUDY THE JOSEPHSON EFFECTS

Zimmerman et al.[113] used a scheme like that shown in Fig. 22a to voltage bias a point contact and to observe the Josephson oscillation at 30 MHz. The shunt resistor R was very much smaller than the dynamic resistance of the point contact, and the current I very much greater than the critical current i_c, so that the point contact was voltage biased at $V = IR$. By displaying the detector output versus bias current on an oscilloscope or X–Y recorder, they determined those bias voltages for which the Josephson oscillation had a component at ν_0, the resonant frequency of the tank. In addition to a strong peak at $V_0 = \Phi_0 \nu_0$ corresponding to the fundamental Josephson oscillation, they found peaks at $V_0/2$ and $V_0/4$ indicating strong harmonics in the supercurrent wave form. The dominant contribution to the line width in this experiment can be from the Johnson noise in the bias resistor R so that measurements of the line width determine the absolute temperature as discussed in Section 5.4.3.4.

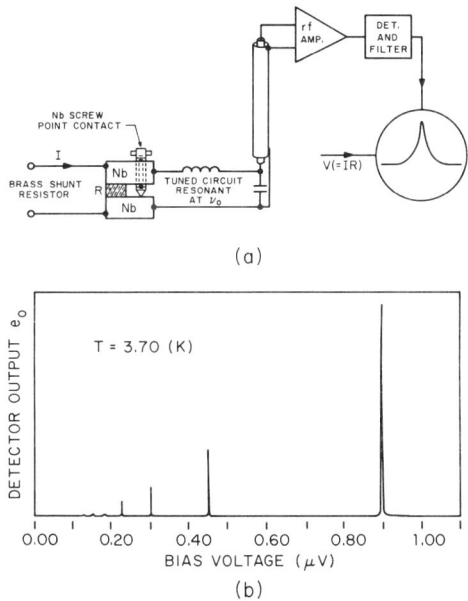

FIG. 22. (a) Method for voltage biasing a point contact and observing the Josephson oscillations at radio frequencies [after J. E. Zimmerman, J. A. Cowen, and A. H. Silver, *Appl. Phys. Lett.* **9**, 353 (1966)]. (b) Detector output versus bias voltage for an experiment like that diagrammed in (a) with $\nu_0 = 60$ MHz, $R \approx 10^{-5}\ \Omega$, and $T = 3.7$ K showing an oscillation corresponding to a multiple quantum transition with $N = 7$ [after R. Sandell, M. Puma, and B. S. Deaver, Jr., *Proc. Int. Conf. Low Temp. Phys. 13th, Boulder, Colorado, 1972*, University of Colorado Associated Press, Boulder, Colorado (to be published)].

Using a very similar experimental arrangement, Zimmerman and Silver[115] discovered oscillation of the type indicated by the data in Fig. 22b giving peaks at $V = (N/n)\Phi_0 \nu_0$, N and n integers. This spectrum corresponds to the coherent motion of N flux quanta across the point contact at each cycle of the oscillation. The sizes of the peaks for $n > 1$ give an indication of the harmonic content and show that the wave form is nearly a sawtooth. These multiple quantum transitions occur when the inductance L of the ring containing the resistor and the point contact is large compared to $\Phi_0/2\pi i_c$. These experiments have been repeated and extended by Kamper et al.[116] and by Sandell et al.[117]

A lumped parameter model has been used by Sullivan et al.[118] to analyze these oscillations, and Sullivan and Zimmerman[119] have developed a mechanical analog which simulates them. Both techniques have proved highly successful in interpreting the extremely nonlinear behavior of circuits containing weak links.

The highest frequencies coupled out of a Josephson oscillator have been obtained by Elsley and Sievers[120] using a point contact in the configuration shown in Fig. 23. The smooth fashion in which the foil matches the point contact to the rest of the cavity was presumed to be important, since if the foil was not present no radiation was observed. They detected up to 10^{-9} W of radiation at frequencies from 60 to 400 GHz. Using a Fourier transform technique they obtained the spectrum of the radiation shown in Fig. 23.

By placing a point contact in a resonant cavity coupled through a small hole to a waveguide, Ulrich and Kluth[121] brought out radiation of about 300 GHz to an external mirror, which directed the radiation back into the waveguide. They detected the presence of the radiation by monitoring

---

[115] J. E. Zimmerman and A. H. Silver, *Phys. Rev. Lett.* **19**, 14 (1967).

[116] R. A. Kamper, L. O. Mullen, and D. B. Sullivan, Superconducting Thin Films, NASA CR-1189, available from Nat. Tech. Information Service, Springfield, Virginia 22151.

[117] R. Sandell, M. Puma, and B. S. Deaver, Jr., *Proc. Int. Conf. Low Temp. Phys.*, *13th, Boulder, Colorado* (1972). Univ. Colorado Associated Press, Boulder, Colorado (to be published).

[118] D. B. Sullivan, R. L. Peterson, V. E. Kose, and J. E. Zimmerman, *J. Appl. Phys.* **41**, 4865 (1970).

[119] D. B. Sullivan and J. E. Zimmerman, *Amer. J. Phys.* **39**, 1504 (1971).

[120] R. K. Elsley and A. J. Sievers, Applied Superconductivity Conference Record, 1972. IEEE Publ. No. 72-CHO682-5-TABSC.

[121] R. B. Ulrich and E. O. Kluth, Applied Superconductivity Conference Record, 1972, IEEE Publ. No. 72-CHO682-5-TABSC.

5.2. EXPERIMENTS THAT STUDY THE JOSEPHSON EFFECTS

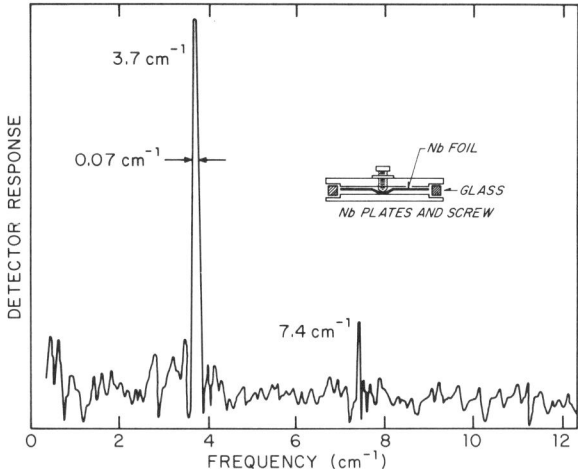

FIG. 23. Spectrum of radiation from a point contact mounted in the configuration shown in the inset. Radiation was coupled out of the dewar through a circular waveguide and the spectrum obtained using a Fourier transform technique (after R. K. Elsley and A. J. Sievers, Applied Superconductivity Conference Record, 1972. IEEE Publ. No. 72-CHO682-5-TABSC].

the voltage across the point contact at constant bias current as the mirror was moved, and found a periodic variation corresponding to a wavelength of about 1 mm.

In a series of papers Clark[122] has reported investigations of generation and detection of radiation by planar arrays of 1-mm diameter tin and niobium spheres pressed together to form two-dimensional arrays of weak links. He found peaks on the I–V curve of the array that were enhanced when plane mirrors were placed on each side of the array. Periodic variation of the sizes of some of the peaks with mirror positions were interpreted as evidence for radiation of wavelength about 1 mm. Similar experiments have been reported by Repici et al.[123] and a theoretical model has been used by Tilley[124] to investigate the possibility of coherent radiation from an array of junctions coupled to a cavity.

[122] T. D. Clark, Phys. Lett. **27A**, 585 (1968); Proc. Int. Conf. Low Temp. Phys., 11th, St. Andrews **1**, p. 686 (1968); Proc. Int. Conf. Sci. Superconductivity, Stanford, 1969 (F. Chilton, ed.), p. 432. North-Holland Publ., Amsterdam, 1971; Proc. Int. Conf. Low Temp. Phys., 12th, Kyoto (Eizo Kanda, ed.), p. 449. Academic Press of Japan, Tokyo, 1971.

[123] D. J. Repici, L. Leopold, and W. D. Gregory, Applied Superconductivity Conference Record, 1972. IEEE Publ. No. 72-CHO682-5-TABSC.

[124] D. R. Tilley, Phys. Lett. **33A**, 205 (1970).

Radiation has also been observed from point contacts comprised of a superconducting point in contact with a normal metal block by Gregory et al.[125] It was postulated that the radiation was due to stimulated emission at the gap frequency.[126] Some controversy has developed over this interpretation and subsequent work[127,128] has not yet answered the question definitively.

5.2.8. Plasma Resonance

Josephson first pointed out the possibility of the existence of plasma oscillations in a tunnel junction.[4] Linearizing Eq. (5.2.28) by setting

$$\sin \theta = \sin \theta_0 + \delta\theta \cos \theta_0 \qquad (5.2.38)$$

yields a wave equation for the small oscillations of the phase

$$\nabla^2 \, \delta\theta + v^{-2} \left(\partial^2 \, \delta\theta/\partial t^2\right) = (\cos \theta_0/\lambda_J^2) \, \delta\theta, \qquad (5.2.39)$$

where $v^2 = 1/\mu_0 \, dC$, C is the capacitance per unit area, $d = \lambda_1 + \lambda_2 + l$ is the effective length of the junction (see Fig. 1), and λ_J is the Josephson penetration depth. The solutions of this equation correspond to electromagnetic waves with dispersion relation

$$\omega^2 = v^2 k^2 + \omega_p^2 \qquad (5.2.40)$$

where

$$\omega_p^2 = v^2 \cos \theta_0/\lambda_J^2 = 2\mu_0 e j_c \cos \theta_0/\hbar C. \qquad (5.2.41)$$

For $k = 0$ there is a longitudinal oscillation with the electric field and current normal to the plane of the junction, and there is a periodic exchange of energy between the electric field and the junction coupling energy. This mode corresponds to a plasma oscillation with frequency

---

[125] W. D. Gregory, L. Leopold, and D. Repici, *Can. J. Phys.* **47**, 1171 (1969).

[126] L. Leopold, W. D. Gregory, and J. Bostock, *Can. J. Phys.* **47**, 1167 (1969).

[127] W. D. Gregory, L. Leopold, D. Repici, and J. Bostock, *Phys. Lett.* **29A**, 13 (1969); L. Leopold, W. D. Gregory, D. Repici, and R. F. Averill, *ibid.* **30A**, 507 (1969); W. D. Gregory, L. Leopold, D. Repici, R. F. Averill, and J. Bostock, *Proc. Int. Conf. Low Temp. Phys.*, 12th, Kyoto, Japan (Eizo Kanda, ed.), p. 445. Academic Press of Japan, Tokyo, 1971; W. D. Gregory, W. N. Mathews, Jr., L. Leopold, J. Bostock, and J. George, in Applied Superconductivity Conference Record, 1972. IEEE Publ. No. 72-CHO682-5-TABSC.

[128] A. A. Fife and S. Gygax, *Appl. Phys. Lett.* **20**, 152 (1972).

ω_p, typically about 10 GHz. This very low frequency occurs because the effective carrier density n_s is small and, in fact, depends on the dc current through the junction since θ_0 is determined by $i = i_\mathrm{c} \sin \theta_0$. Note that the junction capacitance C can be considered to be in parallel with the dynamic inductance

$$\mathscr{L} = (\Phi_0/2\pi i_\mathrm{c})(1/\cos \theta_0) \qquad (5.2.42)$$

and the plasma resonance to be the resonant frequency $\omega_\mathrm{p}^2 = 1/\mathscr{L}C$ of this combination (see Section 5.3.2).

Dahm et al.[129] first noted the dependence of the plasma frequency on the dc current and used this fact for their first experimental observation of the plasma resonance. Essentially, they measured the impedance of a tunnel junction with an extremely low level of applied microwave power. With the applied frequency ω held constant, they varied the dc current (always being certain to be in the zero-voltage condition) to scan the plasma frequency. Because of the experimental advantage of having the signal at the second harmonic rather than the fundamental frequency ω, they used the inherent nonlinearity of the junction to produce second harmonic and then observed the resonant enhancement of the second harmonic signal when $\omega_\mathrm{p} = 2\omega$. Using the Fiske steps to determine the junction capacitance C, they found good agreement between the calculated plasma frequency and the observed one and verified the current dependence of the plasma frequency. Further, they found qualitative agreement with the temperature and magnetic field dependence expected through the functional dependence of the critical current on these quantities.

5.2.9. Subharmonic Structure on the I–V Curve

The I–V curves of tunnel junctions and point contacts are found to exhibit structure at $V = 2\varDelta/ne$, $n = 2, 3, 4, \ldots$.[130–143] These features

---

[129] A. J. Dahm, A. Denenstein, T. F. Finnegan, D. N. Langenberg, and D. J. Scalapino, *Phys. Rev. Lett.* **20**, 859 (1968).

[130] B. N. Taylor and E. Burstein, *Phys. Rev. Lett.* **10**, 14 (1963).

[131] C. J. Adkins, *Phil. Mag.* **8**, 1051 (1963); *Rev. Mod. Phys.* **36**, 211 (1964).

[132] J. M. Rowell, *Rev. Mod. Phys.* **36**, 215 (1964).

[133] I. K. Yanson, V. M. Svistunov, and I. M. Dmitrenko, *Sov. Phys. JETP* **20**, 1404 (1965) [English transl. of: *Zh. Eskp. Teor. Fiz.* **47**, 2091 (1964)].

[134] S. M. Marcus, *Phys. Lett.* **19**, 623 (1966); **20**, 236 (1966).

[135] G. I. Rochlin, *Phys. Rev.* **153**, 513 (1967).

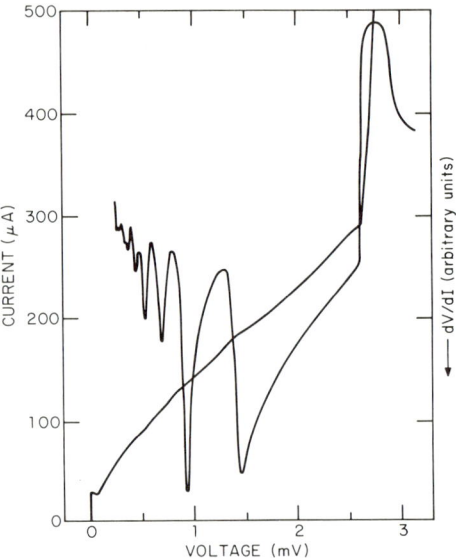

FIG. 24. Subharmonic structure on the I–V curve of a Pb tunnel junction at $T = 1$ K. Also shown is the derivative dV/dI plotted downward so that minima in the plot are minima is conductance. The peak at ~2.7 mV corresponds to $2\Delta/e$ for Pb (after J. M. Rowell and W. L. Feldman, Phys. Rev. **172**, 393 (1968)].

are often called "subharmonic structure." Examples are shown in Figs. 24–26. The precise nature of the features, that is, whether step-like or peak, is not the same for different weak links. However, the fact that some consistent set of features occurs for $V = 2\Delta/ne$ appears well established. For tunnel junctions there is always a large step at $V = 2\Delta/e$ (Fig. 24), due to the onset of quasi-particle conductance. Subharmonic structure is also evidenced by point contacts showing little or no feature at $2\Delta/e$. There is, in fact, evidence that the subharmonic structure is most pro-

[136] I. K. Yanson, Sov. Phys. JETP **26**, 742 (1967) [English transl. of: Zh. Eksp. Teor. Fiz. **53**, 1268 (1967)].

[137] J. M. Rowell and W. L. Feldmann, Phys. Rev. **172**, 393 (1968).

[138] I. K. Yanson and I. Kh. Albegova, Sov. Phys. JETP **28**, 826 (1969) [English transl. of: Zh. Eskp. Teor. Fiz. **55**, 1578 (1968)].

[139] L. J. Barnes, Phys. Rev. **184**, 434 (1969).

[140] A. A. Bright and J. R. Merrill, Phys. Rev. **184**, 446 (1969).

[141] I. Giaever and H. R. Zeller, Phys. Rev. B **1**, 4278 (1970).

[142] M. Puma and B. S. Deaver, Jr., Appl. Phys. Lett. **19**, 539 (1971).

[143] A. S. DeReggi and R. S. Stokes, Proc. Int. Conf. Low Temp. Phys., 13th, Boulder, Colorado, 1972. Univ. Colorado Associated Press, Boulder, Colorado (to be published).

5.2. EXPERIMENTS THAT STUDY THE JOSEPHSON EFFECTS

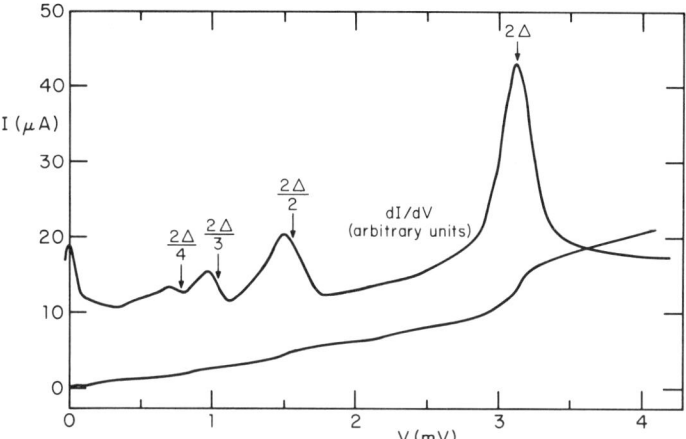

FIG. 25. I–V curve and dI/dV for a Nb point contact at 4.2 K, showing subharmonic structure [after L. J. Barnes, *Phys. Rev.* **184**, 434 (1969)].

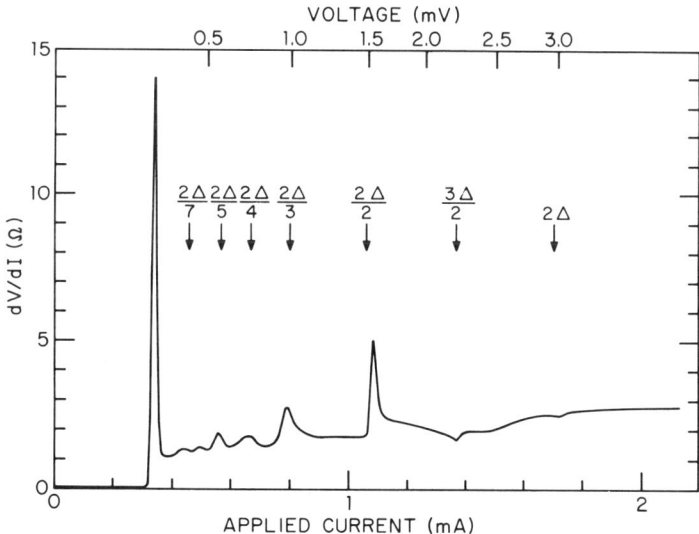

FIG. 26. Differential resistance dV/dI for a Nb point contact at the focal point of a spherical mirror providing broad-band feedback. The data were taken at 3.95 K and the first peak corresponds to the critical current at 0.3 mA [after A. S. DeReggi and R. S. Stokes, *Proc. Int. Conf. Low Temp. Phys., 13th, Boulder, Colorado*, University of Colorado Associated Press, Boulder, Colorado (to be published)].

nounced in junctions containing microshorts.[137] The temperature dependence[136,139,140,142] of all the features follows $\Delta(T)$, the BCS gap parameter, and the structure is only slightly affected by magnetic fields.[134,137,140]

As yet no completely acceptable explanation exists for subharmonic structure. However, most of the mechanisms proposed to account for it involve the Josephson effect. Possible mechanisms have been summarized and critically discussed by a number of authors.[135,137,141] Werthamer[84] has calculated the I–V curve for a tunnel junction assuming that the fundamental frequency of the Josephson oscillation is strongly fed back into the junction. His calculations include both quasi-particle and pair contributions and the effect of the Riedel singularity. The rather formidable expressions that result have not been evaluated in detail, however he discusses the various contributions to the dc current:

(1) Photon-assisted quasi-particle tunneling with the photon being at the Josephson frequency or one of its harmonics can produce dc contributions to the current when

$$eV + h\nu = 2\Delta. \qquad (5.2.43)$$

That is, if

$$\nu = n\nu_J \quad \text{and} \quad \nu_J = 2eV/h,$$
$$V_n = 2\Delta/(2n+1), \quad n = 1, 2, 3, \ldots \qquad (5.2.44)$$

(in essence these are self-induced quasi-particle steps).

(2) The Riedel singularity (Section 5.2.5) can cause an extra pair current contribution when the Josephson frequency or an odd-numbered harmonic of it is $4\Delta/h$. If $(2n+1)\nu_J = 4\Delta/h$, there will be excess supercurrent contributions at

$$V_n = 2\Delta/(2n+1), \quad n = 1, 2, 3, \ldots. \qquad (5.2.45)$$

[A consequence of the singularity in the amplitude of the oscillating supercurrent at $\nu_J = 4\Delta/h$ is that when microwaves at frequency ν are incident on a junction, all of the odd-numbered supercurrent steps will be singular if an odd step falls at $V = 2\Delta/e$, i.e., if $n\nu = 4\Delta/h$. The even steps are singular if an even step falls at $2\Delta/e$. Having the Josephson oscillation strongly fed back into the junction is like having an external oscillator always irradiating the junction at the Josephson frequency ν_J. Thus the dc supercurrent measured at any voltage corresponds to having the junction always biased on the $n = 1$ supercurrent step. When an odd harmonic of v_J falls at $4\Delta/h$, the $n = 1$ step is singular and gives an excess dc supercurrent. These singularities give rise to an odd subharmonic series, Eq. (5.2.45).]

(3) Direct absorption of a photon at the gap energy can give excess quasi-particle current when

$$h\nu = 2\Delta. \tag{5.2.46}$$

With $\nu = nv_J$, these contributions occur at

$$V_n = 2\Delta/2n, \quad n = 1, 2, \ldots. \tag{5.2.47}$$

Werthamer[84] noted that for junctions between different superconductors with gaps $2\Delta_1$ and $2\Delta_2$, the odd series should be given by $V_n = (\Delta_1 + \Delta_2)/(2n + 1)$, and there should be two even series at $2\Delta_1/2n$ and $2\Delta_2/2n$. Using junctions made of lead and a lead indium alloy, Giaever and Zeller[141] report observations of these three series. By using cadmium sulfide as the barrier, they were also able to vary the strength of coupling between the superconductors by shining light on the barrier. Using this technique they were able to determine that the magnitude of all the subharmonic structure increases in proportion to the square of the tunnel matrix element.

If the frequency-dependent surface resistance is the only mechanism satisfying the energetics that produce the even series Eq. (5.2.47), the magnitude of the even and odd series should have different dependence on the strength of coupling.[84] Strässler and Zeller[144] proposed an additional mechanism for the even series, which can explain the Giaever and Zeller results.

Although there is not yet a totally acceptable theory for subharmonic structure, it does appear that strong feedback of the Josephson oscillation and all of its harmonics can account for all the observed effects. The precise shape of the induced structure may vary with the detailed nature of the feedback[122,143] (see Fig. 26).

5.2.10. Phonon Generation by the Josephson Effect

Kinder[145] has observed phonons emitted from a tunnel junction biased at voltages less than $2\Delta/e$. With a constant current through the junction, he used a magnetic field to adjust the voltage across the junction. The

---

[144] S. Strässler and H. R. Zeller, *Phys. Rev. B* **3**, 226 (1971).
[145] H. Kinder, *Phys. Lett.* **36A**, 379 (1971).

energy thus fed into the junction was expected to be converted mostly to phonons because of the severe mismatch for electromagnetic radiation into free space. He used a technique originated by Eisenmenger and Dayem[146] to detect the phonons. Tunnel junctions were fabricated on opposite faces of sapphire single crystals 3–6 mm thick. One junction was biased at $V \leq 2\Delta/e$ and served as a phonon detector. Phonons with energy greater than or equal to 2Δ incident on the detector produce quasi-particles by dissociating pairs, and the resulting increase in tunnel current is a measure of the number of incident phonons.

The other junction was used as a phonon generator. Eisenmenger and Dayem[146] had already shown that when the generator bias V_g was greater than $2\Delta/e$, the phonons received at the detector corresponded to two relaxation processes in the generator junction. The quasi-particles injected across the barrier decayed first to the top of the energy gap and then combined to form a Copper pair emitting a phonon of energy 2Δ. When $V_g \geq 4\Delta/e$, phonons resulting from both processes were detected. However, for V_g between $4\Delta/e$ and $2\Delta/e$ only the recombination phonon was detected.

Kinder used direct quasi-particle tunneling at $V_g = 2\Delta/e$ to calibrate the detector. He then measured the intensity of the phonons emitted from

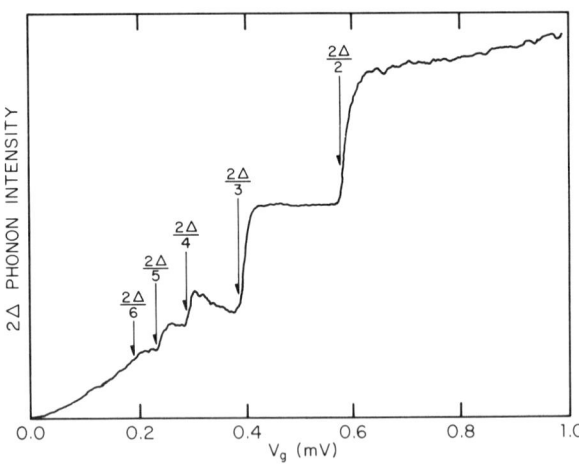

FIG. 27. Intensity of phonons of energy $\hbar\omega \geq 2\Delta$ emitted by a tunnel junction at constant current I_g as a function of voltage V_g (adjusted by an applied magnetic field). Data are for a Sn junction at $T = 1.00$ K [after H. Kinder, *Phys. Lett.* **36A**, 379 (1971)].

[146] W. Eisenmenger and A. H. Dayem, *Phys. Rev. Lett.* **18**, 125 (1967).

the generator as a function of V_g for voltages less than $2\Delta/e$. His results are shown in Fig. 27. Kinder concludes that the dominant process for energy loss in the Josephson junction is production of excited quasiparticles, which subsequently recombine with the emission of 2Δ phonons. The abrupt increases in phonon intensity at the subharmonic voltages $2\Delta/ne$ are approximately proportional to $1/n$. These results suggest a close connection with the subharmonic structure found on the I–V curve (see Section 5.2.9).

5.2.11. Effects of Thermodynamic Fluctuations and Noise

Since the maximum coupling energy between two superconductors joined by a weak link is $\Phi_0 i_c/2\pi$, when $\Phi_0 i_c/2\pi \leq kT$ thermodynamic fluctuations will disrupt the coupling. Thus for $T \approx 4$ K, fluctuation effects will become significant for $i_c \approx 10^{-7}$ A. For an isolated weak link in thermal equilibrium, the power spectrum of the current fluctuations is related to the frequency-dependent loss through the Johnson–Nyquist relation. However, weak links are not usually operated at thermal equilibrium nor are they isolated, since they are intentionally, or by the nature of their structure, coupled to external circuit elements, and they are subject to external noise sources. Several cases have been studied theoretically and experimentally.

The power spectrum of the fluctuations of the current for a tunnel junction operating at finite bias voltage has been calculated by Scalapino[147] and by Stephen[148] and expressed in terms of the separate contributions from the quasi-particle and the pair currents and the experimentally determined I–V curve. The power spectrum $P(\omega)$ gives the frequency distribution of the current fluctuations

$$\langle \delta I^2 \rangle = 1/2\pi \int_0^\infty P(\omega)\, d\omega. \quad (5.2.48)$$

The predicted power spectrum for a tunnel junction biased at voltage V and for $\omega < eV/\hbar$ is

$$P(\omega) = 2eI_q(V)\coth(eV/2kT) + 4eI_p(V)\coth(eV/kT), \quad (5.2.49)$$

[147] D. J. Scalapino, *Proc. Symp. Phys. Superconducting Devices*, Charlottesville (1967) (see Ref. 22); see also D. Rogovin and D. J. Scalapino, *Superconductivity, Proc. Int. Conf. Sci. Superconductivity*, Stanford, *1969* (F. Chilton, ed.), p. 399. North-Holland Publ., Amsterdam, 1971.

[148] M. J. Stephen, *Phys. Rev.* **182**, 531 (1969).

where $I_q(V)$ and $I_p(V)$ are the dc contributions from quasi-particles and from pairs, respectively. This is a spectrum of the "shot noise" type. For $eV \ll kT$ this reduces to

$$P(\omega) = 4kT[I_q(V) + I_p(V)]/V, \qquad (5.2.50)$$

which is of Johnson noise form with an effective conductance $(I_q + I_p)/V$.

The voltage fluctuations are related to the current fluctuations by the dynamic impedance of the junction; at frequencies high enough that the time constant associated with the junction capacitance can be neglected, the spectrum $P_V(\omega)$ of the voltage fluctuations is

$$P_V(\omega) = R_D{}^2 P(\omega), \qquad (5.2.51)$$

where $R_D = dV/dI$, the dynamic resistance of the junction.

Kantor and Vernon[149,150] have measured the noise spectrum for Nb point contacts and find good agreement with these calculations. These results imply a more general validity of the above results, which were derived for tunnel junctions.

The voltage fluctuations across a biased junction frequency modulate the Josephson oscillation. Measurements of the line width of the radiation from Sn and Pb tunnel junctions were reported by Dahm et al.[151] They compared these experimental results with the calculated line width expected from their calculation of the contribution from quasi-particle current fluctuations combined with the contribution from pair current fluctuations calculated by Stephen,[148]

$$\Delta \nu = 4\pi R_D{}^2 (2e/h)^2 (kT/V)(I_q + I_p). \qquad (5.2.52)$$

(Other calculations of the line width have been made by Larkin and Ovchinnikov[152] and by Lee and Scully.[153]) They found satisfactory agreement with the theory but line widths somewhat larger than predicted by the calculations. Vernet and Adde[154] measured the line width of radia-

---

[149] H. Kanter and F. L. Vernon, Jr., *Phys. Lett.* **32A**, 155 (1970).

[150] H. Kanter and F. L. Vernon, Jr., *Phys. Rev. Lett.* **25**, 588 (1970).

[151] A. J. Dahm, A. Denenstein, D. N. Langenberg, W. H. Parker, D. Rogovin, and D. J. Scalapino, *Phys. Rev. Lett.* **22**, 1416 (1969).

[152] A. I. Larkin and Yu. N. Ovchinnikov, *Sov. Phys. JETP* **26**, 1219 (1968) [English trans. of: *Zh. Eksp. Teor. Fiz.* **53**, 2159 (1967)].

[153] P. A. Lee and M. O. Scully, *Phys. Rev. B* **3**, 767 (1971).

[154] G. Vernet and R. Adde, *Appl. Phys. Lett.* **19**, 195 (1971).

5.2. EXPERIMENTS THAT STUDY THE JOSEPHSON EFFECTS 247

tion emitted from Nb point contacts and found good agreement with the theory.

The line width of the Josephson oscillation in thin film bridges has been measured by Kirschman and Mercereau.[155] They find their data consistent with a simple two-fluid model incorporating fluctuations in the supercurrent stimulated by Johnson noise in the normal current. Thiene and Zimmerman[156] have used a vortex flow model to calculate the spectrum of voltage fluctuations for thin film bridges.

The effects of thermal fluctuations on the I–V curves of weak links have been the subject of several theoretical[157–162] and experimental papers.[49,163–165] The calculations usually assume a model of the junction that includes the effects of junction capacitance and quasi-particle current. The qualitative effect of fluctuations is to reduce the critical current and to round the transition to finite voltages. Anderson and Goldman[49] first reported measurements on nearly ideal tunnel junctions, for which they determined the I–V curves at various temperatures very near T_c. They found reasonable agreement with the calculations of Ambegaokar and Halperin.[159] Using thin film bridges which have extremely small capacitance, Simmonds and Parker[163] were able to approximate the conditions of the Ambegaokar and Halperin calculation more closely, and found extremely good agreement with the shape of the I–V curves as a function of temperature (Fig. 28).

Using extremely small (less than 10^{-6} cm^2) tin and lead tunnel junctions, Buchner et al.[164] have measured the I–V curves as a function of external noise coupled into the junction from a Gaussian noise generator (Fig. 29). The observed variation of the dc supercurrent with external

---

[155] R. K. Kirschman and J. E. Mercereau, *Phys. Lett.* **35A**, 177 (1971).

[156] P. Thiene and J. E. Zimmerman, *Phys. Rev.* **177**, 758 (1968).

[157] Yu. M. Ivanchenko and L. A. Zil'berman, *JETP Lett.* **8**, 113 (1968) [*English transl. of: ZhETF Pis. Red.* **8**, 189 (1968)].

[158] Yu. M. Ivanchenko and L. A. Zil'berman, *Sov. Phys. JETP* **28**, 1272 (1969) [*English transl. of: Zh. Eksp. Teor. Fiz.* **55**, 2395 (1968)].

[159] V. Ambegaokar and B. I. Halperin, *Phys. Rev. Lett.* **22**, 1364 (1969).

[160] Yu. M. Ivanchenko and L. A. Zil'berman, *Sov. Phys. JETP* **31**, 117 (1970) [*English transl. of: Zh. Eksp. Teor. Fiz.* **58**, 211 (1970)].

[161] A. C. Biswas and S. S. Jha, *Phys. Rev. B* **2**, 2543 (1970).

[162] L. A. Zil'berman and Yu. M. Ivanchenko, *Sov. Phys. Solid State* **12**, 1530 (1971) [*English transl. of: Fiz. Tverd. Tela* **12**, 1922 (1970)].

[163] M. Simmonds and W. H. Parker, *Phys. Rev. Lett.* **24**, 876 (1970).

[164] S. A. Buckner, J. T. Chen, and D. N. Langenberg, *Phys. Rev. Lett.* **25**, 738 (1970).

[165] H. Kanter and F. L. Vernon, Jr., *Phys. Rev. B* **2**, 4694 (1970).

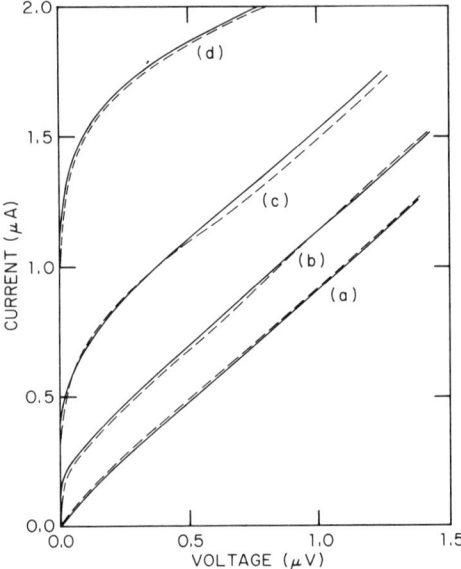

FIG. 28. Current–voltage characteristics of a Sn thin film bridge at several temperatures near T_c. The solid lines are experimental curves and the dashed curves are a one-parameter fit with the theory of Ambegaokar and Halperin. The relevant parameters are (a) $T = 3.873$ K, $\gamma = 2.0$, $I_c = 0.16$ μA; (b) $T = 3.869$ K, $\gamma = 6.2$, $I_c = 0.50$ μA; (c) $T = 3.866$ K, $\gamma = 13$, $I_c = 1.05$ μA; and (d) $T = 3.862$ K, $\gamma = 28$, $I_c = 2.3$ μA, where $\gamma = I_c h/ekT$ [after M. Simmonds and W. H. Parker, Phys. Rev. Lett. **24**, 876 (1970)].

noise is well described by the Ivanchenko and Zil'berman[158] calculation. In contrast to the previous experiments, which observed effects on the I–V curves due to intrinsic fluctuations near T_c, these experiments were well below T_c, but the amplitude of the external noise was large compared to the coupling energy.

Kanter and Vernon[165] have measured the effects of external noise on the I–V curves of point contacts. They find discrepancies with the existing theories and discuss differences between the effects of noise above and below the Josephson frequency.

Another manifestation of thermal fluctuations is found in the tilting and rounding of the microwave-induced supercurrent steps on the I–V curve. Henkels and Webb[166] have reported extensive measurements of step profile as a function of temperature and microwave power. An

---

[166] W. H. Henkels and W. W. Webb, Phys. Rev. Lett. **26**, 1164 (1971).

5.2. EXPERIMENTS THAT STUDY THE JOSEPHSON EFFECTS

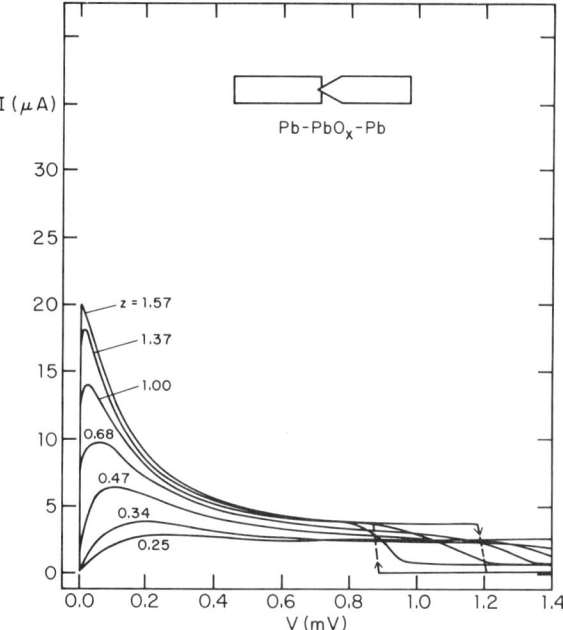

FIG. 29. I–V curves for an extremely small tunnel junction at various noise levels characterized by the parameter $z = hI_c/2ekT$, the ratio of the coupling energy to kT. The junction had the form shown in the inset with area $\sim 10^{-6}$ cm². Data are for $T = 1.6$ K, $I_c = 54$ μA, $H = 0$, and a bandwidth of 0.5 MHz for the external noise generator [after S. A. Buckner, J. T. Chen, and D. N. Langenberg, *Phys. Rev. Lett.* **25**, 738 (1970)].

example of their data and a fit to Stephen's theory[167] for the effect of noise on a driven Josephson oscillator is shown in Fig. 30.

Yanson[168] has reported a study of the effect of thermal fluctuations on the variation of the critical current with magnetic field for tunnel junctions.

In the study of phase transitions an important concept is that of the generalized susceptibility, which relates the order parameter and its thermodynamically conjugate field. In a Josephson junction consisting of two metals with different transition temperatures, when the temperature is near the transition temperature of one and far below that of the other, fluctuations of the order parameter of the one near T_c lead to an

---

[167] M. J. Stephen, *Phys. Rev.* **186**, 393 (1969).

[168] I. K. Yanson, *Sov. Phys. JETP* **31**, 800 (1970) [*English transl. of: Zh. Eksp. Teor. Fiz.* **58**, 1497 (1970)].

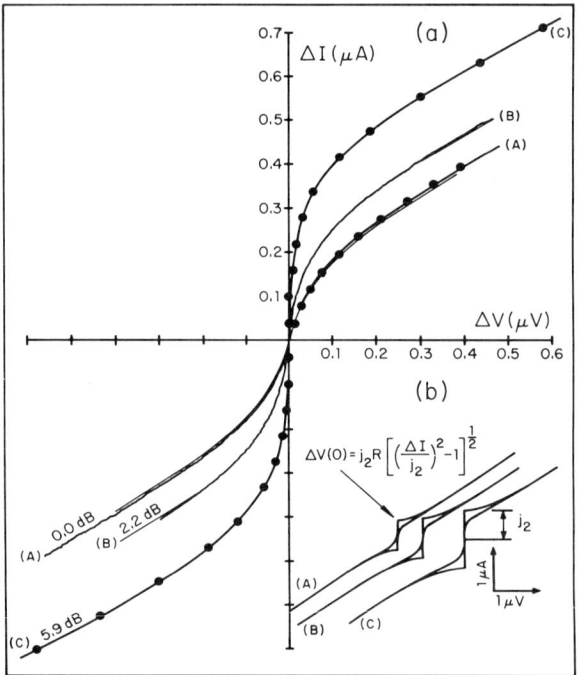

FIG. 30. Development of the $n = 1$ microwave-induced supercurrent step as the microwave power is first applied showing the effects of noise. (a) Experimental data and a one-parameter (j_2) fit to Stephen's theory for the effect of noise on the step shape are shown for three power levels. (b) Lower-resolution plots of the same steps as in (a) along with theoretical curves for the corresponding cases with no fluctuations [$\Delta V(0)$], using the fitted j_2 values. Data taken with a mechanically and thermally stable Nb point contact [R. A. Buhrman, S. F. Strait, and W. W. Webb, *J. Appl. Phys.* **42**, 4527 (1971)] with $T = 3.801$ K and $\nu = 3.14$ GHz [after W. H. Henkels and W. W. Webb, *Phys. Rev. Lett.* **26**, 1164 (1971)].

enhanced conductivity across the junction. This leads to a pair current in excess of the junction quasi-particle current. Ferrell[169] first suggested the connection between pair tunneling and fluctuations and that the generalized susceptibility relating the superconducting order parameter and the pairing field might be measured. Using this effect Scalapino[170] showed that the dc I–V curve measured at various magnetic fields can be interpreted as a measure of the frequency and wave vector dependence of the generalized susceptibility. Equivalent or related theoretical results

[169] R. A. Ferrell, *J. Low Temp. Phys.* **1**, 423 (1969).
[170] D. J. Scalapino, *Phys. Rev. Lett.* **24**, 1052 (1970).

have been obtained by Takayama[171] and Kulik.[172] Quantitative measurements of the excess pair current determined from the I–V curves of Sn–Sn oxide–Pb, and Al–Al oxide–Pb junctions have been reported by Anderson et al.[173] They find good agreement with the predictions of the theory. Their measurements at zero magnetic field constitute a determination of the $k = 0$ frequency-dependent susceptibility for the superconductor. Other measurements[174,175] in qualitative agreement with the theories have been reported and some semiquantitative results on Pb–Sn oxide–Sn junctions have been obtained.[49]

5.3. Characteristics of Various Types of Weakly Linked Superconductors

5.3.1. Types of Weak Links

Numerous types of weakly linked superconductors have been studied. Most fall into one of the general classes depicted in Fig. 4, namely tunnel junctions, SNS junctions, point contacts, Clarke slugs, and thin film bridges. The objective of this section is to give more specific information about each type and to discuss some phenomenological descriptions and models which have been found useful in understanding them.

5.3.1.1. Tunnel Junctions. Tunnel junctions are the most widely studied form of weak links and are best understood both from the point of view of microscopic and phenomenological theory. Matisoo[176] has presented a comprehensive review of tunnel junctions including fabrication, analysis, and application. Early experimenters found junctions capricious, fragile, and short-lived. However, with good technique it is now possible to produce junctions that are stable and have reproducible

---

[171] H. Takayama, *Proc. Int. Conf. Low Temp. Phys.*, *12th* (E. Kanda, ed.), p. 267. Academic Press of Japan, Tokyo, 1971.

[172] I. O. Kulik, *Sov. Phys. JETP* **32**, 510 (1971) [*English transl. of*: *Zh. Eksp. Teor. Fiz.* **59**, 937 (1970)].

[173] J. T. Anderson, R. V. Carlson and A. M. Goldman, *J. Low Temp. Phys.* **8**, 29 (1972).

[174] S. G. Lipson, C. G. Kuper, and A. Ron, *Superconductivity*, *Proc. Int. Conf. Sci. Superconductivity*, *Stanford* (F. Chilton, ed.), p. 269. North-Holland Publ., Amsterdam, 1971.

[175] K. Yoshihiro and K. Kajimura, *Phys. Lett.* **32A**, 71 (1970).

[176] J. Matisoo, Applied Superconductivity Conference Record, 1972. IEEE Publ. No. 72-CHO682-5-TABSC.

characteristics. As many as 400 have been constructed on a single substrate.[177] Detailed information on the fabrication of tunnel junctions is given in Chapter 4.

Tunnel junctions are typically constructed by evaporating or sputtering one thin film superconductor on to a glass substrate, exposing this film to oxygen to produce an oxide about 20 Å thick, then evaporating a second superconducting film on top of the oxide barrier. Many different superconductors and insulators have been used, but aluminum, lead, tin, and niobium with the barrier formed by oxidizing one film are most common. Other techniques for forming the barrier include oxide formation by a glow discharge in oxygen,[178,179] reaction with an absorbed layer of gas,[180] and deposition of various insulating films.[178] Semiconductors including tellurium[181] and cadmium sulfide[141] have been used for barriers, and junctions have been formed on a silicon substrate.[182] Particular attention has been centered on forming tunnel junctions with Nb, and sputtering techniques have been particularly successful.[179,180,183–185]

Several typical configurations for tunnel junctions are shown in Fig. 31. The junction area may range from ~ 1 to 10^{-6} mm². The room temperature resistance usually ranges from 10^{-3} to $1\ \Omega$ and the critical current from a few microamperes to a few milliamperes. A typical I–V curve for a good tunnel junction has the form shown in Fig. 32. When the measurement is made with a voltage source there is a sharp drop at i_c to the very small current at finite voltages along the quasi-particle portion of the curve until the characteristic step $V = 2\varDelta/e$ is reached. When measured with a source of finite impedance, the I–V curve changes from its maximum dc value to the heavy curve along a load line determined by the resistance of the source. When a current source is used, as the current is increased, there is an abrupt switch at the maximum dc supercurrent to the quasi-particle portion of the curve at that current. When the current

[177] D. N. Langenberg, T. F. Finnegan, and A. Denenstein, Electronics, p. 42 (March 1, 1971).
[178] W. Schroen, *J. Appl. Phys.* **39**, 2671 (1968).
[179] R. Graeffe and T. Wiik, *J. Appl. Phys.* **42**, 2146 (1971).
[180] L. O. Mullen and D. B. Sullivan, *J. Appl. Phys.* **40**, 2115 (1969).
[181] J. Seto and T. Van Duzer, *Appl. Phys. Lett.* **19**, 488 (1971).
[182] A. Bruck, *UMIST Solid State Phys. Conf. ONR London Conf.* Rep. C-4-72 by W. Condell, p. 2 (1972).
[183] J. E. Nordman, *J. Appl. Phys.* **40**, 2111 (1969).
[184] J. H. Greiner, *J. Appl. Phys.* **42**, 5151 (1971).
[185] K. Schwidtal, *J. Appl. Phys.* **43**, 202 (1972).

5.3. CHARACTERISTICS

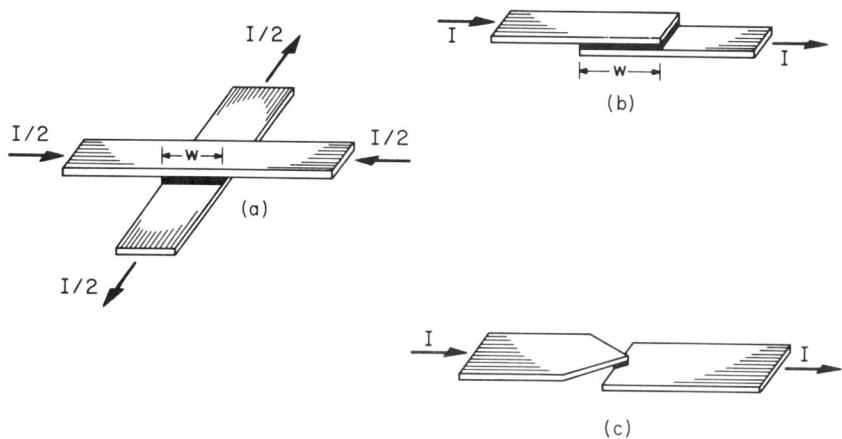

FIG. 31. Typical configurations used for tunnel junctions. (a) The cross-type junction is shown with symmetric current bias; (b) in-line junction; (c) microjunction.

is decreasing from the finite voltage portion of the curve, the path traced is along the lower portion of the I–V curve with a discontinuous switch back to zero voltage at some small value of the current. This hysteretic behavior is discussed more fully in Section 5.3.3.

Standard tests for the quality of a tunnel junction include measurement of the magnetic field dependence and the temperature dependence of the critical current. They should have the forms discussed in Section 5.2.3. The Fiske steps can be used to determine the capacitance of the junction.

An unusual type of I–V curve with a steeply sloped linear portion preceding the discontinuous switch to the quasi-particle branch of the

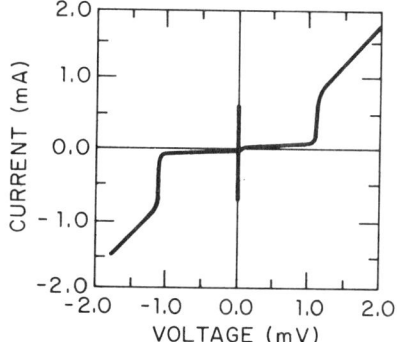

FIG. 32. I–V curve for nearly ideal Sn–Sn oxide–Sn tunnel junction at $T = 2.1$ K.

curve has been reported for wide (~1 mm) tunnel junctions.[186,187] These results have been interpreted in terms of internal flux flow in the junction, and the data appear to be in good agreement with the calculations.

5.3.1.2. SNS Junctions. The properties of SNS junctions of the form shown in Fig. 33 (inset) have been studied extensively by Clarke.[188,189] The junctions were formed by evaporation in an ultrahigh vacuum system with considerable precautions for cleanliness so that the surfaces were not oxidized between evaporations. Typical I–V curves are shown in Fig. 33 for Pb–Cu–Pb junctions. With copper barriers several thousand

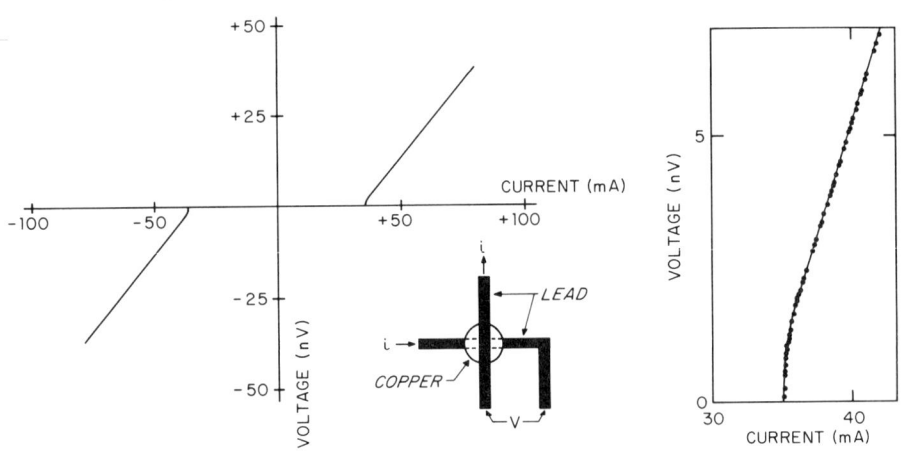

FIG. 33. Voltage–current characteristic for an SNS junction of the form shown in the inset consisting of 7000-Å-thick Pb films 0.2 mm wide separated by a 5520-Å-thick layer of copper at $T = 2.98$ K. An enlarged portion of the curve is shown at the right [after J. Clarke, *Proc. Roy. Soc.* **A308**, 447 (1969)].

Angstroms thick, the normal resistance was approximately $10^{-7}\ \Omega$. The critical current is a rapidly varying function of temperature (see Fig. 8b) and barrier thickness, and the magnetic field dependence of the critical current is similar to that of a wide current-limited tunnel junction (Fig. 7). The supercurrent step structure exhibited by SNS junctions in the presence of an alternating current is extremely sharp and exhibits subharmonic steps (see Fig. 9b).

[186] A. C. Scott and W. J. Johnson, *Appl. Phys. Lett.* **14**, 316 (1969).
[187] A. Barone, *J. Appl. Phys.* **42**, 2747 (1971).
[188] J. Clarke, *Proc. Roy. Soc.* **A308**, 447 (1969).
[189] J. Clarke, *Phys. Rev. B* **4**, 2963 (1971).

Clarke et al.[190,191] have used a simple model to discuss the properties of SNS junctions. They assume that the total current density

$$J = J_\text{s} + J_\text{n} = J_\text{c} \sin \theta + \sigma V, \tag{5.3.1}$$

where the supercurrent density is given by the Josephson equation and the normal current density is related to the voltage V by a conductivity σ, assumed independent of V. This equation together with Eqs. (5.2.30) determine the time and space dependence of the currents. This simple picture explains all of the important differences between the SNS junction and the tunnel junction, at least qualitatively. In the case of a junction wide enough to be limited by its own field, the calculated behavior is that of quantized flux flow across the barrier. The phase locking of this flux flow leads to the step structure in the presence of an applied alternating current. The subharmonic steps are related to the finite impedance of the current sources.

The theory of SNS junctions has been treated from a microscopic point of view by several authors.[192-195]

5.3.1.3. Thin Film Bridges. Several types of thin film bridges (Fig. 34) have been used to study the Josephson effects and are increasingly being used for devices. Nisenoff[196] has reviewed their properties and applications. One type (Fig. 34a), often called a "Dayem bridge," consists of a narrow constriction typically with length and width both less than 1 μ formed in a tin, indium, or aluminum film a few hundred Angstroms thick. Bridges of this type are formed either by evaporating through a very precisely formed mask, or by scribing with a micromanipulator to cut the film and leave a narrow bridge. Large bridges must be operated very near the transition temperature; however, bridges much less than

---

[190] J. Clarke, A. B. Pippard, and J. R. Waldram, *Superconductivity, Proc. Int. Conf. Sci. Superconductivity, Stanford* (F. Chilton, ed.), p. 405. North-Holland Publ., Amsterdam, 1971.

[191] J. R. Waldram, A. B. Pippard, and J. Clarke, *Phil. Trans. Roy. Soc. London* **A268**, 265 (1970).

[192] L. G. Aslamazov, A. I. Larkin, and Yu. N. Ovchinnikov, *Sov. Phys. JETP* **28**, 171 (1969) [*English transl. of*: *Zh. Eksp. Teor. Fiz.* **55**, 323 (1968)].

[193] I. O. Kulik, *Sov. Phys. JETP* **30**, 944 (1970) [*English transl. of*: *Zh. Eksp. Teor. Fiz.* **57**, 1745 (1969)].

[194] C. Ishii, *Progr. Theoret. Phys.* **44**, 1525 (1970).

[195] J. Bardeen and J. L. Johnson, *Phys. Rev. B* **5**, 72 (1972).

[196] M. Nisenoff, Applied Superconductivity Conference Record, 1972. IEEE Publ. No. 72-CHO682-5-TABSC.

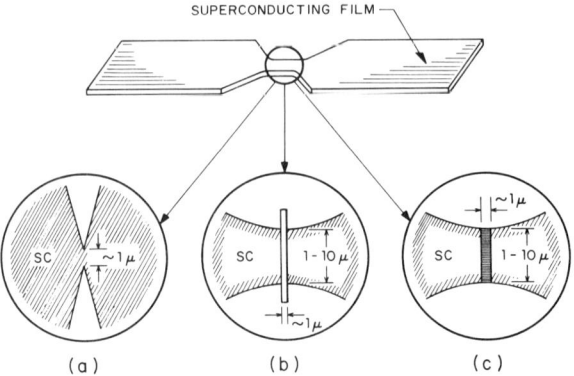

FIG. 34. Types of thin film bridges: (a) narrow neck; (b) normal metal overlay or underlay; (c) anodized strip.

1 μ wide exhibit Josephson effects over a wide range of temperatures. A typical *I–V* curve is shown in Fig. 35a.

Parks and Mochel[197] first observed that bridges of this kind exhibit structure in their resistive behavior. Anderson and Dayem[60] showed that they exhibit supercurrent steps when radiated with microwaves and first related their behavior to the Josephson effect. For wide bridges they

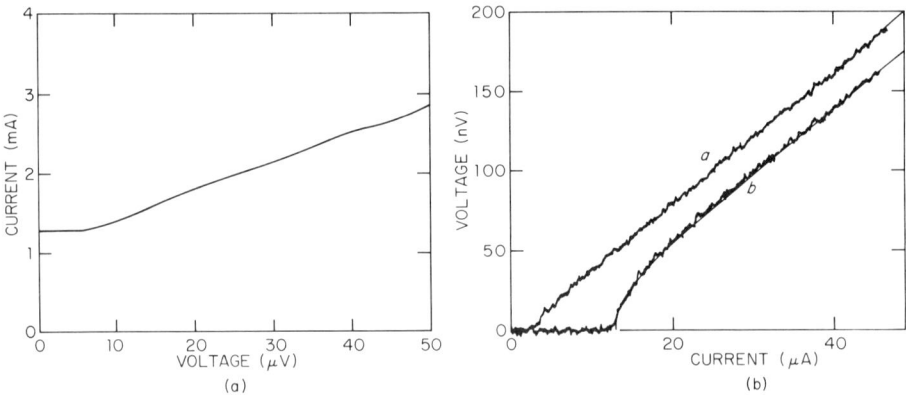

FIG. 35. *I–V* curves for thin film bridges. (a) Dayem bridge in Sn film [after A. H. Dayem and J. J. Wiegand, *Phys. Rev.* **155**, 419 (1967)]. (b) Bridge with normal meta, overlay [after R. K. Kirschman, H. A. Notarys, and J. E. Mercereau, *Phys. Lett.* **34Al** 209 (1971)].

[197] R. D. Parks and J. M. Mochel, *Rev. Mod. Phys.* **36**, 284 (1964); *Phys. Rev. Lett.* **13**, 331 (1964).

5.3. CHARACTERISTICS

described the finite voltage behavior as quantized flux flow and related the subharmonic steps observed in the presence of microwaves to synchronized flow of multiple flux quanta. There have been additional studies of the effects of microwaves on these bridges,[68–72] including the enhancement of the critical current first found by Dayem and Wiegand, which has been related to an increase in phase coherence across the link (see Section 5.2.4.1).

The use of thin film bridges for the study of quantum interference effects was originated by Lambe et al.[198] Mercereau[199] analyzed the behavior of a ring containing a bridge with an applied rf field and described its use as a magnetometer, and, a ring containing two bridges has been used as an interferometer by Fulton and Dynes.[200]

A second type of thin film bridge shown in Fig. 34b is formed by evaporating a narrow strip of normal metal across the superconducting film. Notarys and Mercereau[201,202] have investigated this type bridge extensively, using tin, lead, and indium as the superconductor and gold or copper as the normal metal. (This structure is sometimes called a "Notarys bridge.") In their experiments the length of the bridge ranged from less than 1 to 40 μ, the width from 5 to 10^3 μ, and the thickness of the superconductor from 300 to 3000 Å. Normal state resistances ranged between 10^{-4} and 10^{-1} Ω.

Alternatively, using films of niobium or tantalum, thin film bridges were formed by locally anodizing a small strip across the film creating a locally weakened section as shown in Fig. 34c.[203] In this case the superconductor was coated with a photo resist and a narrow strip cleared across the metal film through the protecting resist. The underlying film was anodized through the cleared strip.

A typical I–V curve for these structures is shown in Fig. 35b. There is an excess current above V/R, where R is the normal resistance. For

---

[198] J. Lambe, A. H. Silver, J. E. Mercereau, and R. C. Jaklevic, *Phys. Lett.* **11**, 16 (1964).

[199] J. E. Mercereau, *Proc. Symp. Phys. Superconducting Devices*, Charlottesville (1967) (see Ref. 22).

[200] T. A. Fulton and R. C. Dynes, *Phys. Rev. Lett.* **25**, 794 (1970).

[201] H. A. Notarys and J. E. Mercereau, *Superconductivity, Proc. Int. Conf. Sci. Superconductivity*, Stanford, 1969 (F. Chilton, ed.), p. 424. North-Holland Publ., Amsterdam, 1971.

[202] J. E. Mercereau, *Proc. Workshop Naval Appl. Superconductivity*, Panama City, Fla. (J. E. Cox and E. A. Edelsack, eds.), NRL Rep. 7302 (Nov. 1970).

[203] R. K. Kirschman, H. A. Notarys, and J. E. Mercereau, *Phys. Lett.* **34A**, 209 (1971).

$V \gg i_c R$ there is a constant excess current approximately equal to $i_c/2$ (where i_c is the critical current) up to voltages 10^3 times those shown in Fig. 35b. The variation of i_c with magnetic field perpendicular to the film is found to be similar to the diffraction pattern found for a Josephson junction.

In these bridges the overlaid or anodized strip produces a narrow band in which the superconducting order parameter is locally reduced and which may be too small to support quantized flux flow. Instead the finite voltage regime is pictured as a phase slip process resulting from a relaxation type oscillation of the order parameter. Using a simple equivalent circuit model, Notarys and Mercereau[201,202] account for this type oscillation and for the excess supercurrent being $i_c/2$. Baratoff *et al.*[204] have used the Landau–Ginsburg equation to calculate the current–phase relation for this type structure and Deaver and Pierce[205] have discussed the relationship between a related equivalent circuit model and the current–phase relation.

Niobium selenide has been used by Consadori *et al.*[206] to form a bridge linking a superconducting ring with which they demonstrated quantum interference effects in a magnetometer configuration. They cleaved a single crystal platelet to a thickness of a few hundred Angstroms and then cut away material to form the desired loop configuration and the narrow bridge.

Bridges of NbN have been fabricated by sputter etching and studied by Janocko *et al.*[207]

5.3.1.4. Point Contacts.
A point contact usually consists of a sharpened superconducting wire or rod with its tip placed in contact with the plane surface of a second superconductor and a mechanism provided for adjusting the pressure between the two. Although point contacts had been used previously for tunneling experiments, their use in the study of Josephson effects was begun by Zimmerman and Silver,[208] who used them for quantum interference experiments. They have become the most widely used form of weak link for magnetometers and high-fre-

---

[204] A. Baratoff, J. A. Blackburn, and B. B. Schwartz, *Phys. Rev. Lett.* **25**, 1096 (1970).
[205] B. S. Deaver, Jr. and J. M. Pierce, *Phys. Lett.* **38A**, 81 (1972).
[206] F. Consadori, A. A. Fife, R. F. Frindt, and S. Gygax, *Appl. Phys. Lett.* **18**, 233 (1971).
[207] M. A. Janocko, J. R. Gavaler, C. K. Jones, and R. D. Blaugher, *J. Appl. Phys.* **42**, 182 (1971).
[208] J. E. Zimmerman and A. H. Silver, *Phys. Lett.* **10**, 47 (1964).

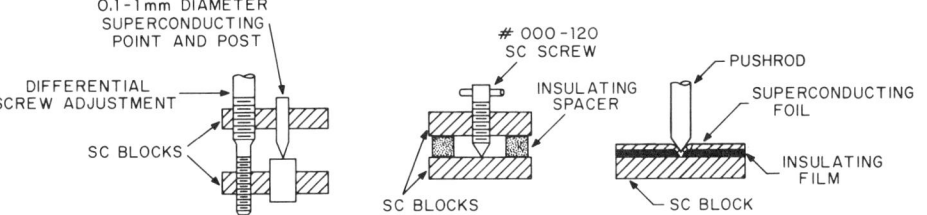

FIG. 36. Some typical configurations for point contacts.

quency applications. A very comprehensive review of the properties and applications of point contacts has been published by Zimmerman.[55]

Several examples of point contacts are shown in Fig. 36. Additional examples of point contacts used for devices are depicted in Figs. 50, 51, 53, and 58. Many combinations of materials have been used but the most common are niobium and Nb alloys. By using glass spacers with an expansion coefficient matched to that of Nb, thermally stable contacts can be achieved.[209] Points are sometimes chemically etched (using a solution of equal parts of hydrofluoric acid, nitric acid, and water for Nb), or simply mechanically polished to a fine tip. Sometimes the tip or the plane are oxidized or anodized to provide an insulating barrier. Alternatively, perfectly clean contacts have been made by breaking a narrow neck in a Nb wire immersed in liquid helium and reestablishing contact between the broken ends.

A variety of different I–V curves can be obtained with point contacts (Fig. 37) and their shape is correlated roughly with the size of the critical current, which can typically be adjusted over about three orders of magnitude. The data in Fig. 37 were obtained by Zimmerman[55] for clean point contacts and show a remarkable constancy of the product $i_c R$ at $\pi \Delta / 2$, the theoretical value for tunnel junctions. This is consistent with the slight variation of i_c with T for $T \ll T_c$ for clean contacts. Most data for point contacts give values of $i_c R$ much less than their theoretical value, and the critical current is often found to vary very rapidly with temperature (see Fig. 8c).

The I–V curves of point contacts frequently have pronounced structures at $V = 2\Delta/e$ and Δ/e, and often at many more of the subharmonic voltages $2\Delta/ne$.[139] Traditionally the feature at $2\Delta/e$ has been associated with tunneling; however, Zimmerman finds it with perfectly clean point contacts as well. Cases are found where the feature at $2\Delta/e$ is absent but

---

[209] R. A. Buhrman, S. F. Strait, and W. W. Webb, *J. Appl. Phys.* **42**, 4527 (1971).

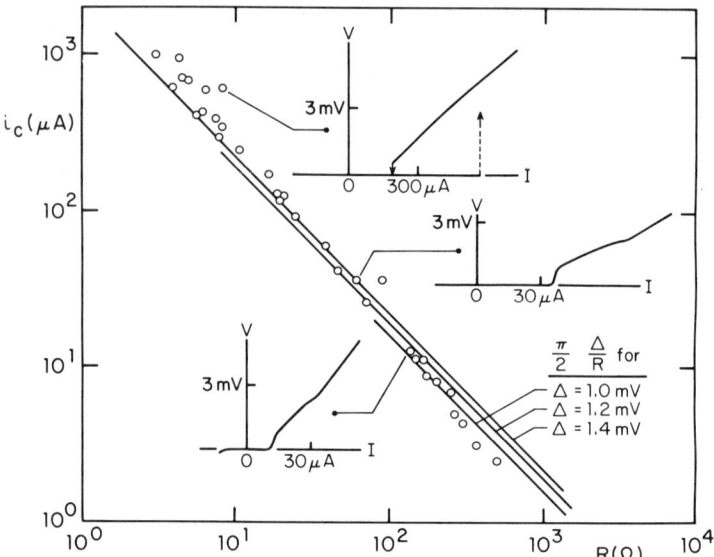

FIG. 37. Critical current versus resistance for many successive adjustments of a particular perfectly clean Nb point contact at 4 K. Some V–I characteristics are shown [after J. E. Zimmerman, Applied Superconductivity Conference Record, 1972. IEEE Publ. No. 72-CHO682-5-TABSC].

a prominent feature occurs at Δ/e. Careful examination of the I–V curve and its derivatives reveals a bewildering array of structure, which includes in addition to the subharmonic structure at $2\Delta/ne$, features that can be identified with antenna modes, external resonances, and impurity molecules coupled to the junction.[142]

Very little theoretical analysis of point contacts has been published; however, lumped parameter models (Section 5.3.3) have been very successful in accounting for their behavior.[74–78,210]

5.3.1.5. Clarke Slug. A blob of Pb–Sn solder on a small Nb wire forms a particularly interesting and useful weak link originated by Clarke[211] and dubbed "slug" (superconducting low-inductance galvanometer). Typical slugs are about 0.5 cm long, use Nb wire 50–100-μ diameter, and have a normal resistance of about 1 Ω. The solder does not wet the wire but is usually weakly linked at several spots within the

---

[210] S. Shapiro, Applied Superconductivity Conference Record, 1972. IEEE Publ. No. 72-CHO682-5-TABSC.

[211] J. Clarke, *Phil. Mag.* **13**, 115 (1966).

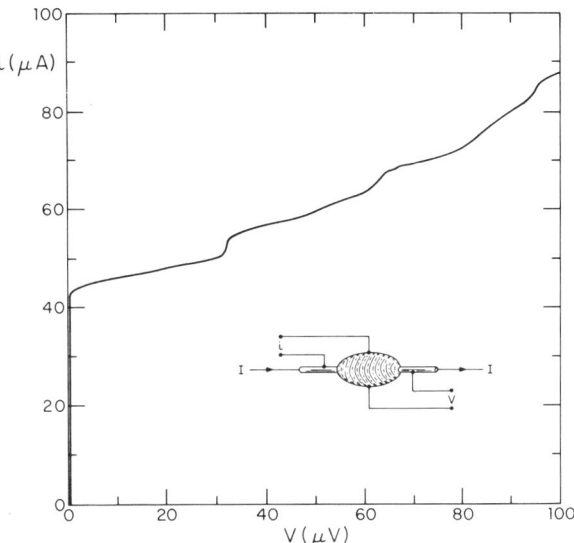

Fig. 38. *I–V* curve for a slug of the type shown in Fig. 60 biased as shown in the inset.

blob. These multiple links give the slug its most useful property, namely a periodic dependence of the maximum zero-voltage current I_c on the current I flowing along the wire (see Fig. 38 inset). This interference phenomena leads to its use as a galvanometer. The property depends on the existence of an area within the blob between two (or more) weakly linked spots into which the magnetic field from the current I penetrates. For some purposes it is convenient to increase this area by using an insulated wire with the insulation notched to permit weak links at two relatively well-defined spots as shown in Fig. 60. The *I–V* curve shown in Fig. 38 is for a slug of this type; the periodic structure on the *I–V* curve is due to the field produced by the bias current i when the slug is asymmetrically biased.[212]

There have been other investigations of interference phenomena with slugs and similar weak links,[213–216] and slugs have been used for ex-

[212] J. Clarke and T. A. Fulton, *J. Appl. Phys.* **40**, 4470 (1969).

[213] R. de Bruyn Ouboter, M. H. Omar, A. J. P. T. Arnold, T. Guinau, and K. W. Taconis, *Physica* **32**, 1448 (1966).

[214] M. H. Omar and R. de Bruyn Ouboter, *Physica* **32**, 2044 (1966).

[215] I. M. Dmitrenko and S. I. Bondarenko, *JETP Lett.* **7**, 241 (1968) [*English transl. of*: *ZhETF Pis. Red.* **7**, 308 (1968)].

[216] O. Jaoul, *Rev. Phys. Appl.* **5**, 885 (1970).

periments on microwave-induced steps.[59] Some of the properties and techniques for using them for various types of measurements are discussed in more detail in Section 5.4.5.

5.3.2. Current–Phase Relation and Various Phenomenological Descriptions of Weak Links

The current–phase relation $i(\theta)$ completely describes the supercurrent properties of a weak link. In most of the preceding discussions, $i(\theta)$ has been assumed to be sinusoidal although, in general, this may not be true. Many of the experimental results do not depend on the precise form of the current–phase relation but only on the fact that it is periodic. Others, e.g., ones that depend on the specific wave form of the supercurrent oscillation or on the current induced in a superconducting loop containing a weak link by an external field, are sensitive to the form of $i(\theta)$.

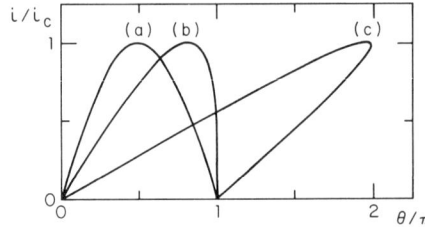

FIG. 39. Some types of current–phase relations showing a progression from sinusoidal to one characterizing a relaxation oscillation.

In the limit of extremely weak coupling the calculated current–phase relation is always sinusoidal.[217] For stronger coupling, the calculated $i(\theta)$ depends on the particular kind of weak link. The current–phase relation has been calculated as a function of a weakness parameter for thin film bridges[72,204,213] using the Landau–Ginsburg theory, and for SNS junctions[193–195] using a microscopic theory. The results are schematized roughly by the curves in Fig. 39, which indicate the sinusoidal case (Fig. 39a) for weak coupling and an initial linear relation between current and phase followed by a rapid variation for stronger coupling (Figs. 39b and c). The highly tipped curve indicates the case for a long link, in which it is possible to have phase shifts greater than π, and is symptomatic of a

[217] P. E. Gregers-Hansen, M. T. Levinsen, and G. Fog Pedersen, *J. Low Temp. Phys.* **7**, 99 (1972).

relaxation oscillation at finite voltage with periodic phase shifts by a multiple of 2π.

Quantum interference effects have been used to measure $i(\theta)$ for point contacts[218-220] and for thin film bridges.[200,217] These measurements show no departure from a sinusoidal relation except near T_c, where it is suggested that fluctuations may have affected the results.[221,222]

There is, of course, a close relationship between $i(\theta)$ and the wave form of the supercurrent oscillation. If the instantaneous voltage V_0 is constant, θ increases linearly in time. Thus the supercurrent wave form $i(t)$ has the same functional form as $i(\theta)$. In practice the instantaneous voltage cannot be kept constant because of the other conduction mechanisms inherent in the junction or due to impedances to which it is coupled, so the wave form deviates from this form.

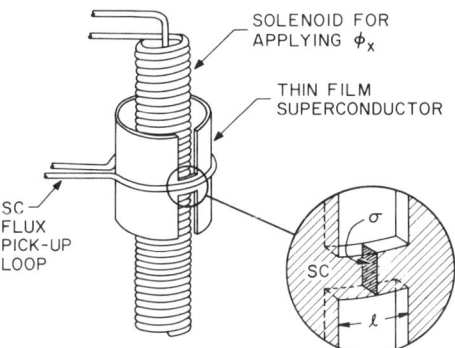

FIG. 40. Superconducting ring containing a weak link. Coil geometry shown is for measuring magnetization of the ring.

In principle, a simple experiment serves to measure $i(\theta)$.[7] Consider a superconducting ring containing a weak link, for example, a short narrow neck (Fig. 40) of length l and cross-sectional area σ with dimensions much smaller than the penetration depth λ. The quantized fluxoid condition

$$\oint (m/2n_s e^2)\mathbf{j}_s \cdot d\mathbf{s} + \oint \mathbf{A} \cdot d\mathbf{s} = n\Phi_0 \tag{5.3.2}$$

[218] J. E. Zimmerman and A. H. Silver, *Solid State Commun.* **4**, 133 (1966).
[219] A. H. Silver and J. E. Zimmerman, *Phys. Rev.* **157**, 317 (1967).
[220] T. A. Fulton, *Solid State Commun.* **8**, 1353 (1970).
[221] J. E. Zimmerman, *J. Appl. Phys.* **41**, 1589 (1970).
[222] J. E. Zimmerman, *J. Appl. Phys.* **42**, 30 (1971).

[see Eq. (5.2.5)] must be true for any closed path in a coherently connected superconductor. If the ring is thick with respect to λ everywhere except at the link, a path can be chosen within the superconductor where $j_s = 0$ everywhere except at the link. Thus the first term in the fluxoid is zero except across the link where the current can be assumed to be uniformly distributed so that $j_s = i/\sigma$, and since the second term is just the total flux ϕ within the ring the fluxoid can be written as

$$(m/2n_s e^2)(i/\sigma)l + \phi = n\Phi_0. \tag{5.3.3}$$

Also the total flux can be expressed as

$$\phi = iL + \phi_x, \tag{5.3.4}$$

where ϕ_x is the externally applied flux (imposed, for example, with a long solenoid through the ring as shown in Fig. 40) and iL is the flux produced by the supercurrent, where L is the inductance of the ring.

Recalling from Eq. (5.2.4) that the terms in the fluxoid are essentially phase contributions whose sum must be $n2\pi$ to ensure a single-valued wave function, we can identify the phase shift θ across the weak link as the first term in Eq. (5.3.3). Thus

$$\theta = (\pi/\Phi_0)(m/n_s e^2)(l/\sigma)i \tag{5.3.5}$$

and Eq. (5.3.3) is then

$$\theta + (2\pi/\Phi_0)(iL + \phi_x) = n2\pi. \tag{5.3.6}$$

Note that this identification is consistent with the definition of θ in Eq. (5.2.10) since

$$\chi_2 - \chi_1 = \hbar^{-1} \int_l (m/n_s e)\mathbf{j}_s \cdot d\mathbf{s} + (2e/\hbar) \int_l \mathbf{A} \cdot d\mathbf{s}, \tag{5.3.7}$$

where the integrals are taken across the link but

$$\theta = \chi_2 - \chi_1 - (2e/\hbar) \int_l \mathbf{A} \cdot d\mathbf{s} = (1/\hbar) \int_l (m/n_s e)\mathbf{j}_s \cdot d\mathbf{s}. \tag{5.3.8}$$

For a ring with zero flux in it initially, a pickup loop surrounding the ring (Fig. 40) can be used to measure (with a superconducting magnetometer) the total flux ϕ as an external flux ϕ_x is applied. By Eq. (5.3.6), ϕ is proportional to θ. Further, if ϕ_x and L are both known, the current can be calculated and therefore the current–phase relation $i(\theta)$ can be determined. This is the type of measurement first carried out by Silver and Zimmerman.[219]

Some insight into the nature of the Josephson oscillation can be gained by noting the relationship between the velocity dependence of the number of superelectrons and the current–phase relation obtained from this simple picture of a weak link. The superelectron density n_s is a function of the drift velocity v_s and can be calculated from the microscopic theory or phenomenological theories. A simple case is the solution of the Landau–Ginsburg equations for a thin bridge with n_s and v_s assumed constant along the bridge.[223,224] Then

$$n_s(v_s) = n_s(0)(1 - v_s^2/v_m^2), \qquad (5.3.9)$$

where $v_m = \hbar/m\xi$ and ξ is the coherence length. The supercurrent density j_s is then

$$j_s = 2n_s e v_s = 2n_s(0) e v_s (1 - v_s^2/v_m^2), \qquad (5.3.10)$$

which has the form shown in Fig. 41 with a maximum at $v_c = v_m/3^{1/2}$. Note that for velocities between v_c and v_m, the current is a decreasing function of velocity; this has been called the "negative inductance" region[72] and, in general, currents in this range are not stable. (If the link is incorporated into a ring in which the phase θ is the fixed variable and if $l < \xi$, currents in this range can be sustained.[72])

In this approximation v_s and θ are linearly related and Eq. (5.3.10) is essentially the current–phase relation. That is, since $j_s = 2n_s e v_s$, Eq.

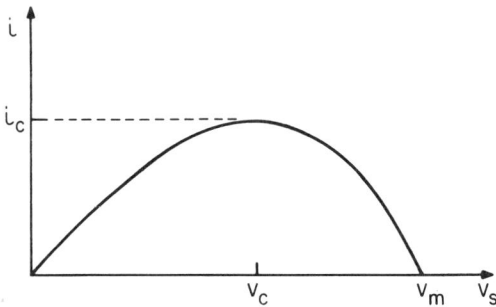

FIG. 41. Velocity dependence of the supercurrent.

---

[223] J. Bardeen, Rev. Mod. Phys. **34**, 667 (1962).

[224] P. G. DeGennes, "Superconductivity of Metals and Alloys," p. 182. Benjamin, New York, 1966.

(5.3.5) is equivalent to

$$\theta = (2ml/\hbar)v_s, \qquad (5.3.11)$$

which together with Eq. (5.3.10) gives the current–phase relation

$$j = n_s(0)ev_m(\xi/l)\theta[1 - (\xi/l)^2(\theta^2/4)]; \qquad (5.3.12)$$

when $l \approx \xi$, this approaches sinusoidal form. A more detailed calculation does, in fact, lead to a sinusoidal current–phase relation for a short bridge.[217]

Mercereau has used the velocity-dependent superelectron density n_s to describe the Josephson oscillation in a thin film bridge as a relaxation oscillation of the order parameter.[202] If a constant voltage is applied to the link, there is a constant acceleration of the pairs. As the current increases, n_s decreases, and when v_c is reached there is a sudden drop in n_s, followed by a transition to a lower current state corresponding to a shift in phase by 2π or some multiple of it. The pairs are accelerated again and this process repeats periodically.

A convenient way to picture this process is to use the representation of the superconducting wave function in a current carrying strip shown in Fig. 42a. The radial distance from the axis to the curve represents the amplitude of the wave function, i.e., $(n_s)^{1/2}$, and the angular position of the vector represents the phase. For a uniform current-carrying strip, the phase advances uniformly along the length giving rise to a helical curve. If at some point the strip contains a narrow neck as shown in Fig. 42b, the order parameter will be reduced and the phase gain per unit length increased. Further increase in the current causes a discontinuous pinching off of one loop of the helix corresponding to a phase slip of 2π between the two sides of the superconductor. For a long bridge it is possible to have several loops of the helix compressed into the small region and all pinched off together corresponding to a phase slip of $n2\pi$. This process corresponds to a highly tipped current–phase relationship.

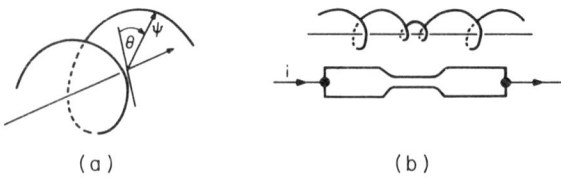

(a)   (b)

Fig. 42. Representations of the macroscopic wave function: (a) for a uniform thin film; (b) for a film containing a narrow neck.

Another useful idea can be illustrated by writing Eq. (5.3.3) as

$$iL_k + iL + \phi_x = n\Phi_0, \qquad (5.3.13)$$

where

$$L_k = (m/2n_s e^2)(l/\sigma) \qquad (5.3.14)$$

is called the "kinetic inductance."[225] By analogy with the ordinary inductance L, which measures the energy E_m stored in the magnetic field of a current by

$$E_m = \int B^2/2\mu_0 \, dV = \tfrac{1}{2}Li^2, \qquad (5.3.15)$$

the kinetic inductance measures the kinetic energy E_k of the current by

$$E_k = \int 2n_s(\tfrac{1}{2}mv_s^2) \, dV = \tfrac{1}{2}L_k i^2. \qquad (5.3.16)$$

The kinetic inductance of a normal conductor is negligible compared to the magnetic inductance. However, for a small superconductor the kinetic inductance can be 10–50 times larger than the magnetic inductance.[225,226]

The phase change across the weak link written in terms of the kinetic inductance is

$$\theta = (2\pi/\Phi_0)iL_k. \qquad (5.3.17)$$

Thus from the Josephson equation (5.2.15),

$$V = d(iL_k)/dt = (i \, dL_k/di + L_k) \, di/dt \equiv \mathscr{L} \, di/dt \qquad (5.3.18)$$

which defines $\mathscr{L}$, the dynamic inductance. The fact that a weak link is inductive was pointed out by Josephson.[2] This can be shown by eliminating θ from Eqs. (5.2.31) leading to

$$V = (\Phi_0/2\pi i_c)[1 - (i/i_c)^2]^{-1/2} \, di/dt, \qquad (5.3.19)$$

where $\mathscr{L}$ can be identified as the coefficient of di/dt. This expression for $\mathscr{L}$ is equivalent to Eq. (5.2.42). For currents $i \ll i_c$, $\mathscr{L} \approx \Phi_0/2\pi i_c$, which is usually 10^{-9}–10^{-10} H. Note that for a tunnel junction with

---

[225] W. A. Little, *Proc. Symp. Phys. Superconducting Devices*, Charlottesville (1967) (see Ref. 22).

[226] R. Meservey and P. M. Tedrow, *J. Appl. Phys.* **40**, 2028 (1969).

capacitance C, the plasma frequency corresponds to the resonant frequency $\omega_p = 1/(\mathscr{L}C)^{1/2}$.

5.3.3. Equivalent Circuit Models

The basic theory of tunnel junctions is well established and can account for pair and quasi-particle contributions to the total current and for effects of junction geometry.[3,84] The theory of SNS junctions has also been treated from a microscopic point of view.[193-195] For other types of weak links, the basic theory is much less well established and other approaches must be used to explain experimental results.

A phenomenological theory[191] of SNS junctions considers the total current as the sum of pair and quasi-particle currents with the pair current satisfying the set of Eqs. (5.2.30) and the quasi-particle current related to the electric field by a normal conductivity. From this theory the time and space dependence of the currents can be calculated and the I–V curve can be determined.

For many purposes only the time dependence of the current is needed and relatively simple equivalent circuit models have proved extremely successful in describing the characteristics of the various types of weak links. These circuits have been useful for calculating I–V curves and the behavior of weak links as circuit elements as well as for giving considerable insight into the physics of these highly nonlinear, active elements.

McCumber[33,227] pointed out that many of the differences among the I–V curves of various types of weak links are due predominantly to the different ac impedances coupled to the junction by nature of its structure or its environment. Stewart,[34] McCumber,[33] and Scott[228] have used equivalent circuit models to calculate I–V curves in good agreement with the diverse curves observed experimentally. Equivalent circuits have been particularly helpful in analyzing the interaction of weak links with external radiation and with other circuit elements.[210]

Most equivalent circuit models use an ideal Josephson element for which the relation between the voltage V and the current i is expressed parametrically by the pair of Josephson equations

$$i = i_c \sin \theta, \qquad \dot{\theta} = 2\pi V/\Phi_0. \tag{5.3.20}$$

A resistor R in parallel with this element provides a second path rep-

---

[227] D. E. McCumber, *J. Appl. Phys.* **39**, 2503 (1968).
[228] W. C. Scott, *Appl. Phys. Lett.* **17**, 166 (1970).

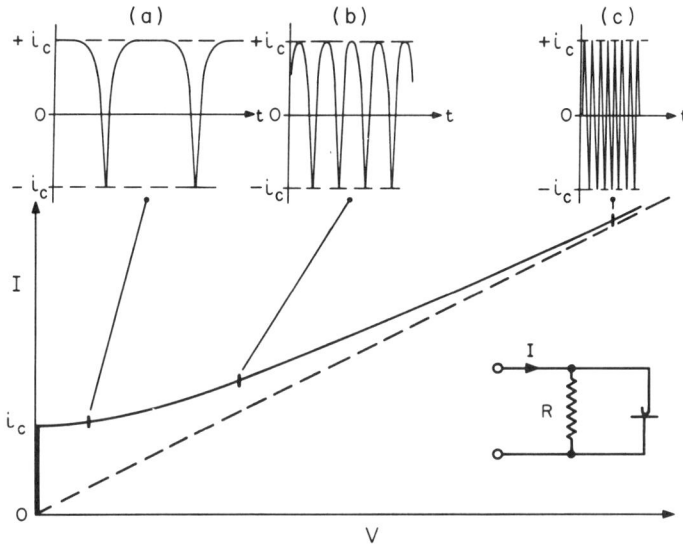

FIG. 43. *I–V* curve for an ideal Josephson element paralleled by a normal resistor. The equivalent circuit and wave forms at several biases are shown.

resenting the quasi-particle current. This combination (Fig. 43) is the simplest equivalent circuit and gives a reasonable representation of some point contacts and thin film bridges.

If this combination is driven with a constant voltage source, the current through the junction oscillates sinusoidally and the current through the resistor is constant. Since the average current through the junction is zero, the dc *I–V* curve is a straight line with slope I/R. However, if a constant current source is used, the current through the junction is sinusoidal only if $I \gg i_c$. For $I \approx i_c$ the wave form of the junction current is extremely nonsinusoidal and there is a large average supercurrent. The equations for this circuit can be solved analytically for the current through the junction and for the average voltage

$$\langle V \rangle = R(I^2 - i_c^2)^{1/2} \tag{5.3.21}$$

giving the *I–V* curve shown in Fig. 43.[33,227,118]

Using this model of a weak link, Hamilton[109] has constructed an electronic analog with which he can obtain the junction current wave form and the *I–V* curve. By including additional circuit elements as shown schematically in Fig. 44, he has studied the effects of applied rf, coupling to an external cavity, and various bias source impedances. Some *I–V*

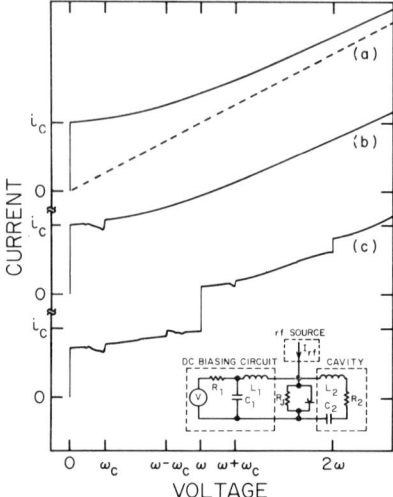

FIG. 44. Calculated I–V curves for (a) the equivalent circuit shown in Fig. 43; (b) with coupling to a cavity resonant at ω_c; (c) with the further addition of external rf at frequency ω [after S. Shapiro, in "Science and Technology of Superconductivity" (W. D. Gregory, W. N. Mathews, Jr., and E. A. Edelsack, eds.), p. 631. Plenum Press, New York, 1972]. The analog model used for these calculations is shown in the inset [after A. Longacre, Jr., Applied Superconductivity Conference Record, 1972. IEEE Publ. No. 72-CHO682-5-TABSC].

curves obtained with this circuit are shown in Fig. 44 and are essentially identical to experimentally observed curves.[107,108,229] Shapiro and Werthamer[106] had previously studied these effects using an analog computer.

This model has also been used to calculate the spectrum of the oscillating supercurrent,[109,230] with applied ac voltage to analyze mixing experiments,[108] and to explain the failure of the critical current of point contacts and thin film bridges to follow the simple Bessel function dependence on microwave power.[75–79]

A tunnel junction is better represented by the circuit shown in Fig. 45, since the intrinsic capacitance of the junction has a profound effect on its behavior. This circuit was used by Stewart,[34] McCumber,[33] and Scott[228] to calculate the I–V curves for a tunnel junction and to explain the hysteresis commonly observed (see Section 5.3.1.1). Scott also in-

[229] S. Shapiro, in "Science and Technology of Superconductivity" (W. D. Gregory, W. N. Mathews, Jr., and E. A. Edelsack, eds.), p. 631. Plenum Press, New York, 1972.

[230] L. G. Aslamazov and A. I. Larkin, JETP Lett. 9, 87 (1968) [English transl. of: ZhETF Pis Red. 9, 150 (1968)].

cluded a voltage-dependent resistor to account for the sharp step in the quasi-particle conductance. If a constant current I is applied,

$$I = i_J + i_R + i_C$$
$$= i_c \sin \theta + V/R + C\, dV/dt$$

or

$$I = i_c \sin \theta + (\Phi_0/2\pi R)\dot\theta + (\Phi_0/2\pi)C\ddot\theta, \tag{5.3.22}$$

which is the equation of a driven penduluum. For $I > i_c$, only a rotating solution exists corresponding to a nonzero voltage across the junction.

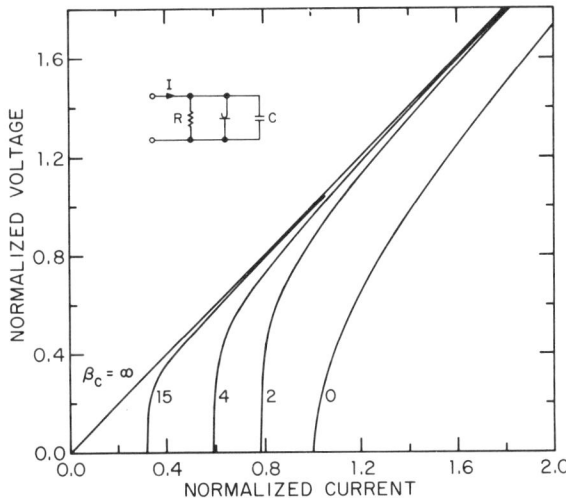

FIG. 45. Voltage–current curves for the equivalent circuit shown, which includes a shunt capacitance. The curves are labeled with values of the parameter $\beta_c = (2\pi/\Phi_0)R^2 C i_c$ [after D. E. McCumber, *J. Appl. Phys.* **39**, 3113 (1968)].

For $I < i_c$, two possibilities exist. A nonrotating solution will obtain if the current is brought up from zero to I corresponding to a dc supercurrent with constant phase angle θ. However, if I is decreased from values greater than i_c, the rotating solution will persist until the energy provided per cycle just equals the energy loss per cycle at which current there is a switch to the nonrotating solution. Thus with the capacitor there are two stable solutions and there is hysteresis in the I–V curve.

The shape of the I–V curve and the amount of hysteresis are determined by the parameter

$$\beta_c = (2\pi/\Phi_0)R^2 C i_c. \tag{5.3.23}$$

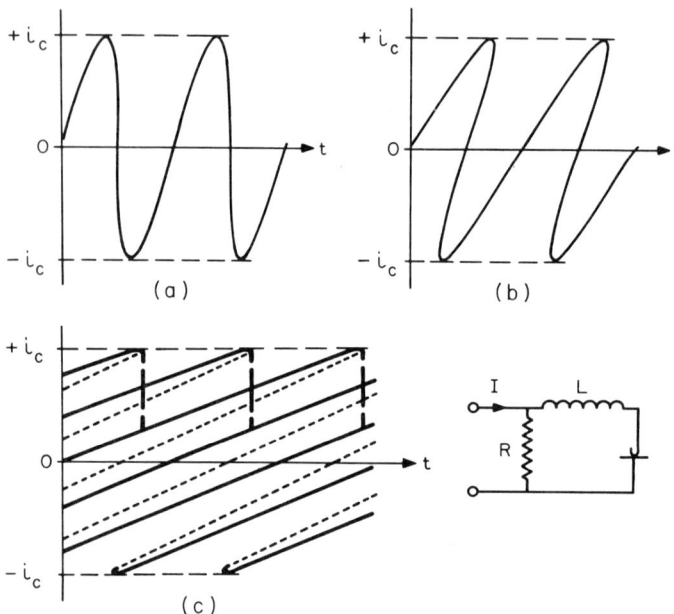

FIG. 46. Supercurrent wave forms for an equivalent circuit (inset) including a series inductor for several values of the parameter $\beta = (2\pi/\Phi_0)i_c L$: (a) $\beta = 1$; (b) $\approx \pi$; (c) $\beta \approx 5\pi$ with a double quantum transition indicated [after D. B. Sullivan, R. L. Peterson, V. E. Kose, and J. E. Zimmerman, *J. Appl. Phys.* **41**, 4865 (1970)].

Some I–V curves calculated by McCumber[33] are shown in Fig. 45. For ideal tunnel junctions $\beta_c \to \infty$, for small SNS junctions $\beta \approx 0$, and for point contacts $0 < \beta_c < 10$. Using small, nearly ideal, tunnel junctions shunted with thin film resistors Hansma et al.[38] have varied β_c over a wide range and beautifully verified the results of these calculations.

Another model (Fig. 46) with an inductance L in series with the ideal Josephson element was suggested by Zimmerman et al.[113] and studied more extensively by Sullivan et al.[118] Some calculated supercurrent wave forms are shown in Fig. 46 for several values of a parameter $\beta = 2\pi i_c L/\Phi_0$. For $\beta > 1$, relaxation oscillations occur, and for very large β, phase slips of $n2\pi$ are possible at each cycle of the oscillation. This model may be a representation of some types of thin film bridges.[205,217] The model has been most useful for discussing superconducting rings containing a weak link, with L representing the ring inductance rather than an inductance intrinsic to the link itself. Sullivan and Zimmerman[231]

---

[231] D. B. Sullivan and J. E. Zimmerman, *Amer. J. Phys.* **39**, 1504 (1971).

have developed a mechanical analog, which has proved extremely valuable for studying this case as well as other time-dependent Josephson phenomena, and have demonstrated cases that were not apparent in analytical solutions.

Nisenoff[196] and Notarys et al.[232] have used an equivalent circuit with an effective voltage generator to describe some thin film bridges and to analyze devices containing them.

5.4. Applications of Weakly Linked Superconductors

In the following sections we attempt to summarize both the fundamental measurements that have been made using the Josephson effects and devices and experimental techniques using weakly linked superconductors. Since most of the individual topics have been described extensively and well in the literature, we will give only a brief introduction to these topics and point to the appropriate papers for details.

5.4.1. Measurements of e/h

Certainly one of the most profound applications of the Josephson effects has been the determination of the fundamental physical constant e/h. As discussed in Section 5.2.4, measurements showed that the Josephson frequency relation $v = 2eV/h$, is precise and independent of material, temperature, magnetic field, and other physical parameters. It was then apparent that this relation provided a way of making a precise determination of the fluxoid quantum $h/2e$. Subsequently, as the precision of this measurement pressed known values of the fundamental constants, the possibility emerged of using the Josephson frequency condition to determine e/h.

In principle, the determination can be made by two methods: (1) direct observation of the Josephson radiation at a given bias voltage; (2) use of the microwave-induced steps by observing the voltage at which a known frequency produces steps in the I–V curve. The latter method has proved more tractable experimentally, using microwave frequencies and high-order steps with voltages of 1 to 10 mV. The first precise measurements were reported by Parker et al.[62,233] After a series of refine-

---

[232] H. A. Notarys, R. H. Wang, and J. E. Mercereau, *Proc. IEEE* **61**, 79 (1973).
[233] W. H. Parker, B. N. Taylor, and D. N. Langenberg, *Phys. Rev. Lett.* **18**, 287 (1967).

ments, these measurements culminated in a reevaluation of the physical constants based on the new value for e/h determined from the Josephson effect.[62,234] The precision of the measurement continues to improve and Finnegan et al.[235] have reported a measurement with a precision of three parts in 10^8. However, difficulties remain in relating their standard to the NBS volt, limiting their accuracy to 12 parts in 10^8. They report the value $2e/h = 483.593718 \pm 0.000060$ MHz/μV$_{\text{NBS}}$. Measurements of e/h have been made at several standards laboratories around the world[236–239] and there have been a number of reviews of these results.[240–244]

5.4.2. Voltage Standard

Taylor et al.[245] pointed out that the Josephson effect provides a very useful technique for maintaining and comparing standards of emf. Voltages are then determined by measuring frequency, which can be done to high accuracy. The problem of transferring from the millivolt range to the 1-V range of a typical standard cell remains. However,

[234] B. N. Taylor, W. H. Parker, and D. N. Langenberg, *Rev. Mod. Phys.* **41**, 375 (1969).

[235] T. F. Finnegan, A. Denenstein, and D. N. Langenberg, *Phys. Rev. B* **4**, 1487 (1971).

[236] T. F. Finnegan, T. J. Witt, B. F. Field, and J. Toots, *Proc. Int. Conf. At. Masses Fundamental Constants*, 4th (J. H. Sanders and A. H. Wapstra, eds.), p. 403. Plenum Press, New York, 1972.

[237] B. W. Petley and K. Morris, *Metrologia* **6**, 46 (1970).

[238] I. K. Harvey, J. C. Macfarlane, and R. B. Frenkel, *Phys. Rev. Lett.* **25**, 853 (1970); I. K. Harvey, J. C. Macfarlane, and R. B. Frenkel, *Metrologia* **8**, 114 (1972).

[239] V. Kose, F. Melchert, H. Fack, and H.-J. Schrader, *PTB Mitteilungen* **81**, 8 (1971).

[240] J. Clarke, *Amer. J. Phys.* **38**, 1071 (1970).

[241] T. F. Finnegan, in "The Science and Technology of Superconductivity" (W. D. Gregory, W. N. Mathews, Jr., and E. A. Edelsack, eds.), p. 565. Plenum Press, New York, 1972.

[242] D. N. Langenberg, in "Precision Measurement and Fundamental Constants" (D. N. Langenberg and B. N. Taylor, eds.), NBS Special Publ. No. 343. U.S. Government Printing Office, Washington, D.C., 1971.

[243] D. J. Scalapino, in "Precision Measurement and Fundamental Constants" (D. N. Langenberg and B. N. Taylor, eds., NBS Special Publ. No. 343. U.S. Government Printing Office, Washington, D.C., 1971.

[244] D. B. Sullivan, Applied Superconductivity Conference Record, 1972. IEEE Publ. No. 72-CHO682-5-TABSC.

[245] B. N. Taylor, W. H. Parker, D. N. Langenberg, and A. Denenstein, *Metrologia* **3**, 89 (1967).

superconducting voltage ratio devices may improve the accuracy of this process.[244] There has been considerable progress toward a practical voltage standard; some of the developments have been reviewed by Petley.[246] It is reported that the U.S. volt may soon be based on the Josephson effect.[244]

In principle, the Josephson effect provides the way of defining the volt. However, since the volt is a derived quantity at present, this would require a major revision of the structure of the fundamental constants.

5.4.3. Superconducting Rings Containing a Single Weak Link

5.4.3.1. Quantized States of the Ring.

Superconducting rings containing a weak link are of great interest both because of the basic physics of the system and because of the numerous superconducting devices based on them. The static behavior expected for the ring is easily deduced from the quantized fluxoid condition. As shown in Section 5.3.2, for a ring with thickness large compared to the penetration depth everywhere except at a single weak link (Fig. 40), evaluating the fluxoid around a path within the ring leads to

$$(\Phi_0/2\pi)\theta + \phi = n\Phi_0, \qquad (5.4.1)$$

where θ is the gauge-invariant phase shift across the link and ϕ is the total magnetic flux in the ring. What is usually of interest is the current i in the ring, or the total flux ϕ in the ring as a function of an externally applied flux ϕ_x (e.g., with a solenoid as shown in Fig. 40.) Thus it is convenient to write

$$\phi = iL + \phi_x, \qquad (5.4.2)$$

where L is the inductance of the ring.

Several simple examples serve to illustrate the behavior as a function of ϕ_x. The simplest case is to assume that there is no phase shift across the link but that the link has a finite critical current i_c. Then from Eq. (5.4.1) we see that the total flux within the ring is always quantized in units of Φ_0. If initially $n = 0$, as ϕ_x is increased from zero a screening current flows always keeping the total flux ϕ in the ring zero until i_c is reached. At i_c there is a discontinuous switch to a different quantized flux state. For the special case $i_c L = \Phi_0/2$, the current in the ring and

---

[246] B. W. Petley, *Contemp. Phys.* **12**, 453 (1971).

the total flux are as shown in Fig. 47a, with the discontinuous switch in quantum state always being by one flux unit.

Actually, there will be a phase shift θ across the link when there is a current flowing and, if the current–phase relationship $i(\theta)$ is known, θ can be eliminated between Eqs. (5.4.1) and (5.4.2), and the current flowing in the ring (or the total flux in the ring) can be calculated as a

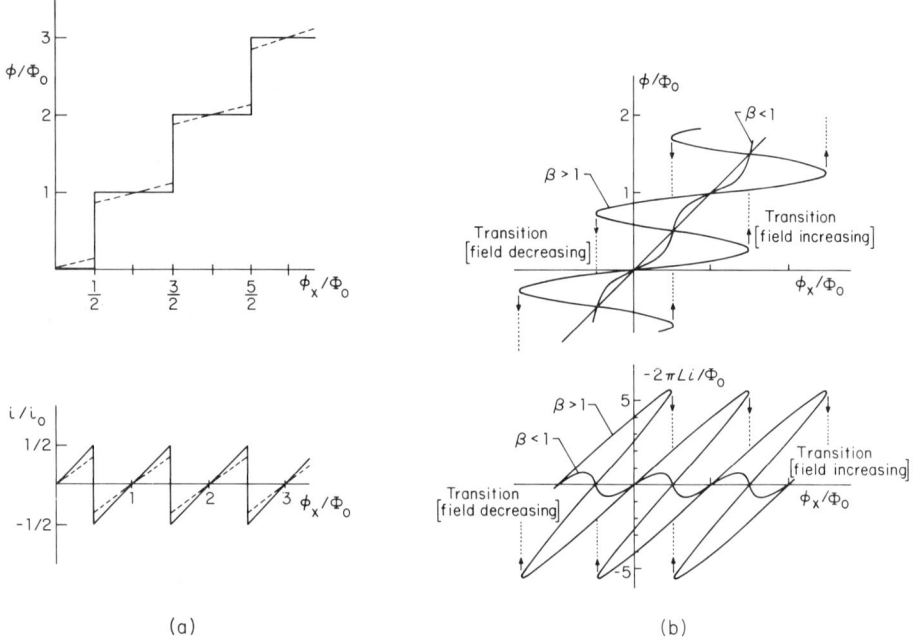

FIG. 47. Total flux ϕ and current i as a function of external flux ϕ_x for a superconducting ring containing a weak link: (a) (Solid line) assuming sharp transition at i_c and no phase shift across link; (dashed line) assuming linear current–phase relation; (b) assuming sinusoidal current–phase relation. Curves for two values of $\beta = 2\pi i_c L/\Phi_0$ are shown [after J. E. Zimmerman, *Cryogenics* **12**, 19 (1972)].

function of ϕ_x. One useful approximation is $i = (\Phi_0/2\pi)(\theta/L_k)$, the so-called linear approximation, where L_k, the kinetic inductance of the link (see Section 5.3.2) is assumed constant. The behavior depends on the value of the parameter $\beta = L/L_k$. For $\beta \ll 1$, the current and flux are indicated by the dashed lines in Fig. 47a. The screening current is less than for the first case considered, and some flux is admitted to the ring continuously between steps because of the phase shift across the link. For $\beta \gg 1$, the possibility for multiple quantum transitions exists and

5.4. APPLICATIONS

the free energy must be calculated to determine what transition minimizes the free energy.

A more realistic case is to assume $i = i_c \sin \theta$ for the weak link. With this current–phase relation, Eqs. (5.4.1) and (5.4.2) lead to

$$i_c \sin(2\pi\phi/\Phi_0) = (\phi - \phi_x)/L \tag{5.4.3}$$

and

$$i/i_c = \sin 2\pi[(iL + \phi_x)/\Phi_0], \tag{5.4.4}$$

from which i and ϕ are plotted in Fig. 47b. In this case the parameter $\beta = 2\pi i_c L/\Phi_0$ determines the behavior. For $\beta < 1$, there is no discontinuous change in current i or in the flux ϕ in the ring, but a continuous transition between states. For $\beta > 1$, when $i \approx i_c$ there is a discontinuous switch in current to a lower quantized state and an abrupt flux change in the ring. These plots of ϕ versus ϕ_x and i versus ϕ_x describe the static response of the ring to an external flux, but should be applicable up to frequencies for which the normal or displacement currents become comparable to the supercurrent and thus up to $\sim 10^9$ Hz for a point contact or thin film bridge. Thus the ac response can also be determined from these characteristics by assuming $\phi_x = \phi_x(t)$.

The first analysis of the quantized states of a ring containing a weak link and of the transitions between these states, and the first measurements of the static and rf response of such rings, was reported in an extremely comprehensive paper by Silver and Zimmerman.[219] Their analysis and experimental techniques have served as the basis for many of the experiments and devices that have followed.

A general technique for studying the response of weakly linked superconducting rings uses the arrangement shown in the lower part of Fig. 48. The ring is inductively coupled to the coil of a tank circuit tuned at some convenient rf frequency, commonly about 30 MHz, and the voltage across the tank measured as a function of rf drive level and static applied field. The response can be calculated[219,247] from the static characteristics (Fig. 47b); however, the essential features can be deduced from a simple model.[248]

Assuming the response shown in Fig. 47a, as the flux ϕ_x is continuously increased, the voltage induced in a coil coupled to the ring will consist of a series of sharp spikes, one for each time a discontinuous flux change

[247] J. E. Zimmerman, *Cryogenics* **12**, 19 (1972).
[248] J. E. Mercereau, *Rev. Phys. Appl.* **5**, 13 (1970).

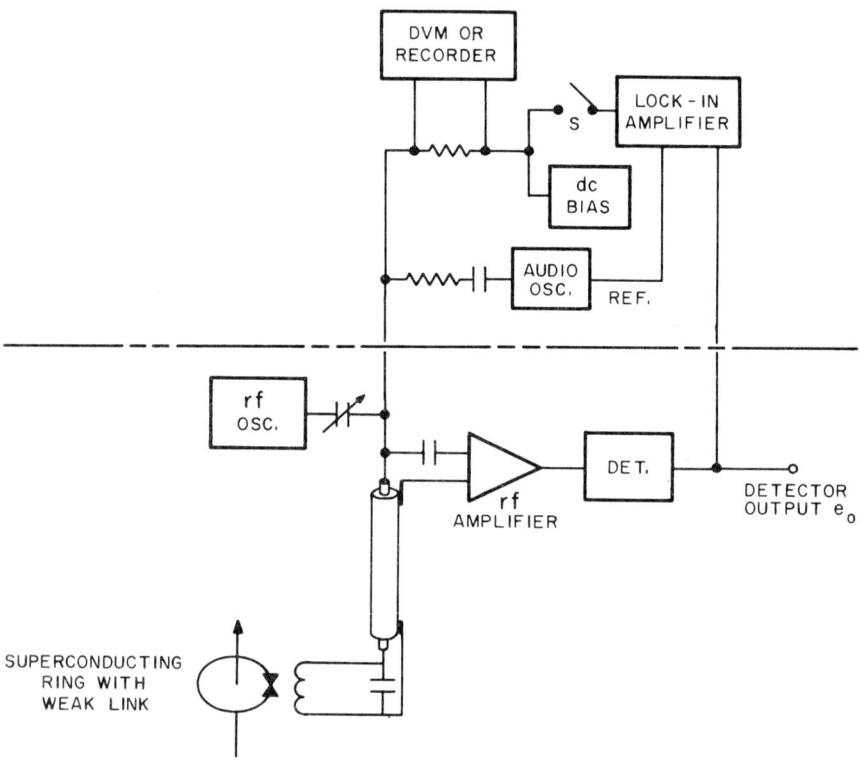

FIG. 48. Circuit for studying the response of a superconducting ring containing a weak link. Only the components shown below the dotted line were used in initial experiments to study the behavior of the ring. With the addition of the components above the line, the circuit becomes a practical magnetometer.

occurs, and these will be periodic in ϕ_x with period Φ_0. This pulse train can be expressed as a Fourier series. Then assuming that

$$\phi_x = \phi_{dc} + \phi_{rf} \cos \omega t, \qquad (5.4.5)$$

and extracting only the component at the fundamental frequency ω (since this is what is observed across a high Q tank circuit), we obtain for the voltage e induced in the tank circuit

$$e \approx \phi_{rf}\omega J_1(2\pi\phi_{rf}/\Phi_0) \cos(2\pi\phi_{dc}/\Phi_0) \cos \omega t, \qquad (5.4.6)$$

where J_1 is the Bessel function of order one. This behavior is nicely exemplified by the plots in Fig. 49 of the amplified and detected rf voltage across a tank circuit coupled to a ring containing a thin film bridge.[248]

This simple analysis would predict no signal at the fundamental frequency for a level of ϕ_{rf} which did not produce a switching between quantized flux states. In fact, there is a variation of the flux in the ring even for smaller amplitudes of the applied flux, and thus there will always be a signal at the fundamental frequency.[219]

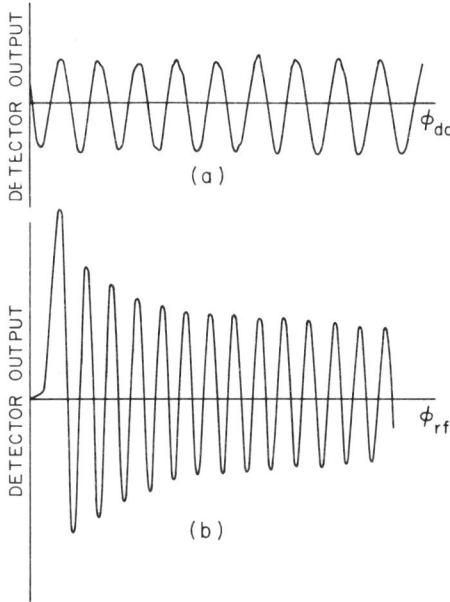

FIG. 49. Detector output (see Fig. 48) as a function of applied flux; (a) showing the periodicity with ϕ_{dc}, ϕ_{rf} held constant; (b) showing the periodic variation with rf level ϕ_{rf} when ϕ_{dc} is constant. The curves are plotted on different scales—each has period Φ_0 [after J. E. Mercereau, *Rev. Phys. Appl.* **5**, 13 (1970)].

5.4.3.2. Magnetometers and Multimeters.

The fact that the amplitude of the rf voltage across the tank is a sensitive function of the flux through the superconducting ring makes possible an extremely sensitive magnetometer. For a ring with an area of 1 cm², the period of variation of the rf amplitude e with ϕ_{dc} is about 10^{-7} G and field changes can be determined to this precision simply by counting cycles of the amplitude variation of the detector output e_0. In fact it is possible to resolve a small fraction of a cycle. A realistic magnetometer circuit (Fig. 48) makes use of a feedback loop to lock the ring to a constant flux condition.

For this purpose, a small audio-frequency flux is also applied and sensed with a lock-in amplifier. The dc output from the lock-in amplifier is fed back as an applied flux to the ring and just cancels the effect of any attempted external flux change $\Delta\phi_x$. Thus the feedback current becomes proportional to $\Delta\phi_x$ and is the quantity that is measured. Sensitivity to flux changes of $10^{-3}\Phi_0$ with a one-cycle bandwidth corresponding to field changes of 10^{-10} G in a 1-cm² pickup loop can then be routinely achieved.

In their initial experiments, Silver and Zimmerman[219] used a niobium ring connected by a niobium point contact essentially like that shown in Fig. 50b. Subsequently many different kinds of weakly linked rings

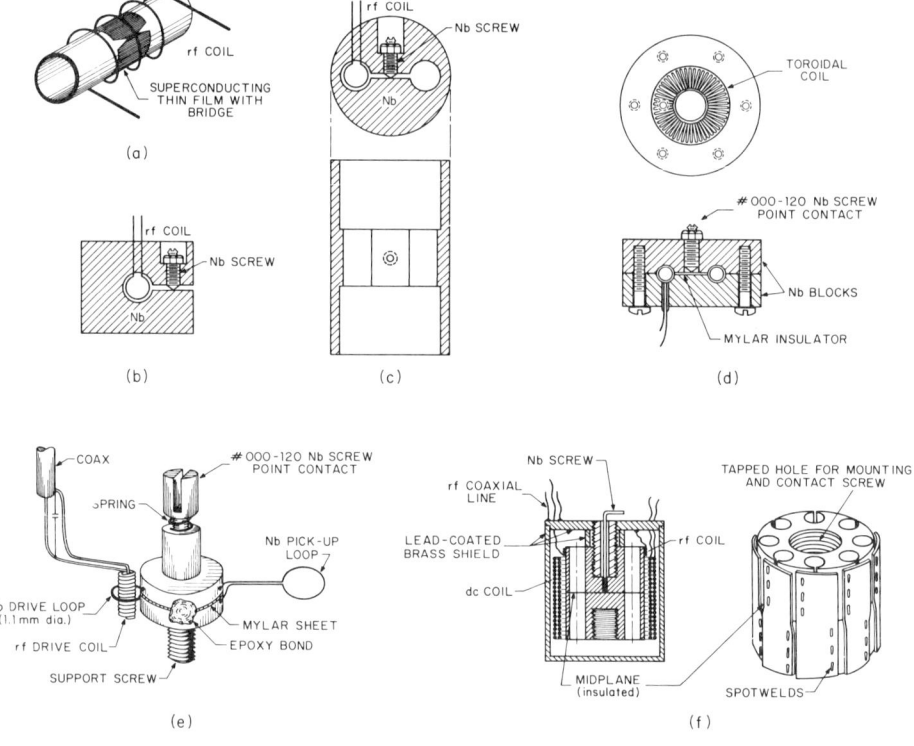

FIG. 50. Magnetometer configurations using rings containing a single weak link: (a) thin film bridge; (b) single ring with point contact; (c) thermally stable double ring [after J. E. Zimmerman, P. Thiene, and J. T. Harding, *J. Appl. Phys.* **41**, 1572 (1970)]. (d) toroid with point contact; (e) double ring with remote pickup loop; (f) multiple-loop squid [after D. B. Sullivan, Applied Superconductivity Conference Record, 1972. IEEE Publ. No. 72-CHO682-5-TABSC].

have been used as magnetometers; some other examples are shown in Fig. 50. A device of this general type is often called a "squid," an acronym for superconducting quantum interference device. All are operated using the general scheme shown in Fig. 48. Forgacs and Warnick[249,250] have described both analog and flux-counting circuits for operating them. Often the rf oscillator is eliminated and the tank is caused to oscillate by providing positive feedback from the output of the rf amplifier to the tank circuit. In either external or self-driven modes, the rf level is adjusted for maximum sensitivity to external flux changes; typically this corresponds to ϕ_{rf} near the first maximum of the Bessel function in Fig. 49b.

Magnetometers using thin film bridges (Fig. 50a) have been studied by Mercereau,[199] Nisenoff,[196,251] and Goodkind.[252,253] Corsadori et al.[206] have demonstrated quantum interference using a bridge formed in a thin flake of niobium selenide. Bridges appear to offer significant advantages of small size and superior stability for device applications, but the temperature dependence of the critical current is often a problem.

So far, the majority of magnetometers have used point contacts. The configuration shown in Fig. 50c has the property of being thermally stable so it can be adjusted at room temperature.[254] It is insensitive to uniform external field along its axis (since it is intrinsically a gradiometer) and is normally operated with the rf tank coil coupled to one hole and a superconducting flux transfer loop coupled to the other. The toroid, Fig. 50d, is also thermally stable and almost insensitive to external fields. It is well suited to application as a galvanometer with both the rf and the current to be measured coupled with coils inside the toroidal cavity. The device[255] shown in Fig. 50e is essentially another form of the two-hole configuration of Fig. 50c. Very high sensitivity to uniform external field can be achieved with the geometry of Fig. 50f, which consists essentially of eight loops in parallel across a single point contact, giving a large cross-sectional area but very small inductance.[256]

---

[249] R. L. Forgacs and A. Warnick, *IEEE Trans.* **IM-15**, 113 (1966).
[250] R. L. Forgacs and A. Warnick, *Rev. Sci. Instrum.* **38**, 214 (1967).
[251] M. Nisenoff, *Rev. Phys. Appl.* **5**, 21 (1970).
[252] J. M. Goodkind and D. L. Stolfa, *Rev. Sci. Instrum.* **41**, 799 (1970).
[253] J. M. Goodkind and J. M. Dundon, *Rev. Sci. Instrum.* **42**, 1264 (1971).
[254] J. E. Zimmerman, P. Thiene and J. T. Harding, *J. Appl. Phys.* **41**, 1572 (1970).
[255] W. L. Goodman, W. D. Willis, D. A. Vincent, and B. S. Deaver, Jr., *Phys. Rev. B* **4**, 1530 (1971).
[256] J. E. Zimmerman, *J. Appl. Phys.* **42**, 4483 (1971).

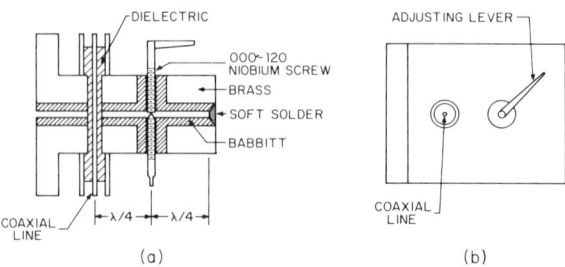

FIG. 51. Broadband magnetometer. Configuration used for squid operating at 9 GHz: (a) longitudinal section; (b) top view [after R. A. Kamper and M. B. Simmonds, *Appl. Phys. Lett.* **20**, 270 (1972)].

Kamper and Simmonds[257] have achieved a broad-band magnetometer by using the configuration shown in Fig. 51 to operate a weakly linked ring at 9 GHz. A section of rectangular X-band waveguide was lined with superconducting metal (babbitt) and the loop was closed with a Nb point contact. The reduced vertical dimension lowered the impedance of the section to give a better match to the point contact. Current at frequencies from dc to 1 GHz passing along the coaxial line were detected via the magnetic field produced in the superconducting loop.

They have also used this device to obtain a quantum-based method for measuring rf attenuation.[258] Since the microwave voltage developed across the weakly linked ring varies as a Bessel function of the microwave drive voltage [see Eq. (5.4.6) and Fig. 49b], the power levels corresponding to minima of voltage across the ring are precisely related by ratios of the zeros of the Bessel function. In their first experiments the precision approached that of the standard plunger-type attenuators.

A crude estimate of the signal power P_s available from one of these magnetometers can be obtained by noting that at each cycle of the rf, one flux unit is switched in and out of a ring of inductance L; thus

$$P_s = (\Phi_0^2/2L)\nu, \qquad (5.4.7)$$

where ν is the frequency. This result indicates the value of making the inductance as small as possible, and shows the motivation for the device shown in Fig. 50f. Zimmerman and Frederick[259] operated a magnetom-

---

[257] R. A. Kamper and M. B. Simmonds, *Appl. Phys. Lett.* **20**, 270 (1972).

[258] R. A. Kamper and M. B. Simmonds, Applied Superconductivity Conference Record, 1972. IEEE Publ. No. 72-CHO682-5-TABSC.

[259] J. E. Zimmerman and N. V. Frederick, *Appl. Phys. Lett.* **19**, 16 (1971).

eter at 300 MHz and found the expected improvement in signal-to-noise over the 30-MHz case. A further improvement was found for the device operated at 9 GHz.[258]

For large rings, the noise limit is determined by thermal fluctuations between quantized flux states. Thus there is a limit on the size of the ring that can be used at a given temperature. For small rings, fluctuations in i_c lead to jitter in the switching between quantized flux states and thus to noise in the magnetometer output. Calculations using various models and some experiments investigating this intrinsic limitation[260-265] give noise levels corresponding to $\Delta\phi_x \approx 10^{-6}$–$10^{-5}\Phi_0$ for a 1-Hz bandwidth. However, the best existing instruments have noise levels which limit the sensitivity to approximately $10^{-4}\Phi_0$, indicating that the limitation is probably due to external noise rather than the intrinsic noise of the device.[266]

The basic magnetometer consisting of a superconducting ring coupled to a tank circuit is essentially a magnetic flux meter. It can be made more sensitive to magnetic field by the use of a superconducting flux transformer. (Fig. 52b). Loops many centimeters in diameter have been coupled to rings of 1 mm diameter. The field amplification varies as the square root of the ratio of the effective volumes of the coils resulting in practical improvement in field sensitivity by approximately a factor of 100.[267] If a pair of pick-up coils in series opposition is used (Fig. 52d), the device becomes a magnetic gradiometer and is insensitive to uniform fields common to both coils.

A zero-resistance ammeter can be achieved by coupling a coil to the superconducting ring (Fig. 52a) or an extremely sensitive voltmeter obtained by connecting a resistor in series with the coil. Johnson noise in the resistor becomes a limitation; however, practical measurements

---

[260] R. E. Burgess, *Proc. Symp. Phys. Superconducting Devices*, Charlottesville (1967) (see Ref. 22).

[261] J. E. Lukens and J. M. Goodkind, *Phys. Rev. Lett.* **20**, 1363 (1968).

[262] D. E. McCumber, *Phys. Rev.* **181**, 716 (1969).

[263] J. Kurkijärvi and W. W. Webb, Applied Superconductivity Conference Record, 1972. IEEE Publ. No. 72-CHO682-5-TABSC.

[264] J. Kurkijärvi, *Phys. Rev. B* **6**, 832 (1972).

[265] R. P. Giffard, R. A. Webb, and J. C. Wheatley, *J. Low Temp. Phys.* **6**, 533 (1972).

[266] J. Clarke, Applied Superconductivity Conference Record, 1972. IEEE Publ. No. 72-CHO682-5-TABSC.

[267] W. S. Goree, Applied Superconductivity Conference Record, 1972. IEEE Publ. No. 72-CHO682-5-TABSC.

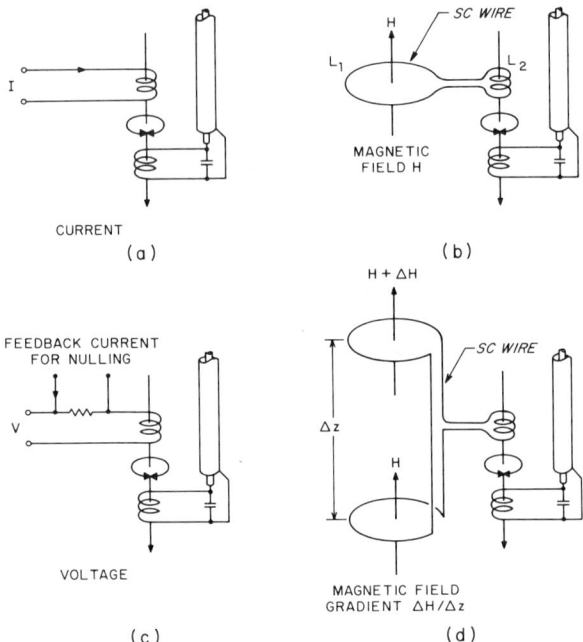

FIG. 52. Devices using an rf operated weakly linked superconducting ring: (a) galvanometer; (b) magnetometer with superconducting (SC) flux transformer; (c) potentiometric voltmeter; (d) magnetic gradiometer.

in the femtovolt range are possible.[268] For voltage measurements, the device can be operated as a self-nulling potentiometer by using current feedback from the lock-in amplifier to the resistor (Fig. 52c).

An impressive variety of devices and instruments based on rf-operated, weakly linked rings have been conceived. In addition to magnetometers,[219] gradiometers,[259] and multimeters,[268] mentioned above, these include inductance bridges,[265,268] an absolute ammeter,[269] a gravimeter,[270] digital current and voltmeters,[244] and systems involving magnetometers with superconducting shields and magnets for magnetization and susceptibility measurements.[267,271] Discussions of the design and operation of devices of this type are available in a number of summary and review

[268] J. E. Lukens, R. J. Warburton, and W. W. Webb, *J. Appl. Phys.* **42**, 27 (1971).
[269] R. Meservey, *J. Appl. Phys.* **39**, 2598 (1968).
[270] W. M. Prothero, Jr., and J. M. Goodkind, *Proc. Symp. Phys. Superconducting Devices*, Charlottesville (1967) (see Ref. 22).
[271] H. E. Hoenig, R. H. Wang, G. R. Rossman, and J. E. Mercereau, Applied Superconductivity Conference Record, 1972. IEEE Publ. No. 72-CHO682-5-TABSC.

papers.[247,248,251,252,254,265,266,268,272–274] These techniques have recently begun to move out of the research laboratory and many of the devices are now available from at least three commercial sources.†

5.4.3.3. Experiments Using Rings with a Single Weak Link.
In their early experiments with weakly linked rings, Silver and Zimmerman[219] originated the rf squid magnetometer and used it to study the quantized states of the ring. Many of the devices and techniques that have evolved from their magnetometer have been discussed above. In the following paragraphs we mention very briefly some experiments that are representative of the diverse types of measurements for which rf-biased weakly linked rings have been used.

When a hollow superconducting cylinder is rotated at constant angular velocity ω about its longitudinal axis, a magnetic field is induced along the axis producing a flux within the cylinder and exactly maintaining the constancy of the fluxoid. The flux enclosed by the cylinder is

$$\phi = \Phi_0(n - 4A\omega m/h), \qquad (5.4.8)$$

where A is the area of the cylinder, m is the electron mass, and n is an integer. Parker and Simmonds[275] have rotated a thin film squid and measured the flux generated by the rotation to determine h/m with an uncertainty of a few parts in 10^4.

The most precise ($\pm 0.1\%$) and the best absolute ($\pm 1/2\%$) measurement of the flux trapped in hollow superconducting cylinders have been made using a point contact squid.[255] The quantized domain structure along the cylinder was mapped and the temperature dependence of the trapped flux measured using a point contact device in the configuration shown in Fig. 50e.

---

[272] M. B. Simmonds and W. H. Parker, *J. Appl. Phys.* **42**, 38 (1971).

[273] W. W. Webb, in "The Science and Technology of Superconductivity" (W. D. Gregory, W. N. Mathews, Jr., and E. A. Edelsack, eds.), p. 653. Plenum Press, New York, 1972.

[274] W. W. Webb, *IEEE Trans. Magn.* **MAG-8**, 51 (1972).

[275] W. H. Parker and M. B. Simmonds, in "Precision Measurement and Fundamental Constants" (D. N. Langenberg and B. N. Taylor, eds.), NBS Special Publ. No. 343. U.S. Government Printing Office, Washington, D.C., 1971.

---

† Develco, Inc., Mountain View, California, S. H. E. Manufacturing Corp., San Diego, California, and Superconducting Technology, Inc., Mountain View, California.

The high sensitivity of these devices has naturally been brought to bear on measurements of thermodynamic fluctuations including fluctuations of the current in the weakly linked ring itself,[261] magnetic fluctuations in conductors[276] observed with a magnetometer, and effects of fluctuations near the superconducting transition on the I–V curves of tin whisker crystals measured with a voltmeter with 10^{-15}-V sensitivity.[277]

Goodkind[270] has observed earth tides using a superconducting gravimeter that employs a magnetometer to detect the position of a magnetically supported superconducting ball. This device has a sensitivity of about 10^{-9} G.

Magnetometers and gradiometers appear to have great potential for medical applications. Time-varying magnetic fields of about 10^{-7} G correlated with the heart beat have been observed with sensors placed near a man's chest,[259,267,278,279] and similar signals have been observed from the excised but still active hearts of small animals.[267] The possible clinical significance of magnetocardiography is being actively investigated.

Systems using superconducting magnetometers in conjunction with superconducting magnetic shields and persistent current magnets make possible measurements of magnetic susceptibility with several orders of magnitude more sensitivity than could formerly be achieved.[252,265,267,271] This system is particularly suited for measuring small changes in susceptibility, for example, as a function of temperature of the sample, and changes of 5×10^{-9} emu/cm^3 corresponding to a magnetic moment of 3×10^{11} electron spins have been detected. The technique has been used for magnetochemical studies of some biologically interesting molecules.[271] It has also been used for thermometry by measuring the temperature-dependent susceptibility of cerium magnesium nitrate.[265]

5.4.3.4. Rings Containing a Resistor—The Noise Thermometer. A superconducting ring containing a very small resistor, e.g., $\sim 10^{-5}$ Ω (see Fig. 53) and a weak link has been used as a means of voltage biasing the link for observing both the fundamental Josephson oscillation and relaxation oscillations corresponding to multiple quantum transitions. These experiments are described in Section 5.2.7. Subsequently Silver and Zimmerman[280] used the voltage biased weak link as an oscillator–

[276] L. Vant-Hull, R. A. Simpkins, and J. J. Harding, *Phys. Lett.* **24A**, 736 (1967).
[277] J. E. Lukens, R. J. Warburton, and W. W. Webb, *Phys. Rev. Lett.* **25**, 1180 (1970).
[278] D. Cohen, E. A. Edelsack, and J. E. Zimmerman, *Appl. Phys. Lett.* **16**, 278 (1970).
[279] J. E. Zimmerman, *J. Appl. Phys.* **42**, 30 (1971).
[280] A. H. Silver and J. E. Zimmerman, *Appl. Phys. Lett.* **10**, 142 (1967).

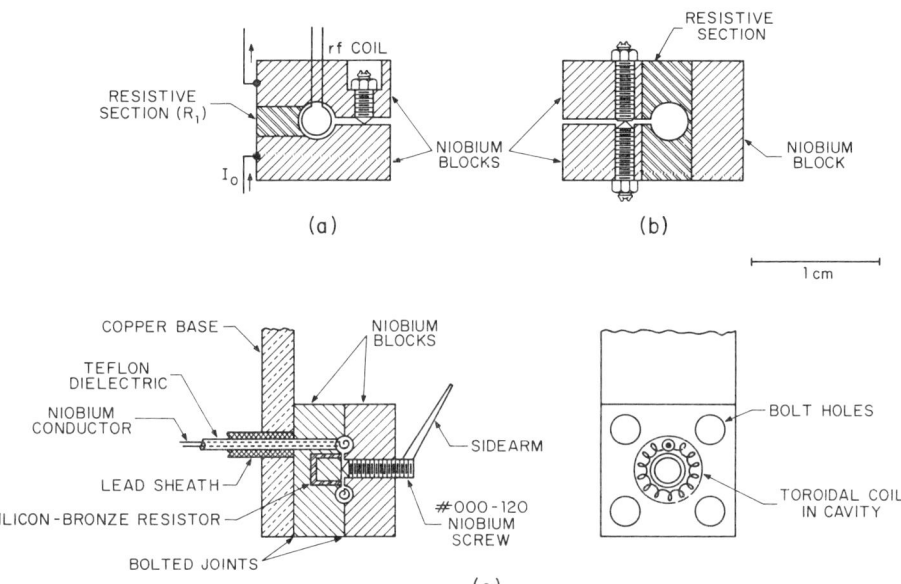

FIG. 53. Weakly linked superconducting rings containing a small normal resistor: (a) Nb ring with brass resistor (typically $\sim 10^{-5}$ Ω); (b) temperature compensated ring with resistor; (c) toroid used for noise thermometry [after J. E. Zimmerman, P. Thiene, and J. T. Harding, *J. Appl. Phys.* **41**, 1572 (1970); J. E. Zimmerman, Applied Superconductivity Conference Record, 1972. IEEE Publ. No. 72-CHO682-5-TABSC].

detector to observe nuclear magnetic resonance by inductively coupling a cobalt sample to the ring and observing the rf voltage across the ring.

In the preceding sections, the behavior of a superconducting ring containing a weak link with a time-varying field was deduced from the static characteristics (Fig. 47), which in turn were obtained from applying the quantized fluxoid condition to the ring. Harding and Zimmerman[281] have shown that rings containing a resistor also exhibit a periodic dependence of their magnetization with period Φ_0 for externally applied flux at low (but not zero) frequency. This behavior can be deduced from a simple calculation. Equating the voltage drops around the ring to the applied voltage when there is a time-varying external flux $\dot{\phi}_x$,

$$-\dot{\phi}_x = L\, di/dt + iR + (\Phi_0/2\pi)\dot{\theta}, \qquad (5.4.9)$$

where L is the inductance of the ring. [This equation is equivalent to

[281] J. T. Harding and J. E. Zimmerman, *J. Appl. Phys.* **41**, 1581 (1970).

differentiating the fluxoid expression Eq. (5.3.6), and adding a resistive voltage term.] Assuming $i = i_c \sin \theta$ for the current–phase relation leads to a periodic dependence of i on ϕ_x, provided $L < \Phi_0/2\pi i_c$ and the low-frequency limit for observing the periodicity is approximately $i_c R/\Phi_0$, not $\sim R/L$ as might be assumed. The crucial feature for periodicity is then not coherence around the ring but the fact that the weak link is a quantized flux "gate."

In the rings containing resistors, Johnson noise causes fluctuations in θ, giving a random walk in the phase as observed and explained by Harding and Zimmerman.[281] The Johnson noise in the resistor also leads to a frequency modulation of the Josephson oscillation of a voltage biased weak link (see Section 5.2.11). Kamper first suggested that a measurement of the line width of the oscillation could be used to determine the temperature and the device would be an absolute thermometer. The calculated line width is[260]

$$\Delta f = 4\pi kTR/\Phi_0^2 = 4.05 \times 10^7 RT \tag{5.4.10}$$

and this variation was verified by Silver et al.[282]

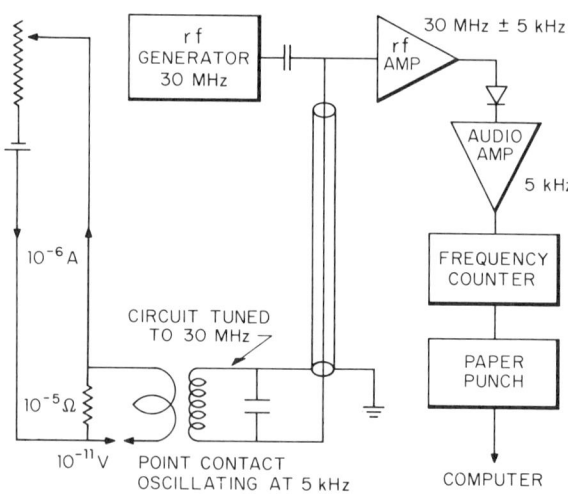

FIG. 54. Circuit for operating a weakly linked superconducting ring containing a normal resistor as a noise thermometer [after R. A. Kamper and J. E. Zimmerman, J. Appl. Phys. **42**, 132 (1971)].

[282] A. H. Silver, J. E. Zimmerman and R. A. Kamper, Appl. Phys. Lett. **11**, 209 (1967).

5.4. APPLICATIONS

A practical noise thermometer[283,284] makes use of the configuration shown in Fig. 53c and the circuit shown in Fig. 54. Rather than the line width, fluctuations in the frequency are determined by counting repeatedly for a fixed time τ. Then

$$\sigma^2 = \langle (f - \bar{f})^2 \rangle = 2kTR/\tau\Phi_0. \quad (5.4.11)$$

Measurements in the millidegree range have already been made,[284] and in principle it may be possible to measure temperatures in the microdegree range.

5.4.4. Superconducting Rings Containing Two Weak Links

5.4.4.1. Interference Effects in Weakly Linked Rings.
Many features associated with the macroscopic quantum nature of the superconducting state were first exhibited experimentally using the configuration shown

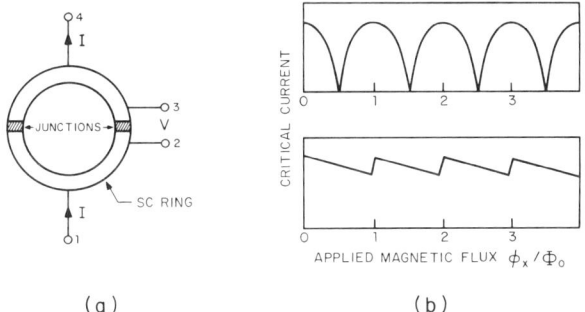

FIG. 55. (a) Superconducting ring containing two weak links—interferometer configuration. (b) Variation of the critical current I_c with external flux ϕ_x through the ring for (upper) symmetric bias and (lower) asymmetric bias (e.g., I between 1 and 3).

in Fig. 55a, a superconducting ring containing two weak links. For most experiments the quantity of interest is the maximum zero-voltage current that can be passed through the parallel combination of weak links. It can be deduced by evaluating the fluxoid, Eq. (5.2.5), around the ring. Using the fluxoid in the form of Eq. (5.4.1) leads to

$$(\Phi_0/2\pi)(\theta_2 - \theta_1) + \phi = n\Phi_0, \quad (5.4.12)$$

---

[283] R. A. Kamper and J. E. Zimmerman, *J. Appl. Phys.* **42**, 132 (1971).

[284] R. A. Kamper, J. D. Siegwarth, R. Radebaugh and J. E. Zimmerman, *IEEE Proc. Lett.* **59**, 1368 (1971).

where θ_1 and θ_2 are the phase differences across the weak links and ϕ is the total flux in the ring. Assuming for convenience two identical weak links, the total current I is

$$I = i_c(\sin\theta_1 + \sin\theta_2), \qquad (5.4.13)$$

which can be written as

$$I = 2i_c \sin[\tfrac{1}{2}(\theta_1 + \theta_2)] \cos[\tfrac{1}{2}(\theta_1 - \theta_2)]. \qquad (5.4.14)$$

The last factor can be expressed in terms of the total flux using Eq. (5.4.12) giving

$$I = 2i_c \sin[\tfrac{1}{2}(\theta_1 + \theta_2)] \cos(\pi\phi/\Phi_0). \qquad (5.4.15)$$

The angle $(\theta_1 + \theta_2)/2$ varies with applied current I; however, the maximum value I_c of I is

$$I_c = 2i_c \, |\cos(\pi\phi/\Phi_0)|, \qquad (5.4.16)$$

which is a periodic function of total flux ϕ with period Φ_0.

The total flux ϕ is the sum of the induced flux due to currents in the two branches of the ring with inductance L_1 and L_2, respectively, and the external flux ϕ_x

$$\phi = i_1 L_1 - i_2 L_2 + \phi_x. \qquad (5.4.17)$$

For a symmetrically biased ring the induced flux is $(L/2)(i_1 - i_2)$, where $L = L_1 + L_2$ and $(i_1 - i_2)$ is the circulating current in the ring. If $i_c L \ll \Phi_0$, the induced flux is negligible and the maximum current is a periodic function of ϕ_x with maximum value $2i_c$ and minimum value zero. If $i_c L > \Phi_0$, the behavior is more complicated. There is a large induced flux giving a more nearly sawtooth shape to the I_c versus ϕ_x curve, and since the maximum change in circulating current is Φ_0/L, the fractional change in I_c is smaller, viz., $\Phi_0/2i_c L$. Asymmetric biasing also produces a sawtooth curve which is useful for some applications.[19] These cases are illustrated in Fig. 55b.

The preceding analysis has neglected any effect of the magnetic field incident on the weak links themselves. If they are tunnel junctions and the field is in the plane of the barrier, then the critical current of each junction has the characteristic diffraction-like behavior [Eq. (5.2.27)] and thus the maximum current has the form

$$I_c \propto i_c \left| \frac{\sin(\pi H/H_0)}{\pi H/H_0} \right| \left| \cos\pi\frac{\phi}{\Phi_0} \right|. \qquad (5.4.18)$$

5.4. APPLICATIONS

Because of the similarities between the behavior of the critical current and optical diffraction and interference phenomena,[3,4] the configuration shown in Fig. 55a is often called a "quantum interferometer."

5.4.4.2. Macroscopic Quantum Interference Experiments.

Jacklevic et al.[285] originated experiments of this type using tunnel junctions in the configuration shown in Fig. 56a, and by measuring the critical current as a function of external flux observed a diffraction modulated interference pattern (Fig. 56b), well described by the above analysis. Subsequently they used two thin film bridges in a ring and verified the

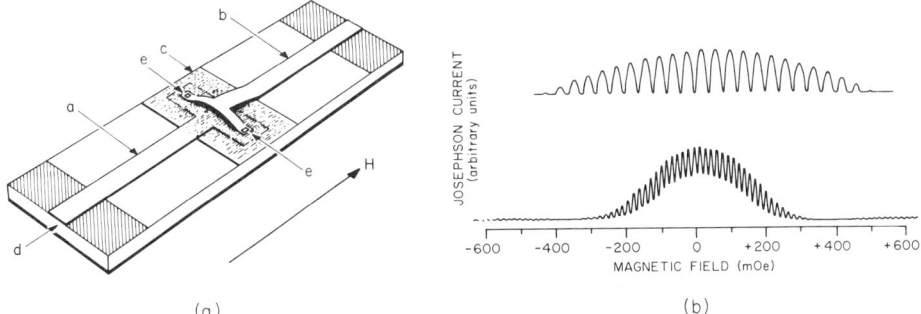

FIG. 56. (a) Interferometer configuration using tunnel junctions. A uniform magnetic field H is applied parallel to the long dimension of the substrate d. The Formvar insulator c is applied over the base tin film a to mask out the junctions e and separate a from the second tin film b. (b) Maximum supercurrent versus magnetic field for configuration of (a). The upper curve has a field periodicity of 39.5 mOe with maximum current 1 mA. The lower curve has field periodicity of 16 mOe with maximum current 0.5 mA. The junction separation is 3 mm and the junction width $w = 0.5$ mm for both cases [after R. C. Jacklevic, J. Lambe, J. E. Mercereau, and A. H. Silver, Phys. Rev. **140**, A1628 (1965)].

periodic dependence on external dc flux as well as the periodic behavior with rf level.[198] In these experiments not the dc critical current but the microwave impedance of the ring was measured. Zimmerman and Silver[286] originated the use of point contacts for Josephson effect experiments by first observing quantum interference effects in rings with two point contacts, and also showed that this circuit made possible an extremely sensitive magnetometer using the changes ΔI_c in the critical

---

[285] R. C. Jacklevic, J. Lambe, A. H. Silver, J. E. Mercereau, Phys. Rev. Lett. **12**, 159 (1964).

[286] J. E. Zimmerman and A. H. Silver, Phys. Rev. **141**, 367 (1966).

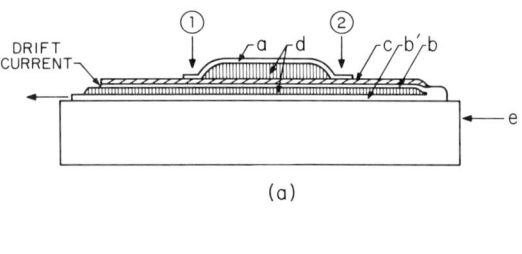

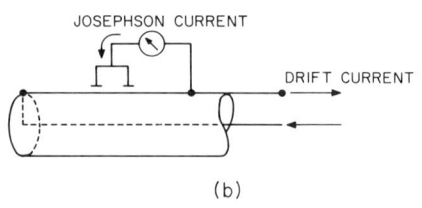

FIG. 57. (a) Cross section of a tunnel junction pair vacuum-deposited on a quartz substrate e. A thin oxide layer c separates thin ($\sim$1100 Å) tin films (a and b). The junctions, (1) and (2), are connected in parallel by superconducting thin films (a and b). A drift current return path b', insulated from b by a Formvar layer d, minimizes the magnetic field at the junctions. (b) The idealized field-free situation. The field-free nature of the experiment was established by an actual measurement of the magnetic field in the junction region [after R. C. Jaklevic, J. Lambe, J. E. Mercereau, and A. H. Silver, *Phys. Rev.* **140**, A1628 (1965)].

current as a measure of flux changes $\Delta\phi_x$ through the loop. Magnetometer applications are discussed in more detail below.

The effects of other quantities on the macroscopic phase in a superconductor were also investigated with interferometers. Using the configuration shown in Fig. 57a, Jacklevic et al.[287] measured the phase shift produced by a current flowing in a thin section of superconductor (b in Fig. 57a) comprising one branch of the interferometer. In the branch b, the contribution of the first term in Eq. (5.2.5) to the phase difference could not be neglected because the film thickness was comparable to a penetration depth λ. The additional phase shift caused by the current, labeled "drift current" in the figure, flowing through this section was observed through the periodic dependence of the critical current on the drift current. In this experiment the effect of magnetic field produced by the drift current was minimized by the geometry, which simulates a coaxial line (Fig. 57b) outside which there should be no magnetic field.

[287] R. C. Jacklevic, J. Lambe, J. E. Mercereau, and A. H. Silver, *Phys. Rev.* **140**, A1628 (1965); see also *Proc. Int. Conf. Low Temperature Phys. 9th* (J. G. Daunt, D. O. Edwards, F. J. Milford, and M. Yaqub, eds.), p. 446. Plenum Press, New York, 1965.

The expected dependence of the maximum current on drift current j_s is given by

$$I_c = 2i_c \left| \cos (e/\hbar) \left(\int \mu_0 \lambda^2 \mathbf{j}_s \cdot d\mathbf{s} + \phi \right) \right|. \qquad (5.4.19)$$

Various measurements verified the dependence of I_c on the drift current and on temperature through the temperature dependence of the penetration depth. Assuming the free electron mass for electrons in the superconductor, the phase shift corresponded to a velocity of $\sim\tfrac{1}{2}$ cm/sec for the supercurrent and to a DeBroglie wavelength of ~ 1 cm.

The same group[288] demonstrated that the phase shift in the ring depends on the total flux inside the ring and is independent of whether the flux is produced by a uniform magnetic field applied across the entire ring or by a long slender solenoid through the center of the ring. Outside the solenoid there is no magnetic field, but only a curl-free vector potential. Ehrenberg and Siday[289] first noted this problem of phase shifts in field-free regions with respect to interference patterns produced in electron microscopes. The concept was discussed extensively by Aharanov and Bohm.[290] The experimentally observed interference effects were the same whether the flux was introduced by the solenoid or through a uniform magnetic field verifying the expected dependence of the phase on $\oint \mathbf{A} \cdot d\mathbf{s}$. However, the diffraction-like effect due to the field on the individual junctions was not present when flux was introduced with the solenoid.

Using the fact that the phase shift θ depends on the supercurrent velocity, Zimmerman and Mercereau[291] used an interferometer to measure the mass of the current carriers in a superconductor. For a circular interferometer of radius r rotated at angular frequency ω, there is a periodic dependence of the phase shift on ω. Using Eq. (5.2.7) with $v_s = r\omega$ gives for the period

$$\Delta \omega = \hbar/m^* r^2, \qquad (5.4.20)$$

where m^* is the effective mass of the carriers [already assumed to be $2m$ in Eq. (5.2.7)]. From the measured variation of the critical current with rotational velocity and the geometry, m^* was found to be the mass of

---

[288] R. C. Jacklevic, J. Lambe, A. H. Silver, and J. E. Mercereau, *Phys. Rev. Lett.* **12**, 274 (1964).

[289] W. Ehrenberg and R. E. Siday, *Proc. Phys. Soc. (London)* **B62**, 8 (1949).

[290] Y. Aharonov and D. Bohm, *Phys. Rev.* **115**, 485 (1959).

[291] J. E. Zimmerman and J. E. Mercereau, *Phys. Rev. Lett.* **14**, 887 (1965).

two free electrons. Also the magnetic field necessary to offset the effect of the rotation was measured. For a long cylinder, the relation

$$B = -(2m/e)\omega \qquad (5.4.21)$$

follows immediately from Eq. (5.2.7). The magnetization induced by rotating a superconductor is often called the "London moment."[292]

Quantum interference effects have been observed to be precisely the same in interferometers constructed of numerous superconductors and with weak links formed in a multitude of different ways. Detailed measurements[293-298] have been made on two weak link and multiple weak link interferometers, and the effect of the ring inductance and asymmetries have been studied.

5.4.4.3. Devices and Measurements Using Superconducting Interferometers.

The interferometer configuration (Fig. 55a), just as the ring containing a single weak link, has been used for numerous superconducting devices. Two interferometers using point contacts are shown in Fig. 58. The dependence of the critical current I_c on the magnetic flux through the ring is the basis for most applications and typically flux changes of $\sim 10^{-3}\Phi_0$ can be detected with a 1-Hz bandwidth.

Changes in I_c rather than its actual magnitude are usually of interest, and several methods have been employed to sense them. One technique applicable to devices with a continuous I–V curve is to bias the device at current slightly greater than I_c and to observe changes in the voltage as a function of applied flux.[299] Another frequently used method involves passing both a dc and a small ac current through the interferometer.

---

[292] F. London, "Superfluids," Vol. I, p. 83. Wiley, New York, 1950; M. Bol and W. M. Fairbank in *Proc. Int. Conf. Low Temperature Phys.*, *9th* (J. G. Daunt, D. O. Edwards, F. J. Milford and M. Yaqub, eds.), p. 471. Plenum Press, New York, 1965; A. F. Hildebrandt, *Phys. Rev. Lett.* **12**, 190 (1964).

[293] A. Th. A. M. De Waele, W. H. Kraan, R. De Bruyn Ouboter, and K. W. Taconis, *Physica* **37**, 114 (1967).

[294] A. Th. A. M. De Waele and R. De Bruyn Ouboter, *Physica* **42**, 626 (1969).

[295] R. De Bruyn Ouboter and A. Th. A. M. De Waele, in "Progress in Low Temperature Physics" (C. J. Gorter, ed.), Vol. VI, p. 243. North-Holland, Amsterdam, 1969.

[296] M. Dmitrenko, S. I. Bondarenko, and T. P. Narbut, *Sov. Phys. JETP* **30**, 817 (1970) [*English transl. of: Zh. Eksp. Teor. Fiz.* **57**, 1513 (1969)].

[297] R. De Bruyn Ouboter and A. Th. A. M. De Waele, *Rev. Phys. Appl.* **5**, 25 (1970).

[298] T. A. Fulton, L. N. Dunkleburger, and R. C. Dynes, *Phys. Rev. B* **6**, 855 (1972).

[299] J. W. McWane, J. E. Neighbor, and R. S. Newbower, *Rev. Sci. Instrum.* **37**, 1602 (1966).

5.4. APPLICATIONS

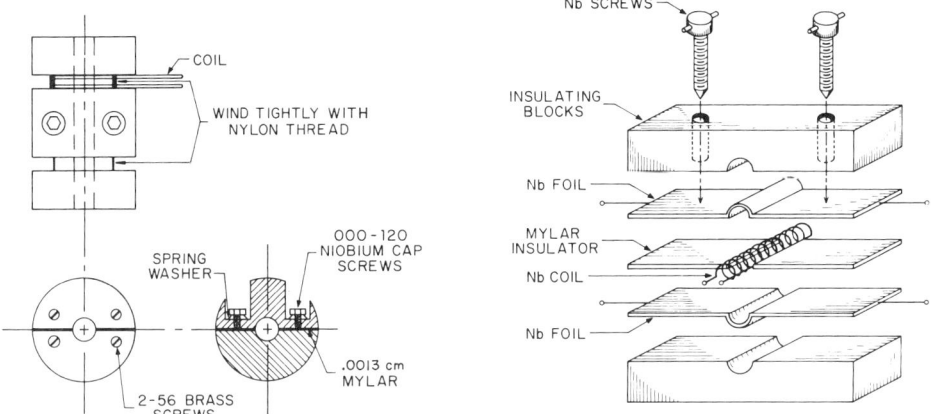

FIG. 58. Interferometer configurations using Nb point contacts [after A. H. Silver and J. E. Zimmerman, *Phys. Rev.* **157**, 317 (1967); J. E. Zimmerman, Applied Superconductivity Conference Record, 1972. IEEE Publ. No. 72-CHO682-5-TABSC].

When the dc level is set slightly below I_c, only the peaks of the ac exceed I_c and produce a series of voltage pulses. The ratio of the pulse width to the space between pulses is then a measure of I_c.[211,300] More commonly the magnitude of the voltage at the fundamental frequency is sensed with a lock-in amplifier and used as a measure of I_c.[287] The ac modulation can also be applied with a coil inductively coupled to the ring and only a dc bias current used.[301]

All these outputs are nonlinear functions of the applied flux, and in most applications are not used directly. Instead they are used as error signals for a feedback circuit (Fig. 59) which keeps the interferometer locked at a fixed value of total flux. The feedback current is then proportional to the change in external flux $\Delta\phi_x$ through the ring. The interferometer can be used directly as a magnetometer or by coupling a coil to the ring, as a low-impedance galvanometer as indicated in Fig. 59. The various techniques for using flux transformers and operating the device as a multimeter are essentially the same as those employed with a ring containing a single link (Fig. 52).

Some experiments that have used interferometers include observations of single quantized flux lines pinned in a type II superconductor,[302] mea-

---

[300] J. Clarke, *Proc. Symp. Phys. Superconducting Devices, Charlottesville* (1967) (see Ref. 22).
[301] D. A. Zych, *Rev. Sci. Instrum.* **39**, 1058 (1968).
[302] J. E. Zimmerman and J. E. Mercereau, *Phys. Rev. Lett.* **13**, 125 (1964).

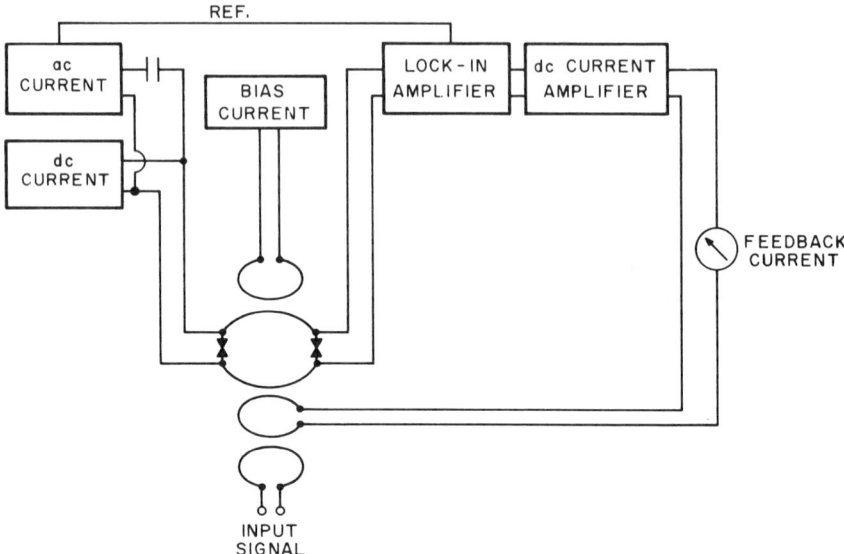

FIG. 59. Typical circuit for operating a ring with two weak links (interferometer) as a magnetometer or galvanometer.

surements of flux creep in applied fields up to 2500 G,[303] observations of phase fluctuations in a superconducting ring near T_c,[304] measurements of fluctuations in the diamagnetism of a superconductor near T_c,[305] a search for quarks,[306] and recording the magnetic field of the heart.[307]

5.4.5. The Clarke Slug

The Clarke slug, which was described in some detail in Section 5.3.1.5, is a particularly simple and useful form of weak-link device. A diagram of an idealized slug is shown in Fig. 60 where weak links are indicated at two discrete notches in the insulation and there is a well-defined area A (surrounded by the dashed line) into which the magnetic field from a

[303] M. R. Beasley and W. W. Webb, *Proc. Symp. Phys. Superconducting Devices, Charlottesville* (1967) (see Ref. 22).

[304] B. T. Ulrich, *Phys. Rev. Lett.* **20**, 381 (1968).

[305] J. P. Gollub, M. R. Beasley, R. S. Newbower, and M. Tinkham, *Phys. Rev. Lett.* **22**, 1288 (1969).

[306] A. F. Hebard and W. M. Fairbank, *Proc. Int. Conf. Low Temp. Phys., 12th, Kyoto* (Eizo Kanda, ed.), p. 855. Academic Press of Japan, Tokyo, 1971.

[307] A. Rosen, G. T. Inouye, A. L. Morse, and D. L. Judge, *J. Appl. Phys.* **42**, 3682 (1971).

5.4. APPLICATIONS

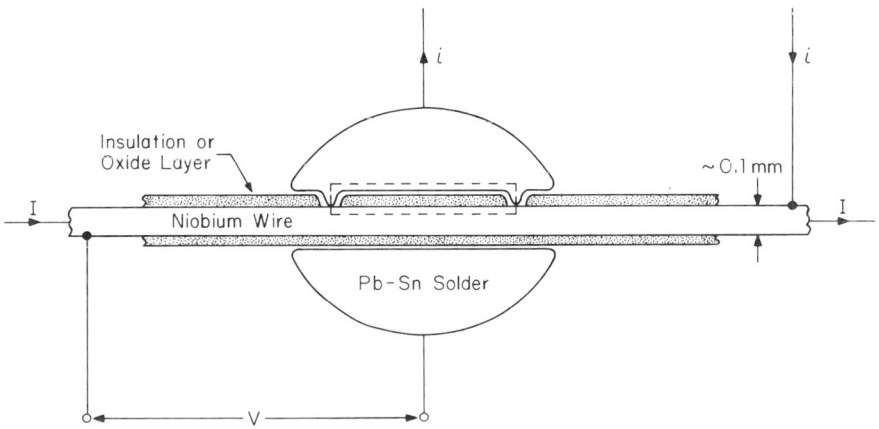

FIG. 60. Cross-sectional view of an idealized Clarke slug showing two discrete weak links and indicating the area (within the dashed line) into which the magnetic field from I penetrates.

current I in the wire penetrates. The critical current I_c is a periodic function of I (Fig. 61a) with period

$$I_0 = 2\pi r \Phi_0 / \mu_0 l(\lambda_1 + \lambda_2 + t), \tag{5.4.22}$$

where r is the radius of the wire, l is the distance between the weak links,

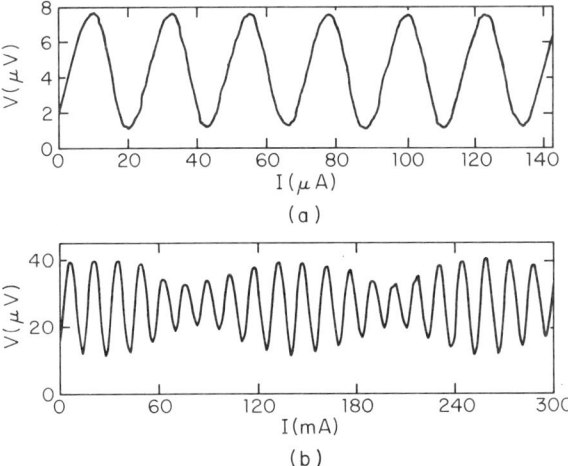

FIG. 61. Periodic variation of the critical current I_c with current I through the wire of a slug (see Fig. 60): (a) for a slug with two weak links with a large area between them; (b) for more than two closely spaced weak links with small areas between them.

t is the thickness of the insulation, and λ_1 and λ_2 are the penetration depths into the wire and into the solder. The current I_0 is that current required to produce one flux quantum in the area A, and there is a clear analogy with the periodic variation of the critical current of an interferometer as discussed in the preceding section.

For slugs made with uninsulated wire there are usually multiple weak-links within the solder drop, giving much more complicated variations of I_c with I (Fig. 61b). The periods correspond to much smaller areas between weak links, since t is negligible, being only the thickness of the oxide on the wire. For most applications the precise form of $I_c(I)$ is unimportant, since the device is normally used as a null detector and operated with feedback as discussed in the previous section.

This device is intrinsically a galvanometer (Fig. 62a) with zero resistance, an inductance of $\sim 10^{-8}$ H, and a current sensitivity of $\sim 10^{-7}$ A.[211] With the addition of a series resistor it becomes a voltmeter. With $R \approx 10^{-8}$ Ω and a current sensitivity of 10^{-7} A, the voltage sensitivity is 10^{-15} V with a 1-sec time constant corresponding to the Johnson noise in the 10^{-8}-Ω resistor at 4 K.[300]

With the Nb wire connected into a continuous superconducting loop (Fig. 62b), the device becomes a magnetometer[308] sensitive to flux

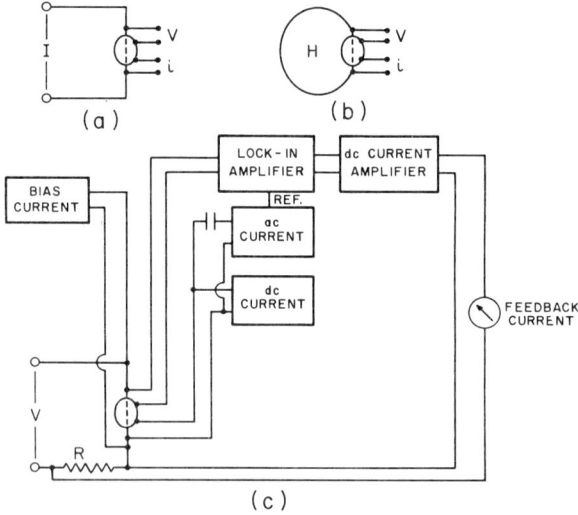

FIG. 62. Devices using Clarke slugs: (a) galvanometer; (b) magnetometer; (c) potentiometric voltmeter.

[308] J. Clarke, *Rev. Phys. Appl.* **5**, 32 (1970).

changes in the loop and can be the basis for all of the devices diagrammed in Fig. 52.

Any of the methods described above (e.g., see Fig. 59) for operating an interferometer can be used with a slug. Alternatively, the slug can be directly coupled as indicated in the diagram for a potentiometric voltmeter, Fig. 62c.[299] A superconducting transformer can be used to match a slug to resistances up to about 1 Ω.[309] Clarke and Paterson[310] have used the positive feedback obtained with asymmetrically biased slugs to obtain current amplification factor of about 100.

Clarke[61] has used slugs to observe ac induced supercurrent steps on SNS junctions and to check the Josephson voltage–frequency relation to high precision.

5.4.6. Detection, Mixing, Harmonic Generation, and Parametric Amplification

5.4.6.1. Response of Weak Links to ac Signals.

The highly nonlinear relation between the supercurrent and the voltage across a weak link, together with the existence of the oscillating supercurrent at finite voltage, makes possible a multitude of extremely sensitive detectors and parametric devices operating from microwave frequencies into the far infrared. Many of these applications have been reviewed by Richards,[311] Shapiro,[210,229] and Clarke et al.[20] This section is intended as a guide to information on these topics.

Equation (5.2.19) expresses the supercurrent that results when both an ac voltage $V_1 \cos \omega t$ and a dc bias V_0 are applied to a weak link. The fact that the maximum zero-voltage current

$$I_0 \propto J_0(2eV_1/\hbar\omega) \tag{5.4.23}$$

is affected by all frequencies is the basis for a broad-band detector. A frequency selective detector is achieved by measuring the maximum dc current on a constant voltage step. For the $n = 1$ step at small ac amplitude, the dc current is a linear function of ac amplitude. This scheme is essentially heterodyne detection with the local oscillator being provided by the Josephson oscillation. Equation (5.2.19) also indicates that a weak link is an effective harmonic generator.

---

[309] J. Clarke, W. E. Tennant, and D. Woody, *J. Appl. Phys.* **42**, 3859 (1971).

[310] J. Clarke and J. L. Paterson, *Appl. Phys. Lett.* **19**, 469 (1971).

[311] P. L. Richards, *in* "Physics of III–V Compounds," Vol. 6. Academic Press, New York (to be published).

If two ac voltages at different frequencies $V_1 \cos \omega_1 t$ and $V_2 \cos \omega_2 t$, as well as a dc bias V_0, are applied to a weak link, the supercurrent is

$$i = i_c \sum_{m,n} (-1)^{m+n} J_m\left(\frac{2eV_1}{\hbar\omega_1}\right) J_n\left(\frac{2eV_2}{\hbar\omega_2}\right)$$
$$\times \sin\left(\frac{2eV_0}{\hbar} t - m\omega_1 t - n\omega_2 t + \theta_{m,n}\right). \quad (5.4.24)$$

This result predicts dc contributions to the supercurrent when $\omega_J - m\omega_1 - n\omega_2 = 0$, where $\omega_J = 2eV_0/\hbar$. Thus there are constant voltage steps not only when $\omega_J = m\omega_1$ and $\omega_J = n\omega_2$, but also when $\omega_J = m\omega_1 \pm k(\omega_1 - \omega_2)$ and $\omega_J = n\omega_2 \pm l(\omega_1 - \omega_2)$, k, l, m, n integers. That is,

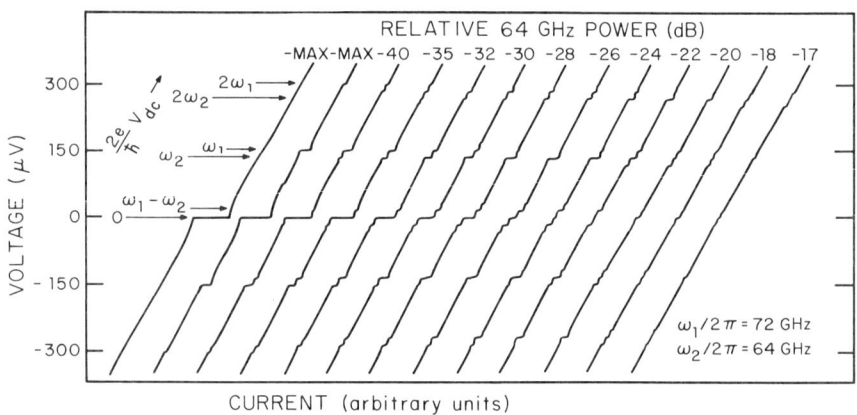

FIG. 63. V–I curves for a point contact showing the effect of simultaneous irradiation with microwaves at two frequencies for Nb–Nb at $T = 4.2$ K. The curve on the left is a V–I curve in the absence of microwave power. The remaining curves are a series of V–I curves with variable power from a source at 64 GHz mixing with fixed power from a source at 72 GHz. The constant–voltage steps are labeled with the Josephson angular frequency corresponding to the voltage at the step. The 72-GHz power produces the steps at ω_1 and $2\omega_1$, while the 64-GHz power produces the steps at ω_2 and $2\omega_2$. Mixing action produces the remaining steps. Note especially the additional steps at the difference frequency $(\omega_1 - \omega_2)$ and at multiples of the difference frequency [after C. C. Grimes and S. Shapiro, *Phys. Rev.* **169**, 397 (1968)].

there are additional steps spaced at multiples of the difference frequency on each side of the usual rf-induced steps resulting from each frequency alone (see Fig. 63). The maximum height for each step occurs for $\theta_{m,n} = \pi/2$ and is obtained by adjusting the bias current. These additional steps are direct evidence for mixing in weak links.

Equations (5.2.19) and (5.4.24) describe only the ideal supercurrent behavior. In an actual weak link it is necessary to consider also the normal and displacement currents. Because of its intrinsicly large capacitance, the typical oxide junction is not very useful for many of the very high-frequency applications which will be discussed below. However, extremely small area junctions can be used successfully. Point contacts have intrinsicly small capacitance, conceivably 0.01-1 pF, and are well suited to high-frequency applications; however, they involve large normal current contributions and inductive effects, which alter the simple picture. These other conduction mechanisms inherent in the weak link itself, as well as the presence of external circuit elements coupled to the weak link, can be quite successfully treated using equivalent circuit models.[210,229,312] Of particular interest for many applications is the effect of coupling to a resonant cavity discussed in Section 5.3.3 (see Fig. 44). In the following paragraphs we summarize some applications of these phenomena.

5.4.6.2. Detection. Detection of radiation by measuring the changes in the maximum zero-voltage current [see Eq. (5.4.23)] has been studied by Grimes et al.[313] They determined the spectral response of several types of point contacts, and for a monochromatic signal in the 4-mm band achieved a noise equivalent power (NEP) of 5×10^{-13} W/Hz$^{1/2}$. This video detection technique has been used by Ulrich[314] in the 1-mm band to observe several astronomical sources.

By measuring the height of the $n = 1$ step of a voltage-biased Nb point contact using a superconducting voltmeter, Taur and Richards[315] have demonstrated linear heterodyne detection at 35 GHz. This form of detection corresponds to observing the dc beat between the Josephson oscillation and the applied radiation. Detection has also been demonstrated with the difference between the Josephson frequency and the applied frequency at some nonzero intermediate frequency, usually that of a resonant cavity coupled to the weak link. Fife and Gygax[316] have detected radiation in the 1-cm and 4-mm bands using a 30-MHz inter-

---

[312] H. Kanter and F. L. Vernon, Jr., *Phys. Lett.* **35A**, 349 (1971).

[313] C. C. Grimes, P. L. Richards, and S. Shapiro, *Phys. Rev. Lett.* **17**, 431 (1966); *J. Appl. Phys.* **39**, 3905 (1968).

[314] B. T. Ulrich, *Proc. Int. Conf. Low Temp. Phys., 12th, Kyoto* (Eizo Kanda, ed.), p. 867, Academic Press of Japan, Tokyo, 1971.

[315] Y. Taur and P. L. Richards, *Bull. Amer. Phys. Soc.* **17**, 45 (1972).

[316] A. A. Fife and S. Gygax, *J. Appl. Phys.* **43**, 2391 (1972).

mediate frequency. Longacre[108] has demonstrated detection for the same bands using a 9.3-GHz i.f.

An extremely sensitive narrow-band detector has been achieved by Richards and Sterling[317] by measuring the change in step height of a cavity-induced step. In analogy with a regenerative receiver they termed this a *regenerative detector*. The effect of the feedback from the cavity is to narrow the bandwidth and enhance the sensitivity. They report an NEP of 5×10^{-15} W/Hz$^{1/2}$ for narrow-band radiation at 200 GHz.

More conventional heterodyne detection with two frequencies incident on the junction is also possible. As indicated above [Eq. (5.4.24)], dc contributions to the supercurrent occur at voltage biases corresponding to all the sum and difference frequencies resulting from mixing in the weak link. These mixing properties have been demonstrated in several experiments[58,64,65] and analyzed extensively by Auracher and Van Duzer.[318] An important case for radiation detection is that when the local oscillator and radiation frequency are only slightly different, in which case the induced step in the I–V curve is modulated at the difference frequency.

In most of the experiments described above, the radiation to be detected was coupled directly to the weak link. Alternatively, electromagnetic radiation can be coupled to a low-inductance ring ($L \approx 10^{-9}$ H) containing a weak link and a very small resistor, typically $\sim 10^{-5}$. Using this configuration Zimmerman[319] has studied heterodyne detection at frequencies from dc to 10 GHz.

5.4.6.3. Mixing and Harmonic Generation.

Weak links have been used for mixing and for harmonic generation from rf to many terahertz. Their efficiency at the highest frequencies appears to exceed that of other techniques.

Grimes and Shapiro[58] made extensive studies of millimeter wave mixing using niobium point contacts. Using signals at 64 and 72 GHz they observed steps spaced by the difference frequency from each of the steps associated directly with the two incident frequencies, and they found that the power dependence of these steps was in general agreement with the product of Bessel functions expected from Eq. (5.4.24). With two nearly identical frequencies they observed steps corresponding to the

---

[317] P. L. Richards and S. A. Sterling, *Appl. Phys. Lett.* **14**, 394 (1969).

[318] F. Auracher and T. Van Duzer, Applied Superconductivity Conference Record, 1972. IEEE Publ. No. 72-CHO682-5-TABSC.

[319] J. E. Zimmerman, *J. Appl. Phys.* **41**, 1589 (1970).

harmonics of the applied frequency and observed directly the intermediate frequency resulting from the difference between the two applied frequencies (Fig. 63). The maximum value of this i.f. occured at maximum slope points between constant voltage steps and was proportional to the differential resistance. This is in contrast to classical detection, where the signal depends on the curvature of the I–V curve. They also observed the beat note between the third harmonic of a signal at 24 GHz and a signal at 72 GHz.

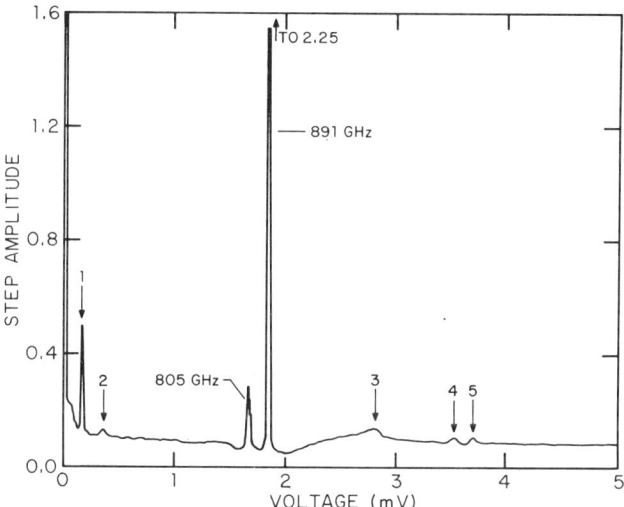

FIG. 64. Amplitude of supercurrent steps on I–V curve of a Nb point contact showing frequency mixing of 805- and 891-GHz laser lines with which it was being irradiated. The peaks are identified as: 1. the diffrence frequency, 2. its second harmonic, 3. Nb energy gap (also present without applied radiation), 4. sum of primary frequencies, 5. second harmonic of 891 GHz [after D. G. McDonald, V. E. Kose, K. M. Evenson, J. S. Wells, and J. C. Cupp, *Appl. Phys. Lett.* **15**, 121 (1969)].

Mixing in weak links has been observed using microwave and laser sources including 805, 891, and 964 GHz from the HCN laser,[64,65,320] 1540 and 1578 GHz from the DCN laser,[65] and 2.5 and 3.8 THz from the water vapor laser[63,65] (see Fig. 64).

An important application of mixing using weakly linked superconductors is to the measurement of the frequency of far-infrared lasers. Mc-

---

[320] D. G. McDonald, A. S. Risley, J. D. Cupp, and K. M. Evenson, *Appl. Phys. Lett.* **18**, 162 (1971).

Donald et al.[63] have used a point contact to mix the 401st harmonic of an X-band oscillator with the 3.8 THz output of the water vapor laser. They have observed the beat note at 9 GHz. Blaney et al.[321] have phase locked X-band klystrons to far-infrared lasers by harmonic mixing in point contacts. By using a point contact tightly coupled to an X-band cavity, Longacre[108] has demonstrated frequency conversion with a single frequency incident on the point contact, the Josephson oscillation serving as the local oscillator. He observed output from the cavity when the difference between the Josephson frequency and a harmonic of the applied frequency was equal to the cavity frequency. In this way he demonstrated direct conversion from 24 to 9.3 GHz. By observing steps on the I–V curve he indirectly demonstrated conversion from 640 to 20 GHz.

5.4.6.4. Parametric Amplification. The current-dependent inductance of a weak link provides the basis for parametric amplification, conceivably at extremely high frequencies. The operation of rf magnetometers and frequency conversion experiments can be interpreted as parametric amplification. Conventional parametric amplifiers using an external pump to vary the inductor have been discussed, as well as self-pumped versions for which the Josephson oscillation of a dc-biased junction serves to vary the inductance at the Josephson frequency.[322-324] Zimmer[325] has demonstrated parametric gain at microwave frequencies using thin film junctions fabricated on the end of a rutile resonator. He used an external pump, and both the pump frequency and the signal frequency were within the band of the resonator. Kanter and Silver[326] have observed parametric amplification at 30 MHz using a voltage-biased point contact to provide a 60-MHz pump frequency.

5.4.7. Digital Devices

So far digital applications have not been widely explored. However, there are several intriguing possibilities. The fact that at a given bias

---

[321] T. G. Blaney, C. C. Bradley, G. J. Edwards, and D. J. E. Knight, *Phys. Lett.* **36A**, 285 (1971).

[322] P. Russer, *Arch. Elek. Übertragung* **23**, 417 (1969).

[323] P. Russer, *Proc. IEEE* **59**, 282 (1971).

[324] A. N. Vystavskiy, V. N. Gubankov, G. F. Leshchenko, K. K. Likharev, and V. V. Migulin, *Radio Eng. Electron. Phys.* **15**, 2121 (1970). See also *Proc. Conf. High Frequency Generation Amplification, Cornell Univ.* (1971).

[325] H. Zimmer, *Appl. Phys. Lett.* **10**, 193 (1967).

[326] H. Kanter and A. H. Silver, *Appl. Phys. Lett.* **19**, 515 (1971).

current a tunnel junction can have two stable states, namely, zero voltage or finite voltage, can be used to produce a logic element for computer use. Matisoo[327] has demonstrated switching times less than 10^{-10} sec. Fast switching times, small size, and newly developed techniques for fabricating stable reproducible junctions may make weak links attractive as computer elements. Anderson et al.[328] have used a number of tunnel junctions in parallel as a quantized flux shuttle and demonstrated its use as a counter or shift register. Fulton[329] has demonstrated similar properties of a very wide tunnel junction.

---

[327] J. Matisoo, *Appl. Phys. Lett.* **9**, 167 (1966); *J. Appl. Phys.* **31**, 2587 (1968).
[328] P. W. Anderson, R. C. Dynes, and T. A. Fulton, *Bull. Amer. Phys. Soc. Ser. II* **16**, 399 (1971).
[329] T. A. Fulton, Applied Superconductivity Conference Record, 1972. IEEE Publ. No. 72-CHO682-5-TABSC.

6. EXPERIMENTAL METHODS IN MÖSSBAUER SPECTROSCOPY*

6.1. Introduction to Mössbauer Spectroscopy

Mössbauer spectroscopy, more formally known as "recoil-free gamma-ray resonance absorption," has become a widely used research tool since its discovery in 1957.[1] As the name implies, it is a spectroscopic technique, in which the absorption of radiation from a monochromatic source induces transitions between the quantum states of an absorbing system. In many ways, Mössbauer spectroscopy yields information similar to that given by the older techniques of EPR and optical resonance, but differs in that the transitions being studied are those between nuclear ground and excited states.

The perturbation of nuclear energy levels by extranuclear fields is sufficiently small to be ignored in many experiments. However, the inherent resolution of Mössbauer spectroscopy is so high that the perturbations of the nuclear energy levels by electronic effects, typically in the range from 10^{-6} to 10^{-10} eV, are readily observed, and, in fact, are the topic of primary interest in Mössbauer spectroscopy research. As in other highly developed spectroscopic techniques, the interesting science lies not in the method itself but rather in what is learned about the physical world from the results of the resonance experiments. The enormous range of Mössbauer effect research, from fundamental studies in relativity through nuclear and solid state physics to studies of biological systems,[2,3] has produced a field that is united only by the techniques used.

---

[1] R. L. Mössbauer, *Z. Phys.* **151**, 124 (1958); *Naturwissenschaften* **45**, 538 (1958) (see also Bibliography).

[2] G. Lang, T. Asakura, T. Yonetani, *J. Phys. C* **2**, 224 (1969).

[3] P. Gilad, S. Shtrikman, P. Hillman, M. Rubinstein, and A. Eviatar, *J. Acoust. Soc. Amer.* **41**, 1232 (1967).

---

* Part 6 is by R. L. Cohen and G. K. Wertheim.

Despite the scope indicated above, the majority of the research is done in the fields of solid state physics and chemistry. Though the details of the experimental technique used vary considerably from one laboratory to another, the basic approach seems at this time to be fairly well standardized. The major goal of this article is to emphasize the standard techniques and approaches that are actually used in perhaps ninety percent of the Mössbauer effect research reported. Toward this end we have emphasized the description of the most widely used systems, and the reasons for their success. More specialized devices and techniques are mentioned briefly. We have also tried to show, when possible, the great extent to which standard equipment can be used in setting up laboratories.

Basic experimental configurations used in Mössbauer spectroscopy are shown in Fig. 1. The radiation consists of gamma rays coming from nuclei that decay from a low-lying excited state to the ground state. This radiation is highly monochromatic, but incoherent. Gamma rays that are emitted without recoil-energy loss[1] from a particular type of nucleus have approximately the right energy to be resonantly reabsorbed by nuclei of the same type. The spectroscopy is normally performed by varying the energy of the gamma rays incident on the absorber and observing the energy and the intensity of the resonant absorption.

The resonant absorption can be observed either directly, by noting the increased attenuation (due to the nuclear absorption) of the gamma rays passing through the absorber, or by observing events following the

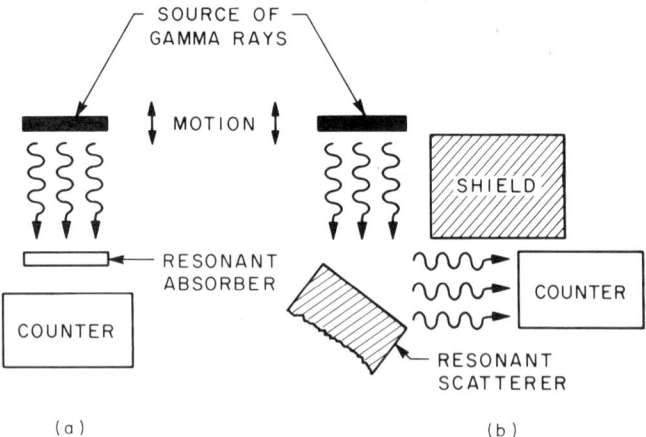

FIG. 1. The two basic configurations normally used in Mössbauer spectroscopy experiments: (a) transmission geometry; (b) scattering geometry.

resonant absorption of the gamma ray, such as the emission of a gamma ray or conversion electron in the subsequent decay of the excited level in the absorber. The first technique, called "transmission geometry," is used in virtually all cases. It would at first appear to be highly advantageous to use the latter technique, "scattering geometry," since this allows, in principle, detection of the resonant absorption events free of background,[4,5] whereas the transmission technique normally allows one to observe only the small increase in absorption caused by the nuclear resonance process. In practice, however, the fact that the counting rate is attenuated by an additional solid angle (the one from the absorber to the detector), the difficulty in providing suitable windows and absorber and detector geometries for the resonantly scattered radiation, and the existence of a strong background from Compton and Raleigh scattered radiation almost invariably make transmission geometry the more effective one. The major research use of scattering geometry at the present time appears to be in the selective excitation of nuclear levels.[6] Scattering geometry has also been used[7] for the study of thick samples, as in metallurgical analysis, where transmission geometry is not realizable.

It is appropriate at this point to mention some of the unusual techniques currently in use. The first of these, and the one most like "normal" Mössbauer spectroscopy, does not use a long-lived radioactive parent but rather a nuclear reaction to populate the Mössbauer state. Coulomb excitation,[8,9] (d, p) reactions,[10] and (n, γ) reactions[11,12] have been used.

[4] J. K. Major, in "Mössbauer Effect Methodology," Vol. 4, p. 89ff. Plenum Press, New York, 1968.

[5] P. Debrunner, in "Mössbauer Effect Methodology," Vol. 4, p. 97ff. Plenum Press, New York, 1968.

[6] N. D. Heiman, J. C. Walker and L. Pfeiffer, *Phys. Rev.* **184**, 281 (1969).

[7] R. L. Collins, "Mössbauer Effect Methodology," Vol. 5, p. 129ff. Plenum Press, New York, 1970.

[8] J. C. Walker and Y. K. Lee, "Hyperfine Structure and Nuclear Radiations" (E. Matthias and D. A. Shirley, eds.), p. 179ff. North-Holland Publ., Amsterdam, 1968; S. A. Wender, L. W. Oberley, and N. Hershkowitz, *Nucl. Instrum. Methods* **89**, 61 (1970); K. A. Hardy, J. A. Hicks, D. C. Russell, and R. M. Wilenzick, *Rev. Sci. Instrum.* **41**, 1652 (1970).

[9] F. E. Obenshain, "Mössbauer Effect Methodology," V. 4, p. 61ff. Plenum Press, New York, 1968.

[10] J. Christiansen, P. Hindennach, U. Morfeld, E. Recknagel, D. Riegel, and G. Weyer, *Nucl. Phys. A* **A99**, 345 (1967).

[11] J. Fink, *Z. Phys.* **207**, 225 (1967).

[12] G. Czjzek and W. G. Berger, *Phys. Rev. B* **1**, 957 (1970).

The primary purpose of such experiments has been to make accessible levels that cannot be reached conveniently from radioactive decay. Such experiments have been used almost exclusively to determine the nuclear parameters of these states. The wider application of these techniques is severely hampered by the fact that the source intensities tend to be 10 to 1000 times lower than those available from radioactive decay, so that extremely long measuring times are required.

A modification of this technique, which was invented to reach substantially different goals,[13] makes use of a nuclear reaction to give a very large recoil momentum to the excited nucleus. This momentum carries it out of the target and implants it into a new host material. The physical separation between the target struck by the beam and the catcher for the radioactive atoms makes it possible to introduce shielding to reduce the intensity of the brehmsstrahlung and other interfering radiations seen by the detector when excitation with charged particles is being performed. The main advantage of this technique, however, is that it makes available for study a number of new physical phenomena. For example, one can hope to determine the site of the implanted ion,[12] and study the implanted ions at essentially zero concentration in the host material. It is also possible, in principle, to investigate the time scale for the slowing down and stopping of the recoiling nucleus.

Another technique makes use of the fact that the nuclear events which precede the emission of the Mössbauer gamma ray generally result in high excitation and ionization of the atom through internal conversion and Auger processes. If the atom is in a metal, it generally regains the lost electrons and returns to its ultimate charge state in time of the order of 10^{-15} sec. If the highly charged ion is in an insulator, however, it is apparent that it may take it a somewhat longer time to reach charge equilibrium, depending on the mobility of electrons in the insulator and the relative electron affinities of the emitting ion and the host. To observe this effect, experiments have been performed[14] in which the radiation feeding the Mössbauer level was detected and used as a timing signal to tell when the excitation of the atom occurred. Then Mössbauer spectra of emitted gamma rays were taken at various time intervals after the preceding transitions. In this way, in principle, the evolution of

---

[13] G. D. Sprouse and G. M. Kalvius, "Mössbauer Effect Methodology," Vol. 4, p. 37ff, Plenum Press, New York, 1968; G. D. Sprouse, G. M. Kalvius, and S. S. Hanna, *Phys. Rev. Lett.* **18**, 1041 (1967).

[14] W. Triftshäuser and P. P. Craig, *Phys. Rev.* **162**, 274 (1967).

6.1. INTRODUCTION TO MÖSSBAUER SPECTROSCOPY

the charged state of the emitting atom after the preceding radioactive decay can be directly observed.

The fourth technique, which has recently come to prominence,[15] is one in which nuclear magnetic resonance is combined with Mössbauer studies of hyperfine levels. In these cases the Mössbauer spectra can be used to observe NMR transitions between hyperfine levels of the nuclear excited state, as well as the effects of various kinds of induced electronic spin flips. When the rf field is applied to magnetic hosts, it is also possible to observe the generation of phonons at the rf frequency, being produced by magnetostriction or domain wall motion.

One application of Mössbauer spectroscopy which has not yet been realized is the study of excited electronic states in solids. For example, the hyperfine structure of an optically excited metastable state could be studied if the ions were continually pumped into that state during the experiment. The difficulty here has been that large amounts of optical power are required to keep enough (approximately 10^{18}) atoms needed to make an absorber in an excited state.

It is also appropriate to mention here that in some favorable cases[16,17] Mössbauer measurements can be made on absorbers that are radioactive, i.e., in which the ground state of the *absorbing* nucleus is not a stable one. Using somewhat specialized techniques[16] the Mössbauer effect has been seen in an absorber with a half life as short as one year.

6.1.1. The Measurables

Mössbauer data generally consist of a series of absorption lines (Figs. 2 and 3) whose area, width, and position contain information in (a) the shift of the centroid, (b) the hfs splitting, (c) the strength of the absorption, and (d) the line width.

(a) The precise value of the energy of a Mössbauer gamma-ray transition depends both on the chemical environment and on the temperature of the emitting or absorbing nucleus. This manifests itself as a shift of the centroid of the Mössbauer effect absorption from zero Doppler velocity whenever source and absorber are chemically distinct

---

[15] N. D. Heiman, L. Pfeiffer, and J. C. Walker, *Phys. Rev. Lett.* **21**, 93 (1968); N. D. Heiman, J. C. Walker, and L. Pfeiffer, *Phys. Rev.* **184**, 281 (1969); *ibid.* **B6**, 74 (1972).

[16] A. J. F. Boyle and G. J. Perlow, *Phys. Rev.* **180**, 625 (1969).

[17] G. M. Kalvius, S. L. Ruby, B. D. Dunlap, G. K. Shenoy, D. Cohen, and M. B. Brodsky, *Phys. Lett.* **29B**, 489 (1969).

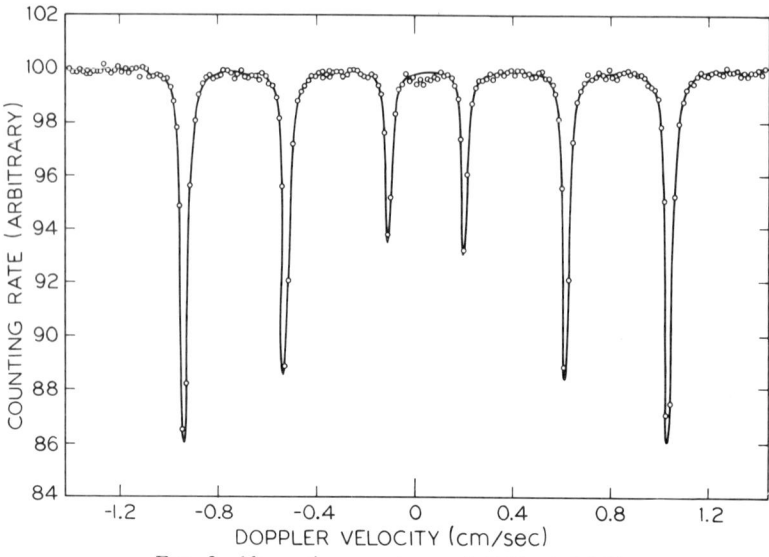

FIG. 2. Absorption spectrum of FeF$_3$ at 4.2 K.

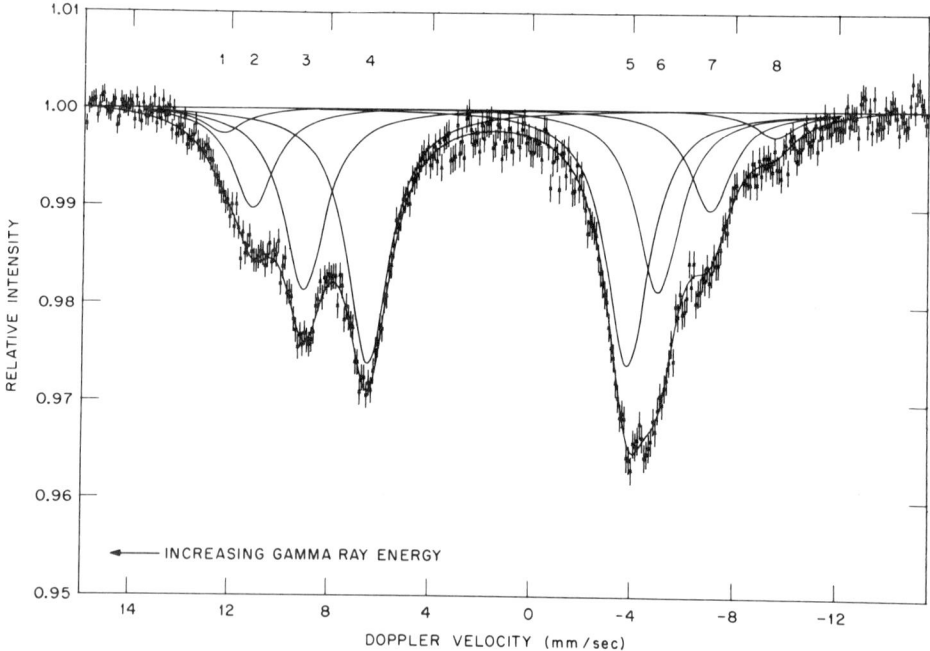

FIG. 3. Mössbauer absorption spectrum of Au^{197} in Au$_2$Mn, showing theoretical least-squares fit to data and individual hf components (Courtesy of D. O. Patterson, L. D. Roberts, J. O. Thomson, and R. P. Levey, Oak Ridge National Laboratory, Oak Ridge, Tennessee).

6.1. INTRODUCTION TO MÖSSBAUER SPECTROSCOPY

or are at different temperatures. Two mechanisms are involved. The first is of electrostatic origin, the second is due to the relativistic, second-order Doppler effect arising from thermal motion.

The first contribution to the centroid shift arises from the electrostatic interaction of nuclear and electronic charges.[18] In general, the effect is due to the s-electron charge density at the nucleus $|\psi(0)|^2$ and to the difference between the mean square nuclear charge radii $\delta\langle r^2\rangle$ of the excited (isomeric) and ground states. Because of this it is generally called *isomer shift* or sometimes *chemical isomer shift* (*IS*). It is given by the equation

$$\delta_{\mathrm{IS}} = (2\pi/3)Ze^2(|\psi_\mathrm{a}(0)|^2 - |\psi_\mathrm{s}(0)|^2)\delta\langle r^2\rangle, \tag{6.1.1}$$

where subscripts a and s stand for absorber and source. The isomer shift provides information not available through other techniques. It can be used to measure $\delta\langle r^2\rangle$, or study chemical changes that affect s-electron charge density. It can be used to identify the chemical state of submicroscopic precipitates or even of isolated impurity atoms.

The second contribution to the centroid shift, δE_T, arises directly from the thermal motion of the atoms.[19] For the most widely used isotopes this shift is smaller than the isomer shift. It is given most generally by

$$\delta E_\mathrm{T}/E = -\langle v^2\rangle/2c^2 = U/2c^2, \tag{6.1.2}$$

where $\langle v^2\rangle$ is the mean square velocity of the emitter atom and U is the internal energy per unit mass. Note that the fractional shift is equal to the ratio of kinetic energy to relativistic rest energy, which is $\sim 10^{-12}$ at room temperature. The thermal shift can provide information on the effective Debye temperature θ_D of impurity atoms in solids. In the Debye approximation this shift is given by

$$\delta E_\mathrm{T} = \frac{3kTE}{2Mc^2}\left[\frac{3}{8}\frac{\theta_\mathrm{D}}{T} + 3\left(\frac{T}{\theta_\mathrm{D}}\right)^3 \int_0^{\theta_\mathrm{D}/T} x^3(e^x - 1)^{-1}\,dx\right], \tag{6.1.3}$$

where M is the mass of the Mössbauer atom.

(b) The ability to resolve the hfs splitting of nuclear levels makes possible some of the most important applications of the Mössbauer

---

[18] See Bibliography. This interaction was first observed in a Mössbauer experiment by O. C. Kistner and A. W. Sunyar, *Phys. Rev. Lett.* **4**, 212 (1960).

[19] R. V. Pound and G. A. Rebka, Jr., *Phys. Rev. Lett.* **4**, 274 (1960), and B. D. Josephson, *ibid.* **4**, 341 (1960).

effect. It competes here with conventional magnetic resonance but provides additional information, and can be used in certain cases where NMR and EPR fail. A major difference is that one always obtains the hfs of both ground and excited nuclear states. The characteristic hfs interactions, both magnetic and quadrupolar, have been useful in identifying the chemical states of atoms produced by nuclear decay.

When the nuclear spin degeneracies of the ground and excited states are lifted by external electric or magnetic fields, the gamma-ray absorption line will be split into components corresponding to transitions between the individual magnetic substates of the nuclear levels. The hfs splitting of ground and excited state levels can be obtained from such spectra.

The nuclear electric quadrupole interaction between the nuclear quadrupole moment Q and an axially symmetric external field gradient (EFG) partially lifts the $2I+1$ fold degeneracy of a nuclear state according to

$$\Delta E_Q = \frac{e^2qQ}{4I(2I-1)} [3m_I^2 - I(I+1)], \quad m_I = I, I-1, \ldots, -I. \quad (6.1.4)$$

The electric field gradient tensor is completely specified (except for its orientation) by two components, which are usually chosen to be the largest (in absolute value) of the $\partial^2 V/\partial^2 x_i$ called eq, and the asymmetry parameter η defined by

$$\eta = (\partial^2 V/\partial x^2 - \partial^2 V/\partial y^2)/eq, \quad (6.1.5)$$

which is zero for axial symmetry. The components have been ordered so that $|\partial^2 V/\partial z^2| > |\partial^2 V/\partial x^2| > |\partial^2 V/\partial y^2|$. As a result, $0 \leq \eta \leq 1$.

The electric field gradient originates from all charges surrounding the nucleus. However, charge distributions with cubic or higher symmetry produce no EFG. Incomplete atomic shells of the atom in which the nucleus resides will also produce an EFG. Both EFG's are further modified by antishielding effects of the closed shells of this atom.

Mössbauer measurements generally give only the product of the EFG and the quadrupole moment, the so-called "quadrupole coupling."

The nuclear magnetic dipole interaction between the nuclear magnetic moment μ and external magnetic fields H completely lifts the spin degeneracy according to

$$\Delta E_H = -\mu H m_I/I, \quad m_I = I, I-1, \ldots, -I. \quad (6.1.6)$$

The magnetic field may be externally applied or may result from the magnetic electrons of the solid itself. In addition to the direct dipolar and orbital fields of unpaired electrons, there are also effective fields arising from the Fermi contact interaction of the nucleus with polarized s-electrons. The latter are especially important in the hyperfine interactions of the transition elements and heavy metals.

(c) The strength of the absorption is measured by the cross section σ, which is a product of the cross section for nuclear resonant absorption σ_0 and the fraction of recoil-free events f. The former is given by

$$\sigma_0 = \frac{\lambda^2}{2\pi} \frac{2I_e + 1}{2I_g + 1} \frac{1}{1+\alpha}, \qquad (6.1.7)$$

where $\lambda = hc/E$ is the wavelength of the gamma radiation, I_e and I_g are the nuclear spins of the excited and ground state respectively, and α is the internal conversion coefficient. (For ^{57}Fe, λ is 0.86 Å.) The recoil-free fraction is given most generally by

$$f = \exp[-4\pi^2 \langle x^2 \rangle / \lambda^2], \qquad (6.1.8)$$

where $\langle x^2 \rangle$ is the mean square thermal vibrational amplitude in the direction of gamma-ray emission. In the Debye approximation it takes the form

$$f = \exp\left[-\frac{E^2}{2Mc^2 k\theta_D}\left(\frac{3}{2} + \frac{\pi^2 T^2}{\theta_D^2}\right)\right], \qquad T \ll \theta_D. \qquad (6.1.9)$$

A graph of f is given in Appendix B. Measurements of the temperature dependence of f are generally used to obtain information on the vibrational properties of the lattice or impurity atoms. A technique utilizing a so-called "black absorber"[20] has proved useful to determine f in sources.

(d) The minimum width of the absorption lines is twice the natural width of the transition, because source and absorber line width are additive, i.e.,

$$\Gamma = \hbar/\tau. \qquad (6.1.10)$$

Experimental line width will generally be greater than this value because of finite-absorption broadening,[21] but this can be corrected for by an extrapolation to zero absorber thickness. Broadening of Mössbauer

---

[20] B. Kolk and B. Harwig, *Nucl. Instrum. Methods* **94**, 211 (1971).
[21] S. Margulies and J. R. Ehrman, *Nucl. Instrum. Methods* **12**, 131 (1961).

lines can arise in paramagnetic ions if the electron spin relaxation time is sufficiently long so that the hyperfine interaction is not averaged out in the Larmor period. There are many other nontrivial effects which can increase the line width, such as unresolved magnetic hyperfine structure, quadrupole splitting, and broadening from isomer shift inhomogenities.

6.2. Mössbauer Spectrometers

The first experiments[1] done by Mössbauer were not, strictly speaking, spectroscopic, but were based on the fact that the recoil-free fraction f vanishes at high temperatures causing the nuclear part of the absorption to disappear. Other experiments have been done using "thermal scanning" to locate Curie temperatures quickly[22] or using the hyperfine interaction arising from an externally applied field to move line positions.[23,24] However, the majority of current Mössbauer spectroscopy is performed with "Doppler effect spectrometers" in which the effective source energy is varied by moving the source or absorber with a known velocity v in the direction of gamma-ray propagation. This changes the effective gamma-ray energy by $\Delta E = vE_0/c$, where E_0 is the gamma-ray energy and c is the velocity of light. Typical values of E_0 are in the range 10–100 keV, so that velocities from 100 cm/sec to $\sim$100 μm/sec are useful values to cover the range of hyperfine interaction energies from 10^{-5} to 10^{-8} eV.

It can be seen from the preceding discussion that the basic elements of a Mössbauer experiment are a radioactive source, an absorber, a radiation detector, a data taking system, and a velocity drive. Below, we emphasize the discussion of velocity drives and the related data collecting systems, which are unique to Mössbauer spectroscopy. Other parts of the systems are widely used in other areas of physics and chemistry, and are discussed primarily in terms of the special restrictions and requirements of the field.

6.2.1. Drives and Data Collection

The required velocities can easily be produced by mechanical or electromagnetic means. Since the velocity applied to the source actually

---

[22] D. G. Howard, B. D. Dunlap, and J. G. Dash, *Phys. Rev. Lett.* **15**, 628 (1965).
[23] C. Sauer, E. Matthias, and R. L. Mössbauer, *Phys. Rev. Lett.* **21**, 961 (1968).
[24] P. P. Craig, D. E. Nagle, and D. R. F. Cochran, *Phys. Rev. Lett.* **4**, 561 (1961).

performs the "tuning" of the spectrometer, it can be seen that the precision with which this velocity can be generated determines the accuracy of the apparatus. The equipment used to produce this velocity is generally called a "drive." Such drives are unique to the field of Mössbauer spectroscopy. They are the furthest outside the experience of most scientists, and are the only component in a Mössbauer spectrometer likely to challenge the experimenter. Unfortunately there are as yet no commercial drives that combine all the virtues of the many arrangements described in the literature.

There are many ways in which drives can be used; the "decision tree" shown in Fig. 4 outlines the multitude of possibilities.

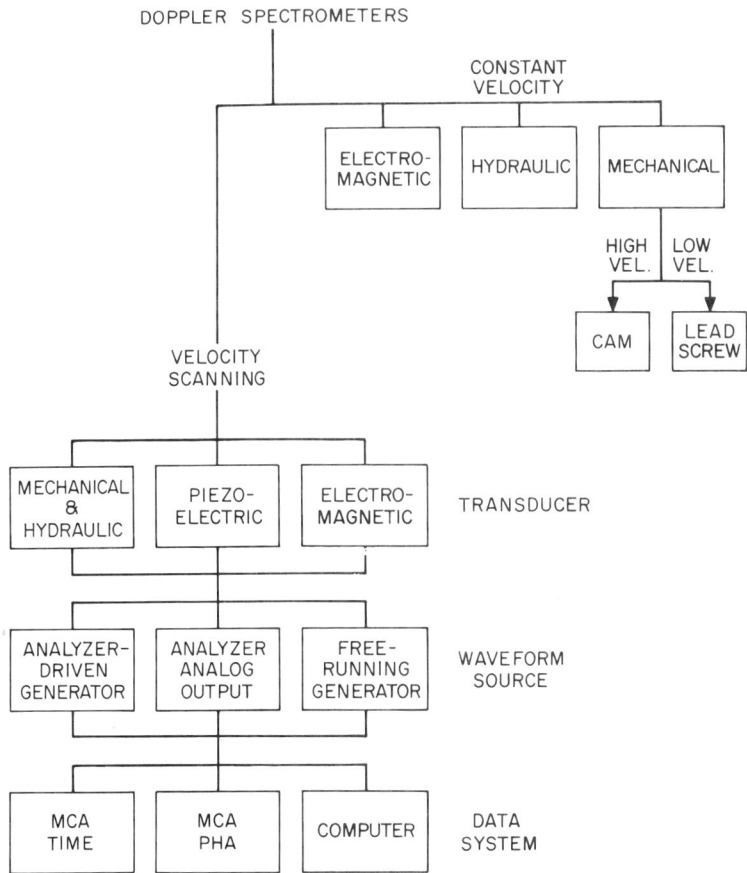

FIG. 4. A "decision tree" showing the frequently used drive and data taking arrangements.

6.2.1.1. Constant-Velocity Drives. Until about 1962, most laboratories used "constant-velocity" systems, in which the source was moved forward with velocity $+v$ and back with $-v$, with the absorption measured separately for the two directions. After a time, a new velocity was chosen, the absorption measured at the new velocity, and point by point, the spectrum was traced out. This system has the advantages that it is relatively easy to obtain an absolute velocity calibration and that the measuring time can be concentrated in those regions of the spectrum that contain the most information.

Most constant-velocity drives reported in the literature have been based on mechanical techniques. One of the simplest embodiments of this approach[25,26] was to make the absorber the rim of a rotating disk or cylinder, tilted with respect to the gamma-ray propagation direction. The absorber then had a component of velocity in the direction of the gamma rays. The most common mechanisms, however, have been lead screws,[27-29] for low-velocity use, and cams,[30] for high-velocity use. Hydraulic mechanisms have also been described for use in special cases.[31,32] Some electromagnetic drives have been developed specifically for constant-velocity operation,[33-35] and it is also possible to use many of the velocity-sweeping drives described below in the constant-velocity mode. Although most mechanical drives are designed for constant-velocity use, constant-acceleration cam drives have been described.[36,37]

---

[25] A. J. Bearden, P. L. Mattern, and P. S. Nobel, *Amer. J. Phys.* **32**, 893 (1964).

[26] P. K. Iyengar and P. S. P. Nathan, *Amer. J. Phys.* **37**, 754 (1969).

[27] K. Cassell and A. H. Jiggins, *J. Sci. Instrum.* **44**, 212 (1967); H. M. Kappler, A. Trautwein, A. Mayer, and H. Vogel, *Nucl. Instrum. Methods* **53**, 157 (1967).

[28] R. H. Nussbaum, F. Gerstenfeld, and J. K. Richardson, *Amer. J. Phys.* **34**, 45 (1966).

[29] R. Booth and C. E. Violet, *Nucl. Instrum. Methods* **25**, 1 (1963).

[30] R. L. Mössbauer, "The Mössbauer Effect," pp. 38–39. Wiley, New York, 1967.

[31] D. E. Nagle and R. D. Taylor, "The Mössbauer Effect" (D. M. J. Compton and A. H. Schoen, eds.), p. 49ff. Wiley, New York, 1965.

[32] J. H. Broadhurst, G. H. Guest, D. A. O'Connor, E. C. Shakespeare, and H. R. Shaylor, *Nucl. Instrum. Methods* **73**, 275 (1969).

[33] P. Flinn, "Mössbauer Effect Methodology," Vol. 1, p. 75ff. Plenum Press, New York, 1965.

[34] P. A. Flinn, *Rev. Sci. Instrum.* **34**, 1422 (1963).

[35] O. C. Kistner, in "Mössbauer Effect Methodology," Vol. 3, p. 217ff. Plenum Press, New York, 1967.

[36] A. J. Bearden, M. G. Hauser, and P. L. Mattern, "Mössbauer Effect Methodology," Vol. 1, p. 67ff. Plenum Press, New York, 1965.

[37] A. H. Schoen, "The Mössbauer Effect," p. 40ff. Wiley, New York, 1967.

All of these constant-velocity systems contain gating and control arrangements to turn off the gamma-ray scalers during the return stroke or to sort out pulses arising during positive and negative velocity scans. An elaborate and extremely sophisticated scheme to return electromagnetic constant velocity drives to the starting position at the end of each stroke with a minimum of disturbance has been described by Chase.[38]

The apparent simplicity, elegance, and intuitive appeal of the constant-velocity spectrometer systems made them very significant in the early day of Mössbauer spectroscopy; but in recent years they have been displaced by the more sophisticated systems to be described below and now are used primarily for a few special experiments,[6,35] for laboratory demonstrations, and for standardization work.

6.2.1.2. Velocity Scanning Spectrometers. The dominance of velocity scanning spectrometers in recent years has been inextricably linked to the development and availability of multichannel analyzers. A few systems involving the use of small on-line general-purpose computers have been described in the literature,[39-41] but these have not yet gained wide acceptance, largely because the characteristics of the commercially available multichannel analyzers and their accessories are very well matched to the requirements of Mössbauer spectroscopy. Multichannel analyzers can be briefly described[42] (see Fig. 5) as small wired-program computers, typically with 100 to 8000 words of storage (1024 words is a common size) usually with a capacity of one million counts per word. Each word of memory is normally called a "channel," and represents a distinct storage location which can be individually accessed, and have its contents incremented, under external control.

These instruments normally include peripherals such as printers and paper tape punches to read out the stored data, cathode-ray tubes for

---

[38] R. L. Chase, *Rev. Sci. Instrum.* **40**, 85 (1969).

[39] R. H. Goodman, "Mössbauer Effect Methodology," Vol. 3, p. 163ff. Plenum Press, New York, 1967; R. H. Goodman and J. E. Richardson, *Rev. Sci. Instrum.* **37**, 283 (1966).

[40] R. H. Voderohe, R. A. Aschenbrenner, and D. H. Jacobsohn, *IEEE Trans. Nucl. Sci.* **NS-17**, 481 (1970).

[41] A. Biran and A. Shoshani, *Nucl. Instrum. Methods* **89**, 21 (1970).

[42] An extremely clear discussion of the problems and techniques of nuclear spectroscopy is given by R. L. Chase, "Nuclear Pulse Spectrometry." McGraw-Hill, New York, 1961. Although the recent developments in the field are not discussed, the emphasis on the fundamental problems, philosophy of counting systems, and analysis of the necessary compromises in such instrumentation make the book of continued value.

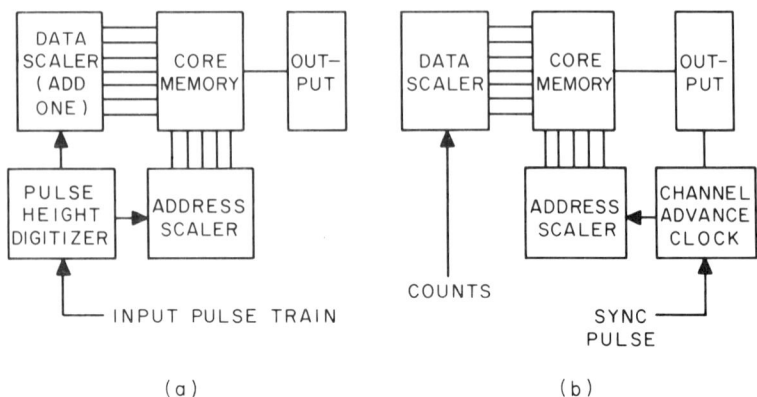

FIG. 5. Functional diagrams of multichannel analyzers (a) in pulse-height analysis mode and (b) in multichannel scaling, or time analyzer mode.

display of memory contents during data taking, and X–Y plotters to graph the stored spectra. Some machines have, in addition, limited capacity for arithmetic manipulation of the memory contents. This operation is not normally used in Mössbauer spectroscopy, unless chemical analysis is being performed.[43] The punched paper tape (or, in some cases, punched card) output is a virtual necessity if the data are to be computer processed.

In data taking, multichannel analyzers normally have two modes available, both shown in Fig. 5. The first method, called "pulse-height analysis," involves the sorting of input pulses by amplitude. Normally, the channel number corresponds linearly to a particular input pulse size. Whenever a pulse of a certain height is sent into the analyzer, the pulse-height digitizer selects the corresponding channel or memory location and increments its contents by one. Thus a histogram of the input pulse size distribution is obtained. In addition to its use in obtaining the Mössbauer spectrum, this pulse-height mode is very useful to check the gamma-ray counting system output for proper detector snd amplifier operation, purity of the observed gamma line, and setting the single channel analyzer.

6.2.1.3. Modulator Mode. The advantage of the pulse-height mode for Mössbauer spectroscopy lies in the fact that it is easy to generate

[43] A. H. Muir, Jr., "Mössbauer Effect Methodology," Vol. 4, p. 75ff. Plenum Press, New York, 1968.

a voltage that is proportional to spectrometer velocity using an electromagnetic generating transducer. This can be either a coil moving in a uniform magnetic field or a small magnet moving within a long coil. The resulting voltage can then be applied to a modulator, which is operated by the output of a single channel analyzer centered on the γ ray of interest. The output of the modulator is then a pulse train (each pulse corresponding to the detection of a gamma ray) with pulse height

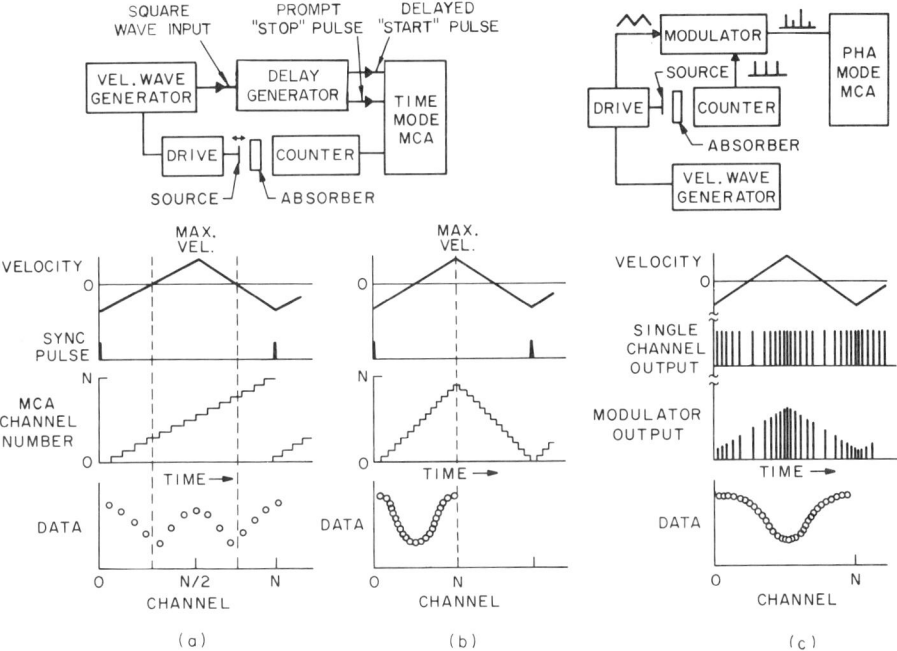

FIG. 6. Mössbauer spectrometer arrangements using multichannel analyzers: (a) time mode with upward multiscaling; (b) time mode with up-down multiscaling; (c) pulse-height mode, with velocity used to modulate single-channel output pulses.

proportional to the instantaneous velocity of the spectrometer when the gamma ray was detected. On routing these pulses to a pulse-height analyzer a histogram is produced, showing how many gamma rays were detected at each spectrometer velocity, i.e., indicating the transmission of the absorber as a function of spectrometer velocity. Since the pulse-height analyzer sorts the input pulses into channel numbers linearly proportional to the input pulse size, the channel number corresponds linearly to the Doppler energy shift. An example of such a system is shown in Fig. 6c.

The main advantage of this type of spectrometer is that since the velocity is detected and measured instantaneously, the velocity produced by the drive does not need to be a very well-established function. In fact, such systems saw widest use in the days before high-quality drives were developed. As competing techniques have improved, this technique has become less popular. One of the most serious difficulties has been that any anomalies in the drive velocity appear as distortions and baseline shifts in the spectrum. For example, if the velocity remains too long near one particular value, a peak will appear at that point in the spectrum. This can be an overwhelming problem when searching for absorptions as small as 0.1% in the experimental spectrum. In an attempt to circumvent this problem, researchers using the modulator technique have normalized their spectra, usually by using another pulse channel sent in through the same modulator. [Most multichannel analyzers have provision for counting pulses (almost simultaneously) on two or more inputs, using a routing pulse to determine in which half or quadrant of the memory each particular pulse shall be stored.] For example, it is possible to store normalizing (clock) pulses in the first half of the memory and the experimental data in the second half, with corresponding channel numbers. At the end of the run the experimenter divides the contents of each channel in the second half by the contents of the corresponding channel in the first half to normalize the data for counting time. In principle this technique can eliminate the variations arising from nonlinearities in the drive and modulator system, at the expense, however, of a substantial loss of accuracy due to the extra statistical variation in the normalizing channel. The experimenter cannot normally circumvent the problem by counting at a much higher counting rate in the normalizing channel than in the data channel, because of the finite deadtime of the pulse-height analyzer system. Typical systems have a deadtime per input count of approximately 10 μsec. Thus if the counting rate in the single channel is 25,000 counts/sec, a not unreasonable value, even with no clock pulses 25% of the input pulses will occur during the deadtime of the analyzer and will not be counted. It can be readily seen that it is effectively impossible to count clock pulses at a substantially higher rate than the gamma ray counting rate in such circumstances.

Another problem arising with the modulator technique is that the channel width variation from channel to channel in the pulse-height analyzer mode is on the order of $\pm 1\%$ in most commercially available instruments. This deviation is far larger than the statistical uncertainty obtained in good data, so that a substantial loss in accuracy can occur.

Standard pulse-height analyzer systems also lack the calibration stability (better than 0.1 channel) that can be routinely obtained with the other systems described below.

Most of these problems could probably be solved by diligent and careful development of the modulator system. A technique combining elements from both the modulator and time mode systems has been described.[44] The decreased use of the modulator system in recent years probably comes not so much from its fundamental problems as from the fact that it is now possible to construct very high-quality drives. Thus the instantaneous velocity monitoring, which is the main advantage of this system, is no longer necessary.

6.2.1.4. Time Mode. The other technique for performing Mössbauer spectroscopy with multichannel analyzers uses the "time mode," also called "multichannel scaling." In this approach a very precise quartz-controlled timer, usually built into the analyzer, opens the channels consecutively, one by one, and allows input counts to accumulate in a channel for a preset time (e.g., 100 μsec). At the end of the preset time that channel is closed, the next channel is opened, and so on until the entire memory has been scanned. When the analyzer reaches the last channel, it can either automatically recycle through the first channel, or halt, awaiting a further command. In this mode of operation the channel number is a linear function of time, and if the velocity produced by the drive is a known function of time this can be directly related to the channel number. An example of this technique is shown in Fig. 6a.

There are three important advantages in the use of the analyzer in this mode. First, since the channel advance is controlled digitally by counting an extremely precise clock signal, the channel number is a linear function of time to one part in 10^4 or 10^5. For the same reason the open time for each channel is very nearly the same as for any other, and channel width inequalities as small as 0.01% are readily obtainable. The third advantage is that essentially all commercial instruments count the input pulses not directly into the core memory but into an intermediate scaler known as the "data register." Thus the input counting rate is not limited by the speed of the magnetic core storage but only by the speed of the data register. This makes the limiting pulse input rate something like 10^7 counts/sec, faster than can be counted at present with most gamma-ray counting systems.

---

[44] R. S. Preston and W. P. McDowell, *Nucl. Instrum. Methods* **81**, 285 (1970).

There are two possible approaches to synchronizing the channel number with the velocity wave form. In one, the drive produces a trigger pulse at some particular point in the velocity wave form, and this is used to start the multichannel analyzer sweep. Alternatively, the analyzer can run freely through the channels repetitively and an analyzer output tied to the channel number can be used to generate or synchronize the drive wave form. These possibilities are shown in Fig. 4 and discussed in greater detail in the section on wave form generators.

The major drawback in taking data in the simple multiscaling mode is that the spectrum is normally accumulated twice, once in the forward and once in the backward direction, with the latter half being a mirror image of the first half (see Fig. 6). This is normally necessary because the analyzer jumps, essentially instantaneously, from the last channel to the first. If the first channel corresponds to the maximum negative velocity, it is impractical for the last channel to represent the maximum positive velocity, since the drive, having mechanical components, cannot follow this sudden velocity change. Consequently the analyzer and drive are usually so arranged that when the analyzer has completed half its sweep, the maximum positive velocity is reached, and in the second half of the analyzer's scan the velocity decreases to the maximum negative value. Thus when the analyzer returns to the first channel, the drive is already at maximum negative velocity, ready to start the next scan. In each scan each velocity is reached twice and, as a result, the spectrum is accumulated twice. In the experience of the authors this has occasionally been advantageous, in that lack of mirror symmetry of the two spectra is a sensitive indicator for most of the drive aberrations normally encountered. Thus it is useful to operate in this mode when initially setting up the apparatus.

There are two disadvantages in operating in this mode: First, computer reduction of the data is required to combine the two halves of the analyzer since the number of points normally taken is so large that hand data reduction is prohibitive. (This operation is called "folding.") Second, data taking requires twice as many channels in the analyzer as will appear in the final spectrum. This can be a disadvantage when very high resolution, requiring many velocity channels, is desired, or if a number of spectra are stored in unused sections of the analyzer while new data is being taken. Two solutions to this problem have been used. The first technique, known as "fast flyback," uses a special wave form generator to provide a high acceleration to take the drive to maximum negative velocity after the maximum positive velocity has been reached.[45-47] Normally, the multichannel analyzer channel advance is inhibited for the duration of the flyback. Then, after the normal stroke starts, the analyzer address advance system is turned on. This technique is limited to low velocities because it is desired to make the flyback time small compared to the measuring time and the acceleration is therefore large. Additionally, since zero velocity inevitably corresponds to an extremum of the displacement, if the source is moved, it is either closest to or furthest away from the counter or collimator near zero velocity, and this results in a distortion of the baseline due to geometric solid angle effects. This effect can be allowed for in computer data reduction, but this imposes an additional inconvenience and uncertainty, especially at high velocities where the geometric effect can be quite large.

The second technique of circumventing the flyback problem is to use the analyzer in what is called the "up–down multiscaling mode" (see Fig. 6b). In this approach, after the maximum channel number is reached, the address scaler is reversed and counts down to lower channel number with each clock pulse until channel zero is reached. This is a capability available in most of the newer multichannel analyzers. The feature has been considered useful enough that there are descriptions in the literature[48,49] of modifications to older machines to provide this capability. Ideally, the velocity may be represented as a symmetric triangular wave form (see Fig. 6b), with its maximum occurring when the highest numbered channel is opened. Then, each channel corresponds to a unique velocity value, even though it is opened twice during each operating cycle. Deviations of the drive from perfect symmetry will be reflected in loss in resolution in the resulting spectrum, imposing severe demands on the drive linearity and stability.

A technique successfully used by the authors, with a free-running triangle generator, involves delaying the triggered start of the multichannel analyzer up–down sweep with a stable variable delay circuit. Comparison of the velocity wave form with the analog address output

[45] H. Brafman, M. Greenshpan and R. H. Herber, *Nucl. Instrum. Methods* **42**, 245 (1966).
[46] Y. Hazony, *Rev. Sci. Instrum.* **38**, 1760 (1967).
[47] T. E. Cranshaw, *Nucl. Inst. Methods* **30**, 101 (1964).
[48] Y. Reggev, S. Bukshpan, M. Pasternak, and D. Segal, *Nucl. Instrum. Methods* **52**, 193 (1967), A Constant Acceleration Mössbauer Spectrometer Utilizing a Multichannel Analyzer Modified for Forward-Backward Address Scaling.
[49] E. Nadav and E. Palmai, *Nucl. Instrum. Methods* **56**, 165 (1967).

of the analyzer (described below) allows adjustment of the variable delay to obtain an exact retrace of the velocity with channel number. This system cannot be used directly in cases where the drive wave form is itself taken from the analyzer analog address output. However, it should be possible to make small corrections (e.g., by adding a small square wave component) to the triangle wave to make the retrace of the true velocity wave form match the channel number of the analyzer in the second half of the multiscaling cycle. Another system which has been described[49] uses separate delays to synchronize the MCA scan in the "up" and "down" directions.

The combination of parabolic motion and up–down multiscaling, as described above, results in a very convenient Mössbauer spectrometer. The spectra produced can be plotted out directly from the multichannel analyzer, or displayed on its cathode-ray screen with a horizontal axis (channel number) that is linearly proportional to energy. The spectrum is inherently free of spurious peaks and valleys arising from aberrations in the electronic system and the data are normally free of first-order effects of geometry changes and slow counting rate drifts.

6.2.1.5. Feedback Velocity Control. One of the great advantages of electromagnetic drives is that they can be made to follow an applied wave form very accurately through the use of inverse or negative feedback.[50] A generalized representation of such a system is shown in Fig. 7. The important basic features of this circuit are standard to servo-mechanisms and appear in slightly different forms in almost all of the electromagnetic drive circuits that have been described in the literature. The first element is a difference circuit that compares the transducer output (presumed to be proportional to spectrometer velocity) with the instantaneous value of a reference input wave representing the desired spectrometer velocity. The output of this difference circuit is called an "error signal" and represents the difference between the actual and the desired instantaneous velocity of the spectrometer. This error signal is amplified, passed through phase compensating networks,[51] and used to supply current to the driving coil to correct the velocity when it deviates from

---

[50] E. Kankeleit, "Mössbauer Effect Methodology," Vol. 1, p. 47ff. Plenum Press, New York, 1965; *Rev. Sci. Instrum.* **35**, 194 (1964).

[51] A brief discussion of the stabilization of feedback systems by means of phase compensating networks is given by D. MacRae, Jr., in "Vacuum Tube Amplifiers" (G. E. Valley and H. Wallman, eds.), Vol. 18, M.I.T. Radiat. Lab. Ser., pp. 333–347. McGraw-Hill, New York, 1948.

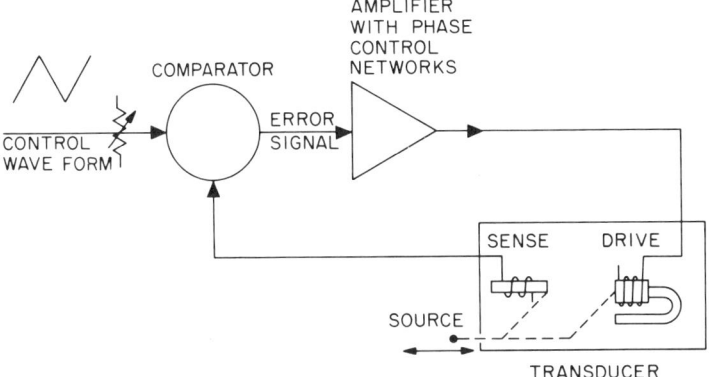

Fig. 7. Essential elements of the feedback velocity control system used in most electromagnet drives.

the desired value. The input wave form, which controls the time dependence of the velocity of the spectrometer, is typically either a triangle, giving constant acceleration, a square wave, giving constant velocity, or a sine wave, giving sinusoidal motion. Many of the electromagnetic systems described in the literature can be used with any of these input wave forms without major changes.

The basic advantages of this system of feedback velocity control are that the velocity can be made to follow an essentially arbitrary wave form and that the accuracy with which the wave form can be followed is very high. (It is possible to obtain conformity within one part in 1000 to an applied wave form under favorable conditions.) The difference between the velocity actually produced and the desired wave form can be easily monitored by observing the error signal. The entire system is relatively immune to perturbations, such as changes in the source mass, vibrations, or room noise, because of the negative feedback. It is also extremely convenient to be able to change the spectrometer scale factor by simply using a potentiometer to adjust the input wave form amplitude.

Although it is, in principle, possible, using the feedback system shown in Fig. 7, to control either the position, velocity, or acceleration of the drive, direct feedback control of the velocity has been by far the most popular choice. A system which has been described as being acceleration controlled[52] differentiates the velocity signal coming out of the trans-

---

[52] J. Pahor, D. Kelsin, A. Kodre, D. Hanzel, and A. Molik, *Nucl. Instrum. Methods* **46**, 289 (1967).

ducer. Arrangements that determine velocity by counting interferometer fringes[53,54] or Moiré[55] fringes from a fine grating are somewhere between the position-controlled and velocity-controlled modes.

The system as described above is extremely simple in concept and is relatively simple in practice. The primary difficulty arises from the fact that the transducer, being an electromechanical device, is very far from an ideal circuit element (see below). To allow for the imperfect characteristics of the transducer, the feedback amplifier is normally built with compensating networks[50,51] to reduce the gain of the feedback system at the resonant frequency of the transducer.

One implementation[56,57] of such a system is shown in Fig. 8. This circuit contains two such compensating networks. One, at the first am-

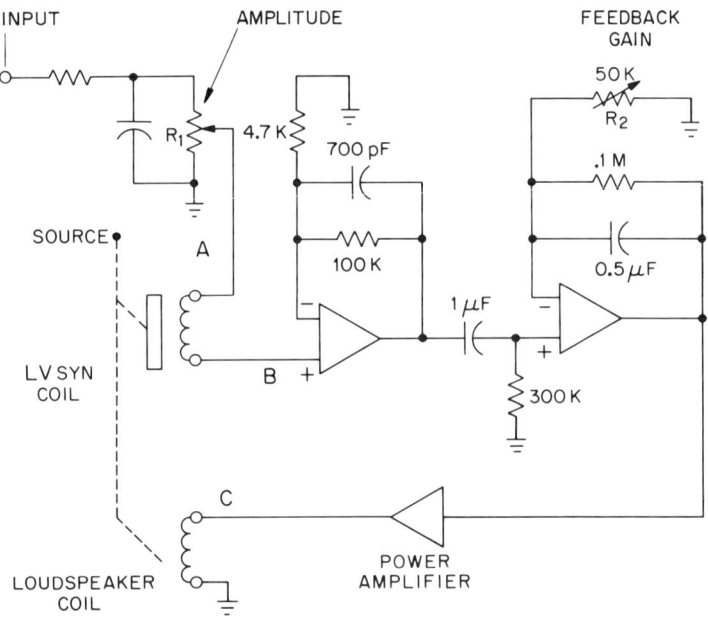

FIG. 8. Circuit diagram of an error signal amplifier.

[53] S. Eylon and D. Treves, *Rev. Sci. Instrum.* **42**, 504 (1971).

[54] F. C. Ruegg and J. J. Spijkerman, Nat. Bur. Stds. Tech. Note 501, p. 89ff. U.S. Govt. Printing Office, Washington, D.C. (Feb. 1970).

[55] J. H. Broadhurst, G. H. Guest, D. A. O'Connor, E. C. Shakespeare, and H. R. Shaylor, *Nucl. Instrum. Methods* **73**, 275 (1969).

[56] R. L. Cohen, P. G. McMullin, and G. K. Wertheim, *Rev. Sci. Instrum.* **34**, 671 (1963).

[57] R. L. Cohen, *Rev. Sci. Instrum.* **37**, 957 (1966).

plifier, decreases the gain by 6 dB/octave above 1.5 kHz. The other makes the second amplifier into an integrator, providing a phase-lag of 90° and an attenuation of 6 dB/octave over the entire frequency domain normally important for velocity control. A brief description of the use of such phase shifting networks for stabilizing feedback systems against oscillation is given in Ref. 51. Typical wave forms generated at various parts of the circuit shown in Fig. 8 are shown in Fig. 9.

This type of velocity feedback control system is very convenient for the experimenter. Once he has set the system up and adjusted the phase compensation networks to provide optimum performance with the transducer being used, the system is essentially free of adjustments and trimming. To a very good approximation, all that is necessary to change the velocity scale is a change in the size of the input voltage wave form. As long as the feedback gain is fairly high, the velocity produced by the spectrometer is determined only by the input wave form and the linearity and calibration constant of the sensing coil. The feedback control almost completely compensates for possible drifts with change in spring constant and amplifier gain, and for moderate changes in source mass. Additionally, the linear nature of the circuit means that the velocity produced will be accurately proportional to the input voltage, so that if a highly linear potentiometer is used to set the amplitude of the wave form, it is unnecessary to calibrate the drive at every scale setting used. As the results of Fig. 9 indicate, existing drives are adequate for the requirements of almost all current research with the Mössbauer effect.

In the system shown in Fig. 7, the entire signal producing the motion of the armature is derived from the error signal. Thus the error signal not only corrects for deviations from linearity of the transducer and amplifier system, but is the source of the entire drive signal. This situation can be improved[52,58] by arranging a series of operational amplifier integrating and shaping networks to produce a current wave form like that shown in Fig. 9c. This current wave form would make the transducer produce a velocity very close to the desired one. Then the feedback would only have to correct for the small deviations in driving motor linearity, variations in spring constant with temperature, and source mass changes. This technique has not been widely used, since existing drives are good enough so that the extra linearity and perfection of this more complex system are unnecessary. For most experimenters, the additional complication, loss of convenience, and requirements for many

[58] R. Zane, *Nucl. Instrum. Methods* **43**, 333 (1966).

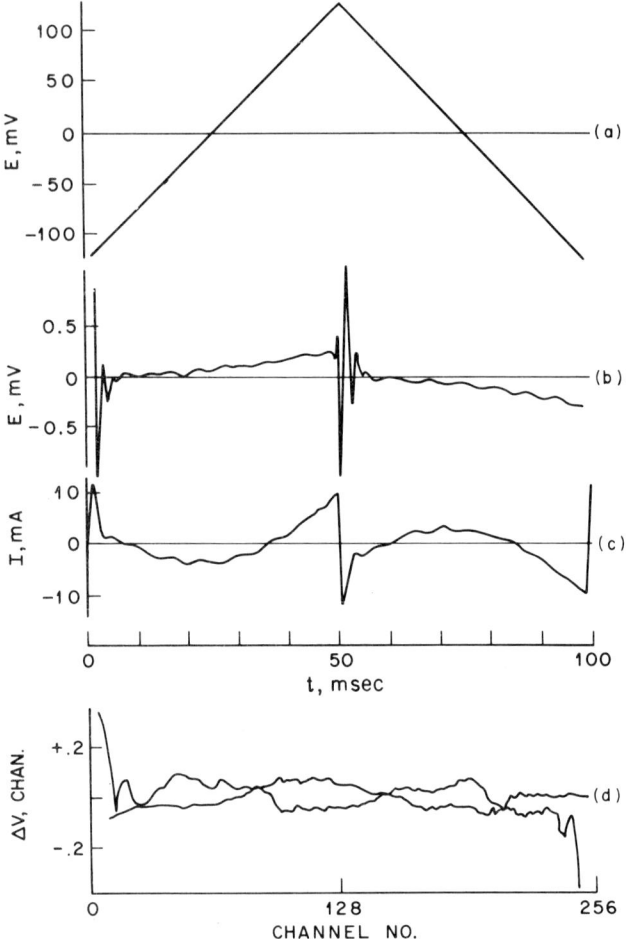

FIG. 9. Wave forms appearing at various labeled points of circuit in Fig. 8 under typical operating conditions (constant acceleration, about 1 cm/sec peak velocity, 10 Hz driving frequency). (a) the input triangle; (b) the error signal; and (c) the current wave form driving the transducer. In up–down multiscaling (Fig. 6b) a good index of the true accuracy of the drive can be obtained by subtracting a fraction of the analyzer analog address output from the signal generated by the velocity transducer. This "deviation signal," shown in (d), is the difference between the instantaneous velocity and the desired velocity, which is proportional to channel number. This type of display facilitates precise adjustment of the timing to make the "up" and "down" sweep directions have the same velocity in a given channel. The mismatch between the two lines shown in (d) (for the "up" and "down" scans) shows the imperfections of the "folding" on a highly magnified scale.

adjustments appear to outweigh the possible gains at this time. In practice, errors due to transducer nonlinearity and pickup of stray 60-cycle magnetic field (from motors and transformers) can create errors in velocity as large as those inherent in the drive.

It is instructive to see what the ultimate limit in velocity accuracy for this type of feedback control system is. If we assume that the velocity transducer has an output of 0.5 V/cm/sec, a typical value, and if the error signal amplifier has an equivalent input noise of 1 µV, then the velocity error signal due to this noise corresponds to approximately 5×10^{-4} mm/sec. This value is negligible in comparison with the observed line widths of Mössbauer resonances. To date these have been limited by the lifetime broadening due to the uncertainty principle, or by inhomogeneous isomer shift or hfs broadening in the source and absorber. Most drive systems in use today noticeably exceed this velocity noise limit of 5×10^{-4} mm/sec.

It should be emphasized that though the attainment of an accurate drive is a necessary step in performing precise Mössbauer spectroscopy, there are other possible instrumental contributions to loss of resolution. If the source is not rigidly connected to the drive, it may not follow the drive wave form accurately. Vibrations and room noise may shake the absorber, resulting in line broadening due to the superimposed noise velocity. If the gamma-ray propagation is not parallel to the drive velocity, the Doppler shift produced will be reduced by the cosine of the angle between them. The existence of this effect, called "cosine broadening," requires that only gamma rays emitted from a cone of limited size[59] (e.g., half-angle 5° or less) be seen by the detector to avoid loss of accuracy in hfs measurements.

6.2.1.6. Electromagnetic Transducers.
Essentially all the electromagnetic transducers which have been discussed in the literature appear to stem from two rather early designs shown in Fig. 10. The basic characteristic of both these designs is the separation of function between the motor, which uses a coil moving in the field of a permanent magnet, and the velocity sensor, which in one case is another moving coil loudspeaker and in the other case is a commercial moving-magnet stationary-coil velocity transducer. It can be seen that there is negligible difference in principle between these two approaches. In practice, the difference is that the commercial transducer has a relatively long stroke over which

---

[59] C. Nistor and T. Tinu, *Rev. Roumaine Phys.* **11**, 551 (1966); R. Riesenman, J. Steger, and E. Kostiner, *Nucl. Instrum. Methods* **72**, 109 (1969).

332   6. MÖSSBAUER SPECTROSCOPY

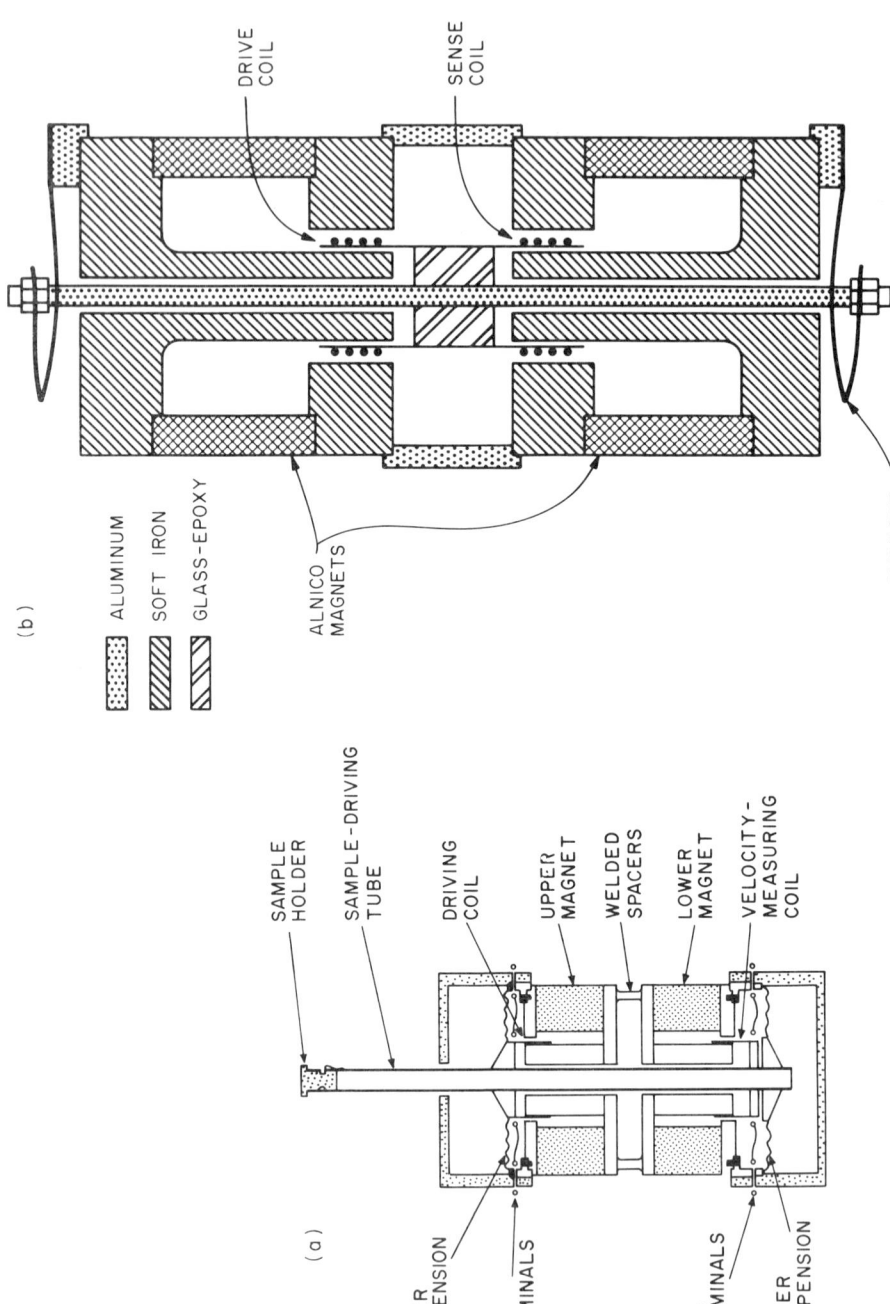

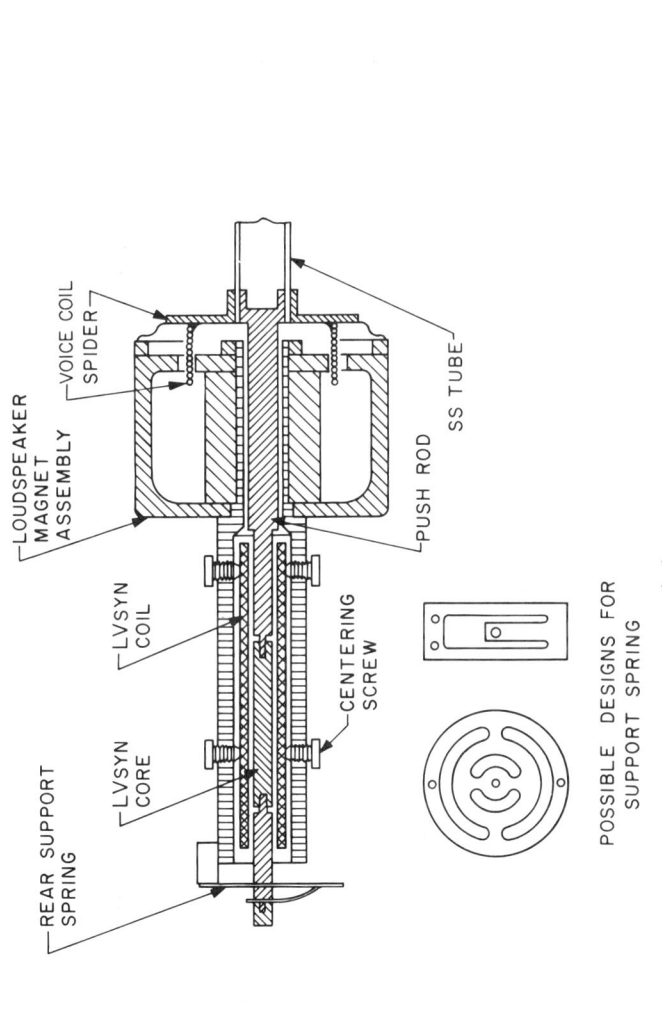

FIG. 10. Examples of electromagnetic transducer design: (a) and (b) double loudspeaker [after F. J. Lynch and J. B. Baumgardner, "The Mössbauer Effect," p. 54ff. Wiley, New York, 1962; E. Kankeleit, "Mössbauer Effect Methodology," Vol. 1, p. 47ff. Plenum Press, New York, 1965; *Rev. Sci. Instrum.* **35**, 194 (1964)]; (c) loudspeaker motor and LVsyn transducer [after R. L. Cohen, *Rev. Sci. Instrum.* **37**, 957 (1966).]

the output voltage is a linear function of velocity. This large stroke is especially useful if low-frequency drive signals or large velocities are desired. However, both types have been used successfully in all velocity ranges, and the advantages of the precision and linearity of the commercial transducer have faded somewhat with the commercial availability of double loudspeaker transducers from a number of suppliers.

To obtain linearity, either the sensing coil must be much longer than the gap or the magnetic gap must be much longer than the sensing coil. In either case, as soon as the displacement becomes large enough that the end of the coil nears the end of the magnetic gap, the ratio between output voltage and velocity will begin to decrease. A clear and detailed discussion of these problems is presented by Kankeleit.[50] The major drawback of the commercially available transducers is that the range of displacement for which the sensing coil output voltage is proportional to the velocity is usually limited to approximately ± 2 mm.

The characteristics of both types of transducers as seen by the feedback control circuit are very similar. There are two important mechanical resonances visible in the amplitude and phase transfer function shown schematically in Fig. 11. The first, typically in the range 10–60 Hz, comes from the mass of the entire moving system (including the load) with the restoring force coming from the support springs shown in Fig. 10. This mode is relatively unimportant in parabolic motion where the properties of the transducer itself are effectively suppressed by the feedback, but is important in most sinusoidal motion drives that are run at or near the frequency of the fundamental mechanical resonance. The next resonance, which typically occurs around 1 kHz with transducers currently in use,[50,60,61] corresponds to the mode in which the voice coils of the driver and sensor move symmetrically toward and away from one another. In this case the "mass" is the reduced mass of the two voice coils and pushrod, and the spring constant is provided by the compressibility of the pushrod connecting the two coils. In this mode the driving and sensing coils move 180° out of phase to each other at the resonant frequency. The feedback, which is negative at low frequencies where the motions of the two coils are in phase, becomes positive feedback at this pushrod compression resonance. This results in oscillation near this resonant frequency if sufficient gain exists in the feedback amplifier system. Thus this resonance severely limits the amount

---

[60] J. Carmeliet and S. Lejeune, *Nucl. Inst. Methods* **62**, 166 (1968).
[61] M. Michalski, J. Piekoszewski, and A. Sawicki, *Nucl. Instrum. Methods* **48**, 349 (1967).

6.2. MÖSSBAUER SPECTROMETERS

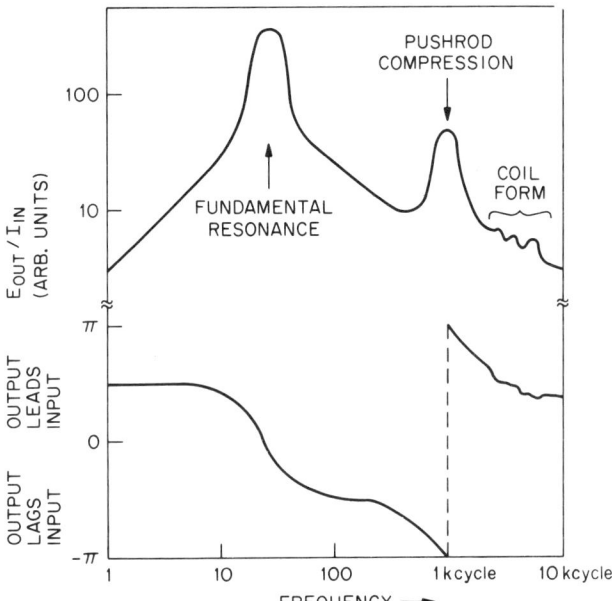

FIG. 11. Sketch of the amplitude and phase transfer function of a typical electromagnetic transducer of the types shown in Fig. 10. Both the resonant frequencies and mechanical Q values vary considerably depending on the details of construction.

of feedback that can be used at high frequencies, and this limits the ability of the drive to attenuate high-frequency room noise and to follow wave forms with high-frequency components. One can envision that a substantial improvement in drive performance might be obtained by making coils as light as possible using aluminum wire and mounting them on a beryllium pushrod and coil form (slotted to reduce eddy current losses) to raise this resonant frequency as high as possible. However, the available drives made with more mundane materials appear adequate for present needs.

It is instructive to examine the compromises involved in the design of these transducer systems and the limitations involved. We have already discussed the difficulty of providing linearity over an adequate displacement range. To make the pushrod compression frequency as high as possible, it at first appears desirable to make the pushrod as short as possible and to place the drive and sensing coils very close together. However, if the two coils are put too close together, the mutual inductance coupling of the driving coil and the sensing coil generates a voltage in the sensing coil when the current in the driving coil changes. This has two deleterious effects. First, it makes the voltage across the sensing coil no longer correspond exactly to the velocity of the moving system; and second, because of its complex phase relationship to the driving current, it can introduce some form of positive feedback into the input of the negative feedback amplifier, producing oscillations or puzzling and highly aberrational drive behavior. The only real cure for this problem is to increase the separation between the coils or to add some magnetic shielding.

The extreme solution to this problem is shown in Fig. 10a, where the loudspeaker magnets have been turned so that their gaps are at the ends of the transducer and the coils are then decoupled by the entire length of the transducer plus the shielding factor due to the two magnets. This arrangement is more subject to problems of interference from stray magnetic fields affecting the magnets and possibly inducing interfering voltages directly in the sensing coil.

It is very desirable to have high magnetic fields (2–8 kOe) in the magnet gaps in these loudspeaker-type transducers. A large field at the motor decreases the size of the drive-coil current required. Indirectly, this reduces the size of the emf introduced in the sensing coil by mutual inductance. A large field at the sensing coil also reduces the relative effects of mutual inductance and stray magnetic fields that may exist in the room. The large sense coil voltage output given by a high magnetic field also makes the velocity signal less critical to handle and less subject to interfering electrical signals.

Within the limitations suggested above and described in the literature mentioned, the design, construction and use of these electromagnetic transducers is relatively straightforward. It is appropriate to say a few words about the choice of support springs. In drives that are working at their resonant frequencies, the requirements on the springs are relatively unexacting and, as long as they are not bowed so that they suddenly snap over into another mode, the operation will probably be satisfactory. For drives being used nonresonantly in parabolic motion, however, the springs chosen are generally the lightest that will provide adequate transverse centering and mechanical support for the moving element. Many spring patterns that will provide large lateral restoring force while producing only small longitudinal restoring force are reported in the literature.[50,57,62] The natural vibration frequencies of the springs themselves,

---

[62] G. Kaindl, M. R. Maier, H. Schaller, and F. Wagner, *Nucl. Instrum. Methods* **66**, 277 (1968).

however, may slightly perturb the drive motion. If this occurs, it can generally be observed as an extra component in the error signal typically in the range 30–90 Hz, well away from any of the fundamental drive resonant frequencies. This can usually be eliminated by placing some mechanical damping material, such as a thin layer of solder or a coating of some rubbery, lossy, polyurethane, or similar resin, on the spring.

The almost universal use of these spring suspension systems is an indication of their effectiveness in comparison with other possible means of support. Systems using sliding or rolling bearing surfaces have not been satisfactory for velocity scanning drives, as the "stick-slip" effect at zero velocity noticeably perturbs the velocity wave form. Slides have been used successfully in constant-velocity drives (where the motion is constant during the measuring cycle); at very low velocities, the "stickiness" becomes a problem even there. Frictionless air-pad bearings have been suggested and used[55]; however, the simpler spring systems have been successful enough to reduce the need for the more complex supports.

It is, in principle, possible to integrate the magnetic circuits for the drive and sense coils, and such designs have been reported.[63] This technique has not been widely used, apparently because the saving (of one magnet) does not make up for the more complex mechanical design required, and because of the increased mutual inductance between the two coils. Some early drive designs used loudspeakers that had two voice coils wound on top of one another, in the same magnetic gap. Though the system was adequate for the drive requirements of the early years, it is not competitive with modern transducers because of the large mutual inductance between the coils. In addition to the problems of stabilization of the feedback system that this may engender, it makes it impossible to ascertain with confidence the size and shape of the true velocity signal.

6.2.1.7. Wave Form Generators. We have so far not discussed the types of velocity wave forms that are used in Mössbauer spectroscopy. Almost all velocity scanning spectrometers reported on to date use either displacement sinusoidal in time, resulting in a velocity sinusoidal in time, or a displacement parabolic in time, which results in a velocity variation that is linear with time. This latter technique is also called "constant acceleration." Each of these modes has its advantages and dis-

---

[63] S. I. Zvarich, L. M. Isakov, and A. I. Shamov, *Ind. Lab. Res.* **34**, 609 (1968); P. J. Ouseph and A. Waheed, *Amer. J. Phys.* **34**, 990 (1966); V. A. Povitskii and A. P. Dodkin, *Ind. Lab. Res.* **34**, 1234 (1968).

advantages, and we will discuss them in turn. (Other useful wave forms are analyzed in Ref. 50.)

The greatest advantage of the sinusoidal drive is that a mechanical oscillatory system, consisting ideally of a mass, a spring, and a small damping element, naturally moves with simple harmonic motion. Thus it is relatively easy to generate a fairly accurate sinusoidal velocity wave form. Normally, sinusoidal drive systems are run at or near the mechanical resonance frequency of the transducer (see below), and thus only low driving power is required and the rate of change of the acceleration is relatively small. In cases where heavy masses, such as large sources or ovens, must be moved, or where the velocity must be carried through a long pushrod, for example, into the bottom of a tall dewar or into the center of a large magnet, these advantages may be overwhelming. The disadvantages of the sinusoidal drive technique are the following: the nature of the sine wave form is such that a relatively large proportion of the measuring time is spent near the maximum velocity, whereas the interesting regions of the spectrum generally tend to lie near zero velocity. For a similar reason, the peak-to-peak displacement required by sinusoidal motion to produce a particular velocity is substantially larger than for, e.g., the linear velocity wave form, and this means that the velocity transducer must be linear over a larger displacement range. This disadvantage is, however, compensated for by the fact that, generally, sinusoidal drives can be used at higher frequencies than the other wave forms, thus requiring smaller displacements. The fact that sinusoidal drives are normally used near their fundamental mechanical resonance makes for other, more subtle, difficulties. For example, one is dependent on the linearity of springs and other mechanical elements that tend to be relatively poorly controlled. Additionally, the drive must be retuned if the source mass or the spring constant are substantially changed. Finally, phase stability of a resonant system being driven at its resonant frequency is extremely poor. Since the correspondence between channel number and drive velocity is normally determined at the sine wave generator, if the phase difference between the drive velocity and sine wave source is not absolutely constant in time, the drive velocity corresponding to a channel will drift. Some of these problems are ameliorated to a certain extent by the use of feedback as described above. One obvious drawback of the sinusoidal drive technique is that the channel number does not correspond linearly to velocity when the "multiscaling" mode is used. Normally the data are either plotted on sine paper or processed by a computer to give plots which are linear in velocity or energy. Sinusoidal drives are normally run with the analyzer triggered from the sine wave generator.

It has been pointed out above that the spectrometer accuracy and stability are determined by the drive, and that the electromagnetic drives can be made to follow voltage wave forms to the order of 0.1%. Thus the responsibility for drive accuracy eventually falls on the wave form generator used to supply the input wave to the drive circuit. When the spectrometer runs in the time mode, the wave form amplitude determines the spectrometer calibration, the wave form purity determines the linearity, and the frequency determines the zero-velocity channel and fold point. Thus all three parameters are critical, and relatively few commercial instruments satisfy all of these requirements. Even if the basic generator is satisfactory, it is very often the case that the resetability is relatively poor, and it may be necessary to break into the circuit and replace potentiometers with fixed resistors. It has been the experience of the authors that though it is possible to check sine wave generators with wave analyzers to the desired low level of distortion, it is more difficult to test a good triangular wave than it is to generate one.

Generally, Mössbauer spectrometers using a sinusoidal velocity wave form take the wave form from commercial, high-purity sinusoidal wave generators. A number of such instruments are commercially available with adequate amplitude and frequency stability and total harmonic distortion less than 0.1%. In multichannel scaling, the experimenter must provide a trigger pulse from this sinusoidal output wave form to start the multichannel analyzer sweep in the same phase relation to the sine wave for each cycle. This trigger pulse can be generated at the time of passage of the sine wave through zero with a "zero-crossing detector." This approach puts zero velocity at the first and last channels of the analyzer as well as in the middle. To avoid the discontinuity and loss of information near zero that this entails, some experimenters[62] integrate or differentiate the velocity sine wave to make a sine wave 90° out of phase with it. Then a detector, sensing when the integrated wave passes through zero, produces the trigger pulse at the time of maximum velocity. An elaborate and sophisticated system designed to measure extremely small isomer shifts to high accuracy is discussed in Ref. 62.

There are five approaches generally used in the generation of triangular waves for constant acceleration drives. Triangle generators with adequate stability and wave form accuracy for Mössbauer spectroscopy use are commercially available. These generators usually provide trigger pulses that can be used directly to initiate the multichannel scaling cycle.

In this case, the multichannel analyzer is slaved to the triangle generator and the entire accuracy and stability of the spectrometer are dependent on the properties of the triangle generator. As an indication of the frequency stability that is desirable, a good spectrometer can exhibit a long-term stability (over weeks) of about ± 0.1 channels or better with data being taken into 256 (folded) channels. This corresponds to a triangle generator stability of one part in 2500. Thus a triangle generator frequency stability of $\sim 0.02\%$ or better is desired. Note that since calibration is normally only performed to $\sim 0.1\%$, the amplitude stability requirements are somewhat less stringent.

A second technique that has been successfully used[50] is to let the analyzer free-run through the channels continuously, returning to channel zero and starting the next sweep immediately. During this sweep through the analyzer, the address register (see Fig. 5), which contains the number of the open channel, is continually changing. The most significant bit of this address register, i.e., the bit having a value 128 in a scan of 256 channels, will be "zero" for the first half of the scan (to channel 127) and then "one" after that. If the output from the flip–flop representing that bit is brought out of the analyzer, amplified, and clipped to a fixed dc level, this voltage can be integrated by a simple operational amplifier circuit to provide a linear ramp. If the clipping circuit is so arranged that it provides an output $+v$ when the address lies in the first half of the memory and a voltage $-v$ when the address lies in the second half of the memory, the integration of this (square wave) voltage will produce a ramp that rises linearly while the address is in the first half, and then falls linearly at the same rate while the address is in the second half. This wave form is exactly what is desired as a reference for constant acceleration spectrometers. The greatest advantage of this approach is that the phase of the velocity wave form is exactly locked to the analyzer channel number with no possibility of drift or deviation. The disadvantages are only that a minor modification to the multichannel analyzer may be required to make this signal available, and that the drive can then not be used or tested without the multichannel analyzer. It is usually also necessary to install some interlocks[45] to prevent the drive from going wild when the multichannel analyzer is in the "stopped" or "read out" mode. A network to prevent the integrator from drifting off to infinity due to inevitable input offsets and inequalities of the positive and negative input voltages has been described in Ref. 50, and is widely used. A modification of this approach has been used to provide the wave form necessary for "fast flyback" operation.[47]

A third approach to the generation of triangular wave forms for constant acceleration drives is the construction of a simple wave form generator using integration of a square wave voltage by operational amplifiers. Such an arrangement would at first appear inferior to the one controlled by the analyzer, since the MCA crystal-controlled channel advance clock is extremely stable. However, an arrangement has been described[64] in which the entire frequency stability is dependent on just two components, and whose overall stability, both short and long term, approaches the analyzer stability. Since the circuit described has the ramp running between two fixed voltages, special networks to prevent the integrator from drifting to infinity are unnecessary, and an extremely linear ramp is obtainable. This triangle generator can be built into the drive chassis, to make a compact arrangement that can be conveniently used with essentially any multichannel analyzer without modification.

A technique that has so far seen relatively little application because of the limitations of available multichannel analyzers is the use of the analog address output voltage as the source of the velocity wave form.[45,46,48,65] Until recently the analog output voltages were inadequate for high-precision spectroscopy because the digital-to-analog converters used had relatively limited differential linearity, excessive switching transients, poorly stabilized full scale voltages, and excessive noise and hum. However, more modern machines are markedly improved in this respect. Linearity and stability figures better than 0.1% are required to equal the performance obtainable with the systems previously described. Used with up–down multiscaling, this approach provides no convenient adjustment to make velocity on the "down" half of the sweep retrace that of the first half, since the analog address output wave form is exactly locked to the channel number. The true drive velocity normally lags the driving wave form slightly, and these deviations of the drive velocity will be of different sign in the "up" and "down" scans, resulting in a slight misfolding of the data. To use the analog address output with normal unidirectional multiscaling, a modification to the analog converter is required that makes the analog voltage decrease in the second half of the memory; this is not easy to do with the required precision.

In an attempt to avoid these problems, some circuits using digitally controlled wave-form generators have been described.[39] These basically use an auxiliary scaling register which is advanced in synchronism with

---

[64] R. L. Cohen, *Rev. Sci. Instrum.* **37**, 260, 977 (1966).
[65] F. C. Ruegg, J. J. Spijkerman, and J. R. DeVoe, *Rev. Sci. Instrum.* **36**, 356 (1965).

the analyzer address, and decode the register output with a digital-to-analog converter. This approach provides a rather flexible wave-form generator, at the cost of considerable complexity relative to the analog generators mentioned above.

A number of systems have been described using a wide variety of techniques to allow scanning only selected regions of the velocity spectrum.[47,61,66]

6.2.1.8. Calibration and Standards. Since most Mössbauer measurements are made with the Doppler spectrometer technique, the calibration is simply a question of establishing the velocity scale. In constant-velocity spectrometers this can be a straightforward procedure. A good calibration can be established by measuring the time for the spectrometer carriage to pass between two points of known separation, giving directly the average velocity for the spectrometer. The timing can be performed either by microscopic observation of two fiducial marks, or by the use of slits of known spacing to interrupt the light falling on a photocell which then gates a timer.

This absolute approach, using basically a meter stick and a clock, can be improved by more sophisticated techniques such as photocell detection of the Moiré fringe pattern of two gratings moved across one another[33,67] and counting of the electrical pulses resulting from the grating motion. In principle, the most precise method, but one fraught with a number of hazards and difficulties, is the use of an optical interferometer,[68,69] with one of the mirrors mounted on the moving part of the spectrometer. A photocell can then be used to count the moving interference fringes resulting from the spectrometer motion. When the light wavelength is known, the number of fringes per second yields directly the spectrometer velocity. All of these optical techniques are made much easier by the use of a laser which provides an intense, highly collimated beam without auxiliary components.

These techniques can also be applied to velocity scanning spectrometers.[68,69] However, most experimenters have found it more convenient to calibrate their spectrometers by measuring a material with

---

[66] N. Gaitanis, A. Kostikas, and A. Simopoulos, *Nucl. Instrum. Methods* **75**, 274 (1969).

[67] H. deWaard, *Rev. Sci. Instrum.* **36**, 1728 (1965).

[68] J. G. Cosgrove and R. L. Collins, *Nucl. Instrum. Methods* **95**, 269 (1971), and S. Eylon and D. Treves, *ibid.* **42**, 504 (1971).

[69] J. P. Biscar, W. Kündig, H. Bömmel, and R. S. Hargrove, *Nucl. Instrum. Methods* **75**, 165 (1969).

6.2. MÖSSBAUER SPECTROMETERS

well-known hyperfine splitting. This method has two advantages over the absolute methods mentioned previously: First, no extra equipment is necessary to perform the calibration; it is carried out with a normal Mössbauer spectrometer. Second, the calibration is highly "operational," i.e., the calibration run is an exact simulation of an actual experiment, and it is likely to be insensitive to the anomalies and aberrations that may be introduced in changing the configuration between calibration runs and measuring runs.

Ever since the early, very precise, measurement of the hyperfine structure of Fe^{57} in iron metal,[70] this particular hyperfine pattern has been the choice of most experimenters for calibration, especially in the range within ± 1 cm/sec. This preference results from the favorable characteristics of Fe^{57} as a Mössbauer isotope, and the well-established splitting and easy availability of iron metal foils. Since the overall splitting of the hyperfine spectrum in iron metal is ~ 10.63 mm/sec at 22°C, a measurement of the hyperfine structure to 0.01 mm/sec or $\sim 1/25$ of a line width provides a calibration good to 0.1%. This calibration accuracy is readily obtained with normal Mössbauer spectrometers. On the other hand, it is rather difficult, using the absolute techniques discussed previously, to calibrate Mössbauer spectrometers to an accuracy much better than this. This fact is probably the major explanation for the widespread use of the Fe^{57} splitting for calibration purposes.

Another "absolute" technique that has been used[71,72] is to vibrate the source (typically, with a piezoelectric crystal) in the direction of gamma-ray emission. If the driving frequency is v (typically 1–100 MHz), the motion generates side-bands at frequencies $\pm nv$ above and below the main line. The position of these sidebands can then be used to calibrate the spectrometer. Accuracies of 0.1% have been obtained by this technique.

Because of the extensive use of the hyperfine structure in Fe^{57} in iron metal as a calibration standard, a precise knowledge of this hyperfine splitting is essential, and the United States National Bureau of Standards has for some years made an effort to measure the splitting more precisely than the result of Ref. 70. The best present result of the National Bureau of Standards and a number of collaborating laboratories is that the overall splitting is 10.627 mm/sec between the outermost lines, and the splitting

---

[70] R. S. Preston, S. S. Hanna, and J. Heberle, *Phys. Rev.* **128**, 2207 (1962).
[71] T. E. Cranshaw and P. Reivari, *Proc. Phys. Soc.* **90**, 1059 (1967).
[72] D. I. Bolef and J. Mishory, *Appl. Phys. Lett.* **11**, 321 (1967).

between the ground state levels is 3.918 mm/sec at 24.1 °C.[73] It has been widely recognized that impurities in iron samples can affect the observed splitting, even at very low concentrations.[74,75] The effects of such impurities on the calibrations have been shown in Ref. 74. It appears that, for calibrations to 0.1% accuracy, iron metal of 99.9% purity generally seems to be adequate, and its state of anneal does not seem to significantly effect the hyperfine structure to this level of accuracy.

Since the hf splitting of the Fe^{57} nucleus in iron metal is only about 10 mm/sec, the velocity range of spectrometers that can be calibrated using this standard is relatively limited. This can be somewhat improved by measuring the complex hyperfine structure shown by a Co^{57} in iron source, measured with an Fe^{57} metal absorber. The resulting complex of lines extends to ± 11 mm/sec, with the outermost lines being fairly intense. For higher velocities, the use of the hyperfine splitting of 26 keV transition in Dy^{161} in magnetically ordered Dy metal at low temperatures has been proposed.[76] Surprisingly, most experimenters who have been working with high-velocity spectrometers have found that if care is taken (especially in ascertaining that the velocity transducer is linear at large displacements), it is quite possible to use extrapolated Fe^{57} calibrations to velocities as high as 50 cm/sec, with 1% accuracy.

The discussion above refers primarily to determining the scale factor for the spectrometer, such as "millimeters per second per channel." An additional factor in calibration which must be considered is where the zero point, or true zero velocity of the spectrometer lies. For constant-velocity spectrometers this is obviously no problem; one can simply turn off the drive. For scanning spectrometers, in the experience of the authors, it is relatively difficult to determine the zero-velocity point exactly using absolute calibration techniques.

Most experimenters have found it convenient to determine the spectrometer zero-velocity point by making a measurement using a source of Co^{57} in iron metal and an absorber of iron metal, both at the same temperature. The resulting spectrum has strong absorption at zero velocity, resulting from the overlap of all six emission lines in the source

---

[73] Noted by R. H. Herber, *in* "Mössbauer Effect Methodology," Vol. 6. Plenum Press, New York, 1971; see also C. E. Violet and D. N. Pipkorn, *J. Appl. Phys.* **42**, 4339 (1971), and J. J. Spijkerman, J. C. Travis, D. N. Pipkorn, and C. E. Violet, *Phys. Rev. Lett.* **26**, 323 (1971).

[74] H. Shechter, M. Ron and S. Niedzwiedz, *Nucl. Instrum. Methods* **44**, 268 (1966).

[75] G. K. Wertheim, *Can. J. Phys.* **48**, 2751 (1970).

[76] R. L. Cohen and G. M. Kalvius, *Nucl. Instrum. Methods* **86**, 209 (1970).

with all six absorption lines in the absorber. The central intense peak in the spectrum then lies at $v = 0$. This peak may be displaced slightly from zero velocity due to chemical (isomer) shifts arising from the fact that the environment of the Co^{57} atoms in the source may not be the same as the environment of the iron atoms in the absorber. Such shifts can generally be expected to be less than 0.02 mm/sec, provided that the spectrum of the source, observed with a single line absorber, shows sharp symmetric lines.

If the zero of spectrometer velocity is independent of the drive amplitude setting, and this is the case for many scanning spectrometers, it is easy to detect such differences between source and absorber isomer shift and make a more accurate determination of the true spectrometer zero. This is done simply by measuring the iron source versus iron absorber spectrum at a number of different velocity settings for the spectrometer. If there is an isomer shift between source and absorber, the center of the observed spectrum will move away from the true spectrometer zero as the spectrometer full-scale velocity is reduced. Simply plotting the observed position of the center line versus the reciprocal of fullscale spectrometer velocity setting should yield a straight line, which, extrapolated to zero reciprocal velocity setting, gives the true spectrometer zero. It should be emphasized again that this technique only works if the spectrometer zero-velocity point is independent of the fullscale velocity setting. Similar approaches have been discussed by Carrell et al.[77]

An alternative to the precise determination of the spectrometer zero-velocity point is to measure and report observed line positions with respect to the line or lines of reference materials that have been measured on the same spectrometer, and that have been widely studied by other experimenters.[62,76] This technique has the advantage that it is independent of the source isomer shift, which generally tends to be relatively poorly controlled. Also, if a standard material is widely used, it is relatively easy to compare results obtained by different laboratories. This technique of measuring and reporting results with respect to a standard absorber greatly reduces the necessity for the precise determination of the spectrometer zero-velocity point. It has the minor disadvantage that the precision of the reported results may need to be reduced slightly to allow for the fact that a difference between two experimental determinations, each with its own uncertainty, is reported.

The selection of reference standard materials has proven to be a dif-

---

[77] J. C. Carrell, R. A. Mazak, and R. L. Collins, *Nucl. Instrum. Methods* **72**, 298 (1969).

ficult problem involving compromises among a number of sometimes conflicting criteria. For the isotope Fe^{57}, both sodium nitroprusside and iron metal have been suggested[78] as comparison standards. A more detailed discussion of this problem and a list of proposed reference materials has been given by Cohen and Kalvius.[76]

In cases where the range of isomer shifts is much greater than the line width, and where the Mössbauer absorption is at least comparable to the photoelectric absorption, precise experiments can be done using two absorbers ("internal standard" technique) in the beam at the same time. The resulting spectrum shows the lines of both absorbers, and the isomer shift difference can easily be determined. For velocity scanning spectrometers, this makes extremely high precision possible. In constant-velocity machines, this technique[79] is vulnerable to drifts occurring between the times the "reference" and "unknown" lines are measured.

6.2.2. Gamma-Ray Detection

There are three types of detectors in general use as radiation detectors for Mössbauer spectroscopy: proportional counters, scintillation detectors, and semiconductor detectors made of silicon or germanium.[80] All of these are commercially available in the configurations required for Mössbauer spectroscopy and there seems little justification for the researcher to make his own detector. Extensive works have been written about proportional counters and about scintillation counters; but semiconductor detectors are so recent that the state of the art is still progressing very rapidly. Good treatments of the principles and applications of semiconductor detectors are given by Ewan[81] and Bertolini and Coche,[82] but at the time of this writing the state of the art had already progressed

---

[78] R. H. Herber, "Mössbauer Effect Methodology," Vol. 6, p. 3ff. Plenum Press, New York, 1971.

[79] W. E. Sauer and R. J. Reynik, *Rev. Sci. Instrum.* **40**, 360 (1969).

[80] K. Siegbahn, ed., "Alpha, Beta, and Gamma Ray Spectroscopy," Vol. 1. North Holland Publ., Amsterdam, 1965, contains a good review of scintillation counting techniques (J. H. Neiler and P. R. Bell, p. 245ff), A brief discussion of gamma-ray detection with proportional counters (S. C. Curran and H. W. Wilson, p. 303ff), and an early very clear review of semiconductor detector techniques and applications (W. M. Gibson, G. L. Miller, and P. F. Donovan, p. 345ff).

[81] G. T. Ewan, in *Progress in Nuclear Techniques and Instrumentation*, Vol. III, 1968, p. 67 (North-Holland Publ. Co.).

[82] E. Bertolini and A. Coche, eds., "Semiconductor Detectors," (North-Holland Publ. Co., 1968).

substantially beyond that described. For detection of the 14-keV gamma ray in Fe^{57}, Xe proportional counters with gamma-ray paths of about 4 cm atm are usually utilized, though both scintillation and silicon semiconductor detectors have been used by some experimenters. The Compton scattering of the intense 122-keV gamma ray that precedes the 14-keV transition tends to produce a large background in the latter types of detectors. Very thin NaI scintillation crystals (~0.2–0.5 mm thick) are usually used to reduce the intensity of this background.

For Sn^{119}, scintillation detectors are almost invariably used. Sn^{119m} sources normally used for research with this isotope produce a strong Sn K_α X-ray complex only about 1 keV above the desired gamma-ray line, and this radiation cannot be resolved from the gammas by scintillation detectors. Normally, a "critical absorber" filter[83] of 40 to 50 μm Pd is used to reduce the number of X-rays incident on the detector. Since the K-shell binding energy of Pd is greater than the 23.8-keV energy of the gamma ray, only the X-rays can be absorbed by K-shell photoeffect, and thus are selectively attenuated. Since the Pd produces secondary X-rays, the Pd filter is usually placed about half way between the source and the detector, preferably between the source and the absorber, to minimize the pickup of these X-rays by the detector. The convenient procedure of using a Pd "cap" on the source or counter involves some loss of efficiency because of the increased number of Pd secondary X-rays detected by the scintillator. As of this writing, Si semiconductor detectors, which are capable of resolving the gamma rays, are not competitive with scintillation counters for Sn^{119} spectroscopy.

From time to time "resonance detectors" have been discussed.[84-86] These generally take the form of a thin film of material, containing Mössbauer absorber nuclei, placed in the center of a proportional counter. When a gamma ray is absorbed by a Mössbauer nucleus in the layer, the after effects of the absorption, such as X-rays, conversion electrons, or Auger electrons, are detected in the proportional counter. This method at first appears to have many advantages; in principle, it detects only Mössbauer gamma radiation and, again in principle,[84] it is capable of providing a resonance of less than 2Γ line width. In actual use, however,

---

[83] B. G. Harvey, "Introduction to Nuclear Physics and Chemistry," p. 229. Prentice-Hall, Englewood Cliffs, New Jersey, 1962.

[84] K. P. Mitrofanov, N. V. Illarionova, and V. S. Shpinel, *Inst. Exp. Technol.* 415 (1963).

[85] K. R. Swanson and J. J. Spijkerman, *J. Appl. Phys.* **41**, 3155 (1970).

[86] Z. W. Bonchev, A. Jordanov, A. Minkova, *Nucl. Instrum. Methods* **70**, 36 (1969).

practical difficulties (primarily the low penetrating power of the secondary radiations and background from Compton and photoelectric processes) that occur in such an arrangement have limited its use to special cases. There appear to be two situations in which the resonant counter has real advantages: The first occurs when the emission spectrum of a source is to be studied, and a standard monochromatic absorber can be used in the detector. The second occurs when the absorber material under study can be incorporated into the counter.

Recently, there has been some work[85,86] using this detection technique for materials analysis problems. In this work, the sample under study was used as the resonance detector, and two advantages were established: an "infinitely thick" sample could be studied,[85] since the scattering geometry was used, and filling the counter with He or argon gas allowed discrimination between conversion electrons[85] (generated in a very thin surface layer) and X-rays from a somewhat thicker layer. Energy discrimination on the emergent conversion electrons has also been used to obtain depth resolution in thin surface layers.[86]

Resonance counters have also been based on parallel-plate spark or avalanche counters.[10] These may be much simpler than standard proportional counters, and have the advantage that time resolution in the nanosecond range can be obtained for use in coincidence experiments. The background may be made small enough that the lack of energy discrimination inherent in avalanche counters is not a problem.

Photographic detection of Mössbauer gamma rays has also been used.[87]

6.2.2.1. Electronic Instrumentation for Nuclear Counters.

The above detectors produce a charge pulse that is proportional to the energy deposited in the detector by the incoming gamma ray. This charge pulse must be amplified, shaped, and put through a pulse-height discriminator[12] to ascertain that it is of the proper height to have originated from a gamma ray striking the detector and not from some unwanted radiation. Since about 1966 the following method[88] has come to be almost universally adopted for the counting of such pulses (see Fig. 12): The charge pulse is integrated by a preamplifier located at or near the detector. The output of the preamplifier is a long-tailed step whose height is proportional to the energy of the incoming gamma radiation. This step pulse

---

[87] E. Kankeleit, *Amer. J. Phys.* **34**, 778 (1966).

[88] An up-to-date, critical, and understandable discussion of contemporary nuclear spectroscopy systems is given in a series of articles by E. Fairstein and J. Hahn, *Nucleonics* **23**, No. 7, 56; No. 9, 81; No. 11, 50 (1965); **24**, No. 1, 54; No. 3, 68 (1966).

6.2. MÖSSBAUER SPECTROMETERS

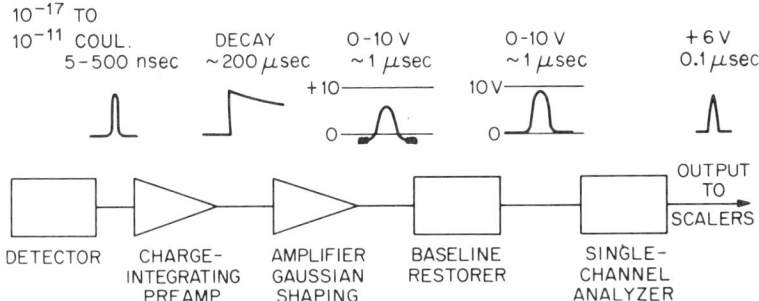

FIG. 12. Basic configuration of electronics commonly used in nuclear pulse spectroscopy. Typical pulse amplitudes and wave forms are indicated.

is normally sent to a main amplifier, which contains complex RC circuits that, by a combination of successive integration and differentiation operations, transform the step function into a pulse of approximately Gaussian shape, typically of approximately 1 μsec length, riding on a flat baseline. This technique is usually called "Gaussian shaping with pole-zero cancellation" and is described in detail Ref. 88. At high counting rates, e.g., above $\sim$10,000 counts/sec, the main amplifier is normally followed by a "baseline restorer," which clamps the output level to zero when no pulse is present. The purpose of this baseline restorer is primarily to prevent shifts in effective pulse height with counting rate due to pulse pileup.

The last element in the counting and detection chain is a single-channel analyzer, which selects pulses within a preset amplitude range corresponding to the gamma-ray energy of interest.

The technique described here has the advantages that the shaping process provides a pulse of close to optimum signal-to-noise ratio, and that high counting rates can be tolerated without significant loss of resolution. Because of these advantages, such spectroscopy systems have almost completely displaced the older CR–RC and delay-line shaping.

For the purposes of Mössbauer spectroscopy, the commercially available preamplifiers, amplifiers, baseline restorers, and single-channel analyzers come so close to the theoretical ideal that there is little advantage to the experimenter in building his own detection systems. Another advantage of the units currently available for this type of nuclear spectroscopy is that with the advent of the NIM system[†] the various manufacturers

---

[†] "Nuclear Instrument Module," a frame (containing a power supply) into which functional modules available from a number of manufacturers can be inserted.

have standardized on power supplies, signal levels, and module geometries so that individual units from different manufacturers can be used together with complete compatibility. Under favorable conditions, counting rates greater than 100,000 counts/sec can be used in the gamma-ray channel if sufficiently strong sources are available.

6.2.3. Radioactive Sources for Mössbauer Experiments

The routine commercial availability of sources such as Co^{57} in copper or palladium for Fe^{57} spectroscopy, and Sn^{119m} in $BaSnO_3$ for Sn^{119} Mössbauer spectroscopy, makes it unnecessary for the experimenter to be concerned about the task of source preparation if these isotopes are being employed. These sources[89,90] for Fe^{57} and Sn^{119} have almost ideal characteristics.

For less widely used isotopes, however, the experimenter must very often make his own sources, and we present below a discussion of criteria and techniques that have proven to be valuable in the past.

By far the most important criterion in choosing a host is that the emitted Mössbauer gamma line shall be unsplit and near the natural line width. This is important not only to simplify the analysis of the resulting spectrum, but because broadening or splitting of the source line results in a loss of the effective resonance intensity. Secondary considerations are: (1) The source should be thin enough so that the gamma rays, which are typically of low energy, can be emitted without excessive photoelectric absorption. (2) None of the Mössbauer isotope in its ground state should be contained in the source, as the resulting "source self-absorption" increases the line width and decreases the observable effect. (3) The host should not produce secondary X-rays of an energy that will interfere with the detection of the Mössbauer gamma rays. (4) The effective Debye temperature should be high to provide a high recoil-free fraction. (5) The source should have good mechanical integrity to avoid the possible accidental escape of radioactive material in case of mechanical failure. (6) Chemical stability under conditions of use is highly desirable.

---

[89] S. M. Qaim, P. J. Black, and M. J. Evans, *J. Phys. C* **1**, 1388 (1968); I. Deszi and B. Molnar, *Nucl. Instrum. Methods* **54**, 105 (1967); J. Kucera and T. Zemcik, *Can. Met. Quart.* **7**, 83 (1958).

[90] M. V. Plotnikova, K. P. Mitrofanov, and V. S. Shpinel, *Inst. Exp. Tech. (USSR)* **4**, 209 (1966); A. V. Kalyamin, B. G. Lur'e, G. V. Popov, Yu. F. Romanov, *Sov. Phys.–Solid State* **10**, 2522 (1969).

6.2. MÖSSBAUER SPECTROMETERS

If the radioactive material can be incorporated as a substitutional impurity in a cubic nonmagnetic metallic host, these criteria are normally satisfied; hence, the wide use of sources such as Co^{57} in copper or[91] Mg_2Sn (metallic surroundings are preferred because they eliminate the possibility of after effects, mentioned previously in Section 1.6). Depending on the particular experiment involved, any of these criteria can be a severe limitation. For example, for Sn^{119} spectroscopy using Sn^{119m} sources, the usable source thickness is limited by Mössbauer self-absorption by the ∼1–3% of stable Sn^{119} normally contained in the source material. The limitation on the useful source thickness from this mechanism can be more severe than that from photoelectric absorption.

If the experimenter can find a compound which appears to satisfy all the previous criteria, there are three alternatives that are normally taken to preparing the source. The radioactive material can be purchased or made, and diffused into the chosen host. In some cases, the desired compound can be made and then irradiated to produce the activity (annealing may be required after the irradiation to remove radiation damage). The most difficult technique is the use of "hot" chemistry or metallurgy to make the compound containing the active material. All of these techniques require at least modest hot lab and health physics facilities since the handling of radioactive materials is involved.

Mössbauer sources typically have high activity in terms of millicuries but nevertheless are not severe radiological hazards because the gamma rays are at relatively low energy and can be easily shielded. On the other hand, experimenters have sometimes been too casual about the physical containment of source material, and instances of contamination of entire laboratories have occurred when unconfined powder was used. There seems to be no justification for the laboratory use of sources, however well contained, composed of loose powder.

Since the running time for experiments varies inversely with source strength, in general the desired strength is the greatest that can be made consistent with radiological hazards, specific activity, and cost, assuming that the resulting count rate is consistent with the capabilities of the counting electronics. Widely used values are 10–50 mCi for Co^{57}, 1–5 mCi for Sn^{119m}, and 10–1000 mCi for Eu^{151}.

---

[91] P. A. Flinn, and S. L. Ruby, *Rev. Mod. Phys.* **36**, 352 (1964); V. A. Bryukhanov, N. N. Delyagin, and R. N. Kuz'man, *Sov. Phys. JETP* **19**, 98 (1964).

6.3. Auxiliary Equipment

Temperature is often an important parameter in Mössbauer experiments. For many isotopes it is essential to operate with a cold source and absorber in order to obtain an appreciable recoil-free fraction. Temperature itself may be the independent variable in the experiment. This is particularly true of experiments concerned with lattice dynamics in which such phenomena as the second-order Doppler shift or the recoil-free fraction are measured. Temperature also enters in the study of such phenomena as spin relaxation, quadrupole splitting, and magnetic hyperfine structure. In the following we will discuss separately means of achieving low and high temperature, although there are certain common problems.

6.3.1. Low-Temperature Techniques

6.3.1.1. Dewars. Low temperature is most conveniently obtained using liquified gases such as nitrogen, hydrogen, and helium.[92] Other materials like methane, ethane, and the various freon compounds have been used less frequently but are particularly useful for constant temperature work above liquid nitrogen temperature. These liquids are generally used in vacuum insulated containers, i.e., dewar flasks, adapted for Mössbauer experiments.[93]

Styrofoam dewars can be used for simple experiments at dry ice or liquid nitrogen temperatures. For long measurements at liquid helium temperatures, glass dewars are very convenient. Probably the most widely used system, though, is the standard commercial all-metal research dewar. Those with indium O-ring sealed removable tail pieces are particularly convenient because they facilitate the change between the various commonly required configurations. These are illustrated in Fig. 13.

In experiments where source and absorber are to be at the same low temperature the vertical configuration has many advantages (Fig. 14). The source and absorber holders and drive assembly are a completely separate unit which can be quickly inserted and removed, facilitating

---

[92] R. B. Scott, "Cryogenic Engineering." Van Nostrand Reinhold, Princeton, New Jersey, 1959, provides a useful compendium of information on low temperature techniques.

[93] G. M. Kalvius, in "Mössbauer Effect Methodology" (I. J. Gruverman, ed.), Vol. 1, p. 163. Plenum Press, New York, 1965.

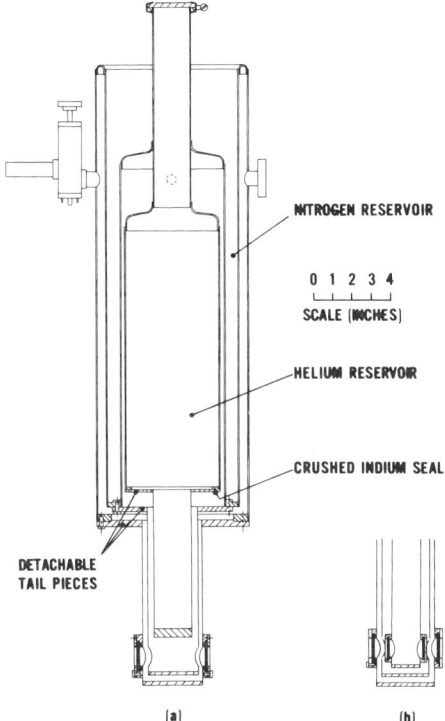

Fig. 13. All-metal helium dewar with indium O-ring sealed interchangeable tail pieces. Two of many possible configurations are shown. (a) Sample mounted in vacuum space. This arrangement is suitable for temperature control applications. (b) Sample immersed in liquid helium. (Courtesy of Janis Research Co., Stoneham, Massachusetts.)

sample changes. Dewar vibrations can be completely decoupled from the experiment.

If only the absorber (or source) is to be cooled, horizontal gamma-ray transmission may be more convenient because the drive can be permanently mounted. In both of the arrangements the sample may be immersed in the coolant, assuring rapid cool-down and a stable sample temperature. Such "immersion" dewars can be designed for rapid sample changing without loss of refrigerant.

The only unusual feature in the design of these dewar tails is the gamma-ray window. The optimum window materials are beryllium and mylar because they provide a minimum of gamma-ray absorption for the thickness required to make a window. The sealing of these dissimilar materials to the dewar is readily accomplished with special low-tempera-

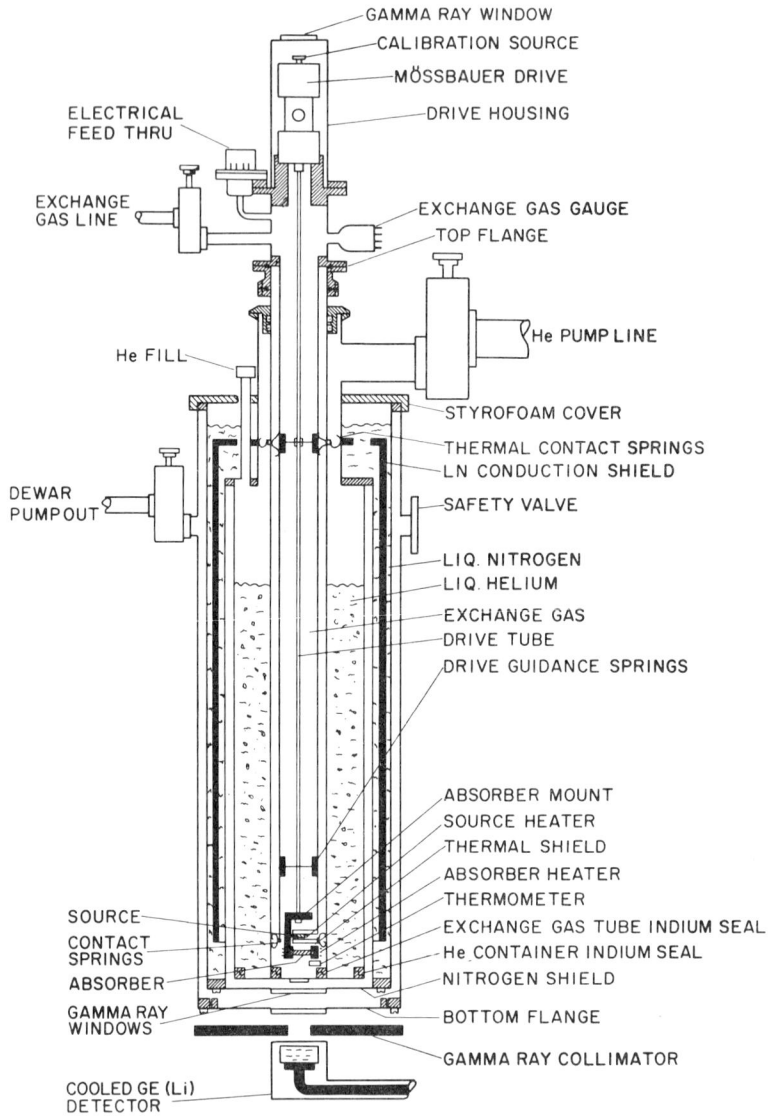

FIG. 14. Vertical drive configuration (Courtesy B. D. Dunlap, G. M. Kalvius, and G. K. Shenoy, Argonne National Laboratory, Argonne, Illinois.)

ture epoxy cement.† Because of their flexibility, mylar windows should have a retainer ring cemented over the bearing surface. Mylar windows of this type have been cycled between room temperature and 4.2 K dozens of times without developing a leak, and can be used even in the pumped helium range. However, mylar of the thickness commonly used, 0.013–0.025 cm, readily transmits helium at room temperature. This may give false indications of trouble during helium leak testing. It may also allow helium to enter the dewar vacuum space before cool-down or through mylar windows in the outer room temperature dewar walls.

Two concerns arise in the use of beryllium windows. Foremost is the well-known toxicity of the metal and its salts,‡ which requires special handling during cutting, machining, or etching. Second is the iron impurity, which is generally present in commercial beryllium and will show up as a spurious background absorption in Fe^{57} Mössbauer experiments. Even if research is planned with an isotope having a high gamma-ray energy, it is advantageous to use beryllium or mylar windows, since calibration and drive checking are usually performed with the 14-keV radiation of Fe^{57}.

Immersion dewars offer only a limited range of temperature which can be obtained by reducing the pressure of the liquid. The ranges are 0.9–4.2 K for He, 14–22 K for H_2, and 65–79 K for N_2. More flexible control generally requires that the sample be physically separated from the liquid. The configuration generally adopted is shown in Fig. 13. The sample is now in an isothermal enclosure provided with a heater, a thermal leak to the main reservoir of liquid gas, and a number of thermometers, all located in the dewar vacuum. Since the sample attains the temperature of its enclosure mainly by conduction, thought must be given to providing a suitable backing material if the cooling and heating times are not to be excessive.

Design of the thermal leak results from a compromise between the need for rapid cool-down and desire to conserve the liquid coolant. Heater power less than 1 W is generally used. The major concern is the thermometry. For precise measurements it is desirable to have a thermocouple or resistance thermometer attached to the Mössbauer sample itself to monitor its response to changes in temperature. The temperature

---

† For example, 7343 Base resin and 7139 Curing agent, manufactured by Narmco Materials Division of Whittaker Corporation, Costa Mesa, California.

‡ "The Merck Index of Chemicals and Drugs." Merck and Co., Rahway, New Jersey, 1960.

of the sample enclosure may be conveniently measured relative to the liquid reservoir. The use of the liquid reservoir as the reference has the disadvantage that the reference temperature changes slightly with atmospheric pressure.

6.3.1.2. Refrigerators and Flow Coolers. There are a number of other techniques for obtaining low temperature in which the cooling is not simply provided by a reservoir of liquified gas. These include thermoelectric junction devices, gas flow coolers, as well as refrigerators based on Joule–Thomson expansion, He^3–He^4 heat of solution, or adiabatic demagnetization.

Thermoelectric devices have a limited temperature range, but will operate from room temperature to well below the ice point. They may be of interest to studies of frozen solutions.

Gas flow coolers range from very simple devices in which the cooling is provided by boil-off from a remote liquid reservoir,[94] to dewars in which liquid He is throttled through a valve into a partially evacuated sample chamber. The latter devices have a number of valuable characteristics. The sample chamber is accessible without breaking the reservoir vacuum, cool-down is rapid, and liquid He temperature itself is readily achieved by filling the sample chamber with liquid. On the other hand, these dewars are sensitive to frozen air and water vapor, which tend to clog the low-temperature valve. Temperature control to better than $\pm 1°$ also appears difficult to achieve in this system.

Another class of devices utilizes the expansion of He or H_2 cooled below the inversion point to achieve liquid gas temperature.† These instruments make it possible to avoid the handling of liquid hydrogen, which is hazardous. On the other hand, they require large volumes of compressed gas or expensive compressors. A closed cycle helium refrigerator has been described by Wiggins *et al.*[95] who also discuss the problem of mechanical vibrations common to these gas flow systems.

6.3.1.3. Cooled Sources. For many isotopes it is necessary to cool both the source and the absorber to obtain the maximum recoil-free fraction. The problem of cooling the source that is being Doppler modulated has been solved in a variety of ways.

[94] W. Wiedemann, W. A. Mundt, and D. Kullmann, *Cryogenics* **5**, 94 (1965).
[95] J. W. Wiggins, J. R. Oleson, Y. K. Lee, and J. C. Walker, *Rev. Sci. Instrum.* **39**, 995 (1968).

---

† Cryotip^R manufactured by Air Products and Chemicals, Inc.

The simplest and perhaps most satisfactory approach is to operate in a vertical configuration as shown in Fig. 14. The source–absorber–drive unit can be removed from the dewar for sample mounting. The source and absorber are cooled by exchange gas to the temperature of the surrounding helium bath. A heater makes it possible to control the temperature of the absorber independently of that of the source.

An alternate approach is to use a horizontal configuration like that shown in Fig. 13a. The source is placed in the dewar vacuum close to the absorber. It is cooled by connecting it to the main coolant reservoir by a flexible metal braid. (Solid metal conductors should not be used because of low-temperature fatigue problems). The Doppler motion is brought into the dewar through a bellows arrangement using a long push rod with low thermal conductivity, e.g., glass or stainless steel. Careful mechanical design is required to make such systems convenient to use.

6.3.1.4. Ultralow Temperature. With liquid helium, temperatures down to ∼1.0 K can be reached by the simple expedient of reducing the pressure on the liquid. Lower temperatures require special techniques, which are described in detail in Chapter 9. Many of these techniques have been applied in Mössbauer spectroscopy.

The use of He^3 in a batch refrigerator makes it possible to reach 0.3 K. Temperatures near ∼0.04 K are obtainable with He^3–He^4 dilution refrigerators adapted for Mössbauer measurements.[96] In these systems it is possible to immerse the sample in the vapor to assure that it is at the same temperature as the coolant. However, it may be preferable to have a sealed refrigeration system to avoid leakage problems. A system that works in either mode has been described in some detail by Enholm et al.[96]

Even lower temperatures can be achieved by adiabatic demagnetization. Shinohara et al.[97] have reached 0.025 K using chromium potassium aluminum and 0.015 K using manganese ammonium Tutton salt. The techniques employed here are not specific to Mössbauer effect except for the use of beryllium windows to provide for gamma-ray transmission. For additional detail see Chapter 9 by W. Weyhman.

[96] G. J. Ehnholm, T. E. Katila, O. V. Lounasmaa, and P. Reivari, *Cryogenics* **8**, 136 (1968); see also G. M. Kalvius, T. E. Katila, and O. V. Lounasmaa, in "Mössbauer Effect Methodology," Vol. 5, p. 231. Plenum Press, New York, 1970.

[97] M. Shinohara, A. Ishiyaki, and K. Ono, *Jap. J. Appl. Phys.* **6**, 982 (1967); **7**, 170 (1968).

6.3.1.5. Temperature Control. Temperature other than those of the liquid gases or the ultra low temperatures described are generally maintained by feedback stabilization. The principles employed are, of course, quite similar to those which apply to feedback systems in general, i.e., an error signal is modified and amplified to generate a control signal.[98] Commercial systems that perform this function are quite widely available, but most are designed for furnaces with rather large thermal mass and slow response time. For accurate control of the small masses required in Mössbauer effect, more sensitive and quicker response is desired.

A system of this type is shown in Fig. 15. As shown, the temperature is measured by a thermocouple with stable reference junction. Alternately, a platinum resistance thermometer or a semiconductor element could be used. The output of this device is measured by a potentiometric microvoltmeter. In these devices the voltage corresponding to the desired temperature can be set and the deviation read on a meter. An electrical

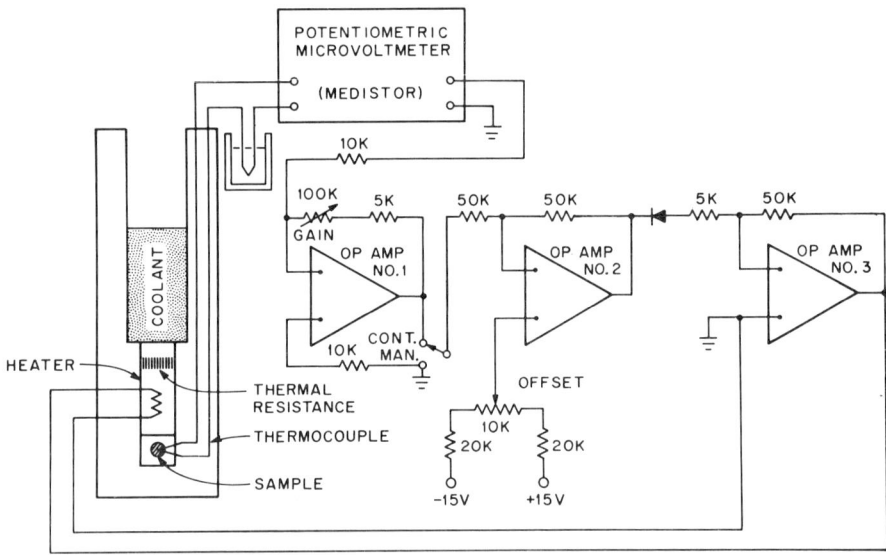

FIG. 15. Proportional feedback temperature control circuit. The output due to unbalance of the microvoltmeter is amplified and used to power the heater. Diode polarity is chosen so that signals corresponding to negative temperature deviations are passed. The system can be modified to provide rate and reset control by substituting suitable RC combinations for the resistors at the first operational amplifier.

---

[98] W. A. Steyert and M. D. Daybell *in* "Mössbauer Effect Methodology," (I. J. Gruverman, ed.), Vol. 4, p. 3. Plenum Press, New York, 1968.

signal proportional to the deviation is also available. It constitutes the error signal, which is then amplified and used to operate a heater. The figure shows additional details such as switch which breaks the feedback loop and allows manual control of the heater current, an offset control which makes possible control around the zero point, and a diode which prevents error signals of the wrong sign from producing heater output.

The system shown has been operated with a Hewlett–Packard dc-coupled amplifier as the output stage. The power available from this unit is adequate for most of the applications described here. If more power is required, a silicon-controlled rectifier ac supply can be substituted for last stage.

6.3.2. Furnaces

The chief concern in the design of furnaces for Mössbauer effect experiments is to keep the physical dimensions small while achieving good temperature homogeneity.[99] The simplest approach, but one which does not meet the above criteria very successfully, is to use a conventional tube type resistance furnace, placing the sample and radiation shields in an evacuated tube running through its bore. Much better results can be obtained by placing the furnace core itself in the evacuated envelope. An example of such a design is shown in Fig. 16. In a vacuum furnace the radiative heat transfer from the core to the outer shell can be effectively suppressed by one or more thermally isolated concentric radiation shields. Water cooling of the outer case is often desirable.

For construction of the high-temperature furnace parts, molybdenum is the best material. At temperatures to $\sim 800°C$, stainless steel has satisfactory properties. The refractory ceramic oxides are also useful but more difficult to handle. Brass must be excluded because of the high vapor pressure of zinc. The vapor pressure of copper is also sufficiently high to cause problems above 600°C. The most serious effect arising from such volatile metals is the contamination of samples and thermocouples. Platinum–platinum 10% rhodium thermocouples are very sensitive to contamination by copper. Contamination of the sample by furnace materials at high temperature is an extremely difficult problem to overcome, and requires careful consideration early in the design.

---

[99] For general background see "High Temperature Technology" (I. E. Campbell, ed.). Wiley, New York, 1956. An interesting furnace for ME work is described by G. V. Vervevkin and Y. V. Vasilev, *Instrum. Exp. Tech. USSR*, 603 (1970).

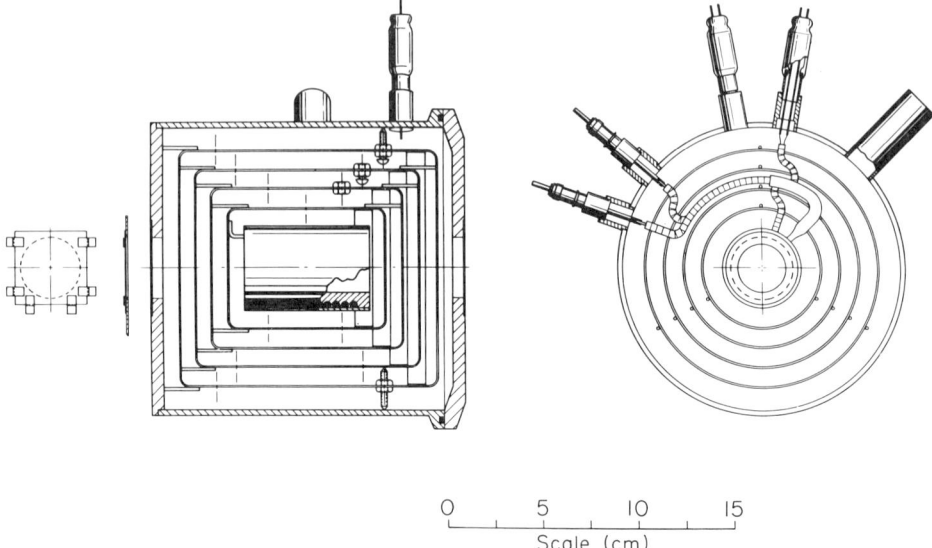

Fig. 16. Radiation-shield insulated vacuum furnace. The furnace core is made of molybdenum, and has a noninductive sheathed heater wound into a machined spiral groove. The radiation shields are spun stainless steel with beryllium windows held by clamps. The Pt–Pt(Rh) control thermocouple is in a deep hole in the molybdenum core.

Other materials that are particularly suitable for the high-temperature parts of Mössbauer vacuum furnaces include graphite and boron nitride. Both have a small gamma-ray absorption and can be machined into sample holders and radiation shields which will cause relatively little gamma-ray scattering. For temperatures up to 800°C, beryllium windows can also be used.

High-temperature thermometry is usually based on thermocouples, although pyrometers may be useful above 1000°C. Suitable thermocouples include Pt–Pt (10% Rh), chromel–alumel, and iron–constantan. The first of these has a relatively small thermoelectric power but is known to be stable and usable to 1600°C. The second has a large thermoelectric power, is usable to 1370°C, but its calibration is less stable. For precise work all thermocouples should be calibrated at fixed points, usually the freezing points of various metals such as lead and gold.

Temperatures to 200°C can be easily reached using commercial button heaters. Platinum, molybdenum, tungsten, and silicon carbide have all been used to produce higher temperatures. Nickel alloys such as Nichrome and Kanthal have high resistance and good chemical and mechanical properties, but cannot reach as high temperatures as the other materials. A particularly simple furnace can be made from silicon carbide tubes with spiral groves to concentrate the power dissipation in the center.

6.3.3. Magnets

Magnetic fields have been used in Mössbauer effect experiments to measure the localization of atomic magnetic moments and to study various aspects of magnetic hyperfine structure. Conventional electromagnets which produce fields of up to 30 kOe have found relatively few applications, in part because the fields obtained with gaps suitable for ME geometries are usually too small.

A great deal of work has been done with water-cooled high-field magnets of the type designed by Francis Bitter.[100] These provide fields of 100 to 250 kOe, but are available only at a few major installations. Mössbauer experiments are difficult to carry out in them because of large stray fields and sizable mechanical vibrations resulting from the enormous flow of cooling water.

Considerably more attractive for Mössbauer experiments are the superconducting solenoids which can now provide fields in excess of 100 kOe with good homogeneity.[101] At present the highest fields have been obtained with Nb_3Sn based material, but these solenoids do not yet provide truly persistent operation because the problems of making superconducting joints remains to be solved. On the other hand, Nb–Zr and Nb–Ti alloy wire solenoids can produce fields up to 70 kOe and can be operated in a persistent mode, i.e., without external power. This results in a considerable reduction in the liquid helium consumption because I^2R losses and thermal conduction losses in the lead-in wires can be entirely avoided.

A number of magnet configurations have been developed for Mössbauer experiments and are commercially available (see Fig. 17). The horizontal magnet with the possibility of room temperature access is the easiest to use, but not the most versatile. The vertical magnet around a central access tube is the simplest to construct but the most difficult to use. The split coil magnet provides both longitudinal and transverse

---

[100] N. A. Blum, *in* "Mössbauer Effect Methodology" (I. J. Gruverman, ed.), Vol. 1, p. 147. Plenum Press, New York, 1965.

[101] P. P. Craig, *in* "Mössbauer Effect Methodology" (I. J. Gruverman, ed.), Vol. 1, p. 135. Plenum Press, New York, 1965; L. J. Swartzendruber, *Nucl. Instrum. Methods* **69**, 101 (1969).

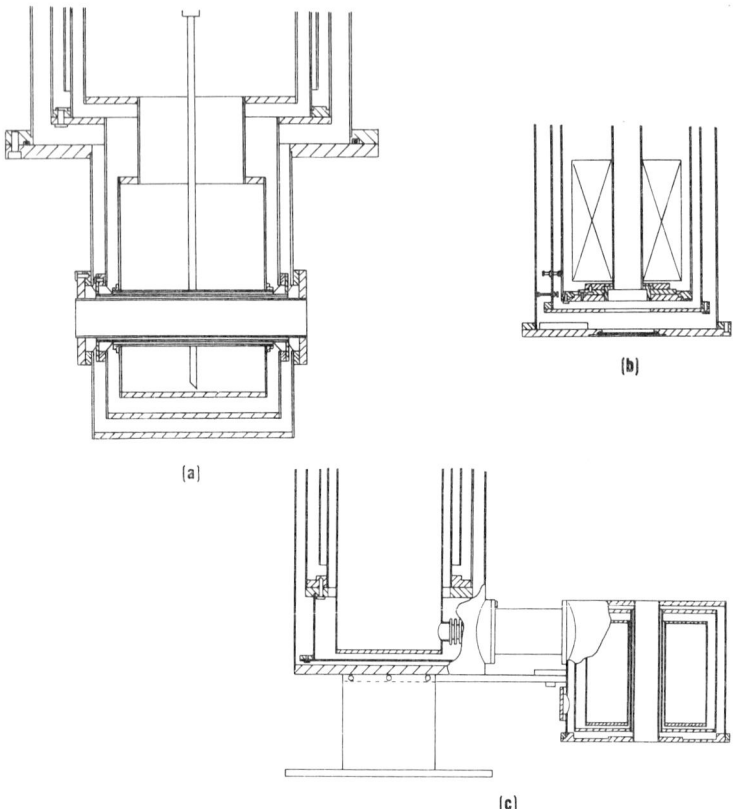

FIG. 17. Superconducting magnet configuration: (a) horizontal with room temperature access. (b) vertical; (c) outboard magnet with room temperature hole. (Courtesy of Janis Research Co., Stoneham, Massachusetts.)

field geometry, but at some cost in maximum field and considerable increase in the complexity of the dewar system. In some respects the vertical outboard magnet with room temperature access offers a combination of attractive features. Samples can be mounted in a completely independent dewar so that the main magnet dewar vacuum need not be broken for sample changes. Cold source and absorber experiments are readily carried out. These advantages are partially offset by the complexity of the dewar itself and by the fact that the full bore of the magnet coil is not available for experiments.

Stray fields of up to 10 kOe have been shown to have negligible effects on proportional counters which are advantageously used in such experiments. Semiconductor particle detectors are satisfactory at 10 kOe at

78 K, but not, in general, at 4 K. Photomultiplier tubes need careful shielding to reduce the stray field below that of the normal earth's field to be useable near superconducting magnets. Electromechanical drive systems can be quite sensitive to stray fields and must be shielded or kept at a sufficient distance to keep fields well below 100 Oe.

The line widths of radioactive sources also can be seriously affected by fringing fields. This can be especially serious for cooled sources. Co^{57} sources in longitudinal magnetic fields are split into four circularly polarized lines. The outer two have 75% of the intensity and are selectively absorbed by the similarity circularly polarized absorber lines. This leads to false line shifts equal to the source splitting *even when the latter is less than the line width*. One source has been reported in the literature (for Fe^{57}) for which the internal field is much smaller than the applied field, greatly reducing the problems of having the source in the fringing field.[102]

6.3.4. High-Pressure Mössbauer Experiment

A number of Mössbauer experiments have been performed to measure the effects of pressure on electronic wave functions. The most popular approach has been the use of the Bridgman anvil technique using tungsten carbide anvils, with the source being put under pressure as shown in Fig. 18. The gamma rays exit radially from the pressure cell, parallel to the plane of the anvils. This technique has been described in detail and has been used to about 100 kbars with unsupported anvils[103,104] (the original Bridgman geometry) and to about 300 kbar using supported anvils.[105-107] Most of these experiments have been done by putting the source under pressure, but a variation of this geometry has been described[106] that allows the use of pressurized absorbers.

---

[102] T. A. Kitchens, W. A. Steyert, and R. D. Taylor, *Phy. Rev.* **138**, A467 (1965); R. B. Frankel, N. A. Blum, B. B. Schwartz, and D. J. Kim, *Phys. Rev. Lett.* **18**, 1051 (1967).

[103] H. S. Möller, *Z. Phys.* **212**, 107 (1968).

[104] V. N. Panyushkin, F. F. Voronov, *Sov. Phys. JETP Lett.* **29**, 7 (1965).

[105] R. Ingalls, "Mössbauer Effect Methodology," Vol. 1, p. 185. Plenum Press, New York, 1965.

[106] P. Debrunner, R. W. Vaughan, A. R. Champion, J. Cohen, J. Moyzis, and H. G. Drickamer, *Rev. Sci. Instrum.* **37**, 1310 (1966).

[107] D. N. Pipkorn, C. K. Edge, P. Debrunner, G. De Pasquali, H. G. Drickamer, and H. Frauenfelder, *Phys. Rev.* **135**, A1064 (1964).

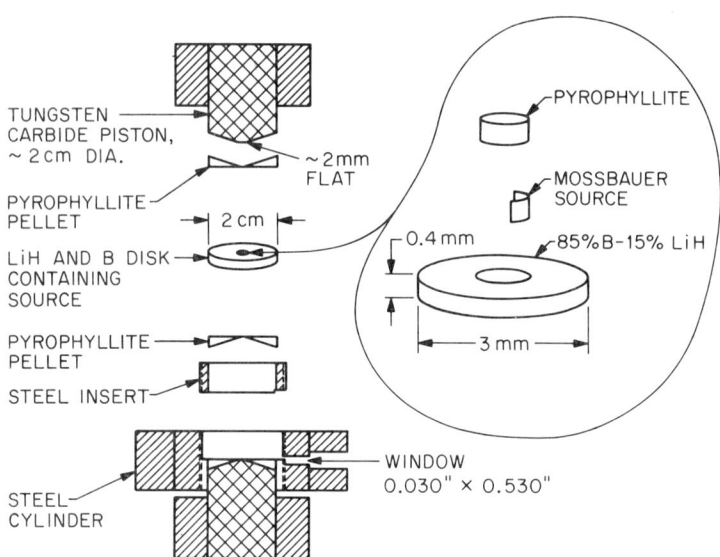

FIG. 18. Design of an advanced apparatus for performing high-pressure experiments (to 200 kbars) using Mössbauer sources. The gamma rays pass radially through the boron–LiH washer, which is used to prevent extrusion of the source [after P. Debrunner, R. W. Vaughan, A. R. Champion, J. Cohen, J. Moyzis, and H. G. Drickamer, *Rev. Sci. Instrum.* **37**, 1310 (1966)].

One group[108] has found success using Bridgman geometry with anvils of the relatively exotic material B_4C. The low gamma-ray absorption of this low-Z material makes it possible to propagate the gamma rays axially through the anvil body, thus making available a much larger absorber area than the arrangements in which the gamma rays propagate along the anvil face. This technique has been used especially for Au^{197} Mössbauer spectroscopy, to about 80 kbars at 4 K. Another experiment in which the gamma rays propagate through the anvil structure was performed using diamond anvils.[109] This technique is relatively simple and straightforward if one has access to the large diamonds required, but the available absorber area is extremely small. A small clamp cell to reach 240-kbar pressures down to 4 K has also been described.[110] The tetrahedral anvil press, which permits larger sample volume than the

[108] L. D. Roberts, D. O. Patterson, J. D. Thomson, and R. P. Levey, *Phys. Rev.* **179**, 656 (1969).
[109] R. H. Herber and J. J. Spijkerman, *J. Chem. Phys.* **43**, 915 (1964).
[110] C. W. Christoe and H. G. Drickamer, *Rev. Sci. Instrum.* **40**, 169 (1969).

above techniques, has been used for Mössbauer spectroscopy at 100 kbars and 600°C.[111]

Experiments at much lower pressures using simple hydrostatic pressure apparatus have been reported. For example, an early experiment at about 3 kbars[112] used a simple pressure cell with an unsupported thick beryllium window. More recently, a pressure cell with a supported beryllium window has been used to do research up to about 10 kbars.[113]

The hyperfine splittings and isomer shift observed with the Mössbauer effect give a great deal of information which is of interest in high-pressure physics and chemistry.[105,107,112] The technique has been used not only to investigate the continuous variations in electronic properties with pressure, but to help map out phase boundaries and determine the properties of new high-pressure phases. The major limitation on high pressure Mössbauer experiments is that the available source area is very small (~ 0.2 mm^2 for the Bridgman anvil technique), limiting the amount of radioactive material that can be used, and the solid angle from which gamma rays can be seen is quite small, further reducing the counting rate. Thus each experiment takes a long time, even with isotopes (such as Sn^{119}) that can provide useful data in just a few minutes in normal equipment. This difficulty is compounded by the fact that pressure-induced changes in isomer shift or hfs tend to be relatively small (unless a phase transition occurs), requiring data with good statistics.

6.3.5. Data Handling

At the end of an experiment the data usually reside in the core memory of a multichannel analyzer, in the form of 0 to 10^6 counts in typically 2^8–2^{10} channels. Thought should be given to the techniques required to handle and process the large volume of digital information that will be obtained. Output must be obtained in a machine compatible form so that further operations can be performed. This output should also provide a permanent record of the experiment. The available options include punched paper tape, punched cards, or magnetic tape, listed in order of increasing expense. The choice depends in part on the

---

[111] W. H. Southwell, D. L. Decker, and H. B. Vanfleet, *Phys. Rev.* **171**, 354 (1968); L. E. Millet and D. L. Decker, *Phys. Lett.* **29A**, 7 (1969).

[112] R. V. Pound, G. B. Benedek, and R. Drever, *Phys. Rev. Lett.* **7**, 405 (1961).

[113] V. N. Panyushkin, *Instrum. Exp. Res.* 478 (1969).

input facilities of the computer used in subsequent data reduction. Printed numerical data are of limited utility and are used mainly as a back-up against failure in the computer-compatible system. Graphical output, on the other hand, turns out to be highly useful because it provides an immediate view of the data, which can be used in planning subsequent stages of data reductions.

Punched paper tape is often the best choice because of its versatility. It can be fed directly into a wide variety of computers and time-sharing systems via a teletype terminal. It can be converted into a page of typed information for compact storage, and is itself compact. Data output is relatively fast and the record is permanent. Conversion to punched cards can be carried out with machines which may be available at a computer center.

Magnetic tape cartridge systems provide a means for high-speed economical readout. They are not computer compatible and generally require that the data be read back into the multichannel analyzer and read out in one of the other forms mentioned above. There appears to be little need for this type of readout in Mössbauer experiments.

APPENDIX A. Data Analysis

There are a number of articles concerned specifically with problems encountered in data analysis. Some offer graphical aids that allow rapid estimation of parameters normally obtained by computer data reduction.

A set of graphs for rapid estimation of component separation in partially resolved doublets has been published by P. B. Moon, *Nucl. Instrum. Methods* **79**, 61 (1970).

Methods of dealing with finite absorber thickness broadening have been discussed by S. Margulies and J. R. Ehrman, *Nucl. Instrum. Methods* **12**, 131 (1961) and D. A. O'Connor, *ibid.* **21**, 318 (1963).

Graphs that allow the evaluation of hyperfine structure parameters of Fe^{57} spectra from experimental data have been given by W. Kündig, *Nucl. Instrum. Methods* **48**, 219 (1967).

Descriptions of computer programs specifically oriented toward analysis of Mössbauer spectra have been given by: S. L. Ruby and J. R. Gabriel, *Nucl. Instrum. Methods* **36**, 23 (1965); J. R. Gabriel and D. Olson, *ibid.* **70**, 209 (1969); D. Agresti, M. Bent, and B. Persson, *ibid.* **72**, 235 (1969); W. Kündig, *ibid.* **75**, 336 (1969); K. A. Hardy, D. C. Russell,

R. M. Wilenzick, and R. D. Purrington, *ibid.* **82**, 72 (1970); Tables of Eigenvalues and Eigenvectors of Hamiltonian Describing the Combined Static Magnetic Dipole and Electric Quadrupole Interaction of a Nuclear Level, by R. M. Steffen, E. Matthias, and W. Schneider, U.S. At. Energy Comm. TID-15749, available from the Office of Tech. Service, Dept. of Commerce, Washington, D.C. is an extensive tabulation for $I \leq 9/2$ for axially symmetric EFG with arbitrary angle between V_{zz} and H.

APPENDIX B. Useful Graphs

Two useful graphs are shown in Figs. 19 and 20.

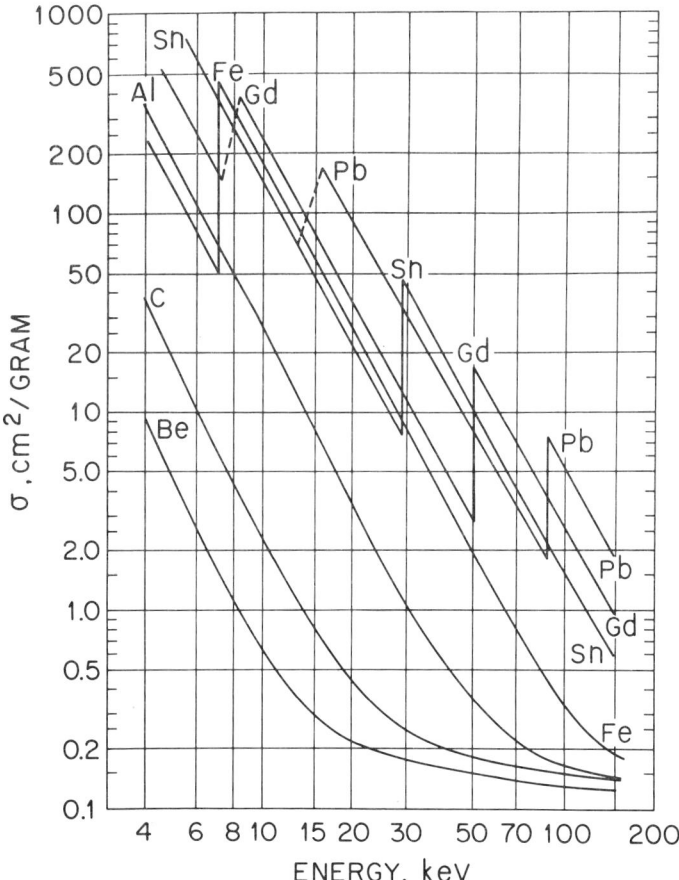

FIG. 19. Gamma-ray absorption cross section versus energy for various materials.

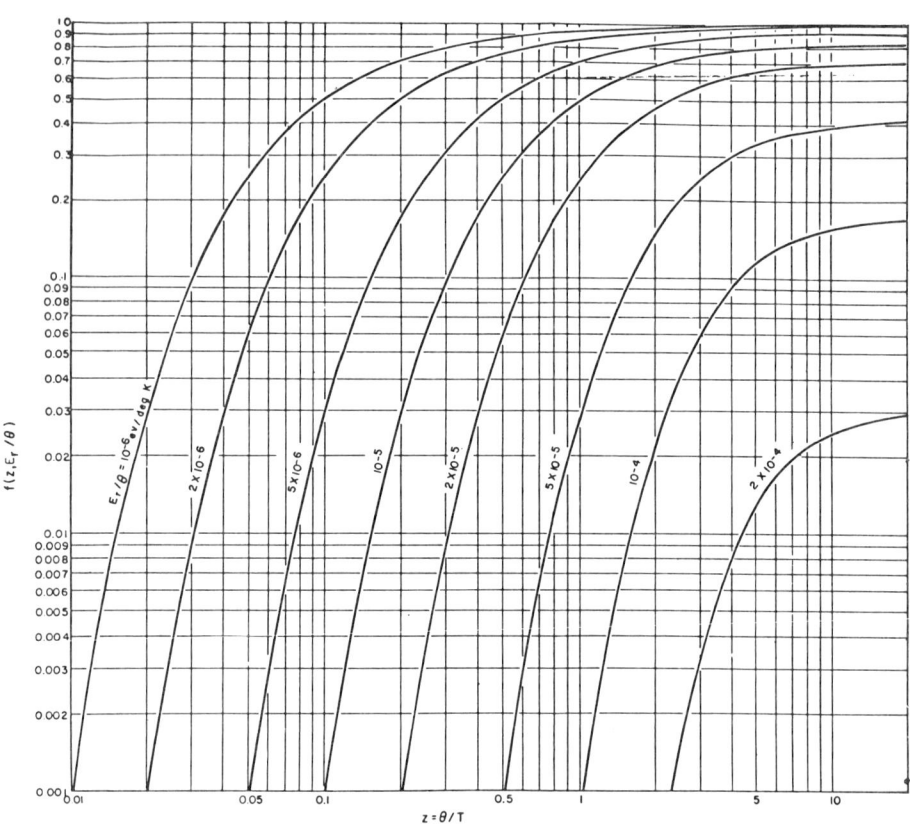

FIG. 20. Recoil-free fraction f as function of recoil energy E_r, Debye temperature θ, and temperature T. Each curve shows the variation of f for a constant value of E_r/θ (units of electron volts per degree Kelvin). (Courtesy of A. H. Muir, Jr., North American Rockwell Corp., Canoga Park, California.)

Bibliography

I. J. Gruverman, ed., "Mössbauer Effect Methodology," Vol. 1-6. Plenum Press, New York, 1965-71, presents the proceedings of a continuing series of conferences emphasizing the techniques of Mössbauer spectroscopy.

Two books: J. G. Stevens and V. E. Stevens, eds., "Mössbauer Effect Data Index" (Covering the 1969 Literature). IFI/Plenum Press, New York, 1970, and A. H. Muir, Jr., K. J. Ando, and H. M. Coogan, eds., "Mössbauer Effect Data Index, 1958-65." Wiley (Interscience), New York, 1966, summarize and tabulate the literature for those periods.

D. M. J. Compton and A. H. Schoen, eds., "The Mössbauer Effect." Wiley, New York, 1962, presents the proceedings of the 1961 Conference on the Mössbauer Effect, and presents some introduction and informal discussion of instrumentation problems.

V. I. Goldanskii and R. H. Herber, eds., "Chemical Applications of Mössbauer Spectroscopy." Academic Press, New York, 1968, contains a series of review articles emphasizing the techniques used to study particular isotopes.

E. Matthias and D. A. Shirley, eds., "Hyperfine Structure and Nuclear Radiations." North Holland Publ., Amsterdam, 1968, presents the proceedings of a 1967 conference in which Mössbauer spectroscopy was a major topic.

7. ULTRASONIC STUDIES OF THE PROPERTIES OF SOLIDS*

7.1. Introduction

Acoustic measurements stand as one of the three primary techniques for the study of the properties of matter: mechanical, electromagnetic, and particle-interaction measurements. The ultrasonic regime is normally regarded as that from about 20 kHz to 500 MHz. The lower audio and infrasonic ranges as well as the higher hypersonic and thermal-vibration ranges generally require quite different experimental techniques, although the physics of the phenomena are often closely related.

The ultrasonic range further subdivides into different ranges with different techniques. The two most common are those employing standing wave measurements used in the kilohertz range (20–1000 kHz) and the pulse propagation technique in the megahertz range (1–500 MHz). The kilohertz range was discussed in the 1959 article in this series by Read et al.[1] Since that time, megahertz pulse techniques have developed enormously and are the principal subject of the present article.

This chapter is intended to provide a practical introduction to the subject, primarily for those who are considering the use of ultrasonic pulse techniques in their studies of the properties of solids. It is not intended to be comprehensive, and in fact it will emphasize those methods in most common use in our laboratory and in others.

The pulse–echo technique developed rapidly after World War II, borrowing heavily from the technology developed for radar. Prominent names in the early developments were Huntington,[2] Mason and McSkimin,[3]

---

[1] T. A. Read, C. A. Wert, and M. Metzger, "Methods of Experimental Physics" (K. Lark-Horovitz and V. A. Johnson, eds.), Vol. 6A, p. 291. Academic Press, New York, 1959.

[2] H. B. Huntington, *Phys. Rev.* **72**, 321 (1947).

[3] W. P. Mason and H. J. McSkimin, *J. Acoust. Soc. Amer.* **19**, 464 (1947).

---

* Part 7 is by E. R. Fuller, Jr., A. V. Granato, J. Holder, and E. R. Naimon.

Arenberg,[4] and Roderick and Truell.[5] Although there are now scores of variations of the basic technique described in the literature, the simple pulse–echo method remains that used most often, and is even commercially available.†

The simplest version of this method provides for an ultrasonic pulse of 1 to 2 μsec duration to be launched into a specimen by means of a piezoelectric transducer bonded to the specimen. The sound pulse propagates through the specimen, is reflected at the other end, and it, as well as subsequent echoes, is detected by the transducer. The echoes are displayed on an oscilloscope and two quantities can be measured, the attenuation and the time of propagation of the sound pulse in the specimen. The latter is used primarily for the study of elastic constants, while the former has been used most in studies of the properties of imperfect crystals. More recently, the importance of simultaneous measurements of both quantities has come to be realized.

In contrast to the standing wave measurements used in the kilohertz range, the pulse–echo technique has the advantages that: (1) polarized shear waves may be used as well as longitudinal waves, (2) measurements can be made as a function of frequency, and (3) the average energy input to the crystal is low, making low-temperature measurements easier. On the other hand, measurements as a function of temperature are generally more difficult in the megahertz range, because of difficulties associated with differential expansion between the specimen and transducer.

The measures most commonly used for losses in the kilohertz range are the dimensionless logarithmic decrement Δ or the inverse Q factor. For megahertz measurements, the attenuation α of the amplitude of a plane wave $u = u_0 \exp[-\alpha x] \exp[i(\omega t - kx)]$ is the standard measure. The relationship between these is $\Delta = \pi Q^{-1} = \alpha \lambda$. Attenuations are sometimes expressed as reciprocal centimeters (nepers per centimeter) or as decibels per centimeter. The number of decibels (dB) is defined as $20 \log_{10}(u_2/u_1)$ so that an amplitude decrease by a factor of 10 corresponds to 20 dB and 1 Np $= 8.68$ dB. Often, attenuation is measured by the amplitude factor $\exp(-\beta t)$ replacing $\exp(-\alpha x)$ and is expressed in units of decibels per second, where $\beta = \alpha v$.

[4] D. L. Arenberg, *J. Appl. Phys.* **21**, 941 (1950).
[5] R. L. Roderick and R. Truell, *J. Appl. Phys.* **23**, 267 (1952).

† Suppliers mentioned most frequently in the literature are Arenberg Ultra-Laboratory, Inc. (94 Green St., Jamaica Plain, Massachusetts 02130) and Matec, Inc. (60 Montebello Road, Warwick, Rhode Island 02886).

7.1. INTRODUCTION

In the earliest measurements, there was insufficient appreciation of the requirements on parallelness of the specimen surfaces. This requirement can be severe at the highest frequencies, leading to optical tolerances, because of the fact that the transducers are phase-sensitive devices. If the phase of the wave front striking the transducer is positive on one half and negative on the other half, then the net response is zero. This effect provides the main limitation on the upper limit of the frequency range for attenuation measurements of low-loss materials.

At low frequencies, diffraction effects occur. The first measurements taking both nonparallelness and diffraction into account to provide absolute attenuation measurements over the megahertz range were by Granato and Truell[6] on germanium in 1956. Because of these corrections, frequency-dependence measurements, although possible, remain among the most difficult.

Measurements of attenuation as a function of other variables are most often made in the 10–30 MHz range. The sensitivity is normally greatest in this range because of an attenuation "window" effect. At lower frequencies the attenuation increases because of diffraction, while at higher frequencies the attenuation increases because of the normally increasing frequency dependence of various loss mechanisms. The optimum frequency depends on the loss mechanism and the diameter of the transducer.

In recent years, techniques described in a later section have been developed which are capable of detecting velocity changes as small as one part in 10^8. The sensitivity is limited by the attenuation of the material, the stability of the electronic circuits, and the temperature control of the specimens. In typical materials, a sensitivity of 10^{-8} corresponds to temperature changes of the order of 1/10 mdeg or of pressure changes of 10^{-2} atm. The high sensitivity permits one to apply nonhydrostatic stresses large enough to produce measurable effects while small enough to avoid plastic deformation of the specimens. This has opened up the field of study of the third-order elastic constants of materials. These constants measure the nonlinear elastic properties just as the ordinary elastic constants (second-order, corresponding to quadratic terms in elastic strain in the elastic energy density) measure the linear elastic properties of solids.

The increased sensitivities now available for velocity measurements also permit one to measure the full response (both in-phase and out-of-

[6] A. V. Granato and R. Truell, *J. Appl. Phys.* **27**, 1219 (1956).

phase components) of an imperfect solid to an elastic stress. This has allowed the fuller identification of mechanisms producing the attenuation. In fact, in many cases, velocity measurements now prove to be superior to attenuation measurements in accuracy for such purposes.

In the following section general characteristics of ultrasonic waves in solids are given, both for small and finite amplitudes. This is followed by a section on experimental techniques and then by one on examples of applications of these techniques to illustrate what one can do with ultrasonic studies. The discussion is incomplete. For useful general references, the reader is referred to books by Truell et al.,[7] Beyer and Letcher,[8] and Mason,[9] to many review articles on specialized topics in the series "Physical Acoustics," edited by Mason and Thurston,[10] and to reviews by Truell and Elbaum[11] and Dignum.[12] A complete bibliography on all articles in ultrasonics is maintained by the Journal of the Acoustical Society of America.

7.2. Ultrasonic Waves in Solids

The discussion will begin with the simple example of a longitudinal wave in an isotropic material. This case can be treated very easily and illustrates many of the principal features of general wave propagation. It therefore serves as a guide for the formal development of the general case which follows. The extension to the general case of small-amplitude waves in an anisotropic elastic solid is considered in Section 7.2.2. The modifications required by finite elasticity theory, needed for determinations of higher-order elastic constants, are discussed in Section 7.2.3. The thermoelasticity theory necessary to interpret measured elastic properties is summarized in Appendix A. Finally, nonelastic behavior

---

[7] R. Truell, C. Elbaum, and B. B. Chick, "Ultrasonic Methods in Solid State Physics." Academic Press, New York, 1969.

[8] R. T. Beyer and S. V. Letcher, "Physical Ultrasonics." Academic Press, New York, 1969.

[9] W. P. Mason, "Physical Acoustics and the Properties of Solids." Van Nostrand-Reinhold, Princeton, New Jersey, 1958.

[10] W. P. Mason and R. N. Thurston, "Physical Acoustics, Principles and Methods," Vols. I–VIII. Academic Press, New York, 1964.

[11] R. Truell and C. Elbaum, "Handbuch der Physik" (S. Flügge/Frieburg, eds.), Vol. X-1/2, p. 153. Springer-Verlag, Berlin, 1962.

[12] R. Dignum, Amer. J. Phys. **32**, 507 (1964).

7.2.1. Longitudinal Waves in an Isotropic Solid

For a simple example we consider an element of volume in a solid, as shown in Fig. 1. Its cross-sectional area normal to x is A, and its thickness is Δx. The left side of the face (at x) is acted upon by a normal stress (force per unit area) of σ, which pulls the face toward decreasing x.

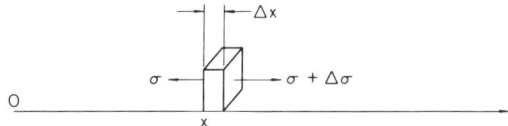

FIG. 1. Stresses acting on a volume element located at x. The cross-sectional area normal to the x direction is A.

On the right face (at $x + \Delta x$) the stress is in general $\sigma + \Delta\sigma$, and pulls toward increasing x. Hence the net force on the element is $A\,\Delta\sigma$. This leads to an acceleration of the element given by

$$A(\Delta x)\varrho\, \partial^2 u/\partial t^2 = A\,\Delta\sigma, \tag{7.2.1}$$

where ϱ is the density of the solid and $u = x - a$ is the displacement of the element from its equilibrium position a. Taking the limit of this as $\Delta x \to 0$, we have the equation of motion for the solid

$$\varrho\, \partial^2 u/\partial t^2 = \partial\sigma/\partial x. \tag{7.2.2}$$

In order to solve for the displacements, a second relationship between stress and the displacement is necessary. In a perfectly elastic solid this relationship is provided by the stress–strain law. The strains, which are displacement gradients

$$\varepsilon = \partial u/\partial a, \tag{7.2.3}$$

are related to the stresses by elastic constants. In most ultrasonic work, it is sufficient to consider only the lowest-order terms in strains so that

$$\sigma = c\varepsilon, \tag{7.2.4}$$

where c is the ordinary, or second-order, elastic constant. Also, no distinction is made between $\partial/\partial x$ and $\partial/\partial a$, or ϱ and ϱ_0 (the equilibrium density) in the infinitesimal development. Thus the equation of motion becomes

$$\varrho_0 \, \partial^2 u/\partial t^2 = c \, \partial^2 u/\partial a^2. \tag{7.2.5}$$

The solution of this equation depends on the boundary conditions involved, but the solution of most interest in ultrasonics is a plane-wave solution of the form

$$u = u_0 \exp[i(\omega t - Ka)]. \tag{7.2.6}$$

This represents a harmonic wave of angular frequency ω and wavelength $\lambda = 2\pi/K$, traveling with a (phase) velocity of ω/K. From Eq. (7.2.5), this velocity is found to be

$$W = \omega/K = (c/\varrho_0)^{1/2}. \tag{7.2.7}$$

Thus measurements of wave velocities provide direct measurements of the elastic constants of the solid. The details of the relationship between the elastic constants of a general, three-dimensional, anisotropic solid and the wave velocities measured ultrasonically are discussed in the next section.

7.2.2. Small-Amplitude Waves in Crystals

In a three-dimensional solid the displacements and position coordinates are vector quantities. In addition, when specifying the stress (force per unit area) acting on the solid, it is necessary to specify both the direction of the stress and the direction of the normal to the element of area on which it acts. These are conveniently expressed in the elements of the second-rank stress tensor $\boldsymbol{\sigma}$, where σ_{ij} is the ith component of the stress acting across an element of area whose normal is in the jth direction ($i, j = 1, 2, 3$). The ith component of the net force on an element of volume is given by the sum of the gradients of all three of the stress elements acting in the ith direction. Hence, the equation of motion is now given by

$$\varrho \, \partial^2 u_i/\partial t^2 = \partial \sigma_{ij}/\partial x_j, \tag{7.2.8}$$

where the Einstein summation convention, implying a sum over repeated indices ($j = 1, 2, 3$), is used.

Only symmetric combinations of displacement gradients are considered in defining the strain parameters

$$\varepsilon_{ij} = \tfrac{1}{2}(\partial u_i/\partial a_j + \partial u_j/\partial a_i), \tag{7.2.9}$$

since antisymmetric combinations, representing rigid rotations, do not affect the elastic strain energy. These strain parameters are also second-rank tensor elements, so the elastic constants relating stress and strain are fourth-rank tensors. The stress–strain law then becomes

$$\sigma_{ij} = c_{ijkl}\varepsilon_{kl} + \cdots, \qquad (7.2.10)$$

where the c_{ijkl} are the second-order elastic constants. In general, there are two types of elastic constants: adiabatic and isothermal. However, for wave propagation at megahertz frequencies, there is insufficient time for heat flow and the measured constants are adiabatic.

There are 3^4, or 81, of these second-order elastic constants. However, this number is considerably reduced by the symmetric definition of these constants. For example, the strains defined in Eq. (7.2.9) are symmetric with respect to the interchange of i and j. The stress tensor is also symmetric.[13] Thus the elastic constants c_{ijkl} are unchanged upon the interchanges $i \leftrightarrow j$ and $k \leftrightarrow l$. These symmetries reduce the number of independent elastic constants from 81 to 36. These 36 constants are customarily written in the reduced Voigt notation.[14] In this scheme each of the symmetric combinations i, j and k, l is replaced by a single index which ranges from 1 to 6 according to

i, j:	11	22	33	23 or 32	13 or 31	12 or 21
Voigt index:	1	2	3	4	5	6

The Voigt subscripts will be denoted by Greek letters, e.g., $c_{\alpha\beta}$.

It is shown in Appendix A that the elastic constants are related to derivatives of the strain energy. As a result, $c_{\alpha\beta} = c_{\beta\alpha}$, so that the number of independent elastic constants is further reduced to 21.

Finally, symmetries of the various crystal structures result in additional reductions. For example, in a cubic crystal there are only three independent second-order elastic constants: c_{11}, c_{12}, and c_{44}. The elastic constant arrays for the various crystal structures have been tabulated, for example, by Brugger.[15]

Much of the literature on elasticity has been concerned with isotropic materials, i.e., those for which all directions are elastically equivalent. For these materials there are only two independent second-order elastic

---

[13] F. D. Murnaghan, "Finite Deformation of an Elastic Solid." Wiley, New York, 1951.
[14] W. Voigt, "Lehrbuch der Kristallphysik." Teubner, Leipzig, 1928.
[15] K. Brugger, *J. Appl. Phys.* **36**, 759 (1965).

constants. The isotropy condition for the second-order elastic constants is seldom satisfied for crystals. Since crystals are necessarily anisotropic on an atomic scale, the isotropy is accidental and of little fundamental significance in those few cases where it is satisfied. For example, Mg is approximately isotropic in second order, but not in third order.[16] The isotropy condition leads to great simplicity in the theory of elasticity, but applies in general only to polycrystalline and amorphous materials. Isotropic elasticity theory is thus of great importance in technical applications, but not in physical discussions focusing on relations between elastic constants and interatomic potentials. As we are primarily interested in the latter, the anisotropic formalism is used here. A useful source on isotropic elasticity theory, including wave propagation, is Landau and Lifshitz.[17]

For the development of the theory of wave propagation in crystals, it is convenient to use and thereby exploit the notational advantages of the full tensor character of the elastic constants in general relations, and to use the reduced notation only for specific examples. The wave equation obtained by using Eq. (7.2.10) in Eq. (7.2.8) is

$$\varrho_0(\partial^2 u_i/\partial t^2) = (\varrho_0/\varrho)(\partial a_j/\partial x_p)c_{ipkl}(\partial^2 u_k/\partial a_j\, \partial a_l) + \cdots . \qquad (7.2.11)$$

For small-amplitude waves, this is linearized by evaluating Eq. (7.2.11) at $\mathbf{x} = \mathbf{a}$ (the undeformed state)

$$\varrho_0\, \partial^2 u_i/\partial t^2 = c_{ijkl}\, \partial^2 u_k/\partial a_j\, \partial a_l . \qquad (7.2.12)$$

A solution of the equation of motion is given by

$$u_i = u_i^0 \exp[i(\omega t - \mathbf{K} \cdot \mathbf{a})], \qquad (7.2.13)$$

where $|\mathbf{K}| = 2\pi/\lambda$. Solutions of this nature represent harmonic plane waves of infinite extent, with planes of equal phase (wave fronts) perpendicular to the $\mathbf{K}$ direction (the propagation direction). The distance between two adjacent planes of equal phase, measured along the $\mathbf{K}$ direction, is λ. The velocity with which a plane of a certain phase of a traveling wave of angular frequency ω travels is the phase velocity, given by

$$W = \omega/|\mathbf{K}|. \qquad (7.2.14)$$

[16] E. R. Naimon, *Phys. Rev. B* **4**, 4291 (1971).
[17] L. D. Landau and E. M. Lifshitz, "Theory of Elasticity." Pergamon, Oxford, 1959.

It is the distance the plane travels in the **K** direction per unit time. Using Eq. (7.2.13) in Eq. (7.2.12) gives the condition

$$(c_{ijkl}N_jN_l - \varrho_0 W^2 \delta_{ik})u_k = 0, \qquad (7.2.15)$$

where $\mathbf{N} = \mathbf{K}/|\mathbf{K}|$ and δ_{ik} is the Kronecker delta. Thus the effective elastic coefficients $\varrho_0 W^2$ in a given direction **N** are given by eigenvalues of the tensor $c_{ijkl}N_jN_l$. The corresponding eigenvectors give the polarization vector or directions of particle displacement. Since this second-rank tensor is symmetric (i.e., with respect to the interchange $i \leftrightarrow k$), the polarization vectors for the three modes of wave motion in a given direction are mutually perpendicular for distinct eigenvalues. In order to have real propagation velocities (i.e., positive eigenvalues), this tensor must also be positive definite. Requirements on the elastic constants in order to satisfy this condition have been discussed by Toupin and Bernstein,[18] Truesdell,[19] and Truesdell and Toupin.[20]

Thus for a given direction in a solid, three small-amplitude plane waves with mutually perpendicular polarizations can propagate. Waves with polarization vectors parallel or perpendicular to the propagation direction are denoted as longitudinal or transverse (shear) waves, respectively. However, in general, a wave is neither purely longitudinal nor transverse. Two special types of directions are commonly of interest in wave propagation experiments: (1) pure mode directions, which have one pure longitudinal wave and two pure transverse waves; and (2) quasi-pure mode directions, which have one pure transverse wave and two mixed waves. Symmetry directions of a crystal are always pure mode directions,[21,22] since the symmetry rotation of any polarization that is not longitudinal or transverse would lead to another independent mode. If the symmetry axis is greater than twofold, similar arguments require that the transverse modes are degenerate and the polarizations are arbitrary.

The elastic constant combinations which determine $\varrho_0 W^2$ are given in Table I for the symmetry directions in cubic crystals. In isotropic materials all directions are pure mode directions, and there is only one longitudinal and one transverse mode. The isotropic condition can be

---

[18] R. A. Toupin and B. Bernstein, *J. Acoust. Soc. Amer.* **33**, 216 (1961).
[19] C. Truesdell, *Arch. Rational Mech. Anal.* **8**, 263 (1961).
[20] C. Truesdell and R. Toupin, *Arch. Rational Mech. Anal.* **12**, 1 (1963).
[21] P. C. Waterman, *Phys. Rev.* **113**, 1240 (1959).
[22] Fedor I. Federov, "Theory of Elastic Waves in Crystals" (translated by J. E. S. Bradley). Plenum Press, New York, 1968.

TABLE I. Elastic Constants for Pure Mode Directions in Cubic Crystals

Propagation direction	Polarization direction	$c = \varrho_0 W^2$
[100]	[100]	c_{11}
	Any direction $\perp$ [100]	c_{44}
[110]	[110]	$(c_{11} + c_{12} + 2c_{44})/2$
	[001]	c_{44}
	[1$\bar{1}$0]	$(c_{11} - c_{12})/2$
[111]	[111]	$(c_{11} + 2c_{12} + 4c_{44})/3$
	Any direction $\perp$ [111]	$(c_{11} - c_{12} + c_{44})/3$

obtained from the cubic crystal results by requiring the two $\langle 110 \rangle$ shear constants to be equal $[(c_{11} - c_{12})/2 = c_{44}]$. The longitudinal and shear constants are then given by c_{11} and c_{44}, respectively.

For general crystal symmetry, the pure mode directions are determined from

$$\mathbf{u}^l = \mathbf{N} \quad \text{or} \quad \mathbf{u}^l \times \mathbf{N} = 0, \qquad (7.2.16)$$

where $\mathbf{u}^l$ is the polarization of a longitudinal wave. Using Eq. (7.2.15), this becomes[15,23]

$$\varepsilon_{ipq} c_{pjkl} N_j N_k N_l N_q = 0, \qquad (7.2.17)$$

where ε_{ipq} is the alternating tensor ($\varepsilon_{ipq} = +1$ if i, p, and q are unequal and in cyclic order, -1 if i, p, and q are unequal and not in cyclic order, and 0 otherwise). The pure mode directions for all the crystalline systems, except monoclinic and triclinic, have been tabulated by Brugger.[15]

The propagation velocity considered thus far has been the phase velocity. However, the magnitude and direction of the flow of energy contained in an element of volume of a plane wave is given by the energy flux density vector.[21,22,24] The velocity with which the energy moves, obtained by dividing the energy flux by the energy density, is given by the group velocity

$$\mathbf{W}_g = (\partial \omega / \partial \mathbf{K}). \qquad (7.2.18)$$

The direction of $\mathbf{W}_g$ determines the path traveled by a packet of waves.

---

[23] F. E. Borgnis, *Phys. Rev.* **98**, 1000 (1955).
[24] T. Hinton and R. E. Green, Jr., *Trans. Met. Soc. AIME* **236**, 439 (1966).

The expression for the group velocity can be found by putting the **K** dependence explicitly in Eq. (7.2.15):

$$\varrho_0 \omega^2 u_i = c_{ipkl} K_p K_l u_k. \tag{7.2.19}$$

Differentiating with respect to K_j, multiplying the result by u_i, and summing [noting that $u_i(\partial u_i/\partial K_j) = 0$] gives

$$(1/W)(\mathbf{W}_g)_j = (1/\varrho_0 W^2) c_{ijkl} N_l u_i u_k. \tag{7.2.20}$$

By taking the inner product of the group velocity $\mathbf{W}_g$ with the propagation direction **N**, the relationship between the group and phase velocities is found to be

$$\mathbf{W}_g \cdot \mathbf{N} = W. \tag{7.2.21}$$

Thus the phase velocity is always less than the group velocity, except when the group velocity is parallel to the propagation direction. Then they are equal. This difference in magnitude results from the difference in path lengths involved in the two velocities. Hence in ultrasonic measurements, where only transit times of waves are measured, phase velocities are always determined.

By arguments similar to those above,[22] it can be seen that the phase and group velocities are equal for evenfold symmetry axes. Furthermore since $c_{ijkl} = c_{jikl}$, it is easily seen from Eqs. (7.2.20) and (7.2.15) that $\mathbf{W}_g = \mathbf{N}W$ for pure longitudinal modes. (An exception to this sometimes occurs for longitudinal waves propagating in a homogeneously stressed solid. This is discussed further in the following section.)

7.2.3. Finitely Strained Solids

7.2.3.1. Introduction.
Thus far, the development has considered only lowest-order terms in the stress–strain relationship. This is the realm of infinitesimal elasticity theory and, in general, is satisfactory for discussing the properties of small-amplitude waves and their relationship to the second-order elastic constants. However, with the advent of more sensitive techniques for measuring velocities, an important aspect of ultrasonics has become the measurement of third-order elastic constants. In order to relate these higher-order elastic constants to velocity measurements, the finite nature of strains must be taken into account.†

---

† The derivations of many of the relations of general thermoelasticity theory used in this section are summarized in Appendix A.

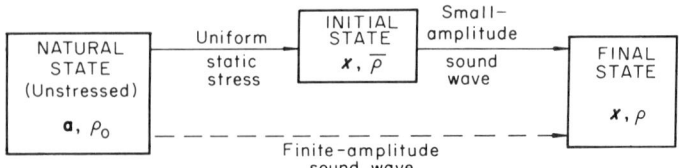

Fig. 2. States of an elastic solid with two possible paths to the finitely deformed final state. Coordinates and densities are given for each state.

Two methods are commonly used for obtaining higher-order elastic constants: (1) measurements of the static-stress dependence of small-amplitude sound waves propagating in a uniformly stressed solid; and (2) measurements of the generation of higher harmonics by the propagation of finite-amplitude sound waves. In both cases, the final state of the solid (i.e., the state of wave propagation) is finitely deformed. In the former case, the finite deformation is due to an initially applied, uniform static stress, and in the latter case due to a finite-amplitude sound wave, as illustrated schematically in Fig. 2. Although the final states for the two cases are not the same, the same formalism can be used for both. Accordingly, the deformation parameters used to relate the different configurations are the transformation coefficients. From the natural to the final configuration they are written as

$$J_{ij} = (\partial x_i/\partial a_j), \qquad (7.2.22)$$

and from the initial to the final configuration as

$$\tilde{J}_{ij} = (\partial x_i/\partial X_j), \qquad (7.2.23)$$

where $\mathbf{a}$, $\mathbf{X}$, and $\mathbf{x}$ are the positions of the same material particle in the natural, initial, and final configurations, respectively. The Lagrangian or finite strain parameters are

$$\eta_{ij} = \tfrac{1}{2}(J_{pi}J_{pj} - \delta_{ij}) \quad \text{and} \quad \tilde{\eta}_{ij} = \tfrac{1}{2}(\tilde{J}_{pi}\tilde{J}_{pj} - \delta_{ij}). \qquad (7.2.24)$$

As in Appendix A, variables referring to the initial state have a tilde over them, and those referring to the natural state, but evaluated in the initial state, have a bar over them.

7.2.3.2. General Wave Equation. The equation of motion of an elastic solid in the absence of body forces is

$$\rho \ddot{x}_i = (\partial \sigma_{ij}/\partial x_j), \qquad (7.2.25)$$

where all quantities are evaluated in the final state. It should be noted that this is the same equation of motion obtained above in Section 7.2.2, and no modifications due to finite elasticity are necessary. In the finite elasticity treatment, the only modification to the infinitesimal development of Section 7.2.2 is in the expression used for the stress σ_{ij}. There the stress was expanded in a Taylor series of infinitesimal strains ε_{ij}, keeping only lowest order. Thus keeping higher-order terms in this expansion would appear to be the simplest approach for considering finite strains. However, in order to identify the resulting higher-order elastic coefficients with the standard Brugger definition of the higher-order elastic constants,[25] it is necessary to use general thermoelasticity theory. So instead, the thermodynamic expression for the stress[13,26,27] [Eq. (A.13) of Appendix A] is used in Eq. (7.2.25). The general wave equation then becomes

$$\varrho_0 \ddot{x}_i = A^S_{ijkl}(\partial^2 x_k / \partial a_j \, \partial a_l), \tag{7.2.26}$$

where all quantities are evaluated at **x** and the superscript S denotes adiabatic processes. The "strain-dependent" elastic coefficients A^S_{ijkl}, which replace c^S_{ijkl} of Eq. (7.2.11) and reduce to them for infinitesimal strains, are given by [Eq. (A.29)]

$$A^S_{ijkl} = t_{jl}\delta_{ik} + J_{im}J_{kn}[c^S_{mjnl} + c^S_{mjnlpq}\eta_{pq} + \tfrac{1}{2}c^S_{mjnlpqrs}\eta_{pq}\eta_{rs} + \cdots], \tag{7.2.27}$$

where the thermodynamic tensions t_{jl} are related to the stress and internal energy by [Eqs. (A.12) and (A.13)]

$$t_{kl} = (\varrho_0/\varrho)J^{-1}_{ki}J^{-1}_{lj}\sigma_{ij} = \varrho_0(\partial U/\partial \eta_{kl}) \tag{7.2.28}$$

and the adiabatic Brugger elastic constants are

$$c^S_{mjnl\cdots} = \varrho_0(\partial^n U/\partial \eta_{mj}\, \partial \eta_{nl} \cdots)|_a. \tag{7.2.29}$$

A similar wave equation can be written in terms of the initial coordinates **X** as

$$\bar{\varrho}\ddot{x}_i = W^S_{ijkl}(\partial^2 x_k/\partial X_j\, \partial X_l), \tag{7.2.30}$$

where the expression for W^S_{ijkl} [Eq. (A.28)] is the same as that for A^S_{ijkl}

---

[25] K. Brugger, *Phys. Rev.* **133**, A1611 (1964).
[26] R. N. Thurston, *in* "Physical Acoustics" (W. P. Mason, ed.), Vol. IA, p. 1. Academic Press, New York, 1964.
[27] D. C. Wallace, *Phys. Rev.* **162**, 776 (1967).

with a tilde over all variables and $c^S_{mjnl}...$ replaced by $C^S_{mjnl}...$, the stress-dependent, adiabatic Brugger elastic coefficients defined about the initial state.

7.2.3.3. Small-Amplitude Waves. For small-amplitude plane waves, linearized wave equations are obtained by evaluating Eqs. (7.2.26) and (7.2.30) in the homogeneously deformed initial state. The elastic coefficients

$$\bar{A}^S_{ijkl} = \bar{t}_{jl}\delta_{ik} + \bar{J}_{im}\bar{J}_{kn}\varrho_0(\partial^2 U/\partial\eta_{mj}\,\partial\eta_{nl})|_X \qquad (7.2.31)$$

and

$$\bar{W}^S_{ijkl} = \bar{\sigma}_{jl}\delta_{ik} + C^S_{ijkl} \qquad (7.2.32)$$

are no longer strain dependent, and indeed reduce to c^S_{ijkl} when the initial stress vanishes. As before, one looks for plane-wave solutions in the form

$$x_i - a_i = u_i \exp[i(\omega t - \mathbf{K}\cdot\mathbf{a})], \qquad (7.2.33)$$

where $\mathbf{u}$ is the vector displacement, $\mathbf{K}$ is the wave vector, ω the angular frequency, and t the time. The propagation direction in the natural state and the natural velocity are given by[28] $\mathbf{N} = \mathbf{K}/|\mathbf{K}|$ and $W = \omega/|\mathbf{K}|$, respectively. Upon substitution into Eq. (7.2.26), the condition for wave propagation is

$$(\bar{A}^S_{ijkl}N_jN_l - \varrho_0 W^2\delta_{ik})u_k = 0. \qquad (7.2.34)$$

Similarly, for a wave description in terms of the initial coordinates, an assumed plane-wave solution in the form

$$x_i - X_i = \tilde{u}_i \exp[i(\omega t - \mathbf{k}\cdot\mathbf{X})] \qquad (7.2.35)$$

gives the wave propagation condition

$$(\bar{W}^S_{ijkl}n_jn_l - \bar{\varrho}v^2\delta_{ik})\tilde{u}_k = 0, \qquad (7.2.36)$$

where $\mathbf{n} = \mathbf{k}/|\mathbf{k}|$ and $v = \omega/|\mathbf{k}|$ are the propagation direction and velocity, respectively.

By comparison of Eqs. (7.2.34) and (7.2.36) with Eq. (7.2.15), it is easily seen that most of the properties of small-amplitude wave propagation in an unstressed solid (natural state) are valid for waves propagating in a homogeneously deformed solid. The exceptions, which are noted below, are due to the fact that the elastic coefficients $\bar{W}^S_{ijkl}$ and $\bar{A}^S_{ijkl}$ do not have complete Voigt symmetry, as does c^S_{ijkl}. (These symmetries

7.2. ULTRASONIC WAVES IN SOLIDS

are discussed in Appendix A.) The pure-mode condition, Eq. (7.2.17), is unaltered, except for being expressed in terms of initial state variables, i.e.,

$$\varepsilon_{ipq}C^S_{pjkl}n_j n_k n_l n_q = 0. \qquad (7.2.37)$$

[The term in the initial stress ($\bar{\sigma}_{jl}$ in $\overline{W}^S_{ijkl}$) vanishes because it is of the form $\varepsilon_{ipq}\tilde{u}^1_p n_q n_j \bar{\sigma}_{jl} n_l = (\tilde{\mathbf{u}}^1 \times \mathbf{n})_i (n_j \bar{\sigma}_{jl} n_l)$.] The group velocity, $\mathbf{v}_g = (\partial\omega/\partial\mathbf{k})$, in the initial state becomes

$$(1/v)(\mathbf{v}_g)_j = (1/v)(\partial\omega/\partial k_j) = (1/\bar{\varrho}v^2)\overline{W}^S_{ijkl}n_l\tilde{u}_i\tilde{u}_k$$
$$= (1/\bar{\varrho}v^2)[\bar{\sigma}_{jl}n_l + C^S_{ijkl}n_l\tilde{u}_i\tilde{u}_k], \qquad (7.2.38)$$

and thus, as before,

$$\mathbf{v}_g \cdot \mathbf{n} = v. \qquad (7.2.39)$$

Also, by taking the cross product of $\mathbf{v}_g$, Eq. (7.2.38), with $\mathbf{n}$ and using the pure mode condition, Eq. (7.2.37), for longitudinal waves, it is easily shown that

$$(1/v)(\mathbf{n} \times \mathbf{v}_g)_i = (1/\bar{\varrho}v^2)[\varepsilon_{ikj}n_k\bar{\sigma}_{jl}n_l]. \qquad (7.2.40)$$

Thus the group velocity is parallel to the propagation direction for all longitudinal waves, whenever the initial stress satisfies $\mathbf{n} \times (\bar{\boldsymbol{\sigma}}\mathbf{n}) = 0$. This condition holds, for example, for an unstressed initial state (the case considered above in Section 7.2.2), for hydrostatic pressure, or for a uniaxial stress either perpendicular or parallel to the propagation direction.

The usual method of experimentally determining third-order elastic constants is by measuring the static-stress dependence of the second-order elastic coefficients. In the laboratory one generally knows the unstressed specimen length L_0 and the propagation direction $\mathbf{N}$ in the natural state, and not the deformed length L and actual propagation direction $\mathbf{n}$. Thus, from the experimental viewpoint, it is usually more convenient to use the natural configuration as the reference state. Accordingly, Thurston and Brugger[28] derived general equations for the stress dependence of $\varrho_0 W^2$ in terms of the second- and third-order elastic constants. To obtain a rotationally invariant form for the equation of motion, Eq. (7.2.34), the polarization vectors $\mathbf{u}$ are transformed according to[18]

$$u_k = \bar{J}_{kq}U_q. \qquad (7.2.41)$$

[28] R. N. Thurston and K. Brugger, *Phys. Rev.* **133**, A1604 (1964).

This eliminates the rotational dependence of $\bar{A}^S_{ijkl}N_j N_l$ for rotations of the initial configuration relative to the natural configuration. Then Eq. (7.2.34) becomes

$$(w_{pq} - \varrho_0 W^2 \delta_{pq})U_q = 0, \qquad (7.2.42)$$

where

$$\begin{aligned} w_{pq} &= \bar{J}^{-1}_{pi}\bar{J}_{kq}\bar{A}^S_{ijkl}N_j N_l \\ &= N_j N_l[\bar{t}_{jl}\delta_{pq} + \bar{J}_{kq}\bar{J}_{kn}\varrho_0(\partial^2 U/\partial\eta_{pj}\,\partial\eta_{nl})_S\,|\mathbf{x}]. \end{aligned} \qquad (7.2.43)$$

From the experimental viewpoint it is sufficient to consider only initial configurations where the stress depends on a single scalar variable P,[28] where P is the applied pressure for hydrostatic stress experiments and the applied force per undeformed area for uniaxial stress experiments. The experimentally measured quantity is

$$(\varrho_0 W^2)' = (\partial[\varrho_0 W^2]/\partial P)_T. \qquad (7.2.44)$$

Differentiation of Eq. (7.2.42) with respect to P at constant temperature and evaluation at $P = 0$ yields[28]

$$\begin{aligned}(\varrho_0 W^2)'_{P=0} &= (U_p w'_{pq} U_q)_{P=0} \\ &= (\bar{t}'_{ab})_{P=0}[N_a N_b + U^0_p U^0_q(2ws^T_{pqab} + s^T_{rsab}c_{pjqlrs}N_j N_l)], \end{aligned} \qquad (7.2.45)$$

where

$$w = (\varrho_0 W^2)_{P=0} = c^S_{pjql}N_j N_l U^0_p U^0_q, \qquad (7.2.46)$$

and $\mathbf{U}^0$ is the limiting polarization vector $\mathbf{U}$ as $P \to 0$ for the corresponding eigenvalue w. The isothermal compliances, $s^T_{rsab} = (\partial\eta_{rs}/\partial t_{ab})_T$, are the inverse of the isothermal Brugger elastic constants, i.e., $s^T_{rsab}c^T_{abmn} = \frac{1}{2}(\delta_{rm}\delta_{sn} + \delta_{rn}\delta_{sm})$. The third-order elastic constants obtained by this analysis are mixed derivatives (i.e., neither purely adiabatic nor isothermal), namely

$$c_{pjqlrs} = \varrho_0[\partial/\partial\eta_{rs}\,(\partial^2 U/\partial\eta_{pj}\,\partial\eta_{ql})_S]_T. \qquad (7.2.47)$$

Conversion of these derivatives to the pure derivatives, which are usually calculated theoretically, involves small corrections which have been given by Brugger.[25]

For initial hydrostatic pressure or uniaxial compression in the direction of the unit vector $\mathbf{M}$, the stress derivative of the thermodynamic tension is[28]

$$(\bar{t}'_{ab})_{P=0} = \begin{cases} -\delta_{ab} & \text{for hydrostatic pressure,} \quad (7.2.48) \\ -M_a M_b & \text{for uniaxial compression.} \quad (7.2.49) \end{cases}$$

7.2. ULTRASONIC WAVES IN SOLIDS 387

For these two types of stress, Eq. (7.2.45) can be written in the form[29]

$$-(\varrho_0 W^2)'_{P=0} = \delta + 2wF + G. \qquad (7.2.50)$$

For hydrostatic pressure, the parameters in Eq. (7.2.50) are

$$\delta = 1, \quad F = s^T_{pqaa} U_p{}^0 U_q{}^0, \quad G = s^T_{rsaa} c_{pjqlrs} N_j N_l U_p{}^0 U_q{}^0, \qquad (7.2.51)$$

and for uniaxial compression they are

$$\delta = (\mathbf{N} \cdot \mathbf{M})^2, \quad F = s^T_{pqab} U_p{}^0 U_q{}^0 M_a M_b, \quad G = s^T_{rsab} c_{pjqlrs} N_j N_l U_p{}^0 U_q{}^0 M_a M_b. \qquad (7.2.52)$$

These parameters as well as w have been tabulated by Brugger[29] for pure mode directions in all of the crystalline systems except monoclinic and triclinic. As an example, consider a transverse wave in a uniaxially

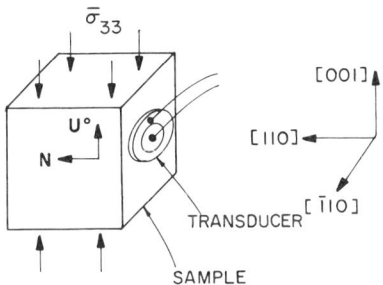

FIG. 3. Typical uniaxial-compression experiment to obtain third-order elastic constants (in this case a combination of c_{144} and c_{166}).

compressed crystal of cubic symmetry with propagation direction $\mathbf{N} = [110]$, polarization direction $\mathbf{U}^0 = [001]$, and loading direction $\mathbf{M} = [001]$ (see Fig. 3). A measurement of the natural velocity (i.e., transit time) versus stress gives the experimentally measured quantity

$$-(\varrho_0 W^2)'_{P=0} = 2c^S_{44} s^T_{11} + s^T_{12} c_{144} + (s^T_{11} + s^T_{12}) c_{166}.$$

7.2.3.4. Finite-Amplitude Waves. In order to treat finite-amplitude waves, one must use the strain-dependent coefficients W_{ijkl} or A_{ijkl} of Eq. (7.2.27). However, it is usually sufficient to consider only the first few terms of the expansion in strain. Thurston and Shapiro[30] have

---
[29] K. Brugger, *J. Appl. Phys.* **36**, 768 (1965).
[30] R. N. Thurston and M. J. Shapiro, *J. Acoust. Soc. Amer.* **41**, 1112 (1967).

obtained Eq. (7.2.27) for A_{ijkl}, expanded to second-order in the displacement gradients $u_{ij} = (\partial u_i/\partial a_j) = J_{ij} - \delta_{ij}$. Some of the features of finite-amplitude waves can be seen by keeping only third-order terms in the equation of motion[30]

$$\varrho_0 \ddot{u}_i = (\partial^2 u_k/\partial a_j\, \partial a_l)[c_{ijkl} + (\partial u_p/\partial a_q)(M_{ijklpq})], \quad (7.2.53)$$

where

$$M_{ijklpq} = c_{ijklpq} + \delta_{kp}c_{ijlq} + \delta_{ik}c_{jlpq} + \delta_{ip}c_{jqkl}. \quad (7.2.54)$$

It is found that pure transverse solutions do not exist for this equation, and the polarizations are no longer mutually perpendicular.[31] An approximate solution to Eq. (7.2.53), for short distances of propagation of a purely longitudinal wave, is given by an expression of the form[31]

$$u_i = A \sin(Ka - \omega t) - A^2 B^2 K^2 a \cos[2(Ka - \omega t)], \quad (7.2.55)$$

where a is the distance measured along the propagation direction and B is a constant involving second- and third-order elastic constants. Thus, if a pure sinusoidal longitudinal wave of angular frequency ω and amplitude A is introduced along a pure mode direction of a crystal, and the amplitude of the longitudinal wave of angular frequency 2ω is measured as a function of distance of propagation (or as a function of the amplitude of the initial wave), the third-order constants in B can be determined. In cubic crystals there are three pure longitudinal mode directions ([100], [110], and [111]), so that three combinations of third-order constants can be measured as shown in Table II. These measurements have been carried out in copper by Gauster and Breazeale,[32] who find

TABLE II. Combination of Third-Order Elastic Constants Appearing in B of Eq. (7.2.55) in Cubic Crystals

Propagation direction	Third-order constants
[100]	c_{111}
[110]	$c_{111} + 3c_{112} + 12c_{166}$
[111]	$c_{111} + 6c_{112} + 12c_{144} + 24c_{166} + 2c_{123} + 16c_{456}$

[31] M. A. Breazeale and J. Ford, *J. Appl. Phys.* **36**, 3486 (1965).
[32] W. B. Gauster and M. A. Breazeale, *Phys. Rev.* **168**, 655 (1968).

results in agreement with other measurements in copper.[33,34] Reviews of finite-amplitude waves have been given by Beyer[35] and Thurston and Shapiro.[30]

Unfortunately, the presence of dislocations in the material also leads to the generation of second harmonics,[36,37] so care must be taken to eliminate these effects if second harmonic generation is to be used to measure third-order elastic constants. There is one further complication in this type of measurement, in addition to the dislocation problem mentioned. When a sinusoidal wave is introduced, it becomes distorted as it progresses through the medium, and a discontinuity in the first derivative of the displacement will eventually occur,[30,31] forming a shock wave. The distance traveled before this discontinuity occurs is called the *discontinuity distance*, and is also a function of the second-order and third-order elastic constants.[30,31] The above analysis is only valid for propagation distances short compared to this discontinuity distance. As an example, the discontinuity distance in germanium is approximately 100–500 cm for an initial displacement amplitude of 10^{-8} cm.[31]

7.2.4. Attenuation and Dispersion

7.2.4.1. Intrinsic Attenuation and Dispersion.

In a real solid there will always be some dissipative mechanisms present such as those involving dislocations, relaxing point defects, or conduction electrons. Hence the solid is not perfectly elastic, and stress is not simply related to elastic strain alone. Additional types of terms must be included in the stress–strain law (such as the dislocation strain for the case of dislocation motion) and the stress is, in general, no longer in phase with the strain motion. These lead to attenuation of sound waves as well as dispersion in the velocity and a further difference between group and phase velocities. The most convenient way of treating this behavior in the wave propagation analysis is to allow K in Eq. (7.2.6) to be a complex number $K_0 - i\alpha$, or to write the plane-wave solution as

$$u = u_0 \exp[-\alpha a] \exp[i(\omega t - K_0 a)], \qquad (7.2.56)$$

---

[33] Y. Hiki and A. V. Granato, *Phys. Rev.* **144**, 411 (1966).
[34] K. Salama and G. A. Alers, *Phys. Rev.* **161**, 673 (1967).
[35] R. T. Beyer, *in* "Physical Acoustics" (W. P. Mason, ed.), Vol. IIB, p. 231. Academic Press, New York, 1965.
[36] A. Hikata, B. B. Chick, and C. Elbaum, *Appl. Phys. Lett.* **3**, 195 (1963).
[37] T. Suzuki, A. Hikata, and C. Elbaum, *J. Appl. Phys.* **35**, 2761 (1964).

where a is the propagation direction and α is the attenuation coefficient. The treatment of attenuation and velocity changes, and the relationship between the two effects, is analogous to that found in dielectric and magnetic susceptibility theory. Detailed models for the attenuation and velocity change are briefly discussed for a few mechanisms in Section 7.4.

7.2.4.2. Apparent Attenuation and Dispersion. In order to measure absolute values of intrinsic attenuation and velocity, other apparent sources of loss must first be identified and corrected for. For example, as will be discussed in Section 7.3.2, the sample faces from which sound waves are reflected must be flat and parallel to within a small fraction of the sound wavelength. Otherwise, interference will occur between different regions of the pickup transducer because of slight variations in the path length traveled by waves in different regions of the sample. This leads to an apparent nonexponential loss. Further, the finite extent of the sound wave can affect both the attenuation and velocity measurement. In the usual megahertz measurement, a transducer having lateral dimensions less than those of the sample is used to introduce pulses of short duration ($\sim$10 cycles). Hence one has the choice of measuring the velocity of the maximum amplitude of the envelope (group velocity) or of a certain phase of one of the cycles inside the envelope (phase velocity). Both these velocities are altered by the finite length of the pulse in a dispersive medium, but fortunately the correction is small in most cases. The correction (typically amounting to less than 0.1%) can be estimated[38] provided the pulse is long compared to the period of oscillation. However, there is a limitation in that a particular phase of a cycle initially inside the envelope will eventually be shifted to the edge of the envelope because of the difference between W_g and W. Detection of that particular point will become impossible. This would only become serious if W_g and W were substantially different.

The finite lateral extent of the wave, limited by the transducer size, is a more serious problem. As the signal diverges from the finite source, interference between elements of the wave traveling different paths is again possible, thereby contributing an additional amount to the measured attenuation and velocity. Velocity corrections are generally small but attenuation corrections can be significant, especially in very low attenuation samples. This apparent attenuation due to diffraction has been

---

[38] G. L. Wire, thesis, Univ. of Illinois (1972).

estimated by Seki et al.[39] to be on the order 1 dB/distance of R^2/λ, where R is the radius of the source and λ is the sound wavelength. This gives a correction on the order of 0.13 dB/cm for $R = 0.635$ cm, $W = 4.8 \times 10^5$ cm/sec, and $f = 15$ MHz. More recent calculations by Papadakis[40] and Cohen[41] take into account effects of anisotropy on diffraction. In any case, diffraction effects can be separated, in principle, by finding the magnitude of the component of attenuation that varies as $1/f$.

A number of other factors also affect the results of an actual measurement. (1) When a pure longitudinal or shear wave is coupled into a sample direction whose allowed polarizations are neither pure shear or longitudinal (when the sample is not exactly oriented, for example), two or more waves of the allowed polarization are created. These waves can interfere to produce apparent attenuation and velocity changes. (2) Energy loss into the pick-up transducer leads to phase shift and attenuation corrections. (3) Because of finite sample size, sidewall reflections can affect the experimental results. A consideration of these matters requires a quantitative discussion of reflection and refraction phenomena, which is not given here. The reader is referred to standard works, such as Mason's "Physical Acoustics and the Properties of Solids,[9]" for details.

7.3. Experimental Techniques

Experimental techniques for measuring ultrasonic velocities, velocity changes, and attenuation are reviewed in this section. The generation and detection of ultrasonic waves in solids, both by piezoelectric transducers and direct rf generation, are discussed in Section 7.3.1. The necessity and techniques for sample preparation are given in Section 7.3.2. Finally, a review of the various experimental methods for measuring velocity, small velocity changes, and attenuation are presented in Section 7.3.3.

7.3.1. Generation of Ultrasonic Waves

7.3.1.1. Transducers. The generation of ultrasonic waves in solids is usually accomplished by means of the conversion of some type of energy (usually electrical) into mechanical energy, and the detection of the waves by the reverse process. A device that performs this conversion is called

---

[39] H. Seki, A. V. Granato, and R. Truell, *J. Acoust. Soc. Amer.* **28**, 230 (1956).
[40] E. P. Papadakis, *J. Acoust. Soc.* **35**, 490 (1963); **36**, 414 (1964); **40**, 863 (1966).
[41] M. G. Cohen, *J. Appl. Phys.* **38**, 3821 (1967).

a *transducer*. Several types of effects can be utilized to accomplish this energy conversion.

The most common is the piezoelectric effect. The application of an external mechanical strain to a piezoelectric crystal produces an electric polarization, and thus, the appearance of electric charges on certain faces of the crystal. The inverse effect is also possible, namely, strains are produced in a piezoelectric crystal when electric charges are put on its faces externally (e.g., by an electric field). Both the direct and inverse effects are linear, in that a change in direction of the stress producing the strain results in a polarity change of the electric charges, and vice versa. Some common piezoelectric crystals used for transducers are quartz, tourmaline, dipotassium tartrate (DKT), ethylene diamine tartrate (EDT), and cadmium sulfide. A second electrical effect similar to the piezoelectric effect is electrostriction. Whereas the piezoelectric effect is linear, this effect is quadratic, and spontaneous electric polarizations can occur. Some commonly used ferroelectrics displaying this property are Rochelle salt, ammonium dihydrogen phosphate (ADP), potassium dihydrogen phosphate (KDP), and ceramic barium titanate and lead zirconate titanate (PZT). For further reference, the reader is referred to the works by Cady[42] and Mason.[43] Cady has made an extensive study of the piezoelectric effect, and Mason deals with piezoelectric materials and their properties as related to transducer application.

Since most ultrasonic studies employ quartz transducers, their properties will be considered in more detail. Quartz is the crystalline form of silicon dioxide SiO_2. Its room temperature density is 2.649 g/cm³, and its melting point is 1750°C. At atmospheric pressure the low-temperature α-phase of quartz undergoes a phase transformation at 573°C to the high temperature β-phase and loses its piezoelectric properties. Increased pressure lowers the transformation temperature. There are a variety of reasons for the popularity of quartz as a transducer material. Its piezoelectric constants are less temperature-dependent than those of other transducer materials. It has great mechanical strength and durability, and can be ground very thin. This is important in order to produce the high frequency sometimes used in ultrasonic applications. Finally, the relatively high transformation temperature (e.g., Rochelle salt melts at 55°C and barium titanate loses its piezoelectricity at 120°C) makes it

[42] W. G. Cady, "Piezoelectricity." McGraw-Hill, New York, 1946.
[43] W. P. Mason, "Piezoelectric Crystals and Their Application to Ultrasonics." Van Nostrand-Reinhold, Princeton, New Jersey, 1950.

convenient for use over a wide range of temperatures. For temperatures above 500°C, tourmaline is a useful transducer material since it retains its piezoelectric properties until approximately 1800°C.

In ultrasonic experiments one usually wishes to propagate either longitudinal or transverse modes. The cuts of quartz useful for the generation and detection of these modes are: X-cut transducers for longitudinal or compressional waves; and Y-cut and AC-cut transducers for the transverse or shear waves. X-cut quartz transducers are constructed from plates cut with their parallel faces perpendicular to the X crystallographic axis. The application of an electric field across these plates (i.e., in the X direction) gives rise to a strain in the X direction. A compressional strain in the Y direction and an ε_{23} shear strain are also generated. However, these modes are usually highly damped when the transducer is cemented to the surface of a solid.[9] A Y-cut quartz plate, whose parallel faces are perpendicular to the Y axis, will generate both a thickness-shear strain ε_{12} and a face-shear strain ε_{13} when an electric field is applied in the Y direction. These modes of vibration are highly coupled and can generate extraneous modes due to cross-coupling.[43] Rotation of the thickness direction (normal to the plate) by $+31°$ about the X axis gives an AC-cut plate. AC-cut quartz transducers generate a very pure shear mode because coupling to the face-shear mode is minimal. Also, for a given thickness, AC-cut quartz plates are less fragile than Y-cut plates and may be driven at higher voltages.[44] [For a description of other various cuts of quartz, see Chapter VI of the book by Mason.[43]]

7.3.1.2. Bonding. The energy generated by a suitable electromechanical transducer must be coupled into the material under investigation. This is usually accomplished by physically bonding the transducer to a face of the specimen with some kind of adhesive. A wide variety of substances have been used for this purpose. They can generally be classified in one of the following groups: (1) low-temperature gases; (2) viscous liquids and greases; (3) supercooling solids and waxes; and (4) high-temperature cements and epoxies. A few representative bonding agents commonly used along with approximate temperature ranges for their usage are listed in Table III. Most of the low-temperature bonding agents listed, as well as techniques for their use, have been discussed by Bateman.[45]

[44] An Introduction to Piezoelectric Transducers. Monograph prepared by Valpey Corp. (1965).
[45] T. B. Bateman, *J. Acoust. Soc. Amer.* **41**, 1011 (1967).

TABLE III. Commonly Used Bonding Agents and Their Approximate Temperature Range of Usage

Bonding agent	Source	Temperature range
Natural gas		1.5–60 K
4-methyl 1-pentene	Matheson, Coleman & Bell	1.5–100 K
Di-2-ethylhexylsebacate (Plexol)	Rohm and Haas Corp.	1.5–200 K
DC 200 series silicone fluid	Dow–Corning Corp.	100–180 K 240–300 K[a]
Nonaq stopcock grease	Fisher Scientific Co.	1.5–300 K
Phenyl salicylate (salol)		Room temperature
Phenyl benzotate		Room temperature
Oil of wintergreen		−50−−5°C
Eastman 910 contact cement	Eastman	20–100°C
Epoxies		100–450°C
Sauereisen (Al$_2$O$_3$ based)	Sauereisen Cements Co.	450–∼1800°C
Aremco ultra-temp 516 (Zirconia base)	Aremco Products, Inc.	450–∼1800°C
Platinum paste	Engelhard Industries	To melting point of Pt

[a] A. G. Beattie, *J. Appl. Phys.* **43**, 1448 (1972).

The choice of a suitable bonding material is usually governed by two experimental variables, the frequency and the temperature. To avoid large losses due to reflection, the bond thickness should be small compared to the acoustic wavelength. Thus, for high frequencies, very thin bonds are required. For the frequency range of 100 to 500 MHz, requirements on the bond thickness and the techniques used to achieve these thin bonds have been discussed by McSkimin.[46] Thick bonds should also be avoided for precision velocity measurements. The phase-shift corrections to the measured transit time for large bond thicknesses make absolute velocity determinations difficult.[47–49]

If the transducer is bonded to a sample at a temperature different from that of the experiment (e.g., room temperature bonding for either cryogenic or furnace temperature studies), the differential thermal ex-

[46] H. J. McSkimin, *J. Acoust. Soc. Amer.* **34**, 404 (1962).
[47] H. J. McSkimin, *J. Acoust. Soc. Amer.* **33**, 12 (1961).
[48] H. J. McSkimin and P. Andreatch, *J. Acoust. Soc. Amer.* **34**, 609 (1962).
[49] H. J. McSkimin, *J. Acoust. Soc. Amer.* **37**, 864 (1965).

pansion between the sample, bonding agent, and transducer can severely strain or damage any one of them. One way to minimize this effect is by the use of viscous bonds. Another is to bond at temperatures closer to the experimental temperature. A technique for bonding at liquid nitrogen temperature for studies at liquid helium temperature is described by Lehoczky et al.[50] Also, for studies as a function of temperature, one should be careful that temperature-dependent properties of the bond do not give rise to a spurious temperature dependence of the velocity.[51]

7.3.1.3. Direct rf Ultrasonic Wave Generation. A method not yet highly developed, but which offers great promise because of the elimination of bonding problems, is that of direct rf ultrasonic wave generation. In the presence of an applied dc magnetic field, it is possible for an rf current in the surface of a metal to couple directly with acoustic waves. The strength of the coupling is directly proportional to the magnetic field intensity. The inverse process is also possible; namely, in a magnetic field, ultrasonic waves in metals produce surface currents whose amplitude is proportional to the magnetic field intensity. Since small coils of wire near the metallic surface can be used to detect the surface current, no contacts or transducers are needed on the sample. Thus there is a direct conversion of electromagnetic energy into mechanical energy and vice versa. Theoretical arguments for this effect have been given by Quinn,[52] and for megahertz frequencies this direct conversion has been observed[53-60] in a variety of metals, as well as in a semimetal and a semiconductor. Recently, this effect has been observed in the absence of an

[50] A. Lehoczky, J. T. Lewis, and C. V. Briscoe, *Cryogenics* **6**, 154 (1966).
[51] A. G. Beattie, *J. Appl. Phys.* **43**, 1448 (1972).
[52] J. J. Quinn, *Phys. Lett.* **25A**, 522 (1967).
[53] P. K. Larsen and K. Saermark, *Phys. Lett.* **24A**, 374, 668 (1967).
[54] J. R. Houck, H. V. Bohm, B. W. Maxfield, and J. W. Wilkins, *Phys. Rev. Lett.* **19**, 224 (1967).
[55] A. G. Betjemann, H. V. Bohm, D. J. Meredith, and E. R. Dobbs, *Phys. Lett.* **25A**, 753 (1967).
[56] R. L. Thomas, G. Turner, and H. V. Bohm, *Phys. Rev. Lett.* **20**, 207 (1968).
[57] W. D. Wallace, J. R. Houck, R. Bowers, B. W. Maxfield, and M. R. Gaerttner, *Rev. Sci. Instrum.* **39**, 1863 (1968).
[58] M. R. Gaerttner, W. D. Wallace, and B. W. Maxfield, *Bull. Amer. Phys. Soc.* **14**, 64 (1969); *Phys. Rev.* **184**, 702 (1969).
[59] R. L. Thomas, G. Turner, and D. Hsu, *Phys. Lett.* **30A**, 316 (1969); *Phys. Rev. B* **3**, 3097 (1971).
[60] W. D. Wallace, M. R. Gaerttner, and B. W. Maxfield, *Phys. Rev. Lett.* **27**, 995 (1971).

applied dc magnetic field for pure metals at low temperatures,[59-61] and theories have been given by Southgate[62] and Alig.[63] For a more detailed treatment of direct ultrasound generation, the reader is referred to reviews by Dobbs[64] and Wallace.[65]

7.3.2. Sample Preparation

Since most ultrasonic studies of crystalline solids involve wave propagation along symmetry axes, it is important that the sample faces be well oriented with respect to these axes to avoid serious errors in measurement. Misorientation errors, when small, are on the order of θ^2,[21] where the misorientation θ is the angle between the propagation direction and the symmetry axis. Misorientations of about 5° can lead to errors on the order of 1%. A simple but adequate means of orienting single crystals to better than $\frac{1}{2}°$ has been discussed by Ochs.[66] This method involves a double-exposure picture of Laue back reflections. After the first exposure, the sample is rotated by almost 180° around an axis parallel to the incident X-ray beam. The second exposure is then taken on the same film. If the sample face is perfectly oriented, the centers of symmetry of the two exposures will be superposed. Misorientation of the face, however, will result in a separation between the centers of symmetry. By geometrical considerations, this separation can be easily related to the misorientation angle. Corrections can be made by use of a goniometer mounted along the axis of rotation.

After the crystal has been properly oriented, its faces can be polished so that they are flat and parallel. As mentioned in Section 7.2.4.2, the nonparallelness of the specimen faces can contribute significantly to the measured ultrasonic attenuation. As a plane wave reflects back and forth between nonparallel faces, its wave front becomes distorted. Thus the part of the wave falling on one side of the transducer can be out of phase with that striking another part. Since the transducer is a phase-sensitive device (i.e., it integrates the signal it receives over its surface), phase variations over the surface can result in interference and apparent at-

---

[61] W. D. Wallace, M. R. Gaerttner, and B. W. Maxfield, *Bull. Amer. Phys. Soc.* **14**, 64 (1969).

[62] P. D. Southgate, *J. Appl. Phys.* **40**, 22 (1969).

[63] R. C. Alig, *Phys. Rev.* **178**, 1050 (1969).

[64] E. R. Dobbs, *J. Phys. Chem. Solids* **31**, 1657 (1970).

[65] W. D. Wallace, *Int. J. Nondestruct. Test.* **2**, 309 (1971).

[66] T. Ochs, *J. Sci. Inst. (J. Phys. E)* **1**, 1122 (1968).

tenuation. This attenuation source has been estimated as[6]

$$\beta \quad (\text{dB/sec}) = 8.68\pi^2 f^2 R^2 \phi^2 \bar{n}/LW, \tag{7.3.1}$$

where f is the frequency, R the transducer radius, ϕ the nonparallelness angle, $\bar{n}$ the number of the echo near which the attenuation is being measured, L the specimen length, and W the sound velocity in the medium. Using the values $f = 300$ MHz, $W = 4.9 \times 10^5$ cm/sec, $L = 1$ cm, $R = 0.635$ cm, and $\bar{n} = 4$, one finds that the nonparallelness angle must be less than 44×10^{-6} in order to ensure that the loss due to nonparallelness will be less than 0.05 dB/μsec. Such tolerances on parallelness can be obtained with proper polishing techniques. A thorough discussion of polishing procedures has been given by Ochs,[66] and the details will not be given here.

7.3.3. Velocity and Attenuation Measurements

7.3.3.1. Basic Techniques.
Velocity measurements can be broadly divided into continuous wave techniques and pulse–echo techniques. Continuous wave systems generally consist of some type of composite oscillator where the sample–transducer system is at a resonant condition. This type of measurement is most common in the kilohertz region, where a sample one-half wavelength long oscillates at the fundamental resonance. Velocity measurements, which correspond to phase velocities, are given by $2f\lambda$. Attenuation measurements can also be made by measuring the input acoustic power required to maintain the oscillations at constant amplitude. In systems of this type the sample composite oscillator is generally an active component of the oscillator circuit. This type of measurement is somewhat inconvenient because of sample geometry and acoustic isolation requirements. For a more detailed discussion of continuous wave systems in the kilohertz region, the reader is referred to the work of Mason[9] and Read et al.[1]

Continuous wave composite oscillator systems are finding increased use in the megahertz region. Here the transducer–specimen combination is used as a two-terminal network, whose impedence varies rapidly near a resonance of the system. The variation in the equivalent resistance and capacitance of a typical system is shown in Fig. 4, which also includes a block diagram of an rf bridge used to measure R and C.[67,68] Generally

---

[67] A. G. Beattie, H. B. Silsbee, and E. A. Uehling, *Bull. Amer. Phys. Soc.* **7**, 478 (1962).

[68] G. A. Alers, *in* "Physical Acoustics" (W. P. Mason, ed.), Vol. IVA, p. 277. Academic Press, New York, 1966.

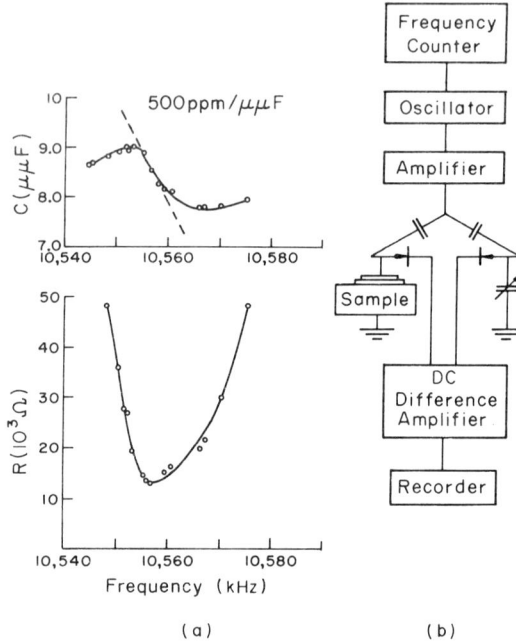

FIG. 4. Under CW excitation the transducer–specimen combination can be described electrically as a capacitor and a resistor in parallel. (a) The graphs show the variation of this capacity and resistance in the vicinity of the frequency at which standing acoustic waves are set up in the sample. (b) Also shown is a block diagram of a simple capacity bridge [A. G. Beattie, H. B. Silsbee, and E. A. Uehling, *Bull. Amer. Phys. Soc.* **7**, 478 (1962)] for detecting changes in the standing wave frequency [after G. A. Alers, "Physical Acoustics" (W. P. Mason, ed.), Vol. 4A, p. 277. Academic Press, New York, 1966].

such a bridge is balanced for a particular resonant frequency, and after small velocity changes, such as those produced by dislocation pinning or magnetoacoustic effects, the frequency is changed to produce a new balance. The relative velocity change is given by the relative frequency shift. Beattie et al.[67] have used this system to measure changes as small as 10^{-7}. This high sensitivity is only obtained for very low-loss specimens, however. This type of measurement is most useful when sample thickness is too small to provide sufficient separation of pulses in a pulse system. Attenuation measurements can be made by analyzing the resistive loss in the system. A comprehensive review of continuous wave megahertz measurements has been given by Bolef and Miller.[69]

[69] D. I. Bolef and J. G. Miller, *in* "Physical Acoustics" (W. P. Mason and R. N. Thurston, eds.), Vol. VIII, p. 96. Academic Press, New York, 1971.

Generally, measurements in the megahertz region are carried out using more convenient pulse–echo techniques. Here pulses of short duration are introduced into a sample by a transducer, and velocity and attenuation measurements are made by measuring the transit time and amplitude loss of echoes of the pulse as it is reflected back and forth between the transducer and opposite face of the sample. Sample size is not restricted as in the resonant oscillator system; a typical sample is a $\sim$1-cm cube. For typical sound velocities and equipment recovery times, the pulses must be about 1 μsec in duration to provide adequate separation between pulses. In order that the pulses enclose enough cycles for plane-wave analysis, frequencies greater than about 5 MHz are required.

There are two basic approaches to the velocity measurements, as noted before: measurement either of the transit time for the envelope of the rectified echoes or of a particular unrectified cycle within an echo. For a nondispersive medium these transit times are equal (the difference between W_g and W resulting solely from the path length difference[22]), so no distinction is necessary. On the other hand, there is a further distinction between phase and group velocities in real dispersive materials, and a distinction between transit times of the pulse envelope and of unrectified cycles must be made. However, the experimental precision with which the envelope velocity can be measured is generally less than the difference between W_g and W, so the distinction is not important in that case. Because of the imprecise nature of envelope velocity measurements, they are generally only used for rough measurements of second-order constants. The more precise absolute velocity and velocity-change systems measure the transit time of a particular cycle either directly or through some sort of interference technique.

In all these velocity-measurement systems some method must be used to correct for the phase shifts due to reflection at the transducer face of the sample. Generally there are two separate effects due to the transducer. First there is a major phase shift resulting from the transmission of a part of the sound wave into the quartz. This can result in the effective loss of an entire cycle in the pulse, so that care must be taken in deciding which cycle in the various echoes is the appropriate one. This problem has been considered in detail by McSkimin,[47] who was able to provide a criterion for deciding. The second effect of the interface is a much smaller phase shift, which is related to the thickness and material of the transducer bond. This problem has been analyzed by McSkimin and Andreatch[48] and McSkimin,[49] who calculate the phase shift for a few materials. The correction is slightly frequency-dependent and is minimum at or near the resonant frequency of the transducer; for very thin bonds this correction can be made very small (corresponding to a velocity correction of $\sim 10^{-4}$). Because of its small frequency dependence, it is generally neglected in velocity change measurements. Nevertheless it is generally the limiting factor in the accuracy of measured absolute transit times.

Pulse–echo attenuation measurements generally utilize the envelope of the rectified echoes. The two most common methods consist of a direct measurement of the relative amplitudes of two different echoes, and a comparison of the decay of a train of echoes with an exponentially decaying curve displayed simultaneously. Both types of systems are available commercially. One of the measurement systems provides for the direct, continuous, automatic recording of attenuation to within an accuracy of 0.001 dB/μsec.

These attenuation measurements must be corrected to account for energy losses not associated with attenuation in the sample, e.g., the energy lost during reflections. Fortunately most measurements of attenuation resulting from a particular mechanism can be deduced from a *change* in attenuation produced in a sample. For example, the dislocation contribution can be determined by measuring the attenuation change produced by immobilizing the dislocations with irradiation-produced defects in the crystal. Hence the (constant) contributions of experimental effects to the attenuation do not affect the final result. However, apparent attenuation contributions from interferences and diffraction are dependent on the sound velocity, which generally also changes during a typical attenuation-change measurement. Hence unless great care is taken to eliminate or minimize these effects, the measured attenuation *change* can be affected.

7.3.3.2. Measurement Systems. The most straightforward phase-velocity measurement involves the direct display of the unrectified echoes on an oscilloscope and the determination of the elapsed time between corresponding cycles in two different echoes. Experiments of this type, generally using a very stable time-mark generator as the time standard, enable the velocities to be determined to within parts in 10^4. The principal advantage of this type of measurement is its simplicity. Since various experimental corrections to measured velocities (such as phase shifts) cannot generally be estimated more accurately than 10^{-4}, this method is very useful in absolute velocity and second-order elastic constant measurements. Although it is not precise enough for most velocity-change measurements, it is useful when changes are fairly large, e.g., those due to dislocations. In this case the method has the added advantage that simultaneous attenuation measurements can be conveniently made.

A further sophistication of this approach is provided by the technique proposed by May[70] and refined by Papadakis,[71] involving the visual superposition of two unrectified echoes. The manner in which this is accomplished is illustrated by the simplified block diagram shown in Fig. 5. The oscilloscope is triggered by an audio oscillator with a period of the round-trip transit time of the ultrasonic pulse in the sample (typically 10 μsec). This frequency is divided by a large number (10–100)

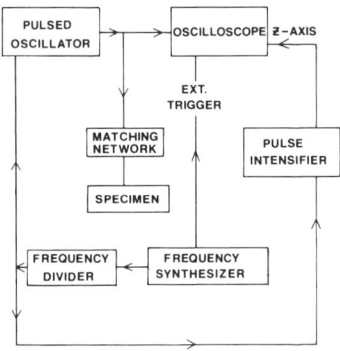

FIG. 5. Simplified block diagram of the Papadakis velocity-measurement system.

and used to trigger the generator of the ultrasonic pulses. Hence all the echoes are visually superimposed on the oscilloscope, whose sweep time is roughly equal to the pulse duration. Then for each echo train generated, a set of two pulses of echo length duration is generated after a delay time corresponding to an integral number of echo transit times. These pulses are connected to the oscilloscope Z axis input to intensify the brightness of the two echoes displayed at those two particular times. By decreasing the oscilloscope brightness it is then possible to see only the two selected echoes, and the sample velocity is given by the inverse of the repetition rate when the corresponding cycle of the two echoes exactly superimpose. Because both echoes are superimposed, the selection of the proper cycle in each echo (mentioned above) is simplified. Thus the method is particularly well suited for absolute velocity and second-order elastic con-

[70] J. E. May, Jr., *IRE Nat. Conv. Rec.* **6**, Pt. 2, 134 (1958).
[71] E. P. Papadakis, *J. Acoust. Soc. Amer.* **42**, 1045 (1967).

stant measurements. Also, because of the increase in sensitivity provided by the ability to measure changes in velocity directly, as well as the ability to make conventional attenuation measurements simultaneously, the system is useful for certain velocity and attenuation change measurements (particularly in dislocation studies). Such a simultaneous velocity-attenuation system is available commercially. However, the velocity measurements are still limited in sensitivity to parts in 10^5, and must be made on a point by point basis.

For a wide variety of ultrasonic studies it is necessary to measure velocity changes of 10^{-6} and smaller. For example, the magnitude of magneto-acoustic effects on velocity is typically a few parts per million for 10-kG fields.[68] The measurement of third-order elastic constants for stress-induced variations of ultrasonic velocities is discussed in Section 7.2.3. There are a number of reasons for maintaining the applied stress at very low levels where velocity changes are on the order of 10^{-6}. For example, nonlinear effects, such as plastic flow, become important for uniaxial stresses above a few kilograms per square centimeter. Furthermore, even in the measurement of some dislocation effects, which generally involve velocity changes of 10^{-5} or greater, the effects can occur very rapidly so that continuous measurement systems are necessary. For all these reasons, a great deal of effort has gone into the development of more sophisticated velocity-measurement systems capable of high resolution as well as continuous detection. The present discussion is not intended to be a complete catalog of velocity-measurement systems; only a few of the more widely used systems will be discussed.

One of the most direct approaches for measurements of phase velocities is the sing-around system.[72,73] In this method a single cycle is selected from one of the echoes near the end of the echo train, and it is used to trigger the generation of the next ultrasonic pulse and echo train, from which a cycle is again selected, and so forth. Hence the repetition rate of the generation of pulses is determined by the transit time between the first pulse and selected echo. Any changes in the ultrasonic velocity directly produce a corresponding change in repetition rate, which is the measured quantity. Velocity changes of parts in 10^7 have been measured in samples with sufficiently low attenuation. The principal disadvantage of this system is that any change in amplitude of the selected cycle results in a significant change in the repetition rate. Hence the velocity-change

[72] N. P. Cedrone and D. R. Curran, *J. Acoust. Soc. Amer.* **26**, 963 (1954).
[73] R. L. Forgacs, *IRE Trans. Instr.* **I-9**, 359 (1960).

measurement is sensitive to amplitude stability of the initial pulse and to attenuation changes in the sample during the measurement.

An instrument described by Blume[74] overcomes most of the difficulties of the sing-around system while maintaining its principal advantage of convenient simultaneous attenuation measurement. In this system, as illustrated in Fig. 6, the ultrasonic driving pulse is obtained by gating and amplifying a portion of a continuous reference wave. Then one of the unrectified echoes is selected from the train and compared in phase to the reference signal. Hence the reference signal serves as a time base for

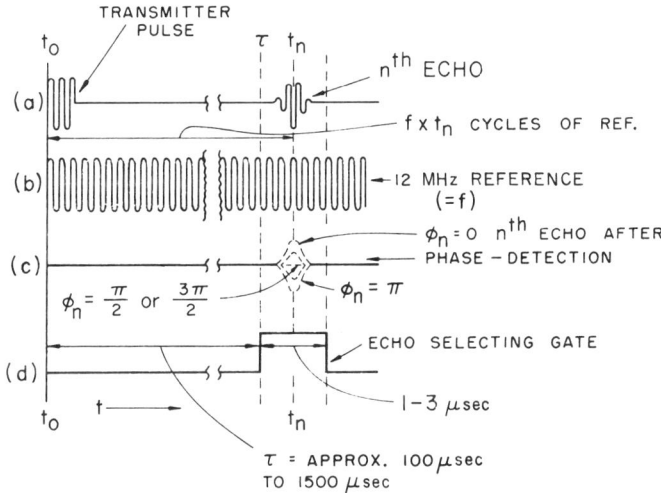

FIG. 6. Basic timing diagram of the Blume detection system [after R. J. Blume, *Rev. Sci. Instrum.* **34**, 1400 (1963)].

measuring the total transit time for the selected echo. For example, if the echo is maintained in phase with the reference signal, the transit time is an integral number of cycles, i.e., n/f. Velocity changes are given directly by the relative frequency change of the reference frequency necessary to maintain the phase equality. The phase relationship between reference and echo can be determined in a number of ways: direct visual observation of an oscilloscope display, or the visual or electronic detection of a maximum or minimum in the algebraic sum of the two, for example. Blume uses the technique of phase-sensitive detection to electronically detect the phase relation. Phase detection is very useful

[74] R. J. Blume, *Rev. Sci. Instrum.* **34**, 1400 (1963).

because of its high signal-to-noise ratio and sensitivity, and is used in other velocity measurement systems (see below) as well as in numerous other measurement systems. Basically the operation of such a detector, as illustrated in Fig. 7, is as follows. The signal is fed into an amplifier whose output polarity relative to input signal is periodically switched from plus to minus. This switching is synchronized with a reference signal so that if the input signal has the same frequency and relative

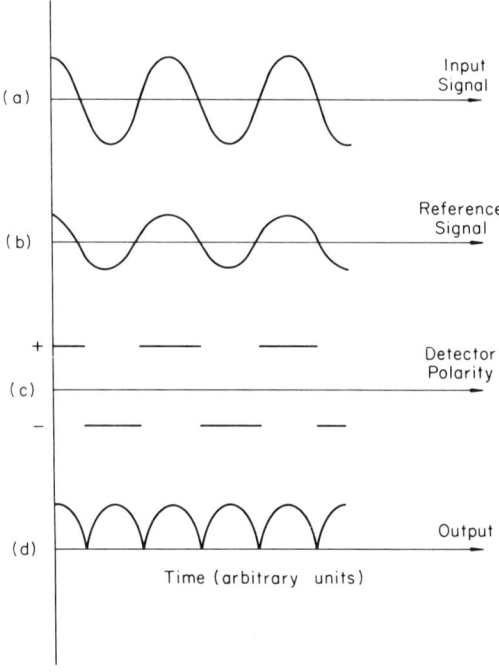

FIG. 7. Timing diagram for the phase-sensitive detector. In the figure the input signal is in phase with the reference signal.

phase as the reference signal, as illustrated, the detector output will always be positive. If the relative phase is changed, the detector output will contain more and more of a negative portion until, when the relative phase of input and reference is 180°, the output is always negative. The integral of the detector output therefore changes from maximum positive to maximum negative as the phase shift varies from 0 to 180°, reaching zero when the signal and reference are in phase quadrature (90 or 270°) (see Fig. 6c). This null at phase quadrature is independent of signal amplitude, and the variation of output polarity with phase enables such

7.3. EXPERIMENTAL TECHNIQUES

a detector to be used to automatically maintain the phase-quadrature condition. As in Blume's system, it is only necessary to connect the phase-sensitive detector output to a servo mechanism which controls the reference frequency. Whenever the quadrature condition is not met, the output of the detector drives the servo in the appropriate direction until the reference frequency is such that quadrature is again satisfied and the detector output is zero.

The Blume system has been used to detect velocity changes to within a few parts in 10^8. Its only real shortcomings lie in the fact that phase-sensitive detection in the megahertz region requires fairly sophisticated electronics, and that the control signal is fairly small because of the low duty cycle during which phase comparisons are made (roughly 1 μsec every 1000 μsec).

A large number of velocity-measurement systems are based on some type of interference of echoes, either with each other or with other reference signals. One of the earliest and most straightforward interference systems is the two-crystal system described by Espinola and Waterman[75] and developed by Hiki and Granato.[33] Here a single ultrasonic pulse is fed simultaneously to two samples nearly identical in geometry as well as in other properties. For any slight difference in transit time between echoes in the two samples, there will be an interference node at or near some echo in the train. Any change in the velocity of one of the samples (caused, for example, by stressing a sample in order to measure third-order constants) will result in a shift of this interference node to a different position. In the Hiki–Granato system, one sample, the reference sample, was carefully controlled in temperature. Then when the velocity of the stressed sample changed, the temperature of the reference sample was changed so that the corresponding temperature-induced change in velocity $[(1/W)\,dW/dT$ is typically 5×10^{-4} K^{-1}] equaled the stress-induced change. This equality of velocity changes was monitored by (visually) determining when the interference node returned to its position prior to the change; the stress-induced velocity change is therefore given by the calibrated temperature-induced change. This system is capable of measuring velocity changes to within parts in 10^6. It is very simple in principle, and inexpensive to construct. However, preparation of two nearly identical samples is difficult and time-consuming, and can be expensive for more exotic crystals.

There are many single-specimen types of interference systems, only a

---

[75] R. P. Espinola and P. C. Waterman, *J. Appl. Phys.* **29**, 718 (1958).

few of which will be mentioned here. The Lamb–Williams technique[76,77] consists of superimposing a second generating pulse on one of the echoes of a first pulse, and establishing a destructive interference condition. The second generated pulse is identical in shape and phase to the first, and its amplitude is adjusted to be equal to that of the echo onto which it is superimposed. The interference condition is determined by observing a null in the remaining echoes, and is related to the delay time between the two pulses. Conventional measurements of this time have provided velocity-change measurements to within a part in 10^5. It is most limited because of the distortions caused by reflections of the echoes, making the determination of the null difficult.

Probably the most widely used interference techniques are variations of pulse-superposition systems developed by McSkimin and co-workers.[47–49,78] In these systems the repetition rate of the generating pulses is very high, so that another generating pulse is superimposed on the pth echo of every preceding pulse (see Fig. 8). The nonsuperimposed echoes between the main pulse and the pth echo then represent the sum of a large number of echoes from preceding pulses. The repetition frequency of the pulse generation in the pulsed oscillator system shown in the figure is then varied until a constructive interference condition is attained where summed echoes are a maximum. The summation of the echoes from a large number of pulses leads to a greater variation in amplitude of the summed echoes near the constructive interference condition, and hence a greater sensitivity in detecting small velocity changes. The usual choices of p are $p = 2$ (every other echo is superimposed, as in Fig. 8) and $p = 1$. For the $p = 1$ mode, the superposition, which occurs on every echo, must be periodically (e.g., every 10 pulses) interrupted briefly (e.g., for two or three echoes) in order to observe the amplitude of the nonsuperimposed echoes.

In another variation the pulses can be gated from a continuous carrier so they are all phase coherent, and the interference condition is determined by this carrier frequency.[49] This gated carrier system is generally more sensitive, primarily because of the higher frequency used to determine the interference condition (10 MHz as opposed to the repetition frequency of about 50 kHz used in the pulsed oscillator system).

[76] J. Williams and J. Lamb, *J. Acoust. Soc. Amer.* **30**, 308 (1958).

[77] A. D. Colvin, Masters thesis, Rensselaer Polytechnic Inst., Troy, New York (1959).

[78] H. J. McSkimin and P. Andreatch, *J. Acoust. Soc. Amer.* **41**, 1052 (1967).

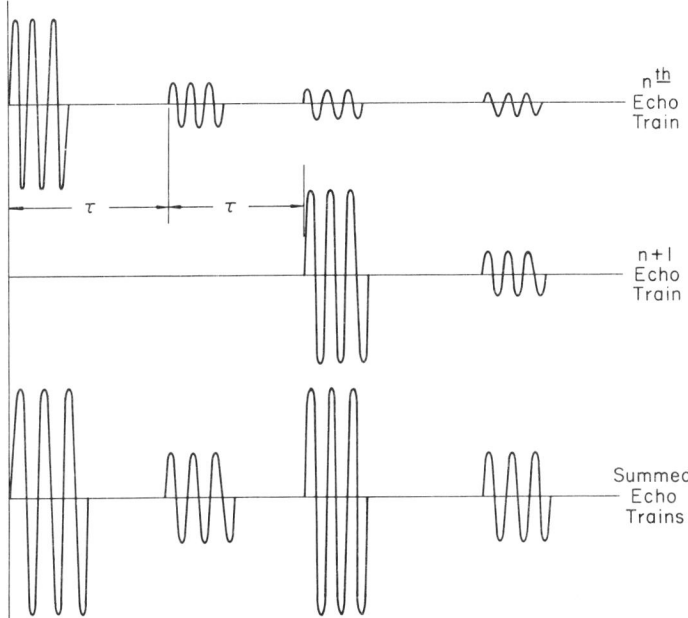

Fig. 8. Illustration of pulse superposition in the $p = 2$ mode of a pulsed oscillator system. Two succeeding driving pulses and their nonsuperimposed echoes are shown, along with the total superimposed pattern. Constructive interference occurs when the time between driving pulses is twice the transit time τ of the pulse in the sample, as shown.

In any of the interference systems, the precision is limited by the sensitivity of the detection of the interference maximum or minimum. The direct visual observation of the interference condition on an oscilloscope display is limited in precision to several parts in 10^6, even in a pulse-superposition system. Hence a great deal of work has been carried out in developing accurate electronic systems for detection of these interference maxima and minima. Several approaches are discussed in an article by McSkimin.[49] One of the more successful approaches involves the sinusoidal modulation of the interference-determining frequency about the interference "resonant" frequency. The amplitude of the echoes then also varies periodically. Furthermore, it can be seen from the shape of the resonant curve of the echo amplitude, shown in Fig. 9, that both the amplitude and polarity of the echo oscillations are determined by the slope of the resonance curve. Hence the echo oscillations will be equally large in amplitude when the interference-determining frequency is at

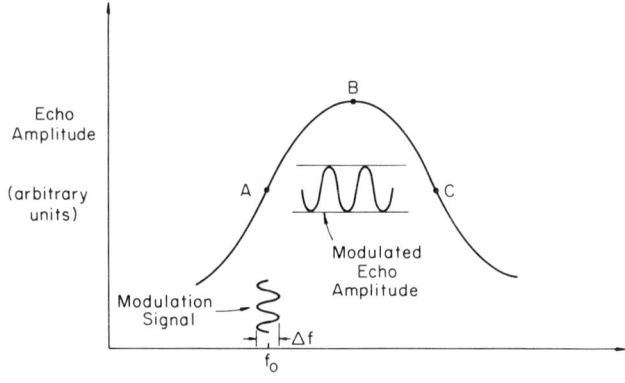

FIG. 9. Summed echo amplitude versus the frequency which determines the resonance for a pulse superposition system. If the interference frequency is modulated an amount Δf about f_0, the echo amplitude is modulated by an amount determined by the position of f_0.

points A and C, but opposite in polarity, and very small (or zero) in amplitude when it is at the resonant frequency, point B.

McSkimin[49] has used this technique in both a gated carrier and a pulsed oscillator superposition system ($p = 2$ mode). The modulated echoes were filtered and fed directly into a servo motor, which in turn controlled the repetition rate or carrier frequency. Because of the polarity and amplitude variation of the echo oscillation, this arrangement automatically maintains the system at the resonant frequency. He was able to determine velocity changes to within parts in 10^6 for the pulsed oscillator system, and parts in 10^7 for the gated carrier.

A detection system designed by Holder[79] combines this technique with the high signal-to-noise features of phase-sensitive detection, noted above. Here the modulation frequency serves as the reference frequency of the phase-sensitive detector, into which the oscillating echoes from a $p = 2$ mode gated carrier superposition setup are fed. The detector output, whose magnitude and sign again depend on the relative positions of carrier center frequency and resonant frequency, is then used to control the carrier frequency—either via a servomechanism[79] or directly through integration of the detector output.[80] Sensitivities of 10^{-8} have been achieved with the system,[79] and it has proven useful in a variety of

[79] J. Holder, *Rev. Sci. Instrum.* **41**, 1355 (1970).
[80] D. Read and J. Holder, *Rev. Sci. Instrum.* **43**, 933 (1972).

measurements[16,81,82] because of its precision, stability, and relative insensitivity to attenuation changes. It is especially useful for samples with fairly high attenuation where, because of the high duty cycle involved, sensitivities of 10^{-5}–10^{-6} are easily attained.

A principal drawback of all the interference type of velocity measurements is that conventional attenuation measurements cannot be conveniently carried out simultaneously because of the superposition involved. However, one method for overcoming this difficulty has been described by Read and Holder.[80] They use the velocity-measurement system just described and interrupt the superposition periodically (roughly 1000-μsec intervals) for a period of between 10 and 100 μsec. During this interval the echo pattern decays and conventional attenuation measurements can be made. Actual measurements show that the simultaneous attenuation measurement has little effect on the sensitivity, accuracy, stability, or response time of the velocity measurements.

There are, of course, many other velocity and attenuation techniques. Most of these are designed for special purposes, however. Some important examples are:

(1) Buffer rod techniques,[83] where the sample is isolated from the transducer by a buffer material, providing a certain amount of isolation of the sample.

(2) By placing a notch or shoulder on the reflecting end of a very long rod, two reflected echoes are produced for each arriving pulse, separated by the transit time between the shoulder and sample end.[84] This is most useful in cases requiring isolation, such as high-temperature measurements where the transducer end of the sample can be kept at room temperature while the notch end is at an elevated temperature. The difference in transit times between the two reflected echoes is a function only of the velocity in the high-temperature region.

(3) In transparent samples, sound waves cause variations in the density, and hence in index of refraction. These variations can be used to set up interference patterns from light traveling through the sample in the region of the sound wave.[85,86] This is most useful where sample

---

[81] W. S. deRosset and A. V. Granato, *J. Appl. Phys.* **41**, 4105 (1970).
[82] K. Kawasaki and A. Ikushima, *Phys. Rev. B* **1**, 3143 (1970).
[83] H. J. McSkimin, *IRE Trans.* **PGUE-5**, 25 (1957).
[84] R. D. Holbrook, *J. Acoust. Soc. Amer.* **20**, 590 (1948).
[85] P. Debye and F. W. Sears, *Proc. Nat. Acad. Sci. U.S.* **18**, 410 (1932).
[86] R. Lucas and P. Biquard, *J. Phys. Radium* **3**, 464 (1932).

preparation is a problem and good echo patterns are difficult to obtain. Such is the case, for example, in rare-gas solids grown in the range of 4 to 100 K.

A fairly comprehensive listing of velocity measurement techniques has been given by McSkimin.[87]

7.4. Applications of Ultrasonic Waves to Measuring Physical Properties

Ultrasonic measurements of physical properties of solids can be basically classified into two groups. The first deals solely with the elastic properties of a solid, i.e., the second- and third-order elastic constants. Here one is not concerned with structural defects (dislocations, interstitials, vacancies, impurities, etc.) or thermal defects (phonons), and the solids under investigation are essentially dispersionless, i.e., the frequency dependence of the velocity and attenuation is too small to be of importance. Although all solids contain defects, in many cases one can still obtain an "acoustically perfect" solid in the respect that the defects present will not contribute to the velocity or attenuation. For example, the effect of mobile dislocations on the measured sound velocity and attenuation can often times be eliminated by the creation of dislocation pinning points, the most common of which are impurities and defects (produced by neutron or gamma irradiation). Thus the first group of ultrasonic measurements is concerned only with the elastic constants of "acoustically perfect" solids. The second group of ultrasonic measurements deals with "real" materials, where one is concerned with dislocation effects (decrement and modulus defect), electronic effects, phonon–phonon interactions, magnetic effects, etc.

7.4.1. Elastic Constants

7.4.1.1. Significance of Elastic Constants. Elastic constants play a dual role of importance in the theory of solids. First of all, a knowledge of only the elastic constants enables one to determine, for example, such properties as the compressibility, the Debye temperature and its pressure dependence, the Grüneisen parameters, and various thermodynamic properties of defects (both structural and thermal) in solids. These

---

[87] H. J. McSkimin, *J. Acoust. Soc. Amer.* **33**, 606 (1961).

7.4. APPLICATIONS TO MEASURING PHYSICAL PROPERTIES

properties can be found without having any information about interatomic forces or potentials. This use of elastic constants has recently been reviewed by Holder and Granato[88] and will not be detailed here. The second important role of elastic constants is a result of their being among the most accurately measurable properties of a solid. Ultrasonic techniques have generally enabled one to determine elastic constants to a greater degree of accuracy than can be obtained from theoretical calculations. Thus the elastic constants can serve as a useful guide in developing any related theory. Since elastic constants describe how the energy density of a solid changes with respect to various volume and shear deformations, they are useful in determining the nature of binding forces and interatomic potentials.

In Table IV, the second- and third-order elastic constants, along with the temperature and pressure derivatives of the second-order constants, have been listed for a number of solids having cubic symmetry. Some of these tabulated values will be used later for illustrating relations among certain elastic constants. For a comprehensive tabulation of existing data, the reader is referred to Landolt and Börnstein[89] and Simmons and Wang.[90]

7.4.1.2. Methods of Calculation.

There are two methods for calculating the elastic constants of a crystalline solid: the method of homogeneous deformation and the method of long waves. The results obtained from the two methods should be in agreement when the same model of a solid is used in both cases.[91] The method of long waves, which has been used to calculate second-order elastic constants, has an advantage in that one can determine phonon dispersion curves. However, second-order elastic constants are easier to calculate by the method of homogeneous deformation, and, more importantly, the method can be extended to the calculation of third- and higher-order elastic constants with little difficulty. For these reasons, the method to be described here is that of homogeneous

---

[88] J. Holder and A. V. Granato, in "Physical Acoustics" (W. P. Mason and R. N. Thurston, eds.), Vol. VIII, p. 237. Academic Press, New York, 1971.

[89] Landolt-Börnstein, "Elastic, Piezoelectric, Piezooptic, Electrooptic Constants and Nonlinear Dielectric Susceptibilities of Crystals," New Series, Vol. III/2. Springer-Verlag, Berlin, 1969.

[90] G. Simmons and H. Wang, "Single Crystal Elastic Constants and Calculated Aggregate Properties: A Handbook," 2nd ed. M.I.T. Press, Cambridge, Massachusetts, 1971.

[91] D. C. Wallace, *Rev. Mod. Phys.* **37**, 57 (1965).

TABLE IV. Room-Temperature Values of Second- and Third-Order Elastic Constants for Various Cubic Crystals[a]

Constant	Cu	Ag	Au	Al	NaCl	KCl	Si	Ge
c_{11}	1.661	1.222	1.929	1.068	0.496	0.409	1.658	1.285
c_{12}	1.199	0.907	1.638	0.604	0.131	0.070	0.639	0.483
c_{44}	0.756	0.454	0.415	0.283	0.128	0.063	0.796	0.668
Ref.	b	b	b	c	d	e	f	g
$(1/c_{11})\, dc_{11}/dT$	-2.4	-2.8	-1.8	-3.29	-7.56	-8.1	-0.53	-0.91
$(1/c_{12})\, dc_{12}/dT$	-1.6	-2.1	-1.5	-1.11	2.14	5.7	-0.75	-0.91
$(1/c_{44})\, dc_{44}/dT$	-3.6	-4.2	-3.3	-5.12	-2.58	-1.9	-0.42	-0.92
Ref.	h	h	h	c	d	i	f	g
dc_{11}/dP	6.36	7.03	7.01	7.35	11.89	13.00	4.33	5.01
dc_{12}/dP	5.20	5.75	6.14	4.11	2.13	1.56	4.19	4.35
dc_{44}/dP	2.35	2.31	1.79	2.31	0.37	-0.56	0.80	1.40
Ref.	j	j	j	k	d	e	f	g
c_{111}	-12.71	-8.43	-17.29	-10.76	-8.64	-7.26	-8.25	-7.10
c_{112}	-8.14	-5.29	-9.22	-3.15	-0.50	-0.24	-4.51	-3.89
c_{123}	-0.50	$+1.89$	-2.33	$+0.36$	$+0.09$	$+0.11$	-0.64	-0.18
c_{144}	-0.03	$+0.56$	-0.13	-0.23	$+0.07$	$+0.23$	$+0.12$	-0.23
c_{166}	-7.80	-6.37	-6.48	-3.40	-0.59	-0.26	-3.10	-2.92
c_{456}	-0.95	$+0.83$	-0.12	-0.30	$+0.13$	$+0.16$	-0.64	-0.53
Ref.	b	b	b	c	d	e	l	l

[a] Values are in units of 10^{12} dyn/cm^2. Also listed are the temperature derivatives (in units of 10^{-4} K^{-1}) and the pressure derivatives (dimensionless) of the second-order constants.

[b] Y. Hiki and A. V. Granato, *Phys. Rev.* **144**, 411 (1966).
[c] J. F. Thomas, Jr., *Phys. Rev.* **175**, 955 (1968).
[d] K. D. Swartz, *J. Acoust. Soc. Amer.* **41**, 1083 (1967).
[e] J. R. Drabble and R. E. B. Strathen, *Proc. Phys. Soc. (London)* **92**, 1090 (1967).
[f] H. J. McSkimin and P. Andreatch, Jr., *J. Appl. Phys* **35**, 2161 (1964).
[g] H. J. McSkimin and P. Andreatch, Jr., *J. Appl. Phys.* **34**, 651 (1963).
[h] Y. A. Chang and L. Himmel, *J. Appl. Phys.* **37**, 3567 (1966).
[i] R. A. Bartels and D. E. Scheule, *J. Phys. Chem. Solids* **26**, 537 (1965).
[j] W. B. Daniels and C. S. Smith, *Phys. Rev.* **111**, 713 (1958).
[k] R. E. Schmunk and C. S. Smith, *J. Phys. Chem. Solids* **9**, 100 (1959).
[l] H. J. McSkimin and P. Andreatch, Jr., *J. Appl. Phys.* **35**, 3312 (1964).

deformation, and the reader is referred to the text by Born and Huang[92] for a discussion of long waves.

In order to calculate the elastic constants of a crystalline solid, one must first have a model for the energy density of the lattice. The calculation then proceeds by determining how the energy density changes with respect to various homogeneous deformations, i.e., deformations for which the resulting structure remains a perfect lattice. The two most widely used types of deformations are those described by the Lagrangian strain parameters and by the Fuchs strain parameters. In the first case, derivatives of the energy density are taken with respect to the Lagrangian strains η_{ij}. The resulting derivatives, when evaluated in the undeformed state, are the Brugger elastic constants[25]

$$c_{ijkl\cdots} \equiv (\partial^n E/\partial \eta_{ij}\, \partial \eta_{kl} \cdots)_{\eta=0}, \qquad (7.4.1)$$

where E is the energy of the crystal per unit undeformed volume. In the other case, one calculates derivatives of the energy density with respect to Fuchs-type strain parameters.[93] These strains consist of a volume deformation, illustrated in Fig. 10a, and of various volume-conserving shear deformations, one of which is illustrated in Fig. 10b. When evaluated in the undeformed state, these derivatives are referred to as *Fuchs elastic constants*. The Fuchs constants are linear combinations

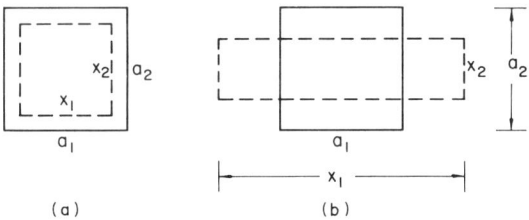

FIG. 10. The Fuchs v and e strains. The dashed and solid lines represent the volume of the deformed and undeformed crystal, respectively. The coordinates of the deformed state are $\mathbf{x}$, and those of the undeformed state are $\mathbf{a}$. (a) For the v strain, $x_i = v^{1/3} a_i$ ($i = 1, 2, 3$). Thus v is the ratio of the volume of the deformed state to that of the undeformed state. For cubic structures, this deformation yields the bulk modulus, i.e., $(\partial^2 E/\partial v^2)_{v=1} = \frac{1}{3}(c_{11} + 2c_{12})$. (b) For the e strain, $x_1 = (1 + e)^{1/2} a_1$, $x_2 = (1 + e)^{-1/2} a_2$, and $x_3 = a_3$. Volume is conserved, and for cubic structures this deformation yields the shear constant $(\partial^2 E/\partial e^2)_{e=0} = \frac{1}{2}(c_{11} - c_{12})$.

---

[92] M. Born and K. Huang, "Dynamical Theory of Crystal Lattices." Oxford Univ. Press, London and New York, 1966.
[93] K. Fuchs, *Proc. Roy. Soc.* (*London*) **A153**, 622 (1936); **A157**, 444 (1936).

of the Brugger constants. The relationships between the Fuchs and Brugger constants have been discussed, for example, by Suzuki et al.[94] for cubic structures and by Naimon et al.[95] for hexagonal structures. The Fuchs constants generally are more convenient to calculate when dealing with noncentral potentials, whereas the Brugger constants are the preferred type for central potentials. In either case, the final results are usually expressed in terms of the Brugger constants to facilitate comparison with available experimental data.

The energy density of a crystal can be expressed as the sum of two terms: a contribution from the static lattice and a contribution from thermal and zero-point vibrations. Most elastic constant calculations consider only the static-lattice energy, since lattice vibrations have a small effect on the calculated results (e.g., elastic constants typically change by $\sim 10\%$ in going from 0 K to room temperature[96]). When comparing static-lattice calculations with experiment, it is not completely correct to use the experimental values at 0 K because of zero-point vibrations. The appropriate experimental values at 0 K are provided by linear extrapolation from the high-temperature region[96] (i.e., temperatures above the Debye temperature). In the next section, static-lattice contributions will be discussed. Since no attempt will be made to obtain rigorous agreement between theory and experiment, the room-temperature values in Table IV will provide sufficient experimental data for comparison with theory. A brief discussion of the temperature dependence of the elastic constants will also follow.

7.4.1.3. Static-Lattice Contributions to the Elastic Constants

7.4.1.3.1. CENTRAL POTENTIALS. For a central interatomic potential $\phi(r)$, where r is the distance between two atoms, the energy of the crystal (per unit undeformed volume) is

$$E = (1/2V_0) \sum \phi(r). \qquad (7.4.2)$$

Here V_0 is the volume of the undeformed state, and the summation is over pairs of atoms. For a homogeneous deformation of a crystal whose atoms are at centers of symmetry, the difference of the square of the

---

[94] T. Suzuki, A. V. Granato, and J. F. Thomas, Jr., *Phys. Rev.* **175**, 766 (1968).

[95] E. R. Naimon, Tetsuro Suzuki, and A. V. Granato, *Phys. Rev. B* **4**, 4297 (1971).

[96] G. Leibfried and W. Ludwig, *in* "Solid State Physics" (F. Seitz and D. Turnbull, eds.) Vol. XII, p. 275. Academic Press, New York, 1961.

7.4. APPLICATIONS TO MEASURING PHYSICAL PROPERTIES

separation of two material particles in the deformed and undeformed states is

$$r^2 - r_0^2 = 2\xi_i \xi_j \eta_{ij}, \qquad (7.4.3)$$

where ξ_i is the difference of the Cartesian coordinates of the particles in the undeformed state, i.e., the ith component of $\mathbf{r}_0$. [When atoms are not at centers of symmetry, it is necessary to consider internal strains. In that case, Eq. (7.4.3) is not strictly valid. For a discussion of internal strains, see, for example, Born and Huang[92]; Naimon et al.[95]; Fuller and Naimon.[97]] From Eqs. (7.4.1)–(7.4.3), it follows that

$$\partial/\partial \eta_{ij} = \xi_i \xi_j (1/r) \, d/dr \equiv \xi_i \xi_j D, \qquad (7.4.4)$$

and that the first-, second-, and third-order Brugger elastic constants are given by

$$c_{ij} = (1/2V_0) \sum \xi_i \xi_j \, (D\phi)_{r=r_0}, \qquad (7.4.5)$$

$$c_{ijkl} = (1/2V_0) \sum \xi_i \xi_j \xi_k \xi_l \, (D^2\phi)_{r=r_0}, \qquad (7.4.6)$$

and

$$c_{ijklmn} = (1/2V_0) \sum \xi_i \xi_j \xi_k \xi_l \xi_m \xi_n \, (D^3\phi)_{r=r_0}. \qquad (7.4.7)$$

The first-order constants vanish for a crystal in equilibrium, e.g., for cubic structures, the lattice parameter is determined by requiring that $0 = \partial E/\partial \eta_{11} = \partial E/\partial \eta_{22} = \partial E/\partial \eta_{33}$, and all other first-order constants are zero by symmetry. For cubic structures, the above equations immediately give the relations

$$c_{12} = c_{44}, \qquad c_{112} = c_{166}, \qquad \text{and} \qquad c_{144} = c_{123} = c_{456}, \qquad (7.4.8)$$

where the Voigt reduced notation[14] has been used. These relations are known as *Cauchy relations*. They hold when the forces are central and all atoms are at centers of symmetry. Among the common cubic materials, the Cauchy relations are expected to hold best in the alkali halides and the rare-gas solids.

For the alkali halides, one of the dominant forces between ions is the long-range Coulombic force. Ghate[98] and Fuller and Naimon[97] have calculated the electrostatic contributions to the elastic constants of alkali

---

[97] E. R. Fuller, Jr., and E. R. Naimon, *Phys. Rev.* B **6**, 3609 (1972).
[98] P. B. Ghate, *Phys. Rev.* **139**, A1666 (1965).

halides. The results for the NaCl-type structure are

$$c_{11}^{es} = -1.569e^2/R^4,$$
$$c_{12}^{es} = c_{44}^{es} = +0.348e^2/R^4,$$
$$c_{111}^{es} = +10.264e^2/R^4,$$
$$c_{112}^{es} = c_{166}^{es} = -1.209e^2/R^4,$$

and

$$c_{144}^{es} = c_{123}^{es} = c_{456}^{es} = +0.679e^2/R^4,$$

where R is the nearest-neighbor distance and e is the electronic charge. For NaCl, $e^2/R^4 = 0.381 \times 10^{12}$ dyn cm^{-2}, which gives $c_{11}^{es} = -0.598$, $c_{12}^{es} = +0.133$, $c_{111}^{es} = +3.911$, $c_{112}^{es} = -0.461$, and $c_{144}^{es} = +0.259 \times 10^{12}$ dyn cm^{-2}. Upon comparison with the experimental results given in Table IV, it is seen that the electrostatic calculations are in good agreement with experiment for c_{12}, c_{44}, c_{112}, and c_{166}, and in fair agreement with experiment for c_{144}, c_{123}, and c_{456}. However, it can be seen that electrostatic forces by themselves cannot account for the observed values of c_{11} and c_{111}. Even though the electrostatic energy provides the main contribution to the binding energy of ionic crystals, it is necessary to consider the closed-core repulsive energy between ions when calculating certain elastic constants. This repulsive interaction is known to be very steep and of extremely short range, and thus it is convenient to consider contributions from different neighbors. In the NaCl-type structure, each atom has as nearest neighbors six atoms of the opposite kind, with coordinates $\xi = (0, 0, \pm 1)R$, $(0, \pm 1, 0)R$, and $(\pm 1, 0, 0)R$. Assuming that the short-range repulsive forces are central, Eqs. (7.4.6) and (7.4.7) can be used to show that the nearest neighbors will only contribute to c_{11} and c_{111}. Thus the repulsive interaction explains somewhat why some of the experimental elastic constants agree with the electrostatic values and others do not. However, a problem which to date still remains unsolved is why the observed value of c_{144} is not in excellent agreement with the electrostatic value. As just seen, the nearest-neighbor short-range repulsive interactions do not contribute to c_{144}. In addition, it can be shown that second-nearest neighbors also contribute nothing to c_{144}.[88] There have been other measurements[99-101] of c_{144} for NaCl, but the

---

[99] Z. P. Chang, Phys. Rev. **140**, A1788 (1965).
[100] J. R. Drabble and R. E. B. Strathen, Proc. Phys. Soc. (London) **92**, 1090 (1967).
[101] R. F. Marshall, thesis, Univ. of Illinois (1972).

7.4. APPLICATIONS TO MEASURING PHYSICAL PROPERTIES

recent one by Marshall,[101] using inertial stresses, seems to be the most direct measurement. He found the experimental value of c_{144} to be $\sim 30\%$ lower than the electrostatic value. The discrepancy could not be explained by van der Waals forces, thermal-vibration effects, central-force repulsion between third-nearest neighbors, or Coulomb effects due to the finite sizes of charged ions. The existence of a weak noncentral force between ions could not be excluded.

When short-range central forces predominate, a neighbor-by-neighbor analysis provides simple relations among the elastic constants. An example is provided by the rare-gas solids. There the interatomic potentials are described fairly well by the Lennard–Jones "6–12" potential,[92] which is both central and short range. Most of the rare gases solidify in the fcc structure. For this structure there are twelve nearest neighbors. Their coordinates are given by $\xi = (\pm\frac{1}{2}, \pm\frac{1}{2}, 0)a$, $(\pm\frac{1}{2}, 0, \pm\frac{1}{2})a$, and $(0, \pm\frac{1}{2}, \pm\frac{1}{2})a$, where a is the lattice parameter. For only nearest-neighbor central force interactions, Eqs. (7.4.6) and (7.4.7) can be used to show that

$$c_{44} = \tfrac{1}{2}c_{11}, \quad c_{112} = \tfrac{1}{2}c_{111}, \quad \text{and} \quad c_{144} = 0, \qquad (7.4.9)$$

where it is understood that the Cauchy relations hold. Data on the second-order elastic constants of rare-gas solids have begun to appear in the last few years. For these materials, the temperature dependence of the elastic constants is quite substantial, and it is difficult to get values extrapolated to 0 K for comparison with static-lattice calculations. Roughly speaking, it is found that $c_{12} \approx c_{44} \approx \tfrac{1}{2}c_{11}$, as expected from Eqs. (7.4.8) and (7.4.9). More realistic calculations for the rare-gas solids include further neighbor contributions and three-body forces, as well as lattice-vibrational effects.

The forces in metals are known not to be central, as the second-order Cauchy relations fail seriously for most metals. However, Hiki and Granato[33] have found that the third-order elastic constants of copper, silver, and gold follow the Cauchy relations much more closely than do the second-order constants, as seen by inspection of Table IV. The interpretation that Hiki and Granato gave to their results was that short-range central forces, in this case arising from d-shell overlap, play a progressively greater role as one progresses from calculations of the energy through the lattice constant to second-, third-, and higher-order derivatives of the total energy. If short-range forces play a dominant role in third- and higher-order elastic constants, then the nearest-neighbor atoms should make the most important contributions. Thus, since copper, silver, and gold have the fcc structure, one might expect [from Eqs. (7.4.8) and (7.4.9)] that

$$c_{112} \approx c_{166} \approx \tfrac{1}{2} c_{111} \quad \text{and} \quad c_{123} \approx c_{456} \approx c_{144} \approx 0.$$

These relations are satisfied to a much greater extent than are the second-order Cauchy nearest-neighbor relations.

7.4.1.3.2. NONCENTRAL POTENTIALS. It should be apparent that the calculation of elastic constants is greatly simplified when central forces predominate. Unfortunately, central-type interactions are unrealistic as far as most metallic and covalent materials are concerned. Still, there has been some success in predicting the elastic constants of various metallic and covalent crystals. For simple metals, i.e., those with small ion cores and no d electrons, pseudopotential methods have been used to calculate elastic constants. (For a discussion of pseudopotentials, see Harrison.[102]) There the energy of the metal is usually written as a sum of three terms: a volume-dependent term E_v, which includes the Fermi, exchange, and correlation energies; an electrostatic term E_c, which is the Coulomb energy of positive point charges arranged on a lattice and embedded in a uniform sea of conduction electrons; and a band-structure term E_{BS}, which represents the deviation of the electron energy from that of free electrons. E_c is, of course, a central-type interaction, but E_v and E_{BS} are not. Thus the Cauchy relations are not expected to hold (e.g., see the values for aluminum in Table IV). When calculating the elastic constants of metals, the preferred strain parameters are usually the Fuchs parameters, the main reason being the complicated dependence of E_{BS} on the Lagrangian strain parameters. For nonprimitive metallic structures, the Fuchs parameters are also preferred because they reduce the number of elastic constants for which internal-strain contributions must be calculated.[95] Complete sets of the second- and third-order elastic constants have been calculated for the alkali metals,[94] for aluminum and lead,[103] and for magnesium.[95] Agreement with experiment was excellent for the alkali metals, good for magnesium, fair for aluminum, and slightly less than fair for lead. However, comparisons with the complete set of experimental third-order constants could only be performed for aluminum[104] and magnesium.[16] It should be noticed that the agreement

---

[102] W. A. Harrison, "Pseudopotentials in the Theory of Metals." Benjamin, New York, 1966.
[103] Tetsuro Suzuki, *Phys. Rev. B* **3**, 4007 (1971).
[104] J. F. Thomas, Jr., *Phys. Rev.* **175**, 955 (1968).

between theory and experiment decreases as the valence increases ($Z = 1$ for the alkali metals, $Z = 2$ for Mg, $Z = 3$ for Al, and $Z = 4$ for Pb). This is most likely due to the fact that the band-structure contributions to the elastic constants increase for higher valence. Harrison[102] has noted that quantitative calculations become more difficult for higher valence, since the results are increasingly sensitive to errors in the band-structure energy.

There has been some theoretical work on the elastic constants of covalent materials, particularly for these having the diamond structure. Since the covalent bond has strong directional properties, simple central-force-type models cannot be used to accurately predict the elastic constants. Keating[105] introduced a phenomenological model with two types of interactions, those due to bond-stretching forces and those due to bond-bending forces. These two interactions yielded two parameters (force constants), which could be adjusted to fit the experimental values of the second-order elastic constants. Thus, as far as the second-order elastic constants of diamond-type crystals are concerned, this is a two parameter model to predict three constants. Keating's results were in very good agreement with experiment for the second-order elastic constants of diamond, silicon, and germanium. Keating also extended this method to the third-order elastic constants of diamond-type crystals. Here there are three parameters (force constants) that are adjusted to fit the six values of the experimental third-order elastic constants. Very good agreement between theory and experiment was obtained for the third-order constants of silicon and germanium. Using force constants similar to those introduced by Keating,[105] and combining them with effective point-ion forces, Martin[106] has presented a method for calculating the second-order elastic constants of zincblende structures. Fuller[107] has recently extended this method to the third-order constants of zincblende structures.

7.4.1.4. Temperature Dependence of the Elastic Constants. The temperature dependence of the elastic constants has been extensively discussed by Born and Huang[92] and by Leibfried and Ludwig,[96] and only the main predictions of the theory will be mentioned. At high temperatures (i.e., those above the Debye temperature), the second-order elastic constants should decrease linearly with increasing temperature. At very

[105] P. N. Keating, *Phys. Rev.* **145**, 637 (1966); **149**, 674 (1966).
[106] R. M. Martin, *Phys. Rev. B* **1**, 4005 (1970).
[107] E. R. Fuller, Jr., thesis, Univ. of Illinois (1973).

low temperatures, however, the second-order constants should decrease as T^4 with increasing temperature. The temperature dependence of a typical elastic constant (c_{11} for BaF_2[108]) is illustrated in Fig. 11. The experimental data for most materials generally satisfy the theoretical predictions, but the T^4 dependence at low temperatures has not yet been unambiguously established.[109] There have also been several empirical equations[109] to represent the temperature dependence of the second-order elastic constants, and these are sometimes very accurate over a large temperature range for a given material.

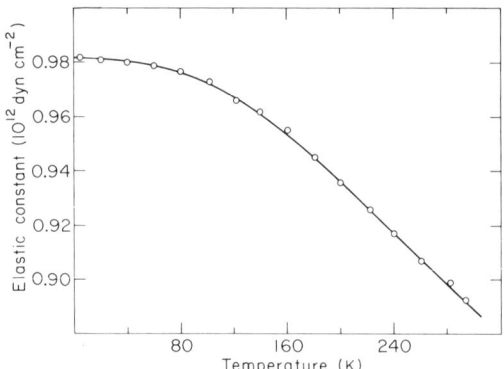

FIG. 11. The temperature dependence of c_{11} for BaF_2 [after D. Gerlich, *Phys. Rev.* **135**, A1331 (1964)].

Using the continuum approximation in the high-temperature limit, Hiki *et al.*[110] have derived expressions for the temperature dependence of the second-order elastic constants in terms of third- and fourth-order elastic constants. Since there are no available data for the fourth-order elastic constants at the present time, it is of interest to try to invert the procedure and determine the fourth-order constants from the experimental values of the temperature derivatives. Such an approach has been used for the noble metals,[110] NaCl,[111] and Cu–Zn.[112] This subject has recently been reviewed by Holder and Granato.[88]

As far as third-order elastic constants are concerned, Ghate[98] has

---

[108] D. Gerlich, *Phys. Rev.* **135**, A1331 (1964).
[109] Y. P. Varshni, *Phys. Rev. B* **2**, 3952 (1970).
[110] Y. Hiki, J. F. Thomas, and A. V. Granato, *Phys. Rev.* **153**, 764 (1967).
[111] K. D. Swartz, *J. Acoust. Soc. Amer.* **41**, 1083 (1967).
[112] K. D. Swartz, thesis, Univ. of Illinois (1966).

calculated their temperature dependence for the alkali halides. In addition, the temperature dependence of the third-order constants of copper has been measured by Salama and Alers[34] and by Peters et al.[113]

7.4.2. Real Materials

In real materials, dislocation effects generally provide attenuation and velocity dispersion. When these effects are removed, for example by irradiation, then phonon–phonon effects remain in insulators. In metals there is, in addition, an electronic loss which is important at low temperatures.

7.4.2.1. Dislocation Effects. In a solid containing no defects, ultrasonic plane waves should propagate with little attenuation. A plane wave traveling in a solid containing dislocations is attenuated, and the attenuation depends sensitively upon the mobility of the dislocations. The mobility, in turn, depends upon the interactions of dislocations with point defects, phonons, and other dislocations. Thus measurements of ultrasonic attenuation or ultrasonic wave propagation in crystals are uniquely suited for the study of dislocation interactions with other defects. On the other hand, if the dislocation interactions are such as to render the dislocations immobile, then no ultrasonic attenuation is obtained. Thus the technique is limited to the study of mobile dislocations.

Advantages of ultrasonic attenuation measurements derive from (1) the sensitivity of the method, (2) the selectivity of the measurements, (3) the fact that the results can be made quantitative, and (4) the fact that the measurements are nondestructive. For example, point-defect concentrations of the order of 10^{11} cm^{-3} can have an important effect on the ultrasonic attenuation caused by dislocations. Also, the ultrasonic attenuation is sensitive only to those point defects which arrive at dislocations, in contrast to properties such as resistivity, which are sensitive to all the defects in the lattice. The method is especially suited to the study of defect interactions, since it is only by these interactions that changes in the ultrasonic wave propagation characteristics are induced. A large amount of detail can be found because of the large number of variables that can be controlled (frequency, strain amplitude, temperature, point-defect concentration, ultrasonic mode orientation, purity, deformation, and others).

---

[113] R. D. Peters, M. A. Breazeale, and V. K. Paré, *Phys. Rev. B* **1**, 3245 (1970).

Measurements of ultrasonic attenuation tend to complement, and not to compete, with measurements made by direct observations and also with other indirect observations. Ultrasonic attenuation measurements differ from other indirect observations in sensitivity and in the linearity of the effects with defect concentrations. For example, properties such as resistivity, lattice parameter, density, and stored energy are usually linear in the defect concentration, and this fact leads to simplicity in interpretation of results. However, it is normally necessary to have defect concentrations considerably in excess of parts per million to obtain measurable effects. On the other hand, ultrasonic attenuation measurements require more interpretation (the ultrasonic attenuation depends on the fourth power of the pinning point density in common cases), but fewer defects are required for measurable effects, and more detail is obtained in the measurements.

There are a number of ways in which dislocations can contribute to ultrasonic attenuation. A dislocation segment oscillating between two pinning points (vibrating string model) gives one characteristic type of damping (low-amplitude resonance). Dislocations that break away from pinning points under stress lead to another type (amplitude-dependent damping). Dislocations which move by overcoming Peierls barriers are supposed to give rise to the low-temperature Bordoni peaks. There are also other mechanisms, models for which have not yet been developed. In particular, there is not as yet a suitable model for the highly deformed state. However, the single dislocation segment model, which neglects interactions between dislocations, seems to work surprisingly well for moderate deformation (up to about 4%). In what follows we discuss only the vibrating string model, since its predictions are definite, and many experimental checks of these predictions are now available.

The basis for ultrasonic attenuation effects lies in the fact that dislocation motion contributes to the total strain developed in a specimen under stress. For a given applied stress, a solid containing dislocations has a larger strain than a perfect crystal, so that the elastic modulus appears to be lower. Under the action of an alternating stress, the dislocation component of the strain may lag behind the applied stress. This leads then not only to a reduction in modulus but also to a damping of the applied stress. Simple estimates of the magnitude of the expected effect for typical dislocation densities lead to much larger values than the observed effects, if it is assumed that the dislocations are perfectly mobile with no restrictions on their motion. It may, therefore, be concluded that there must be impediments to the motion of dislocations. Generally speaking, the same types of obstacles have been assumed as in yield stress theories, where a similar problem is faced. These are, for example, atomic pinning points, network points, jogs, other dislocations, and so on. However, even with such restrictions, one must explain why the dislocation motion lags behind the applied stress. For smooth dislocation motions, one may imagine that the dislocation is viscously damped as it moves through the electron or phonon gas. Also impediments which lead to a jerky motion of the dislocation will lead to a phase lag. Examples of the latter type of effect are provided by motion over Peierls barriers at low stresses, and catastrophic unpinning of dislocations at high stresses.

A model proposed by Koehler[114] and developed further by Granato and Lücke,[115] using an analogy between a vibrating string in a viscous medium and dislocation oscillations, has been proven to be correct in most of its particulars. In this model, advantage is taken of the fact that a dislocation has an effective mass per unit length and an effective tension.

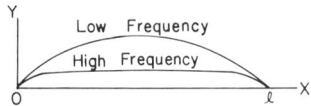

FIG. 12. Schematic dislocation displacement $y(x)$ as a function of coordinate x for low frequencies and high frequencies. At low frequencies, the displacement is limited by tension forces. At high frequencies, the displacement is limited by viscous forces.

Thus an equation of motion for small oscillations of a dislocation may be written as

$$Ay_{tt} + By_t - Cy_{xx} = b\sigma, \qquad (7.4.10)$$

where A is the effective mass for unit length, y is the dislocation displacement measured from the equilibrium position as indicated in Fig. 12, B is the viscous damping constant, C the tension, b the Burgers vector, σ the applied stress, t is the time, x is a coordinate along the dislocation, and subscripts denote differentiation.

The solution of Eq. (7.4.10), together with the boundary conditions $y(0)$ and $y(l) = 0$, where pinning points are placed at $x = 0$ and l, gives the dislocation displacement y as a function of the angular frequency ω of the applied stress $\sigma(\omega)$. Using this the dislocation strain and then

---

[114] J. S. Koehler, "Imperfections in Nearly Perfect Crystals" (W. Shockley, J. Hollomon, R. Maurer, and F. Seitz, eds.), p. 197. Wiley, New York, 1952.

[115] A. V. Granato and K. Lücke, *J. Appl. Phys.* **27**, 583 (1956).

the effective modulus (or ultrasonic wave velocity) and decrement (or ultrasonic wave attenuation) are easily found. The resulting decrement has a typical resonance-type frequency behavior which depends, however, on the magnitude of the viscous damping constant B. Theoretical estimates of the damping constant to be expected for dislocation interactions with phonons and electrons have been made by many authors, and the calculations have recently been reviewed by Brailsford.[116] From these one expects: (1) that the phonon interaction at room temperature is usually much larger than the electron interaction, and (2) that the phonon interaction is so large that the dislocation resonance is overdamped at room temperature. This has the effect of broadening out the resonance and moving it to lower frequencies. Also, when account is taken of the fact that not all dislocations have the same length, the expected maximum is broadened out further. For typical expected dislocation segments of length of order of 1 μm in pure undeformed materials, the expected resonance frequency is at about 1 GHz. However, because of the large damping (we may picture the dislocation as a string moving in heavy molasses), the maximum may be brought down to frequencies as low as 100 kHz.[117] At the same frequencies where the decrement goes through a maximum, a dispersion in the elastic constant is expected. As is indicated schematically in Fig. 12, the displacement of the dislocation is a function of frequency. (The dislocation strain, and thus the modulus reduction, is proportional to the area swept out by the dislocation.) At low frequencies, the velocity of the dislocation is small so that the viscous force is small. The displacement is then limited by the tension forces and is parabolic in shape. At high enough frequencies, however, the viscous forces become dominant, and the displacement of the dislocation cannot achieve its full value. In this case the dislocation moves more like a rigid rod over most of its length, coming down to zero displacement only near the pinning points. Thus we expect the effect of pinning points to be large at low frequencies and negligible at high frequencies.

The dispersion effect is illustrated in Fig. 13, in which the velocity of compressional waves was measured by Granato et al.[118] in a sodium chloride crystal as a function of frequency, before and after a slight deformation. Before deformation there is only a small dispersion of 0.5% centered at about 75 MHz. After the deformation, the magnitude

---

[116] A. D. Brailsford, *J. Appl. Phys.* **43**, 1380 (1972).

[117] A. V. Granato and R. Stern, *J. Appl. Phys.* **33**, 2880 (1962).

[118] A. V. Granato, J. de Klerk, and R. Truell, *Phys. Rev.* **108**, 895 (1957).

7.4. APPLICATIONS TO MEASURING PHYSICAL PROPERTIES 425

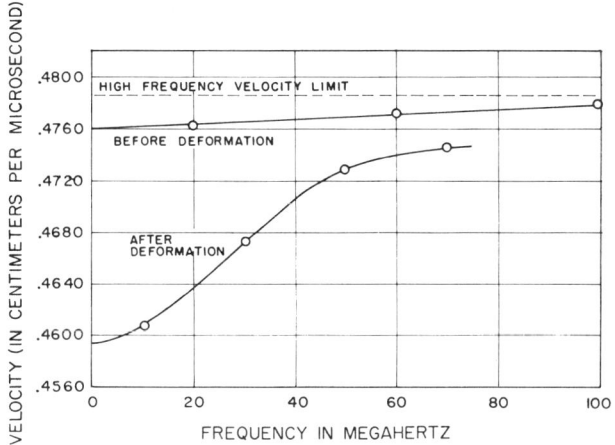

FIG. 13. Velocity dispersion for compressional elastic waves propagating in the [100] direction in NaCl. Deformation increases the magnitude of the dispersion and moves it to lower frequencies [after A. V. Granato, J. de Klerk, and R. Truell, *Phys. Rev.* **108**, 895 (1957)].

of the dispersion has increased to about 4% and has moved to a lower frequency (35 MHz). At room temperature, the dispersion was observed to gradually recover toward the initial condition, presumably as a result of dislocation pinning by deformation-induced defects able to diffuse at room temperature. The interpretation of the effect according to the vibrating string model is as follows. At low frequencies, the dislocations are in phase with the ultrasonic stress. When the stress is applied, the apparent elastic constant (and therefore the ultrasonic velocity) is reduced because the dislocation motion makes the specimen less rigid. However, at high frequencies the dislocations can no longer follow the rapidly changing stress, so that the modulus approaches the true elastic value.

The frequency dependence of the damping arising from dislocation motion is illustrated in Fig. 14. These are measurements by Stern and Granato[119] showing the effect of cobalt gamma irradiation on the damping of high-purity copper. Before irradiation, the damping has a maximum at a few megahertz. The gamma rays produce electrons that are energetic enough to displace lattice atoms, giving interstitials which can be effective as pinning points. After 50 hr of irradiation in a 6000 Ci cobalt source, the height of the maximum decreased, and the location increased to about 100 MHz. The damping at low frequencies is much more sensitive

---

[119] R. M. Stern and A. V. Granato, *Acta Met.* **10**, 358 (1962).

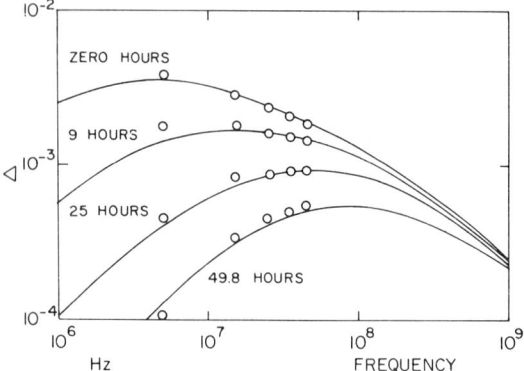

FIG. 14. Decrement in copper as a function of frequency and cobalt gamma irradiation time in a 6000 Ci source. The solid curves are theoretical [after R. M. Stern and A. V. Granato, *Acta Met.* **10**, 358 (1962)].

to the increased number of pinning points than that at high frequencies, as we expected from our previous discussion of Fig. 12.

According to the theory, the height of the maximum should be proportional to ΛL^2 and the location of the maximum proportional to $1/BL^2$, where Λ is the total dislocation density and L is the average segment length between pinning points. At frequencies much lower than that at which the maximum occurs, the damping should be proportional to $\Lambda L^4 B\omega$ and the modulus to ΛL^2.

A characteristic feature of dislocation damping effects is the pronounced dependence of the damping (and modulus) on the amplitude of the external stress. The amplitude dependence of the damping is interpreted in the vibrating string model as resulting from a break-away of the dislocation from weak pinning points as the stress amplitude is increased. The predicted amplitude in the string model given by Granato and Lücke[115] is that the logarithm of the product of the attenuation and strain amplitude should be linear in the reciprocal strain amplitude.

A novel and interesting application of this model was made in 1965 by Tittmann and Bömmel,[120] who measured ultrasonic attenuation in superconducting lead. While studying the phonon–electron interaction in pure lead at 51 MHz, an amplitude dependence, first found by Love and Shaw,[121] was observed. Some of Tittmann and Bömmel's results are

---

[120] B. R. Tittmann and H. E. Bömmel, *Phys. Rev. Lett.* **14**, 296 (1965); *Phys. Rev.* **151**, 178 (1966).

[121] R. E. Love and R. W. Shaw, *Rev. Mod. Phys.* **36**, 260 (1964).

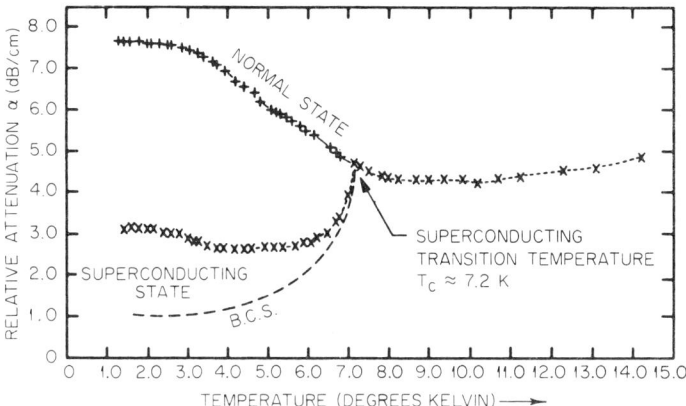

FIG. 15. Temperature dependence of ultrasonic attenuation at medium strain amplitude in pure single crystalline lead with frequency of 51 MHz and longitudinal waves along the ⟨111⟩ direction [after B. R. Tittmann and H. E. Bömmel, *Phys. Rev. Lett.* **14**, 296 (1965); *Phys. Rev.* **151**, 178 (1966)].

shown in Fig. 15. The upper curve shows the typical attenuation versus temperature in the normal state, while the lower (dashed) curve shows that expected in the superconducting state. The measured values are intermediate, and depend strongly on the amplitude of the ultrasonic wave as the temperature is decreased. Tittmann and Bömmel suggest a plausible model for the absence of dislocation damping in the normal state and its presence in the superconducting state. According to their model, the conduction electrons behave as a viscous gas which damps the motion of the dislocations. As the temperature is lowered below the critical temperature, the density of electrons able to damp the dislocation motion decreases, and the breakaway mechanism is able to operate.

A number of experiments were conducted to test these ideas. The amplitude-dependent damping was found to follow the expected strain-amplitude dependence (Fig. 16). Measurements on 99.9% pure lead crystals doped with tin as an impurity showed a strong reduction in amplitude dependence.[120] The frequency dependence of the effect was found to be approximately constant, in agreement with the theory. The ranges studied were 30 MHz–1 GHz for the impure lead and 30–150 MHz for the pure lead. It can be inferred from these measurements that the (undamped) resonance frequencies of the dislocation in the pure lead lie higher than 150 MHz. The measurements demonstrate that the electron gas is an effective source of damping at low temperatures and open up new possibilities for the study of dislocation–electron interactions.

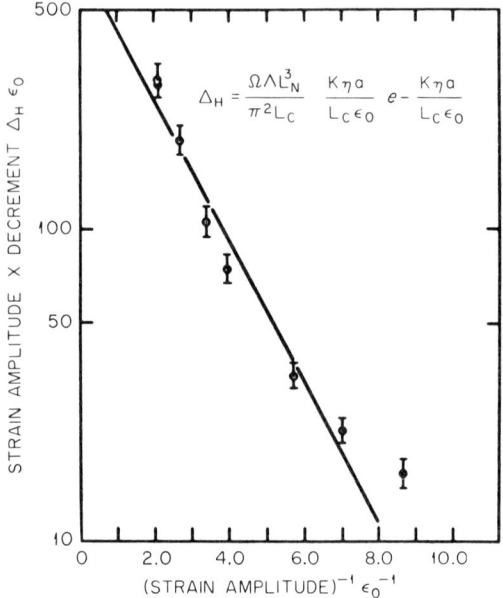

Fig. 16. Amplitude-dependent ultrasonic attenuation in pure lead at 4.2 K with the longitudinal waves along the $\langle 111 \rangle$ direction [after B. R. Tittmann and H. E. Bömmel, Phys. Rev. Lett. **14**, 296 (1965); Phys. Rev. **151**, 178 (1966)].

7.4.2.2. Electronic Effects

7.4.2.2.1. Ultrasonic Attenuation in Metals. When an ultrasonic wave propagates through a metal, some of the acoustic energy is dissipated as a result of the interactions between the wave and the conduction electrons, i.e., electron–phonon interactions. This attenuation mechanism is very dependent upon the parameters k and l, where k is the wave vector of the ultrasonic wave and l is the mean free path of the conduction electrons. A comprehensive review of this subject has been given by Morse,[122] and here we will mention only some of the general features. For the case of $kl \ll 1$, the attenuation varies as the square of the ultrasonic frequency and has the temperature dependence of l. For $kl \gg 1$, the attenuation varies linearly with frequency and, because it is independent of l, has essentially no temperature dependence. In addition, the attenuation for $kl < 1$ is less than that for $kl > 1$ by approximately the factor kl. This accounts for the fact that electronic attenuation of ultrasonic waves can only be observed in reasonably pure materials at very low tempera-

---

[122] R. W. Morse, "Progress in Cryogenics," Vol. I. Heywood, London, 1959.

tures. For example, typical room-temperature values of kl are $\sim 10^{-4}$, and the corresponding attenuation is $\sim 10^{-3}$–10^{-4} cm^{-1}, i.e., an entirely negligible attenuation. The theoretical predictions for the frequency and temperature dependence of the electronic attenuation of ultrasonic waves have generally been confirmed by experiment, e.g., Morse[122] and Fate.[123] An example of the agreement between theory and experiment is illustrated in Fig. 17. By considering the case of a relaxation process of a longitudinal wave in a free electron gas, Morse[124] obtained a linear relation between the electronic attenuation and the electrical conductivity. This relation has also been obtained by Mason,[125] using a viscous electron gas model, and by Kittel,[126] using an electron transfer mechanism. Thus attenuation measurements at low temperatures provide a direct means of determining the residual resistivity of a metal.

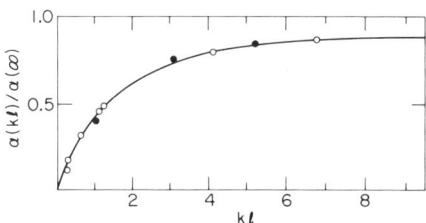

FIG. 17. Observed kl dependence of longitudinal-wave attenuation as compared with theory [after R. W. Morse, "Progress in Cryogenics," Vol. I. Heywood, London, 1959]. Open circles are measurements in indium; solid circles are measurements in copper. The attenuation has been normalized by the asymptotic value as $kl \to \infty$.

In superconductors, the ultrasonic attenuation is found to decrease rapidly as the temperature is lowered through the critical temperature T_c. This effect was first observed in a lead by Bömmel.[127] However, if the superconductivity is destroyed by a magnetic field, no sudden drop in the attenuation occurs at the critical temperature. The difference between the attenuation in the superconducting state and that in the normal state could not be explained until the advent of the BCS theory.[128] It was

[123] W. A. Fate, *Phys. Rev.* **172**, 402 (1968).
[124] R. W. Morse, *Phys. Rev.* **97**, 1716 (1955).
[125] W. P. Mason, *Phys. Rev.* **97**, 557 (1955).
[126] C. Kittel, *Acta Met.* **3**, 295 (1955).
[127] H. E. Bömmel, *Phys. Rev.* **96**, 220 (1954).
[128] J. Bardeen, L. N. Cooper, and J. R. Schrieffer, *Phys. Rev.* **106**, 162 (1957); **108**, 1175 (1957).

found that the temperature dependence of the attenuation in the superconducting state was related to the temperature dependence of the relative number of "superconducting" versus "normal" electrons. For longitudinal waves, the theory predicted that the ratio of the attenuation in the superconducting state to that in the normal state is given by

$$\alpha_s/\alpha_n = 2/\{\exp[\Delta(T)/k_B T] + 1\}, \tag{7.4.11}$$

where $2\Delta(T)$ is the temperature-dependent superconducting energy gap. The theory also predicts that the energy gap at 0 K should be a constant, $2\Delta(0) = 3.5 k_B T_c$, for weak-coupling superconductors. Equation (7.4.11) was derived for an isotropic superconductor with $kl \gg 1$, but Tsuneto[129] later verified its validity as well for $kl < 1$. He also considered the case of transverse waves in superconductors. By treating $2\Delta(0)$ as a variable parameter, Eq. (7.4.11) is commonly used to determine the anisotropy of the energy gap. The agreement between theory and experiment for α_s/α_n is reasonable (e.g., Morse[122]), but small deviations still need to be explained. Mason[130] suggests that these deviations could be due to electron-damping of dislocations, but the measurements by Fate[123] do not support this suggestion.

7.4.2.2.2. MAGNETOACOUSTIC MEASUREMENTS. In reasonably pure metals at low temperatures (i.e., the $kl > 1$ region), an oscillatory variation of attenuation as a function of applied magnetic field can be observed. The periods of these oscillations are directly related to dimensions of the Fermi surface. Depending on the magnitude of the magnetic field, basically three different effects may be observed (e.g., Dignum[12] and Beyer and Letcher[8]). At low fields (~35 G for gallium[12]), a large attenuation increase occurs when the ultrasonic frequency is an integral multiple of the cyclotron frequency. This effect is known as *acoustic cyclotron resonance*. For fields on the order of several hundred gauss, the phenomenon known as geometric resonance can occur. The condition for maximum acoustic attenuation here is that the electron orbit in the direction of the sound wave be an integral number of sound wavelengths. The attenuation will go through a minimum when the orbit size is an integral number of half wavelengths. This effect is illustrated for cesium[131] in Fig. 18. The third effect occurs at still higher fields and is quantum-

---

[129] T. Tsuneto, *Phys. Rev.* **121**, 402 (1961).
[130] W. P. Mason, *Phys. Rev.* **143**, 229 (1966).
[131] J. Trivisonno and J. A. Murphy, *Phys. Rev. B* **1**, 3341 (1970).

7.4. APPLICATIONS TO MEASURING PHYSICAL PROPERTIES

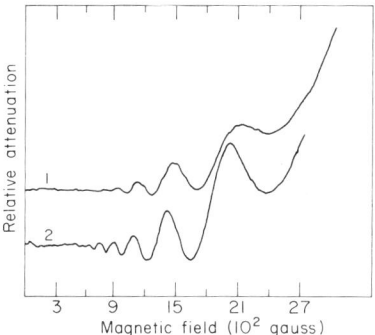

FIG. 18. Relative attenuation of 70-MHz longitudinal waves versus magnetic field with **k** ∥ [001]. The curves are for cesium [after J. Trivisonno and J. A. Murphy, *Phys. Rev. B* **1**, 3341 (1970)]. The upper curve is for **H** ∥ [110] and the lower curve for **H** ∥ [100]. The curves have been shifted for clarity.

mechanical in nature. Here one must consider the fact that the electron orbits are quantized (Landau levels). The oscillations which occur here are of the de Haas–van Alphen type, i.e., they are a result of variations in the density of electron states at the Fermi energy with magnetic field. The de Haas–van Alphen effect measures the periodicity of the diamagnetic susceptibility, whereas the magnetoacoustic effect is concerned with the periodic variations of the ultrasonic attenuation.

The above measurements are dependent upon the electron being able to complete its orbit before being scattered by a defect or impurity. That is why the experiments are only possible in pure materials and at low temperatures, where lattice vibrations are minimum. Because the electrons involved in magnetoacoustic measurements are those at the Fermi surface, useful data on the dimensions of the Fermi surface can be obtained.

7.4.2.3. Other Sources of Loss

7.4.2.3.1. PHONON VISCOSITY. In insulating crystals containing no dislocation losses, there remains an appreciable loss arising from the nonlinear interaction between the ultrasonic wave and thermal phonons of finite temperatures. Examples of such measurements in quartz by Bömmel and Dransfeld[132] are shown in Fig. 19. Quantitative estimates of the loss have been given by Bömmel and Dransfeld[133] and by Woodruff

---

[132] H. E. Bömmel and K. Dransfeld, *Phys. Rev. Lett.* **2**, 298 (1959).
[133] H. E. Bömmel and K. Dransfeld, *Phys. Rev.* **117**, 1245 (1960).

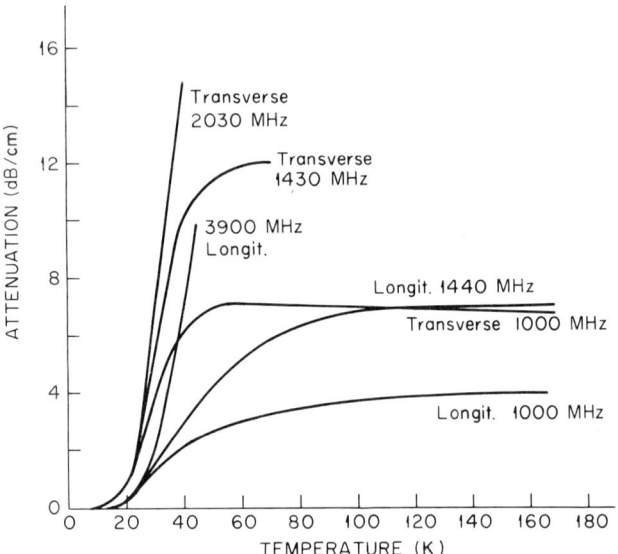

FIG. 19. Temperature-dependent part of ultrasonic attenuation in quartz [after H. E. Bömmel and K. Dransfeld, *Phys. Rev. Lett.* **2**, 298 (1959)].

and Ehrenreich.[134] At high temperatures, the loss is expected to be proportional to ω^2/W^5 and the square of a suitably weighted average third-order elastic constant. The inverse fifth dependence on the velocity leads to larger losses for shear waves than for compressional waves. Below about 40 K the loss becomes strongly temperature-dependent, going to zero with the phonon density. The existence of this loss, which cannot be eliminated as is the case with dislocation losses, has made it necessary thus far to operate gigahertz devices at low temperatures.

7.4.2.3.2. THERMOELASTIC LOSSES. The stresses associated with ultrasonic waves produce temperature changes. The resulting heat flow produces losses which are well known in work at lower frequencies. The losses from flow into specimens from the external media and between grains of polycrystalline specimens have been discussed in detail by Zener.[135] The latter effect can be important in the megahertz region as well. For single crystals, there is in addition a loss resulting from the heat flow between the crests and valleys of the ultrasonic wave. For the

---

[134] T. O. Woodruff and H. Ehrenreich, *Phys. Rev.* **123**, 1553 (1961).

[135] C. Zener, "Elasticity and Anelasticity of Metals." Univ. of Chicago Press, Chicago, Illinois, 1948.

megahertz region, this loss is proportional to ω^2, but is generally smaller than other losses already mentioned. A calculation of the magnitudes of the attenuations has been given by Lücke.[136]

7.4.2.3.3. SURFACE WAVES. A field of study which has grown enormously in the past several years is the study of surface waves on crystals. The availability of the excitations on the surface where they may be tapped readily, as well as the possibilities available for guidance and interactions between waves, leads to many possibilities for circuit devices. Much of this work appears in the electrical engineering literature. A review of the field has recently been given by White.[137]

7.4.2.3.4. LOSSES FOUND IN PARTICULAR MATERIALS. The losses mentioned earlier are common to all metals. There are in addition loss mechanisms for special materials. These include stress-induced ordering, magnetoelastic interactions, interactions with nuclear spin systems, ultrasonic-electron–spin interactions, ultrasonic attenuation (and amplification) in piezoelectric semiconductors, and the acoustoelectric effect. The reader is referred to reviews mentioned earlier for details of these effects. Major attention in the past few years has been given to studies of phase changes in crystals, particularly to the use of ultrasonic studies of scaling laws.[138,139]

Acknowledgment

This work was supported in part by the U.S. Atomic Energy Commission under Contract AT(11-1)-1198, in part by the National Science Foundation under Grant GP-27369, and in part by the Advanced Research Projects Agency under Contract HC 15-67-C-0221.

APPENDIX A. Thermoelasticity Theory and Related Elastic Coefficients

A.1. Thermoelasticity

Thermoelasticity theory provides the framework for interpreting the effective elastic coefficients obtained from experiment. When considering the propagation velocity and its properties in an undeformed elastic

---

[136] K. Lücke, *J. Appl. Phys.* **27**, 1433 (1956).
[137] R. M. White, *Proc. IEEE* **58**, 1238 (1970).
[138] A. Ikushima, *J. Phys. Chem. Solids* **31**, 283 (1970).
[139] B. Lüthi and R. J. Pollina, *Phys. Rev. Lett.* **22**, 717 (1969).

TABLE V. Notation

	Description	Present	R. N. Thurston[a]	D. C. Wallace[b]
1.	Coordinates and densities			
	Natural state	$\mathbf{a}, \varrho_0$	$\mathbf{a}, \varrho_0$	$\bar{\mathbf{X}}, \varrho_0$
	Initial state	$\mathbf{X}, \bar{\varrho}$	$\mathbf{X}, \bar{\varrho}$	$\mathbf{X}, \varrho_1$
	Final state	$\mathbf{x}, \varrho$	$\mathbf{x}, \varrho$	$\mathbf{x}, \varrho$
2.	Deformation and strain parameters			
	Connecting natural to final state	J_{ij}, η_{ij}	$(\partial x_i/\partial a_j), \eta_{ij}$	$\bar{\alpha}_{ij}, \bar{\eta}_{ij}$
	Connecting natural to initial state	$\bar{J}_{ij}, \bar{\eta}_{ij}$	$(\partial X_i/\partial a_j), \bar{\eta}_{ij}$	$a_{ij}, \bar{\eta}_{ij}$ at $\mathbf{X}$
	Connecting initial to final state	$\tilde{J}_{ij}, \tilde{\eta}_{ij}$	$(\partial x_i/\partial X_j), S_{ij}$	α_{ij}, η_{ij}
	Infinitesimal displacement parameters	$\tilde{\varepsilon}_{ij}, \tilde{\omega}_{ij}$	$e_{ij}, \tilde{\omega}_{ij}$	$\varepsilon_{ij}, \omega_{ij}$
3.	Stresses			
	Initial state	$\bar{\sigma}_{ij}$	$\bar{T}_{ij}$	$T_{ij}(\mathbf{X}) = C_{ij}$
	Final state	σ_{ij}	T_{ij}	T_{ij}
4.	Elastic coefficients			
	Brugger constants[c]	$c_{ijkl}...$	$c_{ijkl}...$ at $\mathbf{a}$	$\bar{C}_{ijkl}...$
	Brugger coefficients	C_{ijkl}	$\bar{C}_{ijkl}$	C_{ijkl}
	Huang coefficients[d]	$\bar{W}_{ijkl}$	B_{ijkl}	S_{ijkl}
	Wave propagation coefficients[a]	$\hat{W}_{ijkl}$	$\hat{C}_{ijkl}$	—
	Birch coefficients[e]	B_{ijkl}	S_{ijkl}	B_{ijkl}
		Ω_{ijkl}	R_{ijkl}	$(\partial T_{ij}/\partial \omega_{kl})$
	Voigt coefficients[f]	V_{ijkl}	—	V_{ijkl}

[a] R. N. Thurston, *J. Acoust. Soc. Amer.* **37**, 348 (1965).
[b] D. C. Wallace, *Phys. Rev.* **162**, 776 (1967).
[c] K. Brugger, *Phys. Rev.* **133**, A1611 (1964).
[d] K. Huang, *Proc. Roy. Soc.* (*London*) **A203**, 178 (1950).
[e] F. Birch, *Phys. Rev.* **71**, 809 (1947).
[f] W. Voigt, "Lehrbuch der Kristallphysik." Teubner, Leipzig, 1928.

material, a theory of thermoelasticity for small strains from a configuration of zero stress, as discussed by Voigt,[14] is sufficient. However, to treat a finitely strained elastic material, one must generalize to an initial configuration of applied stress. A large amount of literature has appeared on this subject.[13,18,19,26–28,91,140–145] The use of different strain parameters as independent variables has led to many different types of elastic coefficients. This, along with the various notations which have evolved, can be very confusing when reviewing the literature. Except for notation, which might never be unified, the subject has recently been clarified in articles by Thurston[144] and Wallace.[27] Thurston discusses and interprets the effective elastic coefficients for wave propagation and their symmetries. Wallace presents a general theory of thermoelasticity from which five sets of thermoelastic coefficients are defined and related. The notation to be used here, as well as that of Thurston and Wallace, is listed in Table V for reference.

For every initial temperature T, three states of the elastic solid will be considered (see Fig. 20). The "natural" or unstressed state has coordinates **a** and density $\varrho(\mathbf{a}) = \varrho_0$. An applied uniform stress homogeneously deforms the solid from the natural state of zero stress to the "initial" state of arbitrary uniform stress $\bar{\sigma}_{ij}$. The initial state has coordinates **X** and density $\varrho(\mathbf{X}) = \bar{\varrho}$. The "final" state is obtained from the initial state by the application of an additional stress $\boldsymbol{\sigma} - \bar{\boldsymbol{\sigma}}$. It has co-

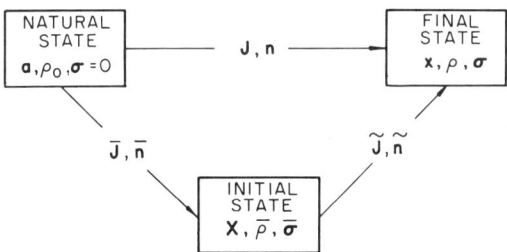

Fig. 20. States of the elastic solid with coordinates, densities and stresses, and the strains connecting them.

---

[140] F. Birch, *Phys. Rev.* **71**, 809 (1947).
[141] K. Huang, *Proc. Roy. Soc. (London)* **A203**, 178 (1950).
[142] C. Truesdell and W. Noll, "Handbuch der Physik" (S. Flügge, ed.), Vol. III/3, p. 1. Springer-Verlag, Berlin, 1965.
[143] A. E. Green, *Proc. Roy. Soc. (London)* **A266** (1962).
[144] R. N. Thurston, *J. Acoust. Soc. Amer.* **37**, 348 (1965).
[145] T. H. K. Barron and M. L. Klein, *Proc. Phys. Soc. (London)* **85**, 523 (1965).

ordinates **x** and density ϱ. Thus if the position of a material particle in the natural state is **a**, the positions of the same material particle in the initial and final states are **X** and **x**, respectively. Usually the initial state corresponds to a static stress, and the final state to wave propagation.

A convenient set of deformation parameters relating the different states are the transformation coefficients

$$\text{Natural to final:} \quad J_{ij} = (\partial x_i/\partial a_j), \quad (A.1)$$

$$\text{Natural to initial:} \quad \bar{J}_{kj} = (\partial X_k/\partial a_j), \quad (A.2)$$

and

$$\text{Initial to final:} \quad \tilde{J}_{ik} = (\partial x_i/\partial X_k). \quad (A.3)$$

The subscripts i, j, k represent components referred to a common Cartesian system, and each runs from 1 to 3. The three sets of transformation coefficients are obviously related by

$$J_{ij} = \tilde{J}_{ik}\bar{J}_{kj}. \quad (A.4)$$

The Lagrangian strain parameters connecting the states are

$$\eta_{ij} = \tfrac{1}{2}(J_{pi}J_{pj} - \delta_{ij}), \quad (A.5)$$

$$\bar{\eta}_{ij} = \tfrac{1}{2}(\bar{J}_{pi}\bar{J}_{pj} - \delta_{ij}), \quad (A.6)$$

and

$$\tilde{\eta}_{ij} = \tfrac{1}{2}(\tilde{J}_{pi}\tilde{J}_{pj} - \delta_{ij}), \quad (A.7)$$

where δ_{ij} is the Kronecker delta. For each deformation there are nine independent transformation coefficients. However, there are only six independent Lagrangian strains, since they are symmetric. However, when taking derivatives, it is convenient to treat all nine Lagrangian strains as independent. This can be done if functions are properly symmetrized first.[13] The infinitesimal displacement parameters are another set of strain parameters sometimes used. For example, from the initial to the final state the six symmetric infinitesimal displacement parameters

$$\tilde{\varepsilon}_{ij} = \tfrac{1}{2}(\tilde{J}_{ij} + \tilde{J}_{ji} - 2\delta_{ij}) \quad (A.8)$$

measure pure infinitesimal strains, and the three antisymmetric infinitesimal displacement parameters

$$\tilde{\omega}_{ij} = \tfrac{1}{2}(\tilde{J}_{ij} - \tilde{J}_{ji}) \quad (A.9)$$

measure pure infinitesimal rotations.

The standard thermodynamic potentials to be used here are the internal energy per unit mass U and the Helmholtz free energy per unit mass F. (The potential U is used for adiabatic processes such as ultrasonic wave propagation, while F is used for isothermal processes.) The thermodynamic equilibrium states of the solid are assumed to be well defined when the configuration and either the entropy per unit mass S or the temperature T are known. Implicit in this is the assumption that the solid is nondissipative (or that only states obtained by elastic deformation from the natural state be considered). This ensures single-valued thermodynamic potentials, and hysteresis effects are ignored. Furthermore, the internal energy and Helmholtz free energy should be invariant for any arbitrary rigid deformation of the solid (i.e., a deformation in which relative positions of material particles do not change). Thus the configuration dependence of the potentials is expressible in terms of the Lagrangian strains

$$U(\mathbf{x}, S) = U(\mathbf{X}, \tilde{\mathbf{\eta}}, S) = U(\mathbf{a}, \mathbf{\eta}, S),$$
$$F(\mathbf{x}, T) = F(\mathbf{X}, \tilde{\mathbf{\eta}}, T) = F(\mathbf{a}, \mathbf{\eta}, T). \tag{A.10}$$

It is convenient to consider cases for which either the initial or the natural state is the reference state. In general, when the reference state is the initial configuration, variables will have a tilde over them. A bar over a quantity will denote a variable referred to the natural state but evaluated in the initial state. Thus, the internal energy has independent variables S and $\tilde{\mathbf{\eta}}/\bar{\varrho}$ (or $\mathbf{\eta}/\varrho_0$) with respect to the initial (or natural) state. The conjugate dependent variables, T and the thermodynamic tension $\tilde{\mathbf{t}}$ (or $\mathbf{t}$), are given by the standard equations

$$T = (\partial U/\partial S) \tag{A.11}$$

and

$$\tilde{t}_{kl} = \bar{\varrho}(\partial U/\partial \tilde{\eta}_{kl}), \qquad t_{kl} = \varrho_0(\partial U/\partial \eta_{kl}). \tag{A.12}$$

Similar equations hold for the Helmholtz free energy when the temperature T is the independent variable. The stresses in the final and initial states are given by[13,26,27]

$$\sigma_{ij} = (\varrho/\bar{\varrho})\tilde{J}_{ik}\tilde{J}_{jl}\tilde{t}_{kl} = (\varrho/\varrho_0)J_{ik}J_{jl}t_{kl} \tag{A.13}$$

and

$$\bar{\sigma}_{ij} = \tilde{t}_{kl}(\mathbf{X}) = (\bar{\varrho}/\varrho_0)\bar{J}_{ik}\bar{J}_{jl}t_{kl}. \tag{A.14}$$

The strains from the initial to the final state usually correspond to wave propagation. Since they are generally small, it is useful to expand the internal energy in a Taylor series in $\tilde{\boldsymbol{\eta}}$

$$\bar{\varrho}U(\mathbf{X}, \tilde{\boldsymbol{\eta}}, S) = \bar{\varrho}\bar{U} + \bar{\sigma}_{ij}\tilde{\eta}_{ij} + \tfrac{1}{2}C^S_{ijkl}\tilde{\eta}_{ij}\tilde{\eta}_{kl}$$
$$+ (1/3!)C^S_{ijklmn}\tilde{\eta}_{ij}\tilde{\eta}_{kl}\tilde{\eta}_{mn} + \cdots, \qquad (A.15)$$

where

$$C^S_{ijkl\ldots} = \bar{\varrho}(\partial^n U/\partial \tilde{\eta}_{ij}\,\partial \tilde{\eta}_{kl}\ldots)\,|_{\mathbf{X}} \qquad (A.16)$$

are the adiabatic thermodynamic elastic coefficients at arbitrary initial configuration. If the static strain from the natural to the initial state is also small, then

$$\varrho_0 U(\mathbf{a}, \boldsymbol{\eta}, S) = \varrho_0 U_0 + \tfrac{1}{2}c^S_{ijkl}\eta_{ij}\eta_{kl}$$
$$+ (1/3!)c^S_{ijklmn}\eta_{ij}\eta_{kl}\eta_{mn} + \cdots, \qquad (A.17)$$

where

$$c^S_{ijkl\ldots} = \varrho_0(\partial^n U/\partial \eta_{ij}\,\partial \eta_{kl}\cdots)\,|_{\mathbf{a}} \qquad (A.18)$$

are the adiabatic thermodynamic elastic constants introduced by Brugger.[25] There is no linear term in Eq. (A.17), since the natural state is unstressed. The isothermal Brugger elastic constants are defined in an analogous way by starting with the Helmholtz free energy. It should be mentioned that other definitions of the third-order elastic constants have been introduced.[13,18,140,146] They usually differ from Brugger's definition by a constant factor.[25] The Brugger elastic coefficients $C_{ijkl\ldots}$ can be related to the Brugger elastic constants $c_{ijkl\ldots}$ by differentiation of Eq. (A.17) with respect to $\tilde{\boldsymbol{\eta}}$ and evaluating the result in the initial configuration. One obtains for the stress and second-order elastic coefficients[27]

$$\bar{\sigma}_{ij} = (\bar{\varrho}/\varrho_0)\bar{J}_{im}\bar{J}_{jn}[c^S_{mnpq}\bar{\eta}_{pq} + \tfrac{1}{2}c^S_{mnpqrs}\bar{\eta}_{pq}\bar{\eta}_{rs}$$
$$+ (1/3!)c^S_{mnpqrstu}\bar{\eta}_{pq}\bar{\eta}_{rs}\bar{\eta}_{tu} + \cdots], \qquad (A.19)$$

$$C^S_{ijkl} = (\bar{\varrho}/\varrho_0)\bar{J}_{im}\bar{J}_{jn}\bar{J}_{kp}\bar{J}_{lq}[c^S_{mnpq} + c^S_{mnpqrs}\bar{\eta}_{rs}$$
$$+ \tfrac{1}{2}c^S_{mnpqrstu}\bar{\eta}_{rs}\bar{\eta}_{tu} + \cdots]. \qquad (A.20)$$

Also, the Brugger coefficients c_{ijkl} and C_{ijkl} have complete Voigt symmetry:

$$c_{ijkl} = c_{jikl} = c_{ijlk} = c_{klij}. \qquad (A.21)$$

---

[146] R. F. S. Hearmon, *Acta Crystallogr.* **6**, 331 (1953).

A.2. Wave Equation and Related Elastic Coefficients

The equation of motion of an elastic solid in the absence of body forces is

$$\varrho \ddot{x}_i = (\partial \sigma_{ij}/\partial x_j). \qquad (A.22)$$

Using Eq. (A.13) and the identity of Euler, Piola, and Jacobi [Thurston,[26] pp. 91–92, or Truesdell and Toupin,[147] p. 226]

$$\frac{\partial}{\partial x_j}\left(\frac{\varrho}{\bar{\varrho}}\tilde{J}_{jl}\right) = 0 = \frac{\partial}{\partial x_j}\left(\frac{\varrho}{\varrho_0}J_{jl}\right), \qquad (A.23)$$

the equation of motion becomes

$$\bar{\varrho}\ddot{x}_i = (\partial/\partial X_l)(\tilde{J}_{ik}\tilde{t}_{kl}) \qquad (A.24)$$

in terms of the initial coordinates, and

$$\varrho_0 \ddot{x}_i = (\partial/\partial a_l)(J_{ik}t_{kl}) \qquad (A.25)$$

in terms of the natural coordinates. Using the Taylor expansions for the internal energy [Eqs. (A.15) and (A.17)] to obtain the thermodynamic tensions, Eqs. (A.24) and (A.25) reduce to

$$\bar{\varrho}\ddot{x}_i = W^S_{ijkl}(\partial^2 x_k/\partial X_j \, \partial X_l), \qquad (A.26)$$

$$\varrho_0 \ddot{x}_i = A^S_{ijkl}(\partial^2 x_k/\partial a_j \, \partial a_l), \qquad (A.27)$$

where

$$W^S_{ijkl} = \tilde{t}_{jl}\delta_{ik} + \tilde{J}_{im}\tilde{J}_{kn}[C^S_{mjnl} + C^S_{mjnlpq}\tilde{\eta}_{pq} + \tfrac{1}{2}C^S_{mjnlpqrs}\tilde{\eta}_{pq}\tilde{\eta}_{rs} + \cdots] \qquad (A.28)$$

and

$$A^S_{ijkl} = t_{jl}\delta_{ik} + J_{im}J_{kn}[c^S_{mjnl} + c^S_{mjnlpq}\eta_{pq} + \tfrac{1}{2}c^S_{mjnlpqrs}\eta_{pq}\eta_{rs} + \cdots]. \qquad (A.29)$$

These are completely general equations of motion and it should be noted that the elastic coefficients W_{ijkl} and A_{ijkl} are strain dependent. For small-amplitude waves linearized equations of motion are obtained by evaluating Eqs. (A.26) and (A.27) in the homogeneously deformed

---

[147] C. Truesdell and R. A. Toupin, "Handbuch der Physik" (S. Flügge, ed.), Vol. III/1, p. 226. Springer-Verlag, Berlin, 1960.

initial state. The elastic coefficients thus obtained are

$$\overline{W}^S_{ijkl} = \bar{\sigma}_{jl}\delta_{ik} + C^S_{ijkl} \tag{A.30}$$

and

$$\overline{A}^S_{ijkl} = \bar{t}_{jl}\delta_{ik} + J_{im}J_{kn}\varrho_0(\partial^2 U/\partial\eta_{mj}\,\partial\eta_{nl})_S\,|_\mathbf{X}. \tag{A.31}$$

The coefficients $\overline{W}_{ijkl}$ are the thermodynamic generalizations of those obtained by Huang[141] and have been discussed in several papers.[18,27,91,143–145] The coefficients $\overline{A}^S_{ijkl}$ are those used by Thurston and Brugger[28] to relate the third-order elastic constants to experiment. (As a matter of notation, our $\overline{A}^S_{ijkl}$ corresponds to what Thurston and Brugger denote as $\tilde{A}^S_{ijkl}$.)

The effective elastic coefficients $\overline{W}_{ijkl}$ (or $\overline{A}_{ijkl}$) have the familiar Voigt symmetry only when the initial stress is zero (i.e., $\bar{\sigma} = 0$). Then both coefficients reduce to c_{ijkl}. In general, the only symmetry they obey is

$$\overline{W}_{ijkl} = \overline{W}_{klij}. \tag{A.32}$$

However, from the equation of motion it is seen that only the symmetric combination $\overline{W}_{ijkl} + \overline{W}_{ilkj}$ is observed in a wave propagation experiment. Thus it is convenient to introduce wave propagation coefficients defined by[144]

$$\begin{aligned}\hat{W}_{ikjl} &= \tfrac{1}{2}(\overline{W}_{ijkl} + \overline{W}_{ilkj}) \\ &= \bar{\sigma}_{jl}\delta_{ik} + \tfrac{1}{2}(C^S_{ijkl} + C^S_{ilkj}).\end{aligned} \tag{A.33}$$

Although these coefficients do not have complete Voigt symmetry, they are symmetric with respect to the interchanges $i \leftrightarrow k$ and $j \leftrightarrow l$. Namely,

$$\hat{W}_{ikjl} = \hat{W}_{kijl} = \hat{W}_{iklj}. \tag{A.34}$$

Accordingly, they can be written in a 6×6 matrix array $\hat{W}_{\alpha\beta}$ using Voigt notation. Unlike the Brugger elastic constants $c_{\alpha\beta}$, the $\hat{W}_{\alpha\beta}$ array is not symmetric, i.e., $\hat{W}_{\alpha\beta} \neq \hat{W}_{\beta\alpha}$ in general. A complete discussion of the coefficients $\overline{W}_{ijkl}$ and $\hat{W}_{ikjl}$ and properties of waves in stressed solids has been given by Thurston.[144]

Since other types of elastic coefficients are sometimes used, it is useful to relate them to the coefficients which enter the equation of motion. The stress–strain equation from the initial configuration is

$$\sigma_{ij} = \bar{\sigma}_{ij} + B_{ijkl}\tilde{\varepsilon}_{kl} + \Omega_{ijkl}\tilde{\omega}_{kl} + \cdots, \tag{A.35}$$

where $B_{ijkl} = (\partial\sigma_{ij}/\partial\tilde{\varepsilon}_{kl})\,|_\mathbf{X}$ and $\Omega_{ijkl} = (\partial\sigma_{ij}/\partial\tilde{\omega}_{kl})\,|_\mathbf{X}$. Using the relation between $\tilde{\boldsymbol{\varepsilon}}$ and $\tilde{\mathbf{J}}$, differentiation of Eq. (A.13) for σ_{ij} yields[27]

$$B_{ijkl} = \tfrac{1}{2}(\bar{\sigma}_{il}\delta_{jk}+\bar{\sigma}_{ik}\delta_{jl}+\bar{\sigma}_{jl}\delta_{ik}+\bar{\sigma}_{jk}\delta_{il} - 2\bar{\sigma}_{ij}\delta_{kl})+C_{ijkl}, \quad (A.36)$$

$$\Omega_{ijkl} = \tfrac{1}{2}(\bar{\sigma}_{il}\delta_{jk} - \bar{\sigma}_{ik}\delta_{jl} + \bar{\sigma}_{jl}\delta_{ik} - \bar{\sigma}_{jk}\delta_{il}), \quad (A.37)$$

where the variations can be either adiabatic or isothermal. These coefficients are the thermodynamic generalization of those given by Birch.[140] They do not have complete Voigt symmetry, B_{ijkl} being symmetric only with respect to the interchanges $i \leftrightarrow j$ and $k \leftrightarrow l$ and Ω_{ijkl} symmetric (antisymmetric) with respect to the interchange $i \leftrightarrow j$ ($k \leftrightarrow l$). The use of Eq. (A.35) in the equation of motion, Eq. (A.22), when evaluated at the initial configuration, yields the effective elastic coefficients $B_{ijkl} + \Omega_{ijkl}$. Thus

$$\hat{W}_{ikjl} = \tfrac{1}{2}(B_{ijkl} + \Omega_{ijkl} + B_{ilkj} + \Omega_{ilkj}), \quad (A.38)$$

as is easily verified from Eqs. (A.33), (A.36), and (A.37). Hence $\overline{W}_{ijkl}$ and $B_{ijkl} + \Omega_{ijkl}$ are on an equal basis, and only the symmetric combination in j and l of either can be measured in a wave propagation experiment.

Other coefficients commonly used in infinitesimal elasticity are the Voigt elastic coefficients.[14] They are the constants which appear in the Taylor expansion of the internal energy in terms of the infinitesimal displacement parameters. Thus three adiabatic Voigt elastic coefficients will be defined as

$$V_{ijkl} = \bar{\varrho}(\partial^2 U / \partial \tilde{\varepsilon}_{ij}\, \partial \tilde{\varepsilon}_{kl})|_{\mathbf{X}}, \quad (A.39)$$

$$V'_{ijkl} = \bar{\varrho}(\partial^2 U / \partial \tilde{\varepsilon}_{ij}\, \partial \tilde{\omega}_{kl})|_{\mathbf{X}}, \quad (A.40)$$

$$V''_{ijkl} = \bar{\varrho}(\partial^2 U / \partial \tilde{\omega}_{ij}\, \partial \tilde{\omega}_{kl})|_{\mathbf{X}}. \quad (A.41)$$

Derivatives of Eq. (A.15) are easily taken, yielding

$$\begin{aligned}V_{ijkl} &= \tfrac{1}{4}[\bar{\sigma}_{il}\delta_{jk} + \bar{\sigma}_{ik}\delta_{jl} + \bar{\sigma}_{jl}\delta_{ik} + \bar{\sigma}_{jk}\delta_{il}] + C^S_{ijkl} \\ &= \tfrac{1}{4}[\overline{W}^S_{ijkl} + \overline{W}^S_{jikl} + \overline{W}^S_{ijlk} + \overline{W}^S_{jilk}],\end{aligned} \quad (A.42)$$

$$\begin{aligned}V'_{ijkl} &= \tfrac{1}{4}[\bar{\sigma}_{il}\delta_{jk} - \bar{\sigma}_{ik}\delta_{jl} + \bar{\sigma}_{jl}\delta_{ik} - \bar{\sigma}_{jk}\delta_{il}] \\ &= \tfrac{1}{2}\Omega_{ijkl} = \tfrac{1}{4}[\overline{W}^S_{ijkl} + \overline{W}^S_{jikl} - \overline{W}^S_{ijlk} - \overline{W}^S_{jilk}],\end{aligned} \quad (A.43)$$

and

$$\begin{aligned}V''_{ijkl} &= \tfrac{1}{4}[-\bar{\sigma}_{il}\delta_{jk} + \bar{\sigma}_{ik}\delta_{jl} + \bar{\sigma}_{jl}\delta_{ik} - \bar{\sigma}_{jk}\delta_{il}] \\ &= \tfrac{1}{4}[\overline{W}^S_{ijkl} - \overline{W}^S_{jikl} - \overline{W}^S_{ijlk} + \overline{W}^S_{jilk}].\end{aligned} \quad (A.44)$$

8. THE USE OF IONS IN THE STUDY OF QUANTUM LIQUIDS*

8.1. Review of the Structure of Ions in Liquid Helium

8.1.1. The Bubble Model for the Negative Ion and the Positronium Atom

The structure of the negative charge in liquid helium, though originally puzzling, has proven to be quite interesting. Commenting on observations of the anomalously long lifetime of orthopositronium in liquid helium,[1,2] Ferrell[3] suggested that, owing to the spin exchange repulsion between positronium and helium atoms, it would be energetically favorable for a cavity or bubble to be formed in the liquid. The outward zero-point kinetic pressure exerted by the positronium atom was supposed to be balanced by the surface tension of the bubble. The positronium atom would thus be confined in a three-dimensional potential well. Ferrell's estimate of the radius, obtained by minimizing the total energy with the assumption that the bubble could be represented by an infinite square well and using a number for the surface tension obtained from measurements on bulk liquid samples, yielded a value of about 22 Å. This may be compared with the average interatomic spacing in liquid helium of about 3.6 Å. The bubble is evidently quite large on an atomic scale, so that one would not expect much overlap between the wave function of the positronium atom and the liquid. The bubble model, therefore, accounts for a reduction in the pickoff rate, and hence the observed increase in lifetime.

Careri seems to have been the first to suggest that this model might also apply to the negative ion in liquid helium, picturing it as an excess

---

[1] D. A. L. Paul and R. L. Graham, *Phys. Rev.* **106**, 16 (1957).
[2] J. Wackerle and R. Stump, *Phys. Rev.* **106**, 18 (1957).
[3] R. A. Ferrell, *Phys. Rev.* **108**, 167 (1957).

---

* Part 8 is by **Frank E. Moss**.

electron confined in a bubble.[4] In this case, the exchange repulsion dominates the Coulomb attraction at short range due to the low polarizability of the helium, and the electron can lower its zero-point energy by creating a bubble.[5]

A considerable amount of research has since demonstrated the validity of the bubble model. In the case of positronium, experimental data on the lifetime as a function of density for both gaseous[6] and liquid helium, as well as measurements of the momentum distribution of positronium atoms confined in bubble states,[7] have provided direct evidence of an exchange-force-induced bubble with a radius of about 21 Å. Experimental work on the excess electron in helium has been much more extensive. Three types of experiments are of significance: (1) studies on the injection of electrons across the vapor–liquid interface[8] and from photocathodes[9] have shown that liquid helium presents a barrier of about 1 eV to electron penetration, thus providing direct evidence of the exchange repulsion; (2) observations of the pressure dependence of the escape probability of negative ions trapped on vortex lines in rotating superfluid helium (see Section 8.5.3) have demonstrated the compressability of the bubble in essential agreement with the model[10–12]; and (3) the depth of the potential well in which the electron is confined has been determined spectroscopically, by observing photoejection of electrons from bubble states.[13,14]

While these experiments confirm the picture of the negative ion in liquid helium as an excess electron trapped in a bubble, the details of the model are less definite. In particular, the spectroscopic experiments yield values for the depth of the potential well of from 0.6 eV near the vapor pressure, increasing to about 1.1 eV at 15 atm, to be compared

---

[4] G. Careri, *in* "Progress in Low Temperature Physics" (C. J. Gorter, ed.), Vol. III, p. 60. North-Holland Publ. Amsterdam, 1961.

[5] J. Jortner, N. R. Kestner, S. A. Rice, and M. A. Cohen, *J. Chem. Phys.* **43**, 2614 (1965).

[6] L. O. Roellig and T. M. Kelly, *Phys. Rev. Lett.* **18**, 387 (1967).

[7] C. V. Briscoe, S. I. Choi, and A. T. Stewart, *Phys. Rev. Lett.* **20**, 493 (1968).

[8] W. T. Sommer, *Phys. Rev. Lett.* **12**, 271 (1964).

[9] M. A. Woolf and G. W. Rayfield, *Phys. Rev. Lett.* **15**, 235 (1970).

[10] R. J. Donnelly and P. H. Roberts, *Proc. Roy. Soc.* **A312**, 519 (1969).

[11] B. E. Springett, *Phys. Rev.* **155**, 139 (1967).

[12] W. P. Pratt and W. Zimmerman, *Phys. Rev.* **177**, 412 (1969).

[13] J. A. Northley and T. M. Sanders, *Phys. Rev. Lett.* **18**, 1184 (1967).

[14] C. Zipfel and T. M. Sanders, *Proc. Int. Conf. Low Temp. Phys., 11th, St. Andrews, Scotland*, **1**, 296 (1968).

with the 1-eV barrier to electron penetration of the liquid measured at the vapor pressure. Interpretations of data from vortex trapping of the ions as well as from the spectroscopic experiments yield values for the bubble radius of about 16 Å† at the vapor pressure, decreasing to about 11 Å near the solidification pressure.

8.1.2. The Electrostriction Model for the Positive Ion

In contrast to the situation applicable to the negative ion, studies on the positive ionic structure have been relatively few. The currently accepted model has been developed by Atkins,[15] based on the idea of electrostrictive compression of a polarizable fluid around a point charge. Treating the fluid as a classical continuum, a straightforward thermodynamic analysis shows that the helium should be solid for radii smaller than about 7 Å. This implies that the positive ion might be pictured as a solid helium "snowball" containing about 30 helium atoms. The effective mass (the extra mass calculated from the density profile plus the hydrodynamic mass associated with motions of the solid sphere) might be as large as $50m_{He}$, where m_{He} is the helium atomic mass.

While the electrostriction model is admittedly based on naive assumptions, Atkins has nevertheless argued that the positive ion in liquid helium is probably a He_n^+ ionic molecule, where n is some small integer. In this connection, it should be noted that He_2^+ ionic molecules have been observed in helium gas and are known to possess the quite high binding energy of about 2.5 eV.

There are very few experimental results that help clarify the structure of the positive ion. The effective mass has, however, been measured by observations of the momentum relaxation time of ions in a microwave cavity.[16] The data, interpreted in terms of the electrostriction model, yield an effective mass of about $50m_{He}$ with a radius of about 6 Å. In addition, trapping and escape of positive ions from vorticity in superfluid helium has been demonstrated,[17] although it is apparently difficult

---

[15] K. R. Atkins, *Proc. Int. School Phys., Enrico Fermi, Course XXI, Liquid Helium* (G. Careri, ed.), p. 401. Academic Press, New York (1963).

[16] A. J. Dahm and T. M. Sanders, *Phys. Rev. Lett.* **17**, 126 (1966).

[17] A. G. Cade, *Phys. Rev. Lett.* **15**, 238 (1965).

---

† See, however, Ref. 12. At least one interpretation of the data from vortex trapping experiments yields a larger value.

to interpret the data in terms of a value for the ionic radius. Clearly, a more definite picture of the structure of the positive ion must await further experimental results.

8.2. Production of Ions in Liquid Helium

8.2.1. Radioactive Sources

The most commonly used method of producing charges in liquid helium has been by ionizing radiation. Very frequent use is made of the isotope Po^{210}, which emits a 5.3-MeV alpha particle with a half-life of 138 days. Less frequently used is 475-year Am^{241} emitting alphas of about 5.5 and 5.4 MeV. These emitters are favored because the alpha range in the liquid is about 0.2 mm, so that the region of ionization is confined to a thin layer near the surface of the source. It is possible to obtain[†] electroplated Am^{241} sources of about 0.05 $\mu Ci/cm^2$, while the short half life of Po^{210} allows the preparation of sources with surface activities near 100 $\mu Ci/cm^2$, from which ionic current densities of the order 10^{-10} A/cm^2 can be drawn in liquid helium. Since this is a small current, normally requiring the use of an electrometer (however, see Section 8.3.3), it is frequently important to maximize the surface activity. Polonium sources are easily prepared by autodeposition on a variety of substrates. The polonium is obtained in a 1 : 3 normal solution of HNO_3 with a specific activity of about 0.2 mCi/ml.[‡] After heating to between 80 and 90°C, a drop of this solution is placed on the substrate, which is also heated to the same temperature range. The polonium will be deposited onto the substrate within about 30 sec. The drop may then be removed and the substrate washed in distilled water. Source making in this way must be carried out under a fume hood to protect against the possibility of accidental vaporization of the radioactive solution.

Suitable substrates must not readily oxidize, otherwise, in operation as an ion source, charges may build up on the oxide layer and cancel or significantly alter the applied electric fields and thus also the ion currents. For this reason, and because it is easy to autodeposit or electroplate on brass, gold is the preferred substrate; however, polonium sources may be autodeposited onto stainless steel and even nichrome as well.

[†] New England Nuclear Corp., 575 Albany Street, Boston, Massachusetts.
[‡] Amersham/Serle Corp. 200 Nuclear Drive, Des Plaines, Illinois.

8.2. PRODUCTION OF IONS IN LIQUID HELIUM

Beta sources have also been used, though less frequently, to produce ions in liquid helium. If the object is to produce a large density of ions in a well-defined region near the source, the 18-keV β from tritium is useful. This particle has a maximum range of about $\frac{1}{2}$ mm in liquid helium. The total ionization per particle is, of course, much less than for alpha radiation. However, this may be compensated by increasing the source activity. Since pure tritium gas has a specific activity of about 2.5 Ci/cm³ STP, multicurie sources are obtainable.[†] If, however, a large ion current is the object, there is no appreciable advantage in using H^3. In the author's experience, a 9-Ci H^3 source with extracting electric fields in the range 100–1000 V/cm produced currents of the order 10^{-10} A, which is approximately what may be obtained from a good Po^{210} source of comparable surface area.

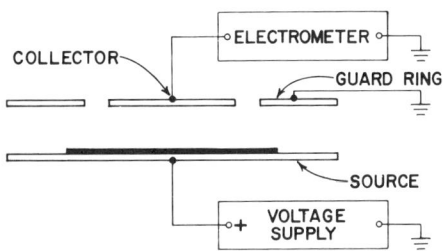

FIG. 1. Arrangement for producing and collecting ions in liquid helium. In this diode, the radioactive isotope is located on the surface of the source shown by the dark area. The guard rings ensure a uniform electric field between source and collector. With the positive terminal of the voltage supply connected to the source, a current of positive ions is measured by the electrometer.

A typical arrangement for producing and collecting ions in liquid helium is shown in Fig. 1. The dc collector current measured by the electrometer I_c is given by

$$I_c = A_c J, \tag{8.2.1}$$

where A_c is the collector surface area and J is the ion current density in the drift space. In the steady state,

$$J = env_d, \tag{8.2.2}$$

[†] New England Nuclear Corp., 575 Albany Street, Boston, Massachusetts.

where e is the charge on the ions, n is the ion density, and v_d is the drift velocity.

An example volt–ampere characteristic of a diode immersed in superfluid helium is shown in Fig. 2. The detailed shape of the characteristic cannot be motivated (except see Section 8.3.3) from Eq. (8.2.2), because of the complicated way in which n depends upon the electric field. It is, however, possible to measure v_d independently (see Section 8.3.1) so that n may be determined from the volt–ampere characteristic.

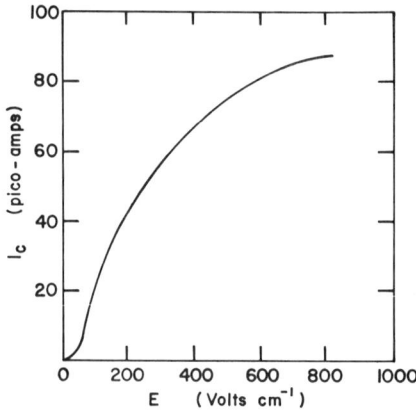

FIG. 2. Characteristic of the Po-210 diode, showing collected current versus electric field in the source–collector region.

8.2.2. Photoelectric Injection of Electrons

This method of producing electrons in liquid helium has been used in only two reported experiments.[9,18] In both cases, cesium–antimony photocathodes were used in a photodiode configuration. Since these cathodes are susceptable to contamination by impurities, they must be prepared and sealed under high vacuum prior to the experiments. The sealed envelope is then immersed in liquid helium and subsequently broken, so that the liquid is admitted to the space between anode and cathode. If the cathode is exposed to light and held at a negative potential with respect to the anode, a current of negative ions is injected into the

---

[18] A. Dahm, J. Levine, J. Penley, and T. M. Sanders, *Proc. Int. Conf. Low Temp. Phys.*, 7th (G. M. Graham and A. C. Hollis Hallett, eds.), p. 495. Univ. of Toronto Press, Toronto, Canada, 1960.

liquid. The light intensities, voltages, and magnitudes of the collected current are not reported. For a given voltage, however, a large attenuation (possibly several orders of magnitude) of the collected current is to be expected when liquid is admitted to the evacuated envelope. This has been explained[19] in terms of the electron–helium repulsive barrier and the formation of the negative ion bubble state (see Section 8.1.1) while still within range of the electron image potential in the photocathode.

8.2.3. Injection of Hot Electrons by Tunnel Diodes

This technique has been developed by Silver[19,20] using Al–Al$_2$O$_3$–Au thin film diodes. The diodes are fabricated by first evaporating an aluminum film of about 2000 Å onto a glass substrate. The Al film is then oxidized, providing an insulating layer of between 60 and 150 Å thickness. Finally a gold film of about 200 Å thickness is evaporated onto the oxide. An electron tunnel current through the oxide barrier results if the gold is biased at a positive driving voltage with respect to the aluminum. For large enough driving voltages, some of the tunnel electrons in the high-energy tail of the distribution are energetic enough to escape the gold and penetrate the liquid. These are termed "hot" electrons, since they are injected at considerably greater than thermal energies in a quasi-free, highly mobile, state. They are thermalized by collisions with the helium atoms, finally decaying into negative ion bubble states some distance from the diode surface. Some of these negative ions may then be collected in an electric field applied to the helium between the tunnel diode and a suitable collector.

An experimental arrangement for injection of electrons by means of a tunnel diode is shown in Fig. 3, and an example of a current–applied field characteristic is shown in Fig. 4. The reduction in collected current, at a given applied field, for the lower temperatures is a reflection of the sensitivity of the injection mechanism to the electron–liquid helium barrier. This barrier increases in magnitude with the liquid density as the temperature is lowered to the lambda point. Since the tunnel diodes typically have active areas of about 0.1 cm², the collected currents shown in Fig. 4 represent injection current densities in the range of 10^{-9} A/cm².

[19] M. Silver, D. G. Onn, P. Smejtek, and K. Masuda, *Phys. Rev. Lett.* **19**, 626 (1967).

[20] D. G. Onn and M. Silver, *Phys. Rev.* **183**, 295 (1969).

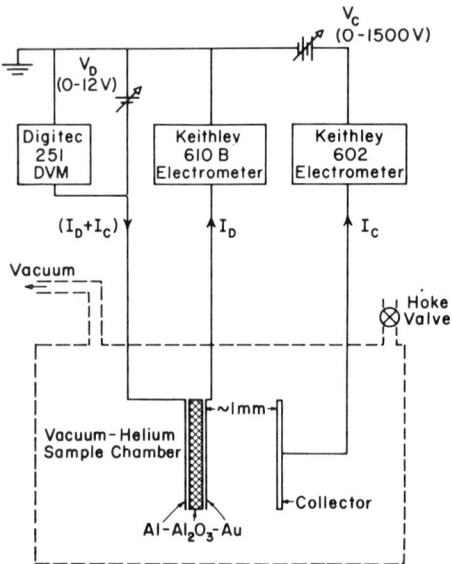

FIG. 3. Experimental apparatus for producing negative ions in liquid helium by injection of hot electrons from a tunnel diode, where V_D is the diode driving voltage and V_c establishes the applied electric field. The chamber represented by dotted lines permits testing at low temperatures in both vacuum and liquid helium [D. G. Onn and M. Silver, *Phys. Rev.* **183**, 295 (1963)].

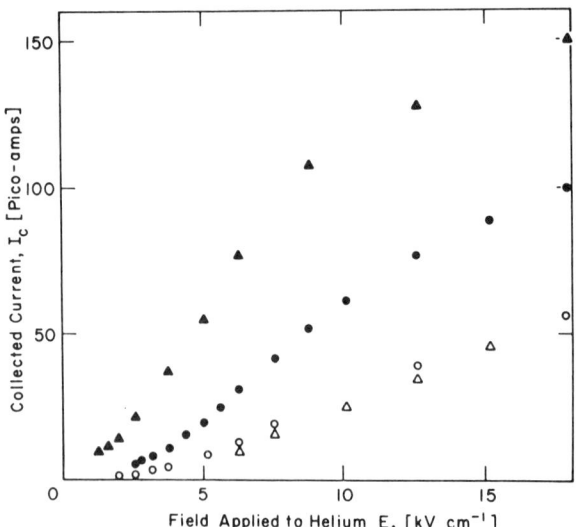

FIG. 4. A collected current–applied field characteristic for tunnel diode injection of electrons into liquid helium at several temperatures (▲: 4.65 K; ●: 4.2 K; △: 2.3 K; ○: 1.3 K) [D. G. Onn and M. Silver, *Phys. Rev.* **183**, 295 (1963)].

8.2.4. Injection of Electrons from Thermionic Cathodes

It has been observed that in HeII, and probably as a result of its unique heat transport properties, it is possible to form stable vapor bubbles around small diameter, heated filaments.[21,22] As a result of the reduced heat transfer to the liquid, the temperature of the filament increased and small enough diameter filaments can achieve incandescence.[23] It is only necessary to apply a positive electric field at the surface of the vapor bubble in order to inject thermionically emitted electrons into the liquid.

The current densities injected into the liquid are evidently quite high. An example current–voltage characteristic for a 5-μm diameter tungsten filament used as the cathode of a 1-cm-long cylindrical diode immersed in superfluid helium is shown in Fig. 5. Data are given for several different filament temperatures and two depths below the free liquid surface at a bath temperature of 1.3 K. These data show that injection current densities several orders of magnitude greater than those produced by ionizing radiation or by tunnel junctions are possible with the thermionic filament. The effects of space charge within the vapor bubble and charge capture by vorticity in the liquid (see Section 8.5.5) on the voltage–current characteristic are quite unknown,† as are the exact bubble size and the details of the heat transfer at the vapor–liquid interface which account for stable bubble formation. In one experiment[23] it has, however, been estimated that only about 30% of the power delivered to the filament is actually dissipated in the liquid; the remainder evidently exiting the dewar as light.

Etched tungsten wires‡ of 3, 5, and 10 μm diameter have successfully been used as thermionic cathodes. They can be easily mounted on gold-plated electrodes using ordinary silver paint. When proper care is exercised in order to avoid contamination either by finger prints or by frozen air, the filaments have proven very rugged. It has been possible to illuminate and extinguish a 10-μm diameter filament up to about 50

---

[21] E. L. Andronikashvili and G. G. Mirskaya, *Sov. Phys. JETP* **2**, 406 (1956).

[22] J. S. Vinson, F. J. Agee, Jr., R. J. Manning, and F. L. Hereford, *Phys. Rev.* **167**, 180 (1968).

[23] G. E. Spangler and F. L. Hereford, *Phys. Rev. Lett.* **20**, 1229 (1968).

† See however, D. M. Sitton and F. Moss, *Phys. Rev. Lett.*, **29**, 542 (1972).
‡ Sigmund Cohn Corp., Mt. Vernon, New York.

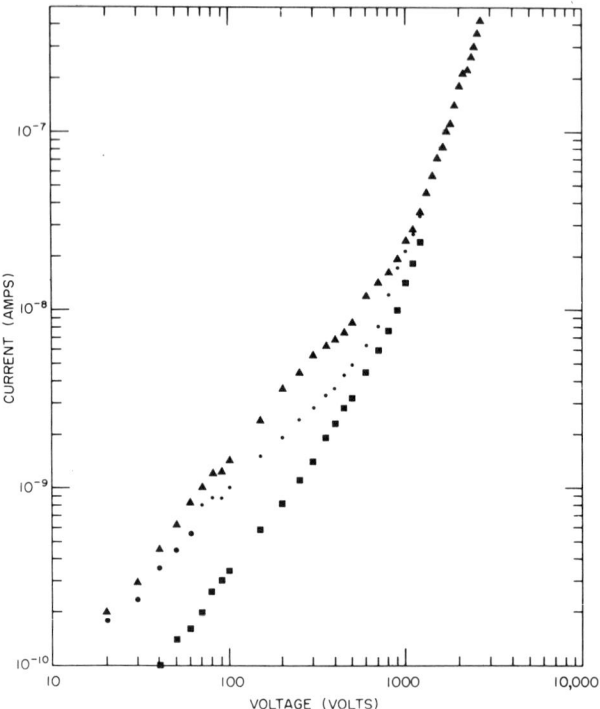

FIG. 5. Current–voltage characteristic of a cylindrical thermionic diode immersed in superfluid helium, at the following values of filament, power temperature, and depth, respectively: ■: 0.088 W, 2000 K, 3.0 cm; ●: 0.13 W, 2110 K, 3.0 cm; ▲: 0.13 W, 2160 K, 1.5 cm [G. E. Spangler and F. L. Hereford, *Phys. Rev. Lett.* **20**, 1229 (1969)].

times under superfluid helium without failure. The same filament was cycled several times between ambient and helium temperatures in the course of the experiment.

8.2.5. Field Emission

Injection of electrons in several simple liquids by field emission from fine points was demonstrated by Gomer[24] in 1965, though no quantitative data were presented for helium. More recently McClintock[25] has presented detailed data on injection of both positive and negative ions in liquid helium from field emission ion sources. The sources were fabricated

---

[24] B. Halpern and R. Gomer, *J. Chem. Phys.* **43**, 1069 (1965).
[25] P. V. E. McClintock, *Phys. Lett.* **29A**, 453 (1969).

8.2. PRODUCTION OF IONS IN LIQUID HELIUM

by mounting an electrolytically etched tungsten tip at a distance d of either 5 or 10 mm from a brass anode. The current to the anode was measured as a function of tip potential V for several temperatures. Currents of negative ions I_e and positive ions I_{ion} were collected for the appropriate polarity of tip potential. The data are shown in Fig. 6, where injection currents as large as those produced by the thermionic emitters are evident, at least for positive ions. Field emission sources have been used by Henson in positive ion mobility experiments in liquid helium[26] and other simple liquids.[27]

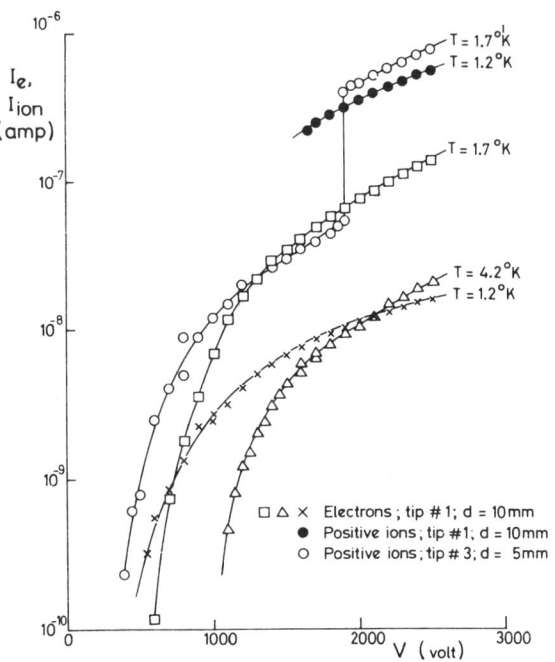

FIG. 6. Injected currents in liquid helium from a field emission ion source [P. V. E. McClintock, *Phys. Lett.* **29A**, 453 (1969)].

8.2.6. Gaseous Discharges and Laser Breakdown

Finally, mention should be made of these two potentially useful methods for ion injection, even though they have not yet been well developed for use in liquid helium. In the gaseous discharge technique,

[26] B. L. Henson, *Phys. Rev. Lett.* **24**, 1327 (1970).
[27] B. L. Henson, *Phys. Rev.* **135**, A1003 (1964).

a discharge ion source is placed in the vapor above a free liquid surface. Either electrons or positive ions may be drawn across the vapor–liquid interface to be collected on a submerged electrode. The method has been used to measure the electron–liquid helium penetration barrier.[28]

The laser-induced breakdown of liquid helium is a very recent accomplishment.[29] A Q-switched ruby laser of 50-MW peak power and 25-nsec pulse duration was focused onto a sample of liquid helium. Threshold intensities for breakdown of the liquid between 10^{10} and 10^{11} W/cm², depending on the temperature, were observed. While no effort has been made to collect the ions produced in the breakdown, the method is mentioned here because of its possible application for production of a short-duration burst of well-localized ions.

8.3. Methods for Measurement of the Ionic Drift Velocity

8.3.1. The Velocity Spectrometer

Although some measurements of the ionic drift velocity in liquid helium had been made previously,[30] the first detailed studies over a wide temperature range were made by Meyer and Reif.[31,32] At a given liquid helium temperature, they used a velocity spectrometer to measure the ionic drift velocity as a function of the electric field in a drift space. A diagram of the spectrometer is shown in Fig. 7. A Po^{210} source S is used for the production of ions which are eventually collected at collector C. The drift space is defined by grids A and B, with the electric field established by a voltage source connected between these grids (indicated by the battery in Fig. 7). Ions of the desired polarity are extracted from the source region by the voltage source connected between A and S. The grid pairs AA' and BB' are gates which are alternately opened and closed simultaneously at some frequency ν, by application of the square-wave voltages shown between A and A' and B and B'. For example, positive ions can pass through AA' only when the square wave as shown is applied to the terminals a_0 and a, i.e., when A' is positive with respect to A. Otherwise the gate is closed (A' negative with respect to A). The

---

[28] W. T. Sommer, *Phys. Rev. Lett.* **12**, 271 (1964).
[29] G. Winterling, W. Heinicke, and K. Dransfeld, *Phys. Rev.* **185**, 285 (1969).
[30] R. L. Williams, *Can. J. Phys.* **35**, 134 (1957).
[31] L. Meyer and F. Reif, *Phys. Rev.* **110**, 279 (1958).
[32] F. Reif and L. Meyer, *Phys. Rev.* **119**, 1164 (1960).

8.3. MEASUREMENT OF THE IONIC DRIFT VELOCITY

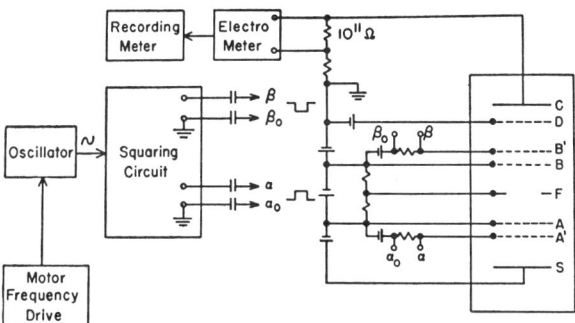

FIG. 7. A schematic diagram of the velocity spectrometer used for the measurement of ionic drift velocities [F. Reif and L. Meyer, *Phys. Rev.* **119**, 1164 (1960)].

gates are phased such that when AA' is open, BB' is closed. As a result, the ion current reaching C will be a maximum when the drift time of the ions between A and B is equal to an integral number of periods ν^{-1} of the gating square waves.

In order to measure the drift velocity, it is sufficient to record the frequency ν_1 at which the first maximum in collected current takes place, and to know the distance d between grids A and B so that $\nu_1 = v_d/d$. In practice, the precision of the measurements can be improved by making a plot of collector current as a function of frequency. This graph will be triangular in shape with maxima occurring at frequency multiples $\nu_m = m\nu_1$ with $m = 1, 2, 3, \ldots$. Thus several determinations of the fundamental frequency ν can be made by recording several successive maxima. At a given frequency, the response of the recording meter is steady even though the ion current arrives at the collector in pulses, because of the long response time of the electrometer compared to the period of the pulses. Likewise, when recording data, the frequency must be driven at a rate slow enough that significant changes in collector current do not occur in times short compared to the electrometer response time. It remains to note that this analysis applies when ideal gates are assumed; a condition approached when the gate grid spacing is very small compared to the drift space distance. An analysis which accounts for nonideal grids can be found in Ref. 32, Appendix B.

In the experiment of Reif and Meyer, the drift space distance was about 1 cm and the gate grid spacing was about 1 mm. The guard electrode F located midway between grids A and B, and maintained at a potential midway between them, served to improve the field uniformity in the drift space. The grid D served to screen the collector from currents induced by the square waves and by ion motions between B' and D. The polarities of the voltage sources as shown in Fig. 7 are correct for experiments with positive ions. A source strength of about 10 μCi was used and ion currents were measured in the range 10^{-13} A or less.

No assertions about the absolute precision of the spectrometer have been made. However, data on the drift velocity were repeatable to within 5% when the drift space distance d was changed by up to 70% if account was taken of the nonideal gates.

8.3.2. The Cunsolo Method

Perhaps because it is simpler, this method of drift velocity measurement has been more widely used than the velocity spectrometer. Invented by Cunsolo[33] in 1961, the apparatus consists essentially of a triode, square-wave voltage generator, filter, electrometer, and recorder, as shown in Fig. 8. The triode consists of a radioactive source for ion production,

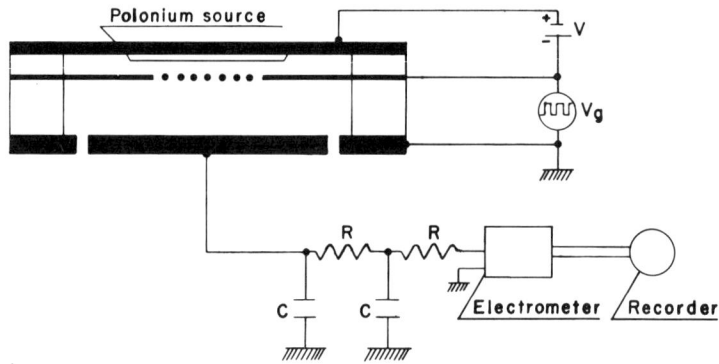

FIG. 8. Apparatus for measurements of the drift velocity of ions in liquid helium [S. Cunsolo, *Il Nuovo Cimento* **21**, 76 (1961)].

an extraction grid maintained at a suitable potential V relative to the source, and a collector and guard ring maintained at ground potential. The polarity of V as shown is correct for the extraction of positive ions. During the half period when the square wave maintains the grid and source at a positive potential relative to the collector, the ions are drawn into the drift space beyond the grid, where they move toward the collector in the electric field established by the peak square-wave voltage and the

---

[33] S. Cunsolo, *Il Nuovo Cimento* **21**, 76 (1961).

8.3. MEASUREMENT OF THE IONIC DRIFT VELOCITY

grid–collector spacing d. During alternate half periods, ions in the drift space are recalled and collected on the grid. Currents induced in the collector by the square wave as well as by the motion of ions in the drift space will be averaged to zero by the filter whose time constant must be long compared to the square-wave period. Only those ions which actually reach the collector can contribute to the dc collector current. Therefore, the electrometer will always read zero unless the square-wave half period $(2\nu)^{-1}$ exceeds the time required for ions to traverse the drift space d/v_d. The principle of the method is that the frequency can be adjusted to a value ν_0, so that the collector current has just become zero, in which case $v_\mathrm{d} = 2\nu_0 d$. In practice the collector current is plotted as a function of the frequency, and ν_0 is determined by extrapolation as shown in Fig. 9.

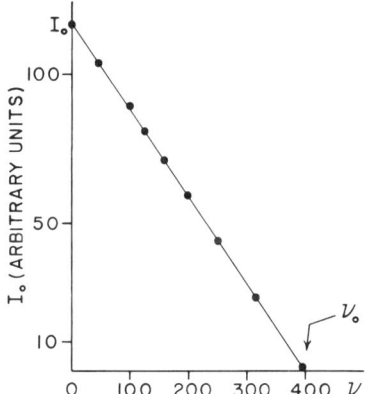

FIG. 9. Plot of average collector current as a function of square-wave frequency showing extrapolation to find ν_0 at $T = 0.85$ K [S. Cunsolo, *Il Nuovo Cimento* **21**, 21, 76 (1961)].

The linear frequency dependence of the collector current can be understood by considering the total charge q deposited on the collector during the positive half period

$$q = \begin{cases} env_\mathrm{d}[(2\nu)^{-1} - d/v_\mathrm{d}] & \text{for } (2\nu)^{-1} > d/v_\mathrm{d}, \quad (8.3.1) \\ 0 & \text{for } (2\nu)^{-1} \leq d/v_\mathrm{d}, \quad (8.3.2) \end{cases}$$

where the quantity in brackets is the amount of time by which the half period exceeds the ion time-of-flight across the drift space. The collector current, averaged by the filter, will just be this charge divided by the full period v^{-1}, viz.,

$$I_\text{c} = \begin{cases} I_0(1 - 2vd/v_\text{d}) & \text{for} \quad (2v)^{-1} > d/v_\text{d}, \\ 0 & \text{for} \quad (2v)^{-1} \leq d/v_\text{d}, \end{cases} \quad \begin{array}{l}(8.3.3)\\(8.3.4)\end{array}$$

and where $I_0 = (env_\text{d}/2)A_\text{c}$ is the collector current for zero frequency, i.e., one-half the current obtained when a dc voltage equal to the peak square voltage is applied between the grid and collector. It is not practical to obtain v_d from measurements of I_0 because of the uncertainties in n.

Cunsolo's method is capable of drift velocity measurements with repeatability better than 1%, while absolute measurements are probably in error by less than 5%, depending on the length of the drift space used and the uniformity of the electric field. The chief disadvantage of the method is that a square-wave voltage must be used to establish the field in the drift space. For experiments requiring high electric fields or large drift space distances (for improved absolute accuracy) the necessary square-wave voltage may be quite high; in fact, nearly 1000 V peak-to-peak. In addition, the high-voltage square wave must be supplied at a relatively low output impedance in order to drive the connecting cable capacitance without undue integration of the wave form. These requirements become more acute as the temperature is lowered, since for a given drift space voltage and separation, v_0 increases exponentially with inverse temperature (see Section 8.4). By contrast most other methods herein described use dc voltage sources to establish the electric field in the drift space.

8.3.3. Velocity Measurements Using Signal Averagers

It has been pointed out that ionic currents in liquid helium are small enough that long time response electrometers are necessary for their measurement. Because of this, in the velocity measurement methods so far described, it has not been possible to directly observe the time dependence of the ionic current. Instead, the time average of this quantity was observed as a function of some gating frequency, and the drift velocity was deduced from these data.

The development of modern signal averaging techniques has, however, made it possible to observe and record directly current wave forms of magnitude as small as 5×10^{-13} A with time resolutions down to 10 μsec.

8.3. MEASUREMENT OF THE IONIC DRIFT VELOCITY

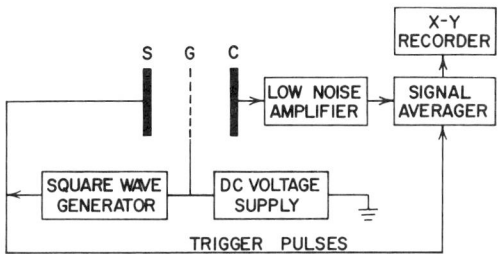

FIG. 10. Apparatus for observing the ion current wave form in a triode [L. Bruschi and M. Santini, *Rev. Sci. Instrum.* **41**, 102 (1970)].

Bruschi and Santini[34] have used a triode similar to that used by Cunsolo, but with the square-wave voltage applied between the source S, and the grid G as shown in Fig. 10. The wave form of the collector current at C is observed by first passing it through a low noise amplifier and then extracting it from the noise with the signal averager, in this case a Waveform Eductor.[†] Sketches of the wave forms to be expected are shown in Fig. 11. At $t = 0$, V_{SG} drives the source positive with respect to the grid, and positive ions drift toward the grid during the time t_1. No collector current flows because of the shielding action of the grid. After time t_1 the first ions penetrate the grid and enter the drift space. During the

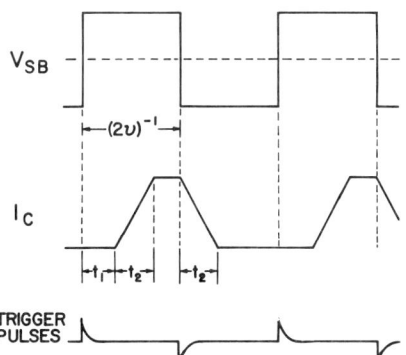

FIG. 11. Wave forms generated by the apparatus of Fig. 10 showing the source–grid square wave V_{SG}, the collector current I_c, and trigger pulses obtained from the square wave [L. Bruschi and M. Santini, *Rev. Sci. Instrum.* **41**, 102 (1970)].

[34] L. Bruschi and M. Santini, *Rev. Sci. Instrum.* **41**, 102 (1970).

† Princeton Applied Research Corp., Princeton, New Jersey.

time t_2 the induced current on the collector rises linearly since the rate at which charge enters the drift space is constant, i.e., the total charge in the drift space $N(t)$ builds up linearly with time. After a time t_2 the first ions reach the collector, the total charge in the drift space, and hence the collector current thus remain constant until the square wave reverses polarity. When V_{SG} is negative, charge cannot traverse the grid and enter the drift space, so that the collector current now decreases linearly as the remaining charge is swept out and, after a time t_2, becomes zero. This cycle is repeated until the noise has been sufficiently averaged out. These results may be summarized as

$$I_c = e\, dN(t)/dt, \tag{8.3.5}$$

which, assuming plane geometry and neglecting diffusion effects, becomes

$$I_c = (ev_d/d)N(t), \tag{8.3.6}$$

where

$$N(t) = \begin{cases} 0 & \text{for } 0 \leq t < t_1, & (8.3.7) \\ nv_d A_c(t - t_1) & \text{for } t_1 \leq t < (t_1 + t_2), & (8.3.8) \\ nA_c d & \text{for } (t_1 + t_2) \leq t < (2\nu)^{-1}, & (8.3.9) \\ nA_c \{d - v_d[t - (2\nu)^{-1}]\} & \text{for } (2\nu)^{-1} \leq t < [(2\nu)^{-1} + t_2]. & (8.3.10) \end{cases}$$

The time $t_2 = d/v_d$ is the time required for the ions to cross the drift space and can be obtained directly from the collector current wave form. Either the part of the wave form corresponding to filling or emptying the drift space with ions can be observed on the signal averager by choosing to trigger it on either the positive or the negative trigger pulses. Two examples of the emptying part of the current wave form corresponding to extreme situations of high and low drift velocities are shown in Fig. 12.

In order for the current wave form to be reproduced with fidelity, it must be true that $RC \ll t_2$, where the response time RC is usually the product of the input resistance of the amplifier and the collector capacitance-to-ground. This time constant should not be too small, however, as the signal-to-noise ratio of the amplifier is unnecessarily degraded. It is usually sufficient to equate the amplifier response time with the minimum resolving time of the signal averager. For example, in Fig. 12a, the 100 channel average is spread over 890 msec, so that the time spanned by each channel (minimum resolving time) is 8.9 msec, and this would be a sufficiently small amplifier response time. On the other hand,

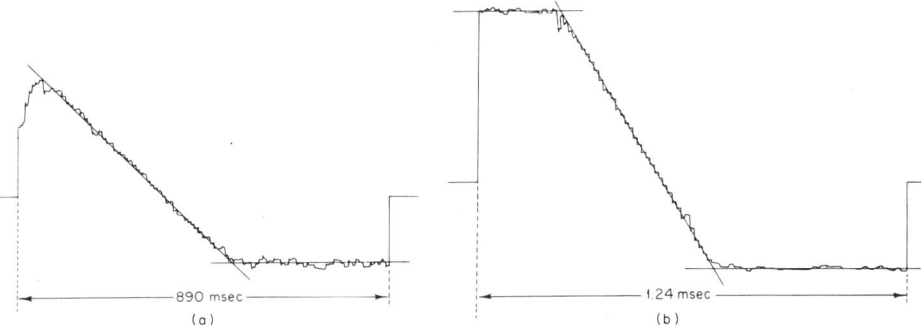

FIG. 12. Examples of a signal averager readout. (a) Liquid nitrogen, $T = 78$ K, maximum ion current: 7×10^{-13} A. (b) Liquid helium, $T = 0.93$ K, maximum ion current: 5×10^{-12} A [L. Bruschi and M. Santini, *Rev. Sci. Instrum.* **41**, 102 (1970)].

the situation shown in Fig. 12b requires the much more rapid response of about 12 μsec.

The characteristics and operating principles of the signal averagers themselves will not be elaborated here. In addition to the Waveform Educator, other commercially manufactured averagers are available.[†]

8.3.4. The Space–Charge Limited Diode

In the preceding paragraphs techniques were described for direct measurement of the ionic drift velocity (more accurately, the time of flight across a well-defined drift space). A more useful concept is, however, the mobility (see Section 8.4) defined by

$$\mu = v_d/E, \tag{8.3.11}$$

where E is the electric field in the drift space. In this section an extremely simple method, which measures the mobility directly, is described.

The technique, applied to ion mobility measurements in liquid helium in 1962 by Bartoli and Scaramuzzi,[35] derives from the operation of a diode, as shown for example in Fig. 1, under the condition of complete space–charge limitation of the current. This condition arises when the space–charge density is large enough that the electric field $E = -\nabla V$

---

[35] A. Bartoli and F. Scaramuzzi, unpublished.

[†] Nicolet Instrument Corp., Madison, Wisconsin; Biomation, Inc. Palo Alto, California.

approaches zero at the source. In this case, Poisson's equation $\nabla^2 V = -4\pi e n/\varkappa$, where $\varkappa$ is the dielectric constant of the liquid in the diode, can be solved.[36] Using Eq. (8.2.1) and the definition of mobility given by Eq. (8.3.11), the result is that

$$I_c = \alpha \mu V^2, \qquad (8.3.12)$$

where α is a constant which depends only upon the geometry of the diode. The simplest procedure is to record the diode collector current as a function of the square of the dc collector voltage. The result is a straight line of slope $\alpha\mu$ so long as the conditions of complete space–charge limitation apply to the diode. The constant α may then be evaluated by comparing the data with results obtained using other methods. The constant can also be calculated[35] with results accurate to at least 5% if care is taken to eliminate the effects of fringing fields in the design of the diode. The method has been successfully used both with plane and cylindrical diodes for ion mobility measurements in liquid He^3 [37] and He^4.[38]

Henson[39] has recently demonstrated that field emission points (see Section 8.2.5) can be operated in a space–charge-limited corona discharge region in liquid He^4. The characteristics of these discharges have been used for ion mobility measurements in the temperature range from 2.18 to 1.7 K by Sitton and Moss.[40] The points were fabricated by electrolytically etching a 0.006-in.-diameter tungsten wire in a solution of 0.5 gm NaOH in 30 ml of triply distilled water. An ac voltage of 1 V rms was applied between the tungsten wire and a copper electrode. The rate of etching was greater near the surface of the solution so that after 2 to 3 min the bottom part of the wire dropped off. The points were then washed in distilled water and mounted on a planar, gold-plated electrode with a drop of silver paint.

The great advantage of the space–charge limited diode is, of course, its extreme semplicity. Grids are unnecessary, and only dc apparatus, a voltage supply, and an electrometer are required. The method is,

---

[36] H. F. Ivey, *in* "Advances in Electronics and Electron Physics," Vol. 6, p. 137. Academic Press, New York, 1954.

[37] P. de Magistris, I. Modena, and F. Scaramuzzi, *Proc. Int. Conf. Low Temp. Phys.*, *9th, Columbus, Ohio*, p. 349 (1964).

[38] D. T. Grimsrud and F. Scaramuzzi, *Proc. Int. Conf. Low Temp. Phys., 10th, Moscow*, p. 197 (1966).

[39] B. L. Henson, *Phys. Lett.* **33A**, 91 (1970).

[40] D. M. Sitton and F. Moss, *Phys. Lett.* **34A**, 159 (1971).

however, limited to measurements in the lower range of mobility in liquid helium between about 1.1 K and the lambda point, otherwise the condition of charge limitation cannot be maintained for very strong sources.

8.4. The Use of Ion Mobility Measurements in the Study of Microscopic Excitations in Liquid Helium

8.4.1. Introduction

In this section the applications of the experimental methods for velocity or mobility measurements to studies on the microscopic structure of liquid helium are described. This is useful because it provides the opportunity to study examples of the actual data and to examine the inevitable variations in choice of method which arise upon applications to particular experiments. First it is necessary to review briefly the theoretical basis for the microscopic structure of the quantum liquids.[41]

In 1941 Landau[42] showed that the unique properties of liquid He^4 below the lambda point could be described by superimposing a set of elementary excitations, each with momentum $p = \hbar k$, on a zero-momentum superfluid ground state. The dispersion curve $\varepsilon(p)$ which he proposed[43] for the excitations of energy ε is composed of two branches: a low-temperature linear branch

$$\varepsilon = c_1 p \tag{8.4.1}$$

and a higher-temperature quadratic branch with a minimum

$$\varepsilon = \Delta + [(p - p_0)^2/2m^*]. \tag{8.4.2}$$

Physically, the linear branch represents states populated by longitudinal phonons traveling at the first sound velocity c_1; and the quadratic branch represents states populated by rotons with effective mass m^* requiring a creation energy Δ. The form of the dispersion curve is shown in Fig. 13.

The thermodynamic properties of HeII for temperatures greater than zero are determined by this "gas of excitations." The number densities

---

[41] For a more detailed development see for example, J. Wilks, "The Properties of Liquid and Solid Helium," Oxford Univ. Press (Clarendon), London and New York, 1967, and references contained therein.

[42] L. D. Landau, *Zh. Eksp. Teor. Fiz.* **11**, 592 (1941); *J. Phys. Moscow* **5**, 71 (1941).

[43] L. D. Landau, *J. Phys. Moscow* **11**, 91 (1947).

of the rotons and phonons can be calculated using Eq. (8.4.1) or Eq. (8.4.2) with the correct statistics by minimizing the appropriate free energy with respect to particle number in a given volume.[44] At about 0.7 K the densities are equal. The liquid properties are then largely determined by the rotons in the range $0.8 \lesssim T \leq T_\lambda$, called the *roton region*; and by the phonons for $0 < T \lesssim 0.5$ K called the *phonon region*. It is thus to be expected that the mobility of an ion in liquid helium is determined by scattering collisions with the roton gas in the higher-temperature range and with the phonon gas at lower temperatures, or by the creation of these excitations. Further elaboration of this point is necessary.

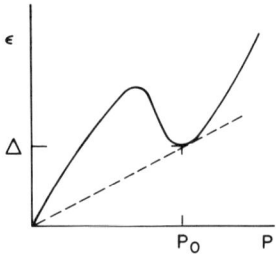

FIG. 13. The form of the dispersion curve $\varepsilon(p)$ for elementary excitations in liquid helium II.

Landau argued that energy may not be exchanged between the superfluid background and any object for relative velocities less than a certain critical velocity v_c defined by the slope of the broken line, shown in Fig. 13, drawn through the origin and tangent to the roton branch. A proper interpretation in terms of an ion traveling in HeII at $T = 0$ is that so long as $v < v_c$ there is no possible mechanism by which the ion can lose energy. When $v \geq v_c$ the ion would be expected to lose energy by roton creation[†] and this has lately been observed.[45] For $T > 0$ and $v < v_c$, the ions can lose energy by collisions with the excitation gas, i.e., the ion mobility is characteristic of roton–ion scattering at high temperatures, and phonon–ion scattering for $T \to 0$.

[44] L. D. Landau and E. M. Lifshitz, "Statistical Physics," p. 201. Pergamon, Oxford, 1958.
[45] G. Rayfield, *Phys. Rev. Lett.* **16**, 934 (1966).

[†] It develops that other excitations (see Section 8.5.4) can be created at smaller critical velocities, so that roton creation by ions was not initially observed.

8.4.2. The Roton Region

In this section the results of several experiments on ion mobility in the roton region are presented. It should be noted that the mobility as defined in Eq. (8.3.11) is dependent on the electric field in the drift space. For this reason, the experimental procedure invariably adopted is to measure v_d as a function of E, for values of the latter quantity as small as possible. The zero-field mobility is then defined by

$$\mu(0) = \lim_{E \to 0}(v_d/E), \tag{8.4.3}$$

and this can be determined from the slope at the origin of a graph of v_d versus E. It should be noted that the space–charge-limited diode inherently measures the zero-field mobility. The results of the experiments are measurements of the mobility as a function of liquid temperature.

Estimates of the functional dependence of $\mu(0)$ on T can be made using the ideas of simple kinetic theory.[32] The drift momentum for an ion of effective mass m_i in an electric field is given by $m_i v_d = eEl(\bar{v}_{ir})^{1/2}$, where l is the mean free path between collisions and $\bar{v}_{ir}$ is the mean relative speed between ions and rotons. Further, $l = (n_r \sigma_{ir})^{-1}$, where n_r is the roton number density, and σ_{ir} is the ion–roton scattering cross section. The mobility is then

$$\mu = e/m_i \bar{v}_{ir} n_r \sigma_{ir}. \tag{8.4.4}$$

Now if v_d is vanishingly small, Eq. (8.4.4) represents the zero-field mobility; and further, since $m_i \gg m^*$, the mean ion–roton relative velocity becomes approximately the roton thermal velocity $(kT/m^*)^{1/2}$, where k is the Boltzmann factor. The roton number density can be calculated as outlined in the preceding section with the result that

$$n_r = [2(m^* kT)^{1/2} p_0^2/(2\pi)^{3/2} \hbar^3] e^{-\Delta/kT}. \tag{8.4.5}$$

The temperature dependence of the mobility can thus be written

$$\mu(0) = (\text{constant}/\sigma_{ir} T) e^{\Delta/kT}, \tag{8.4.6}$$

and this is reasonably well obeyed experimentally.

Examples of data obtained by the experimental methods discussed in the preceding sections are shown in Fig. 14, where $\mu(0)$ is plotted on a logarithmic scale versus T^{-1}, for both positive and negative ions. The data in the roton region lie on straight lines in agreement with Eq. (8.4.6), shown by the solid lines in Fig. 14, if $(\sigma_{ir} T)$ is approximately constant

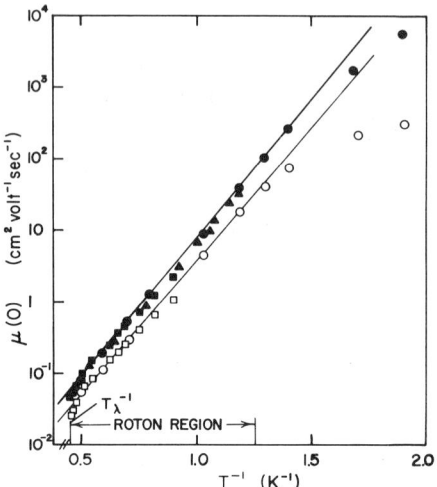

FIG. 14. Zero-field mobility of ions in liquid helium by various methods (●, ○: velocity spectrometer [F. Reif and L. Meyer, *Phys. Rev.* **119**, 1164 (1960)]; ■, □: space–charge limited diode [Scaramuzzi, unpublished]; ▲: Cunsolo method [S. Cunsolo, *Il Nuovo Cimento* **21**, 76 (1961)]. Filled symbols are for positive ions; open symbols, negative ions.

and for values of $(\Delta/k)_+ = 8.8$ K and $(\Delta/k)_- = 8.1$ K for positive and negative ions, respectively. These values can be compared to a more recent value of 8.65 K for the roton energy gap obtained by neutron scattering[46,47] and to 8.9 K obtained by fitting thermodynamic data to functions calculated from Eq. (8.4.2).[48] It is not clear why different values are obtained for the roton energy gap measured with positive and negative ions. It should be pointed out, however, that Eq. (8.4.6) was obtained using simple kinetic theory. In addition, assuming a hard sphere roton–ion interaction would lead to the prediction that σ_{ir} is temperature-independent, in disagreement with the facts. It seems clear that an accurate representation of the scattering process must take into account the difference in positive and negative ion structure. The data depart from the representation of Eq. (8.4.6), as expected, for temperatures below the roton region, since the scattering process there becomes increasingly dominated by the phonons.

[46] J. L. Yarnell, G. P. Arnold, P. J. Bent, and E. C. Kerr, *Phys. Rev.* **113**, 1379 (1959).
[47] D. G. Henshaw and A. D. B. Woods, *Phys. Rev.* **121**, 1266 (1961).
[48] C. J. N. Van Den Meijdenberg, K. W. Taconis, and R. De Bruyn Ouboter, *Physica* **27**, 197 (1961).

8.4.3. The Phonon Region

For low temperatures, the mean free path for ion–phonon scattering becomes quite long, and in order to make meaningful mobility measurements special apparatus is necessary. Schwarz and Stark have constructed a 28.2-cm-long drift space in order to study the drift velocities of negative[49] and positive[50] ions over a range of temperatures including the phonon region down to 0.34 K. The ions were produced by an Am^{241} α source, and negative ions were extracted from the source region by placing a small positive potential on a grid located 1 cm in front of the source. Well-defined pulses of ion current were injected into the drift region by gating the grid potential. In order to obtain a uniform electric field, the drift space was divided by 26 guard rings held at appropriate potentials. At the end of the drift space the ion current was collected and amplified, and the wave form observed by signal averaging. An example of the drift velocity as a function of electric field obtained in this way is shown in Fig. 15. The ion mobility is the slope of the solid line.

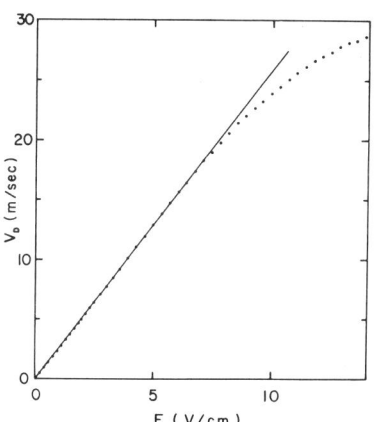

FIG. 15. Drift velocity versus electric field for negative ions in liquid helium at 0.5 K [K. W. Schwartz and R. W. Stark, *Phys. Rev. Lett.* **21**, 967 (1968)].

Below 0.5 K Schwarz and Stark find experimentally that $\mu(0)$ = (constant)T^{-3}. They show that this temperature dependence implies that the ion–phonon momentum transfer cross section must vary inversely with the phonon momentum, and this is exactly what is expected for *inelastic* acoustic scattering processes. This is the only experiment

---

[49] K. W. Schwarz and R. W. Stark, *Phys. Rev. Lett.* **21**, 967 (1968).
[50] K. W. Schwarz and R. W. Stark, *Phys. Rev. Lett.* **22**, 1278 (1969).

which has provided direct information on the negative ion–phonon interaction, and it invites interesting speculations on the structure of the negative ion. By contrast, the positive ion was found to scatter sound like a classical hard sphere.

8.4.4. He^3–He^4 Solutions

The He^3 isotope, when dissolved in liquid He^4, cannot contribute to the superfluid ground state because it must obey Fermi statistics. It therefore behaves like an excitation, and hence can exchange energy with ions by scattering collisions. Dilute He^3–He^4 solutions can be prepared such that at low-enough temperatures, the phonon and roton densities are completely negligible compared to the He^3 concentration. The ion–He^3 interaction can then be studied independently. This experiment has been done by Meyer and Neeper[51] using the double-gate velocity spectrometer. Only preliminary results are reported, in which it was observed that the product of mobility and concentration is only weakly concentration-dependent, but quite strongly temperature-dependent, in the range down to $T = 0.05$ K. Strikingly different temperature dependences are evident for the two species of ion. The results cannot as yet be interpreted, but probably reflect the difference in positive and negative ion structure.

8.4.5. Pure He^3

The behavior of ions in liquid He^3 is expected to be quite different from that in liquid He^4 or He^3–He^4 solutions due to the Fermi statistics obeyed by the He^3 atom. In particular, at low-enough temperatures the effects of degeneracy ought to be observable. The significance of this is that in order to exchange energy with the liquid the ion must scatter a He^3 atom into an unoccupied state, and as the temperature of the Fermi system is reduced toward zero the density of unoccupied states likewise approaches zero. The scattering probability therefore decreases, and as a result the ion mobility is expected to increase in the degenerate retion. Though there seems to be agreement on this point, calculations of the temperature dependence yield diverse results owing to the absence of a consensus on the proper scattering model.[52]

[51] L. Meyer and D. A. Neeper, *Proc. Int. Conf. Low Temp. Phys., 11th, St. Andrews, Scotland* **1**, 287 (1968).

[52] B. D. Josephson and J. Lekner, *Phys. Rev. Lett.* **23**, 111 (1969), and references contained therein.

Recent measurements on the mobility of ions in liquid He³ at low temperature have been reported by Anderson et al.[53] The method used was the double-gate velocity spectrometer, with a tritium β source for ion production. The mobility of the positive ions was indeed found to increase with decreasing temperature in the range between 0.1 and 0.03 K. The negative ions, however, exhibit a mildly decreasing mobility with temperature. The results cannot be justified by current theory.

8.5. Ion Techniques for Studying Macroscopic Quantum Excitations

8.5.1. Rotating Superfluid He⁴

There are excellent theoretical and experimental reasons for believing that rotating superfluid helium is permeated by a linear array of vortex lines parallel to the axis of rotation each with a circulation $\varkappa_1$, quantized in units of h/m_{He}. This can be understood in a simple way by adopting London's idea of the superfluid as a condensate of particles obeying Bose statistics.[54] In this view, a single quantum state—the zero-momentum ground state—becomes macroscopically occupied as the temperature of the system is reduced below some critical temperature T_c. At $T = 0$ K, the ground-state occupation number equals the total number of particles in the system. To the extent that liquid helium can be represented by this model, the superfluid fraction ϱ_s/ϱ, at some temperature $0 < T < T_c$, is to be associated with the fraction of particles occupying the ground state. It is appropriate to ask what results could be expected from imparting some momenta to the superfluid. Both linear and angular momenta can be imparted, as exemplified by fluid flow through tubes and stirring or motion in a rotating bucket, respectively. For a discussion of tube flow the reader is referred to a standard text.[55]

Angular fluid motion—vortex flow—can be considered as a collective motion of all the ground state particles, i.e., the superfluid. Since these particles are all in the same quantum state, their angular momentum about the symmetry axis would be expected to be quantized in the same way as the electron in a Bohr orbit. The Bohr–Sommerfeld quantum

---

[53] A. C. Anderson, M. Kuchnir, and J. C. Wheatley, *Phys. Rev.* **168**, 262 (1968).
[54] F. London, "Superfluids," Vol. 2, p. 40. Dover, New York, 1961.
[55] I. M. Khalatnikov, "Introduction to the Theory of Superfluidity." Benjamin, New York, 1965.

condition can thus be applied to the orbit of any arbitrary helium atom so long as it participates in the superfluid motion. The momentum of the atom is identical with the superfluid momentum p_s, so that

$$\oint \mathbf{p}_s \cdot \mathbf{dl} = nh; \qquad n = 0, 1, 2, \ldots, \qquad (8.5.1)$$

or

$$\varkappa_n \equiv \oint \mathbf{v}_s \cdot \mathbf{dl} = nh/m_{\text{He}}, \qquad (8.5.2)$$

where $\mathbf{v}_s$ is the superfluid velocity, $\mathbf{dl}$ is an element of length along the orbit, and n is a quantum number. It turns out that only singly quantized ($n = 1$) vortices are observed. Further, it has been observed that the motion of liquid helium in a rotating bucket, and even at very low temperatures, approximates the equilibrium solid body rotation observed for classical viscous liquids.[56] In the latter case, the circulation is 2Ω, where Ω is the angular velocity of the bucket. Equating the total vortex and solid body circulations, the vortex line density ϱ_l thus becomes

$$\varrho_l = 2\Omega/\varkappa_1, \qquad (8.5.3)$$

where $\varkappa_1 = h/m_{\text{He}} \approx 0.998 \times 10^{-3}$ cm^2 sec^{-1}. Thus a bucket of superfluid helium rotating at 1 rad/sec is permeated by a density of about 2000 vortex lines/cm^2.

Abundant experimental evidence supports the vortex line theory of rotating superfluid. The original experiments were designed to detect the presence of vorticity through its interaction with second sound.[57] More recently it has proven possible to decorate the vorticity with ions, and these experiments are discussed in the following paragraphs.

8.5.2. Capture of Ions by Quantized Vortex Lines

The interaction of negative ions with rotating superfluid helium was first observed by Careri et al.[58] who observed a strong attenuation of a beam of negative ions directed perpendicular to the axis of rotation. Relatively little effect was noticed when the beam was directed parallel

---

[56] D. V. Osborne, *Proc. Phys. Soc. (London)* **63**, 909 (1950).
[57] H. E. Hall and W. F. Vinen, *Proc. Roy. Soc.* **A238**, 204 (1956).
[58] G. Careri, W. D. McCormick, and F. Scaramuzzi, *Phys. Lett.* **1**, 61 (1962).

to the axis of rotation. The experiment was repeated by Tanner et al.[59] who showed in addition that the ions were captured by the vortex lines and indeed could travel along the lines to be collected at the boundaries of the apparatus. The arrangement of the ion-producing source and the gold-plated collecting electrodes mounted on a plastic frame is shown in Fig. 16a. The axis of rotation[†] is parallel to a line joining the centers of electrodes B and T, so that when S is made negative with respect to C by a suitable dc voltage supply, negative ions are drawn across the mounting frame to collector C. A weak potential between T and B results in a collection of ions traveling along the vortex lines. The currents measured at collectors C and B as a function of angular velocity of rotation, shown in Fig. 16b, indicate that nearly all current lost to collector C is recovered on collector B. Ions are thus continually captured from the transverse beam, and the concept of a capture width σ_c is useful.

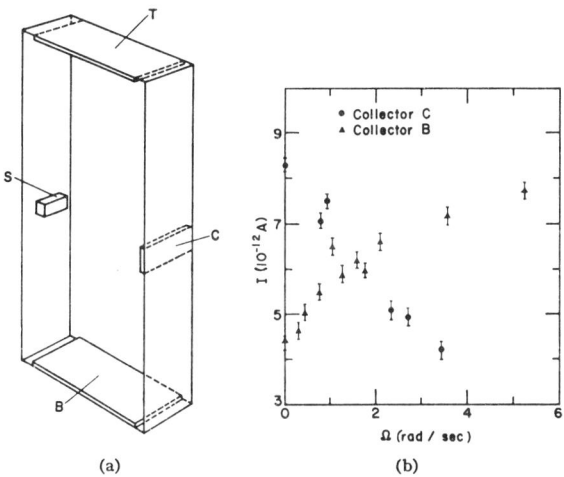

FIG. 16. (a) Source and electrode arrangement for observing ion capture in rotating superfluid helium. The S-to-C distance is 4.2 cm; the T-to-B distance is 5 cm; (b) collector currents as a function of angular velocity (●: collector C; ▲, collector B) [D. J. Tanner, *Phys. Rev.* **152**, 121 (1966)].

[59] D. J. Tanner, D. E. Springett, and R. J. Donnelly, *Proc. Int. Conf. Low Temp. Phys., 9th, Columbus, Ohio, 1964*, p. 346. Plenum Press, New York, 1965.

[†] This was achieved by mounting the liquid nitrogen and helium dewars together with a vibrating-reed electrometer on a large motor-driven turn table. Electrical access was provided through mercury slip rings. The temperature of the helium was altered by pumping on the dewar with a booster diffusion pump through a rotating vacuum seal.

The attenuation of current in the transverse beam can then be written as

$$I = I_0 \exp(-\sigma_c \varrho_1 x), \tag{8.5.4}$$

where x is the transverse distance measured from the source and I_0 is the current collected in the absence of rotation. The dependence on angular velocity results from Eq. (8.5.3). The capture width can in principle be obtained from Eq. (8.5.4) by measuring I_0 and $I(x = d)$, where d is the S-to-C distance, for a given angular velocity. In practice, the method must be altered in order to avoid problems arising from the space charge of the trapped ions. The effect of this space charge is to reduce I_0 below the value measured for zero angular velocity (and hence no trapped ions).

Tanner[60] has resolved the space–charge problem by placing a grid in front of the source shown in Fig. 16a so that the ion beam can be gated and its intensity controlled. The gating was accomplished by a square wave voltage between the source and grid. The triode structure was enclosed by a grounded metal container upon which the vortex lines terminated, so that when the source-grid gate was closed the trapped space charge could leak off. Data are obtained by measurements of I_0 and $I(d)$ for various angular velocities. The capture widths are obtained by matching these data to Eq. (8.5.4).

A somewhat different approach to the measurement of capture width has been taken by Douglass,[61] who measured both the trapped and the free charge resulting from passing a pulse of ions across the vortices. His cylindrically symmetric electrode assembly is shown in Fig. 17. Ions are extracted from the source S by the grid G, and a measured current I is caused to flow to the collector C for a certain build-up time. The ion beam is then turned off by reversing the grid-source voltage. After a brief waiting time, during which the remaining free ions are swept out, the potentials of G and C are switched so that trapped ions can travel along the vortex lines through the screen grid F to be collected on C. Data are obtained by measuring the trapped charge as a function of the build-up time for constant grid–collector C current. The slope of this graph then gives the rate R at which charge is trapped. The capture width can be obtained by integrating

$$dR = env_i \sigma_c \varrho_1 \, dV \tag{8.5.5}$$

[60] D. J. Tanner, *Phys. Rev.* **152**, 121 (1966).
[61] R. L. Douglass, *Phys. Rev.* **141**, 192 (1966).

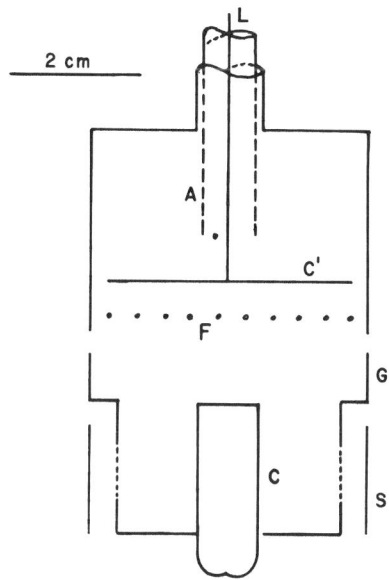

FIG. 17. Cylindrical rotating electrode assembly for capture width measurements [R. L. Douglass, *Phys. Rev.* **141**, 192 (1966)].

over the volume enclosed by the grid and collector C. In this case, v_i is the ion–vortex line relative velocity which is usually taken to be the ion thermal velocity. The ion number density n can be evaluated from Eq. (8.2.2), the grid–collector geometry (cylindrical, in this case), and the measured current I.

As it turns out, the capture width is a strong function of the electric field so that cylindrical geometry is not a good choice. Equation (8.5.5) cannot then be integrated unless the field dependence of σ_c is known. Both Tanner and Douglass find that σ_c goes roughly as the inverse of the field. Tanner's data, obtained using a plane triode with a 3-cm grid–collector spacing, are shown in Fig. 18. The decrease of σ_c toward zero between 1.6 and 1.7 K should be regarded as an artifact of the experimental apparatus, which results from an inability to distinguish between currents of free and trapped ions. This can happen when the mean trapped time (see Section 8.5.3) becomes comparable to either the time constant of the electrometer, or the free ion time-of-flight across the grid–collector drift space. These difficulties can be avoided by replacing the electrometer with a signal averager for direct observation of the collector current wave form (see Section 8.5.5).

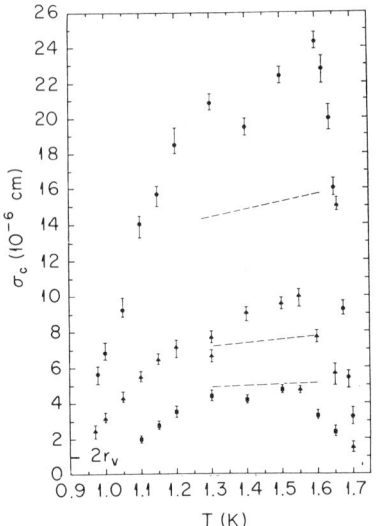

FIG. 18. Capture width for negative ions in rotating superfluid helium. The values for V_{CG} are -50 V (●), -150 V (▲), and -300 V (■). [D. J. Tanner, *Phys. Rev.* **141**, 192 (1966)].

8.5.3. Escape of Ions from Quantized Vortex Lines

A model for ion trapping and escape by quantized vorticity has been developed by Donnelly and Roberts.[62] Central to the model is the fact that an impurity in the vortex flow field must experience a radial force directed toward the vortex core. In this sense, the impurity is any object which excludes the superfluid flow, so that there exists a nonzero pressure gradient over the impurity volume. Donnelly has shown that the potential which derives from this attractive force is approximately the kinetic energy of the excluded superfluid, i.e., the integral of $\frac{1}{2}\varrho_s v_s^2$ over the volume of the impurity. This integration must be performed numerically because of the mixed cylindrical and spherical geometries. A further problem is the uncertain behavior of ϱ_s near the vortex core. To a certain approximation, however, the potential of an ion $u(r)$ near a vortex line and the depth of the potential well $u(0)$ have been calculated.[63]

In the presence of a not-too-large electric field, a saddle point is

[62] P. M. Roberts and R. J. Donnelly, *Proc. Roy. Soc.* **A312**, 519 (1969).

[63] R. J. Donnelly, "Experimental Superfluidity," p. 179. Univ. of Chicago Press, Chicago, Illinois, 1967.

created in the potential of an ion trapped on a vortex line. One mechanism for the escape of ions is by thermal activation over the saddle point. If this is the only escape process, and if a large number of vorticies are decorated with an initial number of trapped ions N_0, the time dependence of the number of ions remaining trapped is

$$N(t) = N_0 e^{-Pt}, \qquad (8.5.6)$$

where P is the escape probability. In an experiment, the current of escaping ions is measured, so that

$$I(t) = -dN/dt = PN_0 e^{-Pt}. \qquad (8.5.7)$$

For thermal activation processes, the escape probability is given by

$$P = Ce^{-\Delta u/kT}, \qquad (8.5.8)$$

where Δu is the activation energy, in this case the potential difference of the saddle point and well minima. Donnelly and Roberts[62] have calculated P using a model considering the ion as a Brownian particle in a potential field $u_T = u(r) - e\mathscr{E}x$, where $u(r)$ is due to the circulation of superfluid and $\mathscr{E}$ is an electric field perpendicular to the vortex line. They found that C varies only weakly with temperature and depends on the radii of curvature of the saddle point and well minima.

Douglass,[64] and more recently Pratt and Zimmerman,[65] have measured the mean trapped time $\tau = P^{-1}$ for negative ions captured on vortices in rotating HeII. Douglass' apparatus is shown in Fig. 17. By making current measurements on collector C' as a function of time he could distinguish between trapped and free ions. The trapped ions move from the space–charge region between C and G along the vortex lines (see Section 8.5.6) in the presence of an electric field perpendicular to the lines. At low-enough temperatures a negligible number of ions escape via thermal activation. This provides a normalization so that at higher temperatures, the mean trapped time could be related to the observed time dependences and magnitudes of the free and trapped currents on C'. The agreement of Donnelly's theory with Douglass' results is shown in Fig. 19.

[64] R. L. Douglass, *Phys. Rev. Lett.* **13**, 791 (1964).
[65] W. P. Pratt and W. Zimmerman, *Phys. Rev.* **177**, 412 (1969).

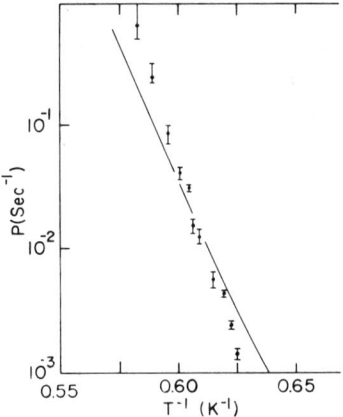

Fig. 19. Escape probability versus inverse temperature for negative ions trapped on vortex lines in rotating HeII, with $R = 1.6$ nm and $E = 25$ V/cm. The line is Donnelly's theory and the points are Douglass' experimental results [P. M. Roberts and R. J. Donnelly, Proc. Roy. Soc. **A312**, 519 (1969)].

8.5.4. Creation of Vortex Rings with Ions

Using apparatus similar to that shown in Fig. 7 (see Section 8.3.1), Rayfield and Reif[66] have made drift velocity measurements on negative and positive ions in the temperature range between 0.28 and 0.7 K. A dc potential V was placed between the ionization source S and grid A'. The ions thus arrive at A' with energy eV if dissipative effects (collisions with phonons and rotons) are negligible. Indeed, Rayfield and Reif found that ions can traverse field-free regions of several centimeters in length at 0.28 K. Velocity measurements were made on ions traversing the region between grids A and B which was free of dc fields. The result was that the velocity was a decreasing function of the energy of the ions at grid A'. This behavior would be expected if the ions were trapped on vortex rings with quantized circulation. The energy and velocity of the rings are given by

$$E = \tfrac{1}{2}\varrho_s \varkappa^2 R[\ln(8R/a) - \tfrac{7}{4}] \tag{8.5.9}$$

and

$$v = (\varkappa/4\pi R)[\ln(8R/a) - \tfrac{1}{4}], \tag{8.5.10}$$

[66] G. W. Rayfield and F. Reif, Phys. Rev. **136A**, 1194 (1964).

where R is the ring radius and a is the core parameter.[67] At constant energy, R can be eliminated between Eqs. (8.5.9) and (8.5.10), yielding the dispersion relation. Matching this relation to the experimental data yielded values for the core parameter and the circulation quantum at 0.28 K, as shown in Fig. 20.

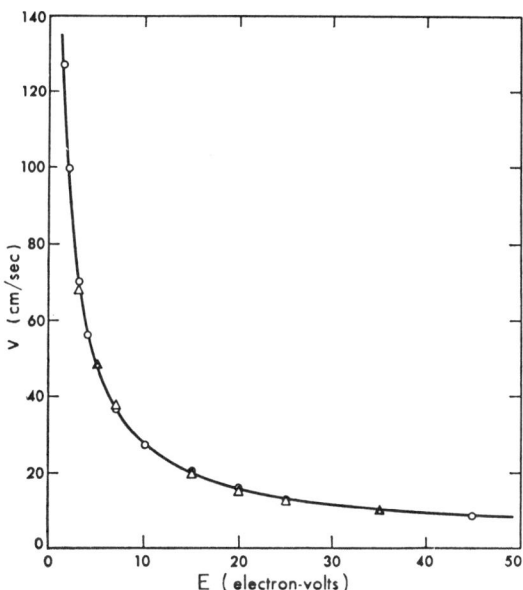

FIG. 20. Velocity as a function of energy for charged vortex rings in HeII. The curve derives from Eqs. (8.5.9) and (8.5.10) for constant R, with $K = (1.00 \pm 0.03) \times 10^{-3}$ cm²/sec and $a = 1.28 \pm 0.13$ Å. The points indicate charge: ○, positive; △, negative [G. W. Rayfield and F. Reif, *Phys. Rev.* **136A**, 1194 (1964)].

At higher temperatures, dissipative effects were studied by measuring the energy loss of rings in the space between A and B. Assuming that the friction force on unit length of vortex line is the product of a temperature-dependent dissipation constant and the line velocity, and neglecting frictional forces on the trapped ion, the total friction force on a vortex ring is

$$F = \alpha(T)(\ln(8R/a) - \tfrac{1}{4}), \qquad (8.5.11)$$

where $\alpha(T)$ is the dissipation constant proportional to the phonon,

---

[67] G. W. Rayfield, *Phys. Rev.* **168**, 222 (1968), has shown the correct numerical factors in Eqs. (8.5.9) and (8.5.10) to be $\tfrac{3}{2}$ and $\tfrac{1}{2}$.

roton, and He³ impurity densities and their respective scattering cross sections with the vortex line.

Careri *et al.*[68] have also observed quantized vortex rings in a higher temperature range near 1 K. Creation of the rings and charge capture were associated with a giant discontinuity observed in data on the ionic drift velocity versus electric field. The velocities were measured using the Cunsolo method (see Section 8.3.2). Examples of the data are shown in Fig. 21a for positive ions.

In the region near 1 K, friction forces on the vortex ring are large and determined by the roton density. One expects a charged ring in an electric field to expand rapidly to an equilibrium radius which results in a circumference and drift velocity such that the electric and friction forces are

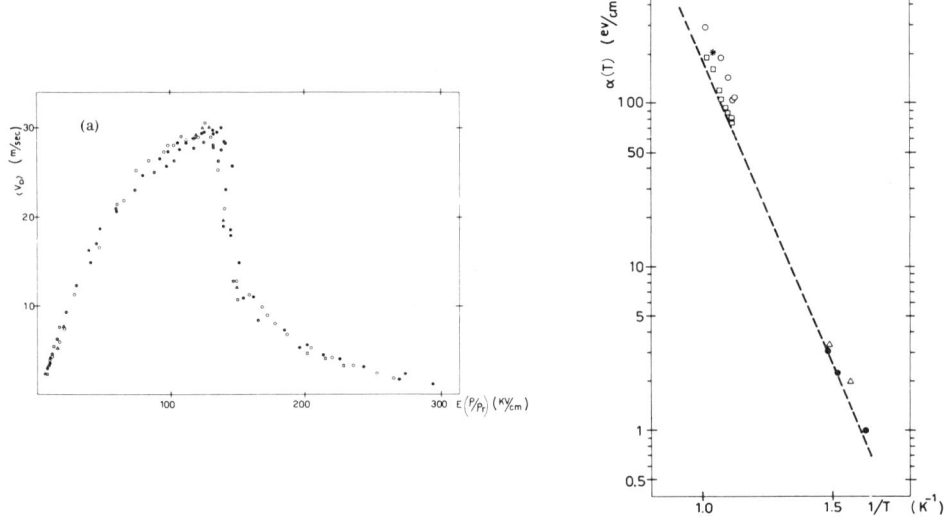

FIG. 21. (a) Drift velocity versus electric field divided by the fractional roton density showing the transition to the vortex ring state. The data points at the indicated pressures and temperatures are: □, S.V.P., 0.910 K; ■, S.V.P., 0.883 K; △, 5 kg/cm², 0.920 K; ○, 6 kg/cm², 0.920 K; ●, S.V.P., 0.988 K; ⊙, S.V.P., 0.975 K. (b) The dissipation constant versus inverse temperature. Open triangles and closed circles are data from the work of Rayfield and Reif [*Phys. Rev.* **136A**, 1194 (1964)]. The broken line exhibits the temperature dependence of the roton density [see Eq. (8.4.5), Section 7.4.2]. The data points are: △, □, negative ions, S.V.P.; ●, ○, positive ions, S.V.P.; ✱, positive ions, 8 kg/cm² [G. Careri, S. Cunsolo, P. Mazzoldi, and M. Santini, *Phys. Rev. Lett.* **15**, 392 (1965)].

[68] G. Careri, S. Cunsolo, P. Mazzoldi, and M. Santini, *Phys. Rev. Lett.* **15**, 392 (1965).

exactly balanced; $F = e\mathscr{E}$. Eliminating R between Eqs. (8.5.10) and (8.5.11) and defining the mobility of the ring by $\mu = v/\mathscr{E}$ results in

$$\ln \mu = \ln 2e\varkappa/\pi a\alpha(T) - e\mathscr{E}/\alpha(T) - \tfrac{1}{4}. \tag{8.5.12}$$

Equation (8.5.12) was matched to the drift velocity data in order to obtain values for a and $\alpha(T)$. The latter quantity is shown in Fig. 21b.

8.5.5. The Use of Ions in Studies on the Structure of Turbulent Superfluid Helium

One of the well-known properties of HeII is its ability to transport a heat flux with very small temperature and pressure gradients. Within the framework of the two-fluid model, a normal fluid (consisting of the phonons and rotons) carries entropy away from a source of heat. For zero-mass current, the normal fluid flow is countered by a noninteracting superfluid flow of zero entropy. Above certain critical values of the heat flux, corresponding to critical velocities of the counterflowing fluids, an interaction develops which leads to the appearance of much larger temperature and pressure gradients. HeII in this state has been called "turbulent" by analogy to the breakdown of laminar flow in classical fluids. The kinetic energy of the counterflowing fluids is sufficient to create vorticity. Normal dissipative effects which account for the decay of vorticity in classical fluids are absent in HeII, and in addition, one expects quantization of the circulation in a superfluid. The structure of turbulent HeII can thus be pictured as a tangled mass of quantized vortex lines.[69]

Sitton and Moss[70] have shown that the structure of turbulent HeII can be studied using negative ions as probe particles. In their experiment a 10-μm diameter, incandescent, tungsten filament was used as the cathode of a 1.6-cm-diameter cylindrical diode, as shown in Fig. 22. The filament was used both for injection of electrons (see Section 8.2.4) and as a source of heat for the generation of turbulence. A rectangular wave voltage was applied between the filament and collector. During a charging period, the filament was held at a negative voltage with respect to the collector and ions were drawn through the diode charging the vorticity. After the charging current reached an equilibrium value the voltage was reversed, and the collector currents during this recall period

---

[69] R. P. Feynman, *in* "Progress in Low Temperature Physics" (C. J. Gorter, ed.), p. 36. North Holland Publ., Amsterdam, 1955.

[70] D. M. Sitton and F. Moss, *Phys. Rev. Lett.* **23**, 1090 (1969).

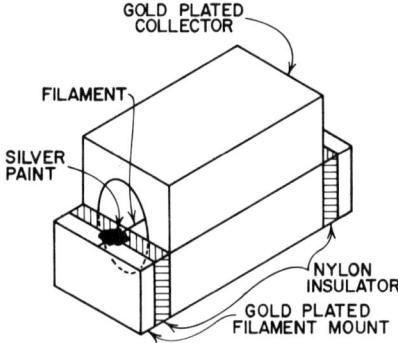

FIG. 22. Cylindrical diode with incandescent tungsten filament [D. M. Sitton and F. Moss, *Phys. Rev. Lett.* **23**, 1090 (1969)].

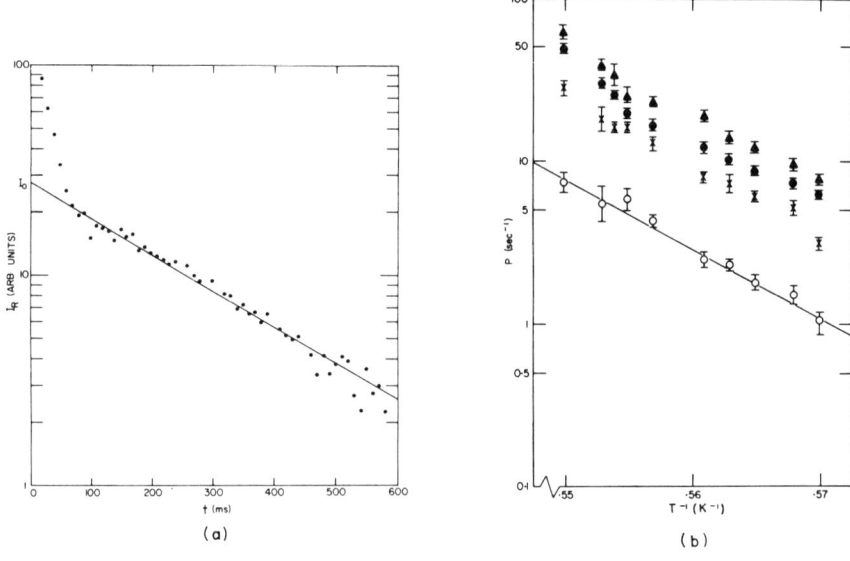

FIG. 23. (a) Diode current during the recall period showing the escape of ions trapped on vorticity in turbulent HeII for $T = 1.740$ K and a recall voltage of 200 V. The line is Eq. (8.5.7) for $P = 4.0 \pm 0.2$ sec^{-1}. (b) The escape probability versus inverse temperature for recall voltages of 400 V (▲); 300 V (●); 200 V (×); and extrapolated results for zero recall voltage (○). The line is Eq. (8.5.8) using the theoretical results of [P. M. Roberts and R. J. Donnely, *Proc. Roy. Soc.* **A312**, 519 (1969)] for C and Δu [D. M. Sitton and F. Moss, *Phys. Rev. Lett.* **23**, 1090 (1969)].

were analyzed with a signal averager. The cycle was repeated until the signal averager displayed an acceptably low noise level. An example of the signal averager readout of the time dependence of the diode current during the recall period is shown in Fig. 23a. Two components of the current are observed. The fast component, shown from 0 to about 60 msec is attributed to the sweep out of ions which were untrapped at the end of the charging period. The slow component represents the escape via thermal activation of ions initially trapped on vorticity, and can be matched to Eq. (8.5.7) in order to obtain P and $I_0 = PN_0$. The escape probabilities obtained in this experiment are shown in Fig. 23b and are in agreement with the results of experiments on ion escape from vortex lines in rotating helium[64,65] and with theory.[62]

8.5.6. Mobility of Ions along Linear Vortex Lines

The mobility of trapped ions traveling along linear vortex lines in rotating helium has been measured by Douglass[71] using the method discussed in Section 8.5.2, and by Glaberson et al.[72] The latter used a time-of-flight method similar in principle to Douglass' apparatus (Fig. 17) except that a signal averager was used to analyze the collector current, and two gating grids were placed above the space charge region. The drift space was 4.93 cm in length, necessitating the use of guard rings spaced at equipotentials. The mobilities of both free and trapped ions were measured simultaneously with the result that the trapped ion mobility was always less than that of the free ions at a given temperature. This result has been explained[71] qualitatively in terms of trapped ion scattering from vortex waves. However, an adequate theoretical treatment must incorporate a suitable model for the vortex core.

8.6. Studies of Superfluid Surfaces and Films

8.6.1. Interaction of Ions with the Free Liquid Surface

Positive and negative ions in liquid helium near a free surface experience an attractive potential which tends to inhibit evaporation of ions across the surface. The potential is accounted for by Coulomb forces induced by the difference in dielectric constants of the liquid and vapor.

---

[71] R. L. Douglass, *Phys. Rev.* **174**, 255 (1968).
[72] W. I. Glaberson, D. M. Strayer, and R. J. Donnelly, *Phys. Rev. Lett.* **20**, 1428 (1968).

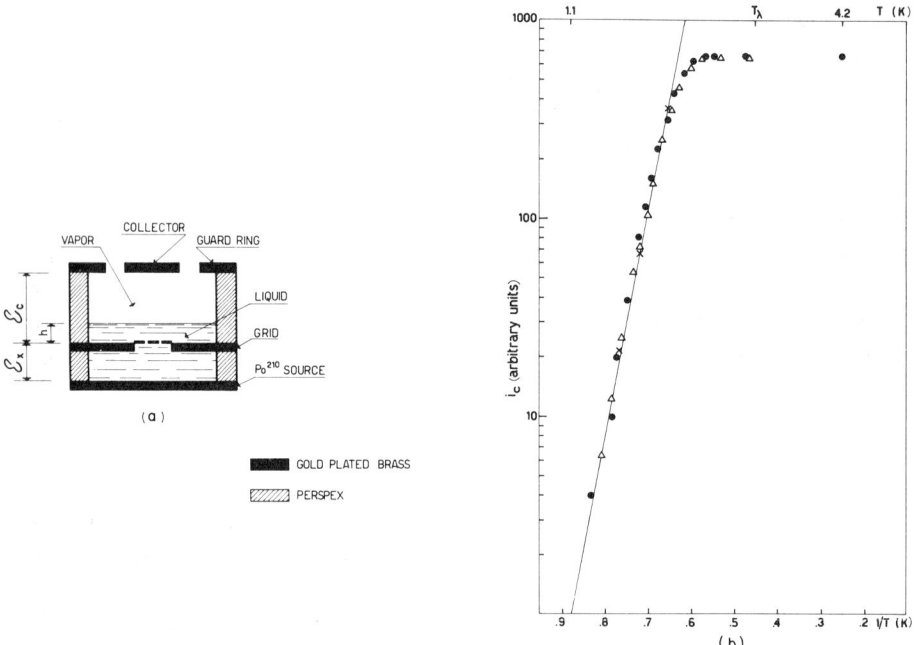

FIG. 24. (a) Experimental cell for measuring the barrier to the evaporation of negative ions across the liquid surface. Ions are extracted from the source in a field $\mathscr{E}_x$ and collected in a field $\mathscr{E}_c$. (b) I_c versus inverse temperature for $\mathscr{E}_c = 20$ V/cm and $\mathscr{E}_x = 400$ V/cm. The line is Eq. (8.6.1) with $E_b/k = 25$ K.

If this potential barrier is small enough, thermionic emission of ions through the barrier is possible. The current collected in the vapor above a free surface is then given by

$$I_c = AT^{1/2} \exp(-E_b/kT), \qquad (8.6.1)$$

where A is a constant and E_b is the magnitude of the barrier. Bruschi et al.[73] have measured E_b for negative ions evaporating from superfluid helium. The experimental cell is shown in Fig. 24a. The guard ring serves the double function of ensuring field uniformity and of preventing ions traveling through the superfluid film (see following section) from reaching the collector. An example of the experimental data is shown in Fig. 24b.

In the same apparatus, it was not possible to observe the evaporation of positive ions. This leads to the conclusion that the barrier for positive ions is much larger than for negatives in spite of the equivalence of the

---

[73] L. Bruschi, B. Maraviglia, and F. Moss, *Phys. Rev. Lett.* **17**, 682 (1966).

polarization forces of the two species. The difference is accounted for by the short-range repulsive forces that lead to bubble formation around an excess electron in liquid helium (see Section 8.1.1), and these results were among the first to clearly indicate the validity of the bubble model for the negative ion.

Subsequent experiments[74] using the same techniques have confirmed and extended these observations. Of particular note is an experiment due to Schoepe and Dransfeld[75] which detected the evaporation of ions trapped on vortex lines intersecting the surface. In this experiment the barrier was not observed.

8.6.2. Motion of Ions in Superfluid Films

Bruschi *et al.*, in the experiment discussed in the previous section, demonstrated that ion currents can flow in the superfluid film. In a subsequent experiment, Bianconi and Maraviglia[76] observed the effect of film flow on this ion current. Their apparatus is shown in Fig. 25. Liquid helium was condensed into the brass outer container to a level between the source S and collector C. The liquid level was monitored by capacitance measurements on the ring electrodes A_1, A_2, and A_3. Liquid also entered the glass-kovar inner beaker which could be raised or lowered with respect to the liquid level in the can by a rod inside the filling line. The liquids inside and outside of the beaker are connected through the film. A displacement h of the beaker results in a change in the chemical potential $\Delta\mu = mgh$ for an isothermal system. This leads to inflow–outflow oscillations, and the levels vary sinusoidally with time after a displacement of the beaker. In a positive electric field between S and C, positive ions are drawn from the source toward the liquid surface. For large liquid levels, there is no appreciable evaporation for positive ions into the vapor, however, due to the large surface potential barrier; instead, they travel through the film and are collected on the gold ring near the rim of the beaker. The ions can be emitted into the vapor if the liquid surface is within 2 mm of the source, and this was attributed to the transport of ions trapped on vorticity through the barrier.[77] The

---

[74] W. Schoepe and C. Probst, *Proc. Int. Conf. Low Temp. Phys.*, 12th, Kyoto, Japan p. 107 (1970).

[75] W. Schoepe and K. Dransfeld, *Phys. Lett.* **29A**, 165 (1969).

[76] A. Bianconi and B. Maraviglia, *J. Low Temp. Phys.* **1**, 201 (1969).

[77] S. Cunsolo, G. Dall'Oglio, B. Maraviglia, and M. V. Ricci, *Phys. Rev. Lett.* **21**, 74 (1968).

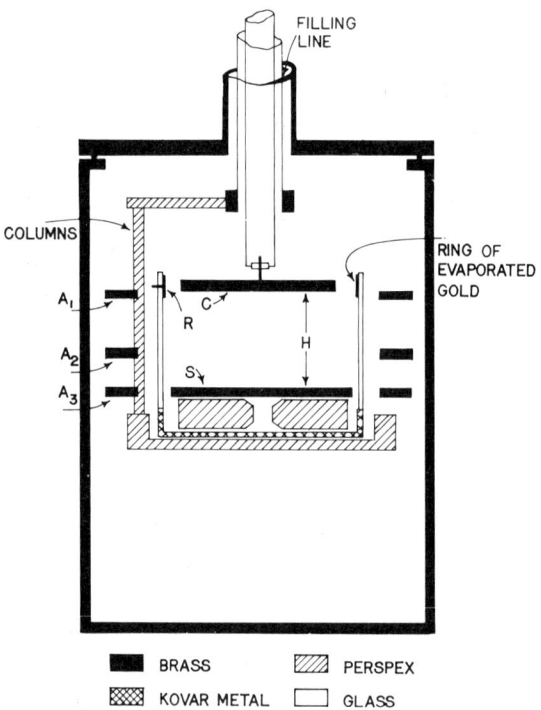

FIG. 25. Arrangement for observing ion currents in superfluid films [A. Bianconi and B. Maraviglia, *J. Low-Temp. Phys.* **1**, 201 (1969)].

current of ions traveling through the vapor was collected on C, and could be observed simultaneously with the film current. The results were that both currents were modulated sinusoidally during oscillating film flow. The current through the vapor was modulated at the frequency of the oscillation of the liquid level, and level changes as small as 10^{-6} cm could be observed. The film current, in contrast, was modulated at twice this frequency. The conclusion was that ion motion in the film was inhibited *both* during inflow and outflow.

Maraviglia has tentatively suggested that this effect results from polarization of the roton momentum in the presence of a superfluid flow of velocity $\mathbf{v}_s$. The excitation energy is then given by $E = E(p) + \mathbf{p} \cdot \mathbf{v}_s$, and is lowered for rotons whose momenta are directed oppositely to $\mathbf{v}_s$. This would result in the observed modulation frequency if the polarized roton–ion scattering cross section is not isotropic.

9. THERMOMETRY AT ULTRALOW TEMPERATURES*

9.1. Introduction

Of increasing importance in solid state physics is the temperature region below 100 mK. Recent development of the ^{3}He dilution refrigerator to a reliable instrument capable of maintaining temperatures below 20 mK continuously has placed a powerful new tool in the hands of the researcher. From this platform, cooling to temperatures below 1 mK will become a major aim for the 1970's. Areas likely to dominate the attention of the solid-state physicist working at ultralow temperatures will be thermal contact between materials, magnetism (especially nuclear magnetic effects and nuclear ordering), quantum fluids (the solid state physicist's interest in superconductors here becomes broadened to include liquid helium), ultralow noise devices, and the zero-temperature limit of many other properties.

Of the measurements that must be performed at ultralow temperatures, thermometry is the most frequent, important, and difficult. The difficulties arise not from any complexity in the equipment used or from operational problems but rather from the great care that must be lavished on the design of the instruments and sensors and in their calibration. Thermal contact being extremely difficult to achieve, temperature measurements by a sensor that is not the sample itself are particularly susceptible to errors. Regardless of the nature of the thermometer, calibration to a true thermodynamic temperature is exceptionally difficult below about 20 mK, since the thermodynamic establishment of a temperature scale requires the measurement of entropy. Truly reversible processes are virtually impossible at ultralow temperatures; so, for example, seldom do measurements of the susceptibility-temperature relation for paramagnetic salts agree at the low end of the useful range for a particular salt when measured in different ways (by different observers). In view of these problems, an important approach has been to compare different

---

* Part 9 is by **Walter Weyhmann**.

phenomena for which reasonably well–established theories exist against the same type of sensor. A prime example of this approach has been the use by Wheatley and collaborators of a CMN (cerium magnesium nitrate) salt pill in measuring the properties of dilute solutions of ^{3}He in ^{4}He, which should exhibit ideal Fermi gas behavior. Recently, increased emphasis has been placed on using properties of materials that are incontrovertible in the temperature range of interest. Nuclear magnetic susceptibility in pure materials such as copper should show a Curie law temperature dependence to the microdegree range and should therefore be trustworthy in the temperature region presently accessible.

Before proceeding, let us recall the terminology used with regard to thermodynamic temperature as opposed to a measured property used for temperature indication. The former we designate T. Some useful function of the observed parameter, which function might be expected to approximate the thermodynamic temperature, we call T^*. Thus, for a paramagnetic salt, the inverse of the susceptibility is used as a measure of T^*, T^* being simply related to T at some high-temperature calibration points. At low temperatures, the salt will no longer exhibit ideal (Curie law) behavior and deviations between T^* and T will occur which must be determined.

Thermodynamic techniques of determining the T^*–T relationship for a thermometer with which adiabatic, reversible processes can be performed are extensively discussed in the standard literature, particularly in references given later on paramagnetic salts. The most frequently used method is a simple (in theory) application of the second law

$$T = dQ/dS = (dQ/dT^*)/(dS/dT^*). \qquad (9.1.1)$$

The requirement of reversibility makes this difficult in practice, especially below 20 mK.

In choosing a thermometer (or thermometers) for a particular measurement, the first decision must be the class of sensor to be used with regard to the determination of the T^*–T relation. Here we might divide thermometers into three classes; absolute scale, fixed scale, and arbitrary scale. In the first of these—absolute—the temperature may be determined by a measurement of one or more observables without recourse to thermodynamic methods or to high-temperature fixed points. The absolute value of nuclear susceptibility would provide such a measurement and such thermometers (particularly with nuclear orientation techniques)

9.1. INTRODUCTION

are available. By thermometers with fixed T^*–T relations, we refer to those sensors whose thermal behavior is known and reproducible from sample to sample (and one hope laboratory to laboratory) but for which an absolute temperature cannot be assigned without reference to a known fixed point. Nuclear susceptibility as measured by nuclear magnetic resonance and susceptibility of paramagnetic salts of specified shape and preparation are examples of this class. An intermediate class between these two exists wherein the relation, once determined, is thereafter available and reproducible without reference to fixed points. These have, then, a "universal" scale. Phase-transition curves and the temperature dependence of nuclear spin–lattice relaxation times in metals may be very useful in the future. Thermometers with arbitrary T^*–T relations such as salt pills of arbitrary preparation and resistance thermometers can provide high resolution when used in combination with thermometers whose calibration is known. Because of their apparent simplicity, resistance thermometers continue to be popular and will not be surpassed for some time to come as indicators perhaps even below 10 mK.

The experimenter must choose his thermometers for their performance with regard to resolution and accuracy. Nuclear orientation can provide absolute temperatures with moderate accuracy and fair resolution and so is not very useful by itself in determining heat capacities where resolution of 0.1% is usually required. Resistance sensors can provide high resolution with simplicity and convenience down to 20 mK if great care is taken, but the absolute temperature–resistance relation must be independently determined.

Finally, the experimenter must determine desired performance in such matters as measurement time, drift, and thermal contact. Nuclear orientation requires equipment and techniques that are familiar primarily to nuclear physicists, and relatively long measurement times (in excess of 1 min/point). On the other hand, the basic techniques of resistance measurement are well known and the measurements can be made in a few seconds and even read out continuously on a strip-chart recorder.

In what follows, we describe in some detail the most popular thermometers using resistance, paramagnetic susceptibility (electronic), and nuclear susceptibility. Less time will be spent on the use of paramagnetic salts than properly reflects their use in practice today, but numerous excellent treatises exist on this elsewhere. In spite of its apparent simplicity, considerable attention will be placed on resistance thermometry because successful use of it depends on attention to a large number of subtle details. Nuclear orientation receives considerably more attention than its present use would seem to merit because of its great promise of providing absolute thermometry against which other thermometers can be calibrated, because no previous review article has dwelt at length with its use as a thermometer for general use (it is always used by experimenters in the field of nuclear orientation), and because the author's major research work in the last few years has been in this field. A brief description of promising techniques now under development is included under "Miscellaneous Thermometers" to alert the experimenter to future possibilities.

Most contemporary textbooks writers consider B to be the fundamental field and always use

$$E = -\mu \cdot \mathbf{B}, \tag{9.1.2}$$

$$\mathbf{T} = \mu \times \mathbf{B}, \tag{9.1.3}$$

etc. However, for a magnetic system with a linear field response,

$$\mathbf{M} = \chi \mathbf{H}, \tag{9.1.4}$$

and so

$$\mathbf{B} = \mathbf{H} + 4\pi \mathbf{M} = (1 + 4\pi\chi)\mathbf{H}. \tag{9.1.5}$$

Thus it is convenient, when discussing the susceptibility of a paramagnetic system in the Curie-law region, to use H and carefully account for all sources of the local field. In all other cases we shall use B. In weakly magnetic systems typically discussed in nuclear resonance work, H is usually used in the literature and, since we use cgs units, no error will be made for such systems by simply replacing our B with H. Our use of B emphasizes that in strongly magnetic materials great care must be taken to account for all sources of local field. Such corrections, unimportant for most temperature regimes, become increasingly important as ultralow temperatures are reached.

A word about the bibliography: We have chosen for reference those articles which, in the author's opinion, provide the most up-to-date information which can prove immediately useful in the construction of laboratory equipment. Usually these will be the latest papers on the subject, and in only a few instances has an effort been made to track down the history of the development of a technique so that proper credit can be given the originator. The purpose of this review is to provide a working paper for the experimenter rather than a comprehensive coverage of the field. A set of useful references on general techniques at ultralow temperatures is included for the reader's convenience.[1-10] After the title of a section there may appear references of general usefulness for that topic.

9.2. Resistance Thermometers[3-6,11]

Resistance thermometers have long played an important role in temperature measurements at low temperatures and continue to do so in the ultralow region. They are very easy to use if proper precautions are taken: the sensors are small and require no magnetic fields, the bridges for making the measurements use standard electronic principles and devices and are highly reliable, measurement is fast and can be continuous and automatic, and precision can be very high at higher temperatures. Competition to these is growing from devices that measure nuclear susceptibility in metals such as nuclear resonance and superconducting quantum interference devices (SQUID's), but for monitoring purposes the resistance sensors will be hard to beat. Reliable performance to 10–15

---

[1] ULT—1970, *A Conf. Ultra-low Temp. Technol.: past, present, future*, NRL Rep. 7133. Sponsored by the Office of Naval Res., Washington, D.C. (April 1970).

[2] "Temperature, Its Measurement and Control in Science and Industry." Based on the *5th Temp. Symp., Washington, D.C.*, Instrum. Soc. Am., Pittsburgh, 1973. (Not to be confused with an earlier edition with this title.)

[3] G. K. White, "Experimental Techniques in Low-Temperature Physics," 2nd ed. Oxford Univ. Press (Clarendon), London and New York, 1968.

[4] F. E. Hoare, L. C. Jackson, and N. Kurti (eds.) "Experimental Cryophysics." Butterworths, London and Washington, D.C., 1961.

[5] C. A. Swenson, Low temperature thermometry, 1 to 30 K, CRC Critical Reviews in *Solid State Sci.* **1**, 99 (1970).

[6] L. G. Rubin, Cryogenic thermometry: a review of recent progress, *Cryogenics* **10**, 14 (1970).

[7] D. de Klerk, Adiabatic demagnetization, *in* "Handbuch der Physik" (S. Flugge, ed.), Vol. XV, Part II, pp. 38ff. Springer-Verlag, Berlin, 1956.

[8] H. M. Rosenberg, "Low Temperature Solid State Physics." Oxford Univ. Press (Clarendon), London and New York, 1963.

[9] W. R. Abel, A. C. Anderson, W. C. Black, and J. C. Wheatley, Thermal and magnetic properties of liquid He^3 at low pressure and at very low temperatures, *Physics* **1**, 337 (1965).

[10] J. C. Wheatley, O. E. Vilches, and W. R. Abel, Principles and methods of dilution refrigeration, *Physics* **4**, 1 (1968).

[11] A. C. Anderson, Carbon resistance thermometry, *in* "Temperature, Its Measurement and Control in Science and Industry." Based on the *5th Temp. Symp., Washington, D.C.*, Instrum. Soc. Am., Pittsburgh, 1973.

mK has been reported for sensors fashioned from radio resistors, and this is likely to be pushed still lower with heavily doped germanium sensors as better thermal contact is achieved.

Two types of resistance elements are now in use: carbon in the form of radio resistors manufactured by particular companies and heavily doped germanium. Metals, both pure and alloys, do not have enough temperature dependence below 1 K to make them useful. For the two materials now popular, the resistance is typical of semiconductor materials and should increase roughly exponentially with $1/T$. Carbon resistors are formed of mixtures of carbon particles and insulating material and one should not expect reliable adherence to some law even empirically derived. At high temperatures, this type of resistor can be very well fit by the equation of Clement and Quinell[12]

$$\log R + K/\log R = B + A/T. \tag{9.2.1}$$

At very low temperatures R becomes very large and the $K/\log R$ term can be ignored. The fact that this logarithmic behavior is not observed in practice below 0.1 K can probably be attributed to the increasing seriousness of self-heating of the resistor by both the measurement power and spurious pickup. Equations have also been proposed by Lounasmaa[13]

$$T = (R + R_1)/(R - R_0), \tag{9.2.2}$$

Penar and Campi[14]

$$\ln R = AT^{-B}, \tag{9.2.3}$$

and Hetzler and Walton[15]

$$1/T = A + BR + CR^{1/2}. \tag{9.2.4}$$

The latter authors show that these two equations fit Speer 470 Ω, $\tfrac{1}{2}$-W, Grade 1002 resistors better than the Clement and Quinell equation in the 0.3–3-K region. Their usefulness is yet to be established at very low temperatures. Other early equations[16] have not been widely used.

[12] J. R. Clement and E. H. Quinell, *Phys. Rev.* **79**, 1028 (1950); *Rev. Sci. Instrum.* **23**, 213 (1952).
[13] O. V. Lounasmaa, *Phil. Mag.* **3**, 652 (1958).
[14] J. D. Penar and M. Campi, *Rev. Sci. Instrum.* **42**, 528 (1971).
[15] M. C. Hetzler and D. Walton, *Rev. Sci. Instrum.* **39**, 1656 (1968).
[16] For examples see D. de Klerk, "Handbuch der Physics" (S. Flugge, ed.), Vol. XV, Part 2, p. 187. Springer-Verlag, Berlin, 1956.

Two brands of resistors are useful below 0.1 K, $\frac{1}{2}$-W, Grade 1002 resistors made by Speer Electronic Components (27–200-Ω values being the most popular)[†17,18] and model ACW 2, $\frac{1}{4}$-W resistors made by Ohio Carbon Company (the 50-Ω value being the only one tested to date to the author's knowledge).[‡19] Other resistors so far tested, such as the popular Allen-Bradley ones used above 1 K, are not useful below 0.1 K. Terry[20] has constructed self-supporting carbon-film thermometers, which he has used in liquid ^{3}He down to 50 mK and which exhibit a nice $1/T$ dependence at very low T ($10 \leq 1/T \leq 25$ K^{-1}). The resistance extrapolates to just below 1 MΩ at 10 mK. It is important to calibrate each thermometer throughout the temperature range in which it is to be used and not to scale published data on nominally similar resistors nor to extrapolate below the lowest calibration temperature. The author has obtained Speer resistors of the same type and value inside one month's time and found them to have considerably different temperature characteristics. Furthermore he has seen no resistor plot much more than one decade as a straight line on the standard log R versus log T plot.

Germanium resistors used down to 0.3 K typically have resistance of about 100 Ω or more at 4.2 K and would be much too high to be convenient below 50 mK. Resistors for use above 1.5 K typically are chosen to have 4.2 K resistances of about 1 kΩ, reach 10 kΩ at 1.5 K, and are uselessly high at 50 mK. Thus heavily doped germanium must be used (bordering on metallic conduction) for which it is not presently possible to obtain spatially uniform doping over large volumes. The author has learned of a chip of germanium produced and used at Bell Telephone Laboratories having a 4.2 K resistance of 10 Ω, an excellent exponential behavior down to a few tens of millidegrees Kelvin, and an extrapolated resistance of 60 Ω at 10 mK.[21] Such pieces of material can be obtained by slicing the heavily doped end of an ingot and testing the pieces individually. Experience with such material is not yet great enough to give any further guidelines here. However, the standard suppliers of ger-

---

[17] W. C. Black, Jr., W. R. Roach, and J. C. Wheatley, *Rev. Sci. Instrum.* **35**, 587 (1964).
[18] A. S. Edelstein and K. W. Mess, *Physica* **31**, 1707 (1965).
[19] Ralph L. Rosenbaum, *Rev. Sci. Instrum.* **41**, 37 (1970).
[20] C. Terry, *Rev. Sci. Instrum.* **39**, 925 (1968).
[21] K. Andres, private communication.

† Airco Speer Electronic Components, Foster Brook Road, Bradford, Pennsylvania.
‡ Ohio Carbon Co., 12508 Berea Road, Cleveland, Ohio.

manium thermometers such as CryoCal, Inc.,[†] Scientific Instruments, Inc.,[‡] and Solitron[§] can supply sensors with 4.2-K resistances in the 20–50-Ω range which must then be tested by the purchaser. Rosenbaum[19] has measured two such units. The development of low-resistance germanium would provide a sensor free of drift and more reproducible in absolute value on thermal cycling.

For semiconductor materials with a single, low-concentration impurity, the conductivity at low temperatures would be dominated by the thermal ionization of the impurity levels and a simple exponential dependence would result. For standard Ge thermometers in the 1–4-K range, this simple limit is not generally reached. For such thermometers, the simple equation

$$R = CT^{-A}e^{-B/T} \qquad (9.2.5)$$

has been proposed as being moderately ($2\frac{1}{4}\%$) accurate.[22] For heavily doped material at very low temperature, a generally accepted equation does not yet exist.

One drawback of resistors presently available for use below 0.1 K is that they are magnetoresistive. For the Speer resistors this amounts to about $-\frac{1}{2}\%/1000$ G to 10 kG at 0.3 K.[17,18] At lower temperatures this rate of change increases considerably. Also the rate of change is higher at fields below 3 to 4 kG. Kirk and Adams[23] find that Speer resistors show a change of resistance proportional to H/T at high fields and very low temperatures. The germanium resistors show the expected H^2 dependence below about 50 kG, above which they approach a linear dependence, and are approximately independent of the direction of the applied field.[¶24,25] The fractional change of resistance with field increases

---

[22] L. T. Clairborne, W. R. Hardin, and N. G. Einspruch, *Rev. Sci. Instrum.* **37**, 1422 (1966).

[23] W. P. Kirk and E. D. Adams, *Phys. Rev. Lett.* **27**, 392 (1971).

[24] J. E. Hunzler, T. M. Geballe, and G. W. Hull, Jr., Germanium resistance thermometers, *in* "Temperature, Its Measurement and Control in Science and Industry," Vol. 3, Pt. 1, p. 391. Compiled by the Amer. Inst. of Phys., Rheinhold, New York, 1962.

[25] D. L. Decker and H. L. Laquer, *Cryogenics* **9**, 481 (1969).

---

† Cryocal, Inc., 1371 Avenue E, Riviera Beach, Florida.

‡ Scientific Instruments, Inc., 632 South F Street, Lake Worth, Florida.

§ Solitron Devices, Inc., Transistor Division, 1177 Blue Heron Blvd., Riviera Beach, Florida.

¶ Cryocal, Inc., 1371 Avenue E, Riviera Beach, Florida.

9.2. RESISTANCE THERMOMETERS

as the temperature decreases, but so does the sensitivity of resistance to temperature. As a result the apparent temperature change is -1 mK/(kG)^2 or less in the 1–10-K interval and is roughly constant in this interval. Measurements at very low temperatures have not yet been published.

On cycling to room temperature and back to 0.3 K or less, carbon resistors shift calibration somewhat. The shift is, however, surprisingly small and typically runs about 0.3% in resistance to below 0.1 K, a temperature error of about $\frac{1}{2}$ mK at 0.3 K and $\frac{1}{4}$ mK at 0.1 K.[18] Specially prepared resistors have been reported to show an order-of-magnitude better stability in the 1–4.2-K range.[26] Johnson and Anderson[27] have investigated the drift of sliced Speer resistors below 0.3 K and found that in optimal cases reproducibility at the 10^{-3} level can be achieved.

The major problem in the use of resistors is the difficulty in achieving thermal contact below 100 mK. This problem is highly exacerbated by the inherent necessity of dissipating power in order to make a measurement and of attaching leads to the resistor in the first place. The power dissipated in the resistor must be kept below the 10^{-12}-W level below 50 mK if the thermometer is to be useful! This power level would shift the apparent temperature at least 1% at 10 mK through self-heating. Thus resistance bridges are typically operated well below the 10^{-13} W level and shielding from power line pickup and radio-frequency interference are essential. This problem can be illustrated with a very simple example: Consider a home FM receiver with a 5-ft antenna, a good approximation to the length of leads in a cryostat. A station in town might easily produce 100 μV rms across the terminals of the antenna which has an impedance of 300 Ω. Thus the power delivered to the receiver is 10^{-10}–10^{-11} W, and this from a single station!

Since the thermal conductivity of the insulating material on the body of a radio resistor is very poor, the first step is to remove some of it to expose the resistive core. Symko[28] has ground away the insulation and attached fine insulated copper wires to the resistor with G.E. 7031 varnish to use for attaching the resistor to the sample. He has also filed the ends flat down to the leads to obtain good contact to the copper and removed the tinning from the leads. In this way he was able to achieve good contact and reproducible measurements down to 10 mK, though the

---

[26] L. P. Mezhov-Deglin and A. I. Shal'nikov, *Cryogenics* **9**, 60 (1969).
[27] W. L. Johnson and A. C. Anderson, *Rev. Sci. Instrum.* **42**, 1296 (1971).
[28] O. G. Symko, *J. Low Temp. Phys.* **1**, 451 (1969).

resistor continued to respond down to 5 mK. Robichaux and Anderson[29] have cut longitudinal slabs from the center of Speer resistors, using the center only in order to keep the copper leads imbedded in the resistor for electrical and thermal contact. The end of the resistor was filed down, thereby eliminating the tinned part sticking out of the body. This procedure has the advantage of giving a larger contact area-to-volume ratio along with a small heat capacity. This treatment yielded resistors having a uniformly increasing sensitivity as the temperature decreased to 20 mK, whereas unmodified resistors inserted into holes in a copper block with Apiezon N grease as the contact agent showed the same sensitivity down to 70 mK but decreasing sensitivity below that. In the author's laboratory, similarly prepared resistors do not show evidence of self-heating until below 11 mK and are used to 7 mK. Using data provided by Anderson et al.,[30] Symko[28] has calculated that the impedance to such resistors as those above would be given by, for small temperature changes,

$$\dot{Q} = \alpha A(T_1^4 - T_2^4) \approx 4\alpha A T^4 (\Delta T/T), \qquad (9.2.6)$$

where A is the contact area, α is very roughly 7×10^{-3} W/cm² K⁴, and the approximate form is written to display the fractional change of temperature with heat input.

Equally important is the reduction of heat input down the leads to the resistors. The leads themselves must be of low thermal conductivity and alloy wires such as Evanohm,† constantan, or manganin serve the purpose quite well if one does not care about an added but constant resistance in series with the sensor. If this is unacceptable, the leads may be tinned with a superconducting solder, most of which will be superconducting at the low fields at which measurements are typically taken. For use in high fields, fine Nb–Zr and Nb–Ti wires might be used. Anderson[31] has devised an ingenious and simple method of grounding leads by passing them through a folded copper foil insulated with cigarette paper and cemented with G.E. 7031 varnish. Alternatively, the leads can be wound around a copper pot or post in the apparatus and thermally anchored with G.E. 7031 varnish or Apiezon N grease. An arrangement

---

[29] J. E. Robichaux, Jr. and A. C. Anderson, *Rev. Sci. Instrum.* **40**, 1512 (1969).

[30] A. C. Anderson, G. L. Salinger, and J. C. Wheatley, *Rev. Sci. Instrum.* **32**, 1110 (1961).

[31] A. C. Anderson, *Rev. Sci. Instrum.* **40**, 1502 (1969).

---

† Wilbur B. Driver Co., 1875 McCarter Highway, Newark, New Jersey.

9.2. RESISTANCE THERMOMETERS

which works well is to use a three-terminal Wheatstone bridge and to run three equal-length wires from room temperature into the experimental chamber through the 4.2-K region. A transition is then made to superconducting wire and two of the wires (Be careful which two!) may then be tied together. Where a complete conduction path exists through the apparatus, the compensating pair may be grounded and a single wire run to one terminal of the resistor, the other terminal being grounded. In this arrangement the bridge must, of course, be floated. The compensating pair can be used common to many resistors. This arrangement carries the additional advantage that the resistor can be tied to the apparatus through a conductor, possibly increasing the thermal contact.

Shielding of leads from the apparatus to the resistance bridge is a well-known instrumentation problem. Here we are more interested in protecting the sensor than the detection amplifiers, but the problem is the same. In order to keep the power below 10^{-14} W, the voltage across the resistor must be less than 0.1 mV rms for a 1-MΩ resistor and 10 μV rms for 10 kΩ. Shielding from power line pickup at this level is a critical problem indeed. Standard procedure calls for the use of a twisted pair or triplet (to the resistor terminals) surrounded by two independent shields, the inner of which is tied at the grounded end of the signal line (only one end of which is grounded) and the outer is tied at the two ends to the signal ground and the ground closest to the other end[32] (see Fig. 1). Inside the cryostat shielding of the leads must be maintained (by use of a metal dewar or by use of independent shields, usually stainless steel for low thermal loss).

Shielding from radio frequencies is an even more difficult problem. The most obvious solution is to place the apparatus inside a screened room. This has several drawbacks, the most obvious being cost, difficulties with spatial freedom including filling with liquid helium, plumbing, etc. More subtle points include the fact that much electronic equipment that one might place *inside* the room close to the cryostat generates rf interference unless expressly shielded and otherwise protected, especially power supplies containing silicon-controlled rectifiers (SCR's), logic circuits such as digital voltmeters, and flashing neon lights. The alternative solution is to place rf bypass circuits on the leads at a point close enough to the sensor that further pickup can be avoided, for example at the entry to the vacuum can or, in the case of a metal dewar, at the top

---

[32] Ralph Morrison, "Grounding and Shielding Techniques in Instrumentation." Wiley, New York, 1967.

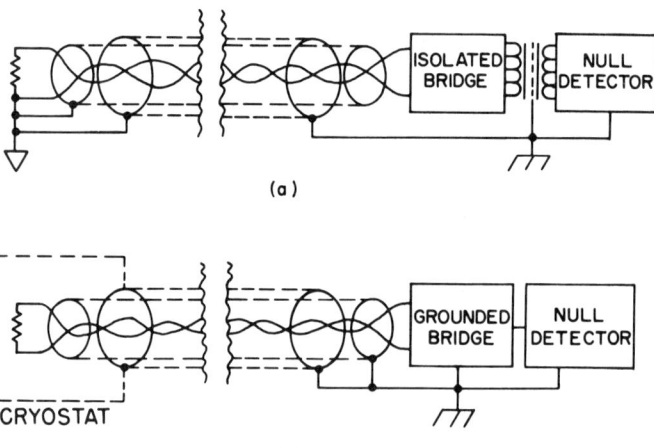

FIG. 1. Grounding arrangements for interconnecting cables between bridge and resistor sensor: (a) grounded source—isolated bridge; (b) floating source—grounded bridge.

flange. At high temperatures, in the liquid helium range, the standard solution has been to simply bypass the resistor with a silver mica capacitor. (Disk ceramic capacitors "freeze out" at low temperatures, their capacitance dropping by over an order of magnitude from their room temperature value. This does not seem to be the case with NPO chip capacitors now popular for use in printed and integrated circuits, and these can be obtained in somewhat smaller sizes than the mica types. These should have better bypass characteristics at high frequencies than capacitors with leads coming out of the body, and they do not appear to crack on cooling if there is no mechanical strain on them. The user would be well advised to test the particular brand in liquid helium before using them.) This solution is probably not adequate at very low temperatures, as capacitor shunts do not make good rf filters. At first sight it would appear that one has a high resistance shunted by the capacitor and thus little of the current should pass through the resistor. In fact one has a transmission line (the twisted pair of leads) of rather low impedance driving the network (more of a voltage than current source) and so the capacitor is not as effective as one requires, though it is certainly a help and may be adequate where large rf interference is not a problem.

A better solution is to place inductance in series with the leads and use capacitor shunts on both sides of these, thus making π section filters. The inductance can be miniature rf chokes or ferrite beads. Chokes

9.2. RESISTANCE THERMOMETERS

should be of the *shielded* variety in order to minimize coupling to neighboring chokes and magnetic pickup with other wires and components. Ferrite beads have the advantage that they are broad-band devices and introduce no rf resonances in the line, though this should be of no importance unless there happens to be a very large pickup in the vicinity of the resonance. The ferrite beads also seem to "freeze out" at low temperatures and become ineffective, though we have not studied this enough to be sure it is generally true; so they should be kept in the room-temperature region unless tested in liquid helium. A filter consisting of 220-μH chokes in each line and bypassed on each end with 560-pF capacitors has a cutoff of about $\frac{1}{2}$ MHz. Compact high-frequency filters can be made by simply running two wires through a ferrite bead with two holes (such as a Ferroxcube No. 56 390 31/4B[†]) and shunting at each end with 270 pF. The cutoff of this is about 10 MHz. Very small commercial filters which can be inserted in connectors are available that have 20-dB attenuation at 10 MHz and improve rapidly as the frequency increases.[‡] Complete connectors with a high density of pins are available with these filters from several of the standard connector manufacturers. These are very convenient but also rather expensive. One connector might be used for over 50 leads, however. Of course, these connectors can be used in the ambient environment only. Still better, filters can be constructed by attaching subminiature, shielded, 47-μH chokes and pin filters in series to a hermetically sealed connector. The filters should pass through and make good contact with a ground plane. Such an arrangement will provide a $\frac{1}{3}$-MHz low-pass filter in each lead and is relatively easy to construct on 18 and 32 pin connectors. The wires must be kept inside a completely sealed tube after passing through the ground plane. It is possible that, in locations where rf interference is exceptionally large, the tube will not adequately shield the wires, especially if only thin-walled stainless steel tubing is used.

Anderson[1] has recently pointed out that the above filters attenuate the normal mode currents in the leads, but the common mode current can still cause problems. The parasitic capacitance between the resistor and the apparatus allows this common mode current to flow through the resistor to the metal near it (see Fig. 2). In fact, the tighter one tries to tie the resistor to the apparatus thermally, the more one increases this

---

[†] Ferroxcube Corp., Saugerties, New York.

[‡] Allen-Bradley, Milwaukee, Wisconsin and AMP, Inc., Capitron Division, Elizabethtown, Pennsylvania.

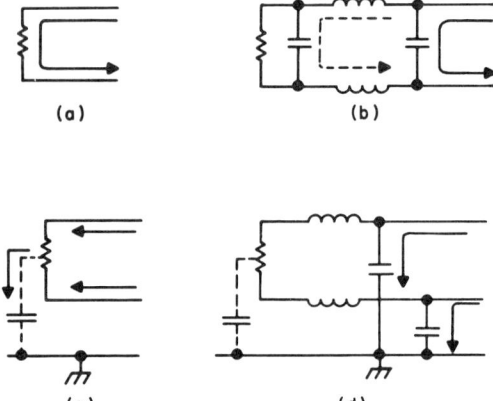

FIG. 2. Filtering rf pickup on lines to resistor sensor. Normal mode rf current: (a) unfiltered; (b) filtered. Common mode rf current: (c) unfiltered; (d) filtered.

capacitance, thereby increasing the pickup. In Anderson's apparatus, placing shunt capacitors to ground as shown in the figure solved the rf pickup problems and eliminated the necessity of using a shielded enclosure.

We shall not dwell at length here on the principles and operation of resistance bridges as these have been adequately covered in the standard literature of experimental low-temperature physics. Rather we shall simply point out considerations the experimenter will want to take into account before choosing his design and beginning construction. The first matter is that of accuracy. At temperatures above 0.3 K, it is possible to achieve very high resolution and accuracy since power dissipated in the resistor is not a serious problem. With a room-temperature reference resistor and 10^{-11}-W measuring power, the noise level of an otherwise perfect, noiseless bridge would limit resolution to 10^{-3} using a 1-Hz detector bandwidth. While somewhat narrower bandwidths could be utilized in the detector, time constants resulting in a 10^{-4} resolution would be much too long. Putting the reference resistor in the liquid helium bath would help somewhat and will be discussed below. Therefore, the bridge need not be overdesigned with regard to resolution if it is not to be used for precision work at higher temperatures.

The frequency of operation is generally chosen to be in the 30–40-Hz range or 400–1000 Hz. Direct current measurements cannot be made with the low noise level required and in any case thermal emf's in the leads and joints could be a nuisance, requiring bipolar measurements and

9.2. RESISTANCE THERMOMETERS

averaging. The advantage of choosing the lower range is that the capacitance placed in the circuit to minimize rf interference causes phase shifts that must be compensated, and this is easier the lower the frequency. At higher frequencies, the transformers used in the bridge are smaller and easier to obtain (usually standard items) and it is easier to filter out power line pickup components in the amplifiers. Neither of these problems is serious, and with field effect transistors (FET's) being available with low noise down to 10 Hz, low-frequency bridges for these purposes are feasible (with a possible exception mentioned below).

Numerous bridge types are available, from the familiar Wheatstone (two-wire) to the Kelvin (four-wire). By keeping the lead resistance low in the helium bath-to-room temperature transition region and by using alloy wires or superconductors in the low-temperature chamber, two-wire bridges will be adequate for most purposes in the ultralow temperature region. The three-wire Wheatstone bridge configuration with a 1 : 1 reference ratio has the advantage of giving a very high accuracy compensation if the leads are of equal length and close together. Of course, any other bridge can be used. Four-wire bridges may be required with low-resistance germanium sensors. (Some references to bridges are given by Swenson.[5] Other bridges are given by Kierstead,[33] Diamond,[34] Ries and Moore,[35] and Ekin and Wagner.[36])

An important word of caution concerning the construction of the bridge: it is easy to generate spurious currents in the bridge which will be transmitted to the resistor. Pickup at the power line frequency and its harmonics is hard to suppress. (Remember that we are trying to keep the power level at the resistor below 10^{-14} W, i.e., 10 μV rms across a 10-kΩ transistor—no trivial task with regard to 60- (or 50-) Hz pickup. Indeed, it is hard enough to observe such spurious sources, and differential instrument amplifiers with high common mode rejection are suggested for this purpose.) Transformers must be magnetically shielded (triply shielded instrument transformers are frequently used and readily available) and the power line should be decoupled with an isolation transformer. This latter transformer and the power supply transformer may need to be magnetically shielded with foil designed for the purpose or placed at some distance from the bridge. Logic circuits, such as digital

---

[33] H. A. Kierstead, *Phys. Rev.* **144**, 166 (1966).
[34] J. M. Diamond, *J. Sci. Instrum.* **43**, 576 (1966).
[35] R. P. Ries and B. K. Moore, *Rev. Sci. Instrum.* **41**, 996 (1970).
[36] J. W. Ekin and D. K. Wagner, *Rev. Sci. Instrum.* **41**, 1109 (1970).

voltmeters and SCR's with their large switching transients, can be used only with the greatest care.

Finally we consider the ultimate capabilities of the bridge. Field effect transistors are available with equivalent input noise voltages of 10 nV/(Hz)$^{1/2}$ to below 30 Hz and input impedances well above 1 MΩ at the frequencies of interest, the optimum source impedance for lowest noise generally being above 10 MΩ. At the 1-MΩ level, a reference resistor at room temperature will have a Johnson noise of about 70 nV/(Hz)$^{1/2}$, far above the capabilities of the FET. Therefore, to realize the full potentialities of available electronics, the reference resistor should be placed in the helium bath and a ratio transformer used to balance the bridge. For lowest noise at lower impedances, one should use junction transistors (optimum impedance about 10 KΩ and $1/f$ noise typically being a problem below 100 Hz) or transformers to match to FET's, 100 kΩ–1 MΩ being an adequate transformed impedance. If a ratio transformer is used, the bridge should operate near the resonant frequency of the transformer, usually about 400 Hz. This minimizes the effect of the transformer shunting impedance. To use a ratio transformer at low frequency, it might be possible to use a low-reference resistance and transform up the resistance with a subsequent transformer. The ultimate resistance bridge may consist of a reference resistor cooled to the temperature of the sensor and a low-temperature parametric amplifier, perhaps with gallium arsenide varactor diodes (which operate in liquid helium) or superconducting elements.

As an example of these points, we discuss a low-cost bridge which has been in use in our laboratory for several years (see Fig. 3). The bridge

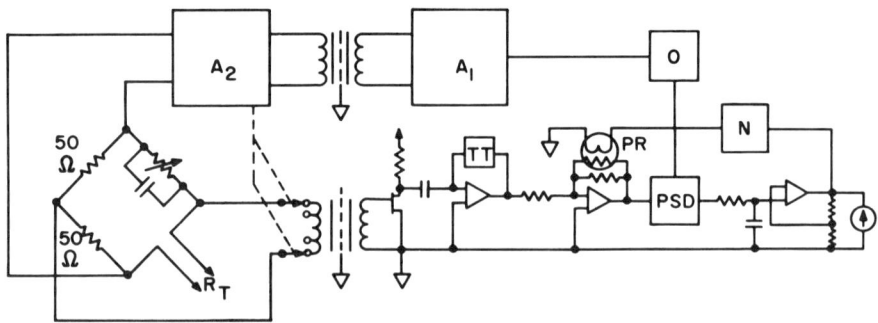

Fig. 3. Schematic diagram of a resistance bridge where O is the oscillator, A_1 is the level set attenuator, A_2 is the resistance level attenuator, R_T is the thermometer resistor, TT is the twin tee, PR is the photo resistor, PSD is the phase-sensitive detector, and N is the frequency compensation.

itself is a simple Wheatstone type with low-value resistors on one side to minimize noise and pickup. Since the variable balancing resistor and the thermometer resistor always have the same ratio at balance, the compensating capacitance can be set once and for all as long as the capacitors in the sensor arm are not temperature sensitive and the highest accuracy is not required.[37] The balancing resistor is a wire-wound decade box with a maximum resistance of 100 kΩ and six decades. A 33-Hz tuning fork oscillator is used for frequency stability. The null amplifier has a FET preamplifier for low noise at low frequency, the low frequency being chosen to minimize capacitance effects. (Several very low-noise FET's are available such as the 2N4867A series. We should point out that the usual, high-quality FET operational amplifiers commercially available do not have the low-voltage noise required for this application as they are constructed for electrometer type applications where low bias current is the prime consideration. Low voltage noise amplifiers can be obtained on special order.[†]) The primary of the coupling transformer is tapped at 1, 3, 10, 30, and 100 kΩ and the secondary is 1 MΩ. The automatic gain control required of the null amplifier is obtained by feeding back the detected output into a tungsten lamp controlling a photoresistor (a commercial unit). This arrangement is used instead of a logarithmic dc amplifier at the output so that saturation of the ac stage cannot possibly occur and instead of a logarithmic ac amplifier so that the gain is set only by the signal level and not by the noise. The shunting resistor across the photoresistor ensures that the center 20% of the meter is of constant gain and linear, the gain being chosen so that the noise level is about 2% of full scale. (Tungsten lamp–photoresistor combinations are very sluggish at low light levels and a frequency compensation network must be used in the loop.) With this bridge, resolution near theoretical is achieved (0.1–0.3% with a measuring power of 10^{-14} W) and the meter stays on scale for any settings of power (up to 10^{-10} W) and any resistance unbalance. For convenience the power level is set with one attenuator and appropriately compensated for the resistance level by a second attenuator, which is ganged with the switch which selects the appropriate transformer tap. The bridge can equally well be operated in the standard three-terminal configuration for lead compensation.

[37] C. Blake, C. E. Chase, and E. Maxwell, *Rev. Sci. Instrum.* **29**, 715 (1958).

† Intech, Inc., 1200 Coleman Avenue, Santa Clara, California.

9.3. Susceptibility Thermometers

The variation of the susceptibility of paramagnetic systems has long been recognized as a powerful tool for the measurement of temperature, particularly in that, in principle, the measurement can be made with very little heat dissipation in the sensor. It was immediately obvious that the susceptibility of the paramagnetic salts used for cooling could itself be used to measure the temperature of the salts. The major problem arose in the establishment of the thermodynamic temperature scale, especially at the lowest temperatures achieved with any particular salt. As the frontier of measurement has been pushed to the millikelvin region, increasing emphasis has been placed on the susceptibility of materials in which the nuclear contribution dominates. This is also partly fostered by the fact that some metals make prime candidates for such thermometers, thermal contact to these is easier to achieve in the majority of cases than to salts. Magnetically ordered materials are useful at higher temperatures, especially insulating antiferromagnetics, and these may be useful at ultralow temperatures as well. They have the advantage that they provide a universal scale to which all observers can refer their measurements. Finally, susceptibility thermometers have a great advantage in that the measurement of susceptibility can be made with very little power dissipation in the sensor or the surrounding apparatus.

An ideal, noninteracting paramagnet obeys the Curie susceptibility law $\chi = C/T$, where $C = (N/V)\{g^2[J(J+1)]\mu_B^2/3k\}$. Curie's law is obtained from the low B/T limit of the exact expression for the magnetization

$$m = M/V = (N/V)g\mu_B J B_J(g\mu_B J B/kT),$$

which correctly describes the saturation limit as well, where $B_J(x)$ is the familiar Brillouin function, so named in spite of the fact that Debye obtained it first.[38] For ideal paramagnetic salts, the Curie equation accurately describes the susceptibility at values of $B/T \ll 2$ kG/K. For nuclear systems, this limit is more than 1000 times larger. The effect of interactions, either from dipole–dipole or exchange forces, is to modify the susceptibility to $\chi = C/(T - \theta)$, the Curie–Weiss law, valid

---

[38] Apparently Van Vleck attached the name "Brillouin function" in order to avoid "confusion with his [Debye's] specific heat function" since Brillouin was the first to use this function in the new quantum mechanics. See J. H. Van Vleck, "The Theory of Electric and Magnetic Susceptibilities," p. 257. Oxford Univ. Press, London and New York, 1932.

9.3. SUSCEPTIBILITY THERMOMETERS

at temperatures high compared to the ordering temperature. In the Weiss molecular field model, θ is the ordering temperature. (In the Néel model of an antiferromagnet, θ is negative; thus the denominator is frequently rewritten with a plus sign.) In real materials, θ has a more complicated behavior in the region just above the ordering temperature because short-range order persists above the point at which magnetization on a macroscopic scale disappears. In practice, a paramagnet obeys the Curie–Weiss law to fairly high accuracy down to temperatures a factor of two or three higher than its ordering temperature. Thus we choose the inverse of the susceptibility as the appropriate measure of the temperature: $T^* = C/\chi$. For example, single-crystal spheres of cerous magnesium nitrate begin to show distinct deviations from the true thermodynamic temperature for $T^*_{\text{sph}} = C/\chi < 5$ mK, the ordering temperature probably being just below 2.0 mK.[39]

Another important property of paramagnets is the rate at which the spins come into thermal equilibrium with the lattice, the spin–lattice relaxation time τ_1; for it is the lattice to which one usually makes thermal contact. (In the case of nuclei in metals, the primary direct contact is to the conduction electrons, but it is also to these that thermal contact is made and the process is still characterized by τ_1. In the case of CMN-to-^{3}He contact which is anomolously large,[40] there is likely direct spin-to-spin coupling at the interface.) If the measurement frequency is low compared to τ_1^{-1}, then the spins will be in thermal equilibrium with the lattice, i.e., we measure the isothermal susceptibility. One can still not be careless about the power level used to make the measurement as losses in the paramagnet will cause heating and a large dB/dt will cause eddy current heating in metals in the sample chamber and vicinity. If $\omega > \tau_1^{-1}$, then the susceptibility measured is the adiabatic susceptibility and the spins can easily be heated above the temperature of the lattice. Thus at high frequencies the greatest care must be taken to ensure that the level of the measuring field is far below that at which deviations would occur (see Section 9.3.1).

Since CMN is the lowest-temperature undiluted paramagnetic salt available and since its region of straightforward application lies above 5 mK, attention at lower temperatures naturally falls on weaker paramagnetic systems. Dilute salts have occasionally been investigated, par-

---

[39] K. W. Mess, J. Lubbers, L. Nielsen, and W. J. Huiskamp, *Physica* **41**, 260 (1969).
[40] W. R. Abel, A. C. Anderson, W. C. Black, and J. C. Wheatley, *Phys. Rev. Lett.* **16**, 273 (1966).

ticularly for cooling below 2 mK. Nuclear systems, however, have the advantage of maintaining high concentration and, in the case of metals such as copper, high thermal conductivity and ease of attachment by soldering or welding.

Because the classical measurements of susceptibility of paramagnetic salts have been adequately and coherently described elsewhere and because the newer techniques are presently in a rapid state of improvement, we shall not dwell on these measurements here. We do hope, however, to point out the more important features of these methods and potential pitfalls as they appear at this time.

9.3.1. Paramagnetic Susceptibility of Localized Atomic Moments[3-7,41-44]

Above we outlined the theory of paramagnetic behavior which applied to localized magnetic moments. Here we need only discuss the measurement of χ and a particular salt useful for ultralow temperatures.

At very low temperatures, the susceptibility of a salt becomes very large and the applied field is no longer the field seen by the local moments. The internal field is the applied field less the field from the "free" poles at the surface of the sample. Obviously the latter is shape-dependent and, therefore, samples of regular shape such as ellipsoids (including spheres) and cylinders are usually chosen. At any particular time, there is also the field of the neighbors, called the *Weiss* (molecular) *field*. For CMN this field is dipolar in origin and the calculation of Lorentz is assumed to be a good approximation. Thus we have

$$H_{\text{loc}} = H_0 + H_{\text{dem}} + H_{\text{W}}, \qquad (9.3.1)$$

where H_0 is the applied field, loc is local, dem is demagnetizing, and W is Weiss. In cgs units, $B_{\text{loc}} = H_{\text{loc}}$. Using standard notation, we write this in terms of the magnetization as

$$H_{\text{loc}} = H_0 - \alpha m + (4\pi/3)m, \qquad (9.3.2)$$

where α is the demagnetizing factor and m is the magnetization per

---

[41] E. Mendoza, Magnetic cooling, in "Experimental Cryophysics" (F. E. Hoare et al., eds.). Butterworths, London and Washington, D.C., 1961.

[42] E. Ambler and R. P. Hudson, Magnetic cooling, *Rep. Progr. Phys.* **28**, 251 (1955).

[43] C. G. B. Garrett, "Magnetic Cooling." Harvard Univ. Press, Cambridge, Massachusetts, 1954.

[44] H. M. Rosenberg, "Low Temperature Solid State Physics," Chapter 9. Oxford Univ. Press (Clarendon), London and New York, 1963.

unit volume. Discussions and tabulations of the demagnetizing factor appear elsewhere.[41,45,46] Using the Curie–Weiss equation for the volume susceptibility

$$\chi = m/H_{\text{loc}} = C/(T - \theta), \tag{9.3.3}$$

one obtains for the externally observed susceptibility, after expressing m in terms of H_0 from the above,

$$\chi_{\text{ext}} = m/H_0 = C/\{T - \theta - C[(4\pi/3) - \alpha]\}$$
$$= C/(T - \Delta). \tag{9.3.4}$$

Thus

$$T = T^* + \Delta. \tag{9.3.5}$$

For a sphere, $\alpha = 4\pi/3$; therefore, $\Delta = \theta$. In any case, the field corrections have the effect of changing the apparent value of θ, so for a particular salt pill α need not be known unless a comparison to other investigators' data is required. In that case it is easiest to use ellipsoids of single crystals.

If a powder sample is used, the correction factor in the denominator can be multiplied by the filling factor f. Since such a sample is not homogeneous, it is not clear how good this approximation is, though it is probably good for reasonably dense powders.[47] For a magnetically anisotropic, powdered salt, the correction factor must be multiplied by another factor n[48] to correct for the spatial averaging of the g factor with respect to the applied magnetic field. If one is comparing the susceptibility measurements with those on an ellipsoid, the g factor is referred to that of the axis along which the ellipsoid is measured, which we shall call g_0. Then

$$n = \bar{g}/g_0, \tag{9.3.6}$$

where $\bar{g}$ is the spatially averaged g factor. The averaged value for a random

---

[45] R. M. Bozorth, "Ferromagnetism," pp. 538ff. Van Nostrand-Reinhold, Princeton, New Jersey, 1951.

[46] W. F. Brown, Jr., "Magnetostatic Principles in Ferromagnetism" (E. P. Wohlfarth, ed.), Vol. 1, Selected Topics in Solid State Physics, p. 187. North-Holland Publ., Amsterdam, 1962.

[47] For a brief discussion of the powder problem see H. Zijlstra, "Experimental Methods in Magnetism" (E. P. Wohlfarth, ed.), Vol. 9, Selected Topics in Solid State Physics, Part 2, p. 83. North-Holland Publ., Amsterdam, 1967.

[48] A. C. Anderson, *J. Appl. Phys.* **39**, 5878 (1968).

distribution of particles is given by

$$\bar{g} = (1/4\pi) \int_0^\pi \int_0^{2\pi} [g_z(\cos\theta)^2 + g_x(\sin\theta\cos\phi)^2 + g_y(\sin\theta\sin\phi)^2] \sin\theta \, d\theta \, d\phi$$

$$= \tfrac{1}{3}(g_x + g_y + g_z), \qquad (9.3.7)$$

where the trigonometric functions such as $(\sin\theta\cos\phi)$ enter once for the projection of H_0 onto the crystal axis and once for the projection of the resulting magnetization back into the field direction. For CMN with $g_x = g_y = 1.84$ and $g_z = 0$, $n = \tfrac{2}{3}$. Thus for powder thermometers, we have

$$\Delta = \theta + \mathrm{nfC}[(4\pi/3) - \alpha]. \qquad (9.3.8)$$

The $C[(4\pi/3) - \alpha]$ term in the expression for χ_{ext} is just the first-order dipole–dipole correction to the susceptibility and higher-order terms must be added for a completely correct expression,[7] just as bR/T^2 is only the first-order dipole–dipole contribution to the specific heat in a dipolar (as opposed to exchange-dominated) system. For an order-of-magnitude calculation, we can guess that the next term might be of order $\delta/T \approx \theta_{\mathrm{d-d}}^2/T$, where $\theta_{\mathrm{d-d}}$ is the dipolar energy in units of temperature (T^{-1} since the term must become more important as $T \to \theta_{\mathrm{d-d}}$ and $\theta_{\mathrm{d-d}}^2$ to preserve the dimensionality required). For chromium methylammonium alum, this term has been observed and agrees with the theoretical calculation.[49,50] This term is not observed when calibrating CMN above $\tfrac{1}{3}$ K.

The susceptibility discussed above is the static or dc value as would be measured with a ballistic galvanometer bridge. This is also thermodynamically the isothermal susceptibility χ_T. At some high frequency, $\omega\tau_1 \approx 1$, where τ_1 is the spin–lattice relaxation time, the spin system will no longer be in equilibrium with the lattice and the adiabatic susceptibility χ_S will be observed.[51–54] The relation between these two

---

[49] M. Durieux, H. van Dijk, H. ter Harmsel, and C. van Rijn, "Temperature, Its Measurement and Control in Science and Industry," Vol. 3, Pt. 1, p. 383. Compiled by the Amer. Inst. of Phys., Rheinhold, New York, 1962.

[50] T. C. Cetas and C. A. Swenson, Phys. Rev. Lett. **25**, 338 (1970).

[51] H. B. G. Casimir and F. K. Dupré, Physica **5**, 507 (1938).

[52] A. H. Cooke, Rep. Progr. Phys. **13**, 276 (1950).

[53] C. J. Gorter, Paramagnetic relaxation, in "Progress in Low Temperature Physics" (C. J. Gorter, ed.), Vol. 2, Ch. IX. North-Holland Publ., Amsterdam, 1957.

[54] G. E. Pake, "Paramagnetic Resonance," Chapter 6. Benjamin, New York, 1962.

susceptibilities is given by

$$\chi_S = \chi_T C_M/C_H, \tag{9.3.9}$$

where C_M and C_H are heat capacities at constant M and H, respectively. For a "dipolar" paramagnet

$$C_M = b/T^2 \tag{9.3.10}$$

and

$$C_H = (b + cH_c^2)/T^2, \tag{9.3.11}$$

where H_c is the constant field in the sample resulting from an applied field and we can use H instead of B as long as the two are linearly related. Thus

$$\begin{aligned}\chi_S &= \chi_T b/(b + cH_c^2)\\ &= \chi_T h^2/(h^2 + H_c^2),\end{aligned} \tag{9.3.12}$$

where h is the rms local field due to the neighboring dipoles. As a function of frequency, the total (or complex) susceptibility is given by

$$\chi = \chi' - i\chi'',$$

$$\chi' = \chi_S + \frac{\chi_T - \chi_S}{1 + \omega^2 \tau_1^2} \xrightarrow{\omega \to 0} \chi_T, \tag{9.3.13}$$

$$\chi'' = \frac{\chi_T - \chi_S}{1 + \omega^2 \tau_1^2} \omega\tau \xrightarrow{\omega \to 0} 0.$$

The power absorbed in an ac measurement per second unit volume is

$$P_{\text{abs}} = (\omega/2)\chi'' H^2, \tag{9.3.14}$$

where $H(t) = H \sin \omega t$. This absorption peaks at $\omega\tau_1 \approx 1$. It is by now obvious to the reader that, to measure χ_T reliably, the frequency and magnetic fields should be kept low. Somewhere in the uhf to microwave range, spin–spin interactions play a role in these effects, but that need not concern us here.

For paramagnets in which the dipolar interactions dominate the exchange forces, the static thermodynamic properties, e.g., susceptibility and heat capacity, are field- and shape-dependent.[55–57] Such effects are

---

[55] P. M. Levy and D. P. Landau, *J. Appl. Phys.* **39**, 1128 (1968).
[56] P. M. Levy, *Phys. Rev.* **170**, 595 (1968).
[57] Heinz Horner, *Phys. Rev.* **172**, 535 (1968).

expected to become important in CMN under the conditions $T < 4$–6 mK and $H_0 > 25$–50 G.[58] Thus it is important that stray magnetic fields, such as residual fields in superconducting magnets after they have been "charged" once,[59] be carefully avoided. This effect has probably not caused problems in past measurements as the magnets are usually removed so as not to disturb the susceptibility bridge or to obtain the lowest possible temperature.

In practice, Δ is determined by calibrating the susceptibility of the salt against the vapor pressure of liquid helium. Since the thermometer is usually quite removed from the liquid ^{3}He pot in evaporation type refrigerators and since the temperature of the mixing chamber in a dilution refrigerator cannot be determined from operational parameters, the vacuum chamber around the salt is filled with ^{4}He gas to achieve thermal contact to the bath. With available mutual inductance bridges, resolution equivalent to a few tenths of a millikelvin in Δ is realizable. Where small vapor pressure thermometers such as ^{3}He bulbs are used, great care must be taken to avoid "cold spots" and to properly correct for the thermomechanical effect in the vapor filled tubes leading out of the apparatus.

The salt almost exclusively used for thermometry at present is cerous magnesium nitrate (CMN), $2Ce(NO_3)_3 \cdot 3Mg(NO_3)_2 \cdot 24H_2O$. It is easy to prepare in high purity from an aqueous solution of cerous nitrate and magnesium nitrate; and of the known concentrated salts, it has the lowest ordering temperature. The salt is hygroscopic and so should be protected from deterioration in the atmosphere. Another important characteristic of this salt is its heat capacity which over much of the temperature range of interest is lower than other salts, thus aiding reasonable thermal response. The lowest crystal field splitting is fairly high, approximately 34 K, resulting in a Schottky anomaly which peaks at 14 K and falls off rapidly below that.[60] At very low temperatures, the dipolar interactions between the moments contribute a T^{-2} term to the specific heat which is given by bRT^{-2}, where $b = 5.76 \times 10^{-6}$ K^2.[61] The lattice and Schottky anomaly contributions are negligible below 0.3 K, but Hudson and Kaeser[61] find that the dipolar contribution is not adequate to completely account for the specific heat below this temperature. This

---

[58] H. Horner, private communication.

[59] A. C. Anderson, W. R. Roach, and R. E. Sarwinski, *Rev. Sci. Instrum.* **37**, 1024 (1966).

[60] C. A. Bailey, *Proc. Phys. Soc.* (*London*) **83**, 369 (1964).

[61] R. P. Hudson and R. S. Kaeser, *Physics* **3**, 95 (1967).

anomalous contribution is less than 10% of the quoted dipolar one below ~30 mK. Theoretically the dipolar specific heat must contain terms higher than T^{-2},[62] and so such deviations are not unexpected. The surprising feature of the Hudson and Kaeser results is that this contribution disappears below 15 mK. Because of low-lying excited states, calibration for highest accuracy should be made below 2.6 K for powder samples and 4 K for single crystals.[50,63] Abel et al.[64] have shown that susceptibility measurements with the ballistic method and a 17-Hz bridge agree.

As mentioned above, $T^* = T$ down to 6 mK for CMN. Below this many different observers obtain many different answers. Wheatley and his collaborators,[65-67] by intercomparing various properties of dilute solutions of ^{3}He in ^{4}He, which should be close to ideal Fermi behavior, obtain $T^* = T$ to below 3 mK with an accuracy of a few tenths of a millikelvin for a powder cylinder of diameter : length ratio of 1 : 1. In this case the solution fills the cylinder to obtain thermal contact. Abraham and Eckstein,[68] using a similar cylinder except containing silver chloride pressed together with wires for thermal contact at high pressures (the AgCl flows under pressure forming a solid cylinder), find a Δ of 1.1 mK for a cylinder and 0.3 mK for a sphere. Using the formulas given above, one calculates $\Delta_{cyl} - \Delta_{sph} = 0.3$ mK as opposed to the 0.8 mK observed. In an earlier experiment, they[69] obtained larger corrections but assumed a T^{-2} specific heat dependence. Several measurements on single-crystal spheres have been performed, three of them by purely thermodynamic techniques[39,61,70] and two by nuclear orientation.[71,72] All of these show increasing deviations of T^* from T below 6 mK. Why

---

[62] See for example, P. H. E. Meijer and D. J. O'Keeffe, Phys. Rev. B **1**, 3786 (1970).

[63] M. J. M. Leask, R. Orbach, M. J. D. Powell, and W. P. Wolf, Proc. Roy. Soc. (London) **A272**, 371 (1963).

[64] W. R. Abel, A. C. Anderson, and J. C. Wheatley, Rev. Sci. Instrum. **35**, 444 (1964).

[65] J. C. Wheatley, Proc. 1963 Low Temperature Calorimetry Conf. Helsinki, 1966 (O. V. Lounasmaa, ed.), Ann. Acad. Sci. Fennicae **AVI**, No. 210, 15 (1966).

[66] A. C. Anderson, D. O. Edwards, W. R. Roach, R. E. Sarwinski, and J. C. Wheatley, Phys. Rev. Lett. **16**, 263 (1966).

[67] W. R. Abel, R. T. Johnson, J. C. Wheatley, and W. Zimmermann, Jr., Phys. Rev. Lett. **18**, 737 (1967).

[68] B. M. Abraham and Y. Eckstein, Phys. Rev. Lett. **24**, 663 (1970).

[69] B. M. Abraham and Y. Eckstein, Phys. Rev. Lett. **20**, 649 (1968).

[70] J. M. Daniels and F. N. H. Robinson, Phil. Mag. **44**, 630 (1953).

[71] R. B. Frankel, D. A. Shirley, and N. J. Stone, Phys. Rev. **140**, A1020 (1965).

[72] James J. Huntzicker and D. A. Shirley, Phys. Rev. **B2**, 4420 (1970).

single crystals should be worse in this respect than powder samples is not understood. Hudson[73] has written a lucid account of this controversy entitled "The Delta Campaign."

Thermal contact to samples of CMN is frequently made by embedding copper wires in a slurry of CMN and Apiezon J oil. This is a rather thick oil that has long been used for making thermal contact in salt pills. When working with liquid helium, for example, in the mixing chamber of a dilution refrigerator, the salt can be placed directly in the helium. Glycerol is also frequently used, but one must be careful as it tends to dehydrate many salts and the salt–glycerol mixture must be kept cold to preserve it. The heat capacity of even small amounts of CMN below 20 mK makes thermal contact a serious problem except, perhaps, when the CMN is in contact with liquid helium-3 (or dilute solutions thereof). Very long thermal time constants of many minutes are frequently observed, in which case thermometry becomes virtually impossible.

Several materials have been measured recently that show some promise. Most of these[74,75] will probably not supplant the more popular materials presently in use but may find specialized applications. Cerium dipicolinate,[76] however, may prove more useful than CMN.

Classically, the susceptibility measurements have been made using ballistic and ac Hartshorn bridges. These are described in detail elsewhere (see, for example, the work of Anderson *et al.*[77]). One gram of sample can provide a resolution of 0.004 K^{-1} in $1/T$. Great care is required in the construction of the coils and surrounding apparatus to ensure that magnetic materials do not reside in the vicinity of the thermometer. Frozen air, metals and insulators with magnetic impurities, and superconducting magnets can all reduce the accuracy of the measurements. In short, all materials used for construction should be thoroughly tested at low temperatures for spurious magnetic behavior, and electrical paths in which eddy currents might be generated should be avoided.

Alternatively, the salt pill can be placed in the coil of an oscillator.[78] The inductance of the coil and, therefore, the frequency of oscillation

---

[73] R. P. Hudson, *Cryogenics* **9**, 76 (1969).

[74] E. C. Hirschkoff, O. G. Symko, and J. C. Wheatley, *J. Low Temp. Phys.* **4**, 111 (1971).

[75] R. C. Sapp and J. F. Tschanz, *Cryogenics* **10**, 498 (1970).

[76] J. C. Doran, U. Erich, and W. P. Wolf, *Phys. Rev. Lett.* **28**, 103 (1972).

[77] A. C. Anderson, R. E. Peterson, and J. E. Robichaux, *Rev. Sci. Instr.* **41**, 528 (1970).

[78] D. S. Betts, D. T. Edmonds, B. E. Keen, and P. W. Matthews, *J. Sci. Instrum.* **41**, 515 (1964).

will be determined in part by the susceptibility of the material inside it. A simple readout with a frequency counter is then possible. By careful construction of the oscillator and rigid mounting of all the components, stabilities approaching 10^{-7} can be achieved over periods of many minutes. Since one is measuring the adiabatic susceptibility, particular care must be taken to keep the power level of oscillation low. Robinson circuits are particularly good in this regard (see Section 9.3.2.1). Tunnel diode oscillators[79-83] completely inside the cryostat and as close to the coil as possible have been used, but the oscillation level is difficult to control and measure.

The development of superconducting quantum interference devices with flux sensitivities of 10^{-9} G cm² (1/100 of a flux quantum) may soon make bridges obsolete. Higher sensitivities may be possible and the readout can be made automatic and continuous. With this type of unit, milligram quantities of CMN might be sufficient for a thermometer. This same high sensitivity will, however, probably make nuclear thermometry more attractive than CMN, particularly below 10 mK.

9.3.2. Paramagnetic Susceptibility of Nuclear Moments

We have seen that CMN becomes hard to use below 6 mK as a thermometer since the T^*-T relation is hard to establish, and in many cases below 20 mK because thermal contact is difficult to achieve. Dilution of the salt lowers the ordering temperature and the magnetization (and thereby, presumably, Δ), but thermal contact will not be improved markedly except through the reduction of the heat capacity. The thermal boundary resistance to the salt will remain high. These difficulties can virtually all be overcome by the use of nuclei in an otherwise weakly magnetic metal. The conduction electrons, of course, give a small net paramagnetic contribution (Pauli paramagnetism less Landau diamagnetism), but this is temperature-independent and is not seen in nuclear resonance or nuclear orientation experiments in any case. The thermal conductivity of metals is high and thermal contact to other metals can be easily achieved by use of normal (not superconducting) solders

---

[79] O. W. G. Heybey, Master's Thesis, Cornell Univ. (1962) (unpublished).
[80] C. Boghosian, H. Meyer, and J. E. Rives, *Phys. Rev.* **146**, 110 (1966).
[81] P. R. Critchlow, R. A. Hermstreet, and C. T. Neppell, *Rev. Sci. Instrum.* **40**, 1381 (1969).
[82] R. B. Clover and W. P. Wolf, *Rev. Sci. Instrum.* **41**, 617 (1970).
[83] R. T. Harley, J. C. Gustafson, and C. T. Walker, *Cryogenics* **10**, 510 (1970).

and various welding techniques. The electron–nuclear contact is reasonably good for many metals.

Thermal contact to the nuclei is controlled by the relaxation rate, the heat capacity of the thermometer, and external thermal resistances. The relaxation time τ_1 between the nuclei and conduction electrons at temperature T_{el} is given by the Korringa relation

$$\tau_1 T_{el} = \text{constant} = K. \qquad (9.3.15)$$

This lies between 10^{-2} and 10 sec K for most metals, being about 1 for copper and $2-3 \times 10^{-2}$ for indium and platinum. This relation is expected to hold for temperatures

$$T_{el} > \mu_N B/k, \qquad (9.3.16)$$

where B is the field seen by the nuclei.[84] The breakdown of the Korringa relation has been observed at about 10 mK for fields of a few hundred kilogauss, the hyperfine field in iron series ferromagnets.[85] We notice that, other things being equal, platinum would be preferred to copper since the relaxation time is considerably shorter. In practice, platinum has proven difficult to use for reasons that are not completely clear but may have to do with the effect of very small amounts of impurities (see also p. 515). The useful isotope ^{195}Pt has $I = \frac{1}{2}$ and an abundance $\sim \frac{2}{3}$.

The lattice heat capacity of the metal is negligible in the range we are interested in and that of the electrons is very small. For copper the electronic and lattice contributions are equal at about 4 K and so the electronic part completely dominates below 1 K and is given by

$$C_{el} = \gamma T, \qquad (9.3.17)$$

where $\gamma = 0.7 \times 10^{-3}$ J/mole K^2 = 0.7×10^{-2} erg/mole (mK)^2 or $C_{el} = 1.1 \times 10^{-3}$ erg/gm mK at 10 mK.[86] The nuclear contribution to the heat capacity is, of course, dependent on the square of the energy splitting of the nuclear m_I states which arises from dipole–dipole interactions between the nuclei, quadrupole interactions, and Zeeman splitting in any applied field. The second of these can be eliminated by using $I = \frac{1}{2}$ nuclei or metals with cubic structure, copper falling in the latter category.

[84] J. A. Cameron, I. A. Campbell, J. P. Compton, and R. A. G. Lines, *Phys. Lett.* **10**, 24 (1964).

[85] W. D. Brewer, D. A. Shirley, and J. E. Templeton, *Phys. Lett.* **27A**, 81 (1968).

[86] E. S. R. Gopal, "Specific Heats at Low Temperatures." Plenum Press, New York, 1966.

9.3. SUSCEPTIBILITY THERMOMETERS

For copper, the dipolar interactions cause an inhomogeneous broadening of the resonance line amounting to about 3 G. Since this is a statistical fluctuation, the resultant energy splitting from this and an applied field is proportional to $B_0^2 + b^2$, where B_0 is the applied field, b is the rms dipolar field, and the resultant field B. Then, using

$$C_N = [g^2 I(I+1)\mu_N^2/3k](B/T)^2, \tag{9.3.18}$$

we find for copper

$$\begin{aligned}C_N &= 3.2 \times 10^{-7} \times (B/T)^2 \quad \text{erg K/mole G}^2 \\ &= 1.2 \times 10^{-5} \times (B/T)^2 \quad \text{erg mK/gm G}^2.\end{aligned} \tag{9.3.19}$$

Thus at 5 G (essentially no applied field) and 10 mK, $C_N = 10^{-6}$ erg/gm mK; and at 100 G and 1 mK, $C_N = 5 \times 10^{-2}$ erg/gm mK.

9.3.2.1. Nuclear Magnetic Resonance Measurements.[87–91] The most direct method of detecting the susceptibility of a particular nuclear species is by nuclear resonance. The theory of magnetic resonance is adequately and amply described elsewhere, and we only need to note here that the resonance condition is given by

$$\omega_0 = \gamma B_0, \tag{9.3.20}$$

where γ is the gyromagnetic ratio and is about 1 kHz/G for copper and platinum. At resonance, the magnetization parallel to the applied field H_0 is given by

$$M_z = M_0/(1 + \gamma^2 B_1^2 \tau_1 \tau_2) \tag{9.3.21}$$

and the transverse component (rotating with frequency ω_0 but $\pi/2$ out of phase with H_1) by

$$\mathscr{M} = \gamma M_0 \tau_2 B_1/(1 + \gamma^2 B_1^2 \tau_1 \tau_2), \tag{9.3.22}$$

---

[87] C. Kittel, "Introduction to Solid State Physics," 4th ed., Chapter 17. Wiley, New York, 1971.

[88] C. P. Slichter, "Principles of Magnetic Resonance." Harper, New York, 1963.

[89] A. Abragam, "The Principles of Nuclear Magnetism." Oxford Univ. Press (Clarendon), London and New York, 1961.

[90] O. G. Symko, Nuclear magnetic thermometry at very low temperatures, in "Temperature, Its Measurement and Control in Science and Industry." Based on the 5th Temp. Symp., Washington, D.C., Instrum. Soc. Am., Pittsburgh, 1973.

[91] M. I. Aalto, P. M. Berglund, H. K. Collan, G. J. Ehnholm, R. G. Gylling, M. Krusius, and G. R. Pickett, Cryogenics 12, 184 (1972).

where τ_1 and τ_2 are the longitudinal (spin–lattice) and transverse (spin–spin) relaxation times, respectively, and B_1 is the circularly polarized rf field component rotating in the same sense as the magnetization and is therefore $\tfrac{1}{2}B_{rf}$.

One of the first things we notice is that, whether one detects M_z or $\mathscr{M}$, the magnetization decreases as the rf field increases. Thus one must keep the quantity $\gamma^2 B_1^2 \tau_1 \tau_2$ low enough that the signal amplitude is not affected to the accuracy desired of the thermometer. Since τ_1 is changing with temperature, this condition is changing. In principle it is possible to decrease B_1 as T decreases since both τ_1 and M_0 are inversely proportional to temperature and therefore signal (see the following paragraphs) is not lost into the noise. In practice this is possible only if great care is taken in the calibration of the receiver gain with each new measurement.

The power per unit volume absorbed by the nuclei is

$$P = 2\omega H_1^2 \chi'' \tag{9.3.23}$$

(where we again use H with χ and remind the reader that $H \neq H_{app}$ in a strongly magnetic material) which becomes, at resonance (at which it is maximum)

$$P_0 = \frac{\omega_0^2 \tau_2 H_1^2}{1 + \gamma^2 H_1^2 \tau_1 \tau_2} \chi_0, \tag{9.3.24}$$

where χ_0 is the static or dc susceptibility. If $\gamma^2 H_1^2 \tau_1 \tau_2 \ll 1$, a necessary condition for accurate thermometry, then at resonance

$$P_0 = \omega_0^2 \tau_2 H_1^2 \chi_0. \tag{9.3.25}$$

(Strictly speaking, we should use B with γ; $H^2 \chi$ is correct since it comes from integrating $H\, dM = \chi H\, dH$.)

The observed signal is frequently proportional to $H_1 \chi''$, that is, to the impressed field strength times the absorptive part of the susceptibility. (In marginal oscillators, the dispersive component of the susceptibility is not observed since it simply shifts the resonant frequency of the LC circuit.) Thus the signal is dependent on H_1 and τ_2. The expressions we have given are for homogeneously broadened lines, i.e., the line width $\Delta\omega_{1/2}$ is $1/\tau_2$. For copper and platinum the line is inhomogeneously broadened by dipole–dipole interactions and in such cases one should replace τ_2 by $\pi g(\omega - \omega_0)$. Because of this broadening, the line shape will not be very susceptible to small perturbations (such as slightly inhomogeneous magnetic fields) and therefore the amplitude of the signal at

9.3. SUSCEPTIBILITY THERMOMETERS

resonance is a good measure of the nuclear susceptibility. If the resonance were susceptible to small perturbations, the integral under the whole line might have to be measured.

Since an rf field must be applied in the case of NMR detection of the susceptibility, eddy-current heating must be avoided by using fine wires. This is also necessary in order to obtain good signals because of finite skin depth. For a cylinder with complete penetration of the ac field, the heating per unit volume is given by

$$\dot{Q} = \frac{\pi^2}{16} \frac{f^2 B_{rf}^2 D^2}{\varrho} \times 10^{-9} \quad \frac{\text{erg sec } \Omega}{G^2 \text{ cm}^4}, \quad (9.3.26)$$

where B_{rf} is the peak rf field, f the frequency, D the diameter of the cylinder, and ϱ the resistivity.[92] In a field of 100 G, the resonant frequency of Cu and Pt is about 100 kHz. For copper magnet wire of 1.3×10^{-3}-cm diameter (A.W.G. # 56) and a resistivity ratio of 70,

$$\dot{Q} = 4 \times 10^{-8} B_{rf}^2 f^2 \quad \text{erg/sec cm}^3 \text{ G}^2 \text{ Hz}^2 \quad (9.3.27)$$

or 4×10^2 erg/sec cm^3 G^2 or 50 erg/sec gm G^2 at 100 kHz. Thus the rf field must be kept low, but this is true in any case to avoid changing the nuclear temperature.

A claim frequently made for the NMR detection of the susceptibility is that it is specific to the nuclear species of interest and not interfered with by impurities such as paramagnetic ions. This may not be the case, particularly if the impurity has Kondo behavior at low temperatures. In this case, the polarized cloud of conduction electrons surrounding the impurity grows spatially as the temperature decreases, in a manner not yet entirely clear. The resonant frequency of nuclei inside this cloud is significantly shifted and the amplitude of the magnetization at resonance (ω_0, B_0) decreased, that is, these nuclei are shifted into the wings of the resonance line and will not contribute to the signal. Nagasawa and Steyert[93] have shown that an Fe impurity in Cu "wipes out" 2000 Cu nuclei from the resonance line at 1.4 K. This wipe out is probably proportional to the susceptibility of the impurity and therefore, for a Kondo system, saturates at a temperature between 1/10 and 1/100 of the Kondo temperature. Below this, the nuclear susceptibility is again expected to

---

[92] Richard M. Bozorth, "Ferromagnetism," p. 778. Van Nostrand-Reinhold, Princeton, New Jersey, 1951.
[93] Hiroshi Nagasawa and W. A. Steyert, *J. Phys. Soc. Japan* **28**, 1202 (1970).

rise as $1/T$. The effect of other types of impurities as they bear on the problem of thermometry has not been investigated below 1 K. From the above it is clear that only materials of the highest purity should be used or, if doubts exist, that the thermometer should be cross-checked against one of another type.

One of two instruments is usually used for detection of the NMR signal; marginal oscillator or free-precession spectrometer. In the former, a Robinson-type oscillator[94] is usually used, a block diagram of which appears in Fig. 4. This oscillator has the advantage that very low levels

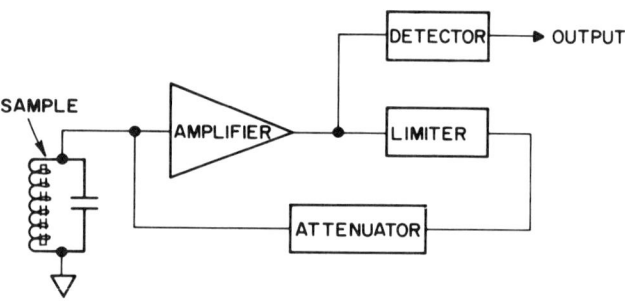

FIG. 4. Block diagram of a Robinson oscillator.

of oscillation can be maintained with good stability. With the integrated circuits having low noise and wide band widths that are presently available, these circuits are now relatively easy to construct and are compact and reliable. Frequency or field sweeping and/or modulation can be used to locate the resonance, the former causing less trouble with eddy currents but being a little trickier to put into practice and maintain constant oscillation levels over wide ranges. Varactor tuning diodes are now frequently used to vary the capacitance electronically.

In the free-precession technique, the magnetization is made to tip out of its equilibrium position and precess about the applied field. The tipping is achieved by applying a resonant, pulsed rf field perpendicular to B_0. The precessing magnetization is then detected by a coil placed orthogonal to B_0 and the driving rf field. The induced voltage in the coil, given by $d\phi/dt$, is linearly proportional to the transverse (to B_0) component of magnetization and the precession frequency. The angle of tilt is proportional to the size of B_1 and the length of the pulse. If t_w is the width of the pulse, then $\theta = \gamma B_1 t_w$. Thus the amplitude of the

---

[94] F. N. H. Robinson, *J. Sci. Instrum.* **36**, 481 (1959).

transverse component of magnetization just after the application of the pulse is $M_0 \sin \theta$, and the flux in the coil is $4\pi A n \eta M_0 \sin \theta \sin \omega_0 t$, where A is the area of one of the pickup coil turns, n is the number of turns, and η is the filling factor. Taking the time derivative, the amplitude of the induced voltage is

$$V = 4\pi C \gamma B_0^2 A n \eta \sin \theta / T \quad \text{(abvolts)} \tag{9.3.28}$$

which can be converted to volts by multiplying the above by 10^{-8} (V/abvolt). Here, C is again Curie's constant. The signal must be measured before it decays in a time τ_2 given by the inverse of the line width. (In fact, in cases such as these the envelope of the free-precession decay is just the Fourier transform of the line shape.) One must be careful to make $t_w \ll \tau_2$. For copper with $\Delta \omega_{1/2} \approx 3$ G, $\tau_2 \approx 100$ μsec so that these conditions are easy to meet, the recovery of the amplifier from saturation after the driving pulse being considerably shorter than this in a well-designed amplifier. The rotation of the magnetization through angle θ reduces the z component to $M_0 \cos \theta$, which is the same as raising the nuclear temperature an equivalent amount. (Symko[28] kept $\theta < 8°$, i.e., $\Delta T/T < 10\%$.) The signal is given by the magnetization just before the application of the pulse, so the effect of this heating of the nuclear system is simply to pump heat into the system which must be removed at a rate given by the spin–lattice relaxation time τ_1. A measurement too close in time to the preceding one would give a correspondingly higher temperature reading depending on all of the factors in obvious ways.

In either of these methods it is necessary to calibrate the amplitude of the signal at known higher temperatures and to extrapolate to lower temperatures assuming the Curie law. If $\mu B_0/kT$ is not considerably less than unity, then the Brillouin function would have to be used for the extrapolation, which would be true for copper in a field of 1 kG at a few tenths of a millikelvin. Since the nuclear signals are much smaller than those from inductance bridges and paramagnetic salts, the calibrations are frequently performed in the ³He range down to $\frac{1}{3}$ K. Thus more care is required in ensuring contact to the reference bath and to the measurement of the reference temperature.

A quite different NMR method of temperature measurement has been used above 1 K, the measurement of the hyperfine frequency in magnetically ordered materials;[2,95] this may find limited application at

---

[95] D. Gill, N. Kaplan, R. Thompson, V. Jaccarino, and H. J. Guggenheim, *Rev. Sci. Instrum.* **40**, 109 (1969).

lower temperatures. The effective field at the nucleus in such a material is approximately proportional to the local atomic moment which in turn depends on the temperature. Since this is a spontaneous moment, no external field need be applied; the frequency dependence being an intrinsic characteristic of the material, the scale is transferable to other laboratories. A major difficulty is the fact that the region of sensitivity is low, running from perhaps 20–80% of the critical temperature. However, the resolution in this region can be extremely high. Thus CMN may again be useful in the 1-mK region in an entirely different way!

9.3.2.2. Nuclear Orientation Measurements.[96–99] Nuclear orientation is unique among the thermometers presently available in that it can provide a measurement of the absolute temperature of the nuclear system without calibration at known, higher temperatures. From the anisotropy of the gamma-ray emission of certain isotopes in particular materials, the temperature can be calculated from independently measured information about the system. Measurements to an accuracy of a few percent are possible in a few minutes of counting, but to obtain the resolution desired in experiments such as heat capacity measurements requires considerably longer times. Thus the main role nuclear orientation is likely to play is

---

[96] R. J. Blin-Stoyle and M. A. Grace, "Handbuch der Physik" (S. Flugge, ed.), Vol. 42, pp. 555ff. Springer-Verlag, Berlin, 1957. A few important errors in this paper occur:

p. 565, $B_q(I) = \cdots [\cdots /(2I + \cdots + 1)!]^{1/2} f_q(I);$

see, for example, the next reference

p. 565, $B_2 = \sum_M \frac{[3M^2 - I(I + 1)]}{\cdots} W(M);$

see B. R. Judd et al., Phys. Rev. **128**, 1733 (1962).

p. 569, $U_k^{(m)} = \sqrt{\cdots} x(-1)^{I_{m-1} + I_m - L_m} \cdots;$

see Brink and Rose, Rev. Mod. Phys. **39**, 306 (1967). The first and third errors may also be found in the corrected copy of Ambler's review article reproduced in M. E. Rose (ed.) "Nuclear Orientation," p. 1. Gordon and Breach, New York, 1963.

[97] S. R. de Groot, H. A. Tolhoek, and W. J. Huiskamp," Alpha-, Beta and Gamma-Ray Spectroscopy" (Kai Siegbahn, ed.) Vol. 2, pp. 1199ff. North-Holland Publ., Amsterdam, 1966.

[98] W. A. Steyert, The Application of Nuclear Orientation to Ultralow Temperature Thermometry, in 5th Temp. Symp. (Ref. 2).

[99] P. M. Berglund, H. K. Collan, G. J. Ehnholm, R. G. Gylling, and O. V. Lounasmaa, J. Low Temp. Phys. **6**, 357 (1972).

9.3. SUSCEPTIBILITY THERMOMETERS

the establishment of the absolute temperature of some other sensor which in turn is used for resolution. Previously well-known materials, e.g., ^{60}Co in Fe, have a useful range of about one decade in temperature over which reasonable accuracy can be easily obtained, this range extending roughly from 5 to 50 mK. There is not much impetus to extend the range upward, as very reliable and highly accurate thermometry already exists above 20 mK; and methods for extending the range downward to at least 0.1 mK and, in principle, lower are now known. In fact, a tunable thermometer useful from below 1 mK to 30 to 40 mK exists (see ^{54}Mn in Cu below) as does a thermometer requiring no applied field (see ^{60}Co in Co below) for the 5–50-mK range. One disadvantage of this type of thermometer is that the count rate at temperature T must be compared with the count rate at $T = \infty$ (≥ 0.3 K in general!), though this is not nearly as great a disadvantage as the calibration that must be made with the previously discussed thermometers. One advantage is that the data is automatically available in digital form. In fact, an excellent complete system can presently be put together entirely from commercially available components for about the price of an automatically balancing cryogenic resistance bridge—and no wires need be run into the cryostat! For about half this price, a simple setup can be purchased.

In order to obtain reasonable orientation and thereby anisotropy in the 10-mK region, the field at the nucleus must exceed 100 kG. While magnets providing such fields are now available, their use for thermometry is obviously out of the question. Nuclear orientation thermometry utilizes the large nuclear polarization that occurs at low temperatures in materials with large hyperfine interactions and the anisotropic radiation pattern inherent in electromagnetic transitions. The hyperfine interaction between a local electronic moment, the nucleus of the same atom, and an applied magnetic field is usually written in the form

$$\mathscr{H} = g_J \mu_B \mathbf{B}_0 \cdot \mathbf{J} - g_I \mu_N \mathbf{B}_0 \cdot \mathbf{I} + a \mathbf{I} \cdot \mathbf{J}, \qquad (9.3.29)$$

where all of the symbols have their usual meanings and $g\mu = \gamma$, the gyromagnetic ratio. If conditions are such that m_I and m_J are good quantum numbers, then the last term above can be replaced by

$$a \mathbf{I} \cdot \mathbf{J} \rightarrow -g_I \mu_N \mathbf{I} \cdot \mathbf{B}_{\mathrm{hf}}, \qquad (9.3.30)$$

where $B_{\mathrm{hf}} = -a m_J / g_I \mu_N$. Thus the complete Hamiltonian becomes

$$\mathscr{H} = g_J \mu_B \mathbf{B}_0 \cdot \mathbf{J} - g_I \mu_N \mathbf{B}_{\mathrm{eff}} \cdot \mathbf{I}, \qquad (9.3.31)$$

where

$$\mathbf{B}_{\text{eff}} = \mathbf{B}_{\text{hf}} + \mathbf{B}_0. \qquad (9.3.32)$$

This replacement would, for example, be valid for free atoms in the Zeeman field region but not at low fields where $\mathbf{F}$ is a good quantum number. If the electronic fluctuations are very slow compared to the nuclear Larmor precession frequency, then the different nuclei in the sample would see different fields and a thermal average over the m_J would be required. This would presumably be true for a dilute gas and for very dilute paramagnetic impurities in nonmagnetic hosts under certain conditions.[100,101] If the electronic fluctuations are very rapid and all members of the system are effectively identical, as in a ferromagnet, then the average over m_J may simple be replaced by J and a single well-defined value of $\mathbf{B}_{\text{hf}}$ exists. For all of the materials we shall discuss below, $\mathbf{B}_{\text{hf}}$ and $\mathbf{B}_0$ are collinear and $\mathbf{B}_{\text{hf}}$ is isotropic. This may not be the case for rare earth atoms. Also, $\mathbf{B}_0$ here refers to the applied field as seen by the nucleus, i.e., corrected for demagnetization effects and all dipoles in- and outside the Lorentz cavity. (For a good discussion of these points, see the work of Shirley.[102])

The radiation pattern of a decaying nucleus (or of an antenna when viewed from a distance large compared to its size) can be expanded in a series of Legendre polynomials (if there is an axis of rotational symmetry). Consider a nucleus with its spin in the z direction. If it decays, say by beta decay, then some angular momentum may be carried off by the emitted particle and the nucleus may "tilt" to a new direction. Thus the polarization of an ensemble of such nuclei will be less after the decay. If another decay now occurs which we detect, usually gamma decay, then the anisotropy of the radiation pattern from the detected decay will be less than what it would have been had the preceding decay not partially reduced the polarization. Further, at any nonzero temperature, the polarization of the ensemble is not 100% and thus the fully developed radiation pattern otherwise characteristic of the decay will not be observed. It is this latter fact that allows us to measure the temperature.

Putting all of these factors together, we can write the radiation pattern as

$$W(\theta) = 1 + \sum_{k=1}^{\lambda} Q_k B_k U_k F_k P_k(\cos \theta), \qquad (9.3.33)$$

[100] D. Spanjaard and F. Hartmann-Boutron, *Solid State Commun.* **8**, 323 (1970).
[101] J. Floquet, *Phys. Rev. Lett.* **25**, 288 (1970).
[102] D. A. Shirley, S. S. Rosenblum, and E. Matthias, *Phys. Rev.* **170**, 363 (1968).

where the F_k are the coefficients describing the radiation pattern of the observed decay and depend on the multipolarity of that decay, U_k are those describing the attenuation of the orientation from preceding decays, B_k are those describing the populations of the m_I levels of the initial state before decay and are the only coefficients that depend on $\mu H/kT$, and the Q_k are coefficients that take into account the attenuation of the anisotropy due to the finite size of the detectors. The highest-order terms in the series will have $\lambda = 2I$ or $2L$, whichever is smaller, L being the angular momentum carried off in the decay and I being the initial nuclear spin. For gamma rays, parity breaking terms in the decay Hamiltonian are extremely small in all cases of interest to us, so only even k terms appear in the series. For the isotopes we shall discuss the second- and fourth-order terms are required. Since the coefficients describing the nuclear decays need not have the same sign, the counting rate at a particular angle need not be a single-valued function of temperature even if the P_k's have the same sign. However, for the isotopes of interest here, the anisotropy is single-valued in T at $\theta = 0$ or $\pi/2$. For completeness we should also point out that the above description is adequate only if the nuclei do not reorient while in one of the intermediate states as could happen if the intermediate state had a long lifetime. Again, for the isotopes of immediate interest, the intermediate state lifetimes are short enough that reorientation does not occur.

We need not discuss the individual parameters above at length in this article. Rather, we shall list the U_k and F_k coefficients in Fig. 7 below for the isotopes discussed. For further information on these matters the reader is referred to the excellent review articles of Blin-Stoyle and Grace[96] and of De Groot et al.[97] (The latter authors use a notation which differs from that of Blin-Stoyle and Grace used here by some normalization factors. These factors are pointed out in the latter article. Most workers in the field are now using the notation used here since the F_k's are used in angular correlation work and tabulations of these exist.[103] Tabulations of all the necessary Clebsch–Gordon (in the B_k's—see below), Racah (in the U_k's), and F_k coefficients needed in the orientation expressions can be found in the work of Appel.[104]) We shall discuss the temperature-dependent B_k's here and refer the reader to information on the Q_k's

---

[103] M. Ferentz and N. Rosenzweig, Table of F Coefficients. Argonne Nat. Lab., No. ANL-5324 (1955).

[104] H. Appel, "Numerical Tables for Angular Correlation Computations in α-, β-, and γ-Spectroscopy: 3j-, 6j-, 9j-Symbols, F- and Γ-Coefficients." Landolt-Börnstein, New Ser., Springer-Verlag, Berlin, 1968.

below. Values for B_k are tabulated in Blin-Stoyle and Grace but the temperature resolution is not very high. The B_k's are given by

$$B_k = (2k + 1)^{1/2} \sum_m C(IkI; m0)W(m), \qquad (9.3.34)$$

where C is a Clebsch–Gordon coefficient and $W(m)$ is the fractional population of the m_I sublevel which, for thermal equilibrium populations, is just given by the properly normalized Boltzmann factors. The arguments of the exponentials in the Boltzmann factors are $g_I\mu_N m_I B_{\text{eff}}/kT = (\mu B/IkT)m_I$, where μ_N is again the nuclear magneton and μ is the dipole moment of the nucleus. The great advantage of nuclear orientation thermometry is that $\mu B/Ik$ is known to high accuracy for the isotope–host combinations of interest.

The apparatus in its simplest form consists of a detector (along the magnetic field axis $\theta = 0$ since the anisotropy is largest in this direction), a high-voltage supply for the detector, a preamplifier mounted on or near the detector, an amplifier which shapes the pulses and has good overload recovery characteristics, a single channel analyzer (SCA), a scaler, and a timer (see Fig. 5). A more sophisticated version would replace the SCA and scaler with a multichannel analyzer (MCA) and, depending on the features of the MCA, would replace the amplifier and timer as well. The high-voltage supply needs to be highly stable and not droop at high current loads as might occur at high count rates. The

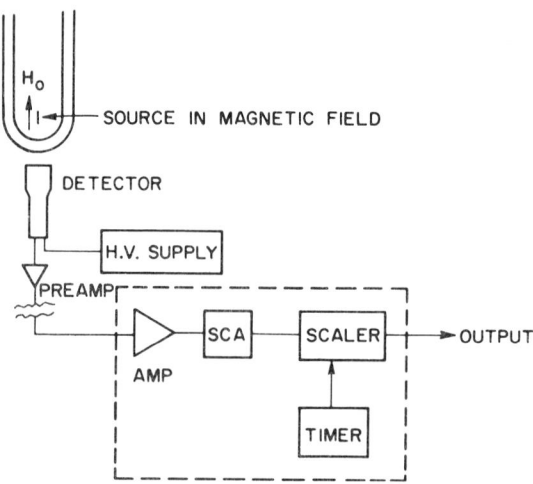

FIG. 5. Block diagram of a nuclear orientation apparatus for determining $W(0)$. Equipment inside dashed block may be replaced by an MCA.

9.3. SUSCEPTIBILITY THERMOMETERS

supplies built into MCAs are not recommended by the author except for monitoring and test purposes as their impedance is usually quite high. The detector for thermometry should be a 7.5 cm × 7.5 cm NaI(Tl) scintillation crystal for most efficient detection. These can be obtained commercially as light-tight units with built-in photomultipliers. The finite size correction factors Q_k have been tabulated by Yates.[105] The high-resolution Ge(Li) detectors presently available make corrections much simpler, but, for reasonably priced units, have very low efficiencies compared to scintillation crystals. For single isotope sources with resolved gammas, the high resolution is not required.

A multichannel analyzer is a great help in setting up the experiment. We recommend a unit with a built-in SCA and with multichannel scaling (MCS) capability. Many such units now contain the necessary amplifier as well. The built-in timer for advancing channels in the multiscaling mode (in which the analyzer accumulates counts from the SCA for a predetermined length of time and then advances to the next channel) frequently offers only short count periods and, therefore, one may still require an external timer. With these capabilities, the experimenter can very simply display the full spectrum of the source and spot any problems in the setup quickly. The SCA can be easily set to produce an output only on pulses in the photopeak. (This can be done with an oscilloscope with the simpler system.) While it may seem statistically better to accept all of the counts, there are two disadvantages in this: the background and electronics noise increase at lower energies, and the scattering from the cryostat appears in the Compton region. Both of these form isotropic backgrounds which would be large if the whole spectrum were used (see Fig. 6). An analyzer has a further built-in feature of considerable value—a logic circuit that cuts off the timer when the analyzer is busy analyzing a pulse. Thus the counting can be done in "live" time and not "clock" time and deadtime corrections for the analyzer need not be made. This feature is operative only in the pulse height analysis (PHA) mode and not the MCS mode. The analyzer time resolution in the latter mode is usually very good, however, and such corrections need be made only for very high count rates. In any case, with either the simpler apparatus or the multichannel analyzer, the deadtime of the system should be measured and the necessary corrections

---

[105] M. J. L. Yates, Finite solid angle corrections, *in* "Perturbed Angular Correlations" (E. Karlsson, E. Matthias, K. Siegbahn, eds.), Appendix 4. North-Holland Publ., Amsterdam, 1964.

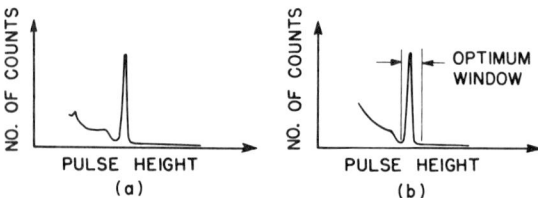

FIG. 6. Optimum window for data accumulation in nuclear orientation thermometer: (a) spectrum of source alone, (b) observed spectrum including low-energy background, noise, and scattering.

applied to the data if required. If only scintillation crystals are to be used, a 256-channel analyzer provides adequate resolution. At 100 sec/channel in multiscaling, this will accumulate data for 7 hr whereas a 400-channel unit for 11 hr, etc. If calibrations of two different thermometers with different radioisotopes are to be performed, a multichannel analyzer is essential.

With regard to deadtime a few precautions should be made. If the amplifiers are frequently overloaded, the system deadtime will be increased by the time it takes these to recover. Thus high-energy backgrounds should be avoided. Any logic circuit contributes deadtime. With scintillation crystal spectrometers the only such circuit usually used is the analyzer (SCA or MCA). With Ge(Li) detectors, sophisticated logic is often used to maintain the high resolution inherent in the detector. Units such as pile-up rejectors and high-resolution biased amplifiers contribute 3–30 or more μsec/pulse deadtime for which the analyzer "live time" switch cannot correct unless custom modifications are made.

The usual procedure is to take "warm" counts for normalization of the data as frequently as possible, whenever the sample is above 0.3 K, but at least at the beginning and end of the run. A very subtle error can creep into the data. If during the warm counts (or at any other data-taking period) the liquid helium level is allowed to drop, the apparatus may expand and the source will move closer to the detector increasing the count rate. This is most likely to happen systematically during the taking of warm counts and would introduce an obvious error. Any apparatus should be checked for this possibility experimentally. For example, an expansion of 1 mm (0.1%/m) with the detector 10 cm from the source will cause a 1% error in $W(0)$. The detector must of course be placed accurately on the axis.

If the experiment lasts for several days, the source half-life may be short enough that the count rates will need to be corrected to a particular time.

9.3. SUSCEPTIBILITY THERMOMETERS

The thermometer will show a time constant at least as long as the nuclear relaxation time, and this should be kept in mind when choosing the frequency with which conditions are changed and the number of count periods before a new change is made.

A final correction applicable at high count rates when two sources are used should be mentioned. Coincidences between two (or more) gammas with energies which sum to the photopeak region will be counted as a signal.[106] This should not be a source of error when one source is used as a thermometer.

Most of the corrections above are either negligible, as in the case of deadtime and coincidence, for the source strengths used in thermometry or are easily avoided, as for expansion, by the careful experimenter. Thus nuclear orientation can provide a straightforward technique for obtaining absolute temperature measurements with an accuracy of a few percent.

If the anisotropy is very large, i.e., $W(0)$ is small, some of the above corrections can lead to fractionally large changes in $W(0)$. This would be particularly true of background subtraction errors and detector placement errors. For low anisotropy, i.e., $W(0) \approx 1$, it becomes difficult to obtain good enough statistics for an accurate temperature determination. We should also notice (see Fig. 8 below) that the sensitivity of $W(0)$ to T decreases at high and low $W(0)$, as it must. Thus the most useful range for thermometry is $0.2 \lesssim W(0) \lesssim 0.9$, though less-accurate determinations can be made outside these limits.

The most popular thermometers used to date have consisted of radioisotopes dissolved in ferromagnetic hosts, usually iron. The first such observations were on ^{60}Co in Co metal.[107,108] Samoilov et al.[109] discovered that impurities dissolved in iron pick up large induced hyperfine fields. These fields range from tens of kilogauss to over a megagauss, making large orientation in the 10-mK region possible. Considerations in the choice of isotope for a particular measurement are the existence of resolved gammas with known nuclear decay coefficients, the value of $\mu B_{\text{eff}}/Ik$, the energy absorbed by the sample from the decay (such as

---

[106] J. A. Cameron, J. P. Compton, R. A. G. Lines, and G. V. H. Wilson, *Proc. Phys. Soc. (London)* **87**, 927 (1966). See the Appendix.

[107] G. R. Khutsishvili, *Zh. Eksp. Teor. Fiz.* **29**, 894 (1955) [*English transl. Sov. Phys.—JETP* **2**, 744 (1956)].

[108] M. A. Grace, C. E. Johnson, N. Kurti, R. G. Scurlock, and R. T. Taylor, *Commun. Conf. Phys. Basses Temp., Paris*, p. 263 (1955).

[109] B. N. Samoilov, V. V. Sklyarevskiĭ, and E. P. Stepanov, *Zh. Eksp. Teor. Fiz.* **36**, 644 (1959) [*English transl. Sov. Phys.—JETP* **9**, 448 (1959)].

beta heating), and the reliability of the metallurgy of making the sample.

Two isotopes are presently popular for thermometry purposes: ^{54}Mn and ^{60}Co. In the range of temperatures presently achieved, one or the other of these can be used. The vital statistics of each of these is given in Fig. 7. First we notice that ^{54}Mn decays by electron capture and therefore causes very little heating, about 5 keV/decay in Auger processes. The heating from a 1.0-μCi source of ^{54}Mn is about 3×10^2 erg/min[110,111]

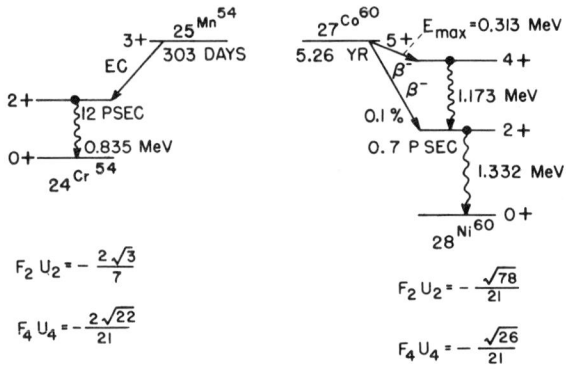

FIG. 7. Decay schemes of ^{54}Mn and ^{60}Co.

and such a source would give a count rate of 2000/min in the photopeak in a 7.5 cm × 7.5 cm NaI detector 10 cm away.[112] Cobalt on the other hand decays by beta decay with an average beta energy of roughly 0.1 MeV, producing a heating rate of about $\frac{1}{2}$ erg/min/μCi.[99] These heating rates do not include γ-ray absorption, which will depend critically on the thickness and subtended solid angle of materials around the thermometer. The count rate would be about the same as for ^{54}Mn in spite of the two gammas per decay since the detector efficiency decreases with energy.

While the two gammas of ^{60}Co will not be fully resolved with a scintillation crystal, this is not important since for both gammas the F_k and U_k coefficients are the same to the accuracy they have been measured. Thus both can be counted and the statistics thereby improved. The lower energy one will lie on top of some of the Compton scatter from the higher

---

[110] W. P. Pratt, Jr., R. I. Schermer, and W. A. Steyert, *J. Low Temp. Phys.* **1**, 469 (1969).
[111] James R. Sites, H. A. Smith, and W. A. Steyert, *J. Low Temp. Phys.* **4**, 605 (1971).
[112] R. Swinehart and W. Weyhmann, unpublished.

9.3. SUSCEPTIBILITY THERMOMETERS

one, but this part of the spectrum is entirely due to scattering in the detector and not from the cryostat. The lifetime of ^{54}Mn is such that decay corrections will need to be made if the experiment lasts more than a day or two, whereas for ^{60}Co this time is longer than a week—assuming about 1% accuracy is required. The nuclear coefficients are listed in the figure.

The measurement of $\mu B/I$ no longer consists of independent measurement of μ and B. In the past, B_{hf} was typically obtained from nuclear resonance on a stable isotope of the desired impurity in the requisite host and μ was obtained from atomic beam or EPR experiments. The errors in the measurement of the moment of a radioisotope are usually somewhat larger than for stable ones because of the quantity of material that can be used. Now very precise measurements of $\mu B/I = h\nu$ are made by finding the frequency at which an applied rf field partially destroys the nuclear orientation.[113] This can be measured to better than 0.1%. For the specific cases we discuss below, the hyperfine field is antiparallel to the applied field, i.e., negative. In these cases $B_{\text{eff}} = B_0 + B_{\text{hf}} = B_0 - |B_{\text{hf}}|$.

Foils are generally used in nuclear orientation because their great surface-to-thickness ratio aids thermal contact. When using ferromagnetic hosts such as iron or cobalt, the applied field (called the *polarizing field* since its purpose is to polarize the domains) should be in the plane of the foil since the demagnetizing factor is very unfavorable normal to the plane. Foils of one to several thousandths of an inch (~ 0.1 mm) thickness are used. With polycrystalline iron, fields of 1.5 kG or greater are required to ensure saturation. Cobalt requires considerably more and is seldom used in polycrystalline form.

The metallurgy of the samples is very important since, if some fraction of the radiospecies does not reside in the proper sites, the orientation would be represented by an unknown admixture of several hyperfine fields. As a case history, the first experiments on ^{54}Mn in iron were performed by diffusing the Mn in at a moderate temperature, about 900°C and the hyperfine field deduced.[114] Later NMR results on the stable isotope became available[115,116] and the measured field from nuclear orientation appeared too low. It is now clear that Mn is rather difficult

---

[113] J. E. Templeton and D. A. Shirley, *Phys. Rev. Lett.* **18**, 240 (1967).

[114] J. A. Cameron, R. A. G. Lines, B. G. Turrell, and P. J. Wilson, *Phys. Lett.* **4**, 323 (1963).

[115] Y. Koi, A. Tsujimura, and T. Hihara, *J. Phys. Soc. Japan* **19**, 1493 (1964).

[116] V. Jaccarino, L. R. Walker, and G. Wertheim, *Phys. Rev. Lett.* **13**, 752 (1964).

to work with and in iron it is best to make the diffusions at greater than or equal to 1100°C[117] or actually melt the sample, though Mn tends to evaporate relatively quickly at these temperatures. We have alluded to the process frequently used in preparing samples for nuclear orientation: diffusion. Onto a piece of foil of the host material a drop of radioisotope solution is evaporated. For ^{60}Co and ^{54}Mn, this is probably a chloride solution. The foil is then placed in an oven and the diffusion process carried out in a H_2 (or dilute H_2 in an inert gas to reduce the hazard of explosions) atmosphere or in a vacuum. The sample is held at the diffusion temperature long enough for considerable penetration (at least several hours at 1200°C and longer as the temperature is lowered) and then frequently annealed at about 1000°C for several more hours. The foil is etched after cooling, the amount of activity being lost being a good indicator of the efficiency of the diffusion process. Melting can be conveniently performed in a rf furnace in a pure alumina crucible or boat. The sample may be held in the molten state for a few tens of minutes or less. After melting, it will be necessary to roll the sample into a foil again and anneal it. It appears that the chloride adequately dissociates at the temperatures used and so electroplating of the radioisotope is not required.

The foil can be attached with solder to a metal piece, usually copper, that acts as a conductor to the part whose temperature is to be measured. When the thermometer is to be used in high fields as required with iron, indium can be used as its superconductivity is quenched at 293 G. There is also evidence that indium remains a good thermal conductor in thin layers after such a field is removed, presumably because of flux trapping. A Cd–Bi eutectic has been found to excel over most other solders,[118] presumably because of the low superconducting transition temperature and because it phase separates into normal Bi and superconducting Cd. It should be noted that this solder has virtually no mechanical strength, but is adequate for mounting a foil onto a flat surface. The action of the flux used in soldering should be mild so that radioisotope is not leached from the foil. In time, activity will accumulate in the solder on the copper as repeated samples are mounted, so as much of the solder as possible should be removed each time and the mounting piece checked occasionally for background contributions.

[117] J. A. Cameron, I. A. Campbell, J. P. Compton, R. A. G. Lines, and G. V. H. Wilson, *Proc. Phys. Soc. (London)* **90**, 1077 (1967).

[118] W. A. Steyert, *Rev. Sci. Instrum.* **38**, 964 (1967).

9.3. SUSCEPTIBILITY THERMOMETERS

The most frequently used thermometer is ^{60}Co in Fe. No metallurgical difficulties have been experienced with the diffusion of the Co into the Fe foil and the calibrations always seem to agree. Along with the ^{60}Co in single-crystal Co discussed below, it can be trusted. A plot of the anisotropy versus temperature for this thermometer appears in Fig. 8. The NMR resonance of the ^{60}Co in Fe occurs at 165.6 MHz in a 675-G applied field.[113] The hyperfine field B_{hf} is negative, as usual, and so increasing B_0 above that needed for saturation decreases B_{eff} ($B_{\text{eff}} = B_0 - |B_{\text{hf}}|$).

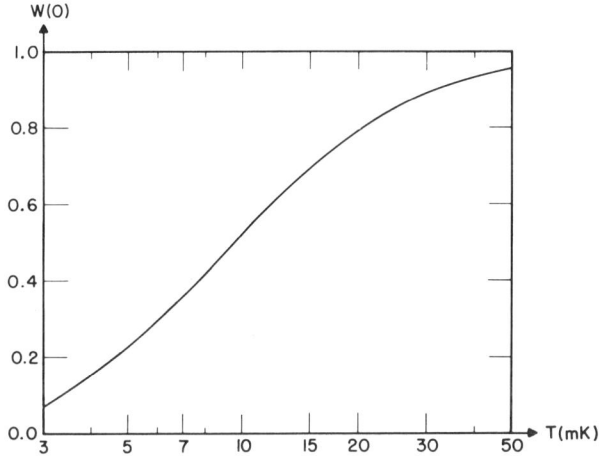

FIG. 8. $W(0)$ versus T for ^{60}Co in Fe with $B_{\text{app}} = 3$ kG.

The magnetic anisotropy of pure hexagonal cobalt is very high. (The stable phase of cobalt is hexagonal below about 390°C. In polycrystalline material some cubic phase can remain frozen in.) The magnetization is directed along the c axis. At the ends of domains there must exist closure domains and these will be perpendicular to the c axis, thereby reducing the gamma-ray anisotropy from the ideal value for all domains collinear. If a slab were cut from a single crystal such that the length L of slab in the c direction were 1 cm, then from information provided by Morrish[119] we calculate that the closure domains occupy only 0.03% of the total volume. This fractional volume is inversely proportional to L. Thus single-crystal cobalt, properly cut, can be used without the application

---

[119] Allan H. Morrish, "The Physical Principles of Magnetism," Chapter 7. Wiley, New York, 1965.

of a magnetic polarizing field. Also, ^{60}Co can be introduced simply by irradiation with neutrons in a pile. After irradiation, the sample should be annealed for several days below 390°C to remove radiation damage. One disadvantage of cobalt as a host is that it is 100% ^{59}Co with a hyperfine splitting of 10 mK; thus its heat capacity is not negligible. Iron has only 2.2% ^{57}Fe with a splitting of 2 mK, the rest being "inert" isotopes. For the nuclear sublevel splitting, one obtains 131 MHz for ^{60}Co in Co(hcp) at $T = 0$.[120]

Manganese-54 can be used in iron where the heating of the cobalt-60 is intolerable. The metallurgical precautions mentioned above need to be taken and the thermometer might be checked against a cobalt one, though this is probably not necessary if the sample has been melted or diffused above 1100°C. The NMR resonance frequency in this case is $\mu B_{hf}/I = 189.9 \pm 0.3$ MHz, again in an applied field of 675 G.[113] The hyperfine field is negative.

The Kondo effect holds out promise of providing a tunable thermometer usable from above 30 mK to below 1 mK. This is a result of the fact that the hyperfine field is strongly dependent on the applied field. For temperatures far below the Kondo temperature, the hyperfine field is temperature-independent and linearly proportional to the applied field for low-field values, $\mu_B B_0 \ll kT_K$. As B_0 becomes very large, the hyperfine field slowly approaches its $T = \infty$ value. At low temperatures and very low fields, there is some conjecture that the effective field approximation breaks down and the full Hamiltonian must be used.[100,101] This has apparently not been observed for Mn in Cu. Once this point is clarified, it will be possible to extend the use of Kondo thermometers to temperatures far below 1 mK. Because of the large spatial extent of the conduction electron polarization clouds surrounding the magnetic impurity, the concentration must be kept very low to preclude interactions between the impurity atoms. The rule of thumb generally used is that the concentration should be kept well below $50T_K$ ppm/K.

At present ^{54}Mn in Cu appears to be the best combination for thermometry purposes. Since ^{54}Mn is obtained carrier free, concentrations in the 10^{-9} range represent completely adequate activity (1–10 μCi) for good counting statistics. The Kondo temperature of this combination is

---

[120] This result is obtained by scaling the NMR result on ^{59}Co of Y. Koi, A. Tsujimura, T. Hihara, and T. Kushida, *J. Phys. Soc. Japan Suppl. B-I* **17**, 96 (1962) by the g-factor ratio $g(60)/g(59)$ obtained from the work of W. Dobrowolski, R. V. Jones, and C. D. Jeffries, *Phys. Rev.* **101**, 1001 (1956).

9.3. SUSCEPTIBILITY THERMOMETERS

near 60 mK and so the field necessary to approach saturation is only a few kilogauss. As mentioned above, complete saturation is very difficult to achieve and the hyperfine field is still rising slowly as the applied field is increased from 20 to 40 kG. The saturation hyperfine field appears to be about -300 kG[121] and at low fields ($B_{app} \leq 140$ G), $B_{eff}/B_0 = 675 \pm 41$.[122] Since the hyperfine field is negative, increasing the field above several kilogauss will actually decrease the anisotropy. Applied fields of 60 to 600 G would be ideal for thermometry in the 1–10-mK range, the anisotropy $W(0)$ being approximately $\frac{1}{2}$ at $T = 1.5$ mK in an applied field of 60 G, 3 mK in 120 G, and 5 mK in 300 G. The metallurgy of this sample is not completely solved. Results on different samples can differ by much more than the statistical error. Manganese readily oxidizes, even when dissolved in another metal like copper; and it may be that this oxidation reduces the amount of Mn in proper sites significantly.[123] Investigations are presently underway in our laboratory to elucidate this point and to find out if preparation in a high vacuum will solve the problem. It may be that cold storage will also be required.

As the temperature is lowered still further, one will want smaller hyperfine fields. This can be achieved by using materials such as Mn in Cu at still lower applied fields, though low fields become harder to set accurately if superconducting magnets are present. Another solution is to use materials with higher Kondo temperatures, thereby requiring higher applied fields. Effective fields at the nucleus of less than 1 kG are required in the 0.1-mK region, so that brute-force orientation (solely the applied field and no hyperfine field) is all that is required.

Notes added in proof:

1. ^{60}Co in Ni has a low hyperfine splitting (69.1 MHz) and may be useful as a thermometer to 2 mK. (See Bacon, Barclay, Brewer, Shirley, and Templeton, *Phys. Rev.* **B5**, 2397 (1972) for listing of the splittings and relaxation times of the various Co isotope thermometers.)

2. Krane, Murdock, and Steyert (*Phys. Rev. Lett.* **30**, 321 (1973)) have reported that various impurities in iron do not align collinearly with the host magnetization at low fields even though the foil is magnetically saturated. The author and co-workers have indeed found a 25% discre-

---

[121] R. J. Holliday, R. Swinehart, and W. Weyhmann, unpublished.

[122] Calculated by the author from the data of W. P. Pratt, Jr., R. I. Schermer, and W. A. Steyert, *J. Low Temp. Phys.* **1**, 469 (1969). It was assumed that the intercept is at zero effective field.

[123] D. H. Howling, *Phys. Rev.* **155**, 642 (1967).

pancy in temperature at 3–5 mK between ^{60}Co in Ni and ^{54}Mn in Fe. The correction is small at temperatures above 15 mK but increasingly serious as the temperature decreases. Either large applied fields must be used or corrections applied when working at very low temperatures.

9.3.2.3. Other Nuclear Susceptibility Measurements. In the Mössbauer effect, the intensity of radiation from different nuclear sublevels is directly observed, as is the energy splitting. Such an observation is to be preferred over a simple measurement of the anisotropy of the gamma radiation as in nuclear orientation thermometry, since it provides several checks on the thermometer. It ensures that the radiation being observed is from nuclei in the proper hyperfine field and it checks that the nuclear system is in internal equilibrium by providing ratios between the populations of several sublevels.[124,125] In practice, however, it is very difficult to use in its present form, mainly because of the high level of activity of the radioisotope required to obtain good counting statistics in reasonable periods of time. It is also necessary to use windows in the dewars transparent to low-energy gamma rays (usually 14-keV gammas from ^{57}Fe) and somewhat more expensive data acquisition equipment (not only is a multichannel scaler required but also a Mössbauer drive unit). Data analysis would be more cumbersome to perform quickly during a run without the help of a small computer, though some moderately accurate indication of temperature could be had by inspection. On the whole, the time required to make a measurement without a hot source prohibits the use of this technique at the present time.

A much more promising technique and one that may rapidly come into dominance over virtually all others where resolution is required is the measurement of the static susceptibility using SQUID's (superconducting quantum interference devices).[90] [The operation of such devices is detailed elsewhere in this volume (Chapter 5).] Since an output can be derived from these devices which changes with changes of the flux in a pickup coil (in which a paramagnetic sample in a magnetic field can be placed), a current can be fed back into a coil coupled to the superconducting circuit to keep the net flux in the circuit constant. In obvious

---

[124] J. G. Dash, R. D. Taylor, P. P. Craig, D. E. Nagle, D. R. F. Cochran, and W. E. Keller, *Phys. Rev. Lett.* **5**, 152 (1960).

[125] R. Dean Taylor, Thermometry below 0.1 K: A comparison of the Mössbauer effect thermometer and the γ-ray anisotropy thermometer, *in* "Temperature, Its Measurement and Control in Science and Industry." Based on the *5th Temp. Symp., Washington, D.C.*, Instrum. Soc. Am., Pittsburgh, 1973.

ways, then, measurement of this current provides a measurement of temperature if the susceptibility is temperature-dependent. Because of the high flux sensitivity of the SQUID's, nuclear paramagnetism is adequately strong in the region below 0.1 K for highly accurate measurements and calibration in the helium-3 vapor pressure range is presently used. Since the total susceptibility of the sample is being measured, extreme care must be exercised to avoid construction materials with temperature-dependent susceptibility in the region of use. The sample itself must also be chosen with the same care and only the highest purity materials used. (The same comments made in the section on NMR apply here.) The superconducting current must also be well shielded from spurious magnetic field fluctuations, and the SQUID itself from rf interference. The measuring field and shielding is obtained by placing the sample in one of the coils of a counter-wound pair and both the coils and sample in an Nb tube with a small trapped flux.

Research on these thermometers is presently intense, and so the reader should consult the current literature for all details on their construction.[1] For example, Hirschkoff et al.[126] have compared measurements on high-purity copper using a SQUID with 17-Hz bridge measurements on CMN, and Bishop et al.[127] have studied several metals for their usefulness in such static magnetization measurements. The importance of these devices is inversely proportional to the brevity of this discussion, and we must repeat that these devices are the most important new tool of the low-temperature physicist for many applications, including thermometry.

Note added in proof:

Giffard, Webb, and Wheatley, *J. Low Temp. Phys.* **6**, 533 (1972), have thoroughly discussed electrical and magnetic measurements using SQUID's as they apply to susceptibility measurements as well as others discussed at the end of this chapter.

9.4. Miscellaneous Thermometers

We group in this section a host of thermometers of varying degrees of promise. We do this not because we think that they are, in any particular case, less important than those discussed above but rather because

---

[126] E. C. Hirschkoff, O. G. Symko, L. L. Vant-Hull, and J. C. Wheatley, *J. Low Temp. Phys.* **2**, 653 (1970).

[127] J. H. Bishop, E. C. Hirschkoff, and J. C. Wheatley, *J. Low Temp. Phys.* **5**, 607 (1971).

their merit is not yet as clear or their development is not yet as advanced as those discussed at length. This section then contains promising devices which may in the future be of considerable importance. For example, one of these, the phase boundary between liquid and solid helium-3, is capable of high resolution and can, in principle, be calibrated to read in absolute temperature. It is not yet clear, however, how well one can make thermal contact to it below 20 mK. Noise thermometry is discussed, perhaps improperly, in the section on superconductors because of the use of Josephson junctions in the detection scheme.

9.4.1. Thermocouples

Surprisingly, one device that has in the past received less and less attention as a thermometer at low temperatures must once again be taken seriously, namely the thermocouple. Two developments account for this new potential: alloys with high thermopower at low temperatures and SQUID's.

Very dilute magnetic impurities in noble metal hosts give rise to giant thermopowers at low temperatures. Some years ago 2.1 at.% Co in Au was popular for use between 4.2 and 77 K because it had about three times the output of constantan when referenced to copper. It proved to be unsuitable in the long run because the cobalt at this concentration is above the solubility limit at room temperature, and so the calibration slowly changed with time. A material with equally large emf is very dilute Fe in Au, 0.02–0.07 at.%, the lower end of that range giving higher emf's at low temperatures.[128] At very low temperatures, the preferred reference material will be a superconductor with its vanishing thermoelectric power and low thermal conductivity. It is also the natural choice for connection to the SQUID voltmeter. In the range above 0.3 K, the thermopower of 0.03 at.% Fe in Au is given by[129]

$$S = -(10.0 + 2.8 \ln T) \quad \mu V/K \qquad (9.4.1)$$

which obviously cannot be extrapolated to ultralow temperatures. It is likely that the thermoelectric power bottoms out just as the resistivity saturates, the logarithmic region extending about $1\tfrac{1}{2}$ decades below the Kondo temperature (which is about 0.24 K for Fe in Au[130]). At lower

---

[128] D. K. C. MacDonald, W. B. Pearson, and I. M. Templeton, *Proc. Roy. Soc.* (*London*) **A266**, 161 (1962).

[129] R. Berman, J. Kopp, G. A. Slack, and C. T. Walker, *Phys. Lett.* **27A**, 464 (1968).

[130] J. W. Loram, T. E. Whall, and P. J. Ford, *Phys. Rev. B* **2**, 857 (1970).

9.4. MISCELLANEOUS THERMOMETERS

temperatures, another alloy may then be chosen; Mn in Cu having a Kondo temperature of about 60 mK would be the next lower choice.

These types of thermocouples are very sensitive to magnetic fields[129,131] and so must either be shielded or placed at an isothermal position outside the field. (The suggestion of Richards et al.[132] is equivalent to the latter.) It will also be difficult to obtain alloys in which the concentration of the "impurity" is well enough determined that a universal thermopower curve can be used.

In these devices, emf's much less than 1 µV must be measured with high accuracy. Using SQUID's, voltage measurements can be made which approach a resolution limited only by the noise due to the resistance of the source. A commercial device manufactured by Keithley Instruments† claims a sensitivity of better than 3 pV p-p for source impedances less than 3×10^{-7} Ω, still over two orders of magnitude larger than the theoretically perfect device at 4 K. Ries and Satterthwaite[133] have constructed a superconducting parametric amplifier capable of near theoretical resolution of 4×10^{-14} V with a 4×10^{-7} Ω source at 4.2 K. Clearly the thermocouple is now a useful device down to 0.1 K with the near theoretical performance superconducting devices.

9.4.2. ³He Melting Curve[134,135]

The melting curve of ³He shows a minimum in the region of 0.3 K and from there to absolute zero the melting pressure rises from roughly 29 to 34 atm. Adams and his co-workers have measured this curve and discussed its usefulness in thermometry.[136,137] (Many other workers have devoted considerable effort to measuring the properties of ³He along this curve and references to their work may be found in the papers mentioned and in reviews on the subject. Recent work below 30 mK is

---

[131] R. Berman, J. C. F. Brock, and D. J. Huntley, *Cryogenics* **4**, 233 (1964).

[132] D. B. Richards, L. R. Edwards, and S. Levgold, *J. Appl. Phys.* **40**, 3836 (1969).

[133] Roger P. Ries and C. B. Satterthwaite, *Rev. Sci. Instrum.* **38**, 1203 (1967).

[134] William E. Keller, "Helium-3 and Helium-4," Chapter 9. Plenum Press, New York, 1969.

[135] J. Wilks, "The Properties of Liquid and Solid Helium" Chapter 22. Oxford Univ. Press (Clarendon), London and New York, 1967.

[136] R. A. Scribner, M. F. Panczyk, and E. D. Adams, *J. Low Temp. Phys.* **1**, 313 (1969); *Phys. Rev. Lett.* **21**, 427 (1968).

[137] R. A. Scribner and E. D. Adams, *Rev. Sci. Instrum.* **41**, 287 (1970).

---

† Keithley Instruments, 28775 Aurora Road, Cleveland, Ohio.

reported in Ref. 138.) Since high resolution can be achieved in pressure measurements, temperature sensitivity of 1 μK and better between 250 and 2.5 mK can be realized. A sample is trapped in a chamber, one end of which forms one plate of a capacitance pressure gauge, by the blocked capillary technique. If the specific volume is properly chosen, the solid and liquid will coexist over a very wide temperature range and the melting curve will be automatically traversed as the temperature changes. Scribner and Adams[137] suggest that a sample of volume about 25 cm³/mole will follow the melting curve for all temperatures below 0.86 K. Continuous readout on a capacitance bridge makes this a particularly convenient device.

The establishment of the temperature scale in this case can proceed from either of two tacks. The classical method of establishing the $T^*(p_m)$–T relation by measuring entropy and heat could be applied using

$$T = (dQ/dp_m)/(dS/dp_m). \qquad (9.4.2)$$

It is also possible to integrate up the Clausius–Clapeyron equation. Confidence in the theory of the entropy of solid and liquid ³He is growing and extrapolations below the temperature at which these are measured may be performed with adequate accuracy in the near future. Thus an accurate p_m–T_m curve may soon exist based on both experimental and theoretical information.

The most promising use of this thermometer is in research on ³He itself. Thermal contact to ³He contained in independent chambers becomes increasingly difficult below 20 mK. The heat capacity of such a thermometer, while relatively small if small quantities of ³He are used, is larger than resistors or metals used in nuclear susceptibility measurements.

9.4.3. Superconductors

Several thermometers have been based on temperature-dependent properties of superconductors. To date, no reliable thermometers have been so made for use below 0.1 K.

The most obvious method is to determine the phase boundary between the normal and superconducting states of a pure superconductor as a

---

[138] R. T. Johnson, O. V. Lounasmaa, R. Rosenbaum, O. G. Symko, and J. C. Wheatley, *J. Low Temp. Phys.* **2**, 403 (1970).

function of applied field.[139] Since the strong diamagnetism associated with the superconducting state can be easily observed, the transition can be readily detected. Because of the well-known parabolic shape of the phase boundary, the sensitivity decreases as the temperature decreases. (Of course, as T approaches zero, the slope of the boundary dH/dT also approaches zero by the Third Law.) Thus the range of any particular superconductor is limited. Black[140] has measured the phase boundary of tungsten below its critical temperature of 15.4 mK and finds that it closely follows ideal BCS behavior. One must be careful to avoid pronounced supercooling effects, particularly if thermal contact is not excellent.[141]

The temperature dependence of the energy gap has also been used in the 0.1–1.0-K region.[142] A tunnel junction between Al and Ag was used with Al_2O_3 forming the insulating layer. Here, of course, some heating is generated by the current and so its usefulness at lower temperatures is not clear. Again, any particular superconductor can be used only over a limited temperature range, about a decade below its critical temperature.

The noise in a Josephson junction shunted by a resistance is a linear function of temperature and its measurement would provide a useful thermometer.[143] Accuracy within the statistical limits has been achieved with this type of device.[144]

9.4.4. Capacitors

Considerable excitement has been generated over the last few years by the development by Dr. W. N. Lawless of Corning Glass Works of a capacitor which exhibits a surprising temperature dependence of capaci-

---

[139] R. P. Ries and D. E. Mapother, Magnetic Thermometry Using a Superconductor, *in* "Temperature, Its Measurement and Control in Science and Industry." Based on the *5th Temp. Symp., Washington, D.C.*, Instrum. Soc. Am., Pittsburgh, 1973.

[140] W. C. Black, Jr., *Phys. Rev. Lett.* **21**, 28 (1968).

[141] R. T. Johnson, O. E. Vilches, J. C. Wheatley, and S. Gygax, *Phys. Rev. Lett.* **16**, 101 (1966).

[142] J. W. Bakker, H. van Kempen, and P. Wyder, *Phys. Lett.* **31A**, 290 (1970).

[143] A. H. Silver, J. E. Zimmerman, and R. A. Kamper, *Appl. Phys. Lett.* **11**, 209 (1967).

[144] Robert A. Kamper, Survey of Noise Thermometry, *in* "Temperature, Its Measurement and Control in Science and Industry." Based on the *5th Temp. Symp., Washington, D.C.*, Instrum. Soc. Am., Pittsburgh, 1973.

tance below 0.1 K.[145-147] Other dielectrics such as doped alkali halide crystals show a monotonically decreasing sensitivity to temperature changes at low temperatures,[148] though they have proven useful in the dilution refrigerator range.[149] The type 1100 Corning capacitor,[†] however, goes through a minimum between 60 and 100 mK. From 30 mK to near the minimum the change in capacitance appears to be proportional to $1/T$,[146] while from 7 to 30 mK the capacitance decreases linearly as the temperature increases.[150] Such a capacitor from the original batch produced by Dr. Lawless has been measured in the author's lab to have a room-temperature capacitance of about 5900 pF, 24,600 pF at 77 K, 16,400 pF at the minimum, and to increase by 50 pF between the minimum and 7 mK. This capacitor has a loss tangent of 2.3×10^{-3} at 7 mK, so with the 5 mV rms of 400 Hz required across the capacitor to obtain a resolution of a few tenths of a picofarad, the power dissipated in the capacitor is roughly 0.5 pW. The major attractiveness of these devices is their apparent insensitivity to magnetic fields. With regard to thermal contact, they will probably prove similar to resistors.

Note added in proof:

Work in the author's laboratory shows that the capacitance of these devices is voltage dependent, so great care must be exercised in their use.

9.5. Conclusions

The giant leap forward that ultralow temperature physics has taken in the last few years has resulted in new demands on thermometric techniques and devices. With the growth of the dilution refrigerator out of its infancy and into a stock tool of the trade and the rapid development of Pomeranchuk and nuclear cooling, measurements in the 1-mK region

---

[145] W. N. Lawless, *Rev. Sci. Instrum.* **42**, 561 (1971).
[146] W. N. Lawless, R. Radebaugh, and R. J. Soulen, *Rev. Sci. Instrum.* **42**, 567 (1971).
[147] L. G. Rubin and W. N. Lawless, *Rev. Sci. Instrum.* **42**, 571 (1971).
[148] A. T. Fiory, *Rev. Sci. Instrum.* **42**, 930 (1971).
[149] R. C. Richardson, private communication.
[150] D. Bakalyar, R. Swinehart, W. Weyhmann, and W. N. Lawless, *Rev. Sci. Instrum.* **43**, 1221 (1972).

[†] Available from Lake Shore Cryotronics, Inc., 9631 Sandrock Road, Eden, New York.

and below may soon be frequent. Thermometers exist which appear to be capable of meeting the challenge. The most promising at this moment appear to be susceptibility measurements using SQUID's, nuclear magnetic resonance, and nuclear orientation.

Acknowledgments

Many members of the low temperature research group at the University of Minnesota have indirectly contributed to this article; but special thanks are due two of my students, Carl Smith and Jim Holliday, and Professors Goldman, Moldover, and Zimmermann. Dr. A. C. Anderson, Dr. N. Brubaker, and Mr. R. Swinehart have read the manuscript with critical and keen eyes. Their comments and corrections are greatly appreciated.

The U.S. Atomic Energy Commission and the Graduate School of the University of Minnesota have supported the author's research programs on which much of the material in this article has been based.

10. SUPERCONDUCTING MICROWAVE RESONATORS*

10.1. Introduction

Resonators are ubiquitous in modern technology. They are used in one or another form for timekeeping, impedance transformation, frequency selection, energy accumulation, and a host of other functions. In general, a resonator is an energy storage device in which energy oscillates back and forth between two distinct forms at a well-defined frequency. There exists a variety of resonator to exploit almost every possible combination of two of the many forms of energy that exist. Some of the more common types of resonators are mechanical, such as tuning forks, resonant reeds, and quartz crystals; or electromagnetic, such as *LC* circuits and microwave cavities. Some resonator functions can be performed in principle by any resonator. Timekeeping is an example, where the choice of resonator type to use is dictated by accuracy requirements and convenience. Other functions require a particular form of resonator. An example is accumulation of microwave energy in order to obtain high rf electric and magnetic fields, which requires an electromagnetic resonant structure.

This article is concerned with one particular type of resonator, the microwave resonant structure. With the application of superconductivity, these devices have recently been developed to an extremely high order of perfection, so high that they are beginning to be used in some applications where entirely different types of resonators have traditionally been used. For example, clock oscillators referenced to superconducting resonators now provide frequency stability better than the best quartz crystal oscillators, and they may well improve by orders of magnitude further in the near future.[1] Superconducting resonators can now be used to

---

[1] S. R. Stein and J. P. Turneaure, *Electron. Lett.* **8**, 321 (1972); *Proc. Int. Conf. Low Temperature Phys.*, *13th*, *Boulder*, *Colorado*, 1972 (to be published by the University of Colorado Associated Press).

---

* Part 10 is by John M. Pierce.

detect mechanical motions less than one fermi (1 F = 10^{-13} cm).[2] This high perfection also makes possible applications for which microwave resonators are uniquely suited, but which are precluded by the high losses in normal metal resonators. Creation of very high rf electric fields on a continuous basis for accelerating charged particles is an example of these. Others are transforming rf impedances by ratios of 10^6 or greater and making efficient antennas which are small compared to a wavelength.

The universal figure of merit for resonators of any type is the quality factor Q, which measures the number of oscillations a resonator will go through before dissipating a substantial fraction of its stored energy. The bandwidth or spectral purity of the oscillation is inversely proportional to the Q. Superconducting microwave resonators with Q's greater than 10^{11} have recently been built,[3-5] and resonators with Q's greater than 10^{10} are no longer uncommon.[6-12] These values are orders of magnitude higher than can be obtained with any other kind of resonator. For comparison, the Q of a high-quality quartz crystal resonator for oscillator use might be as high as 10^7, while the Q values of comparable copper microwave resonators at room temperature might be $\sim 10^4$. Thus the Q's of superconducting resonators can now be a factor of 10^7 greater than their normal metal counterparts.

While it is primarily these high Q values that give superconducting resonators their great potential for application in such a wide variety of laboratory experiments and technological devices, it is also extremely important for many applications that these high Q's can be obtained with high rf fields and a substantial amount of stored energy in the

[2] G. J. Dick and H. C. Yen, *Proc. 1972 Appl. Superconductivity Conf. Annapolis, Maryland*, IEEE Pub. No. 72CH0682-5-TABSC, p. 684.

[3] J. P. Turneaure and Nguyen Tuong Viet, *Appl. Phys. Lett.* **16**, 333 (1970).

[4] M. A. Allen, Z. D. Farkas, H. A. Hogg, E. W. Hoyt, and P. B. Wilson, *IEEE Trans. Nucl. Sci.* **NS-18**, 168 (1971).

[5] P. Kneisel, O. Stoltz, and J. Halbritter, *Proc. 1972 Appl. Superconductivity Conf., Annapolis, Maryland*, IEEE Pub. No. 72CH0682-5-TABSC, p. 657.

[6] J. M. Pierce, *J. Appl. Phys.* **44**, 1342 (1973).

[7] J. P. Turneaure and I. Weissman, *J. Appl. Phys.* **39**, 4417 (1968).

[8] H. Hahn, H. J. Halama, and E. H. Foster, *J. Appl. Phys.* **39**, 2606 (1968).

[9] Ira Weissman and J. P. Turneaure, *Appl. Phys. Lett.* **13**, 390 (1968).

[10] P. Kneisel, *Proc. Joint CERN-Karlsruhe Symp. Superconducting rf-Separators, Karlsruhe, 1969*, Externer Bericht 3/69-19, Kernforchungszentrum Karlsruhe, Karlsruhe, Germany.

[11] Y. Bruynseraede, D. Gorle, D. Leroy, and P. Morignot, *Physica* **54**, 137 (1971).

[12] J. P. Turneaure, *Proc. 1972 Appl. Superconductivity Conf., Annapolis, Maryland* IEEE Pub. No. 72CH0682-5-TABSC, p. 621.

10.1. INTRODUCTION

resonator. Values of Q greater than 10^{10} have been maintained in resonators containing rf magnetic fields greater than 1000 Oe and rf electric fields the order of 70 MV/m.[3,13] To maintain rf fields comparable to these on a continuous basis in a copper resonator requires a prohibitive amount of rf power and produces a prohibitive amount of heat. Fields this high can be produced in such resonators, but only for a few microseconds at a time. These facts indicate the great promise of superconducting resonators for such high-field applications as linear particle accelerators and particle separators.

The art of making these outstanding resonators is largely a product of the last decade. In 1960 the highest Q that had been reported for a superconducting resonator was a few million.[14] Furthermore, most of the work which has produced this great advance has been directed toward building superconducting linear accelerators or rf particle separators, and much of it has appeared in unpublished reports from various accelerator laboratories, especially the High-Energy Physics Laboratory at Stanford University. Thus the major purpose of this article is to bring the results of this recent work to the attention of workers in other fields who may not be familiar with it. The unique properties of these very high quality resonators could be used to great advantage in many fields of experimental physics and technology, and attempts are made in the later sections of this review to point out where these advantages might lie.

The last comprehensive review of rf superconductivity was the exhaustive work in 1963 by Maxwell,[14] which contains a complete summary of the theoretical and experimental work in this field prior to 1963. The anomalous skin effect in normal metals is also covered. No attempt is made to duplicate that work here, and the interested reader is referred to this excellent review for the earlier work. Halbritter[15] has more recently compared experimental results in detail with the current theory of superconductivity. Brief reviews of the theoretical[16] and experimental[12] situation have been given very recently by Turneaure; Schwettman *et al.*[17] have considered specifically the promise of superconductivity for linear accelerators.

---

[13] H. Diepers and H. Martens, *Phys. Lett.* **38A**, 337 (1972).

[14] E. Maxwell, *in* "Progress in Cryogenics," Vol. 4. Heywood, London, 1964.

[15] J. Halbritter, *Z. Phys.* **238**, 466 (1970).

[16] J. P. Turneaure, *Proc. Int. Conf. High Energy Accelerators, 8th. CERN Geneva*, p. 51 (1971).

[17] H. A. Schwettman, P. B. Wilson, J. M. Pierce and W. M. Fairbank, *in* "Int. Advan. in Cryogenic Engineering" (K. D. Timmerhaus, ed.), Vol. 10, p. 88. Plenum, New York (1965).

Because of the publication history in this field, it will often be necessary to refer to unpublished material such as laboratory reports and conference proceedings. However, an attempt is made to reproduce enough details here so that the reader will find it unnecessary to consult this material himself. Where this is not the case, it may be possible to obtain copies of the papers from the laboratories or authors concerned.

This review is divided into four main parts. Section 10.2 is a discussion of the response of a superconducting surface to microwave frequency electromagnetic fields up to the dc critical field in magnitude. Experimental results are given, and the parts of the theory of superconductivity that relate directly to understanding the performance of superconducting resonators are briefly outlined. Section 10.3 is concerned with practical techniques for designing and building high-Q superconducting resonators and for measuring their properties. Suitable materials and methods for preparing high-quality surfaces are discussed. Ways of coupling to high-Q resonators and circuits to measure the Q and resonant frequency are also described. In 10.4 we review briefly a number of applications in several areas of experimental physics and technology where superconducting resonators have already enabled significant advances to be made. In order to be useful to the widest audience, the focus of this review is primarily on small-scale laboratory applications rather than large-scale applications such as accelerators.

10.2. Microwave Properties of Superconductors

10.2.1. The Surface Impedance of Normal and Superconducting Metals

The response of a metal surface to an applied electromagnetic field is usually described in terms of a surface impedance Z. This complex impedance is defined by

$$Z = R + iX = E_t/H_t, \qquad (10.2.1)$$

where E_t and H_t are the tangential electric and magnetic fields at the surface of the conductor. If the response is linear, E_t is proportional to the surface current density, which is equal to H_t.[†] Then the surface density of power flow into the metal from the fields can be written in

---

[†] We shall use mks units in all formulas.

10.2. MICROWAVE PROPERTIES OF SUPERCONDUCTORS

terms of the surface resistance R

$$p_\mathrm{d} = RH_\mathrm{t}^2. \tag{10.2.2}$$

The surface reactance X is related to the phase shift of a wave upon reflection from the surface. It can be expressed in terms of an effective penetration depth δ for the fields into the metal: $X = \mu\omega\delta/2$.

The surface impedance of the walls and their geometry together determine the properties of a microwave resonator, and such resonators are used to measure the surface impedance of materials. Assume first that a particular resonator geometry is constructed of a perfect conductor ($Z = 0$) and that it is resonant in the kth mode at ω_k. Mode calculations for various resonator geometries are covered in numerous references,[†] and they will not be considered here. Now let Z be finite but small (a good conductor). The first-order effect of the surface reactance, which is always positive or inductive for a metal surface, is to shift the resonant frequency downward. The effect of surface resistance is to make the Q finite. The unloaded Q of a resonator is defined by

$$Q_0 = \omega_k U/P_\mathrm{d}, \tag{10.2.3}$$

where P_d is the total power dissipated in the walls of the resonator averaged over a cycle and U is the energy stored in the fields. The combined first-order effects of finite surface impedance can be expressed in the equation[‡]

$$\frac{1}{Q_0} - 2i\frac{\Delta\omega_k}{\omega_k} = \frac{1}{2\omega_k}\int_\mathrm{S}(R + iX)H_k^2\,ds. \tag{10.2.4}$$

---

[18] J. C. Slater, *Rev. Mod. Phys.* **18**, 441 (1946).

[19] S. Ramo, J. R. Whinnery, and T. VanDuzer, "Fields and Waves in Communication Electronics." Wiley, New York, 1965; or S. Ramo and J. R. Whinnery, "Fields and Waves in Modern Radio." Wiley, New York, 1944.

[20] C. G. Montgomery (ed.), "Technique of Microwave Measurements." McGraw-Hill, New York, 1947.

[21] C. G. Montgomery, E. M. Purcell, and R. H. Dicke, "Principles of Microwave Circuits." McGraw-Hill, New York, 1948.

[22] J. C. Slater, "Microwave Electronics." Van Nostrand-Reinhold, Princeton, New Jersey, 1950.

---

† Refs. 18–22 are some general works on microwave theory and practice.

‡ For a derivation of this result, see Ref. 22, p. 67ff.

In this equation, $\Delta\omega_k$ is the (negative) frequency shift, and the field H_k is the unperturbed amplitude of the tangential magnetic field of the kth mode at the surface of the resonator, normalized such that the energy stored in the resonator is 1 J. The integral is over the entire inner surface of the resonator. Equation (10.2.4) includes only the effects of the walls of the resonator. If the medium in the resonator is other than a vacuum, the frequency shifts and loss due to the medium can be considered separately, as can the effects of energy escaping from the resonator through coupling networks or other openings.[†]

In a normal metal at room temperature, Ohm's law adequately describes the microwave response. Thus the current is proportional to the local electric field with a proportionality constant equal to the dc conductivity σ. The surface impedance of such a metal can be written in terms of the classical skin depth $\delta_C = [2/\mu\omega\sigma]^{1/2}$,

$$Z_C = (\mu\omega/2)\delta_C[1 + i]. \tag{10.2.5}$$

Upon cooling a normal metal, the mean free path increases, causing an increase in σ. Thus δ_C decreases, resulting in a decrease in Z_C. However, when the mean free path becomes comparable to δ_C, the relation between the current and the fields becomes nonlocal. Ohm's law is no longer valid, and the current at a point depends upon the electric field in a region around the point of size comparable to the mean free path. The result is that the surface impedance becomes independent of temperature. This phenomenon is called the *anomalous skin effect*,[‡] and it prevents one from obtaining a very great increase in Q by cooling a resonator made of normal metal.

In a superconductor, on the other hand, the situation is dramatically different. At zero frequency, as is well known, the resistance of a superconductor is zero. At finite frequencies, the surface resistance is not zero, but it can be very much less than the normal state surface resistance. It is also very much less than the surface reactance. Unlike the case in a normal metal, the surface impedance of a superconductor is almost purely reactive. For ordinary microwave frequencies, the surface resistance of a superconductor should go to zero exponentially as the temperature is reduced, while at extremely high frequencies (far infrared and

---

[†] These points are discussed in Sections 10.3.1 and 10.3.3.

[‡] See Ref. 14 for a complete description of the anomalous skin effect and references to the fundamental literature.

10.2. MICROWAVE PROPERTIES OF SUPERCONDUCTORS

beyond) the surface impedance of a superconductor is indistinguishable from that of a normal metal.

One can obtain some insight into this behavior in terms of the two-fluid model of a superconductor.[23] In this model it is assumed that the conduction electrons are divided into two classes, superconducting and normal. The fraction of the electrons in the superconducting class varies continuously from zero at the transition temperature to unity at zero temperature. The superconducting electrons move without dissipation; they are accelerated by an electric field; and they also respond directly to a static magnetic field in a way worked out by London.[24] For ordinary microwave frequencies, the electric field effect is unimportant, and it is the magnetic field that causes a supercurrent to flow. This supercurrent tends to exclude the magnetic field from the metal. Thus the penetration of an rf magnetic field into the walls of the superconducting microwave resonator is nearly the same as the penetration of a dc field, i.e., it decays exponentially into the metal with a penetration depth λ, which is much less than δ_C at microwave frequencies. The surface reactance of a superconductor, therefore, is determined by the dc penetration depth. The normal electrons, on the other hand, are assumed to respond to an electric field in the same way as in a normal metal. The oscillating magnetic flux in the penetration layer is $\sim \lambda H_t$, and it induces an electric field parallel to the surface proportional to $\omega \lambda H_t$. This field causes the normal electrons to move and dissipate power. The dissipated power should be proportional to the density of normal electrons ϱ_n and to the square of the electric field. Thus we expect $R \propto \varrho_n \omega^2$, i.e., R should vanish at zero temperature or frequency.

While the two-fluid model is useful in understanding the surface impedance of superconductors at low frequencies and temperatures not too close either to T_c or zero, a detailed understanding of the temperature and frequency dependence comes only through the microscopic theory of superconductivity developed by Bardeen, Cooper, and Schrieffer (BCS).[25] According to this theory, which has been successful in explaining a wide variety of superconducting behavior,[23] there is a gap 2ε in the density of states for the conduction electrons in a superconductor below its transition temperature. This gap is temperature-dependent, varying

---

[23] For an outline of the theory of superconductivity, see for example P. G. DeGennes, "Superconductivity of Metals and Alloys." Benjamin, New York, 1966.

[24] F. London, "Superfluids." Vol. 1. Dover, New York, 1961.

[25] J. Bardeen, L. N. Cooper, and J. R. Schrieffer, *Phys. Rev.* **108**, 1175 (1957).

from zero at T_c to $2\varepsilon(0) \approx 3.5 k_B T_c$ at $T = 0$. The Fermi surface falls in the middle of the gap. Ground state electrons, which are associated in pairs of identical center of mass momentum, are capable of carrying a current, but they cannot scatter inelastically without an energy change of at least 2ε, the energy required to break a pair. In the absence of photons with $\hbar\omega \geq 2\varepsilon$, there is no process which is at all likely to break a pair. Thus there is no dissipation due to this current. These correspond to the "superconducting" electrons of the two-fluid theory. At any finite temperature there will be some electrons thermally excited above the gap. These quasi particles or "normal" electrons do have empty states available at nearby energies, so they do produce dissipation when they carry a current.

The existence of the energy gap has two clear implications for the behavior of the surface resistance. First, the surface resistance may be expected to vanish as $\exp[-\varepsilon(0)/k_B T]$ as $T \to 0$ for any frequency such that $\hbar\omega < 2\varepsilon(0)$. This follows simply from the temperature dependence of the quasi-particle density. Second, for higher frequencies $\hbar\omega > 2\varepsilon(0)$, the microwave photons can directly break a pair and be absorbed, leading to finite dissipation even at $T = 0$. When $\hbar\omega \gg 2\varepsilon(0)$, the gap is unimportant in the interaction between photons and electrons, and the surface impedance is essentially the same as in the normal state. The behavior

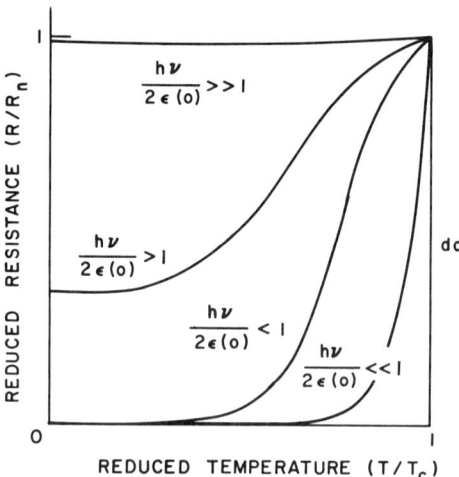

FIG. 1. Qualitative behavior of the superconducting surface resistance as a function of temperature for various frequencies [after H. A. Schwettman, P. B. Wilson, J. M. Pierce, and W. M. Fairbank, *in* "Int. Adv. in Cryogenic Engineering" (K. D. Timmerhaus, ed.) Vol. 10, p. 88. Plenum, New York, 1965].

10.2. MICROWAVE PROPERTIES OF SUPERCONDUCTORS

of the surface resistance of a superconductor with temperature and frequency is shown qualitatively in Fig. 1. Here we imply that the reduced surface resistance is a universal function of the reduced temperature T/T_c and a reduced frequency $\hbar\omega/2\varepsilon(0)$. This is a good first approximation, but it is not valid in quantitative detail.

The implication of the discussion thus far is that, at least for $\hbar\omega < 2\varepsilon(0)$, one can make a resonator of arbitrarily high Q merely by operating at sufficiently low temperature. As one might suspect, this is not the case in practice. The Q of any real resonator is observed to level out and become independent of temperature at some finite temperature. Data for lead[6] and niobium[7] cavity resonators in X-band are shown in Fig. 2. This phenomenon is usually discussed in terms of a residual surface resistance. One assumes

$$R = R_s(T) + R_o, \qquad (10.2.6)$$

where R_s is the highly temperature dependent "superconducting" surface

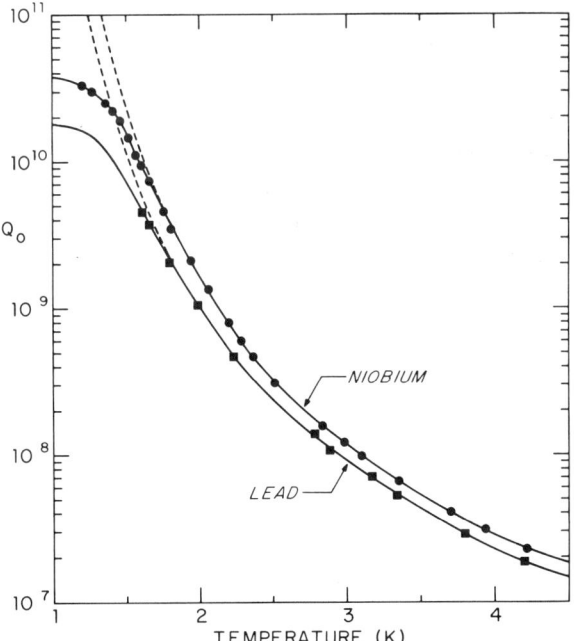

Fig. 2. The unloaded Q of typical lead [J. M. Pierce, *J. Appl. Phys.* **44**, 1342 (1973)] and niobium [J. P. Turneaure and I. Weissman, *J. Appl. Phys.* **39**, 4417 (1968)] cavities in X-band. The dashed curves represent the theoretical behavior of R_s.

resistance discussed above, and R_o is a residual term which is supposed to include all other loss mechanisms. It is usually taken to be temperature-independent. The assumption of temperature independence for R_o is experimentally convenient, since the rapid temperature dependence of R_s makes it difficult to extract the more moderate temperature dependence of likely residual loss mechanisms unless one of them is deliberately enhanced for experimental study. For studies of R_s and for making high-Q resonators it is clearly desirable to reduce R_o as low as possible. Most of the progress of the last decade in the superconducting resonator field has resulted from identifying and eliminating various sources of residual loss. We will return to the problem of R_o below, but first we discuss R_s.

10.2.2. The Superconducting Surface Resistance

The calculation of R_s from the BCS theory is quite complicated. The formal solutions were worked out by Mattis and Bardeen[26] and by Abrikosov et al.,[27] but they require lengthy computer calculation for evaluation. Methods and computer programs for doing this have been worked out by Miller,[28] Turneaure,[29] and Halbritter.[30] Not surprisingly, the results depend upon several electronic parameters of the metal in addition to the energy gap $\varepsilon(T)$, i.e., the electron mean free path, the Fermi velocity, and the London penetration depth of the superconductor λ_L. The latter parameter is related to the penetration depth λ for a dc magnetic field into the superconductor, and it is related to the density of states at the Fermi surface. Estimates of these parameters can be obtained from a variety of measurements on the metal both in the superconducting and in the normal state. The surface impedance also depends upon whether electrons are specularly reflected or diffusely reflected at the surface of the metal.[14] Values calculated for the surface resistance on the basis of the Mattis and Bardeen (MB) theory by Turneaure for lead[29] and

---

[26] D. C. Mattis and J. Bardeen, *Phys. Rev.* **111**, 412 (1958).

[27] A. A. Abrikosov, L. D. Gorkov, and I. M. Khalatnikov, *Sov. Phys. JETP* **35**, 182 (1959).

[28] P. B. Miller, *Phys. Rev.* **118**, 928 (1960).

[29] J. P. Turneaure, Ph. D. Thesis, Stanford Univ. (1967) (unpublished), HEPL Rep. No. 507, High Energy Phys. Lab., Stanford Univ., Stanford, California 94305; see also J. P. Turneaure and I. Weissman, *J. Appl. Phys.* **39**, 4417 (1968).

[30] J. Halbritter, Externer Bericht 3/69-2 and 3/69-6, Kernforschungszentrum, Karlsruhe, Karlsruhe, Germany (1969).

niobium[7] at 11.2 GHz are shown in Fig. 3. These are the two superconductors most commonly used for high-Q resonators. The characteristic exponential temperature dependence at low temperature is clearly evident. Rough values for other frequencies can be obtained by scaling; $R_s \propto \omega^{1.75}$ in both lead and niobium for frequencies between 0.1 and 20 GHz.[12,15]

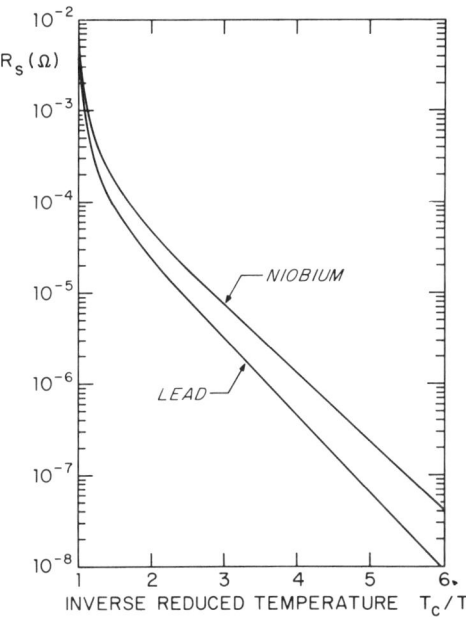

FIG. 3. Calculated values of R_s as a function of reduced temperature obtained by Turneaure for lead [J. P. Turneaure, Ph.D. Thesis, Stanford Univ. (1967)] and niobium [J. P. Turneaure and I. Weissman, *J. Appl. Phys.* **39**, 4417 (1968)]. The respective transition temperatures T_c are 7.18 K for lead and 9.17 K for niobium. At any given temperature, niobium has *lower* R_s than lead.

To compare the MB theory with experimental data such as is shown in Fig. 2, a fitting procedure was developed by Turneaure.[29] The measured surface resistance values are fit by the equation

$$R = AR_{\text{MB}}(T) + R_0. \tag{10.2.7}$$

Here R_0 is a temperature-independent residual resistance and A is a scaling constant which should be nearly unity. It is included to take account of inaccuracies in parameters such as the Fermi velocity and

λ_L which are used to calculate R_MB. In addition, Turneaure leaves $\varepsilon(0)$, the energy gap at zero temperature, as an adjustable parameter. The value of R_MB is extremely sensitive to $\varepsilon(0)$. With this procedure the curves of Fig. 2 were obtained, where the dotted line indicates the behavior of $R - R_0$. The values for $\varepsilon(0)$ obtained in this way are consistent with other means of determining this parameter,[7,29] and the values obtained for A are 0.6–1.0 for lead[29] and 0.97 for niobium.[7] Recently, there have been arguments on the basis of new superheating and supercooling data,[31–33] which are relevant to determining λ_L, and new calculations[33,34] that the accepted values for λ_L in lead, tin, and indium should be changed. Halbritter[15] has shown that if the new values are used in the calculation of R_MB, the surface resistance data for lead, tin, and indium fit Eq. (10.2.7) with $A = 1.0$ within the experimental error ($\sim 5\%$). Thus the correspondence between theory and experiment is good enough so that useful information can be obtained from the surface resistance data on the fundamental parameters of superconductors.

The effects of mean free path and reflection characteristics for electrons at the surface are not overly dramatic for reasonably clean materials, except at rather low frequencies ($f < 1$ GHz). Halbritter[15] has discussed these effects and compared existing experiments on alloys with theory. He finds reasonably good agreement, and we will not consider these effects here.

10.2.3. The Superconducting Surface Reactance

The surface reactance can also be calculated from the MB formulation. As indicated above, it can be expressed as $X = \mu\omega\delta/2$, where δ is an effective skin depth. Turneaure[29] has calculated $\delta(\omega, T)$ for tin using MB, and obtained a very good fit to his experimental data. For low frequencies ($\hbar\omega \ll 2\varepsilon$), δ is not strongly dependent on frequency and roughly equal to twice the dc penetration depth λ. The temperature dependence is complicated, but for most purposes the temperature dependence derived from the two-fluid model is an adequate approximation

$$\delta/2 = \lambda(T) = \lambda(0)[1 - (T/T_\mathrm{c})^4]^{-1/2}. \tag{10.2.8}$$

---

[31] J. Feder and D. S. McLachlan, *Phys. Rev.* **177**, 763 (1969).
[32] F. W. Smith and M. Cardona, *Solid State Commun.* **6**, 37 (1968).
[33] F. W. Smith, A. Baratoff, and M. Cardona, *Phys. Kondens. Mater.* **12**, 145 (1970).
[34] J. C. Swihart and W. Shaw, in "Superconductivity" (F. Chilton, ed.), p. 678. North Holland, Amsterdam, 1971; *Proc. Int. Conf. Sci. Supercond. Stanford (1969)*.

Thus λ is rather sensitive to temperature in the neighborhood of T_c, and it is less sensitive at lower temperatures. For most pure superconductors, $\lambda(0) \sim 5 \times 10^{-6}$ cm. In X-band this implies that $X \sim 5 \times 10^{-3}$ Ω. Thus the surface impedance is almost purely reactive except in the immediate vicinity of T_c.

10.2.4. The Residual Surface Resistance

We turn now to the residual resistance. A number of different effects have been identified that produce residual loss in resonators, and more have been discussed as possible causes. Some of these are entirely extraneous to the superconducting surface, such as lossy material in the resonator, bad joint design, lossy coupling networks, etc. These will be discussed in Section 10.3 on practical resonator design. Here we will only consider effects directly related to the superconducting surface.

The most clearly documented[6] cause of residual resistance in a superconducting surface is magnetic flux trapped through the surface. Figure 4 shows the inverse residual Q of a cylindrical, lead-plated, cavity resonator operated in the TE_{013} mode at 12.2 GHz. The cavity was cooled in various values of magnetic field parallel to the axis of the cavity. The

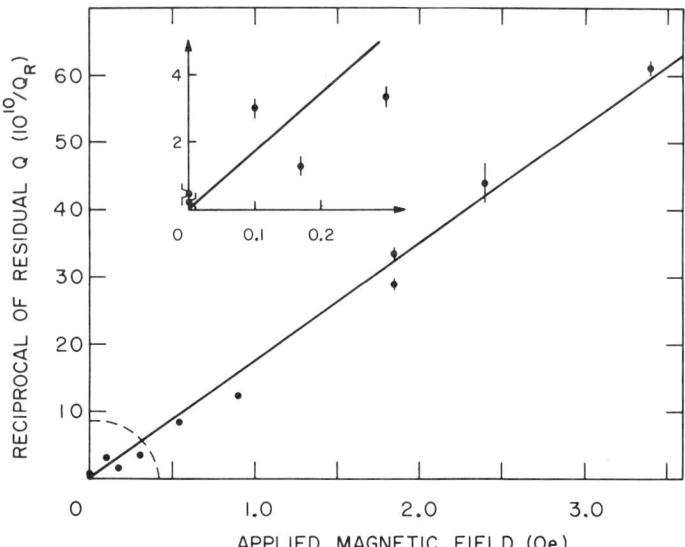

FIG. 4. Inverse residual Q of a lead-plated cavity cooled in various dc magnetic fields applied parallel to the axis of the cylindrical cavity [after J. M. Pierce, *J. Appl. Phys.* **44**, 1342 (1973)].

quantity Q_0^{-1} is proportional to the average R_0 of the cavity walls. The earth's magnetic field was excluded from the vicinity of the cavity by shielding (to $\sim 10^{-4}$ Oe), and a known axial field could be applied with a coil.

When a hollow cavity is cooled through T_c in a field, essentially all the magnetic flux is trapped. There is a tendency for magnetic flux to be excluded from a superconductor by the Meissner effect,[24] but the thin walls of typical cavities have such a large demagnetizing factor that there is rarely an appreciable macroscopic exclusion of an applied field from the cavity. The flux comes through the walls in current vortices,[35] which in a type II superconductor like niobium contain single quanta of magnetic flux.[†] In a type I material, such as lead or tin, vortices containing many quanta are favored.

The size of a vortex in a type I material depends on the number of flux quanta it contains. The dc magnetic field in the core of the vortex, which is a region of normal metal, must be equal to the dc critical magnetic field H_c of the superconductor. Outside the core, the field dies away exponentially with characteristic length λ, which is also the characteristic depth to which the rf fields penetrate the surface of the superconductor. For lead, $\lambda \approx 4 \times 10^{-6}$ cm and $H_c \approx 750$ Oe at 2 K. Calculating the size of the core of a vortex in lead containing n quanta, we obtain $r \approx 10^{-5} n^{1/2}$ cm. Thus a vortex containing even a few flux quanta is larger than λ and in fact may well be larger than the skin depth for rf fields into normal lead at 2 K. At this temperature for microwave frequencies, normal lead is in the anomalous skin effect region, and the skin depth is $\delta_a \approx 2 \times 10^{-5}$ cm. If this is the case, the rf field can be expected to sag into these large vortices, and the rf currents will flow through the normal core rather than detouring it.[6,36,37] This should lead to a residual loss proportional to R_n, the normal state surface resistance of lead. The frequency dependence of the residual loss for the cavity of Fig. 4 was measured[6] by noting the residual Q of two lower modes, TE_{011} and TE_{012}. The result was $R_0 \propto \omega^{2/3}$, which is the characteristic

---

[35] For a discussion of the intermediate state in superconductors and the theories of vortices see P. G. DeGennes, "Superconductivity of Metals and Alloys." Benjamin, New York, 1966.

[36] J. M. Pierce, Ph.D. Thesis, Stanford Univ. (unpublished), HEPL Rep. No. 514, High Energy Phys. Lab., Stanford Univ., Stanford, California 94305 (1967).

[37] J. LeG. Gilchrist, *Proc. Roy. Soc.* (*London*) **A295**, 399 (1966).

[†] $\Phi_0 = h/2e = 2.07 \times 10^{-15}$ Wb.

frequency dependence of a normal metal in the anomalous skin effect region. Furthermore, R_0 should be proportional to the field in which the cavity is cooled, and its magnitude should be just $R_n H_{app}/H_c$. This follows from the fact that the fraction of the cavity ends which is normal is just B/H_c, where B is the macroscopic density of trapped flux. Since in a plated cavity there is little Meissner effect, $B \approx H_{app}$. The slope calculated in this manner is roughly three times the slope of the line in Fig. 4, which is satisfactory agreement, considering that the vortices tend to be located in thin spots in the electroplated lead layer.[6,38] They emerge at the surface in the bottom of depressions on the surface, where the rf field is reduced. The scatter in Fig. 4 can also be explained in this way since the distribution of vortices among depressions depends on the details of the temperature distribution as the cavity is cooled through the transition to the superconducting state.

In a type II material, the way in which residual loss is produced by trapped magnetic flux is more complicated and more interesting. A type II material is one with $\varkappa = \lambda/\xi > 1/2^{1/2}$, where ξ is the coherence length. The coherence length is the characteristic healing length for the energy gap near a boundary between a normal and a superconducting region in the metal. In a type II material there is a negative surface energy associated with such a boundary, so trapped flux tends to break up into flux lines containing single quanta.[35,39] In such a vortex the dc field dies away in a distance $\sim\lambda$ from the center, while there is an essentially normal core $\sim\xi$ in size at the center. To produce normal dissipation in the core, the rf current must penetrate the surface to a depth δ_a from the much shallower depth λ outside the vortex. However the "hole" of size $\xi < \lambda$ is too small to allow the current to sag in this far.[36,37] Thus the rf current can be expected to detour around these small vortices, and we would expect less loss than in the larger vortices characteristic of a type I material. However, these vortices tend to move transversely to a transport current in the material, and if they are not pinned by lattice defects or other inhomogeneities, an electric field is produced and power is absorbed.[40] The flux lines are under tension,[35,39] so motion of the emerging end tends to carry the disturbance deeper into the metal. The driving force is just the Lorentz force $\mathbf{J} \times \mathbf{B}$, and this is opposed by viscous drag and pinning forces. The pinning forces are usually assumed to be quasi

---

[38] J. M. Pierce, *Phys. Rev. Lett.* **24**, 874 (1970).
[39] B. B. Goodman, *Rev. Mod. Phys.* **36**, 12 (1964).
[40] Y. B. Kim, C. F. Hempstead, and A. R. Strnad, *Phys. Rev.* **139**, A1163 (1965).

conservative, i.e., they stem from effective potential wells associated with particular preferred positions in the material for the flux lines.[41] Under this assumption it is clear that a flux line subjected to an ac current of high enough frequency will move subject to the viscous damping only.[42] Pinning will not be important, and the effective mass of a flux line is negligible except at very high frequency ($\sim$50 GHz).[41,43] Thus flux flow will contribute to the loss in a resonator even at low current densities, which at dc would not produce dissipation because they are insufficient to overcome pinning. The depinning frequency at which this crossover occurs can be quite low, especially in reasonably clean materials ($\sim$10 MHz or less).[44,45]

The contributions of this type of flux flow to the surface resistance have been calculated in the case of short mean free path and relatively low frequency, where the skin effect in the normal material is classical.[37,45–47] The calculation also assumes $H_{app} > H_{c1}$, where the flux lines are closely spaced relative to the classical skin depth (spacing $\sim\lambda$). Under these conditions $R \sim R_N [H_{app}/H_{c2}]^{1/2}$. Reasonably good agreement is obtained with experiments on alloy foils and pure niobium with a short mean free path.[44–46] In clean materials or high frequencies where the normal state skin effect is anomalous, the situation is less clear. Rosenblum and co-workers[48,49] observed at 22 GHz in several type II materials a linear dependence on H_{app}. They found $R_H \sim R_N[H_{app}/H_{c2}]$. This is consistent with the assumption that the cores of the vortices act like spots of normal metal, and that the current does not detour around them. The fraction of the area that is normal is just H_{app}/H_{c2}, since at $H = H_{c2}$ the flux lattice spacing is $\sim\xi$. However, as they have since stated,[44] it is likely that flux flow was the operative mechanism here also for the reasons outlined above. Presumably also, the large flux tubes in a type I material can move in the same fashion, but it is not necessary to invoke motion to explain the experimental results discussed above. At very low levels of trapped flux, however, where even in a type I material

---

[41] J. I. Gittleman and B. Rosenblum, *J. Appl. Phys.* **39**, 2617 (1968).
[42] P. G. DeGennes and J. Matricon, *Rev. Mod. Phys.* **36**, 45 (1964).
[43] H. Suhl, *Phys. Rev. Lett.* **14**, 226 (1965).
[44] J. I. Gittleman and B. Rosenblum, *Phys. Rev. Lett.* **16**, 734 (1966).
[45] J. LeG. Gilchrist and P. Monceau, *Phil. Mag.* **18**, 237 (1968).
[46] J. LeG. Gilchrist and P. Monceau, *J. Phys. C (London)* **3**, 1399 (1970).
[47] G. Fischer, R. D. McConnell, P. Monceau, and K. Maki, *Phys. Rev. B* **1**, 2134 (1970).
[48] B. Rosenblum and M. Cardona, *Phys. Rev. Lett.* **12**, 657 (1964).
[49] B. Rosenblum, M. Cardona, and G. Fischer, *RCA Rev.* **25**, 491 (1964).

the vortices contain single flux quanta, it is likely that flux flow is the dominant loss mechanism.

The case of the most interest to users of high-Q, high-frequency resonators is when $H_{app} \ll H_{c1}$ where there are a few isolated lines of trapped flux. One would expect the above flux-flow mechanism to be equally effective here, as these lines will vibrate in their pinning sites just as the relatively dense lattice does when $H_{app} > H_{c1}$. Gilchrist[37] has calculated the absorption of a single line penetrating the surface using low-frequency formulas for the line tension and viscosity. He finds $R_H \propto \omega^{1/2}$, but it is not clear what effect the nonlocal conduction in the anomalous regime will have. Kneisel et al.[50] measured the frequency dependence of loss due to a small amount of trapped flux by exciting several modes of a niobium cavity between 2 and 4 GHz. Their results indicate a frequency dependence more like $\omega^{2/3}$ than $\omega^{1/2}$. If the flux density B is small compared with H_{c1}, the lines do not interact, and we expect $R_H \propto B$. If there is little Meissner effect, we expect $B \approx H_{app}$, but the niobium cavities used[7,51] to measure the dependence of the trapped flux losses on H_{app} have been massive, thick-walled structures of solid niobium. The demagnetizing factors of these walls may well have been small enough so that a partial Meissner effect occurred, causing appreciable distortion of the applied field. Thus B may have differed considerably from H_{app}. Turneaure and Weissman[7] cooled one niobium cavity in an applied field of 1 Oe, and its residual Q was 1.7×10^8 in contrast to 1.3×10^{10} when cooled in very low field. If the loss was uniform on the ends of the cavity, this implies $R_H \approx 4 R_N H_{app}/H_{c2}$. This is reasonable agreement with the results of Rosenblum et al.[48,49] since field distortion could cause nonuniform distribution of trapped flux. On the other hand, Wilson et al.[51] recently cooled a niobium cavity in several values of H_{app} and measured Q's which seem to indicate $R_H \propto H_{app}^2$. This is very difficult to reconcile with an isolated flux line picture unless substantial distortion of the applied field took place. Such a distortion could have a nonlinear dependence on H_{app} because the critical demagnetizing factor for the Meissner effect depends on H_{app}.

Very roughly then, one can take $R_H \approx R_N H_{app}/H_{c2}$ as an estimate of loss due to small amounts of trapped flux passing through the surface

---

[50] P. Kneisel, O. Stoltz, and J. Halbritter, *IEEE Trans. Nucl. Sci.* **NS-18**, 158 (1971).

[51] P. B. Wilson, M. A. Allen, H. Deruyter, V. D. Farkas, H. A. Hogg, E. W. Hoyt, and M. Rabinowitz, *Proc. Int. Conf. High Energy Accelerators, 8th, CERN, Geneva*, p. 237 (1971).

of both type I and type II superconductors. However, in a thick-walled cavity of solid superconductor, considerable deviation from this value may occur.

When $H_{app} < 10^{-3}$ Oe it is very difficult experimentally to characterize the residual loss. It typically varies considerably from surface to surface and even for the same surface upon temperature cycling. Attempts have been made to measure the frequency dependence of the remaining residual loss by measuring the Q of several modes in the same resonator. This method suffers from the unavoidable difficulty that the field distribution is different for different modes. Consequently, if the loss is not uniformly distributed over the surface, confusing results will be obtained. The Karlsruhe group has found residual resistance apparently proportional to ω^2 in both electroplated lead[52] and niobium[50] cavities. However, the author,[6,36,53] working in X-band with lead surfaces that showed considerably less residual loss than theirs, found no consistent frequency dependence and did find strong indication of nonuniform distribution of the loss. The case of niobium is similar. If one compares[12] the lowest values of R_0 measured to date by different groups in different varieties of cavities, one finds values within an order of magnitude of each other at frequencies ranging all the way from 90 MHz to 10 GHz. Thus it seems unlikely that the loss remaining in the best lead and niobium surfaces is simply proportional to ω^2.

Halbritter[54] and Passow[55] have studied processes for direct phonon generation in the metal by the rf fields. They find different effects depending upon whether or not there is a normal rf electric field terminating on the surface. For smooth plane surfaces Halbritter finds a frequency-independent residual loss from an electric field and one proportional to ω^2 from a tangential rf magnetic field. Passow finds a more complicated frequency dependence which in lead peaks near 1 GHz and falls off for higher and lower frequencies. Both of these calculations predict lower residual loss than has been observed experimentally[6,12] to date, and no decision can yet be made between the different frequency dependences predicted. Halbritter also found that surface roughness or cracks in the surface can considerably enhance phonon generation above that for plane surfaces, again producing $R_0 \propto \omega^2$. This could explain

---

[52] L. Szécsi, Thesis, Universität, Karlsruhe, Germany (1970); *Z. Phys.* **241**, 36 (1971).
[53] J. M. Pierce and J. P. Turneaure (unpublished).
[54] J. Halbritter, *J. Appl. Phys.* **42**, 82 (1971).
[55] C. Passow, *Phys. Rev. Lett.* **28**, 427 (1972).

the results of Szecsi[52] in lead and Kneisel et al.[50] in niobium, but it seems unlikely that this is the whole story in the best surfaces for reasons outlined above.

It was argued by the author[6,36] that the remaining residual loss measured in our electroplated lead cavities cooled in very low fields may have been due in part to flux produced by thermoelectric currents generated between the lead plating and the copper substrate as the cavity is cooled. This flux is then trapped as the lead becomes superconducting. The evidence for this effect is not conclusive, but it is a plausible explanation for the extreme variability of the residual loss measured in the same cavity upon repeated cooling from the normal state. The distribution of the flux is sensitive to the details of the cooling process. Cavities cooled very slowly and uniformly did often show a lower residual loss, but this may also have been due to other effects.[6] Even in solid niobium cavities, where the thermoelectric effects are absent, there is always some trapped flux due to unavoidable residual magnetic fields in the region where the cavity is cooled. Rabinowitz[56] has argued that loss due to stationary flux lines will be proportional to ω^2, but it is difficult to see how the pinning frequency could be high enough for the lines to remain stationary at 1–10 GHz.

There are still other sources of residual loss in a superconducting surface. Surface purity, surface crystal size, surface smoothness, work damage, and contamination by surface layers of dielectric have all been shown to affect the residual loss. However, the individual contribution of each of these is not understood. These conditions are difficult to separate experimentally, particularly in the most interesting case, when the residual loss is already at a minimum. The experimental history has shown a steady reduction in residual loss by simultaneous attention to all these surface characteristics. As one would expect, the lowest residual loss has been obtained on surfaces of pure superconductors which are smooth and as free as possible from surface work damage and contamination either on the surface or by interstitial contaminants. It is possible to make arguments as to how some of these surface characteristics can cause residual loss. If there is a layer of dielectric on the surface, an electric field will produce dielectric loss. This will be particularly severe in a resonator mode with a finite normal electric field. Although very good results have been obtained recently with anodized niobium surfaces,[12] other contaminants[6] produce considerable loss. A projection from

---

[56] M. Rabinowitz, *Lett. Nuovo Cim. IV*, 549 (1970).

the surface in the form of a spike causes enhancement of the rf field and an increase of dissipation at its tip, while at the same time it is in relatively poor thermal contact with the rest of the metal. Since the superconducting surface resistance increases rapidly with temperature, there can be a thermal runaway that at some rf field level can keep some metal in the normal state.

By now it seems that ways have been found to reduce the residual loss to the point where any number of mechanisms may explain the remaining loss. Thus it is not clear what one can do to reduce it further in the best surfaces. The best values which have been achieved to date are $R_0 = 3 \times 10^{-8}$ Ω in lead,[6,8,10,52] and $R_0 = 2 \times 10^{-9}$ Ω in niobium.[3-5,57] It does appear to be true, however, that for modes with a normal electric field terminating on the surface, it is considerably harder to obtain low R_0 with lead than niobium. In such modes a typical value for lead is more like 10^{-6} Ω[11,58-60] than the value of less than 3×10^{-9} Ω for niobium.[3]

10.2.5. Dependence of the Surface Impedance on rf Field Level

Of crucial interest to those who wish to use superconducting resonators to produce high rf electric and magnetic fields is the behavior of the surface resistance as a function of rf field strength. We consider first the surface impedance for relatively low fields where the changes are continuous. Breakdown or critical field effects are discussed below in Section 10.2.6.

While the Mattis and Bardeen theory discussed in Section 10.2.2 is a perturbation theory valid only for weak rf fields, there has been some theoretical work done on the behavior of R_s as a function of H_{rf}. By "weak" here, of course, we mean $H_{rf} \ll H_c$, where H_c is the dc critical field of the superconductor. For low frequencies $\hbar\omega \ll \varepsilon(T)$, one expects

---

[57] P. Kneisel, O. Stoltz, and J. Halbritter, *Proc. Int. Conf. Low Temperature Phys., 13th, Boulder, Colorado*, 1972 (to be published by the University of Colorado Associated Press).

[58] D. Gorlé, D. Leroy, P. Morignot, H. Rieder, and Y. Bruynseraede, *Proc. Joint CERN-Karlsruhe Symp. Superconducting rf Separators, Karlsruhe, 1969* (CERN Rep. MPS/MU-SD/69/3, CERN, Geneva).

[59] A. Moretti, J. W. Dawson, J. J. Peerson, R. M. Lill, and M. T. Rebuehr, *IEEE Trans. Nucl. Sci.* **NS-18**, 186 (1971).

[60] A. Carne, R. G. Bendall, B. G. Brady, R. Sidlow and R. L. Kustom, *Proc. Int. Conf. High Energy Accelerators, 8th, CERN, Geneva*, p. 250 (1971).

10.2. MICROWAVE PROPERTIES OF SUPERCONDUCTORS

the superconducting state to respond to the field in much the same way as at zero frequency, i.e., an adiabatic approximation should be valid. Steady fields comparable to H_c cause a slight reduction in ε due to the kinetic energy of the shielding currents in the surface layer. However, the reduction is only about 15% from the zero-field value before the transition to the normal state takes place.[61] The transition at H_c in a magnetic field is first order, and ε jumps discontinuously to zero from nearly its zero field value. However, because R_s depends exponentially on ε, this small reduction in ε can at low temperature cause R_s to increase by several times.[62,63] The magnetic field can also induce surface states, electronic states localized near the surface, which modify the excitation spectrum of the metal in the region of rf field penetration. Halbritter[64] has considered the effects of these states, and he finds a nonmonotonic dependence of R_s on rf magnetic field level that in surfaces perfect enough to have considerable specular reflection for electrons can actually cause a decrease in R_s at higher field levels. Some experimental indications of effects due to these states have been found in high-quality niobium cavities.[16,57]

Direct phonon generation by the rf fields,[54,55] which at low field causes a residual loss, is nonlinear at higher fields. Turneaure[16] has estimated that for $H_{rf} < 1500$ Oe, the extra R due to this effect should be less than 3×10^{-9} Ω.

Finally, there are two loss mechanisms which involve the high *electric* field which accompanies a high rf magnetic field, and do not involve the superconducting state at all. These are electron field emission and multipactoring, which can occur in resonator modes where there is a normal electric field at the surface. For example, in a cavity of Turneaure and Nguyen[3] the peak H_{rf} of 1080 Oe was accompanied by a peak E_{rf} of 70 MV/m at the surface. This field is comparable to those at which field emission becomes significant in dc experiments, particularly if the surface is rough. Turneaure and Nguyen[3] did observe field emission as low as 28 MV/m in one of their niobium cavities which was not as care-

---

[61] See, for example, P. G. DeGennes, "Superconductivity of Metals and Alloys." Benjamin, New York, 1966.

[62] H. A. Schwettman, J. P. Turneaure, W. M. Fairbank, T. I. Smith, M. S. McAshan, P. B. Wilson, and E. E. Chambers, *IEEE Trans. Nucl. Sci.* **NS-14**, 336 (1967).

[63] J. Halbritter, External Bericht 3/68-8, Kernforschungszentrum Karlsruhe, Karlsruhe, Germany (1968); *J. Appl. Phys.* **41**, 4581 (1970).

[64] J. Halbritter, Externer Bericht 3/70-14, Kernforschungszentrum Karlsruhe, Karlsruhe, Germany (1970).

fully protected from surface contamination. Field emission is characterized by the X radiation produced by the accelerated electrons impinging on the walls of the resonator, and since it increases exponentially with field, it can set the limit on the field level that can be reached in a particular resonator. In lead resonators field emission is typically observed at lower field levels ($\sim$10 MV/m).[10,58,65] This can probably be explained by the roughness of typical lead-plated surfaces compared to the extremely smooth surfaces obtainable with niobium. Even in good niobium surfaces, however, field enhancement factors due to roughness of 200 to 400 are typically inferred from the field emission data.[16] Multipactoring[66] is a resonant phenomenon which can occur when the flight time of an accelerated electron across the resonator is such that the field reverses near the time when the electron strikes the opposite wall. Thus electrons ejected by the impact are accelerated back across the resonator to eject more electrons as the field again reverses. The result is a shower effect which can cause severe loading of the resonator.

Changes in the surface reactance due to high rf fields appear to be quite small. However, static radiation pressure can deform a cavity enough to shift its frequency by an observable amount,[3] and it must be taken into account in cavity design.†

10.2.6. The rf Critical Field

When $H_{rf} = H_c$ in a type I material we expect a catastrophic increase in R to something comparable to the normal state surface resistance. Similarly, for a type II material we would expect a drastic increase at H_{c1}, where vortices generated by the rf field can enter the material. For type I materials under certain conditions it is possible to increase a dc magnetic field above H_c to a "superheating" critical field.[33] To observe this experimentally, however, usually requires small samples where there are no nucleation sites, and it is unlikely that much superheating will occur in a macroscopic rf resonator, which inevitably has some flux trapped in its walls.

---

[65] A. Moretti, J. W. Dawson, J. J. Peerson, R. M. Lill, and M. T. Rebuehr, *IEEE Trans. Nucl. Sci.* **NS-18**, 186 (1971).

[66] G. Dammertz, H. Hahn, J. Halbritter, P. Kneisel, O. Stoltz, and J. Vortruba, *IEEE Trans. Nucl. Sci*, **NS-18**, 153 (1971); J. Halbritter, *Particle Accel.* **3**, 163 (1972).

---

† See Section 10.3.1 for further discussion of this point.

Turneaure[29] has measured the rf critical field of tin, a type I material. He finds that the surface resistance remains within a factor of two of its low-field value up to a well-defined critical point when the peak rf magnetic field at the resonator surface is equal to H_{crf}. At this point the Q of the resonator drops catastrophically. Figure 5 is a plot of Turneaure's data[16,29] for H_{crf} as a function of temperature. The solid curve is the dc critical field H_c, measured on samples electroplated similarly to the cavity surface. It is an acceptable fit to the data. The error bars indicate difficulties in estimating the temperature of the cavity surface, which is

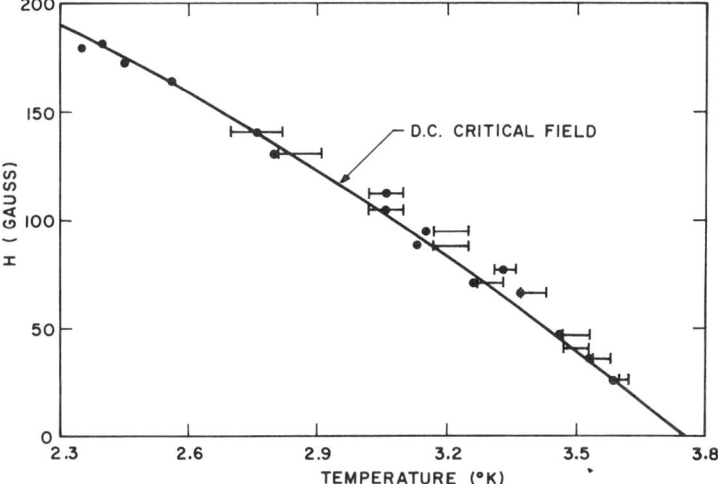

FIG. 5. Critical rf magnetic field measured in a tin-plated cavity. The points are H_{crf}, and the solid curve is the dc critical field. [After J. P. Turneaure, Ph.D. Thesis, Stanford Univ. (1967)].

heated substantially by the high rf power densities involved. Heating of the surface is a common experimental difficulty in experiments involving high rf fields, and we will have more to say about it below. Similar results have been obtained for lead and indium.[11] For these type I superconductors as well, R is not substantially increased until H_{rf} reaches H_c. These measurements were all done in cavities with zero rf electric field at the wall for reasons indicated above.

In type II materials the situation is less clear. We are led by the preceding arguments to expect $H_{crf} = H_{c1}$, and no dramatic increase in R for lesser fields. At 1.2 K in a pure, unstrained, bulk specimen of niobium $H_{c1} \approx 1850$ Oe. This is what makes niobium the most promising material for high-field applications, and almost all the work on H_{crf} in type II materials has been concentrated on niobium. However, all the experiments on niobium[12] indicate values of H_{crf} substantially less than H_{c1}. The highest values of H_{crf} observed to date are 1180 Oe in a mode with no accompanying normal electric field by Diepers and Martens[13] and 1080 Oe in a mode with a high normal electric field by Turneaure and Nguyen.[3] Both these measurements were made in small niobium cavities resonant in X-band. Values of H_{crf} up to 1080 Oe have also been reported very recently in a helical niobium resonator at frequencies ~100 MHz.[67]

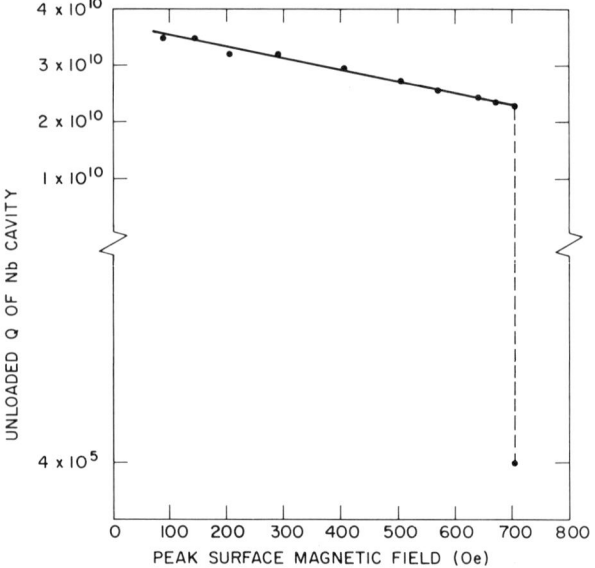

FIG. 6. Unloaded Q of a niobium cavity as a function of the peak of magnetic field at the cavity wall for TM_{010} mode cavity at 8.6 GHz and 1.25 K [after J. P. Turneaure and Nguyen Tuong Viet, *Appl. Phys. Lett.* **16**, 333 (1970)].

A set of data from one of the cavities of Turneaure and Nguyen[3] is shown in Fig. 6. This one does not exhibit the highest H_{crf}, which indicates the variability in H_{crf} obtained by even the most successful experimenters. Figure 6 also show the slight decline in Q as H_{rf} is increased toward H_{crf}. The data of Diepers and Martens[13] show more curvature

[67] A. Citron, J. L. Fricke, C. M. Jones, H. Klein, M. Kuntze, B. Piosczyk, D. Shulze, H. Strube, J. E. Vetter, N. Merz, A. Schempp, and O. Siart, *Proc. Int. Conf. High Energy Accelerators, 8th, CERN, Geneva*, p. 278 (1971).

in this region, which is consistent with $R \propto H_{\rm rf}^2$. More complicated dependences, perhaps due to surface states, have also been observed.[57] However, the total decline in Q shown in Fig. 5 between $H_{\rm rf} = 0$ and $H_{\rm rf} = H_{\rm crf}$ is typical. Thus even in the best niobium surfaces that have been made, $H_{\rm crf}$ is still slightly less than $0.7 H_{\rm c1}$.

The prevailing opinion[16] is that these early breakdown effects are not due to large scale breakdown of the superconducting state by an rf field less than $H_{\rm c1}$. Rather, they are due to a host of effects, most of which quickly cause heating of a small portion of the surface, bringing about a drastic decrease in the Q of a resonator. If for some reason a small local region on the resonator surface develops higher surface resistance than the average, a significant fraction of the total power fed into the resonator will be dissipated in this region, which may raise the temperature there. Because of the extremely rapid temperature dependence of R_s, there is a further increase in the local R, and a thermal runaway can develop. To produce fields comparable to $H_{\rm c1}$ in niobium requires substantial heat flow into the surface, even if the superconducting state does not break down. For example, if $H_{\rm rf} = 1800$ Oe and $R \approx 2 \times 10^{-9}$ Ω, which is the lowest value ever observed in niobium or any other material, the power flow is ~ 1 mW/cm². For a typical cavity in X-band, this implies a total power dissipation ~ 1 W. Furthermore, many of the measurements on niobium have been made on surfaces with 10–100 times higher surface resistance with correspondingly higher heat flow. Thus there is plenty of power available, if focused in a local region in the manner described above, to heat a small region substantially. In addition, if the surface resistance is residual, as it was in the higher resistance measurements mentioned, the low field loss may very well already be localized in a number of relatively small regions on the surface. Similarly, surface roughness or protrusions on the surface will also cause increased local power dissipation due to local field enhancement in regions which are already in relatively poor thermal contact with the bulk.

Rabinowitz[68] has discussed another way breakdown could occur due to local heating. He considers dissipation around lines of trapped flux passing perpendicularly through the surface or parallel to the surface but within the region of rf current flow. In his model breakdown occurs when $H_{\rm rf}$ is equal to the thermally reduced value of $H_{\rm c2}$ at the edge of the core of the vortex.

There are other possible mechanisms to explain the breakdown which

---

[68] M. Rabinowitz, *J. Appl. Phys.* **42**, 88 (1971).

do not initially involve heating. Relatively small amounts of impurities can substantially reduce H_{c1} in niobium. For example, 400 ppm of interstitial oxygen reduces H_{c1} in niobium by a factor of about two,[69] and carbon is about equally effective.[70] However, the impurity concentrations in the material[71] of Turneaure and Nguyen and Diepers and Martens were too small to account for the observed values of H_{crf} unless the impurities were concentrated on the grain boundaries. This kind of concentration could possibly occur in the heat treatment of the material. There are also indications that oxygen resides preferentially on the surface of niobium.[72] A protrusion or bump on the surface will produce an enhancement of the rf field at its tip. Thus it is possible that H_{rf} could approach H_{c1} at such a bump while the average H_{rf} is considerably less. Superheating might explain why this is not a serious problem in type I materials.

In the work that produced the highest values of H_{crf}, great efforts were made to produce a pure, uniform, unstrained niobium surface, and the highest H_{crf} values were first achieved in small X-band cavities.[3,13] Surfaces of comparable quality will clearly be harder to achieve in lower-frequency resonators with larger surface areas exposed to the high fields. As Turneaure has suggested,[12] this may well be the reason that the values of H_{crf} achieved to date in resonators for S-band and below are more typically 600 Oe and below.[12] The 100-MHz helical resonator of Citron et al.,[67] which showed $H_{crf} = 1080$ Oe, also had relatively small area on on the helix exposed to the high field.

In summary then, no fundamental effect has been found which limits H_{crf} to a value less than H_{c1} in type II superconductors. However, they are clearly much more sensitive than type I materials to the quality of the surface. With the intensive development work now going on in a number of laboratories, we can hope that in the not too distant future even large area niobium surfaces will be prepared with H_{crf} approaching H_{c1}.

---

[69] C. S. Tedman, Jr., R. M. Rose and J. Wulff, *J. Appl. Phys.* **36**, 164 (1965).

[70] R. W. Meyerhoff, quoted in J. P. Turneaure and I. Weissman, *J. Appl. Phys.* **39**, 4417 (1968).

[71] This material was similar to that of J. P. Turneaure and I. Weissman, *J. Appl. Phys.* **39**, 4417 (1968). See also J. P. Turneaure, *Proc. 1972 Appl. Superconductivity Conf., Annapolis, Maryland*, IEEE Pub. No. 72CH0682-5-TABSC, p. 621.

[72] M. Strongin, H. H. Farrell, C. D. Varmazis, H. Halama, and O. F. Kammerer, *Proc. 1972 Appl. Superconductivity Conf., Annapolis, Maryland*, IEEE Pub. No. 72CH 0682-5-TABSC, p. 667.

10.3. Methods of Fabricating High-Q Superconducting Resonators and Measuring Their Properties

10.3.1. Design and Fabrication of High-Q Resonators

Possible shapes, sizes, and resonant modes in superconducting resonators are as varied as the experimental uses to which they can be put. Figure 7 is a sample of the variety of superconducting resonators that have been built. Cavities (a) and (b) are for simple TE and TM modes, and they are coupled to waveguide and coaxial line, respectively. Resonators of this sort are used to study the properties of the superconducting surface itself, and they are useful as high-Q circuit elements for oscillators, filters, etc. Structures (c) and (d) are slow-wave structures of the sort used in linear accelerators. They are excited in high-order modes which have an axial electric field that changes in phase by $\pi/2$ or π between adjacent subsections of the structure. Thus a charged particle moving axially along the structure at the correct velocity is continuously ac-

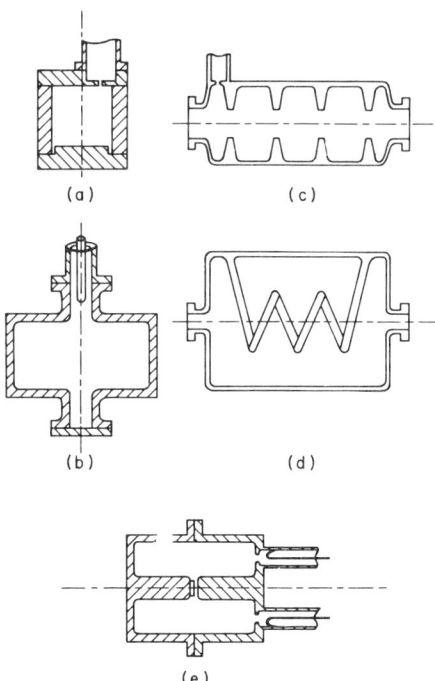

FIG. 7. Several forms of microwave resonators for various purposes discussed in the text.

celerated. Structure (e) is designed for studies of the response of samples to an rf electric field. A relatively uniform and concentrated electric field exists across the gap between the coaxial stubs. Thus a small amount of a sample dielectric substance placed in the gap can have a large effect on the properties of the resonator.

The details of the calculations of mode structure and resonant frequencies for various resonator geometries are beyond the scope of this article. The simpler structures are covered in numerous books,[18-22] and the more complicated cases must be handled with empirical methods or iterative solutions to the boundary value problem on a computer. However, for many purposes an experimental approach, extrapolating from simple geometries that are well understood, may be satisfactory.

Once the required geometry to produce the desired resonant frequency in the desired mode is known, however, there are design criteria which are peculiar to superconducting resonators. These are particularly important if the ultimate in Q is desired. Most of these follow from the fact that a high-quality superconducting surface exhibits much lower loss than almost any other substance. For example, even a small amount of a foreign substance in the resonator such as a dielectric support for a part of the resonant structure or a contamination layer such as oxide on the surface can completely determine the Q of the resonator.

One can take account of several sources of loss in a resonator in the following way: The unloaded Q of a resonator is defined in terms of the total average power dissipated over a cycle and the energy stored in the resonator

$$P_d = \omega U/Q_0. \tag{10.3.1}$$

If P_d is dissipated through several loss mechanisms, we can write

$$P_d = P_1 + P_2 + \cdots = \omega U(1/Q_1 + 1/Q_2 + \cdots). \tag{10.3.2}$$

Thus

$$1/Q_0 = 1/Q_1 + 1/Q_2 + \cdots. \tag{10.3.3}$$

Clearly, the *lowest* Q_i dominates in determining Q_0. In Section 10.2.1 we discussed losses in the superconducting walls of the resonator. These were given in Eq. (10.2.4) as

$$Q_w = G/R, \tag{10.3.4}$$

where the subscript stands for "wall," R is the surface resistance, and G

is a constant determined by the geometry of the resonator, the mode excited, and the frequency. Thus a goal in designing a high-Q resonator is to reduce the extraneous sources of loss to the point where they are comparable to those in the walls. The effect of each loss source can be expressed as an effective Q_i for the resonator.

For a dielectric in the resonator, the calculation of Q_i is rather simple if the loss tangent at the frequency of interest is known, together with the rf electric field in the resonator at the position of the dielectric. The effects of other loss sources are less easy to calculate. One of these is loss due to current flowing across joints in the walls of the resonator. Such joints are almost always necessary in order to fabricate the resonant structure, but it is very difficult in practice to make them in such a way that rf currents can cross them with acceptable loss. Carefully designed and executed electron beam welds in niobium have proven satisfactory, however.[3,12] If at all possible, necessary joints should be located where there is zero rf current in the mode of interest. This has been done in the cavity of Fig. 7a, designed for the TE_{01n} modes. In these modes no current flows from the ends to the sidewalls of the cavity. Alternatively, necessary joints may be "hidden" at the end of narrow tubes leading from the resonator which are waveguides beyond cutoff for the frequency of interest. This has been done in Fig. 7b. Another source of loss is coupling to unwanted modes, which may have lower Q. This problem is actually easier to handle in a high-Q resonator, because the band width of the modes is very narrow. Hence the modes do not overlap unless they are degenerate. The step in the end of the cavity of Fig. 7a is a "mode trap" designed to perturb the frequencies of the degenerate TM_{11n} modes away from those of the desired TE_{01n} modes to avoid this cross coupling.

Another problem which is peculiar to very high Q resonators is that the structure must be extremely rigid mechanically. For example, in an X-band cavity which is ~ 3 cm long and has $Q \approx 10^{10}$, a change in length of 3×10^{-10} cm will shift the resonant frequency by the bandwidth. Such shifts can easily be produced by small pressure changes exterior to the cavity even in quite massive structures. Thus a high-Q cavity makes an extremely sensitive pressure gauge. Vibrations are troublesome for the same reason, particularly in a structure such as Fig. 7d. (See Section 10.4.2 for a further discussion of this point.) Turneaure and Nguyen[3] were able easily to detect the effect of radiation pressure in measurements of frequency shift as a function of excitation level in their cavities, and radiation pressure can couple rf cavity modes to mechanical modes in helical structures.[73] Active feedback techniques can be used to suppress this coupling, however.[67,72]

10.3.2. Materials for High-Q Superconducting Resonators

Niobium and lead are the two superconductors which have been most intensively studied for resonator use. This is because they show the greatest promise for the high-field requirements of accelerator applications, but they are also probably the most suitable materials for a large number of other applications as well. These two superconductors have the highest transition temperatures and critical magnetic fields of all materials which are available commercially in reasonable purity and easily fabricable into the complex shapes often required in resonators while maintaining or recreating a surface which is clean, smooth on an atomic scale, and relatively unstrained. The high critical fields (H_{c1} = 1850 Oe for Nb and H_c = 750 Oe for Pb) are important for high-field applications for the reasons discussed in Section 10.2.5. The high T_c (T_c = 9.25 K for Nb and T_c = 7.19 K for Pb) means that the minimum surface resistance is attained at a higher temperature. This simplifies the cryogenic requirements, and indeed it can be a crucial factor in the economics for a large-scale, high-field application like an accelerator. The power cost of refrigeration is inversely proportional to the temperature at which the heat must be removed. Thus niobium or lead is probably the superconductor of choice unless one desires a particular T_c because he wishes the surface impedance to be strongly temperature-dependent in a particular temperature range or has other specialized requirements.

Superconductors are known with higher transition temperatures and critical fields than Nb and Pb, but they are either alloys of Nb with other metals such as Zr or Ti, or they are intermetallic compounds such as Nb_3Sn. The alloys have short mean free paths which imply relatively low values of H_{c1}.[23] In addition, their metallurgy is difficult, and it is not clear that one can prepare a sufficiently perfect surface for low residual loss. The intermetallic compounds are generally brittle and difficult to fabricate in complex shapes. They too will probably present difficulties in obtaining a good surface. In any case, little work has been done on the application of these materials to high-frequency resonators, and the measured values of R_0 have been quite high.[74]

---

[73] P. H. Ceperley, *IEEE Trans. Nucl. Sci.* **NS-19**, 217 (1972); D. Schulze, H. Strube, K. Mittag, B. Pioszcyk, and J. Vetter, *IEEE Trans. Nucl. Sci.* **NS-18**, 160 (1971).

[74] D. Soumpasis and K. Lüders, *J. Appl. Phys.* **41**, 2475 (1970).

10.3.2.1. Niobium. For the ultimate in low residual loss or for the highest field applications, niobium seems at present to be the material of choice. Techniques for fabricating and processing resonant structures of niobium have been developed first at Stanford University[3,7,9,12,75] in connection with the program to build a superconducting linear accelerator and more recently at Karlsruhe[50,67,76] and Siemens.[13,77,78] This work has resulted in the lowest values of residual resistance ($R_0 \approx 2 \times 10^{-9}\ \Omega$)[3-5,57] and the highest rf critical fields measured to date in any material ($H_{\text{crf}} = 1080$ Oe when there is a normal electric field at the surface[3,67] and $H_{\text{crf}} = 1180$ Oe with no electric field at the surface[13]).

There are two basic processes by which these very-high-quality surfaces have been achieved. Both are aimed at producing a smooth, uniform, unstrained niobium surface of high chemical purity, and both start with electron beam melted niobium with purity slightly better than commercial reactor grade niobium.[7,12] The Stanford process, which was developed first,[3,7,9] uses as a final step an ultrahigh-vacuum firing at about 1900°C. The Siemens process, which has evolved recently,[13,77,78] finishes by applying an anodic layer of amorphous Nb_2O_5, and it does not involve a high-temperature firing. We will outline the processing steps briefly here, but the reader will probably wish to consult the cited references before attempting to follow either of these procedures.

In the Stanford process[3,12,75] the machined or otherwise formed resonator parts are first fired[75] in uhv at about 1900°C. This produces crystals ~1 cm in size and minizes grain boundary length for the chemical polishing step.[4] Next, about 100 μm of material is removed in a chemical polishing bath consisting of 6 parts $HNO_3(70\%)$ and 4 parts $HF(4\%)$ by volume, held at 0°C. This step removes the damaged layer resulting from the machining. Finally, the parts are again fired in uhv at 1900°C to remove chemical contamination from the polishing solution. This leaves a very clean and smooth but highly reactive niobium surface. The parts must be protected from contamination in an inert gas atmosphere until the resonator is evacuated after assembly. Good vacuum practice is essential in the coupling lines to the cavity to avoid cryopumping or gettering contaminants onto the surface when it is

---

[75] J. P. Turneaure, *IEEE Trans. Nucl. Sci.* **NS-18**, 166 (1971).
[76] W. Bauer, A. Citron, G. Dammertz, H. C. Eschelbacher, W. Jüngst, H. Lengeler, H. Miller, E. Rathgeber, and H. Diepers, *Proc. 1972 Appl. Superconductivity Conf.*, Annapolis, Maryland, IEEE Pub. No. 72CH0682-5-TABSC, p. 653.
[77] H. Martens, H. Diepers, and R. K. Sun, *Phys. Lett.* **34A**, 439 (1971).
[78] H. Diepers, O. Schmidt, H. Martens, and F. S. Sun, *Phys. Lett.* **37A**, 139 (1971).

cold, and pumping should be done with ion pumps or turbomolecular pumps to avoid pump oil contamination.

The Siemens process[13,77,78] is simpler in the sense that it does not require an expensive uhv, high-temperature furnace, and the anodic oxide layer may protect the reactive niobium surface and make it less sensitive to contamination.[13] However, degradation has been observed in anodized surfaces kept for long periods at room temperature,[79] perhaps due to conversion of the Nb_2O_5 to NbO at the metal–oxide interface. Also, there is some evidence[16] that anodizing enhances the secondary electron emission involved in multipactoring[66] at high electric fields. In the Siemens process the machined resonator parts are electropolished to remove about 100 μm. The solution is 85 parts H_2SO_4 (conc.) and 10 parts HF (40%) by volume, held at a constant temperature near 30°C. Aluminum cathodes are used. The voltage is adjusted around 10 V to obtain a characteristic current oscillation[78] and is turned off periodically to allow the acid to dissolve the anodic oxide which forms on the niobium surface. This step removes work damage from the machining and smoothes the surface. Electropolishing is superior to chemical polishing for unfired surfaces, which are often small grained, because it does not etch the grain boundaries.[4] After polishing, the surfaces are anodized[77,79] to a depth $\sim$0.1 μm ($\sim$50 V), usually in a solution of NH_3 ($\sim$20%) at room temperature with a current density $\sim$2 mA/cm². This step displaces the surface into the metal to a deeper and presumably purer layer. Thus it performs somewhat the same function as the final firing in the Stanford process.

One can now ask how low a value of R_0 can be obtained with less elaborate surface treatment. Resonators machined of reactor grade niobium with machined or mechanically polished surfaces typically exhibit $R_0 \approx 10^{-6}\ \Omega$.[7,58] Similar results have been obtained with Nb surfaces sputtered onto copper substrates.[80] Surfaces which have been chemically or electrolytically polished to remove work damage as above but are not protected from contamination by good vacuum technique or anodizing will quickly degrade to $R_0 \approx 10^{-6}\ \Omega$.[7,81]

The advent of the Siemens process makes it practical to make high-quality niobium surfaces in many laboratories which do not have access

---

[79] H. J. Halama, *Particle Accelerators* **2**, 335 (1971).

[80] D. Gorle, D. Leroy, H. Rieder, and Y. Bruynseraede, CERN, MPS/MU-NOTE SD69-1, CERN, Geneva (1969).

[81] H. Hahn and H. J. Halama, *IEEE Trans. Nucl. Sci.* **NS-16**, 1013 (1970).

to a uhv furnace. However, there still may be a problem in fabricating the required resonator geometry. The high-purity niobium required is expensive, and the only way which has been found to make satisfactory current-carrying joints is electron-beam welding.[3,12] This may pose a problem if the resonator cannot be designed to avoid such joints. While commercial welding services do exist, the process must be carefully controlled[12] to produce a smooth inner surface without spattering niobium around.[4] Niobium can be electroformed,[82] but the process requires a molten salt bath at high temperature in an inert atmosphere. To do this in the laboratory is a substantial undertaking.

10.3.2.2. *Lead.* Lead may well be the material of choice for many experimental purposes where the ultimate in Q and high electric fields are not required. High-quality lead surfaces have exhibited $R_0 \approx 3 \times 10^{-8}$ Ω,[6,8,10,11] and $H_{crf} \approx 750$ Oe[11] in modes where there is no normal electric field at the walls. In modes with a normal electric field, however, the limit seems to be $R_0 \approx 10^{-6}$ Ω at low field levels.[11,58–60,65] This is probably due to dielectric loss in contaminant films[6] or oxide[83] on the surface, which are difficult to avoid with lead. There are also the indications[52] discussed in Section 10.2.4 that in these modes R_0 is dependent on frequency in some surfaces as strongly as ω^2. Lead surfaces also seem to be quite sensitive to high, normal, electric fields with breakdown occurring at $E \approx 10$ MV/m due to field emission and multipactoring.[10,58,65] This is probably due to the difficulty of obtaining really smooth lead surfaces.

The major advantage of lead is that high-quality resonators can be fabricated in the laboratory with commonly available equipment using ordinary electroplating techniques. The plating technique evolved by the author and colleagues at Stanford University[6,17,29,62,84] is given here. Other laboratories[8,10,11,85] have had comparable results with similar techniques. The substrate we use is OFHC grade copper, and its surface is polished mechanically (600 grit). The piece is then cleaned chemically with acids followed by a KCN dip to remove oxides. While still wet from the distilled water rinse, it is immediately transferred to a lead fluoborate solution (technical grade diluted to 30° Baume). We use 0.4 gm/liter of

---

[82] R. W. Meyerhoff, *J. Appl. Phys.* **40**, 2011 (1969).

[83] T. A. Tombrello and D. A. Leich, *IEEE Trans. Nucl. Sci.* **NS-18**, 164 (1971).

[84] W. M. Fairbank, J. M. Pierce and P. B. Wilson, *Proc. Int. Conf. Low Temp. Phys., 8th* **LT8**, 324. Butterworths, London and Washington, D.C., 1963.

[85] A. Septier and M. Salaün, *C.R. Acad. Sci. Paris, Ser. AB* **269B**, 685 (1969).

animal bone glue as a smoothing additive. The anodes are high-purity lead (99.999% nominal), and several ounces are plated through a new bath to clean it up prior to plating a piece to be used. A layer of lead 5–10×10^{-4} cm thick is applied to the interior surfaces of the resonator parts in about 40 min. This thickness is much greater than the coherence length in lead (~ 800 Å) to avoid proximity effects from the copper substrate. Septier and Salaun[85] found a strong variation of R_0 with the plating current density, which they attributed to impurities from electrodissociation of the organic additive in the plating bath. Therefore, it is worth experimenting with the plating rate in a particular bath. Upon removal from the plating bath, the part is *immediately* rinsed in running distilled water (unaerated), followed *immediately* by a rinse in running absolute ethanol to remove the water. Speed in the last two steps is essential to produce a relatively oxide-free surface, as lead oxidizes rapidly in the presence of water and air. Finally, the ethanol is evaporated by a stream of dry nitrogen. The resulting lead surface is a matte grey-white in appearance with no visible oxide film. The grain size is about 10^{-4} cm. Recently, a study has been made of the effects on the rf properties of electroplated lead layers of process parameters such as copper quality and surface treatment, bath and anode purity, additives, current density, and rinsing techniques. The reader may wish to consult this work.[86] Some workers have successfully used surface treatments with organic agents such as Varsene to retard oxidation.[87]

Lead surfaces must be kept clean if the lowest R_0 is desired. They should be stored in a dry atmosphere until evacuated for use. To avoid cryopumping contaminants onto the surface, the resonator should be sealed prior to cooling. A mild bakeout may also help. For example, one cavity pumped to 10^{-5} Torr without bakeout and sealed showed $R_0 \approx 7 \times 10^{-7}$ Ω. Upon repumping to 10^{-5} Torr with a bakeout at 100°C and resealing, an improvement to $R_0 \approx 7 \times 10^{-8}$ Ω was observed.[6]

It is of interest to consider the residual surface resistance of not so carefully prepared lead surfaces. Solid lead cavities, even though annealed and electropolished have generally shown $R_0 \approx 5 \times 10^{-4}$ Ω or greater.[14]

---

[86] A. Carne, R. G. Bendall, B. G. Brady, R. Sidlow, and R. L. Kustom, *Proc. Int. Conf. High Energy Accelerators, 8th,* p. 262. CERN, Geneva, 1971.

[87] For a detailed description of the Brookhaven plating technique using Varsene, see H. Hahn, H. J. Halama, and E. H. Foster, *Proc. 1968 Brookhaven Summer Study Superconducting Devices Accelerators,* BNL 50155 (C-55), April, 1969, p. 13. Available from Clearing House for Federal Scientific and Technical Information, NBS, Springfield, Virginia, 22151.

On the other hand, cavities plated as above but used for a long period and allowed to oxidize seemed to maintain $R_0 < 10^{-4}\,\Omega$.[88] The increase of R_0 due to corrosion and oxidation may be expected to be sensitive to normal electric fields and the exact nature of the corrosion fissuring of the surface.[54]

It should also be mentioned that lead surfaces with $R_0 \approx 10^{-7}\,\Omega$ have been produced by evaporation at Karlsruhe.[10] The vacuum was $\sim 10^{-9}$ Torr and the rates were 5–10 Å/sec. Samples with $\sim 4\%$ Bi in the lead showed comparable values of R_0.

10.3.3. Measurement Techniques for High-Q Resonators

In order to measure the properties of a resonator, one must couple at least one transmission line to it. Methods for doing this are discussed in Section 10.3.5. With a single line, one measures the reflection coefficient or input impedance of the resonator. With two, one measures transmission through the resonator.

We shall assume for simplicity that the coupling network and transmission line are lossless. This condition can usually be approximated sufficiently well in practice. With energy U stored in the resonator and no incident power, a certain power $P_h \propto U$ will be radiated out the transmission line h. This can be expressed in terms of coupling Q_h, one for each line

$$Q_h = \omega U / P_h. \tag{10.3.5}$$

The total or loaded Q of the resonator can then be written

$$Q_L^{-1} = Q_0^{-1} + \sum_h Q_h^{-1}, \tag{10.3.6}$$

where Q_0 is the unloaded Q of the resonator, the Q with no lines coupled to it. One can then define the coupling coefficients β_h

$$\beta_h = Q_0/Q_h. \tag{10.3.7}$$

Thus

$$Q_0 = Q_L\left[1 + \sum_h \beta_h\right]. \tag{10.3.8}$$

If there is no incident power, the stored energy U, and hence P_h, will decay exponentially with a time constant $\tau = Q_L/\omega$. Thus one can de-

---

[88] W. J. Trela and W. M. Fairbank, *Phys. Rev. Lett.* **19**, 822 (1967); W. J. Trela, Thesis, Stanford Univ. (1967).

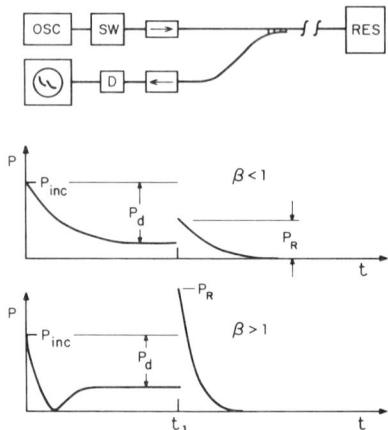

FIG. 8. Microwave circuit for the decrement method of measuring resonator Q, and wave forms observed on the oscilloscope for two values of the coupling parameter β. The microwave switch SW is turned on at $t = 0$ and off at $t = t_1$.

termine Q_L by exciting the resonator in some way and measuring the decay time for free oscillations. The decrement method for measuring Q_0 takes this approach. A single transmission line is usually used, and the reflected wave from the resonator is detected by means of a directional coupler as shown in Fig. 8. The oscillator must be stable enough to stay well within the resonator bandwidth ω/Q_L, and it is turned on and off by a microwave switch. The isolators are to stop any multiple reflections in the system, and the lines from the isolators to the resonator should be well matched with low loss. If the incident power P_{inc} is turned on with the switch at $t = 0$ and off at $t = t_1$, the curves of Fig. 8 show the behavior of the power P at the detector for two values of the single coupling factor β. The oscillator is assumed to be tuned to ω_0, which condition is obtained by maximizing P_R. At $t = 0$, provided $Q_L \gg 1$, P_{inc} is almost totally reflected from the coupling network at the resonator. Hence $P(0) \approx P_{\text{inc}}$. As the resonator fills and the radiated wave from the resonator builds up, it begins to cancel the reflected wave from the coupling network. Thus P drops. If $t_1 \gg Q_L/\omega$, P will approach a constant value $P(t_1) = P_{\text{inc}} - P_d$, where P_d is the power dissipated in the resonator. This follows if the line and coupling network are lossless. At t_1, after P_{inc} is turned off, P consists only of the radiated wave $P_R = \omega U/Q_h$. Since $P_d = \omega U/Q_0$, it follows that

$$\beta = P_R/P_d. \tag{10.3.9}$$

One then determines Q_L from the decay time after t_1. Note that t_1 must be long enough so that a steady state is reached and P is constant. Otherwise $P_{\text{inc}} - P(t_1) = P_d + dU/dt$. Note also that once Q_h is known, a varying Q_0 may be monitored continuously merely by monitoring P with constant P_{inc}. This method is sensitive over a range $0.2 < \beta < 2$, and one must know whether $\beta < 1$ or $\beta > 1$. Also the variations of Q_0 must be slow, $dQ_0/dt \ll \omega$.

The frequency sweep method is a variant of the decrement technique, which is useful when a highly stable oscillator is not available. The oscillator frequency is swept through resonance rapidly, and some energy gets into the resonator. As above, Q_L is determined from the decay of the free oscillations after the oscillator passes through resonance. These may be observed in a reflection setup such as that of Fig. 8 as a beat envelope formed by the radiated power at frequency ω_0 mixing with the reflected signal from the incident power sweeping away from ω_0 in frequency. However, it is better to apply this method with a transmission circuit in which two lines are coupled to the resonator. The swept frequency $\dot{\Omega} = \dot{\omega}/\omega_0^2$ is applied to line 1 with coupling factor β_1, and the decay is observed at line 2 with coefficient β_2. For an adiabatic fast passage through resonance ($\dot{\Omega} \ll Q_L^{-1} \ll \dot{\Omega}^{1/2}$), the expression for the power detected at the output is[89]

$$\frac{P_{\text{out}}}{P_{\text{inc}}} = \frac{\pi}{2\dot{\Omega}Q_L^2} \frac{4\beta_1\beta_2}{(1 + \beta_1 + \beta_2)^2} \exp\frac{-\omega_0 t}{Q_L}. \qquad (10.3.10)$$

In this context "adiabatic" ($\dot{\Omega} \ll Q_L^{-1}$) means that the passage is slow enough so that the cavity tracks the phase of the input; and "fast" means that the energy dissipated in the resonator is not appreciable during the passage ($\dot{\Omega}^{1/2} \gg Q_L^{-1}$). The extraction of Q_0 is simplest if $\beta_1 = \beta_2 = \beta$ by symmetry. Then

$$P_{\text{out}}(t=0)/P_{\text{inc}} = 2\pi\beta^2/\dot{\Omega}Q_0^2 = 2\pi/\dot{\Omega}Q_h^2, \qquad (10.3.11)$$

and

$$Q_0 = Q_L(1 + 2\beta). \qquad (10.3.12)$$

---

[89] This result may be derived using standard references, e.g., E. L. Ginzton, "Microwave Measurements." McGraw-Hill, New York, 1957. It was derived for this particular case by H. Hahn, H. J. Halama, and E. H. Foster, *Proc. 1968 Brookhaven Summer Study Superconducting Devices Accelerators*, BNL 50155 (C-55), April 1969.

Note that P_0/P_1 is independent of Q_0 for a given coupling geometry (given Q_h).

The classic method for measuring Q and ω_0 for a resonator is to map the resonance curve as a function of ω under steady state conditions. This can be done either by reaction or by transmission. The method is discussed in many texts, but it is rather difficult to apply to a resonator of extremely high Q. The signal source must be stable to a small fraction of the resonator bandwidth,[†] and the transmission lines must be very carefully matched to avoid errors in the measurement of the reflection or transmission coefficient.

Finally, one can let the resonator itself determine the frequency of an oscillator. This is particularly useful if the resonant frequency is changing rapidly and must be measured continuously, and it does not require a highly stable microwave signal source. We discuss here methods of using a resonator for frequency control which are particularly useful for measuring resonator parameters. In Section 10.4.3 we discuss the use of superconducting resonators in frequency standards.

Two approaches are possible, which are indicated in Fig. 9. First, one can use an amplifier with sufficient gain at ω_0, such as a traveling wave tube, and feed back power either reflected from or transmitted through the resonator to produce a self-sustaining oscillation. Transmission is probably more satisfactory than reflection since light coupling can be used ($\beta \ll 1$). This approach is shown in Fig. 9a. Automatic gain control is necessary to limit the oscillation, and information on β can be derived from AGC level, thus providing information on Q. Alternatively, the frequency shift associated with a known change of the phase shifter can be measured to determine the resonator bandwidth ω_0/Q_L. As in all microwave systems, liberal use of isolators greatly simplifies the matching requirements on the transmission lines and components. If a Q measurement is a primary objective, one can insert a microwave switch in the loop[90] to stop the oscillation and allow the resonator decay time to be measured. Another variant[91] of this scheme useful for measuring Q_L over a wide range involves putting a low-frequency amplitude modulation ($\omega \ll \omega_0/2Q$) on the carrier and measuring the relative phases of the detected i.f. from the incident and transmitted

---

[90] Nguyen T. Viet, *C.R. Acad. Sci. Paris* **258**, Group 5, 4218 (1964).
[91] M. Salaün, *C. R. Acad. Sci. Paris* **269B**, 347 (1969).

[†] $\Delta\omega/\omega = Q_L^{-1}$ for the full width at half power.

10.3. HIGH-Q SUPERCONDUCTING RESONATORS

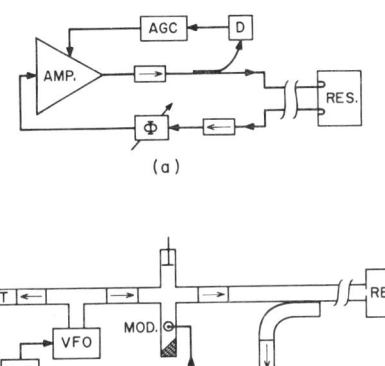

FIG. 9. Two methods to make a resonator determine the frequency of an oscillator (see text for details).

waves. For purposes of making a stable oscillator, this simple feedback circuit suffers from the difficulty that variations in electrical length of the lines between a room-temperature amplifier and the resonator due to varying temperature gradients will cause variations in the loop phase shift and shifts in the operating point of the oscillator.

As a second basic approach to making an oscillator from a resonator, one can lock a constant-amplitude, voltage-tunable oscillator (VFO) to the cavity. An adaptation of a method devised by Pound[92] is shown in Fig. 9b. The power incident on the resonator is modulated at i.f. to produce sidebands far enough from ω_0 so that they are well off resonance. They are simply reflected from the coupling network. These sidebands are produced in the magic tee modulator, and the movable short allows adjustment of the relative phase of the sidebands to that of the carrier. The sidebands thus provide a phase reference for the reflected power, and the i.f. signal produced when they mix with the reflected wave at ω_0 in the detector contains information on both the amplitude and phase of the reflection coefficient. If the modulator is properly adjusted, the i.f. signal is proportional to the imaginary part of the reflection coefficient, which for $\beta \approx 1$ changes sign as the oscillator tunes through the resonance. This i.f. signal is mixed with a properly phased signal from the i.f. oscillator to produce a dc discriminator voltage, which controls

[92] R. V. Pound, *Rev. Sci. Instrum.* **17**, 490 (1946).

the VFO. The dc component of the detected microwave signal is a direct measure of the total reflection coefficient, and hence is a measure of Q_0. Alternatively, one can derive a more sensitive measure of Q_0 at low values of β by measuring power transmitted through the resonator by means of a second transmission line. This system is relatively insensitive to line length variation, since the sidebands traverse the same path to the cryogenic section as the carrier. The modification of Pound's original design to achieve this insensitivity was used by Trela and Fairbank,[88] and it has been developed into a highly stable oscillator by Stein and Turneaure.[1] Their oscillator is discussed in Section 10.4.3.

10.3.4. Frequency Measurement and Control

Direct measurements on extremely high-Q resonators, such as the resonance curve or decrement method described above, put severe demands on the stability of the microwave source. The short-term stability is particularly important, since the sharp frequency characteristic of the resonator will convert FM noise to AM noise. Alternatively, measurements of the resonant frequency to the limits of stability obtainable with an oscillator locked to a resonator require extreme accuracy in the measuring equipment.

The short term stability ($\sim 10^{-2}$ sec) of available quartz frequency standards is about one part in 10^{10}, which is adequate for most purposes with resonator Q's up to 10^{11}. These fixed-frequency oscillators can be made continuously tunable without loss of stability by means of frequency synthesizers, and the signal may be translated to microwave frequencies from the few megahertz operating frequencies of these oscillators by various techniques. For low microwave frequencies (S-band and below), direct frequency multiplication and amplification are practical. For higher frequencies, the multiplied phase noise of the standard becomes prohibitive, and one must phase lock a fundamental microwave oscillator to a harmonic of the standard. This presents difficulties if a wide tuning range is desired, since the i.f. sections of most phase lock systems must be sharply tuned to select a single harmonic. A system devised by L. V. Knight and used by Turneaure[29] and the author,[36] which combines continuous tunability with quartz standard stability is shown in Fig. 10. The phase-lock module adjusts the klystron reflector voltage to keep its rf input in phase with a harmonic of the 5-MHz standard. By interposing an amplitude modulator driven by the tunable frequency synthesizer between the sampling coupler and the

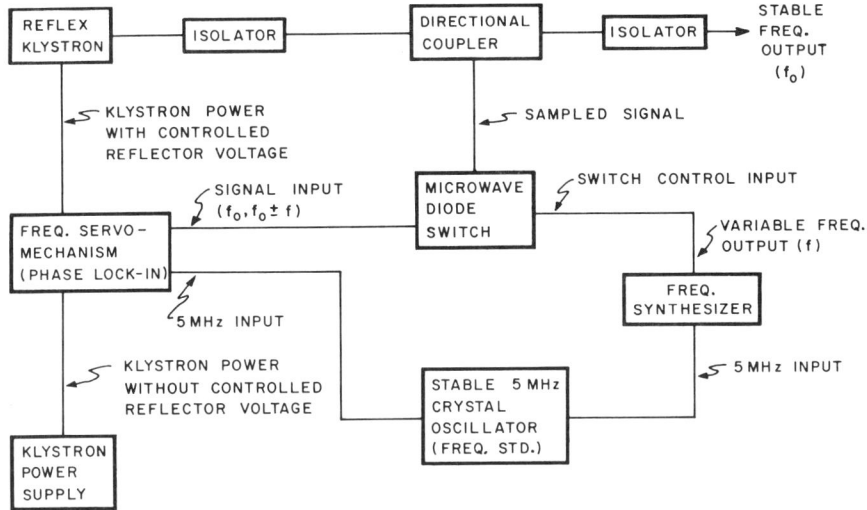

Fig. 10. A highly stable source of microwave frequency power which is continuously tunable.

phase lock circuit, the system is induced to lock onto the sideband rather than the carrier frequency. Thus the carrier frequency is continuously tunable with the synthesizer. Some difficulty arises as $f \to 0$, but in this range one can vary the internal i.f. signal of the lock circuit slightly using the synthesizer.

A source such as that of Fig. 10 is obviously complicated and expensive. Thus there is a real advantage if the experimenter can use the resonator itself to determine the frequency. This is difficult, however, if wide variations in Q, and hence β, are encountered in the experiment.

A number of techniques are available to measure accurately the frequency of a microwave signal. Commercial frequency counters now cover much of the microwave range. The source of Fig. 10 is accurate absolutely to a few parts in 10^9, the long term stability of the standard. If only relative frequency measurements are needed, two oscillators stabilized by superconducting resonators may be beat together.[88] Frequency shifts of small fractions of the resonator bandwidths are easily detectable.

10.3.5. Coupling Networks for Superconducting Resonators

A coupling network is required to couple a superconducting resonator satisfactorily to a normal-metal transmission line. One usually wants to produce coupling factors β comparable to unity without introducing appreciable loss into the system. The coupling required to do this is exceedingly weak by normal standards. We have seen that $\beta \approx 1$ implies $Q_h \approx Q_0$, and Q_0 can be exceedingly high. In addition, since in many experiments Q_0 will vary by several orders of magnitude, it is often highly desirable to make the network adjustable to maintain reasonable coupling.

Roughly speaking, if the coupling Q is Q_h, the fields in the resonator are $Q_h^{1/2}$ times those in the transmission line. It is this fact that allows the construction of a system in which the dominant losses are those in the superconducting resonator rather than those in the transmission line and coupling network, even though the surface resistance of the normal metal may be 10^6 times that of the superconductor. The first design requirement for a coupling network with negligible loss is that this field reduction be accomplished before any normal metal is exposed to the fields. It is usually most convenient to do this by exploiting the dissipationless attenuation which occurs in a waveguide beyond cutoff or by using a geometrical mismatch such as a small aperture or iris in the resonator wall. This part of the network must be constructed with a superconducting surface. Figure 11 shows a number of coupling networks which incorporate one or another of these features.

The second design criterion is that the network couple efficiently to the desired mode of the resonator and inefficiently to undesired modes. As we have mentioned, the extremely sharp resonances of a high-Q resonator help to alleviate the problem of coupling to unwanted modes, but efficient coupling is still desirable to minimize the field levels necessary in the coupling networks. In general, efficient coupling is accomplished by ensuring that the fields excited in the network by the transmission line have as nearly as possible the same geometry as the fields excited by the desired resonator mode. Essentially, Q_h is determined by the overlap integrals between these two waves, one for the electric fields and one for the magnetic.

Perhaps the simplest network is that of Fig. 11d, which couples a coaxial line on axis to the TM_{01} modes of a cylindrical cavity. Figure 7b is an example. A simple (superconducting) extension of the outer wall of the coax forms a waveguide beyond cutoff. The fields excited by the exposed tip of the center conductor have the same symmetry as the desired cavity mode, and Q_h is determined by the length of the cutoff section. This provides a simple means to adjust Q_h as indicated in Fig. 11d.

10.3. HIGH-Q SUPERCONDUCTING RESONATORS

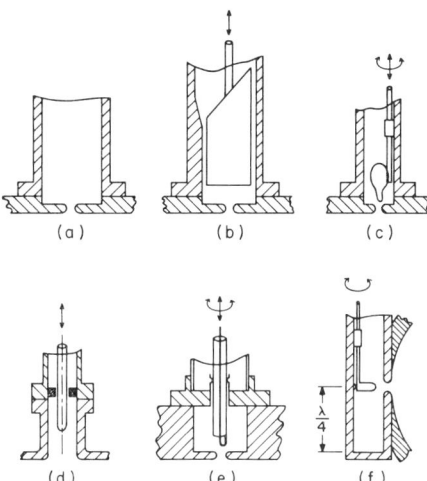

FIG. 11. Several coupling networks suitable for coupling high-Q superconducting resonators to normal metal transmission lines. Some provide variable coupling by means of motions indicated by arrows in the figure (see text for details).

The networks of Fig. 11a–c,e all involve an iris in the shorted end of a waveguide, and they couple by means of the magnetic field only. Thus the iris should be placed in the resonator wall at a position of maximum tangential magnetic field for the desired mode. Figure 7a is one example. The Q_h can be calculated for an iris[93–95] either in the end of a guide or in the sidewall as in Fig. 11f, where the coupling is by the electric field. These calculations are useful as a first approximation, but the coupling is extremely sensitive to the size and shape of the iris, and it is likely that experimentation will be necessary. We emphasize again that the iris itself must be superconducting. The networks of Fig. 11b,c are adjustable versions of 11a. In Fig. 11b the wide dimension of the guide is narrowed so that it is beyond cutoff except where filled with a movable plunger of low-loss dielectric such as Teflon.[96] For a wide range of coupling variation, 11b requires a larger iris than 11c, which increases the possibility of loss. In 11c the magnetic field is enhanced in the immediate vicinity of the hole by means of a wire loop. If the tip of the loop is inserted through the hole, very low Q_h results. On the other hand 11c

---

[93] H. A. Bethe, *Phys. Rev.* **66**, 163 (1944).
[94] I. G. Wilson, C. W. Schramm, and J. P. Kinser, *Bell. Syst. Tech. J.* **25**, 408 (1946).
[95] J. Van Bladel, *Proc. IEE* **117**, 1098 (1970).
[96] R. B. Lewis and T. R. Carver, *Phys. Rev.* **155**, 309 (1967).

is essentially equivalent to 11a when the loop is withdrawn and rotated flush with the side of the guide. This system worked well for the author[17,36,84] for low-power work, but at high power it is difficult to cool the loop. In Fig. 11f a similar approach is used to enhance the electric field with a conducting vane. The network of Fig. 11e couples magnetically to a loop on the end of a coaxial cable.[8,87] The circular section behind the iris is beyond cutoff, and a wide variation in Q_h can be achieved by moving the coax. The outer tube is a vacuum wall.

10.3.6. Special Cryogenic Microwave Techniques

We conclude this discussion of measurement techniques with a few remarks concerning special problems involved in transmitting microwave power from room temperature into a cryogenic environment. The requirements of low attenuation and low thermal conduction are conflicting to a large extent. If moderate attenuation is acceptable in the microwave design, waveguides or coaxial cables of stainless steel can be used. If minimum attenuation is essential, a copper waveguide must be used. Unacceptable heat leak down a copper guide can be avoided by breaking the guide into three sections encased in a stainless steel vacuum jacket. The microwaves can be transmitted efficiently across the gaps between sections without mechanical contact by means of carefully designed choke flanges.[20,21] Radiation heat leak can be suppressed by heat sinking the middle section with liquid nitrogen, and inserting matched radiation baffles in it.

Finally, for demountable vacuum joints which are also tight to microwave radiation to greater than 170 dB,[96] one can use indium wire gaskets crushed between flat flanges. Wire ~0.05 cm in diameter crushed down to ~0.075 cm works well.

10.4. Brief Review of Experiments Using Superconducting Resonators

In the preceding sections we have discussed many experiments designed to elucidate the response of superconducting surfaces to rf fields ranging up to the critical field in magnitude and up to ε/h in frequency. These experiments constitute the bulk of the work which has been done with superconducting resonators, and they have resulted in an understanding of their behavior which is adequate for most design purposes. Superconducting resonators can now be designed with predictable properties for all but the most demanding applications where the extreme of high Q or high fields are required.

In this section we review briefly the experiments which have been done using the unique properties of superconducting resonators for other purposes than studying the behavior of superconductors. The uniquely high Q's obtainable with these resonators even in the presence of rather high rf fields suggest that they should be extremely useful devices in a wide range of experimental and practical applications. Only a few of the many possible applications have been explored, and many opportunities are left for future experimental ingenuity.

Perhaps the most obvious applications for extremely high Q resonators are those which exploit the correspondingly narrow bandwidth of the resonance. The full width $\Delta\omega$ at half maximum power of the resonant absorption curve is given by $\Delta\omega/\omega_0 = Q_L^{-1}$, where ω_0 is the resonant frequency and Q_L is the loaded Q of the resonator. Thus it is apparent that bandwidths less than 1 Hz are obtainable at microwave frequencies. This immediately suggests that these devices should be useful as filters and as the frequency determining elements in stable oscillators. The fact that these resonators can be designed to be tunable over an appreciable range of frequencies makes them even more promising.

10.4.1. Frequency Stability in Superconducting Resonators

To exploit fully the narrow bandwidth of a high-Q resonator, one must first ensure that ω_0 is stable enough. Certainly the center frequency of a filter, for example, must be stable to a small fraction of the bandwidth. The resonant frequency of a resonator is determined by its geometry, the mode excited, the dielectric constant of any material in the fields, and the surface reactance of the walls. All of these are sensitive in some degree to external variables such as temperatures, pressure, or mechanical force. Discussions of the stability problem have been given by Stone and Hartwig,[97] Biquard,[98] and Stein and Turneaure.[1]

Consider first the effect of surface reactance variation on the frequency. It depends upon the geometry of the resonator and the mode in use, but an estimate for the fundamental mode in a simple cavity resonator will illustrate the problem. The surface reactance can be expressed in terms

---

[97] J. L. Stone and W. H. Hartwig, *J. Appl. Phys.* **39**, 2665 (1968).
[98] F. Biquard, *Rev. Phys. Appl.* (*France*) **5**, 705 (1970).

of the penetration depth λ for the field, and λ is a function of temperature

$$\lambda \approx \lambda_0(1 - t^4)^{-1/2}, \qquad (10.4.1)$$

where $t = T/T_c$. If λ changes, then the effective length of the cavity changes by the same amount, and the resonant frequency is inversely proportional to the length. Thus

$$f^{-1} df/dt \approx -L^{-1} d\lambda/dt, \qquad (10.4.2)$$

where L is the length. There is a proportionality constant of order unity which depends upon the mode and the shape of the cavity.[99] Using (10.4.1) in the region $t \ll 1$, we obtain

$$\frac{1}{f}\frac{df}{dt} \approx \frac{\lambda}{L}\frac{(-1)}{\lambda}\frac{d\lambda}{dt} \approx \frac{-2\lambda t^3}{L}. \qquad (10.4.3)$$

As an example consider a 3-GHz lead cavity operated at 2 K. Then $L \approx 10$ cm, and $t \approx 0.3$. For most superconductors $\lambda_0 \approx 5 \times 10^{-6}$ cm. In this case then $f^{-1} df/dt \approx 3 \times 10^{-8}$. To maintain stability to one part in 10^{12} thus requires temperature regulation to 200 μK. One advantage of a cryogenic environment is that this sort of temperature stability or much better is quite easy to maintain over the short term.[100,101] Difficulties[102] are encountered for times longer than a few hours, and if much better than this order of stability is required for very long times, it may be necessary to correct the resonator frequency through one of the tuning methods discussed in Section 10.4.2.

Mechanical motions also must be carefully controlled. In the above cavity a frequency shift of one part in 10^{12} corresponds to a length change of 10^{-11} cm. Thus to obtain a stable resonant frequency, one must pay careful attention to minimizing any vibration of the structure, and one must minimize the effects of radiation pressure which can deform the structure. By using a relatively massive and rigid structure, isolated reasonably well from high-frequency external vibrations and accurately

---

[99] See, for example, J. C. Slater, "Microwave Electronics." Van Nostrand-Reinhold, Princeton, New Jersey, 1950, p. 80.

[100] See, for example, A. C. Rose-Innes, "Low Temperature Techniques." English Univ. Press, London, 1964; or A. J. Croft, "Cryogenic Laboratory Equipment." Plenum Press, New York, 1969.

[101] C. P. Pickup and W. R. G. Kemp, *Cryogenics* **9**, 90 (1969).

[102] W. L. Johnson and A. C. Anderson, *Rev. Sci. Instrum.* **42**, 1296 (1971).

10.4. EXPERIMENTS USING SUPERCONDUCTING RESONATORS

controlling the power level in the resonator, resonant frequencies stable to the order of one part in 10^{14} for times $\sim$1 hr have been achieved.[1] Very-long-term stability may again pose a problem, however, since effects of mechanical creep, which are not understood, could become important. Again it is possible that one will want to use long-term frequency correction. As we mentioned previously, the differential pressure between the interior and exterior of the resonator must be carefully controlled. Changes in this pressure which cause flexing of the walls can easily dominate the other effects discussed above. This same sensitivity to mechanical motion can be exploited to measure displacements smaller than 10^{-13} cm.[1,2]

It is often desirable to have dielectric materials in the resonator for mechanical support or for other reasons, some of which are discussed below. Changes in the dielectric constant will affect the resonant frequency in a way that is easily calculated, but the dominant effect is usually a serious reduction of the attainable Q of the resonator. Even a small amount of almost any dielectric will contribute to the dominant loss in the resonator. This reduction in Q may be acceptable, however, in return for whatever other advantages are obtained. Often the Q of a high-quality superconducting resonator is higher than can profitably be used in a particular application, and trade-offs can be made.

10.4.2. Tuning Superconducting Resonators

Superconducting resonators, like others, can be tuned to a desired frequency by a variety of means, some obvious and some novel. We have seen that tuning may be necessary if the long-term stability of a resonator is too match the attainable short-term stability, and tuning is also necessary to produce a desired resonant frequency to greater accuracy than initial machining tolerances afford. One can obviously tune a resonator by means of a movable wall or piston. Commercial frequency meters are tuned in this way, and very wide tuning ranges can be achieved. However, serious mechanical vibration problems arise because the resonant frequency is usually quite sensitive to the piston position. A variation on this method, designed to trade a narrow tuning range for less sensitivity to vibration, involves a section of the resonator wall which can be distorted inward by application of external pressure. Reducing the tuning range by making the flexible section smaller or stiffer yields any desired degree of stability. Such a tuner can be made to respond to an electrical signal from an automatic control circuit by means of a piezoelectric[103] or other transducer. Another way of reducing sensitivity and increasing stability involves "hiding" the variable tuning network in a tube branching from the resonator which is small enough in diameter to be beyond cutoff. Its length may be varied by a movable short or bellows arrangement.[104] A novel tuning method has been explored by Hartwig and co-workers,[97,105] which involves a material of variable dielectric constant in the resonator. In certain semiconductors the dielectric constant is affected by optical radiation. This is called the photodielectric effect, and it allows a resonator to be tuned dynamically by means of a variable light intensity. This method does, however, severely limit the Q attainable in the resonator.

10.4.3. Frequency Standards Referenced to Superconducting Resonators

While only a few superconducting resonators have been used in filter applications,[97,106] several groups have used superconducting resonators to make highly stable oscillators. We have seen above that the short-term stability of a superconducting resonator can be made very good indeed (one part in 10^{14}). Long-term stability is a more difficult problem. Thus superconducting resonators may be most useful in oscillators which do not require long-term stability but must be tuned easily or in short-term "flywheel" oscillators for atomic frequency standards. These standards presently use quartz oscillators to provide short-term stability, and the frequency is continuously corrected from the atomic oscillator by a phase-lock control loop with a time constant of the order of seconds. The short-term stability of quartz oscillators is limited primarily by noise in the amplifiers as a consequence of the fact that the power level in the quartz crystal must be held at a very low level (10^{-6} W) to avoid problems of nonlinearity. Cutler and Searle[107] and Hafner[108] have discussed in detail the problem of oscillator stability. The effects of amplifier noise on the short-term stability of an oscillator can be reduced by operation at higher fundamental frequency,[108] higher power level in the primary resonator,[107]

---

[103] I. Ben-Zvi, J. G. Castle, Jr. and P. H. Ceperley, HEPL Rep. No. 660, High Energy Phys. Lab., Stanford Univ., Stanford, California 94305 (1971).

[104] D. Gorle, D. Leroy, and P. Morignot, CERN Rep. CERN/MPS-MU/SD 70-2 DL/ld-17.8. CERN, Geneva (1970).

[105] J. L. Stone, W. H. Hartwig and G. L. Baker, *J. Appl. Phys.* **40**, 2015 (1969).

[106] D. G. Arndt, T. W. Eggleston, and C. R. Haden, *Proc. 1972 Appl. Superconductivity Conf., Annapolis, Maryland*, IEEE Pub. No. 72CH0682-5-TABSC, p. 679.

[107] L. S. Cutler and C. L. Searle, *Proc. IEEE* **54**, 136 (1966).

[108] E. Hafner, *Proc. IEEE* **54**, 179 (1966).

10.4. EXPERIMENTS USING SUPERCONDUCTING RESONATORS

and by passing the output through a narrow bandwidth filter. The microwave operating frequencies of superconducting resonators are much higher than the few megahertz operating frequencies of quartz resonators. The dominant nonlinearity of properly designed superconducting cavities is radiation pressure,[1,3] and much higher power levels can be used than in quartz crystals. Finally, the high Q of a superconducting resonator could be exploited in an output filter. Thus superconducting resonators offer a potentially very attractive alternative to quartz crystals in precision frequency standards.

Several ways to make an oscillator with a superconducting resonator were discussed in Section 10.3.3 in connection with Fig. 9. The approach of Fig. 9b, in which an oscillator is locked to the cavity frequency by means of modified Pound stabilizer, is probably the most promising approach to a highly stable oscillator, because it is insensitive to electrical length variations of the transmission lines into the cryostat. Stein and Turneaure[1] are following this route to a very high degree of stability indeed. They have built two X-band oscillator systems in which Gunn-effect oscillators are locked to two cavities with $Q \approx 10^9$. They intercompared the two and measured an rms fractional frequency fluctuation spectrum which had a value of 1.5×10^{-14} at 100 sec averaging time τ and rose approximately as τ^{-1} for averaging times down to 10^{-4} sec. This short-term stability is considerably better than the best quartz crystal oscillators which have been made. They measured a relative drift of 1.6×10^{-14}/hr. They also estimate that stability on the order of 10^{-16} for $\tau = 1000$ sec may be achieved with this technique,[1] which would be far and away better than has yet been achieved by any method.

Nguyen[90,109] has used the direct feedback approach of Fig. 9a, and measured stability of $\sim 6 \times 10^{-11}$ for a 1-sec averaging time τ. The τ dependence of the rms fractional frequency deviation was $\tau^{-1/2}$, which is a consequence of the filtering effect of the cavity in this configuration. The drift was about 3×10^{-10} for 6 min. This basic approach suffers from the difficulties with line-length variations discussed previously, but these variations are likely to be rather slow and may not be too important in a flywheel application where the frequency is continuously corrected. A group at Orsay[110] has simply coupled a cavity closely to a

---

[109] F. Biquard, Nguyen T. Viet and A. Septier, *Proc. 1968 Summer Study Superconducting Devices Accel.* Brookhaven Report BNL50155(C-55). (Available from Clearing House for Federal Scientific and Tech. Inform., Springfield, Virginia 22151.)

[110] J. J. Jimenez, P. Sudraud and A. Septier, *Electron. Lett.* **7**, 153 (1971).

klystron and achieved a stability of $\sim 3 \times 10^{-12}$ for 1 sec with a drift $\sim 10^{-10}$ for 30 min. This is a very simple method to apply if the ultimate in stability is not required. Since noise and line lengths are important in the performance of an oscillator, it may ultimately be desirable to design an oscillator where the amplifier and feedback system is entirely within the cryogenic environment. McAshan[111] has built an oscillator using a degenerate parametric amplifier located immediately adjacent to a resonator and cooled to the same temperature. It was as stable as his measuring equipment (probably $\sim 10^{-10}$), but he has not pursued it further or published his results. Another possibility along this line is to use a cooled tunnel diode as an amplifier. Still another direct method of making an oscillator has been explored by the group at Orsay.[98,109] This is the monotron configuration in which an electron beam passing through a suitably designed cavity interacts to set up an oscillation in the cavity. Their oscillator[109] had a stability of 2×10^{-10} for $\tau = 0.1$ sec with $\tau^{-1/2}$ behavior for shorter times and stability independent of τ for τ up to several minutes. They concluded that the short-term stability of this oscillator was limited by noise from field emission and multipactoring[66] in the cavity which was operated at power levels the order of 1 W.

10.4.4. High-Field Applications

The desire to use high rf electric fields to accelerate elementary particles in linear accelerators or particle separators has been the major impetus for most of the work that has been done on superconducting resonators in the last decade. Most of the work discussed in Section 10.2, particularly that concerned with residual losses and high-field effects, has been done by groups interested in accelerator or separator applications and associated with various high-energy physics laboratories. The most active group has been that at the High-Energy Physics Laboratory at Stanford University and much work has been done at Karlsruhe in Germany and Brookhaven National Laboratory. The construction of an accelerator or particle separator is a large scale project, and any reader interested in pursuing one of these applications will undoubtedly wish to consult directly with these groups. In any case, a detailed review of accelerator or separator design is beyond the scope of this article, and we will merely give a brief summary of work in progress.

The project which is nearest to reality is the 200-ft-long, 2-GeV elec-

---

[111] M. S. McAshan, Stanford Univ., private communication (1969).

tron linear accelerator under construction at Stanford.[12,112] It has niobium accelerating cavities which will operate at 1300 MHz at a temperature of 1.8 K. It will produce 100 μA of 2-GeV electrons with an energy spread of 10^{-4} in long pulses lasting the order of 1 sec. These specifications are impossible to attain in conventional linear accelerators operating at room temperature. At Illinois a superconducting microtron consisting of a 30-MeV superconducting linac through which the beam is recirculated to produce 600-MeV electrons is under development.[113] Proton linacs are under construction at Karlsruhe[67] and under consideration at the California Institute of Technology[114] and at Stanford. At Stanford one aim is to use a superconducting linac to produce pions for cancer therapy at a cost which is feasible for a large hospital. This too is impossible with conventional techniques. Particle separators are under construction at Karlsruhe[115] and under study at Brookhaven,[79,116] Rutherford Laboratory,[60] and CERN.

Insofar as the author is aware, no attempt has yet been made to use the high rf electric field attainable in superconducting resonators for purposes other than accelerating particles. Similarly, no use has been made of the high rf *magnetic* field other than for studying the critical field behavior of the superconductors themselves. It would seem that this offers a fruitful opportunity for experimental ingenuity.

10.4.5. Material Property Studies

Superconducting resonators can be used to study with great sensitivity the dielectric and magnetic properties of materials. The basic properties measured, of course, are the complex dielectric constant or permeability of the material at the frequency of resonance. By using a resonator geometry in which a number of modes can be excited, information on the frequency dependence of ε or μ can be obtained. The real part of ε (μ) shifts the resonant frequency, and the imaginary part lowers the Q. For almost any substance, the loss tangent is much greater than for a super-

---

[112] L. R. Suelzle, *IEEE Trans. Nucl. Sci.* **NS-18**, 146 (1971).

[113] A. O. Hanson, *IEEE Trans. Nucl. Sci.* **NS-18**, 149 (1971); R. A. Hoffswell, *ibid.*, **NS-18**, 177 (1971).

[114] A. J. Sierk, *IEEE Trans. Nucl. Sci.* **NS-18**, 162 (1971).

[115] W. Bauer, G. Dammertz, H. Eschelbacher, H. Hahn, W. Jüngst, E. Rathgeber, and J. Vortruba, *IEEE Trans. Nucl. Sci.* **NS-18**, 181 (1971).

[116] H. J. Halama, and H. Hahn, *IEEE Trans. Nucl. Sci.* **NS-14**, 350 (1967); H. J. Halama, *ibid.* **NS-18**, 188 (1971).

conductor, so the Q is generally lowered much more than the frequency is shifted. However, the resonant frequency can be measured much more accurately than the Q. Thus even a small amount of material exposed to the electric field in a superconducting resonator can effectively dominate the resonator parameters. We have noted this fact before in the sections on resonator design.

Dielectric studies of materials using superconducting resonators have been pioneered mainly by the group at the University of Texas under W. H. Hartwig. They have discussed the theory of this kind of measurement[117] and studied many materials, including low-dissipation construction materials,[117] ionic crystals,[118] and a number of interesting semiconducting materials.[119] Carrier trapping effects in CdS were studied[120] along with the photoelectric effect.[105,121] They have even used a semiconductor wafer in a superconducting resonator as a nuclear radiation detector.[122] Thus a superconducting resonator makes a quite versatile "material property analyzer."[119]

10.4.6. Miscellaneous Applications

In addition to the applications discussed above, upon which most work has been done, there are a number of other possible uses for superconductors at microwave frequencies which have only begun to be explored. Trela and Fairbank[88] used the frequency of a cavity partially filled with liquid helium as a level detector in a helium flow experiment. Helium is one dielectric with an extremely small loss tangent,[117] and it does not seriously degrade the Q of the cavity. It does, however, affect the frequency and cause stability problems. Thus highly stable cavities are normally evacuated. The apparatus of Trela and Fairbank[88] could detect liquid level shifts of 10^{-8} cm, but vibrations and waves in the liquid limited useful sensitivity to $\sim 10^{-6}$ cm. Similarly Dick and Yen[2] are developing a motion detector with which they hope to detect displacements of a gravity wave detector as small as 10^{-17} cm.

---

[117] W. H. Hartwig and D. Grissom, *Proc. Int. Conf. on Low Temp. Phys., 9th*, **LT9**, 1243. Plenum Press, New York, 1964.

[118] D. Grissom and W. H. Hartwig, *J. Appl. Phys.* **37**, 4784 (1966).

[119] J. J. Hinds and W. H. Hartwig, *J. Appl. Phys.* **42**, 170 (1971).

[120] W. H. Hartwig and J. J. Hinds, *J. Appl. Phys.* **40**, 2020 (1969).

[121] G. D. Arndt, W. H. Hartwig, and J. L. Stone, *J. Appl. Phys.* **39**, 2653 (1968).

[122] C. W. Alworth and C. R. Haden, *J. Appl. Phys.* **42**, 166 (1971).

10.4. EXPERIMENTS USING SUPERCONDUCTING RESONATORS

The extremely low loss in superconducting microwave structures makes it possible to contemplate using circuit elements that are quite impractical if built with normal metals. Walker and Haden[123] have made efficient loop antennas for 400 MHz which are only 1 cm² in area. A normal metal antenna this small compared to the wavelength has a loss resistance much higher than the radiation resistance, and it is therefore quite inefficient. With a superconducting loop the loss can be made negligibly small. In addition, a high-Q superconducting resonator was used to transform efficiently the 10^{-3}-Ω radiation resistance of the loop to normal transmission line impedance $\sim 50\ \Omega$. To obtain efficiently a transformation ratio this large requires a comparably high Q, obtainable only with a superconducting resonator.

Similarly, thin film microstrip and strip-line rf circuits made with superconductors offer great promise. The loss in normal-metal microstrips is dominated by loss in the metal, and one cannot obtain particularly high Q. Since a dielectric is an integral part of a microstrip line, one cannot obtain in a superconducting microstrip as high Q as one can in a cavity resonator, but great improvement over normal-metal microstrips is possible. DiNardo et al.[124] have made resonators at 14 GHz which consist of 1-cm loops or straight strips on alumina substrates. They measured Q's as high as 5×10^5. The small size and high Q of these resonators makes them extremely interesting circuit elements. Mason and co-workers[125,126] have studied superconducting microstrip lines for delay line and pulse storage applications. The high attenuation of normal-metal strip lines makes it impractical to obtain large delays in a small space, but superconducting lines are not so limited. Mason[126] has made a superconducting microstrip delay line with comparable delay, attenuation, and risetime to a normal 2.2 cm coaxial delay line and accomplished a size reduction of 30,000 to 1. In addition, because of the inductive nature of the superconducting surface impedance, a line with thickness comparable to the penetration depth shows a marked reduction in phase velocity below the free space value. Furthermore, the phase velocity is variable through the dependence of the penetration depth on temperature and mean free path. Further reduction can be obtained with a high-dielectric-constant substrate. Values of phase velocity as low as

---

[123] G. B. Walker and C. R. Haden, *J. Appl. Phys.* **40**, 2035 (1969).
[124] A. J. DiNardo, J. G. Smith, and F. R. Arams, *J. Appl. Phys.* **42**, 186 (1971).
[125] P. V. Mason and R. W. Gould, *J. Appl. Phys.* **40**, 2039 (1969).
[126] P. V. Mason, *J. Appl. Phys.* **42**, 97 (1971).

$c/40$ have been obtained.[125] Some work along this line has recently been done by the Purdue group[127] as well.

Finally, Ekstrom and co-workers[128] have investigated the response of a 278-m-long coaxial line made of superconductors. They found a rise time of 140 psec for pulses, but the attenuation as a function of frequency was rapidly oscillatory. This was attributed to resonances from minor variations in the electrical parameters of the line, the effects of which are not damped out because of the low attenuation.

In this section we have reviewed very briefly applications of superconducting microwave circuits which have so far been made. From the wide variety of applications we have discussed, it seems clear that microwave superconductivity offers promising answers to a great number of experimental problems, only a few of which have been explored in any detail.

Acknowledgments

The author wishes to thank the many workers in this field who have patiently and promptly sent him works as submitted for publication and unpublished reports. This cooperation has been invaluable during the long task of preparing this review. Conversations with Professor J. P. Turneaure were particularly helpful, and thanks are due to Professor R. V. Coleman and Dr. J. Halbritter for critical readings of the manuscript.

Financial support for the preparation of this article was provided in part by the Office of Naval Research.

[127] V. L. Newhouse, C. A. Passow, and R. L. Gunshor, *Proc. 1972 Appl. Superconductivity Conf.*, *Annapolis, Maryland*, IEEE Pub. No. 72CH0682-5-TABSC, p. 673.
[128] M. P. Eckstrom, W. D. McCaa, Jr., and N. S. Nahman, *J. Appl. Phys.* **42**, 106 (1971).

11. SUPERCONDUCTING DEVICE TECHNOLOGY

11.1. Superconducting Magnets*

11.1.1. Introduction

Conventional laboratory electromagnets can supply dc magnetic fields up to about 25 kOe (2.5 T) in a volume of about 15 cm^3. If higher fields are required, there are three possibilities: the water-cooled, high-power solenoid (Bitter magnet); the cryogenic, but nonsuperconducting, solenoid; or the superconducting solenoid (or hybrids between these). All these magnets generate a field by passing a dc current through an electric conductor arranged in a helical path, and differ only in the way in which they deal with the problem of heat production and dissipation. In most cases no iron is used in the magnetic circuit, since the extra field obtained is not worth the increased weight and cost.

The Bitter magnet produces a maximum magnetic field of 70 to 250 kOe, depending on the size of the working space provided, and on the available power and cooling. Its main advantages are the magnitude of the maximum field, the rapid field sweep obtainable (1 min or less to maximum field), and the room-temperature working space. The disadvantages are: the very high cost, both in investment and in operation, of the megawatt dc generator and cooling system that are required; the mechanical vibration level resulting from the forced cooling; the relatively high level of ripple in the magnetic field; and the relatively poor field uniformity (0.1–1% variation over the working volume). There are probably fewer than a dozen Bitter magnet installations in the world.

Nonsuperconducting cryogenic solenoids can be as simple as a coil of copper wire immersed in a liquid nitrogen bath, or as complicated as a reinforced high-purity aluminum hollow conductor cooled by supercritical helium under several atmospheres pressure. The incentive for using cryogenic solenoids is the great decrease in electrical resistivity

---

* Chapter 11.1 is by C. D. Graham, Jr.

that accompanies decreasing temperature in pure metals. To take full advantage of the reduced resistivity, the temperature needs to be near 10 K or lower, and furthermore the conductors need to be of very high purity and are consequently very soft and weak. As a result, nonsuperconducting solenoids usually offer no advantage over superconducting solenoids, and are rarely used.

Superconducting solenoids capable of generating fields significantly greater than 25 kOe became possible following the discovery of high-field superconductors.[1] They are readily available to produce fields up to 125 kOe, and have been built to provide fields of 150 kOe. Hundreds of superconducting magnets are in regular operation, and they are generally quite reliable with few maintenance problems. The principal advantages of superconducting solenoids are relatively low cost and small size, almost complete freedom from mechanical vibration, low field ripple, and reasonably good field uniformity. Disadvantages are fairly slow field sweep rates (at least several minutes to maximum field except in the case of multifilament windings), and the complication, inconvenience, and expense of working with liquid helium. This last disadvantage loses most of its sting if, as is frequently the case in solid state physics, the experiment itself requires the use of liquid helium. All these points are discussed in more detail below.

11.1.2. General Considerations

The usual arrangement of a laboratory superconducting magnet is shown in Fig. 1. The solenoid itself, weighing 10–40 kg, is suspended in a stainless-steel dewar vessel. The solenoid is immersed in liquid helium, so that the working space in the bore is filled with liquid helium. If the experiment requires a temperature other than 4.2 K, an insert dewar is required; this can have a single vacuum wall if temperatures no higher than about 100 K are needed, but requires a double vacuum wall if the working space must go to room temperature or above. Temperatures below 4.2 K can be attained by pumping on the magnet space and lowering the temperature of the entire solenoid. This is costly in time and in helium, but is sometimes done to increase the maximum field attainable by the solenoid. If lowering the sample temperature is the principal goal, it is usually better to use a pumped insert dewar. A

---

[1] J. E. Kunzler, E. Buehler, F. S. L. Hsu, and J. H. Wernick, *Phys. Rev. Lett.* **6**, 89 (1961).

11.1. SUPERCONDUCTING MAGNETS

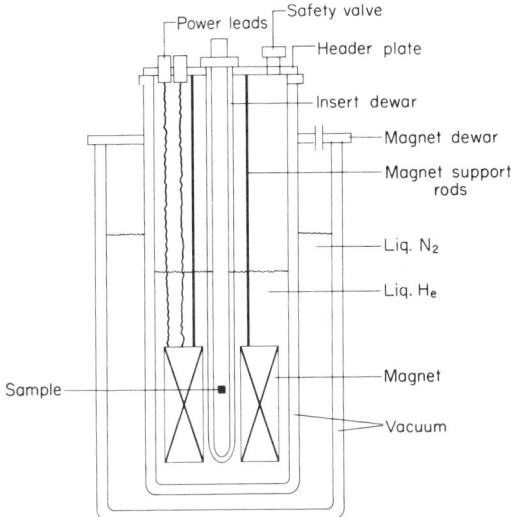

FIG. 1. Typical construction of superconducting magnet and dewar.

variable-temperature insert dewar using helium pumped from the main solenoid reservoir is possible, but the plumbing becomes rather complicated. More elaborate dewars can be built to provide optical access, free access from the bottom of the solenoid, horizontal orientation of the solenoid, etc.

Power is normally provided by a rectifier-type dc power supply, with the current fed through the header plate and led down to the solenoid. Maximum currents are generally in the range 20–100 A, and with reasonable care the liquid helium losses are less than 1 liter/hr. Often the power leads are cooled with the escaping helium gas, in which case the solenoid space must be kept sealed during operation.

Some kind of safety valve must be provided to allow for the rapid escape of helium gas in the event the solenoid "quenches," i.e., reverts from the superconducting to the normal state, with consequent rapid boil-off of liquid helium. This is less traumatic than one might expect, since the helium is boiled off over several tens of seconds, and a 5-cm diameter clear opening is sufficient for any reasonable size laboratory magnet.

Superconducting magnets often provide for "persistent mode" operation; at the desired operating current, a superconducting short circuit link is placed across the solenoid terminals so that the supercurrent flows continuously through the coil without input from the external power supply. Sometimes the main power leads are then removed to eliminate that source of thermal conduction to the helium reservoir.

Some means of measuring the level of liquid helium above the solenoid is required, and automatic warning, refilling, or shut-down equipment can be provided as necessary. The operation of a superconducting magnet, and conducting the experiment in it, requires competent supervision, so that elaborate safety devices may not be worth the cost in money, complication, and unreliability.

It is possible to buy superconducting wire or ribbon and construct a solenoid with no more than ordinary machine-shop equipment. However, there are enough tricks of the trade to make this procedure unwise. It is quite likely to cost more in damaged and worthless superconductor than it saves in purchase price. Superconducting solenoids are available from a number of companies in the United States and elsewhere, and this chapter is written on the assumption that the user will purchase his solenoid rather than build it. A potential buyer should look for advertisements in recent issues of *Physics Today*, *Cryogenics*, *Review of Scientific Instruments*, or the Guide to Scientific Instruments issue of *Science*. There is almost no such thing as a standard model, and virtually all solenoids are built to order.

11.1.3. Materials

Two classes of superconducting materials are in regular use: the *ductile alloys* and the *high-field* or *compound superconductors*. The first ductile superconducting alloys were niobium–zirconium, but these have been replaced by niobium–titanium, usually near 50% Ti by weight. Other alloying elements may be added. For use in magnets, the alloy is in the form of one or more strands embedded in copper. The composite conductor may be round, square, or rectangular in cross section, and the ratio of copper to superconductor is rarely less than 1 : 1. Much higher ratios are used in large magnets (see the discussion of stabilization). The advantages of ductile alloys are ease of handling and relatively low cost; the low cost in turn results largely from the fact that they can be manufactured using conventional metallurgical equipment and methods. The disadvantage is a limitation in maximum field. These alloys are not superconducting in magnetic fields above about 100 kOe at 4.2 K, which means that they have a useful upper limit of about 70 kOe; operation at lower temperature can raise this limit substantially.

The only commercially available material for higher-field applications is Nb_3Sn, which remains superconducting in fields above 200 kOe, and is practically useful up to 150 kOe.† It is a brittle material, and must be made in the form of thin ribbon to be sufficiently flexible for magnet construction. It may be electroplated with copper, or soldered to copper ribbon, or embedded in aluminum, for stabilization (see below).

Superconducting material is tested by subjecting a sample length to a magnetic field applied perpendicular to the direction of current flow, and increasing the current until a measurable voltage drop is detected across the sample. This is known as the *short-sample* test; a series of such tests at different fields define the short-sample characteristics of the material. An example is shown in Fig. 2. In rectangular conductors, the

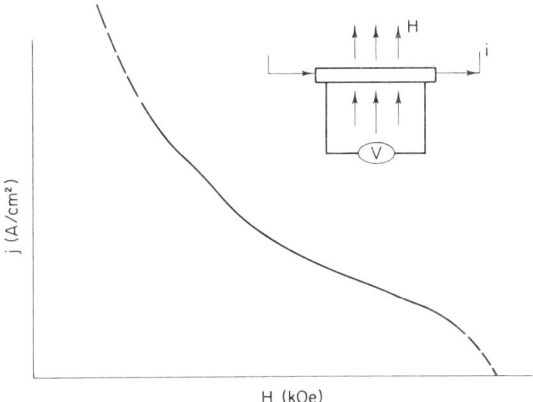

FIG. 2. Short-sample test arrangement and results.

short-sample characteristics may be significantly different depending on whether the field is applied parallel or perpendicular to the wide edge of the conductor. Performance with the field parallel to the current is not usually of interest. The low-field data are often unavailable because the current becomes too high to handle experimentally, and the high-field data are sometimes unavailable because the necessary fields are not conveniently obtainable. If a magnet or other superconducting device performs so that the current density through the conductor, at the point where the field seen by the conductor is highest, corresponds to a point on the short-sample characteristic curve, the device is said to "meet

---

† *Note added in proof*: V_3Ga superconductor is now available; it should be useful for fields above 150 kOe to perhaps 175 or even 200 kOe.

short-sample performance." Failure to meet short-sample performance may be due to quality variations in the material (100% testing is not practicable), but is most often due to instabilities or flux-jumps in the superconductor.

11.1.4. Stabilization

Very large superconducting magnets, such as those used for bubble chambers, are normally designed to be "fully stabilized," that is, there is a large enough cross section of high-conductivity copper in parallel with the superconductor so that even if the superconductor were to fail completely, the current could be carried in the copper without excessive power dissipation, and the system could be operated safely at least long enough to be shut down. The copper is usually bonded directly to the superconductor so that both are wound into the solenoid together.

Small laboratory magnets are not fully stabilized, because the penalty in increased size and weight is too high and because the current densities become too low to obtain high fields. Some copper is bonded to the superconductor, however, and such magnets are said to be "partially stabilized." The copper can improve magnet performance in several ways; for a discussion see the excellent review paper by Chester.[2] Recent theoretical and experimental work has led to the development of "multi-filament" conductors, containing many very small parallel superconducting wires embedded in a normal metal matrix.[3] Properly designed multi-filament conductors are essentially free from instability problems; they are available only with ductile superconductors. The minimum stabilization requirement for a laboratory magnet is that the magnet should not be damaged by a quench from the highest operating field. Increasing degrees of stabilization will allow faster sweep rates and more reliable operation, usually at the cost of increased solenoid size and weight.

11.1.5. Field Uniformity

Field uniformity or homogeneity is usually specified in per cent over a particular diameter spherical volume (DSV); for example, 0.5% over a 2.5 cm DSV. Interpreted strictly, this would mean that the maximum field difference between any two points in a 2.5 cm sphere would be no

---

[2] P. F. Chester, *Rep. Progr. Phys.* **30**, 561 (1967).

[3] Superconducting Materials Group, Rutherford Laboratory, *J. Phys. D* (*Brit. J. Appl. Phys.*) **3**, 1517 (1970).

11.1. SUPERCONDUCTING MAGNETS

greater than 0.5%. In practice, the specification is usually interpreted to mean a field difference of 0.5% (or sometimes $\pm$ 0.5%) between the maximum field on the central axis and a point 1.25 cm away, also on the axis. The Maxwell equation div $B = 0$ requires that field changes going in a radial direction should be opposite in sign and smaller in magnitude than field changes going in an axial direction, so measurements are usually made only on the axis. Furthermore, measurements are usually made only of the axial component of field, although there must be a radial component at all points off the central axis.

The field uniformity specification applies at the maximum field, but not necessarily at lower fields. In particular, there is normally some trapped residual field when the solenoid current is reduced to zero, and the uniformity, if expressed as a percentage of central field, may be very bad at and near zero field.

A magnet built without regard to field uniformity, i.e., built to use the minimum amount of superconducting material, will have a field uniformity of $\frac{1}{2}$–1% over a diameter equal to the bore. Higher uniformity is obtained by adding extra turns (or current density) at the ends of the magnet, or removing turns near the center. For a discussion of the various techniques, see Montgomery.[4] Field uniformity of 0.01% or even 0.001% can be obtained over a 1-cm DSV, although this may require a magnet bore substantially larger than 1 cm.

The previous paragraphs have been concerned with spatial uniformity. One must also consider temporal uniformity, or field ripple. This depends primarily on the power supply, since the solenoid is normally a passive load. A superconducting solenoid is a rather unusual load, however, since it is a pure inductance, and dc power supply specifications are usually written for resistive loads. In particular, when holding a constant current through the solenoid, the power supply will be operating at very low voltage. For a further discussion, see Section 11.1.7; in general, it is not difficult to keep the time variations in field smaller than the space variation.

In persistent mode operation, if available, field ripple is unmeasurably small. However, since magnets normally contain internal connections that are not perfectly superconducting, there is a measurable field decay due to the power dissipated in these connections. The field decay may be as high as 1%/hr, but is usually less than 0.1%/hr; it tends to be considerably worse at very high fields than at moderate fields. For

[4] D. B. Montgomery, *Solenoid Magnet Design*. Wiley (Interscience), New York, 1969.

high-field magnets (over 100 kOe), the long-term stability is usually better during operation with a power supply than in the persistent mode.

Some superconducting solenoids undergo measurable "flux jumps" as the field is increased or decreased. These are combined magnetic and thermal instabilities which lead to more-or-less local field redistributions. If sufficiently large, they cause the coil to quench; otherwise they appear as sudden jumps in field. Increasing stabilization decreases the incidence of flux jumps. If measurements are to be made while the field is sweeping, flux jumps may be a major problem, although in recently built magnets this problem seems to have been largely overcome.

11.1.6. Field Measurement

For many purposes it is sufficient to measure the current through the solenoid and take the field to be proportional to the current, just as is done with a copper solenoid. However, the diamagnetic properties of the superconducting winding influence the field in the magnet, so that the field versus current curve is somewhat nonlinear and hysteretic. For ductile alloy magnets, these effects are generally less than 1% and are usually not significant except near zero field. In high-field (Nb_3Sn) magnets, the problem is often more serious. If exact knowledge of the field is important, it can be measured directly with a search coil and integrator, or a commercial Hall probe or magnetoresistance element, all of which will work satisfactorily at either room temperature or liquid-helium temperature. Commercial rotating-coil gaussmeters are also available to measure axial fields, but must be operated at or near room temperature. Calibration of any field-measuring device presents some difficulty. The search coil and rotating-coil methods can reasonably be assumed to be linear with field (if allowance is made for the increase in coil resistance with field at low temperature), so that a calibration against a known standard at low field can be extrapolated to high fields. The same is not necessarily true of Hall-effect and magnetoresistance measurements.

It is possible to build a field-measuring probe into the magnet, so that it does not occupy any of the working space. The simplest and most convenient method is to use a bifilar winding of ordinary copper wire in the high-field region of the magnet. The resistance increases by roughly a factor of three between 0 and 100 kOe at 4.2 K, and is nearly linear at high field, as shown in Fig. 3. The nonlinear region can be reduced by using high-purity copper. The probe readings must be calibrated in

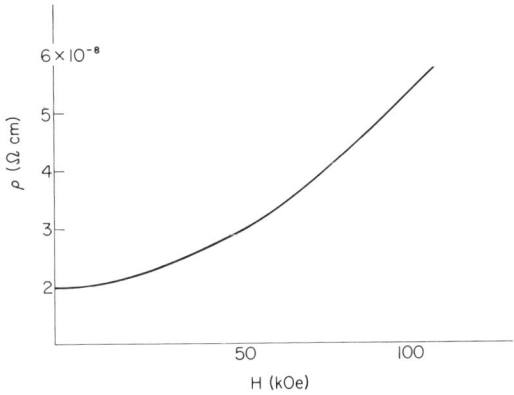

Fig. 3. Magnetoresistance of electrical conductor-grade copper at 4.2 K.

terms of the central field. Such a probe is also useful in measuring the temperature of the magnet during cool-down and warm-up, since the probe resistance is a strong function of temperature.

11.1.7. Power Supply Considerations

Figure 4 shows in schematic form the usual connection of a superconducting magnet and power supply. Since the magnet is essentially a pure inductance, the rate of increase of the magnet current is inversely proportional to the inductance: $E = -L\, di/dt$, where L is the magnet inductance in henries. A small 60-kOe magnet may have a maximum operating current of 40 A and an inductance of 8 H; to reach maximum

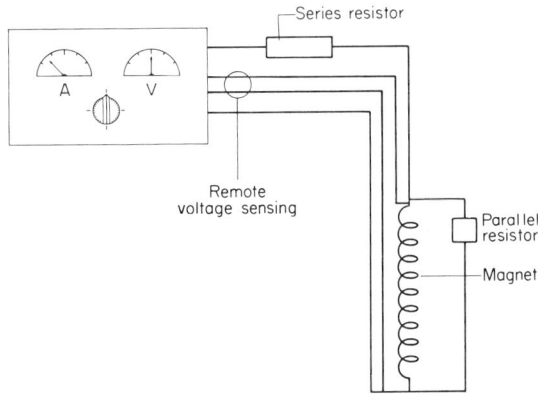

Fig. 4. Power supply connections for superconducting magnet.

field in 10 min then requires an applied potential of 3.2 V. If the power supply can operate either in constant-current or constant-voltage mode, a field sweep is then conveniently obtained by operating in the constant-voltage mode. Steady field is obtained by operation in constant-current mode; however, some power supplies do not control well at very low output voltages into an inductive load. Often a resistance is put in series with the magnet to increase the load on the power supply at constant current; a more elegant method replaces the resistance with a high-current diode, so that the voltage drop is independent of current.

The energy stored in the magnet is given by $\frac{1}{2}Li^2$, which in the case considered above is 6.4 kJ at maximum field. If the power leads are opened while there is field in the magnet, very high voltages can appear at the break, and the stored energy of the magnet can be dissipated in the resulting arc. For protection against this possibility, a shunt resistance is often placed across the magnet terminals, either inside the dewar or just above the header plate.

When the current is decreased, the magnet acts as a source of power, since the stored magnetic energy is being removed. If the power supply voltage is simply set to zero (but the circuit not broken), the magnet current will decay with a time constant $\tau = L/R$, where L is the inductance of the magnet and R is the parallel combination of the shunt resistance (if any) and the lead resistance plus the internal resistance of the supply. The value of R is generally 0.1–0.01 Ω, so that τ is several minutes. The time to reach substantially zero current is of course several times longer.

11.1.8. Operation in Swept Field

Some superconducting magnets are built with internal shunts, so that the current distribution with a voltage applied is not the same as that at constant current. This can lead to significant changes in field uniformity during a rapid sweep. The time constant for attaining steady state conditions after a field sweep on some early magnets was many minutes, but it is now no more than a few seconds or possibly as long as a few minutes, and is generally not serious.

Some magnets undergo a series of small flux jumps while the field is changing, which may be large enough to interfere with sensitive measurements.

None of these effects is significant unless it is necessary to make measurements while the field is changing.

11.1.9. Reversed Field

Superconducting coils are more subject to flux jumps and to premature quenches if they are run with a negative field *after* a previous positive run. The seriousness of the problem seems to vary from magnet to magnet or from manufacturer to manufacturer. There may be no problem, or it may be necessary to run at a slower-than-normal sweep rate in the negative direction, or it may even be impossible to reach full negative field.

11.1.10. Liquid Helium Consumption

It is difficult to specify precise figures for liquid helium consumption, largely because the helium losses depend quite strongly on the depth of liquid in the dewar and on the efficiency of the radiation shields and gas baffles below the header plate. Generally speaking, a small 60-kOe magnet should operate with an average total loss of less than 0.5 liter/hr. A 100-Oe magnet will use perhaps 0.5 liter/hr when held at 4.2 K in zero field, with the losses doubling at maximum current. An insert dewar which holds the sample temperature above 4.2 K will naturally increase the losses, as will optical access to the working space from below, frequent changes of sample, etc.

As a rule of thumb, it will require $\frac{2}{3}$–1 liter of liquid helium per kilogram of magnet weight to cool the magnet from 77 to 4.2 K. Thus a small 60-kOe magnet might weigh 5 kg and require 5 liters of He for cooling, while a large 100-kOe magnet weighing 25 kg would require about 25 liters of liquid helium. Normally an additional volume of liquid helium equal to perhaps half the cool-down volume is added to the dewar as an operating reserve. The helium used in preparing the magnet for operation is generally comparable to the volume consumed in about 24 hr of continuous operation. Unless the magnet is operated continuously for more than a day each time the magnet is cooled to 4.2 K, the helium use during operation will be less than the helium use during cool-down. Some saving in helium can be attained by precooling to 65 K instead of 77 K (see Section 11.1.11).

A persistent current switch consists of a length of superconductor connected across the terminals of the magnet. When persistent operation is not desired, as when the magnet current is increased or decreased, the superconductor in the switch must be driven into the normal state. This can be done either with an applied field or by an increase in temperature.

Use of magnetic field is not generally practical because of the large field from the magnet itself, so a local heater is provided to raise the temperature of the superconducting switch. This heater naturally dissipates energy and boils off liquid helium; the energy dissipation is generally 150–600 mW, and is therefore not negligible in comparison to the total helium losses (1 W boils off 1.4 liter/hr). If the magnet is to be operated most of the time at constant current, the persistent current switch can save helium because the current through the main leads can be dropped to zero; if the field is frequently swept, the losses in the persistent current heater may outweigh the savings in helium at constant current. As remarked above, operation in the persistent mode is electrically very quiet, but may give an objectionable long-term field decay.

11.1.11. Operating Procedure

We include here a brief description of the procedure for cooling and running a superconducting magnet, with an estimate of the time involved in the various steps.

Assuming the magnet is at room temperature, the first step is to precool to liquid-nitrogen temperature. The simplest procedure is to pour or pump liquid nitrogen into the space containing the magnet. When the magnet is at 77 K, the liquid nitrogen is removed by extending a tube to the bottom of the liquid and pressurizing the space over the liquid to about 0.1 atm. with nitrogen or helium gas. The liquid which is removed can be pumped directly into the outer (liquid nitrogen) jacket of the dewar. This step may take 10 min for a small magnet, 45 min for a large one.

If it is necessary to use the minimum amount of liquid helium for cooling the magnet, the temperature can be lowered from 77 to about 65 K by partially evacuating the space above the liquid nitrogen. Pressure should not go lower than about 100 Torr, since nitrogen solidifies at 63 K and 95 Torr. This procedure can save about 30% in the volume of liquid helium needed to cool the magnet to 4.2 K. The cooled liquid nitrogen is transferred out of the dewar in the same way as described above.

If the magnet is of closed construction, so that no liquid nitrogen enters the windings, liquid helium can be transferred in as soon as the liquid nitrogen is removed. If the magnet is of open construction, so that liquid penetrates the winding, it is usually necessary to wait for an hour or more for the liquid nitrogen to drain out. After this time, the liquid nitrogen will have collected in the bottom of the dewar, and must be transferred out as described above. It is possible to accelerate the removal of the trapped liquid nitrogen by lowering the pressure in the magnet space and boiling the nitrogen away, but it is necessary to be careful to avoid freezing any nitrogen in the windings. The magnet may also be precooled by filling the outer dewar jacket with liquid nitrogen and waiting for radiative cooling; this takes many hours.

Before transferring liquid helium, the magnet space should be thoroughly flushed with helium gas, or evacuated and backfilled with helium gas. The liquid helium transfer is accomplished in the usual way, taking care that the transfer tube reaches to the bottom of the magnet. In this way the escaping helium gas will cool the magnet with maximum efficiency. The helium transfer should be reasonably slow; perhaps 10–20 min for a small magnet and 45–90 min for a large one. When the magnet is cold, and covered with liquid helium, the power supply is turned on and the magnet operated as desired. Additional liquid helium can be added as required while the magnet is operating, taking care not to blow out the remaining liquid from the magnet. It is important to remember that stainless steel transfer tubes are slightly magnetic and will experience a significant force if the magnet is energized.

Superconducting coils are generally unshielded, and the stray field in the room can be quite large. In particular, one must keep the dewar well away from steel walls, lab benches, dollies, lift trucks, and relay racks; meters and electronic gear should also be kept at least one meter away.

At the end of a run, it is best to leave the magnet under a slight positive pressure of helium gas, or (less desirable) under vacuum, to avoid condensation of ice or water on the magnet. Unless the magnet is to be used every day, it is probably not advisable to try to keep it at liquid helium temperature. If it is to be used fairly frequently, it may be worthwhile to keep it at liquid nitrogen temperature more or less permanently to reduce preparation time before operation.

11.1.12. Costs

The following cost figures are approximate and were obtained in 1972. A minimum superconducting magnet, with 60-kOe maximum field, 2.5 cm diameter bore, and 1% field uniformity, sells for $ 4000–5000. An appropriate dewar will be about $ 1500, and a power supply about the same. A minimum complete system would then cost $ 7000–8000.

Increasing the magnet size to 5 cm bore and improving the field homogeneity to 0.1% (over 2.5 cm) approximately doubles the price of the magnet; the dewar and power supply are not significantly more expensive.

A 100-kOe magnet is generally 2–3 times the price of a 60-kOe magnet of similar size and uniformity; the dewar and power supply costs may increase by a factor of 1.5 or 2. A minimum complete 100-kOe system is therefore likely to cost about $ 12,000. The price can increase rapidly if special dewars, unusually high uniformity, or other special features are required.

Magnet manufacturers do not normally build dewars, but purchase them from other makers. Magnet manufacturers may or may not make their own power supply units, but will supply all necessary accessories such as connecting cables, calibration curves, etc.

11.2. Superconducting Shielding*

11.2.1. Introduction

It has long been recognized that superconductors should provide the possibility of exceptionally fine shielding of low-temperature experiments against external electromagnetic interference. Workers in low-temperature labs have used superconducting shields with varying degrees of success. Many of the cases where the experimenter feels that superconducting shielding has not lived up to expectations have not been due to a failure of the shielding but have been due to excessively high expectations for the shield. In most shielding designs we can estimate performance by considering the superconductor to be a perfect conductor. However, if we pay inadequate attention to such details as mutual capacitance between input leads and the shield or proper grounding techniques, the shield will not be effective. A description of various solutions to grounding problems is given by Morrison.[1] The factors discussed there should be considered before the performance of a shield is estimated.

The properties of a superconductor which must be considered for the shielding problem are the appearance of an infinite electrical conductivity and the Meissner effect. The Meissner effect, however, is very seldom large enough in a shield made of commercial grade materials to affect the design, so the zero-resistance property is of prime importance. Clearly, the superconductor must be in an environment that the magnetic field at all points on the surface of the shield is below the critical field or resistance will appear at the point in the shield where the critical field is exceeded and the shield will behave no differently than one made of a normal metal.

The existence of the critical field creates a very important design limitation on a superconducting shield. The field at the surface of the superconductor is due both to the external magnetic field and to any currents

[1] R. Morrison, "Grounding and Shielding Techniques in Instrumentation." Wiley, New York, 1967.

* Chapter 11.2 is by W. O. Hamilton.

which are flowing in the shield. Thus, in designing a shield, one must consider the size of the ambient magnetic field plus the fields created by any shielding currents, whether due to changing external magnetic fields or to demagnetizing currents in the shield. This is particularly important at the shield edges.

11.2.2. Shielding against Time-Varying Fields

Because of the infinite conductivity of the superconductor we expect it to be a very effective screen against time-varying fields. Lenz's law ensures that time-varying magnetic fields should be screened from the shield interior by induced surface currents. Electric fields will not exist beyond a depth in which currents can flow. The London equations predict that in the stationary state the magnetic field will decrease exponentially from the surface of the material, the decrease being due to the screening currents. The characteristic length for this decrease is the London penetration depth. In lead this is between 370 and 500 Å and the currents flow in a layer of approximately this thickness.

Thus, for low frequencies, we expect the screening currents to flow in a surface sheath of approximately 500 Å thickness and to exclude the field from the material inside the sheath. Of course, any cavity inside the material will then be perfectly shielded from changing fields outside. Thus a superconductor which is many penetration depths thick should be a much more efficient shield than a normal metal against time-varying fields. The frequency at which the shielding by a superconductor becomes no better than that from a normal metal will be the point at which the normal skin depth becomes roughly comparable to the penetration depth. This is somewhat above microwave frequencies.

We have tacitly assumed in the above discussion that our superconducting shield is made from a type I material: one where magnetic flux does not penetrate the shield until the critical field is reached. Shields can also be made from type II superconductors; however, if the fields against which the shield is used are greater than Hc_1 for the material, a decreased shielding effectiveness may be expected from flux motion in the shield. The conditions governing flux flow are complex and reference should be made to the literature before a shield is designed if the physical situation indicates that flux flow might occur.[2] However, we describe below work which demonstrates that the zero-resistance property of the

---

[2] Y. B. Kim and M. J. Stephen, *in* "Superconductivity" (R. D. Parks, ed.), p. 1107. Dekker, New York, 1969.

type II superconductor will shield against time-varying fields that are not energetic enough to move the trapped magnetic flux. The shielding effectiveness of type I shields has been measured by Cabrera and is described in Section 11.2.3.

One use of a type II shield against time-varying fields has been reported by Day.[3] The experiment to be shielded required a high ambient magnetic field. A type II superconducting cylinder was built around the experiment and means were provided for keeping a section of the shield above its transition temperature while the superconducting magnet which provided the ambient field was energized. The shield was then allowed to become superconducting. The relative location of the shield and magnet was such that the field was about the same inside and outside of the shield and no currents were induced in the shield. Thus there was no pressure for flux jump or flow.

No detectable change in magnetic field inside the shield could be observed at the 10^{-5}-G level even though the ambient field was nearly 10 kG. The field variations which were being shielded were those of the typical laboratory environment and the detection apparatus was sensitive to field variations of 10^{-6} Oe.

Another shield design for fluctuating fields which can be used when smaller magnetic fields are present is made by putting the shield outside the magnet and far enough removed that the critical field Hc_1 is not exceeded. Such a design is described by Hebard.[4]

11.2.2.1. Shield Design. The shield design in such experiments is not critical as long as the magnetic pressure on the shield at no time becomes great enough to cause flux motion. An engineering problem concerns the method used in experiments like Day's to keep the cylinder normal while the external field is being applied. Hildebrandt and Elleman[5] have described a nonelectrical switch whereby they circulate warm helium gas through a tube which is fastened to the shield. The principle is to drive an area of the shield normal so as to allow flux to penetrate and equalize the flux pressure inside and outside the shield. Deaver and Goree[6] have described a system in which the superconducting shield is kept in a separate dewar and cooled separately.

[3] E. P. Day, *Phys. Rev. Lett.* **29**, 540 (1972); Ph.D. Thesis, Stanford Univ. (1972).
[4] A. H. Hebard, Ph.D. Thesis, Stanford Univ. (1972).
[5] A. F. Hildebrandt and D. D. Elleman, *Proc. Symp. Phys. Superconducting Devices*, Univ. of Virginia, Charlottesville (1967).
[6] B. S. Deaver and W. S. Goree, *Rev. Sci. Instrum.* **38**, 311 (1967).

A factor to be considered is the effect of penetrations on the shielding effectiveness. As long as there is no actual wave propagation through the penetration, the shielding effectiveness can be estimated by considering the field outside the shield to be slowly varying and applying the usual solutions to LaPlace's equation to calculate the field inside the shield. For a cylinder, open at the top, we find that a transverse field should decrease at least as $e^{-1.8z/R}$ as one goes down into the cylinder. The field components along the axis of the cylinder should decrease more rapidly, as $e^{-3.8z/R}$. This shielding factor has been measured by Deaver and Goree.[6] The propagation of electromagnetic energy along the wires penetrating the shield can be controlled by filters on the wires and by the techniques discussed by Morrison.[1]

There has been a wide variety of materials used for superconducting shields. Any material should work consistent with the limitations described earlier. Hebard's shield was made of niobium sheet 0.050 in. thick, while that used by Day was plasma-sprayed NbTi.† Lead foil is commonly used and there have been reports of the use of tin, indium, 50–50 Pb–Sn solder, and Woods' metal. Plasma-sprayed Nb has also been used.

11.2.3. Shielding against dc Fields

We have previously discussed shielding against time-varying fields because the only property of a superconductor which is required for shielding is the perfect conductivity. We now need to examine the Meissner effect to see how it can be used to reduce or completely eliminate the effects of the ambient magnetic field on an experiment. It is clear that the perfect conductivity of the superconductor will also shield an experiment against external electric fields within the limitations described by Morrison,[1] so our main concern from this point should be shielding against constant external magnetic fields.

We discussed earlier the penetration of a magnetic field into a surface layer of a type I superconductor. Obviously then, to be an effective shield the superconductor must be many penetration depths thick. This is not a serious design problem unless the experiment dictates that the shield be an evaporated film.

If we had a perfect Meissner effect we should have no difficulty ob-

---

† NbTi can be plasma sprayed by several processes. Most of the work described here was done by Union Carbide Corp., Tarrytown Technical Center, New York.

11.2. SUPERCONDUCTING SHIELDING

taining shielded regions with zero magnetic field. We would make a shield of superconducting material and simply cool it through its transition temperature. The Meissner effect would cause the superconductor to exclude the magnetic flux which penetrated it and the experiment inside would then be in a region of zero magnetic field. Unfortunately, a perfect Meissner effect is never observed.

We know that magnetic flux trapped in a superconductor must be quantized. Trapped flux quanta are what causes an imperfect Meissner effect. The flux lines are pinned in the material at grain boundaries or dislocations. One can decrease the number of pinning sites by using very pure material for the shield and by careful annealing of the shield after fabrication. Experimentally, however, the Meissner effect is still far from complete. Means must be devised to decrease the amount of flux which the Meissner effect must remove from the material. The ambient field must be made as small as possible before the shield is made superconducting. After the shield is superconducting, the ambient field can be changed to any value which does not exceed the critical field at the surface of the shield.

Several techniques have been used to reduce the ambient field. The experimental area can be surrounded by conventional high-permeability magnetic shields. The materials commonly used for these shields are mu-metal (Jones and Laughlin), Hypernome (Westinghouse), Netic and Co-netic (Perfection Mica), and similar materials by other manufacturers.[†] Most experiments using shielding have started by reducing the field with iron shields. Field levels of 10^{-5} Oe can be obtained by degaussing a properly designed double magnetic shield of high-permeability materials. Shields of higher performance have been reported,[7,8] but many of these have been more elaborate than required for a laboratory experiment.

Vant-Hull and Mercereau[9] and Deaver and Goree[6] have attempted to reduce the field penetrating a cylindrical superconducting shield by spinning the shield rapidly while it was cooled through its transition temperature. Any ambient magnetic fields that do not have axial symmetry

---

[7] D. Cohen, *J. Appl. Phys.* **38**, 1295 (1967).
[8] D. Cohen, *Rev. Phys. Appl.* **5**, 53 (1970).
[9] L. L. Vant-Hull and J. E. Mercereau, *Rev. Sci. Instrum.* **34**, 1238 (1963).

[†] Jones and Laughlin Steel Corp., 3A Gateway Center, Pittsburgh, Pennsylvania 15230; Westinghouse Electric Corp., Special Metals Div., Blairsville, Pennsylvania 15717; Perfection Mica Co., 742 Thomas Dr., Bensenville, Illinois 60106.

with respect to the cylinder will, when seen from the frame of the rotating cylinder, appear to be a time-varying field of frequency ω, where ω is the rotation frequency. Eddy currents will, therefore, be set up in the rotating cylinder to counteract these time-varying fields.

If the cylinder is started rotating while warmed above its superconducting transition temperature, these eddy currents will tend to exclude or screen the resulting time-varying field, but because of the finite conductivity, the currents will not screen it completely and will decay if the rotation is stopped. Measurements of the electrical resistivity of lead show that the resistivity of the metal drops abruptly above the transition temperature becoming very small before the metal becomes superconducting. Thus if the cylinder is slowly cooled while rotating, the eddy currents will approach those which would be induced if the material had zero resistance, i.e., they will approach the superconducting screening currents.

As the cylinder makes the transition to the superconducting state, there will be very little flux penetrating the cylinder, since it has already been excluded by the eddy currents. Thus the density of trapped flux lines should be much lower than it originally would have been.

It should be noted that this technique results in a field in the cylinder which has small components perpendicular to the rotation axis. It will not shield against a field which has axial symmetry with respect to the axis of rotation. The technique thus results in a very uniform field along the axis of the cylinder. If this component must also be made small, it is necessary to reduce it by using supplementary external coils or high-permeability magnetic shielding. The reduction must be done while the shield is being cooled, because once it is superconducting the flux is trapped in the cylinder and cannot be removed without driving the cylinder normal.

We have mentioned before that the Meissner effect has never been seen to be perfect. Bol[10] discovered that the amount of flux excluded from metal cylinders which were being spun depended on such things as the strain in the metal. He found that this strain dependence appeared even when the annealed cylinders were not spun at all until reaching helium temperatures and, hence, were probably due to the strain induced by the centrifugal forces of spinning. Because of Bol's measurements, it is probably not sufficient to depend on the Meissner effect itself to obtain an extremely low field.

---

[10] M. Bol, Ph.D. Thesis, Stanford Univ. (1965).

11.2. SUPERCONDUCTING SHIELDING

It has also been found that when shields are made by electroplating the superconductor onto another metal, e.g., Pb plated onto Cu, an increase in field will sometimes be obtained on cooling the shield through the superconducting transition.[11] Pierce[12] has estimated that the thermoelectric potentials associated with the plated materials and the temperature gradients obtained in cooling could result in currents of several amperes circulating in the shields. Once the shield superconducts these currents cannot decay and result in field trapped in the shield.

There are, however, other methods to reduce the field. If a folded flexible superconducting bladder is cooled in a given magnetic field, it will exclude some of the flux incident on the bladder and trap the remainder. If the bladder is then unfolded, the density of flux lines inside the bladder can be much lower, hence, the field is lower. If another bladder is then inserted inside the first and cooled slowly, it should trap less flux than the first because it was cooled in a lower magnetic field. The second bladder can then be expanded thus decreasing the field still further. A third bladder is then inserted inside the second, and so on. Thus, in principle, it should be possible to decrease the field to an arbitrarily small value.

There will come the point where the field on an unopened, nonsuperconducting bladder will be low enough that the flux linked by the bladder is less than one-half of a flux unit. At this point, when the bladder becomes superconducting, it is energetically favorable for it to exclude all the flux and thus yield a zero magnetic field inside.

The first published work on the use of superconducting bladders was that of Brown.[13] He used cylindrical bladders made of lead foil and folded so that upon opening they would decrease the field along the axis of the cylinder. The field transverse to the cylinder axis was decreased by a geometrical factor depending on the folding pattern. Brown obtained field reductions by a factor of 500 in each bladder expansion, but did not check the field below 10^{-6} G because his flux gate magnetometer would not respond to lower fields. Brown also checked the attenuation of external constant fields and found the decrease to be exponential as we would expect (Section 11.2.2.1).

Probably the most complete study of the properties of superconducting

---

[11] W. M. Prothero and J. M. Goodkind, *Proc. Symp. Phys. Superconducting Devices*, Univ. of Virginia, Charlottesville (1967).

[12] J. M. Pierce, Ph.D. Thesis, Stanford Univ. (1967).

[13] R. E. Brown, *Rev. Sci. Instrum.* **39**, 547 (1968).

bladders has been performed by Cabrera and Hamilton.[14] They performed a series of experiments whereby they obtained a field at least as low as 6×10^{-8} G and demonstrated that a shielding factor of at least 10^8 can be obtained with lead bladders.

Cabrera and Hamilton's experiments were performed with bladders made of lead foil of thickness 0.0063 cm. The foil is soldered without flux to make a cylindrical bladder of 10-cm diameter and is soldered closed at the bottom. Cabrera and Hamilton control the unfolding of their shields by gluing the foil to cotton cloth and using the cloth to guide the bladder over a former. As long as the cloth is kept dry before cooling, it remains flexible at helium temperatures. The lead foil at helium temperatures behaves approximately like household aluminum foil at room temperature. Braided nylon fish line has also proved useful. It remains flexible enough to thread over pulleys and transmits at least its test break strength at low temperatures.

The amount of field reduction depends on the folding pattern and the aspect which the bladder offers to the ambient magnetic field. If it offers a large surface perpendicular to the field, it will trap more flux than in the case that a very small surface area was presented to the field. The folding pattern will also control the field inside the bladder because, at seen from inside, each region of trapped flux looks like a current dipole and by proper folding some degree of cancellation can be obtained.

Cabrera and Hamilton measured the shielding factor by passing a complex of superconducting coils along the outside of the balloon and examining the response of a magnetometer inside. Alternating and constant fields of up to 50 G could be induced at the outer surface of the lead. They found that no signal could be detected inside the shield except in the case where there was a pin hole in the lead. At that point careful examination showed that the field reaching the inside of the shield was greatly attenuated because the nature of the shield is such that any magnetic flux line entering the shield through a hole must also go out through the same hole. For a shield without holes they were unable to detect, at the 10^{-7}-G level, any effect of a 50-G field. A 50-G field did not measurably change the field configuration in the bladder, so it did not move trapped flux lines.

The field inside a superconducting shield can be further reduced by "heat flushing," a technique which is designed to reduce the amount of

---

[14] B. Cabrera and W. O. Hamilton, in "Science and Technology of Superconductivity," p. 587. Plenum Press, New York, 1973.

11.2. SUPERCONDUCTING SHIELDING

flux which becomes trapped in a superconducting balloon. If, as we cool a folded superconducting bladder through the transition, we keep the area of the balloon which is in the vicinity of the transition temperature small, then it will take a large field to create a flux unit in this intermediate state region. The small transition region is then moved over the entire balloon, flushing out the flux. Some degree of heat flushing can even be obtained by transferring the first helium very slowly into the dewar containing the balloon.[5,14]

Cabrera and Hamilton used heat flushing to obtain their lowest fields. They constructed a length of double-walled pipe where the space between the walls could be evacuated. A folded superconducting bladder was placed inside the pipe which was then inserted vertically into a previously expanded balloon. Helium gas could be admitted at the upper end of the pipe to allow the level of the helium liquid to rise in the pipe, thus cooling the unfolded balloon from the bottom. They measured the temperature profile in such an experiment and found that a very sharp gradient, on the order of several degrees per centimeter, was maintained across the unfolded balloon. The level difference between the helium outside the pipe and that inside was observed with a differential oil manometer. The level was allowed to rise over the period of several hours. The rate of rise must be slow to avoid the generation of trapped flux due to thermoelectric currents induced by the changing temperature difference in the lead. It was in a series of heat flushing experiments that the low field of 6×10^{-8} ($\pm 6 \times 10^{-8}$) G was measured. The uncertainty in the measurements arose because of some magnetic impurity in the measuring Josephson magnetometer.

11.2.4. Conclusions

The experiments we have described indicate that it is possible to use superconducting shielding to obtain very low electric and magnetic fields and excellent shielding against field changes. The low electric fields come about as a result of the perfect conductivity of the superconductor and the very low magnetic fields because of the quantization of magnetic flux. It should be emphasized, however, that even without striving for low magnetic fields the superconducting shields make sensitive electrical measurements at low temperatures extremely easy. Pickup and interference are eliminated if the sensor is surrounded by a superconducting shield and if proper attention is paid to grounding and the effects of penetrations of the shield.

Acknowledgments

The influence of Professor William Fairbank on much of the work reported here cannot be overestimated. The author gratefully acknowledges the support of the Air Force Office of Scientific Research and the National Science Foundation during the preparation of this article.

12. EXPERIMENTAL METHODS IN THE PREPARATION AND MEASUREMENT OF THIN FILMS*

The use of thin solid films in physical investigations has grown rapidly during the past decade. In solid state physics a rapidly growing volume of experimental and theoretical work is appearing concerning the electronic, optical, magnetic, and superconducting properties of thin films. The use of thin films, however, is not restricted to studies of thin films per se or of phenomena uniquely characteristic of the thin film geometry. Thin films are often convenient to use. It is much easier to prepare a series of alloys in the form of thin films than to prepare a corresponding series of bulk alloys. A flat uncontaminated surface for optical property studies can be produced easily on a thin film in contrast to the meticulous polishing required to produce such a surface on a bulk crystal. With reactive metals such as the alkalis it is impossible to prepare clean surfaces with other than thin film techniques. Also, thin film techniques allow the study of many abnormal metastable structures not present in bulk samples. Most of the growth in the use of thin films, however, has been stimulated by the increasing application of thin films to practical devices. Technological applications of thin films include their use as mirrors, antireflection coatings, interference filters, decorative coatings, passive and active electronic components, photoemissive surfaces, radiation detectors, piezoelectric transducers, and computer memory elements.

In this chapter we will provide a summary of the experimental techniques for producing thin films and a discussion of some characteristic thin film measurements. Emphasis will be placed on the study of the film properties which are characteristic of their unique geometry and of the growth conditions employed in their preparation.[†] The reader

---

[†] The applications of thin films to the study of electron tunneling in solids and to the study of defects in solids are considered in Chapters 4 and 16, respectively.

* Part 12 is by D. C. Larson.

is referred to the general literature for more comprehensive treatments and for more complete bibliographies relating to thin film phenomena.[1-6]

12.1. Preparation of Thin Films

12.1.1. Introduction

Thin films may be formed by rolling, electropolishing, or, in some instances, by cleaving a bulk sample. Generally, however, thin films are formed on a substrate by a vapor or chemical deposition process. The growth of thin films proceeds through several stages. Initially, isolated nuclei are formed on the substrate. These nuclei grow and coalesce, holes are filled in, and eventually a continuous film is formed. The resultant film structure and properties are determined primarily by the structure and temperature of the substrate although other factors such as the deposition rate, the pressure and composition of the residual gases, and the kinetic energy and angle of incidence of the impinging vapor are also known to be important. Glasses and polycrystalline ceramics are commonly used substrate materials for polycrystalline films. To assure adhesion of the film to the substrate the substrates must be cleaned prior to use. Surface contaminants are generally removed by use of detergents and other solvents in conjunction with ultrasonic agitation. After drying, substrates may be heated to elevated temperatures in vacuum or subjected to ionic or electron bombardment for further cleaning. The crystallite size generally increases with increasing substrate temperature and with decreasing deposition rate. In order to produce single-crystal films single-crystalline metal or dielectric substrates are employed. Alkali halide, mica, MoS_2, and MgO substrates may be prepared for use by cleaving in air or in a vacuum system. The oriented growth of a film

---

[1] H. Mayer, "Physik Dünner Schichten," Vol. 2. Wissenschaftliche Verlagsgesellschaft, Stuttgart, 1955.

[2] L. Holland, "Vacuum Deposition of Thin Films." Wiley, New York, 1956.

[3] L. Holland (ed.), "Thin Film Microelectronics." Wiley, New York, 1965.

[4] J. C. Anderson (ed.), "The Use of Thin Films in Physical Investigations." Academic Press, New York, 1966.

[5] K. L. Chopra, "Thin Film Phenomena." McGraw-Hill, New York, 1969.

[6] R. I. Maissel and R. Glang (eds.), "Handbook of Thin Film Technology." McGraw-Hill, New York, 1970.

on a substrate is termed *epitaxial growth* and has been the subject of numerous investigations.[7,8]

The structural order of deposited films can be varied from a completely disordered state (amorphous) to a well-ordered (single-crystal) state. The film structures thus produced are often quite different from the bulk structures of the same materials. Some illustrative examples showing the diversity of possible film structures are presented.

(a) If the conditions at the substrate surface are such that the mobility of the incident atoms or molecules is very low, a highly disordered microcrystalline or amorphous structure will result. Such amorphous structures will result from vapor deposition onto room-temperature substrates of such materials as the elements Ge,[9] Si,[10] C,[11] B,[12] S, and Se,[13] inorganic compounds such as GaAs,[14] GeTe,[15] SiC,[16] and SiO, chalcogenide glasses,[17] polymers, and organic solids such as tetracene[18] and copper phthalocyanine. Vapor deposition onto substrates cooled to liquid-helium temperatures will produce essentially amorphous films of Bi and Ga,[19] Sb,[20] Fe and Cr,[21] Be,[22] and most nonmetallic compounds; the same deposition conditions will result in microcrystalline films of most metals such as Cu, Ag, Au, Al, Pb, Sn, In, Zn, and Hg.

(b) A variety of abnormal metastable thin film structures have been obtained; these structures are stabilized through the influence of the substrate or by impurities. Abnormal fcc structures have been obtained

---

[7] D. W. Pashley, *Advan. Phys.* **5**, 173 (1956); **14**, 327 (1967).

[8] J. W. Matthews, *Phys. Thin Films* **4**, 137 (1967).

[9] G. Breitling, *J. Vac. Sci. Technol.* **6**, 628 (1969).

[10] H. Richter and G. Breitling, *Z. Naturforsch.* **13a**, 988 (1958).

[11] B. T. Boiko, L. S. Palatnik, and A. S. Derevyanchenko, *Sov. Phys.—Dokl.* **13**, 237 (1968) (English transl.).

[12] C. Feldman and W. A. Gutierrez, *J. Appl. Phys.* **38**, 2474 (1967).

[13] O. S. Heavens and C. H. Griffiths, *Acta Cryst.* **18**, 532 (1965).

[14] J. E. Davey and T. Pankey, *J. Appl. Phys.* **35**, 2203 (1964).

[15] K. L. Chopra and S. K. Bahl, *J. Appl. Phys.* **40**, 4171 (1969).

[16] C. J. Mogab and W. D. Kingery, *J. Appl. Phys.* **39**, 3640 (1968).

[17] S. R. Ovshinsky, *Phys. Rev. Lett.* **21**, 1450 (1968).

[18] A. Szymanski, D. C. Larson, and M. M. Labes, *Appl. Phys. Lett.* **14**, 88 (1969).

[19] W. Buckel, *Z. Phys.* **138**, 136 (1954); W. Buckel and R. Hilsch, *ibid.* **138**, 109 (1954).

[20] H. Richter, H. Berckhemer, and G. Breitling, *Z. Naturforsch.* **9a**, 236 (1954).

[21] S. Fujime, *Jap. J. Appl. Phys.* **5**, 1029 (1966).

[22] S. Fujime, *Jap. J. Appl. Phys.* **5**, 59 (1966).

in such normally bcc metals as Nb, Ta, Mo, and W[23-25] and in such normally hcp metals as Hf, Zr, and Re.[25] Abnormal structures have also been observed in thin films of the alkali halides and a number of inorganic compounds.[26]

(c) Single-crystal films of a wide variety of materials can be obtained by vapor deposition onto heated single-crystalline substrates. For example, single-crystal films of elements such as Cu,[27] Ag,[28] Al,[29] Au and Fe,[30] and Ni[31] and compounds such as Ni–Fe[32] and Cu–Au[33] can be prepared by vacuum deposition onto heated {100} surfaces of NaCl. Single-crystal films of a large number of semiconducting and insulating materials may also be prepared by epitaxial growth on suitable substrates.[34] Often several different crystalline orientations are produced so that a textured polycrystalline film rather than a single-crystalline film is produced. Single-crystal metal films of near monolayer thicknesses may be prepared by vacuum deposition onto single-crystal metal surfaces.[35]

12.1.2. Thermal Evaporation

In order to form a thin film by thermal evaporation the source material must be heated in vacuum to a temperature sufficiently high to allow a reasonably high rate of evaporation or sublimation. The vapor is condensed onto a cooler substrate to form a solid film. The pressure in the vacuum system must be sufficiently low so that the vapor atoms or molecules travel in a straight line from the source to the substrate without colliding with the ambient gas molecules. This requires pressures lower

---

[23] P. N. Denbigh and R. B. Marcus, *J. Appl. Phys.* **37**, 4325 (1966).

[24] T. E. Hutchinson and K. H. Olson, *J. Appl. Phys.* **38**, 4933 (1967).

[25] K. L. Chopra, M. R. Randlett, and R. H. Duff, *Appl. Phys. Lett.* **9**, 402 (1966); *Phil. Mag.* **16**, 261 (1967).

[26] See, for example, Chopra (Ref. 5), p. 200.

[27] L. O. Brockway and R. B. Marcus, *J. Appl. Phys.* **34**, 921 (1963).

[28] B. W. Sloope and C. O. Tiller, *J. Appl. Phys.* **32**, 1331 (1961).

[29] C. Sella and J. J. Trillat, *in* "Single Crystal Films" (M. H. Francombe and H. Sato, eds.), p. 201. Pergamon, Oxford, 1964.

[30] J. W. Matthews, *Appl. Phys. Lett.* **7**, 131, 255 (1965).

[31] O. S. Heavens, R. F. Miller, G. L. Moss, and J. C. Anderson, *Proc. Phys. Soc. (London)* **78**, 33 (1961).

[32] R. D. Burbank and R. D. Heidenreich, *Phil. Mag.* **5**, 373 (1960).

[33] H. Sato and R. S. Toth, *Phys. Rev.* **124**, 1833 (1961).

[34] See, for example, I. H. Khan, *in* "Handbook of Thin Film Technology" (R. I. Maissel and R. Glang, eds.), p. 10-1. McGraw-Hill, New York, 1970.

[35] E. Grünbaum, *Proc. Phys. Soc. (London)* **72**, 459 (1958).

12.1. PREPARATION OF THIN FILMS

than 10^{-4} Torr, but even lower pressures are desirable to minimize the adsorption of the ambient gases on the substrate. For this reason ultrahigh vacuum systems which produce pressures lower than 10^{-9} Torr are often employed.

12.1.2.1. Resistive Heating. The temperatures required for evaporation or sublimation are most commonly produced by resistive heating. The source material or evaporant is supported by a filament, boat, or crucible of a refractory material which does not react chemically with the

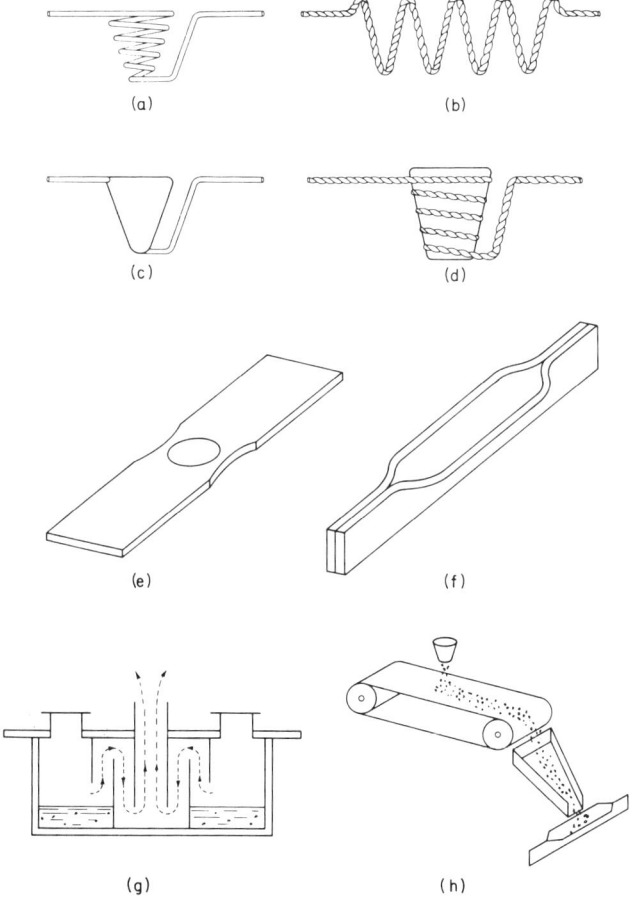

FIG. 1. Thermal evaporation sources: (a) wire basket; (b) wire helix; (c) ceramic-coated wire basket; (d) ceramic crucible with wire-coil heater; (e) dimpled boat; (f) canoe-type boat; (g) baffled chimney. (h) flash-evaporation source [L. Harris and B. M. Siegel, J. Appl. Phys. **19**, 739 (1948)].

TABLE I. Temperatures and Support Materials for Thermal Evaporation

Material	Temperature (°C)		Support materials			Other techniques[a]	Remarks
	mp	$vp = 10^{-2}$ Torr	Wire	Boat, crucible			
Ag	961	1030	W, Mo, Ta	Mo, Ta, C		SP, ED	Does not wet W
Al	660	1220	W, Ta	C, BN, TiB$_2$–BN		SP	Alloys with W, reacts with C and oxide crucibles
Al$_2$O$_3$	2030	~1800	W, Mo, Ta	W, Ta		EB, SP(rf), AN	Some decomposition
As	820	280		C, Al$_2$O$_3$, BeO			Sublimes, polyatomic vapor, toxic
Au	1063	1400	W, Mo	W, Mo, C		EB, SP	Reacts with Ta, wets W and Mo
B	~2200	2000		C		EB, SP	Deposits from C supports probably impure
Ba	720	610	W, Mo, Ta, Nb, Ni, Fe	W, Mo, Ta		SP	Wets refractory metals without alloying, reacts with most ceramics
BaO	1925	1540		Pt, C, Al$_2$O$_3$		EB	Some decomposition
Be	1280	1230	W, Mo, Ta	W, Mo, Ta, C, BeO		EB	Wets refractory metals, vapors and oxides, toxic
Bi	271	670	W, Mo, Ta, Nb, Ni	W, Mo, Ta, C, Al$_2$O$_3$			Vapors are toxic
C	~3800	~2600		C		EB, SP	Carbon arc evaporation, sublimes
Ca	840	600	W	Al$_2$O$_3$			Deposit corrodes in air
Cd	321	265	W, Mo, Ta, Nb, Fe, Ni	Mo, Ta, oxides			Sublimes

12.1. PREPARATION OF THIN FILMS

Material	Temp	Support	Other	Method	Comments
Cds	1750		670	W	Mo, C	EB, SP(rf)	Stoichiometry depends on substrate temperature
Co	1490	1520	W, Nb	Al_2O_3, BeO	EB, SP	Sublimes slowly, alloys with W
Cr	~1900	1400	W, Ta		EB, SP, ED	Sublimes
Cu	1083	1260	W, Mo, Ta	Mo, C, Al_2O_3	SP	Does not wet W, Mo, Ta
Fe	1535	1480	W	W, Al_2O_3, BeO, ZrO_2	EB, SP	Sublimes slowly, alloys with all refractory metals
Ga	30	1130		Al_2O_3, BeO		Alloys with refractory metals
Ge	940	1400	W, Mo, Ta	W, Mo, Ta, C, Al_2O_3	EB, SP, CVD	Wets refractory metals, purest films by electron-beam evaporation
In	156	950	W, Mo	W, Mo, C		Wets W, Mo boats preferred
Li	181	537		Fused quartz, steel	EB	Oxide destroys fused quartz
LiF	870	~1200		Mo		Easily evaporated from Mo boat
Mg	650	440	W, Mo, Ta, Nb, Ni	Mo, Ta, C		Sublimes
MgF_2	1263	1130	W, Mo, Ta	Mo, C	EB	Very little dissociation
Mn	1244	940	W, Mo, Ta	Al_2O_3	EB	Wets refractory metals
Mo	2620	2530			EB, SP	Sublimes slowly, electron-beam evaporation preferred
NaCl	801	670		W, Mo, Ta		Evaporation from oven
Nb	~2500	~2600	W	W	EB, SP	Reacts with W, electron-beam evaporation preferred
Ni	1450	1530	W	Al_2O_3, BeO	EB, SP	Sublimes slowly, reacts with W, electron-beam evaporation preferred

[a] EB, electron beam; SP, sputtering; SP(rf), rf sputtering; ED, electrodeposition; CVD, chemical vapor deposition; AN, anodization.

TABLE I (continued)

Material	Temperature (°C)		Support materials		Other techniques[a]	Remarks
	mp	$vp = 10^{-2}$ Torr	Wire	Boat, crucible		
NiFe	1395	~1600		Al_2O_3	EB, SP	Ni content low in films
Pb	327	715	W, Mo, Ni, Fe	W, Mo, Ta, Fe, Al_2O_3	EB	Does not wet refractory metals, toxic
PbS	1112	675	W, Mo	Quartz		Purest films from quartz furnace
Pd	1550	1460	W	Al_2O_3, BeO	EB, SP	Alloys with refractory metals, sublimes slowly
Pt	1770	2100	W	ThO_2, ZrO_2	EB, SP, ED	Alloys with refractory metals, electron-beam evaporation preferred
Rh	1970	2040	W	ThO_2, ZrO_2	EB, SP, ED	Sublimes slowly, electron-beam evaporation preferred
Sb	630	530	Mo, Ta, Ni	Mo, Ta, C, Al_2O_3, BN		Polyatomic vapor, toxic
Se	217	240	Mo, Ta, 304 stainless steel	Mo, Ta, C, Al_2O_3		Wets all source materials, contaminates vacuum system, toxic
Si	1410	1350		C, BeO, ThO_2, ZrO_2	EB, SP, CVD	Sublimes slowly, SiO contamination with oxide crucibles, electron-beam evaporation gives purest films

Material						
SiO		1025	W, Mo, Ta	Mo, Ta	EB	Evaporation from furnace
SiO$_2$	1730	1250		Al$_2$O$_3$	EB, SP(rf), CVD	Reacts with refractory metals
Sn	232	1250	W, Mo, Ta	Mo, Ta, C, Al$_2$O$_3$	EB	Wets Mo and Ta, attacks Mo
Ta	3000	3060			EB, SP	Ta arc evaporation, electron-beam evaporation preferred
Te	450	375	W, Mo, Ta	Mo, Ta, C, Al$_2$O$_3$		Wets all refractory metals without alloying, contaminates vacuum system, toxic
Ti	1670	1750	W, Ta	W, C, ThO$_2$	EB, SP	Sublimes slowly, reacts with refractory metals, getters gases
TiO$_2$	1840		W, Ta		EB, SP(rf)	Decomposes into TiO upon heating
V	1920	1850	W, Mo	Mo	EB	Sublimes slowly, wets Mo without alloying, reacts with W
W	3380	3230			EB, SP	W arc evaporation, electron-beam evaporation preferred
Zn	420	345	W, Mo, Ta, Nb, Ni	Mo, Fe, C, Al$_2$O$_3$		Sublimes, wets refractory metals without alloying
ZnS	1830	~1000	Mo, Ta	Mo, Ta, C		Partially decomposes, evaporation from furnace
Zr	1850	2400	W	W	EB, SP	Wets and slightly alloys with W, electron-beam evaporation preferred
ZrO$_2$	2700		W	W, Ta	EB, SP(rf)	Yields some ZrO

evaporant. Typical resistance heated evaporation sources are shown in Fig. 1. Some of these sources are easily constructed in the laboratory and others may be purchased commercially. Both single-wire and multi-strand filaments are used with the latter preferred due to the greater surface area. A large number of elements and compounds can be evaporated from filaments or boats of W, Mo, or Ta. More reactive materials such as Al, Fe, Ni, Si and SiO_2 require the use of ceramic-coated baskets or ceramic crucibles. Crucibles of C, Al_2O_3, BeO, BN, and quartz are commonly employed. In order to produce a reasonably high deposition rate the temperature of the source material should be raised to a temperature where the source material has a vapor pressure of 10^{-2} Torr. Approximate source temperatures required for a variety of elements and compounds are listed in Table I. Also listed are recommended support materials for thermal evaporation and suggested alternative techniques. Materials with evaporation temperatures less than their melting temperatures such as Cr, Fe, Zn, and numerous inorganic and organic compounds may be evaporated by sublimation. Those materials which can be fabricated in the form of wires may then be evaporated without using a contaminating source support.

A schematic of an experimental setup used by Garcia et al.[36] to produce Bi films is shown in Fig. 2. This system is basically an evaporator–cryostat combination which allows control of the substrate temperature over the range of 10 to 400 K. Cleaved mica was used as a substrate and gold contacts were deposited onto the mica prior to the bismuth deposition. Six contacts were provided—two for current leads, two placed longitudinally for voltage measurements, and two placed transversely for Hall voltage measurements. Manganin wire with poor thermal conductivity was used for leads to minimize thermal leakage during the low-temperature measurements. The substrates were then placed in the evaporator and the Bi film was deposited, annealed, and then measured without removal from the vacuum system. Once the measurements for a particular thickness were completed, the film was reannealed and an additional layer of Bi was deposited on top of the previous film. In this manner the temperature and thickness dependence of the electrical resistivity, Hall effect and transverse magnetoresistance of the Bi films were obtained. A nitrogen-cooled solenoid yielding a 2-kOe field was provided for the latter measurements.

Another experimental setup used by the author for thin film deposition

---

[36] N. Garcia, Y. H. Kao, and M. Strongin, *Phys. Lett.* **29A**, 631 (1969).

12.1. PREPARATION OF THIN FILMS

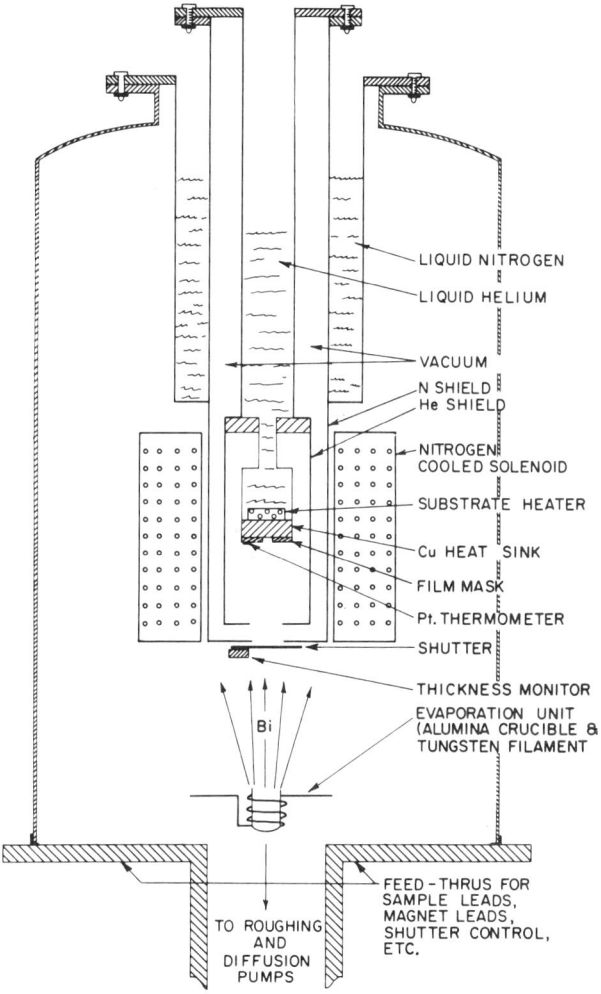

FIG. 2. Schematic of an evaporator with provision for in situ low-temperature measurements [N. Garcia, Y. H. Kao, and M. Strongin, *Phys. Lett.* **29A**, 631 (1969)].

and *in situ* measurements is shown in Fig. 3. This system utilizes a conventional liquid nitrogen trapped diffusion pump with a 2000-liter/sec pumping speed. Three thermal evaporation sources and three separate masks are employed to produce three-layer sandwich film structures as shown in Fig. 4. Microscope glass slides with predeposited Cu–Cr contact pads are used as substrates. The glass slides are attached to a copper block which in turn is attached to a reservoir which can be filled

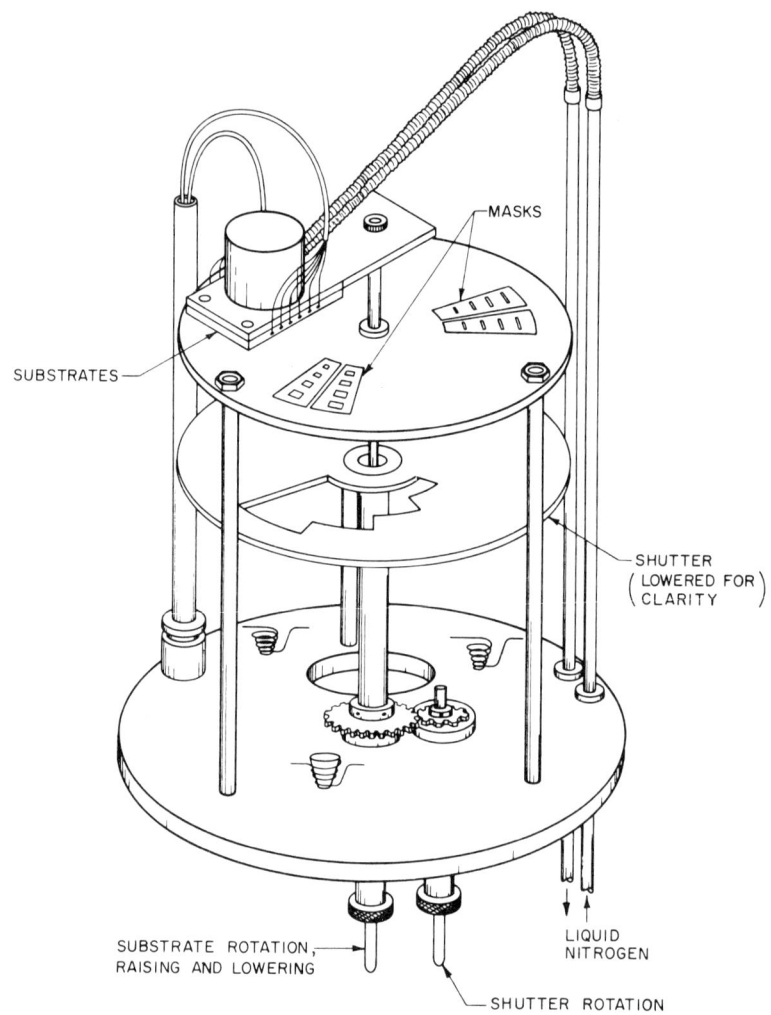

FIG. 3. Vacuum evaporation system used for three-layer film depositions.

with liquid nitrogen or other cryogenic fluids through flexible stainless steel hoses. A substrate heater is also provided. Lead wires are soldered to the Cu–Cr pads prior to the multilayer film depositions. The sandwich structure shown in Fig. 4 was produced by first depositing a 2000-Å-thick metal film through a mask. The thickness of this film can be monitored electrically. The substrate is then moved into position over the second mask and an inorganic or organic insulating film is deposited. Eight different insulator thicknesses are prepared through the step-by-step

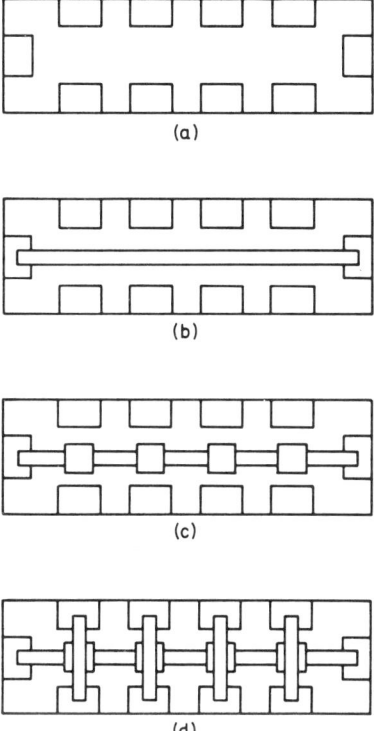

FIG. 4. Preparation of metal–insulator–metal sample: (a) glass microscope slide with predeposited Cu–Cr contact pads; (b) a metal strip has been deposited across the end contacts; (c) an insulating film has been deposited; (d) another metal film has been deposited across the insulating film.

rotary motion of a shutter. Finally the substrate is moved into position over the third mask where 2000-Å-thick metal counter electrodes are deposited. Tungsten wire filaments are generally used for the metal evaporations whereas ceramic crucibles are utilized for the dielectric film depositions. To prevent spattering of the dielectric material, quartz wool is often placed above the charge in the crucible. After deposition, *in situ* measurements of the current–voltage characteristics of the insulator films are made at temperatures ranging from 77 to 500 K. More elaborate multilayer film deposition systems have been described by Behrndt[37] and Randlett *et al.*[38] The latter authors describe a system which

---

[37] K. H. Behrndt, *in* "Thin Films," p. 1. Amer. Soc. for Metals, Metals Park, Ohio, 1964.

[38] M. Randlett, E. Stroberg, and K. L. Chopra, *Rev. Sci. Instrum.* **37**, 1378 (1966).

allows the preparation of films sequentially from any one of five vapor sources, onto any one of five substrates maintained at any temperature from 23 to 450°C, using any one of five masks.

The evaporation of alloys and compounds presents special problems because the components usually differ in their vapor pressures and hence do not evaporate congruently. Consequently the composition of the condensate may differ substantially from the composition of the evaporant. For example, direct evaporation of Cu–Zn or Cu–Cd alloys will produce films with Zn and Cd concentrations, respectively, much higher than the source. On the other hand permalloy (85% Fe, 15% Ni) and such alloys as Cu–Au, Cu–Ag, and Cr–Ge can be deposited with only minor compositional changes. Even when compositional changes occur, however, the source material may be evaporated completely and the film annealed to produce the desired composition in the alloy. Similar problems are encountered with compound evaporations when the compound dissociates during evaporation. For example many III–V compounds such as InSb, GaAs, and AlP dissociate and the resulting films are rich in the relatively volatile group V constituents. For these compounds two-source or flash-evaporation techniques are preferable to single-source depositions. Many of the II–VI and IV–VI compounds can be evaporated without significant compositional changes. Stoichiometric single-crystal films of such compounds as PbS, PbSe, and PbTe,[39] and CdS[40] have been prepared by vacuum deposition.

The use of two or more sources to produce multilayer films was discussed previously. The use of two or more sources simultaneously can be used to produce multicomponent films. The sources may be operated at different temperatures and the problems associated with noncongruent evaporation from a single source are avoided. Two-source evaporations have been used to produce alloys such as $PbTe_xSe_{1-x}$,[41] stoichiometric compounds such as CdS,[42] AlSb,[43] GaAs,[44] GaP,[45] and $BaTiO_3$,[46] meta-

---

[39] J. N. Zemel, in "Solid State Surface Science" (M. Green, ed.), Vol. 1, p. 291. Dekker, New York, 1969.

[40] N. F. Foster, *J. Appl. Phys.* **38**, 149 (1967).

[41] R. F. Bis, A. S. Rodolakis, and J. N. Zemel, *Rev. Sci. Instrum.* **36**, 1626 (1965).

[42] R. J. Miller and C. H. Bachman, *J. Appl. Phys.* **29**, 1277 (1958).

[43] J. E. Johnson, *J. Appl. Phys.* **36**, 3193 (1965).

[44] R. F. Steinberg and D. M. Scruggs, *J. Appl. Phys.* **37**, 4586 (1966).

[45] J. E. Davey and T. Pankey, *J. Appl. Phys.* **40**, 212 (1969).

[46] A. E. Feuersanger, A. K. Hagenlocher, and A. L. Solomon, *J. Electrochem. Soc.* **111**, 1387 (1964).

stable alloy phases, and multiphase structures such as cermets. The use of two-source evaporations to produce stoichiometric semiconducting compounds has been analyzed by Günther[47] and has been developed into the experimental technique known as the "three-temperature technique." This technique exploits the fact that a suitable substrate temperature may be found for compound growth for a wide variation of constituent vapor pressure ratios. The conditions required for compound formation are defined by a "stoichiometric interval" which specifies ranges of vapor pressures and substrate temperatures which allow stoichiometric compound formation. The three-temperature technique has been applied with particular efficacy to the preparation of the III–V and II–VI semiconductor compound films. Metastable phases can also be prepared by codeposition onto cooled substrates (vapor quenching). Phases can be prepared in this manner which violate the equilibrium phase diagrams. Mader[48] has extensively studied a number of metastable alloy systems.

The two-source or three-temperature techniques require rather careful control of the source temperatures. An alternative approach to alloy and compound depositions is that of flash evaporation. With this technique fine particles of the compound are fed continuously into a heater which is hot enough to vaporize all the constituents of the particles. Alternatively, a mixture of the constituents is fed into the heater. The flash evaporation mechanism used by Harris and Siegel[49] to produce metal alloy films is shown in Fig. 1h. The particles are transported from a hopper to the Mo boat by a metal belt. The temperature of the boat is adjusted so that the particles melt and then disappear almost instantaneously. A variety of feed mechanisms have been developed and the technique has been applied to the preparation of thin films of metal alloys, cermets, perovskites, and the III–V semiconducting compounds,[50] as well as to the preparation of chalcogenide glass films.[51]

12.1.2.2. Methods for Refractory Materials.
With resistive heating the source support material is in general heated to a higher temperature than

---

[47] K. Günther, in "The Use of Thin Films in Physical Investigations" (J. C. Anderson, ed.), p. 213. Academic Press, New York, 1966.

[48] S. Mader, in "The Use of Thin Films in Physical Investigations" (J. C. Anderson, ed.), p. 433. Academic Press, New York, 1966.

[49] L. Harris and B. M. Siegel, *J. Appl. Phys.* **19**, 739 (1948).

[50] J. L. Richards, in "The Use of Thin Films in Physical Investigations" (J. C. Anderson, ed.), p. 71. Academic Press, New York, 1966.

[51] H. K. Rockstad, *Solid State Commun.* **7**, 1507 (1969).

the evaporant itself. The possibility of source material–evaporant reaction or direct source support material evaporation presents the possibility of film contamination. Furthermore, it becomes very difficult to supply enough energy by resistive heating to evaporate the more refractory materials. The use of electron bombardment heating overcomes the power limitations imposed by resistive heating and source temperatures greater than 3000°C may be obtained. The electron beam may be directed at the evaporant itself or may be used to heat a conducting refractory crucible such as C or W. The best approach, however, is to utilize direct heating of only a portion of the source material by using a focused electron beam. When the source material is placed on a water-cooled pedestal, a portion of the source material will serve as its own crucible.

The electrons are emitted from a hot tungsten filament and are accelerated by a potential difference of 4 to 10 kV. The electron gun systems may be either work-accelerated or self-accelerated. In the former the source material (work) serves as the anode whereas the latter type of system employs a separate anode with an aperture through which the electron beam passes. Examples of both types of sources are shown in Fig. 5. The pendant-drop source (Fig. 5a) requires a rod or wire source material which forms a molten drop at its end due to the electron bombardment. The drop is held by its surface tension and therefore cannot be heated too far above its melting temperature. Satisfactory evaporation rates have been obtained with such metals as Fe, W, Ni, Ta, Ag, Mo, Pt, Ti, and Zr. The work-accelerated sources shown in Figs. 5b and c provide electrostatic and magnetic focusing, respectively. The focusing of the electron beam allows the melting of only the surface of the evaporant and the water-cooled crucible presents no problems of contamination. The proximity of the vapor source and filament can create problems of glow discharge, however, which may be alleviated with a self-accelerating source design (Fig. 5d). These sources, however, require voltages of 10–20 kV and present X-ray radiation problems.

An electron-beam evaporation system used by Gould et al.[52] is shown schematically in Fig. 6. The vacuum system is bakable and pressures as low as 10^{-11} Torr can be attained prior to an evaporation. A pending drop evaporation source is used and a carrousel mechanism permits the selection of any one of six rods for the evaporation. Films are prepared simultaneously on substrates placed on the central diffrac-

---

[52] G. Gould, C. Grahman, E. Grünbaum, L. Moraga, J. Müller, and D. C. Larson, *Thin Solid Films*, **13**, 61 (1972).

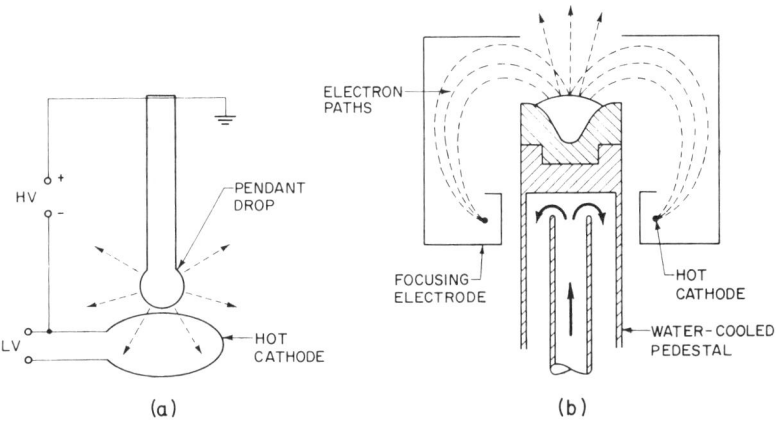

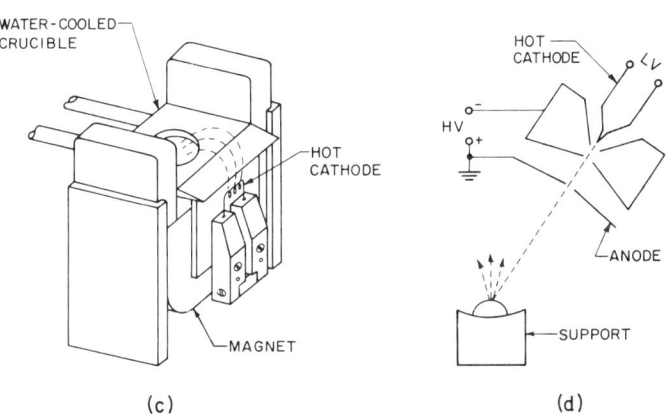

Fig. 5. Electron-beam sources for thermal evaporation: (a) pendant-drop source; (b) source with electrostatic focusing [B. A. Unvala and G. R. Booker, *Phil. Mag.* **9**, 691 (1964)]; (c) source with magnetic focusing (Varian Associates, Palo Alto, California); (d) self-accelerated electron gun.

tion support and in the oven. Heated single-crystal substrates such as cleaved NaCl are employed to allow epitaxial growth of single-crystal films. *In situ* studies of the film growth processes are possible with reflection high-energy electron diffraction (RHEED) or low-energy electron diffraction (LEED). For RHEED observations the substrate may be rotated $\pm 10°$ about the vertical axis and up to 170° about an axis normal to the specimen; the specimen may also be moved continuously by ± 10 mm in a direction normal to the specimen surface. For LEED observa-

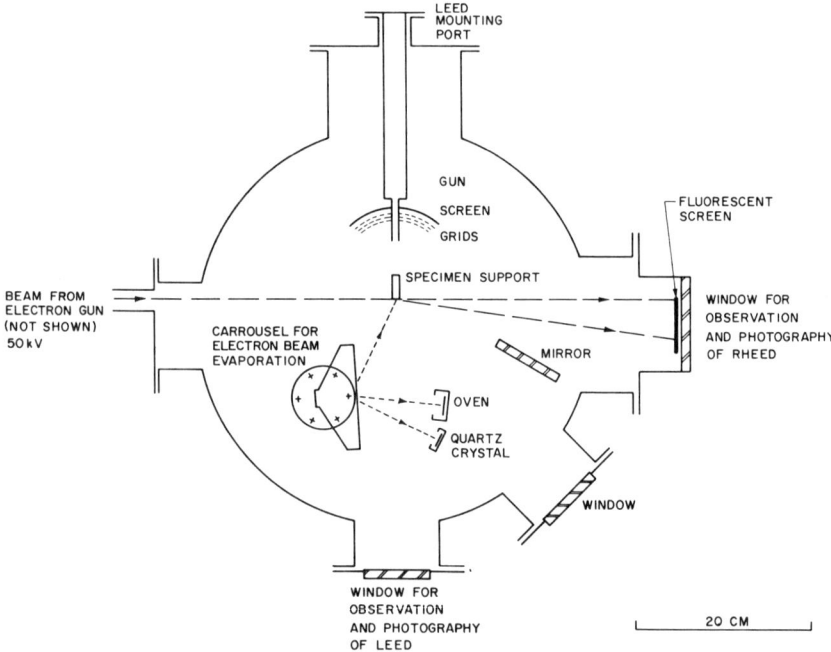

Fig. 6. Schematic of electron-beam evaporation system with provision for *in situ* LEED and RHEED observations [G. Gould, C. Grahman, E. Grünbaum, L. Moraga, J. Müller, and D. C. Larson, *Thin Solid Films*, **13**, 61 (1972)].

tions the specimen must be rotated by 180° around a vertical axis. Epitaxial films for electrical property studies are prepared in the oven. A shutter in front of the oven allows the preparation of six different film thicknesses. The film thicknesses are monitored with a quartz crystal oscillator.

Refractory alloy and compound films can be prepared with two electron-beam heated sources. Two separate sources may be employed or the electron beam can be oscillated back and forth between two sources. With the latter approach the dwelling time of the electron beam on each source can be controlled to produce the desired ratio of vapor pressures above each source.[53] This technique has been used to produce homogeneous Cu–W and Cu–Mo alloy films.[54]

Additional thermal evaporation techniques, which are applicable to refractory materials and which are capable of producing very high

[53] M. Oron and C. M. Adams Jr., *J. Sci. Instrum.* (*J. Phys. E*) **2**, 183 (1969).
[54] M. Oron and C. M. Adams Jr., *J. Appl. Phys.* **40**, 4218 (1969).

evaporation rates, utilize electric arcs, high-power lasers, and exploding wires. Electric arcs are produced between conducting electrodes with standard dc arc-welding generators. Thin films of C for electron-microscope replica studies are commonly produced in this manner.[55] High-intensity laser beams have been used to evaporate fine grain powders of Cr, W, Ti, C, Sb_2S_3, ZnS, $SrTiO_3$, and $BaTiO_3$.[56] The laser source is placed outside the vacuum system and consequently no contamination is introduced from the heating source. Deposition rates as high as 10^6 Å/sec can be achieved. Such high deposition rates can also be achieved by passing a large current from a bank of capacitors through a metallic or semiconducting wire. This exploding-wire technique has been used to prepare Cu, Au, and Ge,[57] InSb and InAs[58] films.

12.1.3. Sputtering

When a target is bombarded with energetic particles, atoms are ejected from the surface of the target. This process, which is called *sputtering*, may be used to produce thin films when the ejected atoms are allowed to condense on a substrate. The sputtering process itself has been extensively studied[59-61] and the application of sputtering to thin film preparation has been reviewed by Maissel.[62]

The sputtering process is characterized by the sputtering yield which is defined as the average number of atoms ejected from the target surface per incident particle (ion). No sputtering is observed for incident ion energies of less than 20–30 eV, but the yield rises rapidly, approaches a flat maximum, and then decreases with increasing incident ion energy. The maximum yield occurs at higher energies for more massive incident

---

[55] G. Thomas, "Transmission Electron Microscopy of Metals," p. 133. Wiley, New York, 1962.

[56] H. Schwarz and H. A. Tourtellotte, *J. Vac. Sci. Technol.* **6**, 373 (1969).

[57] D. M. Mattox, A. W. Mullendore, and F. N. Rebarchik, *J. Vac. Sci. Technol.* **4**, 123 (1967).

[58] L. N. Aleksandrov, E. I. Dagman, V. I. Zelevinskaya, V. I. Petrosjan, and P. A. Skripkina, *Thin Solid Films* **5**, 1 (1970).

[59] E. Kay, *Advan. Electron. Electron Phys.* **17**, 245 (1962).

[60] M. Kaminsky, "Atomic and Ionic Impact Phenomena on Metal Surfaces." Academic Press, New York, 1965.

[61] G. K. Wehner and G. S. Anderson, *in* "Handbook of Thin Film Technology" (L. I. Maissel and R. Glang, eds.), p. 3-1. McGraw-Hill, New York, 1970.

[62] L. I. Maissel, *in* "Handbook of Thin Film Technology" (L. I. Maissel and R. Glang, eds.), p. 4-1. McGraw-Hill, New York, 1970.

ions; energies for maximum yield range from less than 5 keV for H and He ions to 50 keV or more for Xe and Hg ions. The yield varies much more strongly with variation in incident ion than with variation of target material species, but there is a periodicity of yield related to the location of the target material species in the periodic table. The yield also increases with increasing angle of incidence with respect to the normal, except for single-crystal targets where yield maxima and minima are observed. The yield minima are correlated with the crystalline directions in which the ions can penetrate most deeply. The yield, however, is relatively insensitive to target temperature and yields from liquid metals are in many cases similar to the yields obtained with the corresponding solids.

The energies and velocities of the sputtered atoms are typically much higher than those of evaporated atoms. Ejection velocities of sputtered atoms range from 4 to 8×10^5 cm/sec, whereas for comparable deposition rates the average velocity of evaporated atoms is an order of magnitude lower.

12.1.3.1. Glow-Discharge Sputtering. Energetic ions for sputtering are most easily produced in a glow discharge. A glow discharge is formed when a voltage greater than the breakdown voltage is applied between two electrodes in a gas at low pressure. The glow discharge consists of a number of visually distinguishable regions the most important of which is the cathode or Crookes dark space. Most of the applied voltage occurs across this region (cathode fall) and the ions and electrons created in the breakdown are accelerated across this region. The energetic ions strike the cathode and produce sputtered atoms and secondary electrons which in turn produce additional ionization. Under "normal glow" conditions with the voltage close to the breakdown value, the number of secondary electrons produced is just sufficient to sustain the discharge. For effective sputtering "abnormal glow" conditions are required with higher operating voltages and higher current densities.

The optimum pressure range for glow-discharge sputtering is between 25 and 75 mTorr. The length of the cathode dark space is 1–2 cm at these pressures and the optimum cathode (target)–substrate distance is roughly twice the length of the cathode dark space. If the substrate is too close to the cathode, it will interfere with the ion current. A film of uniform thickness will be produced over a substrate area of about one-half that of the cathode area. The cathode may be in the form of a plate, foil, or powder compact which is attached to a suitable support.

12.1. PREPARATION OF THIN FILMS

The vacuum equipment used for glow-discharge sputtering is similar to that used with thermal evaporation techniques. Due to the relatively high operating pressures, diffusion pump systems rather than ion pump systems are generally used. Since much more heat is generated during sputtering than during thermal evaporation, cooling of the system walls, the cathode, and the substrate is desirable, but not essential. The sputtering system used by Francombe[63] for preparing semiconducting compound films is shown in Fig. 7. The electrodes are in the form of flat plates and the substrates are supported by the anode. Three substrates can be coated sequentially and substrate heating facilities are provided. The system is initially pumped down to a vacuum of about 10^{-7} Torr and then the glow discharge is formed in an atmosphere of flowing inert gas such as Ar or Kr. The cathode is maintained at a negative potential of 1–4 kV with respect to the grounded anode and a current-limiting resistor

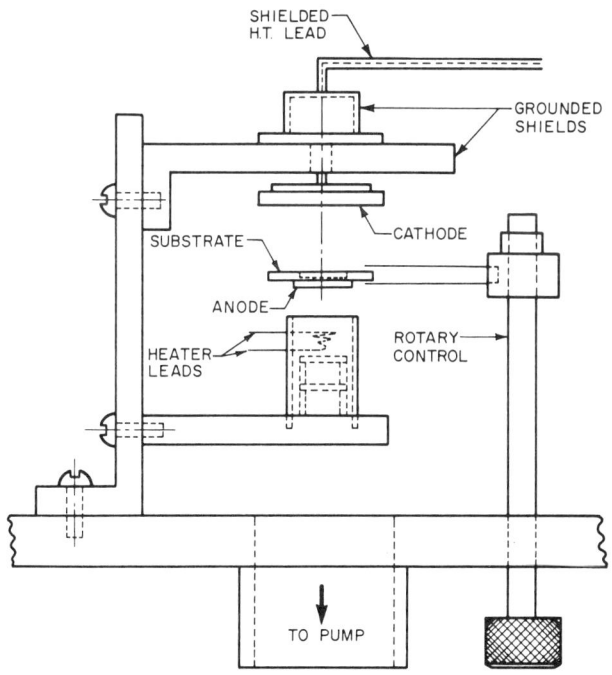

FIG. 7. Schematic of glow-discharge sputtering system [M. H. Francombe, *Trans. 10th Nat. Vacuum Symp.* p. 316. Macmillan, New York, 1963].

[63] M. H. Francombe, *Trans. 10th Nat. Vacuum Symp.*, p. 316. Macmillan, New York, 1963.

is provided to prevent arcing. The rate of deposition depends on the power supplied and highly reproducible film thicknesses can be obtained by control of the power and deposition time alone. Contamination, however, is a serious problem and may result from either outgassing from the vacuum system walls or from decomposition of the diffusion pump oil due to the high operating pressures.

Several electrode arrangements have been used in glow-discharge sputtering. With normal diode sputtering the substrate is attached to the anode and kept at anode potential or the substrate is allowed to float and may acquire a negative potential relative to the anode. If the substrate is held at a large negative potential, the growing film will be subjected to a sputtering action which will preferentially remove impurity atoms leading to "cleaner" films. Another approach which is particularly useful in the preparation of chemically reactive films is getter-sputtering. This method utilizes the gettering action of sputtering material to purify the inert gas before it reaches the part of the system where the substrate is located. The getter-sputtering apparatus used by Hauser et al.[64] for preparing two-film sandwiches is shown in Fig. 8. The cathode is made of the material to be sputtered and is surrounded by an anode can located far enough from the cathode so that it does not penetrate the cathode dark space at any point. The cathode sputters in all directions toward the anode and getters the inert gas which enters from the top of the anode can. The substrate is situated on a cold table and the deposition process is controlled by a tantalum shield. Prior to coating, the cathodes are sputtered for 10 min or longer to bring them to a steady state condition. Typically films are deposited using 1000 V and 2 mA in an atmosphere of 25 mTorr of Ar gas. For the deposition of the second film, the second cathode is brought into place and presputtered. The substrate is maintained at liquid-nitrogen temperatures throughout the procedure to minimize interdiffusion between the two layers of the sandwich. A double-can–double-cathode getter-sputtering apparatus has also been developed to minimize the delay between the deposition of the first and second films.[65] The getter-sputtering technique has been used to prepare superconducting thin films of Nb, Ta, V_3Si, V_3Ge, and V_3Ga,[66] films of rare-earth alloys in the Co–Cu–Sm system,[67] superconducting–normal film

---

[64] J. J. Hauser, H. C. Theuerer, and N. R. Werthamer, *Phys. Rev.* **136**, A637 (1964).
[65] J. J. Hauser, H. C. Theuerer, and N. R. Werthamer, *Phys. Rev.* **142**, 118 (1966).
[66] H. C. Theuerer and J. J. Hauser, *J. Appl. Phys.* **35**, 554 (1964).
[67] H. C. Theuerer, E. A. Nesbitt, and D. D. Bacon, *J. Appl. Phys.* **40**, 2994 (1969).

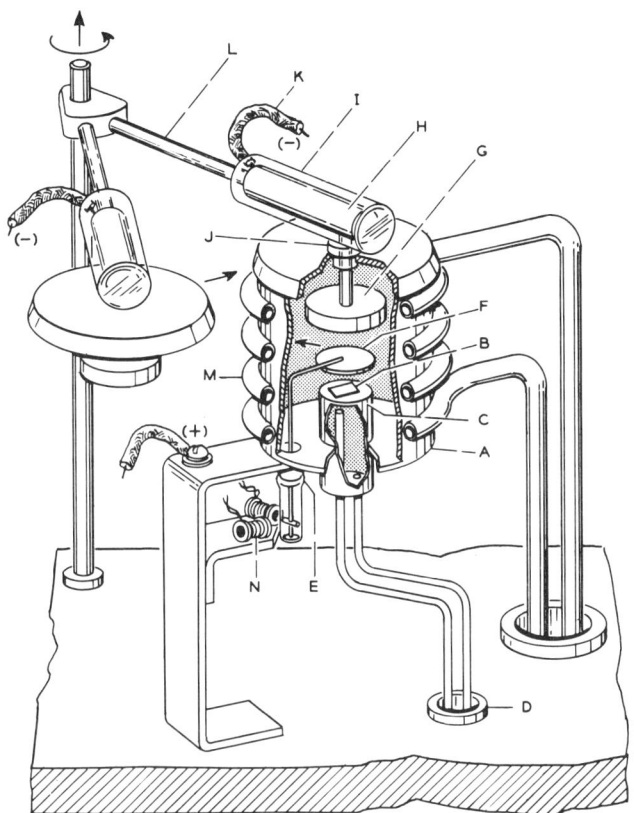

FIG. 8. Getter-sputtering system used for two-layer sandwich depositions. A: anode can; B: substrate; C: liquid-nitrogen cold table; D: liquid-nitrogen feed-through; E: Covar-glass seal; F: tantalum shield; G: electrode; H: electrode holder; I: quartz shield; J: quartz bushing; K: shielded negative lead; L: quartz supported rod; M: liquid-nitrogen cooling coil. N: electromagnet [J. J. Hauser, H. C. Theuerer, and N. R. Werthamer, *Phys. Rev.* **136**, A637 (1964)].

sandwiches of Pb/Pt and Pb/Cu,[64] Pb/Fe, Pb/Ni, Pb/Gd, Pb/Cr, Pb/Mo–1%Fe, and Pb/Pb–2.9%Gd,[65] and other multilayer sandwich structures.[68]

Getter-sputtering techniques are employed to purge the inert gas of undesirable impurities. Reactive gases, however, such as O_2, N_2, NH_3, CH_4, and H_2S may be premixed with the inert gas and deliberately introduced into the sputtering chamber where they react with the sputtered material to form compound films. This process is termed *reactive*

---

[68] J. J. Hauser, *Phys. Rev. Lett.* **23**, 374 (1969); *Phys. Rev* B **1**, 3624 (1970).

sputtering and has been used to produce compounds such as the oxide, hydride, carbide, nitride, or sulfide of the metal or semiconducting cathode. The process of reactive sputtering has been reviewed by Schwartz,[69] and representative applications of its use include the preparation of the oxide and nitride of Ta,[70] the preparation of superconducting films of NbN with transition temperatures close to the bulk value,[71] the preparation of films of AlN with approximately bulk properties,[72] and the preparation of epitaxial layers of the refractory semiconducting material SiC.[73] Dopants may also be introduced during the sputtering process. For example, p-type CdS films have been prepared by sputtering CdS in an argon atmosphere containing small amounts of PH_3.[74]

12.1.3.2. Low-Pressure Methods. Useful sputtering rates are attainable with a self-sustained glow discharge only when the inert gas pressure is greater than 15 or 20 mTorr. Even lower pressures would be desirable to minimize the incorporation of inert gas or impurity atoms into the growing film. Sputtering at lower pressures is possible when electron and/or ion concentrations are increased or when the ionizing efficiency of the electrons is increased.

The electron supply may be increased by introducing a thermionic filament and a third electrode to accelerate the electrons supplied by the filament. This process, called *triode sputtering*, can be used to produce films at ambient pressures as low as 1 mTorr. A triode sputtering apparatus used by Foster[75] to produce piezoelectric ZnO films is shown in Fig. 9. The quartz or sapphire substrates are mounted in a heated aluminum holder 2 cm away from the target which is a pressed and sintered ZnO disk. The orientation of the substrate with respect to the target can be used to control the direction of incidence of the vapor beam which in turn affects the orientation of the deposited film. To obtain this directivity a low-pressure sputtering system is essential as the mean free path of the condensing material must be greater than the target to substrate distance. ZnO films with the c axes of the crystallites perpen-

---

[69] N. Schwartz, *Trans. 10th Nat. Vacuum Symp.* p. 325. Macmillan, New York, 1963.
[70] E. Krikorian and R. J. Sneed, *J. Appl. Phys.* **37**, 3674 (1966).
[71] J. R. Gavaler, J. K. Hulm, M. A. Janocko, and C. K. Jones, *J. Vac. Sci. Technol.* **6**, 177 (1969).
[72] A. J. Noreika, M. H. Francombe, and S. A. Zeitman, *J. Vac. Sci. Technol.* **6**, 194 (1969).
[73] A. J. Learn and K. E. Haq, *Appl. Phys. Lett.* **17**, 26 (1970).
[74] M. Lichtensteiger, I. Lagnado, and H. C. Gatos, *Appl. Phys. Lett.* **15**, 418 (1969).
[75] N. F. Foster, *J. Vac. Sci. Technol.* **6**, 111 (1969).

dicular to the substrate (which produce longitudinal waves only) are produced by deposition under relatively clean vacuum conditions. By introducing organic contaminants (xylene) and applying a bias of -150 V to the substrate, films with the c axes of the crystallites parallel to the substrate and with an average direction toward the target are produced. Such films yield good piezoelectric shear mode transducers.

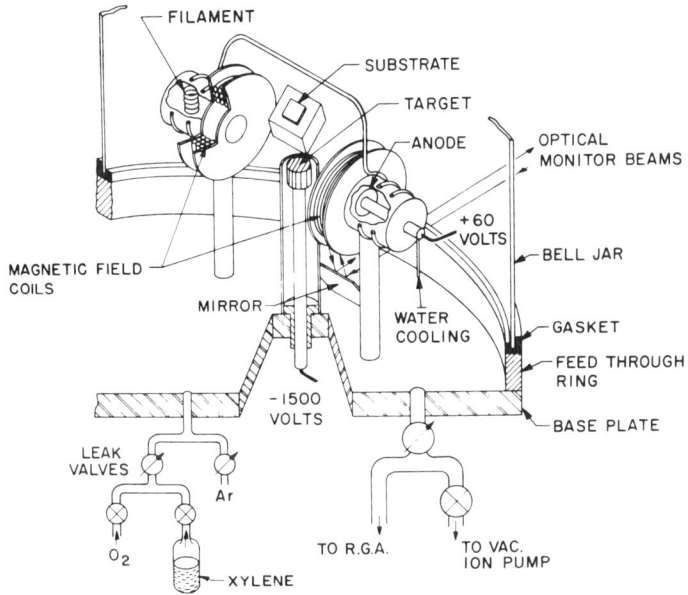

FIG. 9. Triode sputtering apparatus [N. F. Foster, *J. Vacuum Sci. Technol.* **6**, 111 (1969)].

The ionizing efficiency of the electrons may be increased in a triode sputtering system by applying a magnetic field parallel to the hot filament–anode axis. Such a system has been used by Ivanov *et al.*[76] to prepare magnetic films and by Edgecumbe *et al.*[77] to prepare superconducting films of Nb with a single target and films of Nb–Zr, Nb–Sn, Nb–Ti, and Ti–V using two targets. With the magnetic field parallel to the sputtering target the discharge tends to concentrate at one side of the target and reduces the uniformity of the sputtering rate over the target surface.

---

[76] R. D. Ivanov, G. V. Spivak, and G. K. Kislova, *Izv. Akad. Nauk SSSR Ser. Fiz.* **25**, 1524 [*Transl. Bull.* **25**, No. 12 (1961)].

[77] J. Edgecumbe, L. G. Rosner, and D. E. Anderson, *J. Appl. Phys.* **35**, 2198 (1964).

This can be avoided by using a hollow cylindrical target (cathode) with an axial magnetic field. The anode can be either inside or outside the cathode cylinder with the sputtering occuring inside. Three-dimensional substrates such as wires may be uniformly coated when placed inside the hollow target. Gill and Kay[78] have used a hollow cathode with a coaxial anode (inverted magnetron configuration) to sputter films at pressures as low as 10^{-5} Torr.

Low-pressure sputtering can also be induced by a high-frequency electromagnetic field. The high-frequency potential may be applied directly to one of the electrodes, to a coil in the discharge chamber, or to a coil outside the discharge chamber. This technique is especially useful in the preparation of insulating films which cannot be sputtered with dc techniques as the positive charge which accumulates on the target surface during ion bombardment cannot be easily neutralized. When a high-frequency field is applied, the target surface will acquire a negative bias since the electron mobilities in the plasma are orders of magnitude higher than the positive ion mobilities. This negative bias is essential for sputtering to occur and requires frequencies greater than 10 kHz. At higher frequencies a larger negative bias will develop resulting in a higher sputtering rate; low megacycle frequencies are generally needed for best results. Davidse and Maissel[79] have developed an rf sputtering system which has been used to prepare thin films of fused quartz, aluminum oxide, mullite, boron nitride, and various glasses. The rf field (13.56 MHz from a conventional amateur radio transmitter) is applied to a metal electrode behind the dielectric target. A grounded metal shield is placed close to the metal electrode to prevent sputtering of the metal electrode. The glow discharge can be sustained in a 10^{-2}-Torr atmosphere of Ar and at 10^{-3} Torr with a 110-Oe magnetic field normal to the target. The technique of rf sputtering has been shown to yield insulator films of high quality and excellent stability.[80] The method has been successfully applied, for example, to the preparation of Si,[81] SiO_2,[82]

---

[78] W. D. Gill and E. Kay, *Rev. Sci. Instrum.* **36**, 277 (1965).

[79] P. D. Davidse and L. I. Maissel, *J. Appl. Phys.* **37**, 574 (1966).

[80] W. A. Pliskin, P. D. Davidse, H. S. Lehman, and L. I. Maissel, *IBM J. Res. Develop.* **11**, 461 (1967).

[81] M. H. Brodsky, R. S. Title, K. Weiser, and G. D. Pettit, *Phys. Rev. B.* **1**, 2632 (1970).

[82] H. R. Koenig and L. I. Maissel, *IBM J. Res. Develop.* **14**, 168 (1970); J. S. Logan, *ibid.* **14**, 172 (1970); L. I. Maissel, R. E. Jones, and C. L. Standley, *ibid.* **14**, 176 (1970); J. S. Logan, F. S. Maddocks, and P. D. Davidse, *ibid.* **14**, 182 (1970).

12.1. PREPARATION OF THIN FILMS

ZnO,[83] and ferrimagnetic garnet films.[84] Metal films can also be produced by rf sputtering if the rf field is coupled to the target electrode through a capacitor. The capacitor prevents a dc current from flowing and allows a negative bias to develop on the metal electrode. Schrey *et al.*[85] have studied the structure and properties of rf sputtered, superconducting Ta films in comparison with the results obtained with other deposition techniques. A general review of rf sputtering and the application of rf sputtering techniques to the preparation of thin films has been prepared by Jackson.[86]

12.1.4. Chemical Deposition

Thin films may also be formed by a variety of chemical methods. These methods are used to prepare films of thickness ranging from less than 100 Å to greater than 100 μm and with structures ranging from single-crystalline to amorphous. Techniques such as electroplating, anodization, and chemical vapor deposition (CVD) are of great commercial importance and are discussed in a number of standard reference volumes.[87-89] Review articles have been devoted to such topics as the gas phase deposition of organic and inorganic insulating films,[90] oxide layer deposition from organic solutions,[91] chemical vapor deposition of single-crystal films,[92] electrodeposition of thin films,[93] and the preparation of anodic oxide films.[94]

12.1.4.1. Chemical Vapor Deposition.
The process by which the nonvolatile products of a gas phase reaction are allowed to deposit onto a substrate is known as *chemical vapor deposition*. These processes may be further classified by the type of chemical reaction used such as thermal

---

[83] D. L. Raimondi and E. Kay, *J. Vac. Sci. Technol.* **7**, 96 (1970).
[84] E. Sawatzky and E. Kay, *J. Appl. Phys.* **39**, 4700 (1968).
[85] F. Schrey, R. D. Mathis, R. T. Payne, and L. E. Murr, *Thin Solid Films* **5**, 29 (1970).
[86] G. N. Jackson, *Thin Solid Films* **5**, 209 (1970).
[87] F. A. Lowenheim (ed.), "Modern Electroplating." Wiley, New York, 1963.
[88] L. Young, "Anodic Oxide Films." Academic Press, New York, 1961.
[89] C. F. Powell, J. H. Oxley, and J. M. Blocher, Jr. (eds.), "Vapor Deposition." Wiley, New York, 1966.
[90] L. V. Gregor, *Phys. Thin Films* **3**, 131 (1966).
[91] H. Schroeder, *Phys. Thin Films* **5**, 87 (1969).
[92] B. A. Joyce, *in* "The Use of Thin Films in Physical Investigations" (J. C. Anderson, ed.), p. 87. Academic Press, New York, 1966.
[93] K. R. Lawless, *Phys. Thin Films* **4**, 191 (1967).
[94] C. J. Dell'Oca, D. L. Pulfrey, and L. Young, *Phys. Thin Films* **6**, 1 (1971).

decomposition, reduction, disproportionation, and polymerization. The large number of possible chemical reactions allows the formation of films of a wide range of materials on metal or insulating substrates of variable geometry. In contrast to thermal evaporation and sputtering, chemical vapor deposition is possible at atmospheric pressures.

In a thermal decomposition (pyrolysis) process the substrate is heated to sufficiently high temperatures to cause the vapor phase compound to decompose at the substrate surface and leave behind the nonvolatile components. Compounds such as the metal hydrides, metal carbonyls, and organometallic compounds decompose at temperatures below 500°C whereas higher temperatures are required for the decomposition of the metal halides. Representative applications include the production of permalloy films by the thermal decomposition of iron and nickel carbonyls,[95] the production of epitaxial Si and Ge films by the thermal decomposition of silane (SiH_4)[96] and germane (GeH_4),[97] and the production of SiO_2 films by the thermal decomposition of an alkoxysilane.[98] Nonthermal decomposition processes which allow deposition onto unheated substrates have also been employed. Electron beam irradiation of adsorbed molecules of tetramethyl silane and $SnCl_2$ has been used to produce films of Si[99] and Sn,[100] respectively. Amorphous Si films of unusually high resistivity have been prepared by an electrodeless rf glow discharge in silane gas.[101,102]

Lower substrate temperatures may be used if a decomposition process is replaced by a reduction process. The reducing agent is generally hydrogen, but metal vapors are also employed. The material to be deposited is introduced to the reaction chamber in the form of a compound such as a halide. This method is widely used in the preparation of epitaxial films of Si and Ge.[103,104] The appropriate chloride ($SiCl_4$, $GeCl_4$, or $SiHCl_3$) is mixed with dry hydrogen and passed over a substrate heated to 1000 to 1200°C. Dopants may be introduced into the films

[95] T. Matcovich, E. Korostoff, and A. Schemeckenbecher, *J. Appl. Phys.* **32**, 935 (1961).
[96] B. A. Joyce and R. R. Bradley, *J. Electrochem. Soc.* **110**, 1235 (1963).
[97] M. Davis and R. F. Lever, *J. Appl. Phys.* **27**, 835 (1956).
[98] E. L. Jordan, *J. Electrochem. Soc.* **108**, 478 (1961).
[99] E. S. Faber, R. N. Tauber, and B. Broyde, *J. Appl. Phys.* **40**, 2958 (1969).
[100] R. W. Christy, *J. Appl. Phys.* **33**, 1884 (1962).
[101] R. C. Chittick, J. H. Alexander, and H. F. Sterling, *J. Electrochem. Soc.* **116**, 77 (1969).
[102] P. G. Le Comber and W. E. Spear, *Phys. Rev. Lett.* **25**, 509 (1970).
[103] H. C. Theuerer, *J. Electrochem. Soc.* **108**, 649 (1961).
[104] E. F. Cave and B. R. Czorny, *RCA Rev.* **24**, 523 (1963).

12.1. PREPARATION OF THIN FILMS

by the introduction of controlled amounts of group III or V halides or hydrides. Films of the refractory metals have been prepared at relatively low temperatures (500–600°C) by the hydrogen reduction of compounds such as $CrCl_2$, $TaCl_5$, $MoCl_5$, VCl_4, and WF_6.[105]

Chemical transport or transfer processes are also widely used in the preparation of thin films. In such processes material is transported from the source to the substrate by means of a volatile chemical vapor. One type of transport reaction of use in the preparation of epitaxial layers of Si and Ge involves the disproportionation of the diiodides of Si and Ge.[106] To prepare Si films, SiI_2 is formed by passing I_2 over source Si held at 1100°C. The SiI_2 is then passed over a substrate held at 950°C where the deposition of Si and the production of SiI_4 vapor occurs. The SiI_4 may be recycled into the hot zone so that a continuous closed-tube process is possible. Compound semiconductor films such as GaAs, GaP, GaN, CdS, CdSe, ZnSe, and ZnS may also be prepared by chemical transport processes. A single source of the compound may be used or the components may be generated separately from two sources. The system used by Tietjen and Amick[107] to prepare epitaxial films of $GaAs_{1-x}P_x$ on Ge and GaAs substrates is shown schematically in Fig. 10.

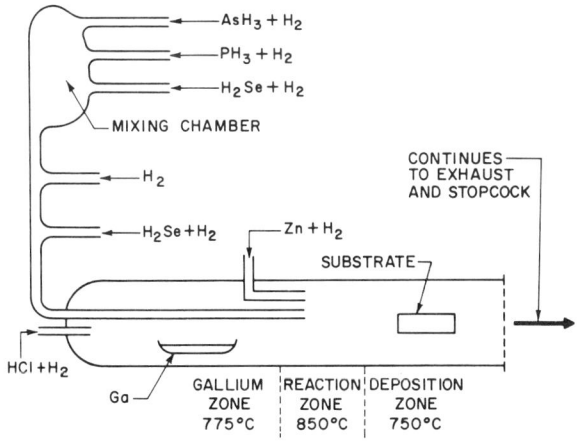

FIG. 10. Schematic of chemical transport vapor deposition apparatus [J. J. Tietjen and J. A. Amick, *J. Electrochem. Soc.* **113**, 724 (1966)].

---

[105] W. M. Feist, S. R. Steele, and D. W. Readey, *Phys. Thin Films* **5**, 237 (1969).
[106] E. S. Wajda and R. Glang, in "Metallurgy of Elemental and Compound Semiconductors" (R. O. Grubel ed.), p. 229. Wiley (Interscience), New York, 1961.
[107] J. J. Tietjen and J. A. Amick, *J. Electrochem. Soc.* **113**, 724 (1966).

The reagent gases AsH_3, PH_3, and H_2 are premixed in a ballast vessel and are carried in H_2 gas into the reaction zone through a long tube which extends beyond the Ga. A mixture of HCl and H_2 is introduced through a shorter tube, reacts with the Ga forming a gallium subchloride, and then passes into the reaction zone. The deposition occurs in an adjacent zone onto substrates held at 750°C. Doping is accomplished by adding H_2Se (*n*-type) into the ballast vessel or by introducing Zn vapor (*p*-type) into the reaction zone. Single-crystal layers of $GaAs_{1-x}P_x$ with high electron mobilities were obtained for $0 < x < 1$. Vapor deposition processes have also been used to prepare for example epitaxial films of the small energy gap semiconductor $Hg_{1-x}Cd_xTe$,[108] the wide energy gap semiconductor SiC,[109] the magnetic semiconductor $CdCr_2S_4$,[110] and the ferrimagnetic oxide $Y_3Fe_5O_{12}$ (YIG).[111]

Polymerization processes may also be employed to prepare films of both organic and inorganic polymers. In electron microscope studies the formation of undesirable polymer films by the interaction of the electron beam with silicone vacuum pump vapors has long been observed. Thin useful pinhole-free insulating films with very good mechanical and chemical properties, however, can be prepared in this manner. In addition to electron bombardment, ion bombardment, ultraviolet radiation, and glow discharge in the presence of a monomer vapor will produce polmer films on a surface. Electron bombardment has been used to produce silicone polymer films as thin as 50 Å[112] and polyepichlorohydrin films as thin as 100 Å.[113] Ultraviolet irradiation techniques, widely used in photoresist etching, have been used to produce insulating films as thin as 50 Å from butadiene vapor.[114] Insulating films of polybutadiene[115] and polydivinylbenzene[116] as thin as 100 Å have been prepared in a glow discharge.

12.1.4.2. Electrodeposition. The preparation of single-crystalline and polycrystalline metal films can be simply accomplished by an electrodeposition process. The apparatus consists of an anode and cathode im-

---

[108] O. N. Tufte and E. L. Stelzer, *J. Appl. Phys.* **40**, 4559 (1969).
[109] R. B. Campbell and T. L. Chu, *J. Electrochem. Soc.* **113**, 825 (1966).
[110] S. B. Berger and L. Ekstrom, *Phys. Rev. Lett.* **23**, 1499 (1969).
[111] R. C. Linares, R. B. McGraw, and J. B. Schroeder, *J. Appl. Phys.* **36**, 2884 (1965).
[112] R. W. Christy, *J. Appl. Phys.* **35**, 2179 (1964).
[113] L. V. Gregor and L. H. Kaplan, *Thin Solid Films* **2**, 95 (1968).
[114] P. White, *Electron. Reliability Microminiaturizat.* **2**, 161 (1963).
[115] N. M. Bashara and C. T. Doty, *J. Appl. Phys.* **35**, 3498 (1964).
[116] L. V. Gregor, *Thin Solid Films* **2**, 235 (1968).

mersed in a suitable electrolyte. Metallic ions in the electrolyte migrate under the influence of the applied electric field and are deposited onto the cathode. According to the laws of electrolysis, the mass of material deposited is proportional to the charge passed through the electrolyte; one gram-equivalent of material requires 1 F (96,490 C) of charge. The structure and properties of the deposit depend sensitively on the rate of deposition, the temperature, and the composition of the electrolyte. The addition of small amounts of certain inorganic and organic substances to the electrolyte can profoundly affect the structure of the electrodeposit. The growth processes involved are often very similar to those encountered in vacuum deposition. The presence of the aqueous medium and the electric field, however, lead to important differences including the possibility of epitaxial growth at room temperature. An extensive survey of the experimental techniques used in the electrochemical deposition of metals and alloys has been given by Lowenheim.[87] The preparation and structure of single-crystal metal films prepared by electrodeposition has been reviewed by Lawless.[93] Wolf[117] has discussed the electrodeposition and application of magnetic alloy films of the Ni–Fe–Co system.

An electrodeposition process of wide use in the preparation of oxide films is anodization. An anodizing treatment causes the natural oxide present on most metals to increase in thickness. The oxide forms on an anode which is placed in an electrolytic cell containing an aqueous solution or molten salt along with a Pt or Ta cathode. Anodization processes which utilize gaseous or solid electrolytes have also been employed.[118] Materials that can be anodized successfully include Al, Bi, Cu, Cr, Nb, Ta, Ti, Zr, Si, and Ge. The films may be prepared under constant current or constant voltage conditions. Under the normally preferred constant current conditions, the film thickness and voltage (overpotential) increase linearly so the voltage developed is a rough measure of the thickness; Al_2O_3 and Ta_2O_5 thicknesses are roughly 13 Å/V and 16 Å/V, respectively. The upper limit of thickness is determined by the breakdown potential; breakdown occurs by sparking or through the formation of recrystallized regions. The anodic films are generally amorphous, but thicker films often recrystallize and single-crystal oxide films have been observed to form on single-crystal surfaces of Al.[119] The dielectric properties and corrosion resistance of anodic oxide films have long been the subject of

---

[117] I. W. Wolf, *J. Appl. Phys.* **33**, 1152 (1962).

[118] J. L. Miles and P. H. Smith, *J. Electrochem. Soc.* **110**, 1240 (1963).

[119] P. E. Doherty and R. S. Davis, *J. Appl. Phys.* **34**, 619 (1963).

much interest; recently many other interesting electronic phenomena such as bistable switching and negative resistance have been observed in anodic oxides such as Al_2O_3, Ta_2O_3, ZrO_2, TiO_2, and Nb_2O_5.[120,121] A comprehensive survey of the formation processes and the experimental procedures for producing anodic oxide films has been given by Young.[88] More recent developments have been reviewed by Dell'Oca et al.[94]

12.1.4.3. Other Chemical Methods. Although an anodization process is required to produce a thick oxide film, thinner oxide films form on most metals when they are exposed to air or heated in an air or oxygen ambient. On many metals the oxide layers which form are pinhole free and resilient and serve as ideal tunneling barriers. According to Giaever[122] the best oxide barriers are formed on Al, Cr, Ni, Mg, Nb, Ta, Sn, and Pb. Tunneling barriers can be produced with difficulty on Cu, La, Co, V, and Bi and not at all on Ag, Au, and In. Insulating films of the oxides of Fe, Co, and Ni, which exhibit many interesting conduction phenomena, have also been produced by a thermal oxidation process.[123] Insulating films of SiO_2 are very important in silicon semiconductor device technology and are often formed by thermal oxidation in dry oxygen or steam.[124] The resulting films have chemical and physical properties very similar to those of fused quartz.

Oxide layers may also be produced from liquid films. For example, stable SiO_2 films can be obtained by dipping a substrate into a colloidal solution of silicic acid and upon removal heating the coated substrate to 200 to 500°C. This process has been used to prepare single-layer and multilayer dielectric film structures with useful optical properties. The technique can be used to prepare oxide films of the higher valence metals and semiconductors such as Al, In, Pb, Si, Sb, Sn, Ta, Ti, and Zr.[91]

Another liquid-film technique which has been used to prepare thin films of wide variety of organic materials is the Langmuir–Blodgett

[120] K. L. Chopra, *J. Appl. Phys.* **36**, 184 (1965).

[121] T. W. Hickmott, *J. Appl. Phys.* **33**, 2669 (1962); *J. Vac. Sci. Technol.* **6**, 828 (1969).

[122] I. Giaever, *in* "Tunneling Phenomena in Solids" (E. Burstein and S. Lundqvist, eds.), p. 19. Plenum Press, New York, 1969.

[123] J. E. Christopher, R. V. Coleman, A. Isin, and R. C. Morris, *Phys. Rev.* **172**, 485 (1968).

[124] W. A. Pliskin, D. R. Kerr, and J. A. Perri, *Phys. Thin Films* **4**, 257 (1967).

technique.[125-127] If a benzene solution of a long-chain fatty acid is placed on the surface of a dilute salt solution in a trough, the benzene will evaporate leaving a monomolecular layer of the fatty acid salt on the surface. By exerting a small surface pressure on the film the molecules will stand upright on the surface with their polar hydrophilic groups sticking into the water. Lowering a glass microscope slide or other solid

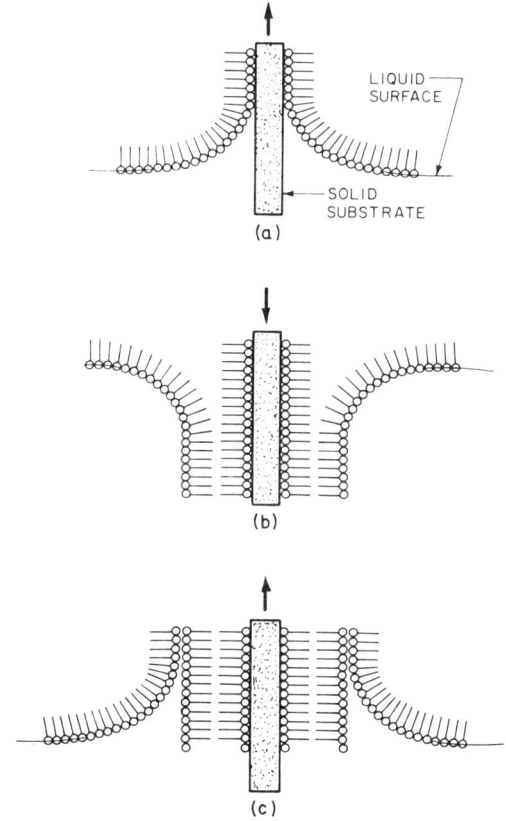

FIG. 11. Preparation of thin films utilizing the Langmuir–Blodgett technique: (a) withdrawal of substrate from solution produces a monomolecular fatty-acid layer; (b) second immersion of substrate into solution produces a second layer; (c) second withdrawal of substrate from solution produces a third layer.

[125] I. Langmuir, *Trans. Faraday Soc.* **15**, 62 (1920).
[126] K. B. Blodgett, *J. Amer. Chem. Soc.* **57**, 1007 (1935).
[127] G. L. Gaines, Jr., "Insoluble Monolayers at Liquid–Gas Interfaces." Wiley (Interscience), New York, 1966.

12. PREPARATION AND MEASUREMENT OF THIN FILMS

substrate into the trough produces no coating, but on withdrawal, the hydrophilic groups adhere to the solid surface and a monomolecular layer is formed. As shown in Fig. 11, a second immersion and withdrawal of the substrate will result in a three-layer film. This procedure can be continued and more than 100 monomolecular layers can be transferred to the substrate. The thickness of each monolayer is simply the length of the fatty-acid molecules and films of thickness ranging from about 10 to 10,000 Å can be prepared with an accuracy of a few tenths of a percent. Multilayer films of a large number of polar–nonpolar compounds such as heavy metal soaps, alcohols, and esters can be built up in this manner. Composite films with regularly spaced ionic dye layers separated by fatty acid salt layers have been used to study light wave fields with very high resolution.[128]

Metal films can be deposited directly from chemical solutions without the application of electrode potentials. Chemical silvering processes are still widely used for back surface reflectors and mirrors.[129] Metal films of Ni, Co, Cu, Pd, and Au can also be produced by chemical plating processes.[130]

12.2. Thin Film Measurements

12.2.1. Introduction

Apart from the obvious geometrical differences between thin films and bulk materials, film structures vary widely and, in general, are quite different from the structure of bulk materials. Both geometrical and structural differences lead to important differences in their respective chemical and physical properties. Many measurement techniques can be applied equally to thin film and bulk samples, making allowance only for the smaller mass of the thin film specimen. Other measurement techniques have been developed specifically for thin film measurements or are applied with particular advantage to thin film specimens. This latter class of measurement methods will be emphasized in the following sections of the chapter.

[128] K. H. Drexhage and H. Kuhn, in "Basic Problems in Thin Film Physics" (R. Niedermayer and H. Mayer, eds.), p. 339. Vandenhoeck and Ruprecht, Göttingen, 1966.
[129] J. Strong, "Procedures in Experimental Physics." Prentice-Hall, Englewood Cliffs, New Jersey, 1946.
[130] A. Brenner, in "Modern Electroplating" (F. A. Lowenheim, ed.), p. 698. Wiley, New York, 1963.

12.2.2. Thickness Measurements

The thickness of a thin film is its most characteristic feature and thickness measurements are therefore of primary importance in any thin film investigation. Since the film surfaces and substrate surfaces are never ideally flat, the thickness is usually specified as a mean thickness. This concept allows one to define an average thickness for even a discontinuous film consisting of isolated islands. Such films, however, are probably best characterized by their mass per unit area. Although many thickness-measuring techniques allow a direct thickness measurement, other techniques utilize physical property measurements that depend on thickness. Such indirect measurements are often employed due to their simplicity and convenience. More detailed discussions of thin film thickness measurements are available in the literature.[131-135]

12.2.2.1. Microbalance Methods.

If the density and area of a film are known, the thickness may be calculated from the weight of the film. This latter quantity can be determined by weighing the substrate before and after the deposition. Weighing can be performed outside the vacuum system, but *in situ* vacuum microbalance techniques are preferred. Microbalances can be categorized by the mechanism employed such as torsion fiber, helical spiral, magnetic suspension, or pivotal suspension. The most versatile balances employ some form of automatic null-balance using electromagnetic or electrostatic forces to counter-balance out-of-balance forces. Microbalances which are easy to use and which have sensitivities of 0.1 μg are available commercially. Such sensitivities correspond to a 0.5 and 4 Å thickness of a 1-cm² area of Au and Al films, respectively. A quartz fiber microbalance described by Mayer *et al.*[136] achieves sensitivities of 10^{-2} μg. The balance employs a 40-μm diameter

---

[131] K. H. Behrndt, *Phys. Thin Films* **3**, 1 (1966).

[132] W. A. Pliskin and S. J. Zanin, *in* "Handbook of Thin Film Technology" (R. I. Maissel and R. Glang eds.), p. 11-1. McGraw-Hill, New York, 1970.

[133] W. Steckelmacher, *in* "Thin Film Microelectronics" (L. Holland, ed.), p. 193. Wiley, New York, 1965.

[134] O. S. Heavens, "Optical Properties of Thin Solid Films," p. 96. Dover, New York, 1965.

[135] K. L. Chopra, "Thin Film Phenomena," p. 83. McGraw-Hill, New York, 1969.

[136] H. Mayer, R. Niedermayer, W. Schroen, D. Stünkel, and H. Göhre, *in* "Vacuum Microbalance Techniques" (K. Behrndt, ed.), Vol. 3, p. 75. Plenum Press, New York, 1963.

quartz torsion fiber. A crossbeam is fused to the torsion wire and carries at one end the balance pan and at the other a magnet. A solenoid above the magnet provides the restoring force. The balance is constructed entirely of quartz and Mo and can be baked at 500°C to achieve pressures below 10^{-9} Torr. Recent developments in microbalance techniques and applications are described in the proceedings of the annual conferences on vacuum microbalance techniques.[136a]

The most popular thin film thickness monitor is the quartz-crystal oscillator. Crystal oscillators are more convenient to use than the microbalances discussed above and at the same time provide high sensitivity. Such monitors generally utilize AT-cut quartz crystal plates in which thickness shear oscillations are excited. The AT-cut (the plane parallel to the X axis rotated 35°20′ about the X axis) is chosen because of its low-temperature coefficient for the resonant frequency. The crystal surfaces are antinodal and mass, added to one or both sides of the crystal plate, will produce a shift in the resonant frequency which is proportional to the added mass. If the mass of the deposited material is sufficiently small, the oscillator will respond only to the added mass and will be unaffected by the film thickness, density, and elastic constants. Film densities must therefore be known in order to calculate the film thicknesses from the measured added masses per unit area. For a typical resonant frequency of 5 MHz, an added mass of 1.8×10^{-8} gm/cm² ($\sim$0.1-Å-thick Au film) will produce a 1-Hz frequency shift. The frequency will decrease linearly with added mass (within 1%) up to an added mass of about 4.5×10^{-3} gm/cm² (25,000-Å-thick Au deposit). When the linear range has been exceeded, the deposited film must be removed or a new crystal plate must be used.

Prior to use, electrodes are generally deposited onto each face of the crystal plate to allow perpendicular excitation. Connections to the oscillator are provided by spring clips or by soldering. The crystals can be baked out and used in ultrahigh vacuum systems. Although the AT-cut is relatively temperature-insensitive, greater stability is attained when the crystal is held in a water-cooled holder. In addition to an oscillator circuit, a frequency-measuring instrument is required. Since small frequency changes must be measured, it is convenient to compare the monitor frequency with some reference frequency to provide an intermediate frequency for the counter circuit. With optimal electronics and operating conditions, sensitivities as high as 10^{-12} gm/cm² have been

---

[136a] "Vacuum Microbalance Techniques." Plenum Press, New York, 1960.

obtained.[137] Crystal oscillators may also be used to monitor and control the rate of deposition from a single source[133] or from two sources to produce alloy films of controlled composition.[138]

12.2.2.2. Optical Methods. The earliest thin film thickness measurements were made by Newton. He observed that the color of a thin transparent film illuminated by white light could be used to determine the film thickness when the refractive index of the film is known. The principle involved is illustrated in Fig. 12. Light incident on a film of thickness d will be partly reflected at the two film surfaces. If both the film and the substrate are nonabsorbing (or only slightly absorbing), a phase change of π will occur at both interfaces if $n_2 > n_1 > n_0$, where n_2, n_1, and n_0 are the refractive indices of the substrate, film, and incident medium, respectively. An interference maximum will then be produced when $m\lambda = 2n_1 d \cos \theta_1 = 2d(n_1^2 - \sin^2 \theta_0)^{1/2}$, where m is an integer, λ is the wavelength, θ_0 is the angle of incidence, and θ_1 is the angle of refraction. The interference maximum for the wavelength λ will produce a characteristic color which must be compared to a color-thickness gauge for the thickness determination. Detailed color charts for SiO_2 on Si are available for this purpose which allow thicknesses ranging from 500 to 15,000 Å to be determined with an accuracy of ± 100 Å for films with indices close to SiO_2.[139] The method can also be applied to films deposited onto absorbing substrates, but complicated phase-shift

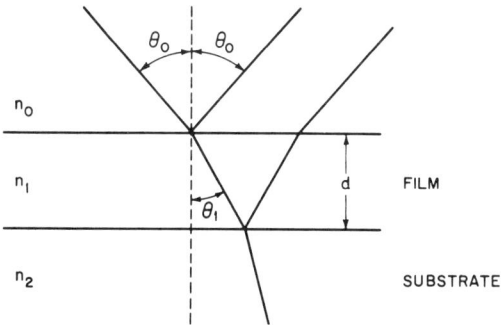

FIG. 12. Schematic diagram showing reflection and refraction of light by a transparent film on a transparent substrate (drawn for $n_2 > n_1 > n_0$).

[137] A. W. Warner and C. D. Stockbridge, in "Vacuum Microbalance Techniques" (K. Behrndt, ed.), Vol. 3, p. 55. Plenum Press, New York, 1963.
[138] R. Gerber, *Rev. Sci. Instrum.* **38**, 77 (1967).
[139] W. A. Pliskin and E. E. Conrad, *IBM J. Res. Develop.* **8**, 43 (1964).

corrections are necessary. More precise methods based on the same interference principles have also been developed.[140] One technique utilizes the change in interference fringes produced by a change in angle of incidence and is called VAMFO (variable angle monochromatic fringe observation).[141] Another technique called CARIS (constant angle reflection interference spectroscopy) utilizes spectrophotometric techniques.[142]

Multiple-beam interferometric techniques are widely used for film thickness determinations as well as for surface topographic studies.[143] These techniques utilize the interference fringes produced by two highly reflecting surfaces placed in close proximity. The experimental setup used for observing Fizeau fringes is shown in Fig. 13. One of the reflecting surfaces is a plate which is coated with a uniform, highly reflecting, semi-transparent film (Fizeau plate). The other reflecting surface is produced by depositing a Ag or Al film over a flat substrate partially covered with the film whose thickness is to be measured. The Fizeau plate is brought into contact with the film surface at a slight wedge angle so that an air-gap no more than a few wavelengths thick is formed. When the interferometer is illuminated by an extended monochromatic light source, dark Fizeau fringes are observed against a bright background. These fringes represent contours of equal air-gap thickness and are separated by a half-wavelength of the monochromatic light. When the interferometer is adjusted so that the fringes are perpendicular to the film step, the fringe displacement provides a direct measure of the film thickness. The thickness of films 30–20,000 Å thick can be measured routinely with an accuracy of ± 30 Å. Film thicknesses can also be measured using fringes of equal chromatic order (FECO). This technique employs two silvered parallel plates (one of which is partially coated with the film whose thickness is to be measured) illuminated by white light. The reflected light is focused on the entrance slit of a spectrograph and dark fringes are observed which are displaced at the film step. By measuring the wavelengths of the fringes in two regions, the film thickness can be accurately determined. Under optimum conditions film thicknesses can be measured with an accuracy of 1 or 2 Å.[144]

---

[140] W. A. Pliskin, *in* "Physical Measurement and Analysis of Thin Films" (E. M. Murt and W. G. Guldner, eds.), p. 1. Plenum Press, New York, 1969.

[141] W. A. Pliskin and R. P. Esch, *J. Appl. Phys.* **36**, 2011 (1965).

[142] F. Reizman, *J. Appl. Phys.* **36**, 3804 (1965).

[143] S. Tolansky, "Multiple-Beam Interference Microscopy of Metals." Academic Press, New York, 1970.

[144] H. E. Bennett and J. M. Bennett, *Phys. Thin Films* **4**, 1 (1967).

12.2. THIN FILM MEASUREMENTS

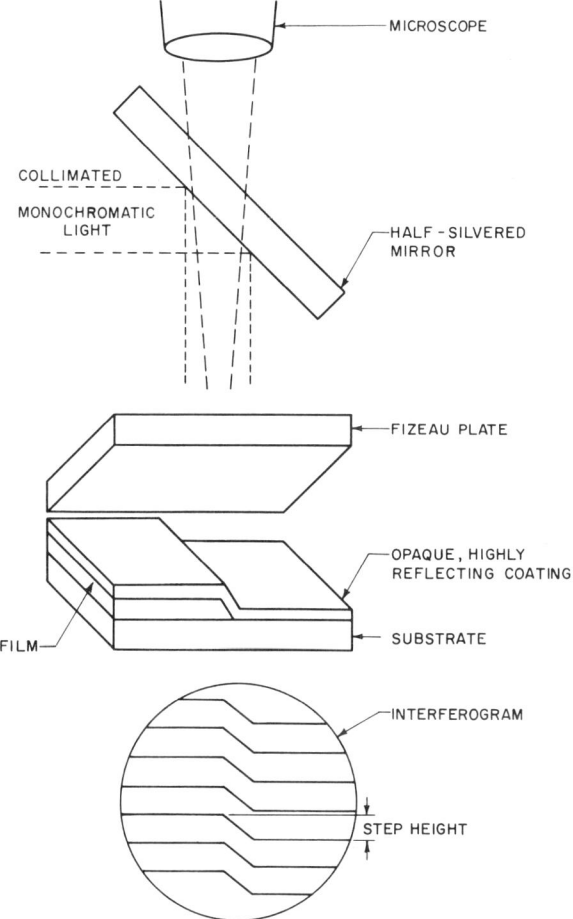

FIG. 13. Schematic diagram of optical interferometric system for thin film thickness measurements.

Thin film thicknesses have also been measured with a Michelson interferometer, a Nomarski polarization interferometer,[145] and by ellipsometry.[146] A number of other optical thickness measurement techniques have been described in standard texts and review articles.[131,132,134,147]

---

[145] M. Francon, "Optical Interferometry." Academic Press, New York, 1966.

[146] E. Passaglia, R. R. Stromberg, and J. Kruger (eds.), "Ellipsometry in the Measurement of Surfaces and Thin Films." Nat. Bur. Std. Misc. Publ. 256, 1964.

[147] O. S. Heavens, *Rep. Progr. Phys.* **23**, 1 (1960); *Phys. Thin Films* **2**, 193 (1964).

12.2.2.3. Electrical Methods. The thickness of metal films deposited onto insulating substrates may be conveniently determined by resistance measurements when the resistivities of the films are known. The bulk resistivities may be assumed when the films are pure, coherent, and thicker than the electron mean free path ($\sim$500 Å at room temperature). Thin film resistivities, however, are usually somewhat higher than the bulk values and may be orders of magnitude higher in ultra thin partially discontinuous films. Furthermore, structural changes during growth as well as mean free path effects can cause the resistivity of a film to vary with thickness. Although it is often difficult to use resistance measurements for absolute thickness determinations, they are still very useful for monitoring film growth. Resistance measurements made during film deposition allow the preparation of films of predetermined thicknesses and afford some control of the rate of deposition. Resistance measuring systems which have been used to monitor and control thin metal film depositions have been described in the literature.[148-150]

The thickness of dielectric films may be determined by capacitance measurements when the dielectric constants of the films are known. This type of measurement, however, suffers from uncertainties with regard to the dielectric constants similar to the resistivity uncertainties encountered with resistance measurements of thin film thicknesses. In addition, the film must be pinhole-free, uniform, and highly insulating to allow use of the simple parallel plate capacitor formula. Capacitance measurements, however, are often useful for measuring the thicknesses of very thin (less than 100 Å) insulating films, where optical measurements are not sufficiently accurate. An optical thickness determination of thicker films, however, is necessary for the determination of the dielectric constant. Capacitance measurements have been used, for example, to determine the thicknesses of thin films of Al_2O_3 and BeO,[151,152] ZnO,[153] SiO,[154] and SiC.[155] The thicknesses of monolayer thick fatty acid salt layers

[148] E. C. Crittenden Jr. and R. W. Hoffman, *Rev. Mod. Phys.* **25**, 310 (1953).

[149] J. A. Bennett and T. P. Flanagan, *J. Sci. Instrum.* **37**, 143 (1960).

[150] J. A. Turner, J. K. Birtwistle, and G. R. Hoffman, *J. Sci. Instrum.* **40**, 557 (1963).

[151] T. W. Hickmott, *J. Appl. Phys.* **33**, 2669 (1962).

[152] D. Meyerhofer and S. A. Ochs, *J. Appl. Phys.* **34**, 2535 (1963).

[153] C. A. Mead, *in* "Basic Problems in Thin Film Physics" (R. Niedermayer and H. Mayer, eds.), p. 674. Vandenhoeck and Ruprecht, Göttingen, 1966.

[154] R. R. Verderber, J. G. Simmons, and B. Eales, *Phil. Mag.* **16**, 1049 (1967).

[155] T. E. Hartman, J. C. Blair, and C. A. Mead, *Thin Solid Films* **2**, 79 (1968).

(24.6 Å for cadmium stearate) have also been determined by capacitance measurements.[128]

12.2.2.4. Other Methods. Film thicknesses can be measured using X-ray absorption, emission, and reflection techniques. One technique relies on measurements of the attenuation by the film of the characteristic radiation of the substrate. The attenuation is an exponential function of the film thickness and the mass absorption coefficient of the film material. Another technique utilizes the characteristic radiation emitted by the film material. The emitted radiation increases in intensity with increasing film thickness. The use and relative merits of these two techniques in the measurement of the thicknesses of Al films deposited onto Si substrates have been discussed by Cline.[156] The emission technique allows thickness measurements on the elements from Al to U with accuracies of ±6 Å and also allows thickness measurements on completely covered films. X-ray reflection measurements have also been used to measure film thicknesses. When the incident angle is in the region of the critical angle for total reflection and the film surface is sufficiently smooth, X-ray interference is observed between the beams reflected from the air–film and film–substrate interfaces. The spacing of the interference fringes is used to measure the film thickness. Sauro et al.[157] have used this technique to measure the thickness of 250 to 1000-Å-thick Cu films on glass substrates with 5% accuracy. Experimental arrangements for X-ray reflection measurements have been reviewed by Bertin.[158]

Stylus instruments such as the "Talysurf" are often used to measure surface irregularities and film thicknesses. The stylus is lightly loaded to prevent penetration into the surface and is moved across a film step for a thickness measurement. The vertical motion is converted by a transducer to an electrical signal which is amplified and recorded. Film thicknesses from 20 A to 10 μm can be measured very accurately and conveniently.[159] The film thicknesses of relatively soft materials, however, are best measured using other techniques.

---

[156] J. E. Cline, in "Physical Measurement and Analysis of Thin Films" (E. M. Murt and W. G. Guldner, eds.), p. 83. Plenum Press, New York, 1969.

[157] J. Sauro, I. Fankuchen, and N. Wainfan, *Phys. Rev.* **132**, 1544 (1963).

[158] E. P. Bertin, in "Physical Measurement and Analysis of Thin Films" (E. M. Murt and W. G. Guldner, eds.), p. 35. Plenum Press, New York, 1969.

[159] N. Schwartz and R. Brown, *Trans. 8th Nat. Vacuum Symp.* p. 836. Pergamon, Oxford, 1961.

12.2.3. Mechanical Measurements

Large internal stresses (as high as 10^{10} dyn/cm^2) develop in thin films during growth. These stresses are tensile or compressive and can produce, respectively, film cracking or buckling. Internal stresses also can produce anisotropies in the electrical, magnetic, and optical properties so that an appreciation of the state of stress in a thin film becomes important for whatever application is envisaged. Thin films also exhibit high tensile strengths due largely to their high defect densities. Since thin films are readily amenable to electron microscopic observation, their mechanical characteristics can be readily correlated with their microstructural characteristics. A comprehensive discussion of the mechanical properties of thin films and the measurement techniques involved has been given by Hoffman.[160] Campbell[161] has also reviewed mechanical property measurements emphasizing techniques employed to measure the adhesion between a thin film and its substrate.

12.2.3.1. Internal Stresses. The stresses that arise in thin films can be separated into two categories: thermal stresses and intrinsic stresses. Thermal stresses are produced when the film and substrate have different thermal-expansion coefficients and when the deposition and measuring temperature differ. The thermal-expansion coefficients of Au, glass, and NaCl, for example, are about 14, 8, and $40 \times 10^{-6}/°C$, respectively. If Au is deposited onto heated glass or NaCl substrates, tensile or compressive stresses, respectively, will develop on cooling to room temperature. Such thermal stresses are readily calculated and can be largely eliminated by a judicious choice of substrate material. Intrinsic stresses are produced during film growth or on annealing in metal, semiconducting, and dielectric films and can be very large and either tensile or compressive in nature. Their presence can be attributed, for example, to contaminants, lattice mismatch between the film and substrate, recrystallization processes, or void formation. The stresses, therefore, are a function of the film thickness and the growth conditions such as the deposition rate, substrate material and temperature, and residual gas pressure.

The internal stress in a thin film will produce a bending of the film–substrate sandwich. A tensile stress will produce a convex film surface and a compressive stress will produce a concave film surface. Measure-

---

[160] R. W. Hoffman, *Phys. Thin Films* **3**, 211 (1966).
[161] D. S. Campbell, in "Handbook of Thin Film Technology" (R. I. Maissel and R. Glang, eds.), p. 12-3. McGraw-Hill, New York, 1970.

ments of such bending during or after film deposition provide the most common methods of film stress determinations. The substrates are in the form of a long, thin, cantilever beam, where the free end deflection is measured, or a beam supported on knife edges, where the center deflection is measured, or a circular plate, where interference fringe patterns are measured. The substrates must be sufficiently thin (50–100 µm) for the deflections to be measurable and substrate materials such as glass, mica, quartz, and Al have been employed. The deflections of the beams may be measured optically with a microscope,[162] mechanically with a stylus instrument,[163] through a capacitance change between the substrate and a fixed plate,[164] or through a capacitance change with the substrate as the moving plate of a differential capacitor.[165] Other methods rely on an electromechanical[166] or magnetic[167] restoration of the null position in combination with a measurement of the restoration force. Some of these methods are extremely sensitive and allow stress measurements to be made during the initial stages of film growth. The Talysurf stylus instrument can detect a beam deflection corresponding to a stress of 10^6 dyn/cm^2 in a 1000-Å-thick film[163] while the highest sensitivity of the capacitance method is 0.2×10^6 dyn/cm^2 for a 1000-Å-thick film.[164] The deflection of a circular plate may be measured interferometrically and used for stress measurements. A schematic drawing of the system used by Finegan and Hoffman[168] is shown in Fig. 14 with the deposited film in a state of tension. The change in the fringe pattern between the substrate and an optical flat is observed. Due to the limited flatness of the available substrates, the substrate profile is remeasured after the film has been dissolved and then used as a reference profile. The circular plate method allows a more convenient measurement of stress anisotropies than the beam methods, but only the beam methods allow direct *in situ* stress measurements during film deposition. Additional interference methods for measuring substrate bending include laser interferometry[169] and holography.[170]

[162] H. P. Murbach and H. Wilman, *Proc. Phys. Soc. (London)* **B66**, 905 (1953).
[163] H. Blackburn and D. S. Campbell, *Phil. Mag.* **8**, 823 (1963).
[164] J. D. Wilcock and D. S. Campbell, *Thin Solid Films* **3**, 3 (1969).
[165] A. E. Hill and G. R. Hoffman, *Brit. J. Appl. Phys.* **18**, 13 (1967).
[166] H. S. Story and R. W. Hoffman, *Proc. Phys. Soc. (London)* **B70**, 950 (1957).
[167] J. B. Priest, *Rev. Sci. Instrum.* **32**, 1349 (1961).
[168] J. D. Finegan and R. W. Hoffman, *Trans. 8th. Nat. Vacuum Symp.* p. 935. Pergamon, Oxford, 1961.
[169] A. Taloni and D. Haneman, *Surface Sci.* **8**, 323 (1967).
[170] P. J. Magill and T. Young, *J. Vac. Sci. Technol.* **4**, 47 (1967).

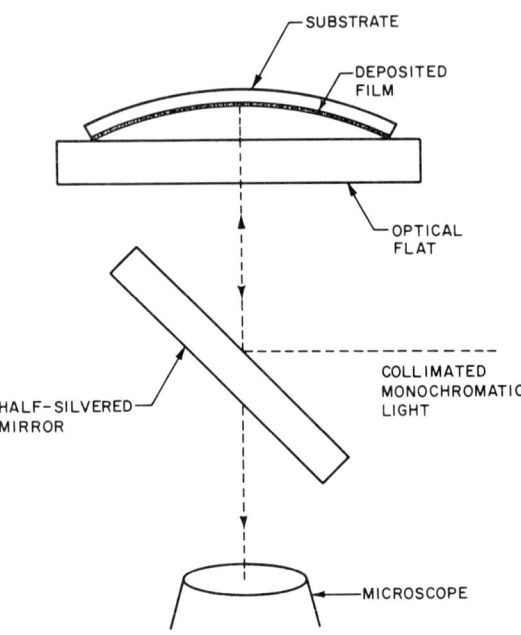

FIG. 14. Schematic of apparatus for observing fringe patterns produced by the variable air-gap between a stressed circular substrate and an optical flat [J. D. Finegan and R. W. Hoffman, *Trans. 8th Nat. Vacuum Symp.* p. 935. Pergamon, Oxford, 1961].

For the cantilever beam configuration, the stress is commonly calculated from the deflection of the free end using an approximate equation derived by Stoney[171]

$$\sigma d_f = E_s d_s^2 \delta / 3L^2 \tag{12.2.1}$$

where σ is the stress in the film, d_f is the film thickness, d_s is the substrate thickness, E_s is Young's modulus for the substrate, δ is the deflection of the free end, and L is the length of the substrate beam. Since this formula gives directly the force per unit width exerted by the film on the substrate, the film thickness must be independently measured in order to calculate the film stress. This expression can be used only if the film is very thin relative to the substrate, if the deflection is less than the substrate thickness, and if the substrate length is at least twice its width. In addition, the formula gives only an average stress when the stress is nonuniform throughout the film thickness and gives uncertain stress values for discontinuous films. A more exact expression incorporating the Poisson

[171] G. G. Stoney, *Proc. Roy. Soc. (London)* **A82**, 172 (1909).

term for the substrate and the elastic constants of the film has been given by Davidenkov.[172]

Stresses in thin films may also be determined by X-ray diffraction techniques by comparing the position and shape of thin film diffraction maxima with the corresponding bulk diffraction maxima. Line-broadening effects due to particle size and defects, however, are hard to separate from the strain broadening so that a careful analysis is required. Borie and Sparks[173] have used a line-broadening analysis to determine the strains and strain gradients in epitaxial films of Cu_2O on Cu. Shifts of the diffraction maxima can be observed in films as thin as 100 Å and can be employed to determine lattice strains directly. Generally the change in lattice parameter and strain perpendicular to the film plane is measured and elasticity theory is used to calculate the stress in the plane of the film. X-ray diffractometer techniques have been employed to measure line shifts and corresponding strains and stresses in such film–substrate systems as Au and Cu on glass,[174] Ni on NaCl,[175] Al on Si,[176] and YIG on YAG.[177] Electron diffraction measurements have also been employed for stress determinations. Although not as accurate as X-ray diffraction methods, they have been used with advantage to measure lattice strains in the very small particles (nuclei) formed during the earliest stages of film growth.[178]

12.2.3.2. Tensile Properties. The extremely high tensile strength of metal whiskers led many investigators to search for comparable properties in thin films. Although films were not found to be as strong as whiskers, they nonetheless are stronger than bulk, and the correlation of their mechanical properties with their microstructural characteristics has been a subject of continuing interest.

The first measurements of the tensile strength of thin films were made by Beams *et al.*[179] The films were electrodeposited onto the cylindrical surfaces of steel rotors. The rotors were suspended magnetically in a vacuum and spun around a vertical axis by a rotating magnetic field.

---

[172] N. N. Davidenkov, *Sov. Phys.-Solid State* **2**, 2595 (1961) (English transl.).

[173] B. Borie and C. J. Sparks, Jr., *in* "Thin Films," p. 45. Amer. Soc. for Metals, Metals Park, Ohio, 1964.

[174] R. W. Vook and F. Witt, *J. Appl. Phys.* **36**, 2169 (1965).

[175] J. F. Freedman, *IBM J. Res. Develop.* **6**, 449 (1962).

[176] G. A. Walker, *J. Vac. Sci. Technol.* **7**, 465 (1970).

[177] R. Zeyfang, *J. Appl. Phys.* **41**, 3718 (1970).

[178] H. J. Wasserman and J. S. Vermaak, *Surface Sci.* **22**, 164 (1970).

[179] J. W. Beams, J. B. Breazeale, and W. L. Bart, *Phys. Rev.* **100**, 1657 (1955).

Both the tensile strength of the films and the adhesion of the films to the rotor surfaces are obtained by determining the rotor speeds necessary to throw off the films for rotors of varying radii. This method eliminates problems associated with gripping the films, but allows only an ultimate strength determination rather than a complete stress–strain relation. Another technique which avoids gripping problems is the bulge method where the thin film is mounted at the end of a hollow cylindrical tube or hole and then pressurized.[180] The film may be mounted by first mounting a plastic film on the end of the tube or hole, then depositing the film to be tested onto the plastic substrate, and finally dissolving the plastic substrate. An alternative mounting method, which allows tensile measurements to be made on films grown epitaxially on rocksalt, utilizes a water jet to drill a hole in the substrate under the film. In either case a pressure differential across the unsupported film causes the film to bulge. The biaxial stress can be calculated from the pressure differential; the strain is calculated from the deflection of the center of the unsupported film which is measured interferometrically. Stress–strain relations for films as thin as 250 Å can be determined in this manner. In addition, the initial strain in the film may be measured by observing the initial bulging of the film after the removal of the substrate material.[181]

Microtensile testers have also been used to measure the tensile properties of thin films and whiskers. Either "soft" machines, where the load is prescribed and the elongation measured, or "hard" machines, where the strain rate is prescribed and the load measured, are used. Marsh[182] has designed a soft apparatus in which a null-torsion balance is used to apply the load and an optical extension detector is used to measure the elongation. Loads ranging from 1 mg to 400 gm can be applied and extensions from a few angstroms to 15 mm can be measured with an error of less than 5 Å. Other soft machines employ solenoid magnets to apply the load and use direct microscope observation or interferometry to detect the elongation.[183,184] Various hard machines have also been described in the literature.[184–186] With all microtensile techniques

---

[180] J. W. Beams, in "Structure and Properties of Thin Films" (C. A. Neugebauer, J. B. Newkirk, and D. A. Vermilyea, eds.), p. 183. Wiley, New York, 1959.
[181] A. Catlin and W. P. Walker, J. Appl. Phys. **31**, 2135 (1960).
[182] D. M. Marsh, J. Sci. Instrum. **38**, 229 (1961).
[183] C. A. Neugebauer, J. Appl. Phys. **31**, 1096 (1960).
[184] J. M. Blakely, J. Appl. Phys. **35**, 1756 (1964).
[185] D. Kuhlman-Wilsdorf and K. S. Raghaven, Rev. Sci. Instrum. **33**, 930 (1962).
[186] A. Lawley and S. Schuster, Rev. Sci. Instrum. **33**, 1178 (1962).

it is difficult to mount the film in such a way as to avoid nonuniform loading leading to tearing of the film at the edges. Attachment of the films to the grips is also difficult and often the films are glued to the grips prior to their removal from the substrate. Much of the scatter in the stress–strain relations obtained with microtensile techniques can be attributed to the handling and mounting procedures employed as well as to flaws, nonuniformities and residual stresses present in the films.

12.2.4. Optical Measurements

Thin films find many applications in optical systems as, for example, mirrors, beam splitters, antireflection coatings, and interference filters. Such applications often require the production of complex multilayer composites with carefully controlled thicknesses and refractive indices.[187,188] These stringent requirements have motivated detailed studies of the optical properties of thin metal and dielectric films. In addition, thin film measurements are useful in determining the bulk optical properties of solids. Thin film techniques allow the preparation of clear large area optically flat surfaces. Reflectivity or photoemission measurements may be made immediately after film deposition or during film deposition to assure that the results are not influenced by contaminants. It is also possible to make transmission measurements of highly absorbing metal films or of semiconducting films above the fundamental absorption edge. Optical transmission (absorption) measurements of ionic materials in the ultraviolet and vacuum ultraviolet also require the use of thin film specimens. If bulk optical properties are being deduced from film measurements, great care must be exercised in film preparation to ensure optical properties representative of the bulk. In general, the best optical spectra are obtained by using epitaxially grown films. Even with careful film preparation, however, the optical absorption behavior of films often departs widely from the bulk behavior; the study of such departures is useful in ascertaining the effect of structural and compositional influences on the optical behavior. The study of the optical characteristics of amorphous materials or of metastable phases (polymorphs) is possible only with thin film specimens.

In this section we will discuss methods for determining the optical constants of thin films. Methods applicable to thin films, however, are

---

[187] A. Thelen, *Phys. Thin Films* **5**, 47 (1969).
[188] P. H. Lissberger, *Rep. Progr. Phys.* **33**, 197 (1970).

often applicable to bulk specimens and vice versa. In general, bulk methods are applicable when the effects of multiple reflection may be neglected. The use of optical property measurements for the determination of the electronic band structures of bulk materials is treated elsewhere.† Interference methods which are used for thickness measurements of transparent films when the refractive indices are known can also be used to determine the refractive indices when the thicknesses are independently measured. These methods have been discussed in Section 14.2.2.2. More comprehensive reviews of thin film optical constant measurements are available in the literature.[134,144,147,189,190]

12.2.4.1. General Reflection and Transmission Methods. The optical constants n_1 and k_1 (complex refractive index $n_1 - ik_1$) of a homogeneous, isotropic film resting on a substrate (see Fig. 12) can, in principle, be derived from two independent reflection and/or transmission measurements if the film thickness and the optical constants of the substrate are known. For example, the reflectance (R) and transmittance (T) can be measured (RT method), the reflectance at two angles of incidence can be measured, or the transmittance of two films of different thickness can be measured (TT method). The latter type of measurement, however, can lead to serious errors as the optical properties are not necessarily independent of thickness. All of these methods require considerable calculation when either the film or substrate are absorbing as the expressions for the reflectance and transmittance become too complicated to be inverted in closed form. Detailed expressions for R and T and analyses of the above methods and many other similar methods have been given by Heavens.[134]

The RT and TT methods have been critically analyzed by Grant[191] and Grant and Paul.[192] These methods were used to determine the optical constant of 100- to 3000-Å-thick epitaxial films of Ge on CaF_2 substrates which are transparent for the 2000- to 6000-Å wavelength radiation used. A schematic diagram of the optical system is shown in Fig. 15. Light from the lamp L passes through the filter F and is focused by a mirror system

[189] F. Abelès, *Prog. Opt.* **2**, 249 (1963).
[190] P. Rouard and P. Bousquet, *Progr. Opt.* **4**, 145 (1965).
[191] P. M. Grant, Gordon McKay Lab. of Appl. Sci., Harvard Univ. Tech. Rep. No. HP-14 (1965).
[192] P. M. Grant and W. Paul, *J. Appl. Phys.* **37**, 3110 (1966).

† See Chapter 1.6.

12.2. THIN FILM MEASUREMENTS

M_1, M_2 onto the entrance slit of the monochromator. Monochromatic light emerges and passes through the chopper C and is then focused by the "mirror lens" system M_4, M_5, M_6 onto the sample S. If the mirrors M_R and M_T are identical and the optical paths S–M_T–PM and S–M_R–PM are equal, then both transmittance and reflectance can be measured by a simple sample-in–sample-out technique. The average angle of incidence was 7° which is essentially normal incidence. A photomultiplier detector (PM) and phase-sensitive detection was employed. Using the appropriate

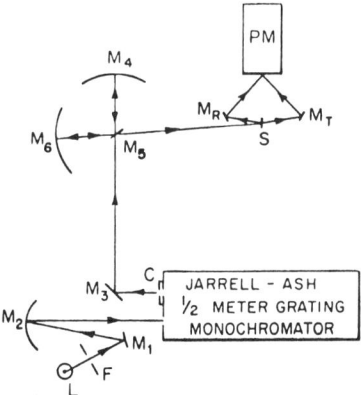

FIG. 15. Schematic of optical system used for thin film reflectance and transmittance measurements [P. M. Grant and W. Paul, *J. Appl. Phys.* **37**, 3110 (1966)].

theoretical expressions for R and T, the optical constants were obtained through a Newton–Raphson iteration procedure. They found that with the RT method there were certain regions in which the derived values of n_1 and k_1 were very sensitive to small changes of R and T. With the TT method they found it possible to choose a pair of films which would reduce the error for all wavelengths used. The reflectance and transmittance data of Grant and Paul have also been analyzed by Verleur.[193] He employed an automatic curve-fitting procedure to obtain a classical oscillator fit and obtained results in better agreement with bulk Ge data than was obtained with the RT or TT methods.

It is sometimes possible to make separate measurements of n_1 and k_1. For metal films, where k_1 is much larger than n_1, a measurement of the transmittance or of the phase change accompanying reflection can be used to determine k_1. For this method to be applicable the metal film thick-

---

[193] H. W. Verleur, *J. Opt. Soc. Amer.* **58**, 1356 (1968).

nesses must be greater than 300–400 Å so that the effects of multiple reflections are suppressed and the substrate must be transparent with known refractive index. With k_1 determined a measurement of the reflectance allows a determination of n_1. This technique has been used by Schulz and Tangherlini[194,195] to determine the optical constants of Ag, Au, Cu, and Al. In order to test the homogeneity and surface roughness of the films they also measured the reflectances R_p and R_s of the p and s (electric field parallel and perpendicular to the plane of incidence) polarizations of the radiation at an angle of incidence of 45°. The expected relation $R_s^2 = R_p$ was obtained with films of Ag, Au, Cu, and Al, but deviations were obtained for Sn and Pb indicating that the latter two metals were relatively unsuitable for reflectivity measurements.

The optical constants of a thin film can also be measured using critical angle methods. When the substrate and film are transparent and the substrate has a higher refractive index n_2 than the film n_1, light incident on the film from the substrate side will be totally reflected at angles of incidence greater than $\theta_c = \sin^{-1}(n_1/n_2)$. Measurements of the critical angle and the refractive index of the substrate allow a determination of the refractive index of the film. This method is insensitive to the presence of surface contamination or oxide layers since the measurements are made on the protected substrate side of the film. When the extinction coefficient k_1 of the film is nonzero, attenuated total reflection occurs; the change of reflectance with angle is then not a step function at θ_c, but it exhibits a maximum slope at an angle θ_m. When k_1 is small, θ_c and hence n_1 can be obtained with good accuracy from the measured θ_m.[196] Alternatively, the film optical constants can be obtained by internal reflectance measurements for different angles of incidence or for different polarizations. The advantages and applications of internal reflection measurements have been discussed at length by Harrick.[197] In the ultraviolet and vacuum ultraviolet spectral regions, the refractive index is less than one for many materials so that a critical angle exists for reflection from the exposed side of the film. Such critical angle measurements have been made by Hunter[198] to determine the indices of refraction of Al, Mg, and In in the far ultraviolet.

[194] L. G. Schulz, *J. Opt. Soc. Amer.* **44**, 357 (1954).
[195] L. G. Schulz and F. R. Tangherlini, *J. Opt. Soc. Amer.* **44**, 362 (1954).
[196] W. R. Hunter, *J. Opt. Soc. Amer.* **54**, 15 (1964).
[197] N. J. Harrick, "Internal Reflection Spectroscopy." Wiley (Interscience), New York, 1967.
[198] W. R. Hunter, *J. Opt. Soc. Amer.* **54**, 208 (1964).

The refractive indices of transparent films may be conveniently determined by Brewster angle measurements. The Brewster angle is the angle of incidence where the reflectance of the p component is zero and is given as $\theta_B = \tan^{-1}(n_1/n_0)$, where n_1 and n_0 are the refractive indices of the medium and the incident medium, respectively. If a film-coated substrate is illuminated by p component light at an angle of incidence θ_B, no light is reflected from the air–film interface and the reflectance of the film-coated substrate becomes the same as the reflectance of the uncoated substrate. This phenomenon forms the basis of the Abelès method,[199] which consists of a comparison of the reflectances of coated and uncoated regions of the substrate. The angle of incidence for matching reflectances may be determined visually or photoelectrically and then used to obtain the refractive index of the film through the Brewster angle expression. The Abelès method is very convenient and is applicable to homogeneous, isotropic, and nonabsorbing films on either metal or dielectric substrates; the method is most sensitive when the optical film thicknesses are near odd multiples of a quarter-wave length at the Brewster angle. When the thin film refractive index is within ± 0.3 of the substrate refractive index, an accuracy of ± 0.002 is possible.[189] An alternative approach to the determination of the Abelès equal reflectance condition has been proposed by Hacskaylo[200] which allows accuracies of about ± 0.0002 to ± 0.0006 for a corresponding index range of 1.2 to 2.3.

12.2.4.2. Spectrophotometric Methods. For a transparent film of refractive index n_1 and thickness d on a transparent substrate of refractive index n_2, the reflectance R and transmittance T ($T = 1 - R$) at normal incidence are easily calculated and are found to exhibit maxima and minima as a function of wave number $1/\lambda$. When $n_0 \lesseqgtr n_1 \lesseqgtr n_2$, the reflectance is a maximum for $1/\lambda = m/2n_1 d$ and a minimum for $1/\lambda = (2m+1)/4n_1 d$, where m is an integer. When $n_0 \lesseqgtr n_1 \gtreqless n_2$, the conditions for maxima and minima are reversed. By measuring the reflectances as a function of wave number with a spectrophotometer, the reflectance values and wave numbers at the turning points can be determined. The refractive index n_1 is first obtained from the reflectances at the turning points and the film thickness is then obtained from the wave numbers at the turning points. If the film thickness is independently measured, the refractive index can be obtained from the turning point

---

[199] F. Abelès, *J. Phys. Radium* **19**, 327 (1958).
[200] M. Hacskaylo, *J. Opt. Soc. Amer.* **54**, 198 (1964).

wave numbers alone. For the method to be applicable the film thickness must be greater than $\lambda/4$ over the range of wave numbers used. In addition, the order m of the interference maxima or minima must be known and the effects of dispersion must be taken into account. If, for example, the film thickness is known, the wavelengths λ_m and λ_{m+1} corresponding to two successive maxima or minima may be used to first determine the order of interference. The order of interference will be given by the integer closest to $\lambda_{m+1}/(\lambda_m - \lambda_{m+1})$ when the dispersion is not too large; for no dispersion the above expression will yield an integer. The orders of interference are then used to calculate first approximations to the refractive indices. These values along with approximations of the dispersion $dn_1/d(1/\lambda)$ may then be used to recalculate second approximations to the refractive indices using more exact dispersion-dependent expressions for the turning point wavelengths.[147,189]

Another technique for measuring the refractive indices of transparent films utilizes the interference patterns produced in light transmitted through films at various angles of incidence.[201] By increasing the angle of incidence the position of a given maximum or minimum can be shifted to the same wavelength as the minimum or maximum which was immediately adjacent to it at a lower wavelength at normal incidence. By determining the interference order numbers as above, the angle of refraction of the light in the film can be obtained and the refractive indices then calculated from Snell's law of refraction. It is also possible to determine the refractive indices of transparent films with a technique called *immersion spectrophotometry*.[202] This technique utilizes reflection measurements of the peak heights of interference maxima as a function of the index of refraction of the fluid in which the film and substrate are immersed. When the known refractive index of the fluid matches that of the film, the peak height goes to zero.

The optical constants of weakly absorbing films on transparent substrates can also be determined with spectrophotometric measurements. Hall and Ferguson[203] have measured the spectrophotometric reflectance and transmittance of CdS and ZnS films at normal incidence to determine their optical constants. Their method is applicable to films with $k_1^2 \ll (n_1 - 1)^2$, where to a first approximation the phase shifts at the air–film and film–substrate interface may be neglected and the refractive

[201] J. C. Banter, *J. Electrochem. Soc.* **112**, 388 (1965).
[202] W. P. Ellis, *J. Opt. Soc. Amer.* **53**, 613 (1963).
[203] J. F. Hall Jr. and W. F. C. Ferguson, *J. Opt. Soc. Amer.* **45**, 714 (1955).

indices calculated from the wavelengths of the transmittance maxima or minima and the known film thickness. The extinction coefficient k_1 is then obtained from the measured transmittance amplitudes at the maxima through use of a small k theoretical expression for the transmittance. These values are then used to calculate the refractive index from measured amplitudes of the reflectance maxima and of the transmittance minima using appropriate theoretical expressions. The values for the refractive index obtained by independent methods agreed within 1%. Graphical methods have also been applied to optical constant measurements of absorbing films. For a transparent substrate of given refractive index n_2, curves may be computed giving the variation with d/λ of the reflectance R and transmittance T at normal incidence for a range of values of film constants n_1 and k_1. Hadley[204] has prepared a set of curves for this purpose for a substrate refractive index of 1.5. Each set of curves gives plots of R and T as a function of d/λ for one value of k_1 and for a range of values of n_1. To use the curves the film thickness and the reflectance and transmittance are measured at a particular wavelength. From each set of curves for a particular k_1 two values for the index of refraction $n_1(R)$ and $n_1(T)$ are obtained. Curves for a different k_1 yield two additional values of $n_1(R)$ and $n_1(T)$. From a plot of $n_1(R)$ and $n_1(T)$ versus k_1 the point of intersection yields the correct values for n_1 and k_1. An optimum accuracy is obtained when the film thickness is an odd multiple of $\lambda/4$. Similar spectrophotometric methods utilizing successive approximations to determine the optical constants of absorbing films have been described by Koehler et al.[205] and by Hass et al.[206]

12.2.4.3. Polarimetric Methods. Optical constant determinations which are derived from the reflectance and/or transmittance of thin films rely on photometric intensity measurements. Polarimetric or ellipsometric methods, however, are based on measurements of angles and phase differences that characterize the change of the state of polarization of light upon reflection at nonnormal incidence from the surface of a thin film. The state of polarization is specified in terms of the phase and amplitude relations of the p and s components (parallel and perpendicular to the plane of incidence, respectively) of the electric field vector. In general, reflection causes a change in the relative phases of the p and s waves and a

---

[204] L. N. Hadley, cited by Heavens (Ref. 147).

[205] W. F. Koehler, F. K. Odencrantz, and W. C. White, *J. Opt. Soc. Amer.* **49**, 109 (1959).

[206] G. Hass, J. B. Ramsey, and R. Thun, *J. Opt. Soc. Amer.* **49**, 116 (1959).

change in the ratio of their amplitudes. The change of phase is designated by the angle $\varDelta$ and the change in the ratio of the amplitudes is designated tan ψ, i.e., $R_p/R_s = \tan \psi \exp(i\varDelta)$, $R_p = E_{rp}/E_{ip}$, and $R_s = E_{rs}/E_{is}$, where R_p and R_s are the amplitude reflectances for the p and s waves, E_{ip} and E_{is} are the electric field amplitudes for the incident p and s waves, and E_{rp} and E_{rs} are the electric field amplitudes for the reflected p and s waves. The quantities $\varDelta$ and ψ are measured experimentally and used to determine the optical constants.

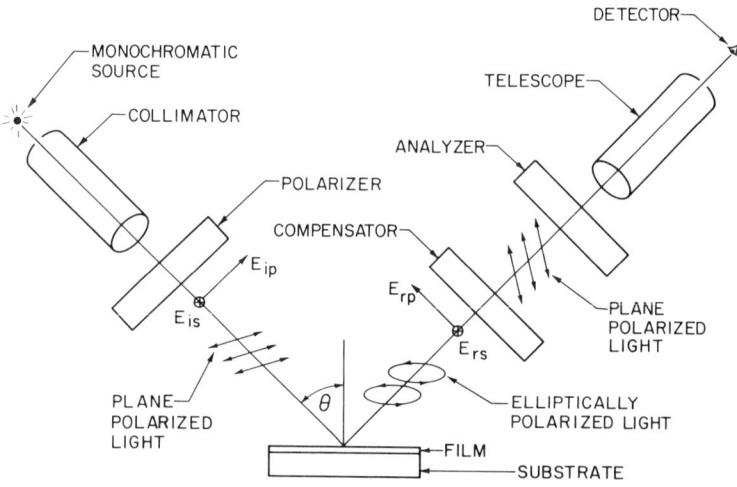

FIG. 16. Schematic of apparatus for polarimetric measurements (ellipsometer).

An experimental arrangement (ellipsometer) for polarimetric measurements is shown in Fig. 16. Light from the source is collimated and polarized with a Glan–Thompson or Nicol prism mounted in a graduated circle. The compensator is a birefrigent plate usually of quarter-wave thickness and is mounted either between the sample and the analyzer as shown or between the sample and the polarizer. The analyzer is a second prism mounted on a graduated circle which is rotated until a minimum of transmitted intensity is indicated by the detector. Detection may be visual or photoelectric. The determination of $\varDelta$ and ψ can be achieved in many ways. One method[207] employs plane-polarized incident light with the plane of polarization rotated 45° from the plane of incidence. The compensator and analyzer are then rotated to produce extinction.

[207] A. Vašiček, *J. Opt. Soc. Amer.* **37**, 145, 979 (1947).

The measurement may be made at any angle of incidence, but angles near the principal angle (the angle at which the differential phase change on reflection for the p and s waves is 90°) usually give the greatest sensitivity. If the angles between the fast axis of the compensator and the analyzer and the plane of incidence are γ and χ, respectively, the angles Δ and ψ are obtained from the relations

$$\tan \Delta = \tan 2\gamma / \sin 2\chi \quad \text{and} \quad \cos 2\psi = \cos 2\gamma \cos 2\chi. \quad (12.2.2)$$

In another method,[208,209] the fast axis of the compensator is set at 45° to the plane of incidence and the polarizer and analyzer are rotated to produce extinction. The compensator need not be exactly a quarter wavelength and can be mounted on either side of the sample. If the polarizer and analyzer angles measured from the plane of incidence are P and A, respectively, and δ is the retardation of the compensator, the angles Δ and ψ are obtained from the relations

$$\tan \Delta = \sin \delta \tan(90° - 2P), \quad \tan \psi = \cot L \tan(-A),$$
$$\text{and} \quad \cos 2L = -\cos \delta \cos 2P. \quad (12.2.3)$$

Since two ellipsometric quantities Δ and ψ are obtained experimentally, both the refractive index n_1 and thickness d of a transparent film may be determined when the optical constants of the substrate are known. The optical constants of the substrate may also be determined by polarimetric methods. Vašiček[207] has developed a method for determining the refractive index and thickness of transparent films on glass substrates. To facilitate computation he presents tables of calculated values of Δ and ψ as well as of γ and χ for thicknesses up to 1.5 μm and refractive indices ranging from 1.0 to 2.75 for a glass substrate of index 1.5163 and incident light of wavelength 5893 Å. An uncertainty of 0.5' in the values of Δ and ψ permits accurate measurements of films as thin as 10 Å. Archer[209] has developed a computer-generated graphical representation of the dependence of Δ and ψ on the refractive index and thickness of transparent surface films on Si. The optical constants of Si were also measured and used in the computation along with the wavelength and angle of incidence employed. This method allows a rapid determination of n_1 and d from the experimental values of Δ and ψ with an accuracy of ± 0.004 in index

---

[208] F. L. McCrackin, E. Passaglia, R. R. Stromberg, and H. L. Steinberg, *J. Res. Nat. Bur. Std.* **67A**, 363 (1963).
[209] R. J. Archer, *J. Opt. Soc. Amer.* **52**, 970 (1962).

of refraction and ± 1–5 Å in thickness for film thickness ranging from 15 to 3000 Å. For very thin films ($d \leq 100$ Å) simple approximations exist for the dependence of $\varDelta$ and ψ on n_1 and d[210]

$$\varDelta = \bar{\varDelta} - \alpha d \quad \text{and} \quad \psi = \bar{\psi} + \beta d, \quad (12.2.4)$$

where $\bar{\varDelta}$ and $\bar{\psi}$ denote the values of $\varDelta$ and ψ for an uncoated substrate, d is the film thickness, and α and β are functions of the wavelength of light, the angle of incidence, and the refractive indices of the film and substrate. The quantities n_1 and d may be obtained from the experimentally determined quantities $\varDelta - \bar{\varDelta}$ and $\psi - \bar{\psi}$.

The optical constants and thickness of an absorbing film cannot be determined from a single measurement of $\varDelta$ and ψ. If the thickness is known, however, the optical constants are, in principle, obtainable by graphical methods.[209] Bulk ellipsometric methods may be employed with films sufficiently thick for the effect of multiple reflections to be negligible. If the substrate is transparent, measurements of the transmittance allow a test of the applicability of the bulk methods. If the film thickness is small compared to the wavelength of light, approximations exist for the dependence of $\varDelta$ and ψ on n_1 and k_1[147]

$$\varDelta = \bar{\varDelta} + bd/\lambda \quad \text{and} \quad \psi = \bar{\psi} + (ad/2\lambda) \sin 2\bar{\psi}, \quad (12.2.5)$$

where $\bar{\varDelta}$ and $\bar{\psi}$ refer to the uncoated substrate, λ is the wavelength, d is the film thickness, and a and b are functions of the angle of incidence and the optical constants of the film and substrate. The optical constants are not directly calculable from the observed $\varDelta - \bar{\varDelta}$ and $\psi - \bar{\psi}$, but are determined by trial and error. It is also important to select an angle of incidence which maximizes $\varDelta - \bar{\varDelta}$ and $\psi - \bar{\psi}$.

Photometric methods, in general, are very sensitive to the presence of very thin layers on substrate surfaces. Oil and Hg films of 5 Å thicknesses can be easily detected[211] and a precision in the determination of sub-mono-layer coverage of a few hundredths of an atomic layer of O on Si has been claimed.[210] A useful review of submonolayer ellipsometry has been provided by Bootsma and Meyer.[212]

---

[210] R. J. Archer and G. W. Gobeli, *J. Phys. Chem. Solids* **26**, 343 (1965).
[211] R. W. Fane, W. E. J. Neal, and R. M. Rollason, *Appl. Phys. Lett.* **12**, 265 (1968).
[212] G. A. Bootsma and F. Meyer, *Surface Sci.* **14**, 52 (1969).

13. THE OBSERVATION OF MAGNETIC DOMAINS*

13.1. Introduction

A magnetic domain is a region in which the spontaneous magnetization is uniform in magnitude and direction. The existence of domains readily explains how a crystal can be demagnetized without loss of the spontaneous magnetization and also underlies the extremely high susceptibilities which can arise in ferro- or ferrimagnetic crystals. Figure 1 shows an exceptionally simple example, a crystal of cubic shape and presumed uniaxial anisotropy which is divided by a single division into just two

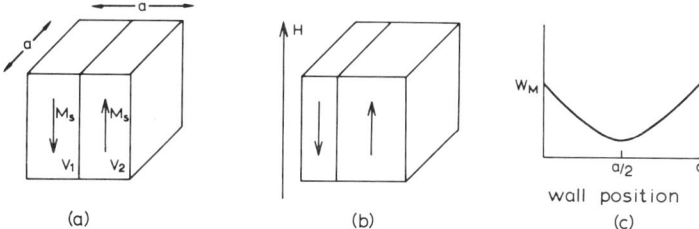

FIG. 1. (a) A simple structure of two 180° domains, of equal volume (demagnetized state), and (b) in the presence of a magnetic field. The variation of the magnetostatic energy indicated by (c) is equivalent to a driving force toward the demagnetized state.

domains. When $v_1 = v_2$, the crystal is demagnetized. When a field is applied, the energy of interaction with the field

$$W_H (= vE_H) = -\mathbf{H} \cdot \mathbf{M}_s v = -HM_s(v_2 - v_1) \qquad (13.1.1)$$

clearly decreases as the wall moves to enlarge v_2 as shown in Fig. 1b. This motion will be limited because the magnetostatic energy is minimum when the wall is central (the demagnetized state), the variation of E_M being as shown schematically in Fig. 1c. Now if the crystal becomes more and more elongated in the direction parallel to the magnetization in the

---

* Part 13 is by D. J. Craik.

domains E_M falls, and in principle the susceptibility can approach infinity for a perfect crystal.

In practice the domain wall, the boundary between the domains, will be seen to interact with crystal imperfections and this will decrease the susceptibility as well as give a finite coercivity and hysteresis: the magnetization curve for the idealized example above would be reversible and anhysteretic. Also, of course, the domain structures of most specimens will be much more complex than that in the example but the principle remains that the magnetic behavior, as represented by the magnetization curve and hysteresis loop, will be largely influenced by demagnetizing effects and domain wall interactions with imperfections. Thus there is a threefold interest in domain structures. They form a representation of the basic properties of the material, their study may give useful information on these basic or intrinsic properties, and most certainly their study can help to explain the "technical" properties such as coercivity, permeability, remanence, and losses (including microwave losses).

Quite recently it has come to be appreciated that domains are valuable not only for what they explain but also simply as domains. The presence (or absence) of a domain at a particular point, or in a particular sequence, can represent an item of information and very great efforts are currently being devoted toward the realization of practical bulk stores of this type as well as of logic devices in which domains may be created, moved, and caused to interact in a controlled manner.

13.2. Principles Governing Domain Formations

Domain structures correspond to the bulk magnetic properties, regardless of whether the spontaneous magnetization is a consequence of ferromagnetism, ferrimagnetism, or weak (canted-spin) magnetism, or whether the material is a metal, oxide, or other compound.

The chief intrinsic properties of any magnetic material are the spontaneous magnetization $\mathbf{M}_s$, magnetocrystalline anisotropy, and magnetostriction. $\mathbf{M}_s$ can, in turn, be considered to result from the existence of ionic magnetic moments or free electron spins, which are aligned by exchange interactions approximately represented by an exchange constant A which may be derived from the exchange integral J. $\mathbf{M}_s$ usually falls slowly as the temperature rises, with a rapid fall to zero at the Curie temperature T_C, but there may be an intermediate compensation temperature at which $\mathbf{M}_s$ falls to zero before rising once more. The magneto-

13.2. PRINCIPLES GOVERNING DOMAIN FORMATIONS

crystalline anisotropy constants K are also strongly temperature-dependent and generally this anisotropy disappears at significantly lower temperatures than does $\mathbf{M}_s$. The domain structure of any specimen corresponds to the combined effects of these intrinsic properties and the shape and size and microstructure of the specimen.

The shape is particularly important since it affects the magnetostatic energy involved. For uniformly magnetized ellipsoids, this energy can be represented as

$$E_M = -\tfrac{1}{2} N M_s^2 \qquad (13.2.1)$$

per unit volume, where the demagnetizing factor N is readily calculated from the eccentricity, or obtained from tables compiled by Stoner[1] or Osborne.[2] E_M has its maximum value ($N = 4\pi$) for thin sheets magnetized normal to their surface, and $E_M \to 0$ for long needles magnetized along their major axes: for a sphere $N = 4\pi/3$, and $N = 2\pi$ for a cylinder magnetized normal to its axis. Calculations of E_M for specimens subdivided into domains are much more difficult but have been made with considerable success. The important point to note is that the subdivision, which breaks up the surface change $\mathbf{M}_s \cdot \mathbf{n}$ into regions of different sign, effects a considerable reduction in E_M.

Consider a spherical crystal of uniaxial material with $\mathbf{M}_s$ constrained to lie along one of the two easy directions by the anisotropy energy which has the approximate form

$$E_K = -K \sin^2 \theta \qquad (13.2.2)$$

with θ the angle between $\mathbf{M}_s$ and the easy direction. (This neglects high-order terms such as $K_2 \sin^4 \theta$.) The magnetostatic energy is $W_M = \tfrac{1}{2}(4\pi/3) \mathbf{M}_s^2 (4\pi r^3/3)$. If now $\mathbf{M}_s$ is supposed to be oppositely directed in two hemispheres, it is readily estimated that W_M is approximately halved, and it is further reduced by more complex arrangements. However, this reduction in W_M is only achieved by introducing the energy which must be associated with the domain boundary itself. Between one domain and the next $\mathbf{M}_s$ must rotate through 180°, and the spatial distribution of the spin directions can be investigated by minimizing the sum of the crystalline anisotropy energy [Eq. (13.2.2)] and the exchange energy which is

$$E_A = A(d\theta/dx)^2 \qquad (13.2.3)$$

---

[1] E. C. Stoner, *Phil. Mag.* **36**, 816 (1946).
[2] J. A. Osborne, *Phys. Rev.* **67**, 351 (1945).

when the angle is supposed to change with respect to translations along a single axis $0X$. It is found[3] that the 180° domain wall has a specific width

$$\delta = \pi(A/K)^{1/2} \qquad (13.2.4)$$

and energy density per unit area

$$\gamma = 4(AK)^{1/2}. \qquad (13.2.5)$$

It seems that $A \approx 10^{-6}$ for most materials, but K can range from zero (in principle) to 10^7 erg cm^{-3} so that limits for δ and γ are about 10^{-6} cm and 10 erg cm^{-2}, and generally K can be taken as the controlling factor. Since the total wall energy for the two-domain sphere is

$$W_\gamma = \pi r^2 \gamma, \qquad (13.2.6)$$

it is easily seen that introducing the domain wall reduces the total energy so long as

$$r > 9\gamma/4\pi M_s^2 = r_c. \qquad (13.2.7)$$

Thus if we take the energy minimization as the sole criterion, we see that domains should always arise in crystals above a certain size (i.e., a certain "magnetic size" or total magnetization).

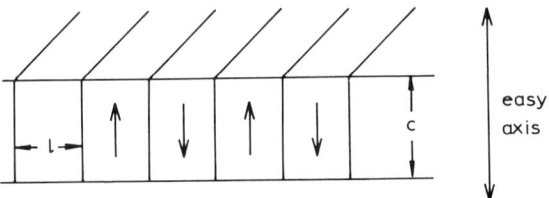

FIG. 2. 180° domains in a uniaxial crystal plate of thickness c.

With the same criterion it is possible to determine, for example, the equilibrium domain spacing and thus energy for a model structure such as that of simple 180° domains shown in Fig. 2. The only variable energy terms are those for the domain wall energy, which is a question of simple geometry, and the magnetostatic energy. To obtain the latter it is necessary to integrate the product of magnetic potential ϕ and surface charge

---

[3] F. Bloch, Z. Phys. **74**, 295 (1932); L. Landau and E. Lifshitz, Phys. Z. Sowjet. **8**, 153 (1935).

13.2. PRINCIPLES GOVERNING DOMAIN FORMATIONS

or pole density $\sigma = \mathbf{M} \cdot \mathbf{n}$, where ϕ is a solution of Laplaces equation with appropriate boundary conditions and has the form of a series while σ can also be expressed in series form since the charge distribution is periodic, giving[4]

$$W_M = \frac{1}{2l}\int_{-l}^{+l} \sigma\phi \, dx = \sum_{\substack{1 \\ n \text{ odd}}}^{\infty} (16lM_s^2/n^3\pi^2)\{1 - \exp(-\pi nc/l)\}. \quad (13.2.8)$$

The minimization of $W_\gamma + W_M$ by standard computer programs gives l and thus the equilibrium energy density as a function of M_s, γ, and c.

An analytical expression can be obtained as an approximation for large values of c/l since the exponential term disappears, $\sum_1^\infty 1/n^3 = 1.0518$ and $W_M = 1.7 M_s^2 l$, and differentiation of $(W_M + W_\gamma)$ gives[5]

$$l = (\gamma c/1.7 M_s^2)^{1/2} \quad \text{cm.} \quad (13.2.9)$$

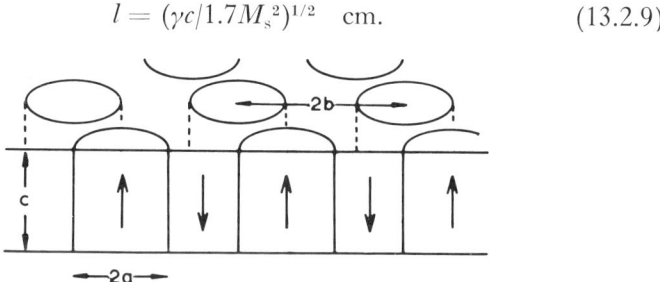

FIG. 3. An alternative structure of 180° domains (cf, Fig. 2) in the form of cylinders. The two directions of magnetization are no longer equivalent, and even a perfect crystal has a finite residue due to the tendency of the cylindrical walls to contract.

Other simple structures may be similarly analyzed. For example, Fig. 3 shows a hexagonal array of cylindrical 180° ("bubble") domains with a magnetostatic energy obtained by double Fourier methods as

$$W_M = M_s^2\left[2\pi c\left\{\frac{a^2}{b^2}\left(\frac{\pi}{3^{1/2}}\right) - 1\right\}^2 + \frac{2a^2}{b}\left(H + \frac{2G}{3\pi}\right)\right], \quad (13.2.10)$$

where

$$H = \sum_{\substack{-\infty(m \text{ even})}}^{+\infty}{}' \left[\frac{J_1^2(a\pi m/b)\{1 - \exp(-c\pi m/b\}}{3m^3}\right.$$
$$\left. + \frac{3^{1/2}J_1^2(a\pi m/3^{1/2}b)\{1 - \exp(-c\pi m/3^{1/2}b)}{m^3}\right]$$

[4] Z. Malek and V. Kambersky, Czech. J. Phys. **8**, 416 (1958).
[5] C. Kittel, Rev. Mod. Phys. **21**, 541 (1949).

and

$$G = \sum_{-\infty}^{\infty}{}' \sum_{-\infty}^{\infty}{}' \frac{J_1^2\{(a\pi/b)(m^2+n^2/3)^{1/2}\}[1-\exp\{(-c\pi/b)(m^2+n^2/3)^{1/2}\}]}{(m^2+n^2/3)^{3/2}} \quad (13.2.11)$$
$(m+n)\text{even}$

and J_1 is the first-order Bessel function: the primes indicate that $n = m = 0$ is omitted. This again reduces, for large c/b to[6]

$$W_M = 1.345(b/c)M_s^2 \quad \text{erg cm}^{-3}. \quad (13.2.12)$$

Having calculated the magnetostatic energy terms, it is relatively easy to compare the total energies of different types of structure and, with this criterion, decide which structures are most likely to occur in a given specimen.

Some structures which include closure domains with 90° domain walls (Fig. 4) give zero demagnetizing energy corresponding to complete flux closure. If this applied to a uniaxial material such as cobalt, a very

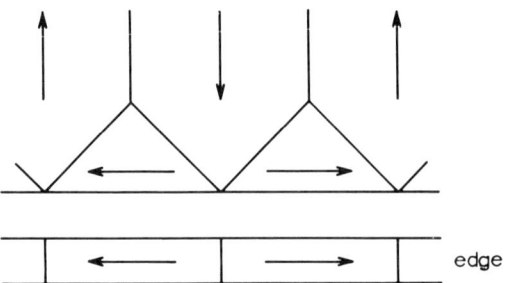

FIG. 4. 180° domains with flux closure structures at the edge of a crystal, as found in iron (easy axes $\langle 100 \rangle$).

high magnetocrystalline anisotropy energy would be involved, of value equal to K per cubic centimeter [Eq. (13.2.2); $\theta = \pi/2$]. The equilibrium spacing is readily computed as $l = (2\gamma c/K)^{1/2}$ cm and for most uniaxial materials which, in fact, have generally high values of K/M_s, the total energy corresponding to this would be found to be greater than that for the simple 180° domains (also with the equilibrium spacing). For this reason 180° domains are said to be characteristic of uniaxial materials, with closure more likely to apply to crystals such as iron with $\langle 100 \rangle$ easy

[6] D. J. Craik and P. V. Cooper, *Phys. Lett.* **33A**, 411 (1970); D. J. Craik, P. V. Cooper, and W. F. Druyvestein, *Phys. Lett.* (1971).

13.2. PRINCIPLES GOVERNING DOMAIN FORMATIONS

axes. For these the anisotropy energy may also be zero, but the magnetoelastic energy must be taken into account.

Turning to materials with cubic structure but with ⟨111⟩ easy axes (K_1 negative), closure is again to be expected but this cannot now be complete. This can be illustrated by the simple example of a thin rectangular block with the crystallographic orientation shown in Fig. 5. This is expected to behave largely as a uniaxial crystal due to shape effects, with 180° domains magnetized parallel to the single [111] axis

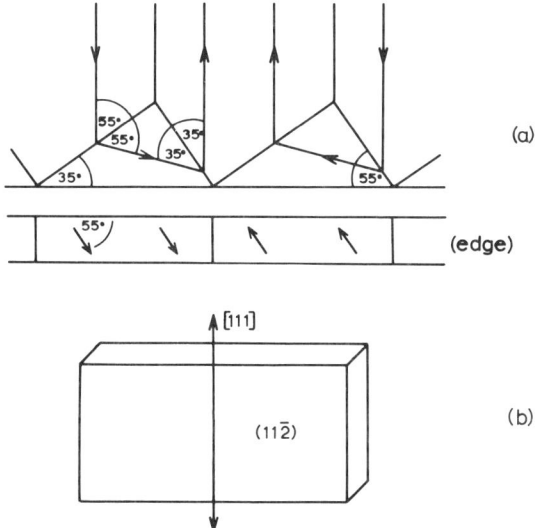

FIG. 5. Incomplete closure as observed in (a) crystals of ferrimagnetic garnets with negative K_1, easy axes ⟨111⟩, cut with the orientation shown in (b).

which lies in the plane of the crystal. Viewing the crystal in plane we might expect closure domains at the edges with 70 and 110° walls across which the normal component of the magnetization is zero, and these are in fact formed in crystals of rare earth iron garnets. However, a magnetostatic energy of a very complex form still arises due to (a) the component of $\mathbf{M}_s$ across the edges and (b) the component across the surface of the partial closure domains themselves as illustrated.

Magnetization curves can be computed simply by including a term for the interaction of the magnetization with the field $W_H = -\mathbf{M} \cdot \mathbf{H}$, in the total energy to be minimized. This gives an equilibrium structure, for any value of **H** in which one set of the 180° domains, for example, is wider than the intervening domains. The computed curves are anhysteretic, usually have zero remanence and coercivity although finite M_r and H_c arise for structures in which the oppositely magnetized regions are inequivalent, such as the arrays of cylinders. Hysteresis loops can be built up by shifting the anhysteretic curves along the positive and negative field axes to an extent which corresponds to the field required to move the domain walls past any imperfections which are present. This distinction, effectively between large-scale and microscopic magnetostatic effects, is somewhat arbitrary but in some cases at least it can be justified by comparing anhysteretic curves with measured loops.[7]

Defining the induced magnetization as that component which is parallel to the applied field implies a scalar susceptibility, although for anisotropic material the more general susceptibility is a tensor.

So far it has been assumed that the magnetization has been uniform within the domains; div $\mathbf{M} = 0$. This is an approximation valid for high values of the ratio K/M_s, since K can be considered to control the alignment with respect to an easy direction and M_s to control the stray fields arising from surfaces or discontinuities which favor deviations from the easy axes. The effects of finite rotational susceptibiliting can only be estimated by rather complex treatments, but they are always such as to reduce the energy to an extent indicated by the factor for simple 180° domains[8]

$$\frac{2}{1 + [(1 + k^*)(1 + k^* \sin^2 \phi)]^{1/2}}, \quad k^* = \frac{2\pi M_s^2}{K} \quad (13.2.13)$$

with ϕ the angle between the easy direction and the normal to the surface. When k^* or μ^* is really high, the effects can no longer be considered as a correction to the uniform magnetization model, which must be abandoned. This emphasizes the role of direct observation in the study of domains, since the structures can scarcely be predicted at all for "soft" materials with $K \to 0$, for example.

The shortcomings of the approach outlined above are quite apparent. One postulates a likely structure with regard to the given principles and then carries out an analysis to see if in fact it has a low energy in comparison with reasonable alternatives. However, there is always the possibility of other structures existing, with even lower energy, which have simply not been thought of.

Clearly it would be more satisfactory actually to derive equilibrium

---

[7] D. J. Craik and D. A. McIntyre, *IEEE Trans. Magn.* **MAG-5**, 378 (1969).

[8] M. Fox and R. S. Tebble, *Proc. Phys. Soc.* **72**, 765 (1958); **73**, 325 (1959).

structures from first principles by energy minimization without the postulation of models, and in such a treatment the domain walls would arise naturally as part of the structure. (Thus exchange energy should be included directly and not simply as affecting the walls.) The principles have long been recognized and applied to states of nonuniform magnetization in single-domain particles, but they have not yet been applied successfully to predict domain structures.[9]

The situation is far more satisfactory if the observation of domain structures is given the primary role in the subject. A large body of empirical information gives a clear impression of what structures are to be expected in any particular specimen with given intrinsic and geometrical parameters, and observations also indicate which quantitative analyses are worth attempting. Systematic observations and measurements may then verify the analytical work, give specific data on domain wall parameters and thus on exchange effects, and help to elucidate the processes of magnetization and the wide range of technical magnetic properties.

13.3. Survey of Methods

The oldest method of observation, devised by Bitter,[10] is based on the visualization of stray field configurations arising from the domain structures by applying magnetic particles to the surface. At a surface containing an easy direction of a specimen with a 180° domain structure, for example, there will be components $\mathbf{M}_s \cdot \mathbf{n}$ across the surface only where it is intersected by the domain walls. An obvious approximation is to treat this as a strip of charge $\sigma = M_s$ of width δ on a surface of zero charge, and the fields and field gradients are then readily calculable (calculations are described in Section 13.5.3).

It is apparent that any field-measurement technique with high resolution is potentially applicable to the study of domains. High sensitivity is not specifically required since the fields are substantial; locally $\sim \pi M_s$. Miniature Hall probes and magnetoresistive elements have been used, while other important classes of method depend upon the magnetic forces on particles applied to the surface and the electromagnetic Lorentz forces on electrons passing near to the surface. Although the Bitter or powder-pattern method continues to be used very extensively, the ob-

---

[9] W. F. Brown Jr., *Phys. Rev.* **105**, 1479 (1957); E. H. Frei, S. Shtrikman, and D. Treves, *ibid.* **106**, 446 (1957).

[10] F. Bitter, *Phys. Rev.* **38**, 1903 (1931); **41**, 507 (1932).

vious drawback to the stray-field methods is that they detect the fields outside the specimens and can only relate to superficial structures which may or may not be clearly related to the domains within the specimen. For example, the author noted that patterns on all four faces of a prismatic cobalt crystal, cut parallel to the easy direction, indicated that the 180° walls disappeared at a stage of the measured magnetization for which the crystal must still have contained many 180° domains which simply did not intersect such surfaces.

Moreover there is a basic objection to any method that detects a consequence of the magnetization distribution rather than detecting the magnetization directly. Direct interactions of light with the magnetization are utilized in the magnetooptic methods (Section 13.6), and Lorentz electron microscopy (Section 13.7) is based on the interaction of an electron beam with the magnetization rather than the stray fields. The Faraday (magnetooptical) and Lorentz methods involve transmission and are not limited to superficial structures, but, of course, the ability to transmit light or an electron beam represents a severe limitation on the specimen. The use of X rays is indicated for the internal examination of really thick specimens, and X-ray topography has been found to image domain structures due to an indirect interaction with the magnetization via the magnetostriction (Section 13.9).

With regard to the discrimination between superficial and internal structures, a major problem in domain studies is the preparation of specimens with strain-free surfaces. Magnetostatic effects may enforce the formation of closure structures or patterns of reverse magnetization near to surfaces, which generally have some fairly definite relation to the internal structures, but a heavily strained layer may so alter the magnetic properties of the material that the strain domains formed, and visualized by powder patterns, for example,[11] have little relation to the structures typical of the underlying strain-free material. Transmission methods can distinguish between superficial and internal structures even when both exist simultaneously.

13.3.1. Specimen and Specimen Surface Preparation

Crystals of rare earth orthoferrites, for example, can be grown in the form of cubes or prisms with natural faces exactly normal and parallel

---

[11] W. C. Elmore, *Phys. Rev.* **51**, 982 (1937); **53**, 757 (1938); **54**, 309 (1938); **62**, 486 (1942); S. Chikazumi and K. Suzuki, *J. Phys. Soc. Japan* **10**, 523 (1955); L. F. Bates, H. Clow, D. J. Craik, and P. M. Griffiths, *Proc. Phys. Soc.* **72**, 224 (1958).

13.3. SURVEY OF METHODS

to the easy direction: barium ferrite and related compounds can be grown as thin platelets which are also suitable for domain studies without further preparation and the same applies to polycrystalline or (epitaxial) single crystal metal films, and to epitaxially grown ferrite single crystals. More generally it is necessary to cut, grind, and polish the specimens to produce surfaces suitable for study, or thin slices which transmit light or electron beams. This produces strains and amorphous surface layers which at best obscure the underlying structures, may also give rise to spurious superficial structures, and in extreme cases (yttrium and rare eath iron garnets) completely control the domain structures formed. Thus the domain structure of polished slices of garnet crystals (unless they are exceptionally finely polished) is one of $180°$ domains magnetized normal to the surface of the slice regardless of the disposition of the normal easy axes, the stress-induced anisotropy overriding the magnetocrystalline anisotropy. One way of investigating the state of a surface is to determine whether observed domain rearrangements correspond to measured magnetization changes; the author noted that domain patterns on a mechanically polished polycrystalline ferrite remained virtually unchanged while the specimen was magnetized to saturation.

There are several ways to remove such strains. The most apparent, for the metals, is electropolishing either using commercial equipment or a bath assembled for the purpose. As well as applying to surface preparation, special methods of electropolishing can be used for preparing metal foils suitable for Lorentz microscopy, these methods having been developed intensively for routine electron microscopy. The requirements are particularly stringent when the surfaces are being prepared for Kerr effect studies, since any surface pits scatter light to an extent which may seriously reduce the contrast obtained, and Houze[12] obtained particularly good results by using commercial equipment which combines mechanical and electropolishing to give very *flat* strain free surfaces.† Even when electropolishing is used it is also advisable to anneal the specimens carefully prior to polishing, taking care not to use conditions so extreme as to risk altering any crystallographic texture.

Electropolishing is not applicable to insulating oxide materials such as spinel ferrites or garnets. Annealing at temperatures quite close to the

---

[12] G. L. Houze, Jr., *J. Appl. Phys.* **38**, 1089 (1967).

† Reinacher polishing equipment supplied by Struers Scientific Instruments, Copenhagen, Denmark.

melting points, e.g., 1450°C for YIG, is of value but only after very fine mechanical polishing. The effectiveness is estimated simply by the results obtained, i.e., the gradual replacement of strain domains by larger in-plane domains with reduced contrast (Section 13.6.1). The author found that holding polycrystalline ferrites at a temperature at which they would normally be sintered was apparently very effective; it may be supposed that considerable mobility of the material in the surface layers exists at the same temperature at which sintering is appreciable, but the treatment must be brief or the specimen parameters are appreciably altered. The use of furnace atmospheres appropriate to the particular material (i.e., the atmosphere which would normally be used for sintering) is obviously essential to prevent decomposition or second-phase formation. Garnets appear to be quite stable in air up to 1450°C.

A chemical polishing technique[13] may be used for garnets and some other oxides. This consists of immersing a polished slice, for example, in a bath of phosphoric acid held at a temperature of 300°C for (211) faces of YIG (see Table I). This is also of value for polycrystalline hex-

TABLE I. Temperatures of Phosphoric Acid Baths for Chemical Polish[13]

Material	Orientation	Temperature (°C)
$Y_3Fe_5O_{12}$	(110)	300
	(211)	300
	(111)	400
$Y_3Ga_5O_{12}$	(211)	280
$Y_3Al_5O_{12}$	(110)	280
$MgFe_2O_4$	(111)	270

agonal ferrites (C. Tanasoiu, private communication). The specimen may be supported in a platinum mesh container, but it must be moved repeatedly to prevent high spots developing at the points of contact. One drawback of chemical polishing is that, although the surfaces can be shown to be strain-free, they appear rarely to be flat but have a slight "orange-peel" texture which can alter the magnetic behavior significantly.

The final method to be noted is one that seems to have received little recognition but holds great promise: the use of ion bombardment equip-

[13] J. Basterfield, *Brit. J. Appl. Phys.* **2**, 1159 (1969).

ment, or ion beam polishing. Commercial equipment† is available in which a beam of inert gas ions with a spatially uniform energy distribution (produced by guns with multiple apertures) is directed at a glancing angle at the surface of a specimen which is continuously rotated. Material is stripped off at a rate typically around 10^{-4} cm/hr to leave, eventually, a surface which is optically smooth (featureless by optical microscopy) and virtually strain-free. Presumably there must be some surface damage, but the depth of this is probably of the order of the lattice spacing. The applicability of this method has been demonstrated by Paulus (who originated the equipment) by studies of powder patterns on single crystals of magnetite,[11] while A. Harrison in the author's laboratory has demonstrated the transformation of strain patterns to in-plane domains in (211) slices of gadolinium iron garnet due to ion bombardment, and has investigated the applicability of the method to polycrystal garnets. The great advantage of this method is its universality with respect to the material under study.

Whatever final methods are used it is essential that the mechanical cutting and polishing be carried out with great care. Devices which present the crystal to a slowly rotating cutting wheel with only a gentle pressure minimize the initial damage, and most of the strained or deformed material can be removed by prolonged and carefully graded diamond powder polishing on a solder, lead, or indium lap. Polishing may terminate with $\frac{1}{4}$-μm diamond or, preferably, a suspension of silica flowing continuously over a special plastic lap. A commercial preparation suitable for this is Syton, a milky liquid which, rather surprisingly, is used and sold by the textile trade and a suitable lap is hard polyurethane foam (marketed, for example, by H. V. Skan, Solihull, England). The foam must be cemented to a stainless steel lap which rotates within a corrosion-resistant dish from which the liquid can be drained away for recirculation. Examples of this type of polishing will be illustrated later.

13.4. Powder Pattern (Colloid) Technique

The basic powder pattern method consists simply of applying a thin layer of a magnetic colloid to a strain-free surface and observing the

[11] M. Paulus, *Z. Angew. Phys.* **17**, 216 (1964).

† Manufactured by ALBA, 92-Asnieres, France; Edwards, Crawley, Sussex, England.

pattern formed by the action of the stray fields, using a metallurgical microscope. The layer is formed under a coverslip. Dark field viewing may be advantageous, particularly if the patterns are faint. This may be the case when dilute colloids are used, to delineate finely spaced structures with high resolution, or when the field gradients are small. For highest resolution a water immersion lens may be used in contact with a drop of the colloid.

The use of magnetite, Fe_3O_4 is virtually universal. This is produced as aggregates of microcrystals about 100 Å in diameter by coprecipitation from a suitable mixture of ferrous and ferric chlorides (e.g., $2gFeCl_2 \cdot 4H_2O$, 5.4gm $FeCl_3 \cdot 6H_2O$ in 300 cm³) by adding NaOH (5 gm in 50 cm³) drop by drop to the heated (70°C) and stirred solution. The floccular precipitate is thoroughly washed by decantation or in a vacuum filter and then dispersed; some tendency to dispersion occurs automatically as the salts are removed. The production of a stable colloid may be achieved by peptization using HCl or by rapid stirring or ultrasonic agitation in the presence of a dispersing or stabilizing agent or by a combination of the two. Dilute solutions (1% or less) of the commercial Aerosol OT (USA) or Manoxol OT (U.K.), i.e., sodium-*di*-2-ethyl hexyl sulphosuccinate, appear to be most useful. Empirical control of the concentration of HCl and the stabilizing agent and of the mechanical treatment can give a wide range of particle sizes and densities, with colors ranging from black to pale amber.

The colloid must be stable and must not attack or stain the surfaces. Beyond this the most important parameter influencing the behavior of the colloid is particle size (the particles being aggregates of the microcrystallites). The magnetic forces on the particles, assumed to be single-domain particles and thus to behave as dipoles, are proportional to their size and to the field *gradients* in which they move. Wide walls, which occur in low-anisotropy crystals, give rise to low field gradients at the intersection of the domain boundaries with the surface and a large particle size is then required. Adequate gradients can exist even when M_s is very low and it is quite easy to produce patterns on rare earth orthoferrites with $M_s < 10$; presumably the walls are very narrow. For studies of thin films it is important to note that the field configurations above Néel walls are much more favorable for pattern formation than those above Bloch walls, and the latter can only be imaged with some difficulty by this method. Bergmann[15] calculated that, to form patterns, the particle

---

[15] W. H. Bergman, *Z. Angew. Phys.* **11**, 559 (1956).

13.4. POWDER PATTERN (COLLOID) TECHNIQUE

diameters should be within the limits

$$3[kT(1 + \mu_w{}^*)/2\pi M_p M_w S_w]^{1/2} < d < 3(2kT/\pi^2 M_p{}^2)^{1/3}, \quad (13.4.1)$$

where M_p and M_w are the saturation magnetization of the particles and the specimen and $\mu_w{}^* = 1 + 2\pi M_w{}^2/K_w$. If the particles are too large, there is a tendency to form closed chains of crystallites in which the flux closure reduces the net dipole moment, and beyond the lower limit the magnetic energies are too small compared with kT. For the highly anisotropic hexagonal ferrites (barium ferrite or magnetoplumbite), this gives 15 Å $< d <$ 99 Å, i.e., indicates dispersion down to the ultimate microcrystallites.

Garrood[16] noted that the usefulness of colloids tended to improve with time, presumably due to crystallite (or aggregate?) growth, and found that the process could be hastened by boiling for an hour or two. A further refinement was to precipitate and boil the colloid in a magnetic field of several kilooersteds to produce particles with larger dipole moments (i.e., fewer flux loops). Jones[17] certainly found this to be significant with respect to difficult studies on NiFe and produced relatively coarse polarized suspensions in Teepol ($\frac{1}{2}$%) with the addition of HCl to pH 5.0–5.2. Patterns on presumably wide Bloch walls in nickel platelets are shown in Fig. 6a, and the colloid also gave the high-resolution patterns of stripe domains in (b). It was not, however, very stable and unsuitable for studies in which the walls were moved.

Unpublished data by J. C. Hendy (Plessey Research Laboratories) gives an example of colloid stability and the control of particle size. Closely spaced stripe domains can form the basis of a display system, using diffraction from small areas in which the direction of the stripes can be controlled by applied fields, by sealing a layer of colloid between an evaporated NiFe film and a cover glass. In-use stability of Manoxol OT stabilized colloids is demonstrated by the mobility of the patterns (as shown in Fig. 7) over periods of several months at least. Chemical passivity is also apparent. It was found in this work that particles of the order of 1000 Å contributed most to the diffraction efficiency; finer particles, down to 100 Å, contributed very little in this particular application. Suspensions containing up to 30 gm magnite in 125 cm³ liquid can be prepared, although 11.5 gm in 150 cm³ is more suitable.

---

[16] J. R. Garrood, *Proc. Phys. Soc.* **79**, 1252 (1962).
[17] M. E. Jones, Thesis, Imperial College, London (1970).

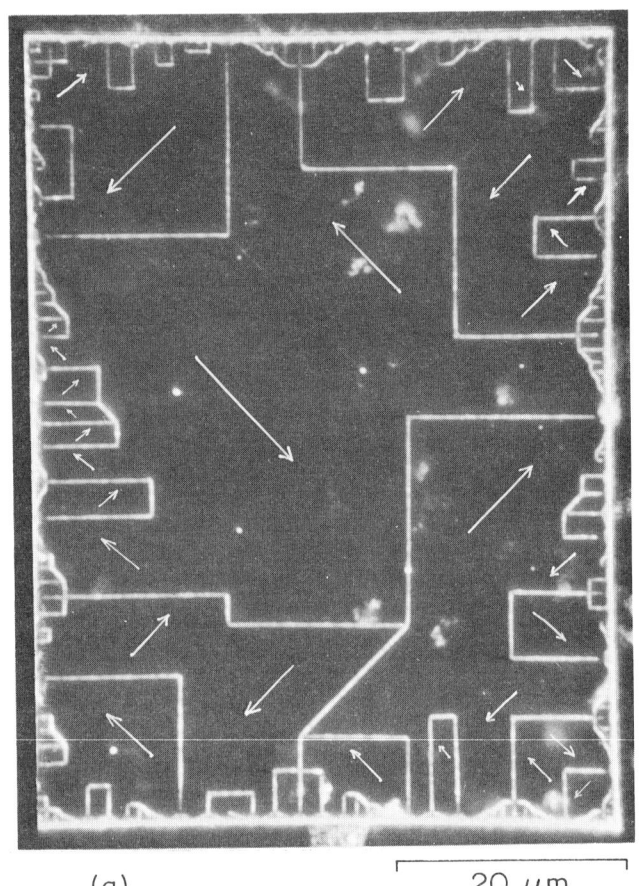

(a) 20 μm

20 μm

13.4. POWDER PATTERN (COLLOID) TECHNIQUE

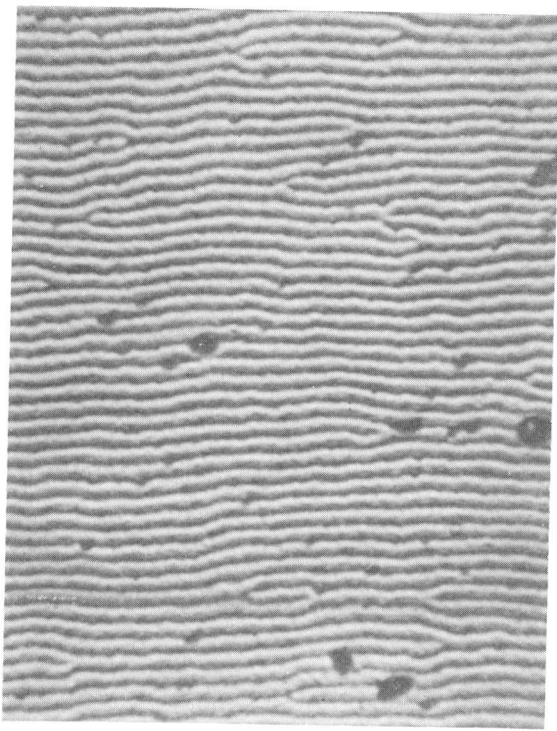

FIG. 7. Uniform stripe domains in a nickel–iron film giving patterns in a Manoxol stabilized colloid which show good long-term mobility and can be used in display systems [J. C. Hendy, Plessey, Ltd., private communication].

Manoxol or Aerosol OT stabilized colloids certainly appear to be easiest to prepare and the most reliable for studying wall motion and domain rearrangements. The obvious temperature restrictions for aqueous colloids can be relaxed somewhat by preparing special suspensions

---

FIG. 6. (a) Powder pattern showing the domain structure of a nickel platelet 2550 Å thick with (001) surface bounded by ⟨100⟩. Due to the large shape anisotropy (and low value of K/M_s), M_s is constrained to lie along ⟨110⟩ directions. Walls of 180° join the corners of the 90° structures and lie between the antiparallel arrows, but are shown only faintly or not at all due to the different stray field distributions [M. E. Jones, Thesis, Imperial College, London (1970)]. (b) Powder patterns of finely spaced stripe domains which form in thicker platelets (9 μm) in which M_s tends to lie closer to ⟨111⟩; dark field microscopy [M. E. Jones, Thesis, Imperial College, London (1970)].

for low[18,19] and high temperatures,[20-22] but once the essential simplicity of the technique is lost it would seem advisable to turn to alternatives.

13.5. Specialized Techniques and Pattern Formation

13.5.1. Replica Method for Electron Microscopy

The dimensions of some domain structures are close to the limit of resolution of the optical microscope, and it is advantageous to examine powder patterns in the electron microscope. For this the patterns must be in the form of replicas which can be stripped from the specimen and mounted on a grid for examination by transmission. The first requirement is for an extremely finely dispersed colloid, ideally a suspension of single crystallites, and the second is that the colloid should dry to form a thin solid film in which the pattern becomes set without the aggregation which normally accompanies the drying of a colloid.

A high degree of dispersion may be obtained by peptization with HCl, after forming and washing the precipitate in the usual way. Unfortunately the concentration required appears to vary around 0.1 N and the best results can only be obtained by preparing a series of samples and testing by their ability to pass a hard filter paper. The pH must be lower than that at the isoelectric point, i.e., pH 6.5. A stabilizing agent which has good film-forming properties is found to be Celacol, a commercial synthetic cellulose derivative. After the addition of the Celacol to the charge stabilized colloid the HCl can be completely removed by dialysis. A little of the resulting colloid is allowed to dry on the specimen surface, a grid is dipped in a solution of nitrocellulose in amyl acetate or in a diluted commercial adhesive (Durofix) and laid on the Celacol film and, after the adhesive has dried, the grid can be stripped off taking the film with it. After reinforcement by carbon evaporation the adhesive is dissolved away before examination in the electron microscope.[23]

[18] L. F. Bates and S. Spivey, *Brit. J. Appl. Phys.* **15**, 705 (1964).
[19] B. Gustard, *Proc. Roy. Soc.* **A297**, 269 (1967).
[20] W. Andra, *Ann. Phys.* **17**, 233 (1956).
[21] W. Andra, *Ann. Phys.* **3**, 334 (1959).
[22] M. Rosenberg, C. Tanasoiu, and V. Florescu, *IEEE Trans. Magn.* **MAG-6**, 207 (1970).
[23] D. J. Craik, *Proc. Phys. Soc.* **B69**, 647 (1956); D. J. Craik and P. M. Griffiths, *Brit. J. Appl. Phys.* **9**, 279 (1958).

13.5. SPECIALIZED TECHNIQUES AND PATTERN FORMATION

Typical results are shown in Fig. 8. Part (a) demonstrates the two-phase structure of a partially ordered PtCo alloy and shows domains within ordered lammellae less than 10^{-4} cm across while 8(b) shows patterns on the basal plane of a crystal of barium ferrite with a near-saturation field applied normal to the surface. The domain boundaries are exceptionally sharply resolved.

13.5.2. Vapor Condensation Method

Evaporated magnetic films are prepared by heating the film material in vacuo with the substrate directly exposed to the source. If the ferromagnetic metals or alloys are melted in a crucible or by hanging loops of the wires over a stout tungsten wire, in a substantial atmosphere of an inert gas, the vapor formed condenses to give a "smoke" of fine particles which undergo Brownian motion and eventually impinge on exposed surfaces. If a prepared specimen is included in the enclosure, the particles may condense upon it to give a domain pattern.[24,25] Such patterns are generally similar to those produced by the powder pattern technique, in both clarity and contrast, and exceptionally good results may be obtained. A great advantage of this particular process lies in the ease of adjusting the temperature of the specimen. Al-Bassam,[26] for example, produced interesting powder patterns on crystals of gadolinium and terbium down to 130 K, using iron particles in a 0.2-Torr helium atmosphere. (The Curie point of terbium is 222 K.)

It is possible to remove the powder deposits from the specimen surfaces in order to examine them in the electron microscope, the technique then being comparable to the replica method of the previous section. An example is shown in Fig. 9a for comparison with the Kerr technique.[27] It has also been noted that the powder deposits show up with fair contrast in the scanning electron microscope, and this might be developed into a technique which, with the great depth of focus, would be particularly useful for specimens with uneven or faceted surfaces.

---

[24] R. I. Hutchinson, P. A. Lavin, and J. R. Moon, *J. Sci. Instrum.* **42**, 885 (1965).

[25] U. Essmann and H. Träuble, *Phys. Status Solidi* **18**, 813 (1966).

[26] T. S. Al-Bassam, Thesis, Durham (1969).

[27] A. Hubert, *Phys. Status Solidi* **24**, 669 (1967).

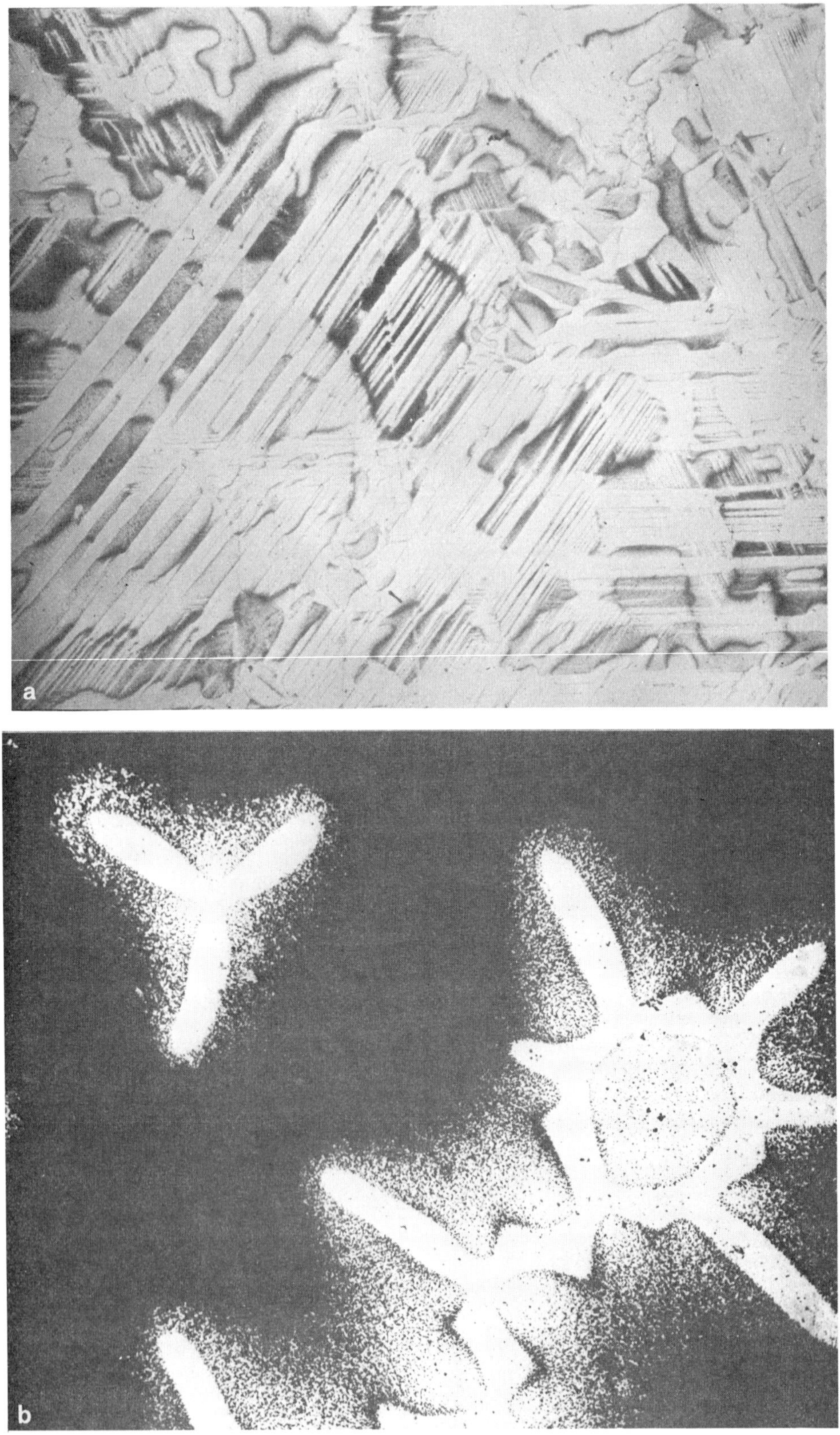

13.5. SPECIALIZED TECHNIQUES AND PATTERN FORMATION

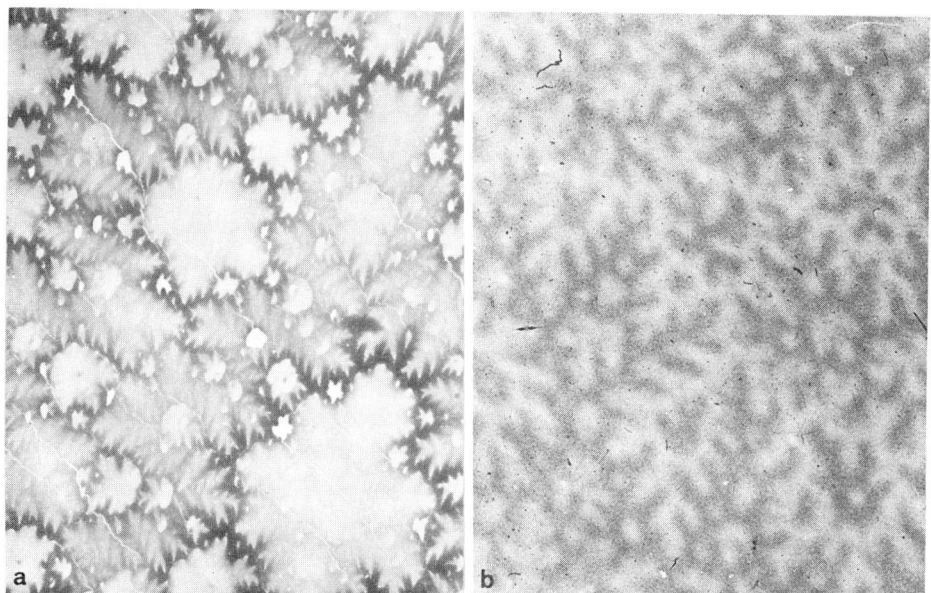

FIG. 9. Domains intersecting a basal plane of cobalt as imaged by (a) the powder deposition technique [R. I. Hutchinson, P. A. Lavin, and J. R. Moon, *J. Sci. Instrum.* **42**, 885 (1965)] and (b) the polar Kerr effect [A. Hubert, *Phys. Status Solidi*, **24**, 669 (1967)]. The first gives an image dependent on the stray field distribution while the second indicates the components of $\mathbf{M}_s$ normal to the surface.

13.5.3. The Formation of Powder Patterns

In forming patterns the colloid particles move in the forces due to the field gradients to a configuration of minimum energy ($W = -\mathbf{m} \cdot \mathbf{H}$, where $\mathbf{m}$ is the particle dipole moment) corresponding to their aggregation in regions of maximum field strength. Since the dipoles are also partially aligned by the fields there are mutual repulsive forces unless aggregation into chains can occur. Comparing the simple models for Bloch and Néel walls in Fig. 10 one sees that the fields and gradients are similar, but since the fields from the Néel walls are parallel to the surface

---

FIG. 8. Colloid replica patterns imaged in the electron microscope: (a) indicates the clear resolution of structures of the order of 10^{-4} cm in spacing in PtCo and illustrates the information which may be obtained on both magnetic and crystallographic structure; (b) shows the precision with which domain edges may be delineated (barium ferrite surface normal to easy axis). The resolution corresponds to the crystallite size of the colloid used, i.e., about 200 Å.

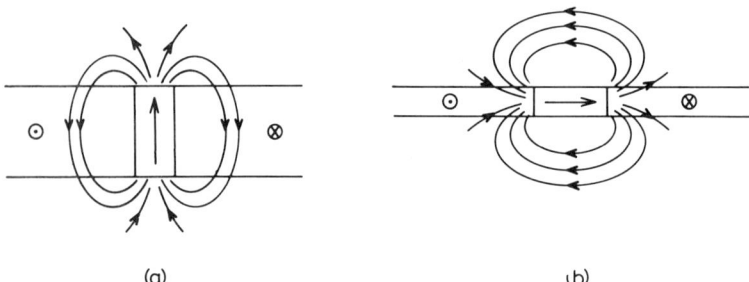

FIG. 10. A simple representation of (a) Bloch and (b) Néel walls in thin specimens, illustrating that the stray fields are more favorable for powder pattern formation above the latter.

chains can form to close the flux from these walls (in so doing they may make small but appreciable differences in the wall energy and structure). A familiar feature of work on films is the relative ease of imaging Néel walls as compared with Bloch walls in the thicker films.

At first sight it is not very apparent that patterns should form at all, in zero applied field on surfaces normal to $\mathbf{M}_s$, as on the basal planes of uniaxial crystals with high K/M_s. The stray field components normal to the surface of such a crystal containing 180° domains (Fig. 11) will have the

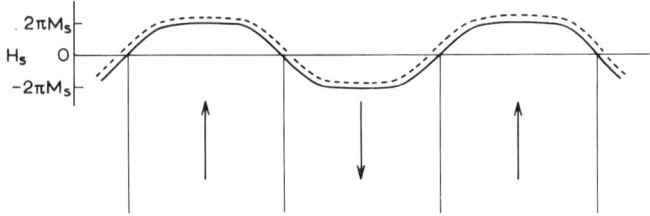

FIG. 11. The normal component of the stray field above 180° domains with $\mathbf{M}_s$ normal to the surface. The broken line shows the effect of a field applied normal to the surface.

form of a square wave at the surface itself, with amplitude $2\pi M_s$, and of a "rounded off" square wave some distance above the surface. Applying a field normal to the surface shifts this up or down, so that the colloid is expected to collect over alternate domains and particularly toward their centers. In practice the boundaries are still sharply delineated (Fig. 8b) and, in zero field, patterns form at the boundaries themselves as shown in Fig. 12a.

To explain this it is necessary to take account of the total stray field magnitude, including the components parallel to the surface and normal to the walls. A straightforward integration of fields from small elements of charge $M_s \, dx \, dy$ gives

$$H_\| = -M_s \log\left[\frac{\{(c+a)^2 + z^2\}^{1/2}}{\{(c+a)^2 + z^2 + b^2\}^{1/2} + b}\right.$$
$$\left.\times \frac{\{c^2 + z^2 + b^2\}^{1/2} + b}{(c^2 + z^2)^{1/2}}\right], \qquad (13.5.1)$$

$$H_\perp = M_s \sin^{-1}\left[\frac{(c+a)b}{\{(c+a)^2 + z^2\}^{1/2}(b^2 + z^2)^{1/2}}\right]$$
$$-M_s \sin^{-1}\left[\frac{bc}{\{(c^2 + z^2)(b^2 + z^2)\}^{1/2}}\right] \qquad (13.5.2)$$

for the components from a strip of charge $a \times b$, at a distance z above the surface as illustrated by Fig. 12b.[28] Clearly $H_\| \to \infty$ as $z \to 0$ at the boundary, and in spite of the logarithmic form the author indicated that with certain assumptions the colloid distributions observed in practice could be reasonably explained: it was shown that the total fields could have maxima in the centers of those domains in which $\mathbf{M}_s$ was parallel to the applied fields, minima near to the walls, and secondary maxima over the walls themselves. Figure 12c shows clearly that, in moderate applied fields, the colloid particles collect primarily in the centers of the domains magnetized parallel to the field, but they are also concentrated at the domain boundaries. Some appreciation of the stray field distributions above surfaces is vital if results are to be properly interpreted.

13.6. Optical and Magnetooptical Properties

Two of the most elegant and useful methods of domain observation are based on the intrinsic optical properties of crystals. Before introducing these magnetooptical (Faraday and Kerr) methods it is useful to survey briefly the methods of producing polarized light and the principles governing the propagation of light through nonmagnetic crystals (by analogy, the electrooptic effects).

The vibrations of a light ray (i.e., the $\mathbf{E}$ and $\mathbf{H}$ vectors) are always parallel to the wave front, and in isotropic media are also normal to the

---

[28] D. J. Craik, *Brit. J. Appl. Phys.* **17**, 873 (1966).

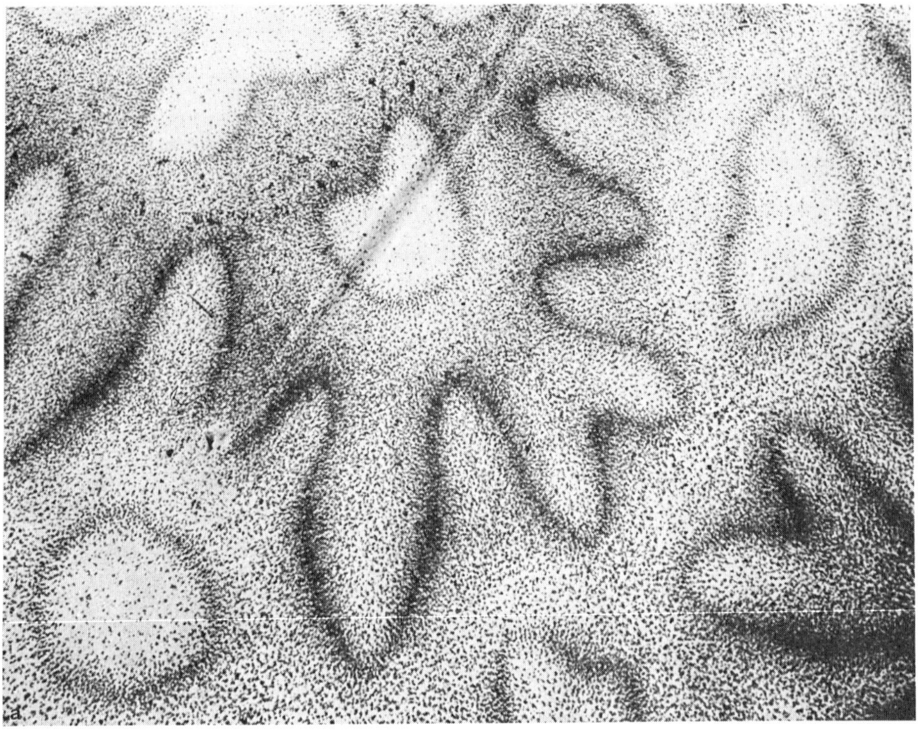

698   13. THE OBSERVATION OF MAGNETIC DOMAINS

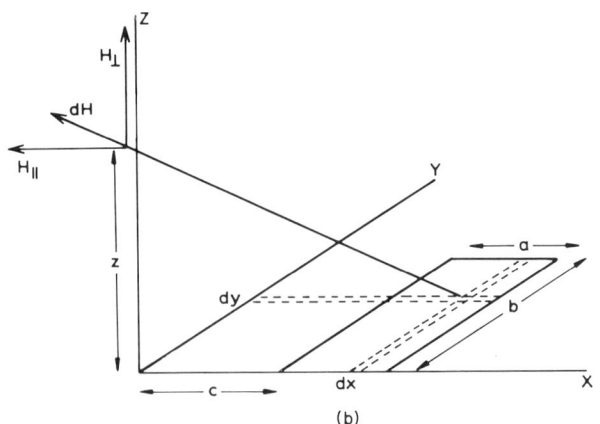

(b)

13.6. OPTICAL AND MAGNETOOPTICAL PROPERTIES

FIG. 12. (a) Electron micrograph of patterns on 180° domain wall intersections at a surface normal to M_s of barium ferrite. Note the formation of chains of particles normal to the wall intersections indicating that the field components parallel to the surface are of primary importance in pattern formation. (b) Diagram for computing stray field components from sheets of charge, with parameters relating to expressions in the text. (c) As for (a), with a field of approximately 500 Oe applied normal to the surface, the colloid particles aggregate in the centers of the domains and at the walls.

path of the ray itself. Light is said to be unpolarized if the vibrations have any orientation in the plane of the wave front, as is the case with light from all sources except the laser. If the vibrations lie in only one plane, the light is plane-polarized.

Birefringent crystals are those in which the velocity of propagation, and thus the refractive index, depends upon the orientation of the plane of polarization with respect to one or more crystallographic directions. For the sake of demonstration just one optic axis will be assumed to exist as in the classic example of calcite, which is strongly birefringent. Furthermore, while passing through a birefringent crystal the vibrations are confined to planes either normal or parallel to the axis. Thus an unpolarized beam incident at an angle to the optic axis can be treated as an "ordinary" ray with a plane of polarization normal to the axis and an "extraordinary" ray with plane of polarization parallel to the axis. The first has a refractive index w and the second a refractive index ε.

First suppose a crystal to be cut with its surfaces normal to the axis and consider a ray at normal incidence which is thus parallel to the axis (Fig. 13a). Whatever the state of polarization of the incident beam the vibrations are all normal to the axis and thus equivalent and the behavior is as for isotropic materials. Now suppose the surfaces are cut parallel to the axis and a normally incident beam is traveling in a direction normal to the axis. This will propagate as an ordinary and extraordinary ray, but these will be coincident (Fig. 13b). A plane-polarized ray is propagated as such if the plane of polarization coincides with the axis or is normal to the axis, but for incident light plane-polarized at an arbitrary angle to the axis two rays with different refractive indices will again be formed; again these will be coincident but due to the different velocities the two rays with orthogonal planes of polarization and initially in phase will emerge with a phase difference Δ. If the thickness and the difference in refractive indices $\varepsilon - w$, is such that $\Delta = \lambda$, the emergent ray will still be plane polarized with the original incident plane. If $\Delta = \lambda/2$, the emergent ray is plane-polarized but effectively rotated through $\pi/2$. If $\Delta = \lambda/4$, the two plane-polarized emergent rays can be seen to continue in such a way that the resultant describes a spiral with constant amplitude and this is known as *circularly polarized light*. For other values of Δ elliptically polarized light emerges, and the description of this can readily be imagined.

To a certain extent, i.e., strictly for monochromatic light due to the dispersive nature of the effects, the birefringence of a specimen can be compensated by a number of devices including the quartz wedge, which imposes a retardation on one of the plane-polarized rays according to the thickness inserted, and the mica quarter wave plate; a sheet of thickness 0.0184 mm giving a path difference of $\lambda/4$ for sodium light ($\lambda = 589$ mμ) and thus, for example, converting incident plane-polarized to circularly polarized light and vice versa. This will be seen to be of real practical value.

If a ray is incident to a birefringent crystal at an angle other than 0 or 90° to the axis, it is split into two rays with different paths and, in a suitably constructed double prism of calcite (a nicol prism), one of the rays can be totally reflected at the interface so that a single ray with plane polarization emerges. If a cone of light passes through a nicol, which is

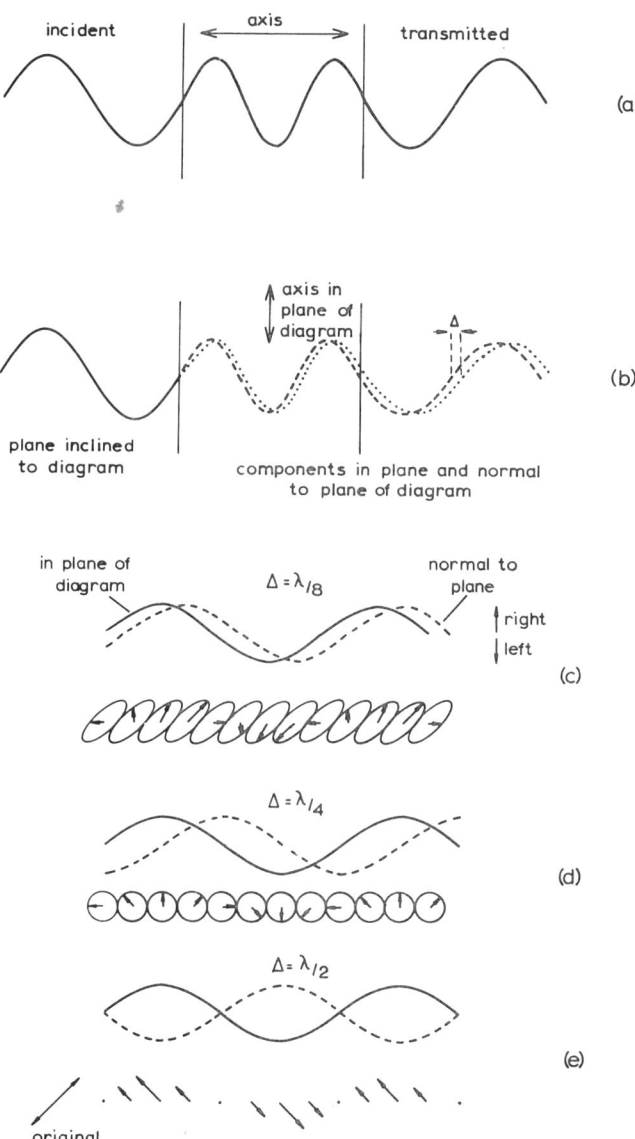

FIG. 13. Propagation of light in anisotropic (uniaxial) crystals: (a) crystal surfaces perpendicular to axis, ray parallel to axis, effective isotropic behavior; (b) surfaces parallel to axis, ray perpendicular to axis, giving two coincident internal rays which emerge with a phase difference $\varDelta$. The dashed and dotted lines represent plane-polarized vibrations in orthogonal planes (i.e., in the plane containing the crystal axis and normal to it). The emergent ray is generally elliptically polarized (c) but will be circularly polarized (d) if $\varDelta = (n/4)\lambda$ and plane-polarized (e) if $\varDelta = (n/2)\lambda$ with an effective rotation of the plane with respect to that of the incident light (n integral).

the configuration most economic of material, a small range of planes of polarization is produced instead of a single well-defined plane and the inclination of the end faces displaces the incident ray laterally. Both these disadvantages are overcome by the Glan–Thompson design, in which the end faces are normal to the axis of transmission and the optic axes are also normal to the incident ray. This design is more expensive and one pays heavily for large lateral dimensions, i.e., for large apertures. Wollaston prisms are also of value in magneto-optic studies. These are again double calcite prisms with the optic axis normal to the beam and normal to each other in the two halves: two superposed incident rays with differently oriented planes of polarization are refracted in different directions and thus separated, to fall on the two halves of a split photocell, for example.

The second practical method for producing plane-polarized light depends on dichroism, first studied in tourmaline crystals. This is the strong absorption of one of the two plane-polarized rays propagated, so that only plane-polarized light emerges if the thickness is adequate. The modern dichroic filters consist of assemblies of aligned microcrystallites, e.g., Polaroid, and in these only about 20% of the preferred polarization is absorbed so that about 30% of the incident light is transmitted. Nicols transmit about 86% of the preferred plane and have the higher extinction ratio of about 10^{-6}, if free of scratches and internal defects; the extinction ratio is the proportion of light transmitted through crossed polars, ideally zero.

Returning to the basic properties, crystals which exhibit mirror isomerism are optically active, i.e., have the property of rotating the plane of plane-polarized light throughout the crystal. It has been seen that birefringence can effect a rotation, but the phenomena are quite distinct since (a) the rotation is progressive throughout an optically active crystal and (ideally) plane-polarized light is always transmitted, (b) rotation is only effected for prescribed thicknesses of a birefringent crystal and this only occurs on recombination of the two rays, i.e., there is no rotation of a single plane-polarized ray within the crystal. Crystals may be both optically active and birefringent, in which case the activity may be studied without complications by using crystals cut normal to the optical axis (ray parallel to the axis) since these will show apparently isotropic behavior. Cubic crystals may be optically active but not birefringent (in principle every noncubic crystal should be birefringent), so in this case the beam may have any orientation. In the general case, plane-polarized light incident on an optically active birefringent crystal will be transmitted as a rotated and elliptically polarized ray and the results only analyzed with some difficulty.

Fresnel envisaged rotation as being due to the difference in the velocity of propagation of right- and left-handed circularly polarized rays, since a plane-polarized ray may be considered as the resultant of two oppositely rotating circularly polarized rays. Thus rotation can also be described as circular birefringence.

There are magnetic counterparts of both birefringence (i.e., magnetic birefringence) and optical activity (i.e., Faraday rotation). Magnetic birefringence is a difference in the refractive index for polarizations perpendicular and parallel to the direction of the spontaneous magnetization, and since $\mathbf{M}_s$ normally lies along a crystal axis this may be considered as an addition to crystal birefringence. It is important to appreciate that cubic crystals can be magnetically birefringent, since the symmetry is altered by the presence of $\mathbf{M}_s$ (i.e., within a domain there is a single axis specified by $\mathbf{M}_s$).

Faraday rotation is a rotation of the plane of polarization for rays propagated parallel to the magnetization, the sense of the rotation depending on the direction of propagation with respect to $\mathbf{M}_s$. There is thus a marked contrast with optical activity in that light passing through a crystal and being reflected back along its own path would undergo opposite senses of rotation as viewed along the magnetization axis and thus a "double rotation," whereas in the case of natural rotation both directions along the crystal axis are equivalent and the total rotation for the double passage would be zero. Faraday rotation was originally discovered in paramagnetics polarized by an applied field. The Fresnel picture has clear significance here since one sense of circular polarization coincides with the sense of the precession of the spins.

The Faraday rotation produced depends upon the thickness traversed and the Faraday constant F for the material, i.e., $\theta = Ft°$ compared with $\theta = KMt°$ for paramagnetics, where K is Kundt's constant. A similar law is often quoted for ferro- and ferrimagnetics, but F does *not* in fact depend on M_s for ordered materials in any direct way, but rather on the ordering itself; Faraday rotations can readily be observed while M_s passes through zero at compensation points of garnets and magnetooptical effects arise in antiferromagnetics. Clearly a high value of F is not a sufficient specification for producing a high rotation, since the material must be adequately transparent and a quality factor may be defined as $Q = F/\alpha$, where α represents the opacity $[I = I_0(1 - \varrho)^2 \exp(-\alpha t)$, where ϱ is the reflectivity]. Values of F and α are given in Table II.

TABLE II. Faraday Rotation and Absorption Coefficients[a]

Material	T (K)	(Å)	F (deg/cm)	α (cm^{-1})
Fe	300	5460	3.5×10^5	7.6×10^5
Co	300	5460	3.6×10^5	8.5×10^5
Ni	300	4000	7.2×10^5	2.1×10^5
MnBi	300	6300	1.4×10^5	3.4×10^6
YIG	300	12,000	2.4×10^2	0.069^b
	300	6000	1.9×10^3	1.2×10^{3b}
GdIG	300	5200	4×10^3	3×10^3
CoFe$_2$O$_4$	300	4000	3.7×10^4	1.7×10^5
EuO	60	12,000	2×10^5	10^2
	5	6600	5×10^5	5×10^5
EuSe	4.2	6600	2.6×10^5	1.7×10^2

[a] M. J. Freiser, *IEEE Trans. Magn.* **MAG-4**, 152 (1968); C. D. Mee and G. I. Fan, *ibid.* **MAG-3**, 72 (1967).

[b] Note the great dependence of α on wavelength: the "infrared window."

The applicability of Faraday rotation to the observation of domains should be apparent: if a specimen is placed between crossed polarizers and the analyzer rotated until extinction is restored for the light passing through one set of domains, the other domains will appear bright (Fig. 14). Due to the dispersion of F, different wavelengths will be more or less extinguished and thus domains with different directions of M_s will appear in different colors and this can be of real value. Domain observation associated with birefringence is also clearly feasible and will be described in the next section. A most important general distinction between the two is that the Faraday effect requires a component of $\mathbf{M}_s$ parallel to the light beam, while domains magnetized normal to the beam may give birefringent effects.

The state of polarization may also be affected by reflection and this must be taken into account in experimental configurations in which beams are turned through 90°, for example. At the so-called *Brewster angle of incidence* natural light is almost completely transformed into plane-polarized light, polarized perpendicular to the plane of incidence, while the refracted ray is polarized in the plane of incidence. If the incident beam is plane-polarized, it becomes elliptically polarized after reflection unless the plane of polarization is parallel or perpendicular to the plane of incidence. Kerr found in 1887 that, with this latter condition, the plane of polarization was rotated by reflection from a material with

13.6. OPTICAL AND MAGNETOOPTICAL PROPERTIES

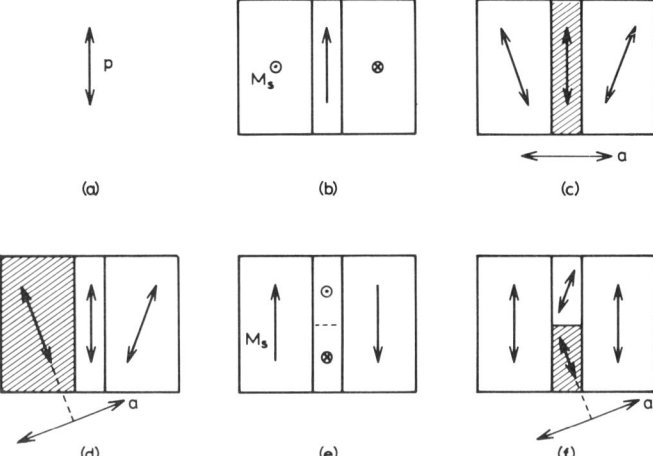

FIG. 14. (a) Plane of polarization of incident beam; (b) crystal containing two domains magnetized normal to the surface (the wall is simplified in structure and exagerated in width), (c) with analyzer crossed both domains appear lighter than the wall due to rotation away from extinction in the domains (with the analyzer parallel to polarizer the wall is lighter than the domains); (d) rotation of the analyzer to restore extinction of light through one domain also makes the other domain lighter; (e) crystal with domains in plane and reversal in the sense of rotation in the wall; (f) the appearance when the analyzer is rotated to give extinction for part of the wall.

a spontaneous magnetization, this again being of obvious significance here. Unfortunately for the terminology Kerr also found that the plane could be rotated by passing light through a dielectric across which an electrostatic field was applied; a quite different effect forming the basis of the Kerr cell.

Three Kerr magnetooptical effects may be distinguished. The first, for plane-polarized rays normal to the surface which is also normal to $\mathbf{M}_s$ (or has a component of $\mathbf{M}_s$ across it), is a rotation analogous to the Faraday effect in transmission (Fig. 15a). This is called the *polar effect*. The reflected beam becomes elliptically polarized as well as rotated. The transverse Kerr effect is the reflection counterpart of magnetic birefringence and is observed when $\mathbf{M}_s$ is parallel to the surface and normal to the plane of incidence of the light as shown in Fig. 15b. This effect has been used for magnetization measurements in which a laser beam, for example, is modulated on reflection from a specimen in which the magnetization is cycled.

The effect most generally used for domain observation is the longitudinal effect, since it applies when $\mathbf{M}_s$ lies in the surface (as is most

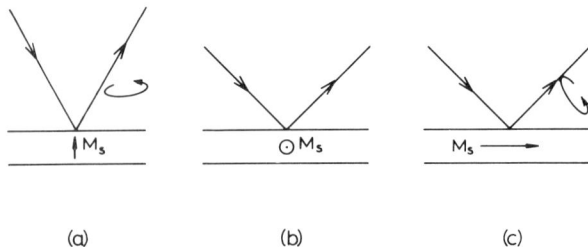

Fig. 15. (a) Polar Kerr effect, generally used at normal incidence; (b) transverse Kerr effect; (c) longitudinal Kerr effect of most general usefulness.

usual) and in this case is parallel to the plane of incidence as in Fig. 15c. This may again be described as the introduction of a small component normal to the incident plane of polarization introducing ellipticity and an effective rotation. Both the ellipticity and the rotation depend on the angle of incidence, as measured by Treves,[29] for example (Fig. 16), the rotation being zero at normal incidence for this effect. The rotation obtained is always low, about 10 min for nickel and 20 min for iron or cobalt but can be considerably enhanced, to about 1° for iron, by coating the surface with an antireflection dielectric layer such as ZnS or SiO.[30,31] According to Lissberger[32] this can give an optimum rotation magnification of 25. The low rotation calls for the use of Glan–Thompson prisms or good quality filters for both polarizer and analyzer.

The ellipticities are inhomogeneous in the usual case in which convergent light is used and the rays have a range of angles of incidence.

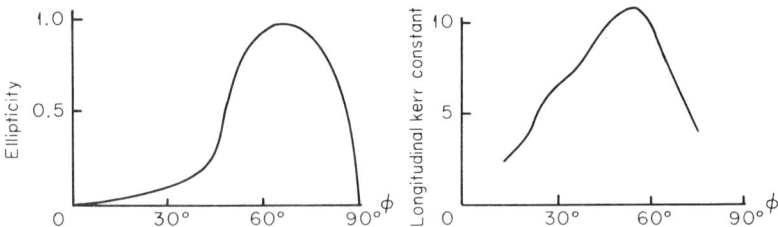

Fig. 16. The ellipticity and Kerr rotation as functions of the angle of incidence [D. Treves, J. Appl. Phys. **32**, 358 (1961)].

[29] D. Treves, J. Appl. Phys. **32**, 358 (1961).
[30] J. Kranz and W. Dreschel, Z. Phys. **150**, 632 (1958).
[31] A. Green, M. Prutton, and W. Carter, J. Sci. Instrum. **40**, 490 (1963).
[32] P. H. Lissberger, J. Opt. Soc. Amer. **51**, 948 (1961).

Thus they can be reduced but not eliminated by the use of a quarter wave plate, suitably oriented, and it follows that the greatest domain contrast is not necessarily obtained at the angle of incidence which gives the greatest rotation but by a compromise between maximum rotation and minimum ellipticity.

13.6.1. Faraday and Birefringence Methods

The application of the Faraday effect is basically very simple and may be achieved by a standard polarizing microscope or simple optical bench. The full resolution of the optical microscope may be used. The chief limitations are, of course, on specimen thickness and the necessity for a component of the magnetization parallel to the light beam; this latter, however, is more apparent than real. Typical thicknesses for reasonable transparency (visible range) are 100 μm for garnets or orthoferrites, 10 μm for spinel ferrites, and 0.1 μm for metals.

Figure 17a shows an optical bench including a heat filter and quarter wave plate and using an intense quartz-halogen lamp (or xenon arc) to aid the examination of thicker specimens.* Long working distance objectives are valuable to facilitate working at controlled temperatures for which the specimen may be enclosed in a gas tight enclosure through which heated or cooled gases may be circulated. Some heating and cooling may also be achieved by thermoelectric elements. Koehler illumination is valuable, particularly for small and rather thick specimens which must be precisely illuminated to prevent relatively intense light passing the specimen and being scattered into the viewing or measuring system. This is shown separately in Fig. 17b and described in the legend.

In practice, there is much to recommend a vertical or near-vertical system for the apparently trivial reason that the specimen does not then require any special mount. The specimens are usually very fragile and susceptible to strain and no adhesive system of mounting is acceptable; certainly nothing could be simpler than laying the specimen flat on a glass slide.

Garnets provide excellent examples of the technique. Diamond polishing usually introduces sufficient strain to define a strain-induced easy axis normal to the surface of the slice, whatever the crystal orientation.

---

* *Note added in proof*: G. Myers, in the author's laboratory has achieved outstanding results, with very high resolution and contrast, by using an argon ion laser for thin garnet specimens ("bubble domain" films).

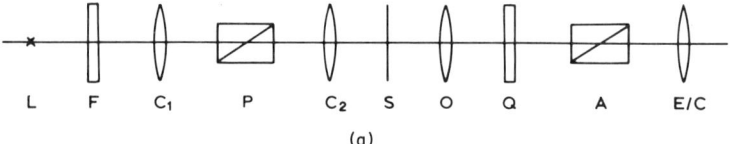

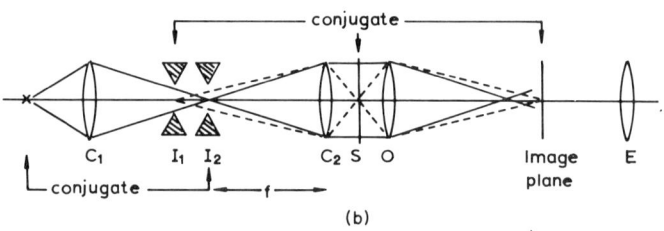

FIG. 17. (a) Components of Faraday effect equipment. L: lamp, F: filter (heat filter and/or monochromatic filter), C: condenser, P: polarizer, S: specimen, O: objective, A: analyzer, Q: quarter wave plate or compensator (quartz wedge), E/C: eyepiece or camera lens. Polaroid is better placed after C_2. (b) Koehler illumination. Since the iris I_1 is conjugate with the image plane it defines the illuminated area, i.e., its image is seen superimposed on the specimen. Since I_2 is conjugate with the filament it defines the effective filament area used, i.e., the intensity of the illumination.

FIG. 18 Strain domains in GdIG by the Faraday effect (a) domain contrast (cf., Fig. 14d); (b) wall contrast (cf. Fig. 14c). The structures are influenced by polishing scratches.

Since M_s is then parallel to the light beam, with the beam normally incident on the specimen surface, a very high contrast can be achieved as shown in Fig. 18a. If the analyzer is oriented so as to give equal light intensity in both sets of domains, then domain walls will present their own contrast, since, in the walls, M_s is oriented normal to the beam (Fig. 18b). When the strain is relieved the normal easy axes are occupied. This may be achieved by chemical polishing, ion polishing, annealing or by very fine mechanical polishing using a colloidal suspension. (Sometimes a combination of these may be required!) Figure 19a,[†] for example, shows a (211) slice of GdIG at room temperature, in which the stress has been partly relieved (cf., Fig. 18) and 19b shows a (211) slice of YIG (one easy axis in the plane) for which a largely strain-free condition is indicated by the obvious occupation of the normal easy axis. Contrast between in-plane 180° domains is now only obtained by tilting the specimen appropriately to give components parallel to the light beam, but the domains may be clearly outlined by the domain wall contrast: the light and dark segments are due to the periodic reversal of the sense of rotation and thus of the direction of M_s in the center of the wall. It is rather difficult to distinguish between strain due to the polishing and strain due to discontinuities occurring during growth of the crystals.

What cannot be shown here are the striking colors due to the dispersion of the Faraday rotation throughout the visible spectrum. These are of more than aesthetic value since, for any setting of the polarizer and analyzer, the directions of M_s corresponding to different colors can be determined by applying magnetic fields appropriately. For a (211) slice of GdIG cut into the shape of a rectangle and chemically polished to a strain-free condition, in-plane 180° domains were light and dark red in conditions for optimum contrast while closure domains were in two different shades of yellow and green.

It is important to remember, in relation to specimen thickness, that transmission may be greatly increased by departing from the visible region, most strikingly by utilizing the infrared "window" for garnets. This necessitates the use of infrared detectors or image converters whose spectral response may combine with the absorption curve to give a very restricted range of operation as indicated by Fig. 20 (R. D. Enoch and E. A. D. White, private communication). Enoch[33] has set up an infrared

---

[33] R. D. Enoch and R. M. Lambert, *J. Phys. E* **3**, 728 (1970).

[†] Crystals kindly supplied by Mullard Research Laboratories.

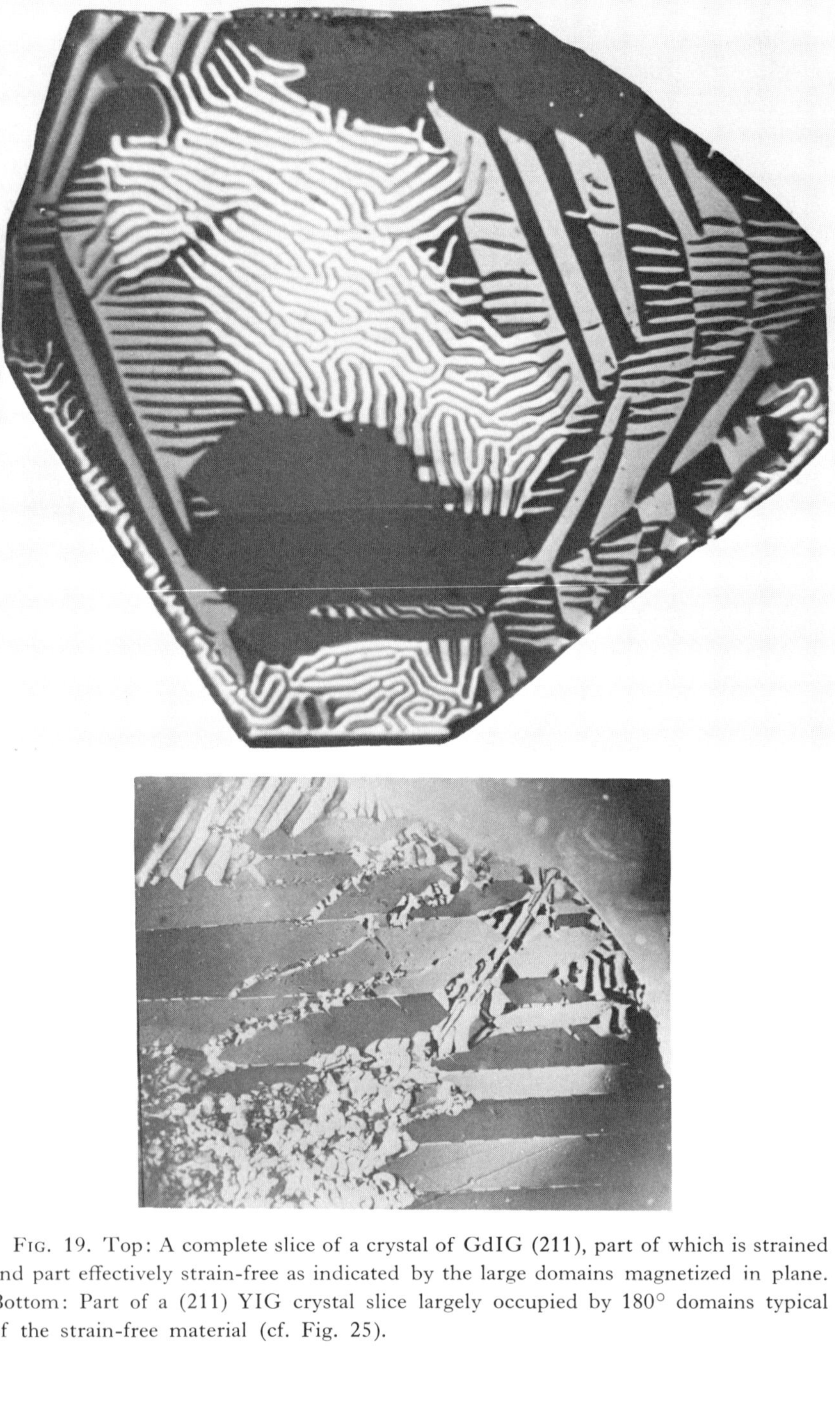

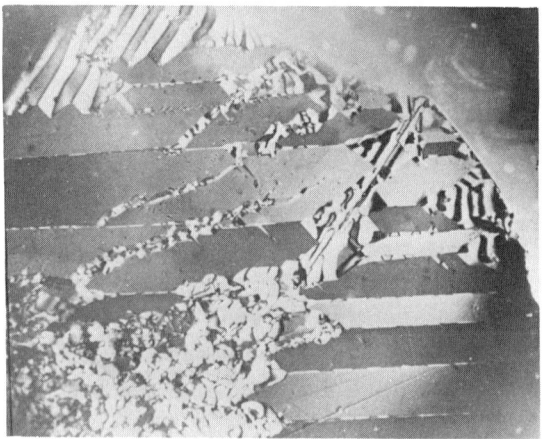

Fig. 19. Top: A complete slice of a crystal of GdIG (211), part of which is strained and part effectively strain-free as indicated by the large domains magnetized in plane. Bottom: Part of a (211) YIG crystal slice largely occupied by 180° domains typical of the strain-free material (cf. Fig. 25).

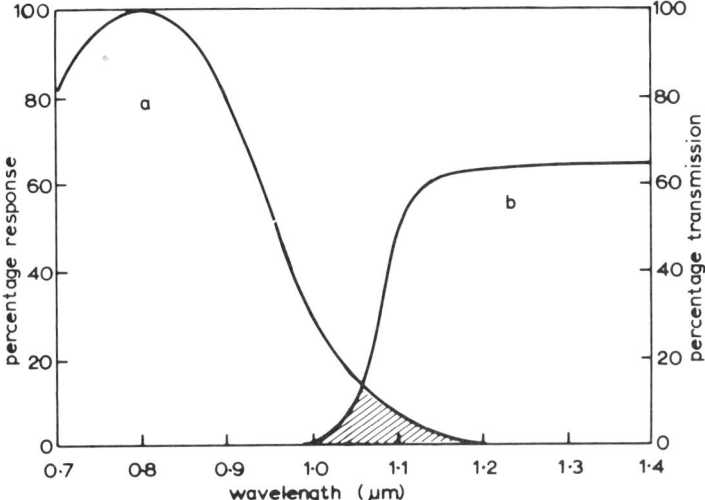

FIG. 20. Restriction of the region of operation of an infrared microscope by the combined effects of the relative response of the image converter (a) and the relative transmission of YIG (b) as functions of wavelength (R. D. Enoch and E. A. D. White).

system with television display which gives striking results for garnet crystals several millimeters thick. In particular, Fig. 21a and b compare the true internal structure of a crystal of YIG 3 mm thick and the impression given by powder patterns on the surface.

The power of the Faraday method for the study of thin evaporated films, of nickel–iron in particular, has been clearly demonstrated by Boersch and Lambeck;[34] special reference should be made to Lambeck's review.[35] In particular, it is noted that tilting a specimen with respect to the light beam to obtain components of M_s parallel to the beam need not lead to a limited region of sharp focus, so long as the specimen remains normal to the axis of the objective, which must have a large aperture (Fig. 22a). Placing the analyzer before the objective avoids the depolarizing effects due to transmission through the lens and the consequent restriction to polaroid rather than prisms is not of great consequence. Lambeck obtained the remarkable micrograph of cross-tie walls and magnetization ripple shown in Fig. 22b. It was also noted that when the analyzer was crossed with respect to the polarizer there would

[34] H. Boersch and M. Lambeck, Z. Phys. **177**, 157 (1964); M. Lambeck, Z. Angew. Phys. **15**, 272 (1963).
[35] M. Lambeck, IEEE Trans. Magn. **MAG-4**, 51 (1968).

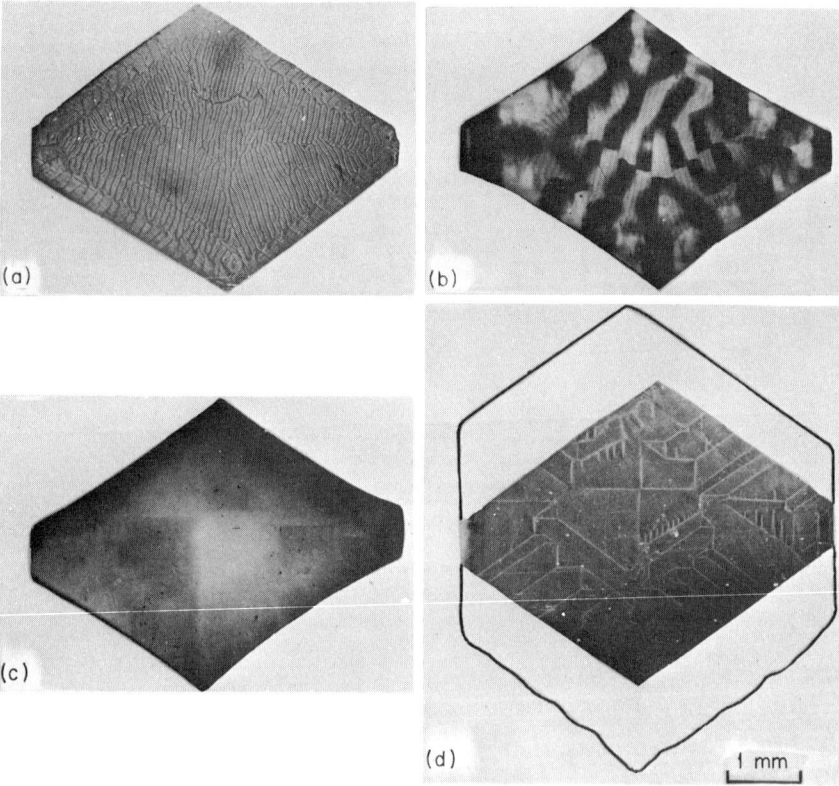

FIG. 21. (a) Powder patterns on the growth face of a thick (uncut) YIG crystal indicating a very fine domain spacing; (b) the large domains in the bulk of the crystal by Faraday microscopy in the infrared, confirming that the fine structures are associated with the growth surface itself; (c) the image obtained after cutting a slice parallel to growth face and finely polishing, with polars crossed and a quarter wave plate between specimen and analyzer to show domains parallel to surface by birefringence. In this case powder patterns (d) correspond to the internal structure and show the details, with the types of wall expected for $-ve\ K_1$. The outline shows the shape of the cut face (R. D. Enoch, Private Communication).

of course be no contrast between the domains but they would still act as a 180° phase object. The reason for this is that since both rays are originally in phase, after rotation is opposite senses the components transmitted by the analyzer have opposite phase (Fig. 23a). The phase difference leads to Fresnel diffraction giving fringes at the domain boundaries as in Fig. 23b, if the aperture of the illumination is sufficiently low. Also, Fraunhofer diffraction spots are formed in the focal plane of

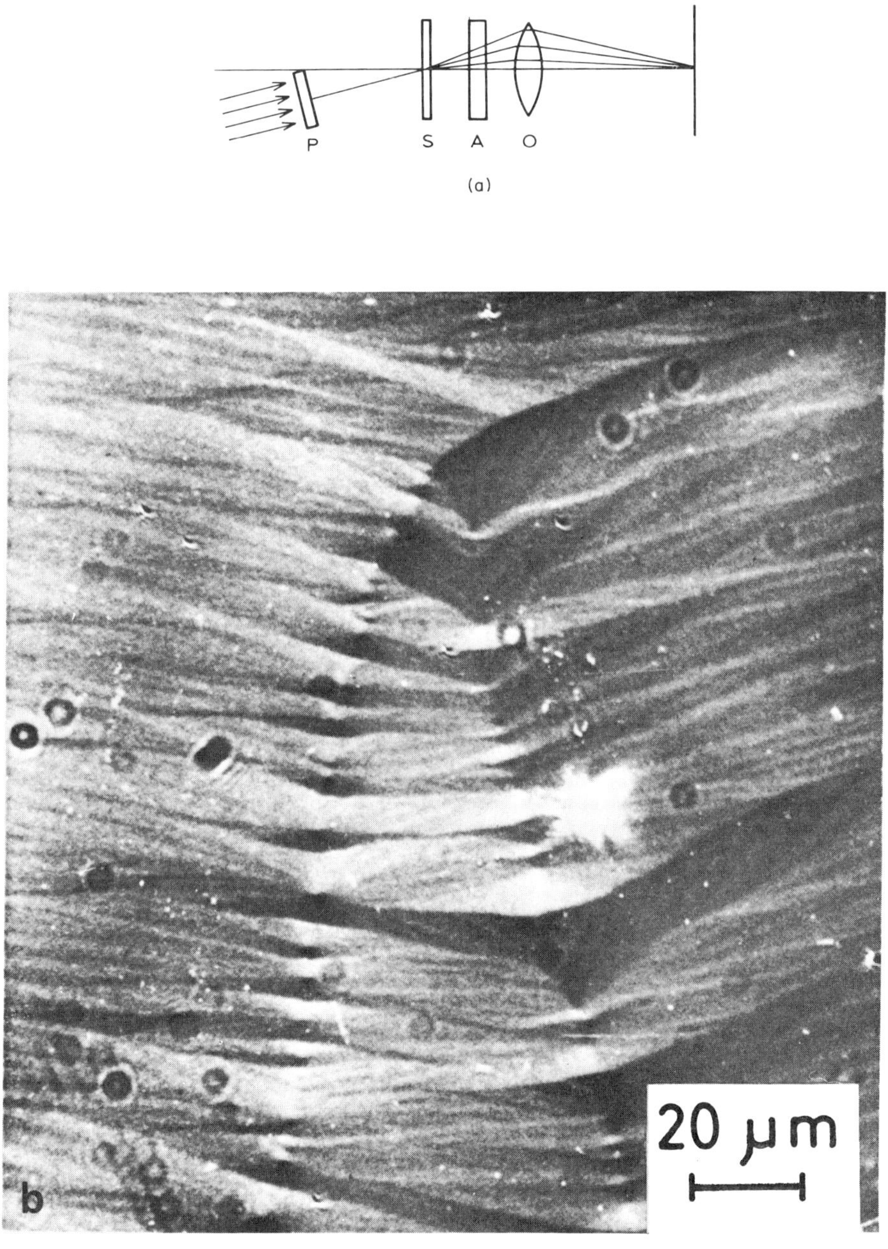

13.6. OPTICAL AND MAGNETOOPTICAL PROPERTIES

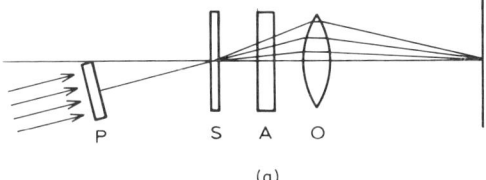

(a)

FIG. 22. (a) Lambeck's arrangement for which all the specimen is in focus (normal to axis of objective) but still gives Faraday contrast with in-plane domains. (b) High-resolution micrograph of cross-tie walls and magnetization ripple in an iron film obtained by this equipment.

714   13. THE OBSERVATION OF MAGNETIC DOMAINS

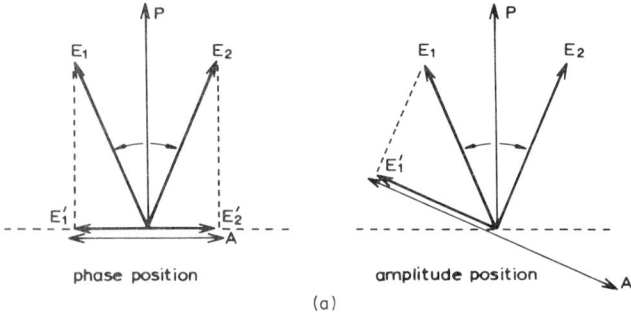

(a)

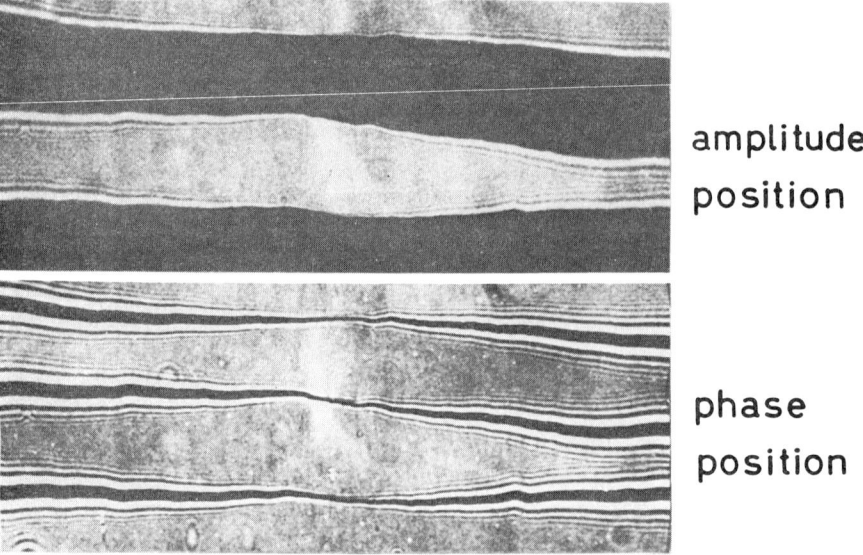

Fresnel diffraction
(b)

13.6. OPTICAL AND MAGNETOOPTICAL PROPERTIES

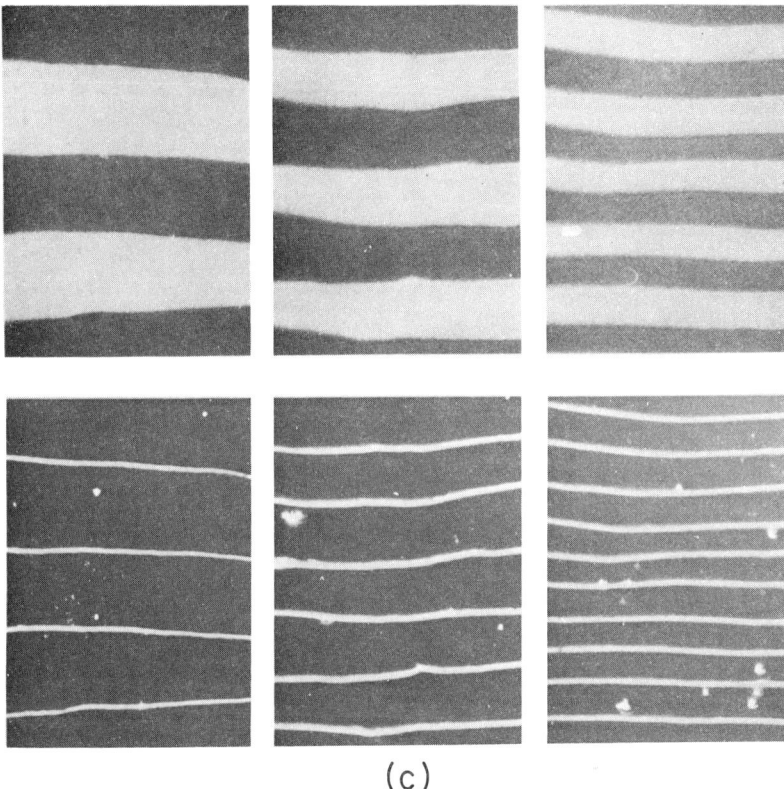

(c)

FIG. 23. When an analyzer is set crossed to a polarizer (a) the components of the light (electric vectors) E_1' and E_2' which are transmitted after rotation by passage through antiparallel domains are equal in amplitude but opposite in phase, thus this may be called the *phase position* as compared to the conventional *amplitude position* giving contrast between the domains as illustrated by (b) for an iron film [M. Lambeck, Z. Angew. Phys. **15**, 272 (1963)]. (c) Domain walls in an iron film revealed by Fraunhoffer diffraction (dark field), as opposed to the Fresnel diffraction in (b) [M. Lambeck, Z. Angew. Phys. **15**, 272 (1963)].

the objective (cf., section 13.7) and if the zero-order spot is stopped out dark field images can be obtained; the walls show up as bright lines (Fig. 23c) and this effect must be distinguished from the sources of wall contrast described earlier.

13.6.2. Magnetic Birefringence

Birefringence may be considered either as a complication of Faraday observations or a method of portraying domains as such. Ytterbium or

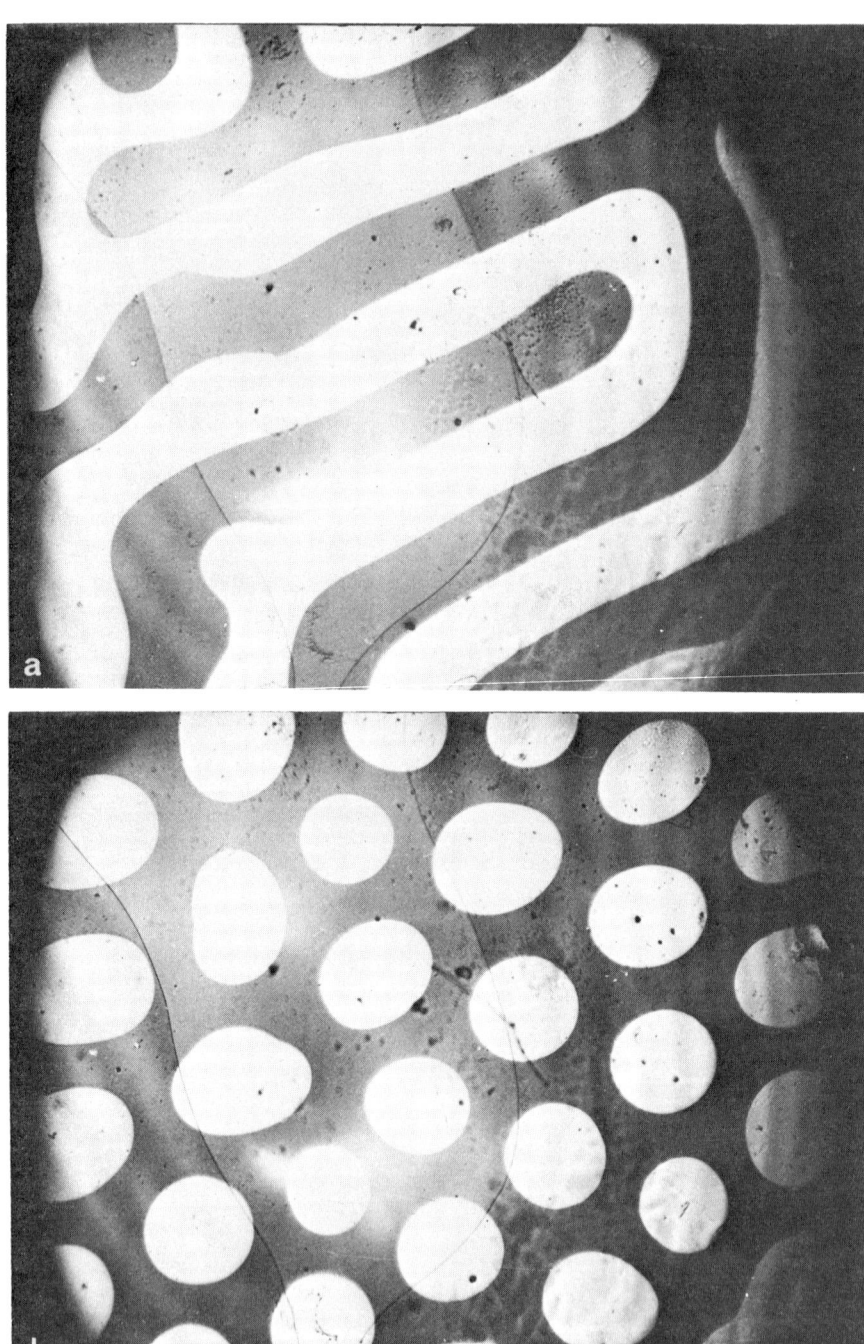

FIG. 24. 180° domains in YbFeO$_3$ with strong Faraday contrast which is only obtained by rotating the specimen and/or a quarter wave plate to an optimum orientation with respect to the plane of polarization of the incident beam: (a) strip domains in zero field; (b) and (c) cylindrical or "bubble" domains in zero field and with 40 Oe applied in the

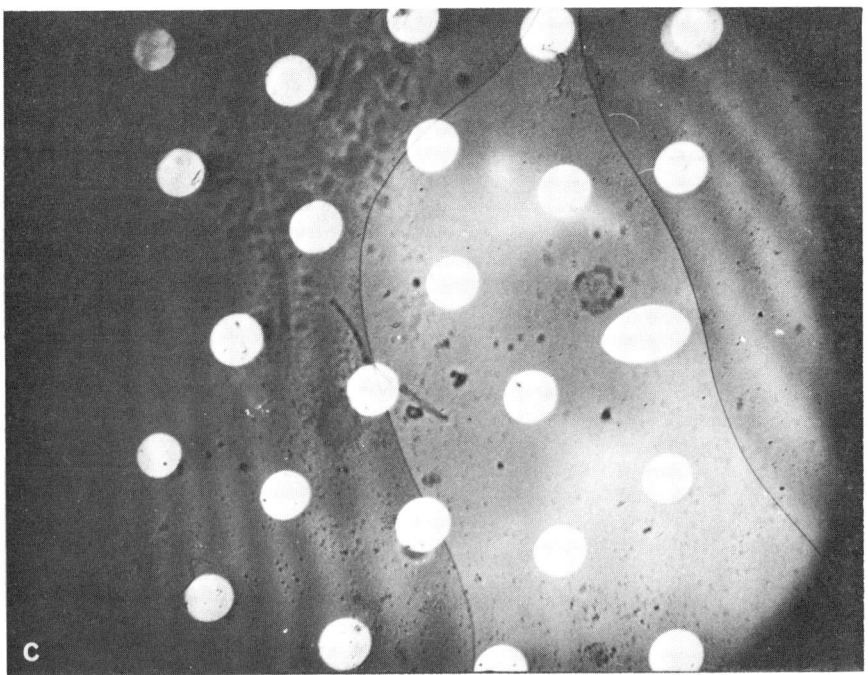

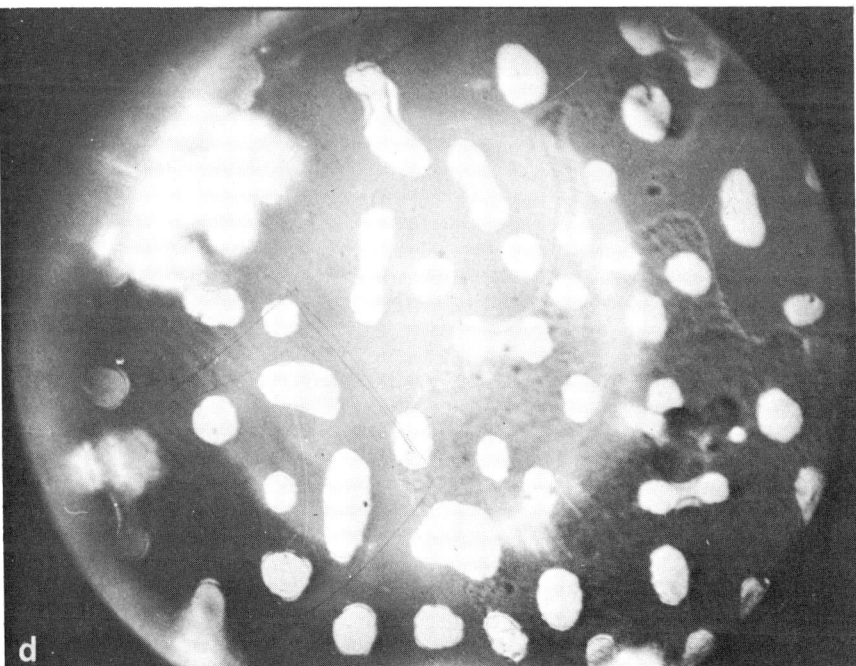

easy direction (parallel to the beam). The regularity of the structures and their rearrangement in applied fields is indicative of good sample preparation (measured $H_c < 1$ Oe) in contrast to (d), which represents an intermediate stage of polishing. Note that a thin layer of grease, used to secure the specimen affected the contrast and this should be avoided.

rare earth orthoferrites are strongly uniaxial and quite transparent at about 100-μm thickness. Plates cut normal to the c axis should thus give high contrast, with $\mathbf{M}_s$ parallel to the light beam, but using simply a polarizer and analyzer it is usually impossible to approach extinction for either set of domains and thus to obtain any great contrast. The lack of extinction clearly indicates the destruction of the plane-polarized state by strong birefringence. This may be compensated by a quarter wave plate in a rotatable mount which, as indicated in the preceding section, superimposes a phase difference, according to its orientation, which compensates for that giving the ellipticity; in practice this condition is readily found by trial. If the polarizers were set closest to extinction for one set of domains before inserting the quarter wave plate then a further rotation of the analyzer is needed. Figure 24 shows 180° domains in ytterbium orthoferrite viewed with the use of a quarter wave plate. A second consequence of the birefringence is the dependence of the contrast obtained, without a compensator, on the orientation of the plane of polarization of the incident beam with respect to the b and c axes of the plate. For a purely uniaxial crystal rotations around the axis of the light beam could have no effect.

Figure 25 shows domains in the plane of a garnet plate which is normal to the light beam, with the polarizer and analyzer crossed. The contrast is supplied by the birefringence itself, since there can be no Faraday rotation with this arrangement. Also it is noted that the light and dark shades cannot be interchanged by rotating the analyzer, as for the Faraday effect (assuming that there might be a small component of $\mathbf{M}_s$ along the beam), but light and dark *are* interchanged by rotating the specimen. This is an excellent demonstration of magnetic birefringence since the crystal structure is cubic and only the magnetization gives the anisotropy. Such effects were first noted by Dillon[36] and Tabor and Chen[37] have given quantitative analyses for crystals which are both "magnetically active" and birefringent.

13.6.3. The Polar Kerr Method

The polar Kerr method involves apparently formidable difficulties, since the incident and reflected rays coincide. However these have been long recognized and overcome in standard metallurgical microscopes of

---

[36] J. F. Dillon, *J. Appl. Phys.* **29**, 1286 (1958).
[37] W. J. Tabor and F. S. Chen, *J. Appl. Phys.* **40**, 2760 (1969).

FIG. 25. A single crystal disc of YIG observed between crossed polarizer and analyzer. There are two easy axes in the plane of the disc (110) and the substantial occupation of these, giving 180, 110, and 70° domain walls, indicates that the specimen is largely strain-free although prepared by careful mechanical polishing only. Most of the contrast is due to magnetic birefringence; the darker and lighter areas interchange on rotating the specimen and there is no contrast between 180° domains. Faraday contrast occurs *within* the 180° walls where M_s is normal to the surface and between the irregularly shaped strain domains near to the edges showing that they are also magnetized normal to the surface and parallel to the light beam. Note the differing senses of rotation in different 180° walls, and the periodic changes of sense for some walls. (Presumably the polarizer and analyzer were not *perfectly* crossed).

the Bausch and Lomb type as originally used by Williams et al.[38] for domain studies. The essential feature is the Foster prism.[39] As seen from Fig. 26 this acts as both polarizer for the incident ray and permanently crossed analyzer for the reflected ray.

[38] H. J. Williams, F. G. Foster, and E. A. Wood, *Phys. Rev.* **82**, 119 (1951).
[39] L. V. Foster, *J. Opt. Soc. Amer.* **28**, 124 (1938).

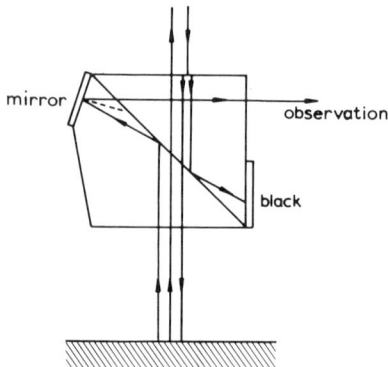

FIG. 26. Foster prism as used for polar Kerr studies [L. V. Foster, *J. Opt. Soc. Amer.* **28**, 124 (1938)]. The ordinary ray is reflected at the cemented interface and absorbed at the blackened surface while the extraordinary ray is transmitted, since the refractive index of the cement matches that of the crystal for this ray. If there is no rotation, the ray reflected from the specimen passes back through the prism. Rotation at the specimen introduces an ordinary component in the prism which is reflected as shown to the eyepiece or camera.

With an inactive specimen the resulting transmission would be negligible (extinction). With rays rotated equally in opposite senses by "up" and "down" domains there would be more transmission but no contrast. However, if an extraneous rotation is introduced by a compensator such that the reflected rays from the "up" domains are brought back to extinction, the "down" domains will give rays off extinction and appear bright and vice versa. The microscope mentioned achieves this rotation by a quarter wave plate but it could also be achieved by a Faraday compensator as used by Fowler et al.[40] for measurements rather than observations. The latter also used a post analyzer (Glan–Thompson prism) to improve the sensitivity of the measurements of rotation; whether this would be advantageous for domain structure observations might be investigated.

The ellipticities induced at normal incidence are small and means for their compensation are not strictly necessary.

Polar rotations are higher than those obtained by the longitudinal Kerr effect (without using dielectric layers) and there are no limitations on objective aperture so the full resolution of the optical microscope may be achieved as evinced by Hubert's micrograph in Fig. 9b. (Figure 9a,

[40] C. A. Fowler, E. M. Fryer, B. L. Brandt, and R. A. Isaacson, *J. Appl. Phys.* **34**, 2064 (1963).

which has already been referred to, shows the same structure by the method of powder deposition in an inert gas and stresses the difference between stray field and more direct methods; they should be considered as complementary rather than contradictory.) Good results may also be obtained on MnBi and magnetoplumbite, for example, but cases in which a large component of the magnetization occurs across a surface must always be considered as exceptional due to the magnetostatic energies involved, and this is the greatest limitation on the scope of the method.*

The studies by Fowler et al. of domains in orthoferrites[40] are doubly interesting since few reflection studies have been made on oxide materials and, since the rotations obtained vary little over about 20° from normal incidence, they realized that polar observations could in fact be made with the type of equipment generally used for the longitudinal method. Values of the rotation at normal incidence were found by the above technique (Faraday compensation) to vary smoothly from 1 min at $\lambda \approx 6000$ Å to 5 min at $\lambda \approx 4500$ Å for both $YFeO_3$ and $HoFeO_3$. The relative rotation for antiparallel domains is double this value.

13.6.4. Longitudinal Kerr Method

The longitudinal Kerr effect gives an effective rotation due to the addition of a "Kerr component" of polarization normal to the original plane whether this is in or normal to the plane of incidence, which can be used to detect domains magnetized in the plane of incidence.[41] The inherent contrast is much lower than that obtained by the Faraday effect, with the more suitable samples at least, and Fowler and Fryer[42] devised a double photographic technique for the reduction of noise. The superimposition of a negative of the domain structure with a reversed negative of the saturated surface prior to printing cuts out extraneous light due to surface asperities, etc. This is not really necessary, however, if the surfaces are of a good optical quality (as is automatically obtained with evaporated metal films but otherwise requires very careful techniques) and particularly if the surfaces are bloomed.

[41] C. A. Fowler and E. M. Fryer, *Phys. Rev.* **86**, 426 (1952).
[42] C. A. Fowler and E. M. Fryer, *J. Opt. Soc. Amer.* **44**, 256 (1954).

---

* The method is clearly suitable for materials with high values of the ratio K/M_s^2 and very striking results have been obtained on the new rare-earth-cobalt permanent magnet materials.

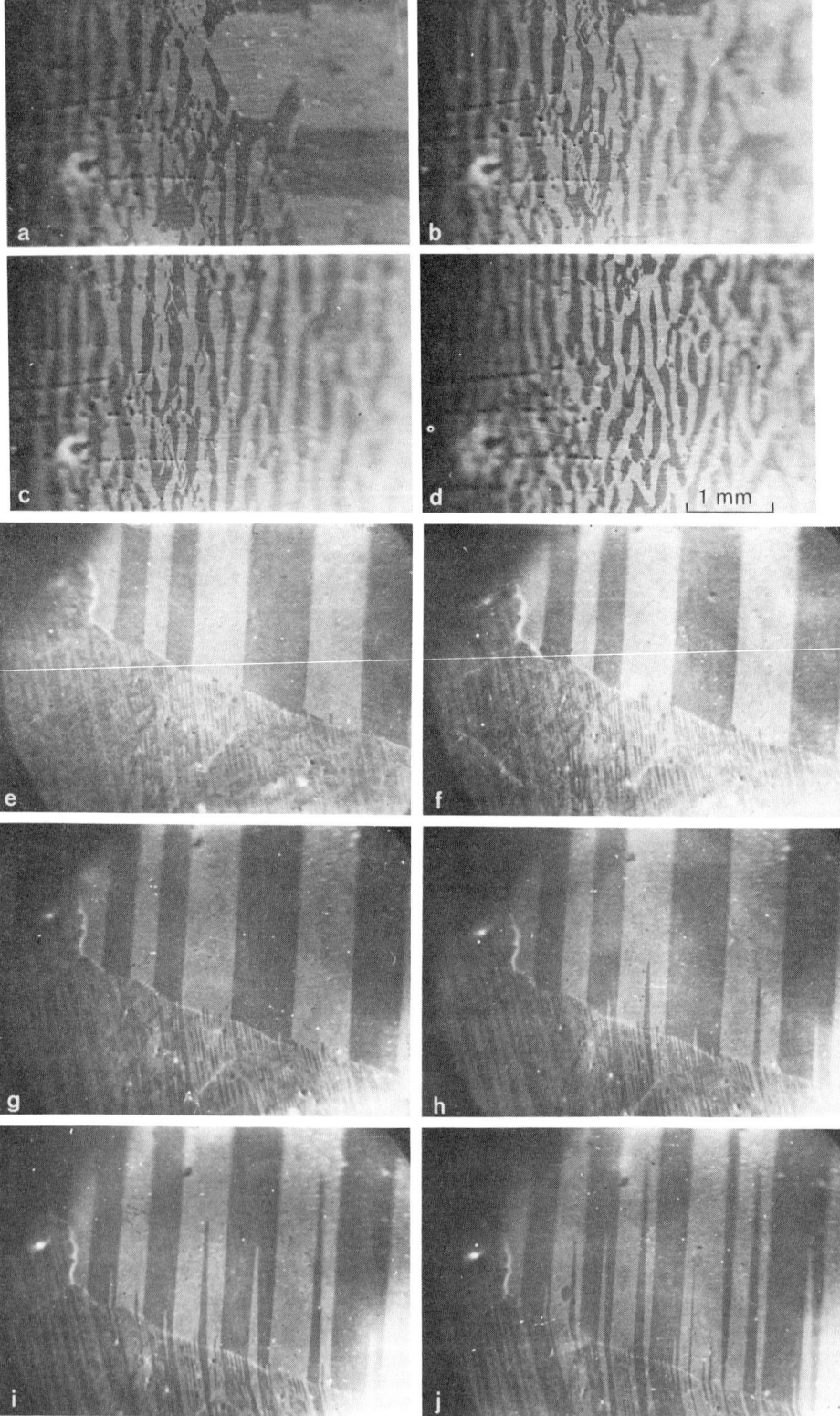

13.6. OPTICAL AND MAGNETOOPTICAL PROPERTIES

Blooming, or coating with a dielectric layer such as ZnS or SiO to reduce the reflectance, is of enormous practical value, as already noted.[30-32] For most purposes the coating need not be very strictly controlled. Having evaporated a cobalt film onto a 3-in slide, a very nonuniform coating was formed by evaporating SiO from a tungsten basket placed near to one end of the slide and, although the interference colors varied through a wide range, more than adequate contrast was always obtained from the coated film although the uncoated film gave very little contrast.

Figures 27a–d show 180° domains in grain-oriented 3% SiFe which break up into more complex structures under the influence of stress.[43] This is chosen to emphasize the relative simplicity of the method since the quality is quite adequate to represent this process, although the particular equipment used was very simply set up. It is also a particularly suitable application for the Kerr method since the process can be studied repetitively and, indeed, powder-pattern studies gave a fair representation of the initial and final structures but not of the transition. A second appropriate application (for any magnetooptical method) is to magnetization dynamics as illustrated by Fig. 28. Houze has shown in this way that attempts to explain 50-cycle losses, for example, in terms of static domain studies are quite inappropriate, since the number of domains is a function of frequency. Both the Kerr and Faraday effects may be used for measurements of domain wall velocity.[44]

Micrographs such as Fig. 27 obtained with a very simple arrangement can only be in precise focus along one line and efforts to overcome this by approaching normal incidence would lead to greatly reduced contrast. Stein and Feldtkeller[45] demonstrated the great value of the simple procedure of rotating the objective lens with respect to the film plane with regard to Scheimpflugchen's rule, that the object plane, lens plane, and

---

[43] D. J. Craik and R. J. Fairholme, *J. Phys. (Paris)* **32**, 46 (1971).

[44] F. C. Rossol, *J. Appl. Phys.* **40**, 1082 (1969); F. C. Rossol and A. A. Thiele, *ibid.* **41**, 1163 (1970).

[45] K. U. Stein and E. Feldtkeller, *Z. Angew. Phys.* **23**, 100 (1967).

FIG. 27. Domains in singly oriented silicon iron with a carefully electropolished and bloomed surface, demonstrating the effects of stress. (a)–(d) shows the progressive destruction of 180° domains due to compression along the easy axis initially occupied, and (e)–(j) show rotation toward the strip axis of M_s in a poorly oriented crystallite, and consequent modification of the structure in the well-oriented crystallite, due to tension up to 3.9 kg/mm². The processes can be followed repeatedly by this method.

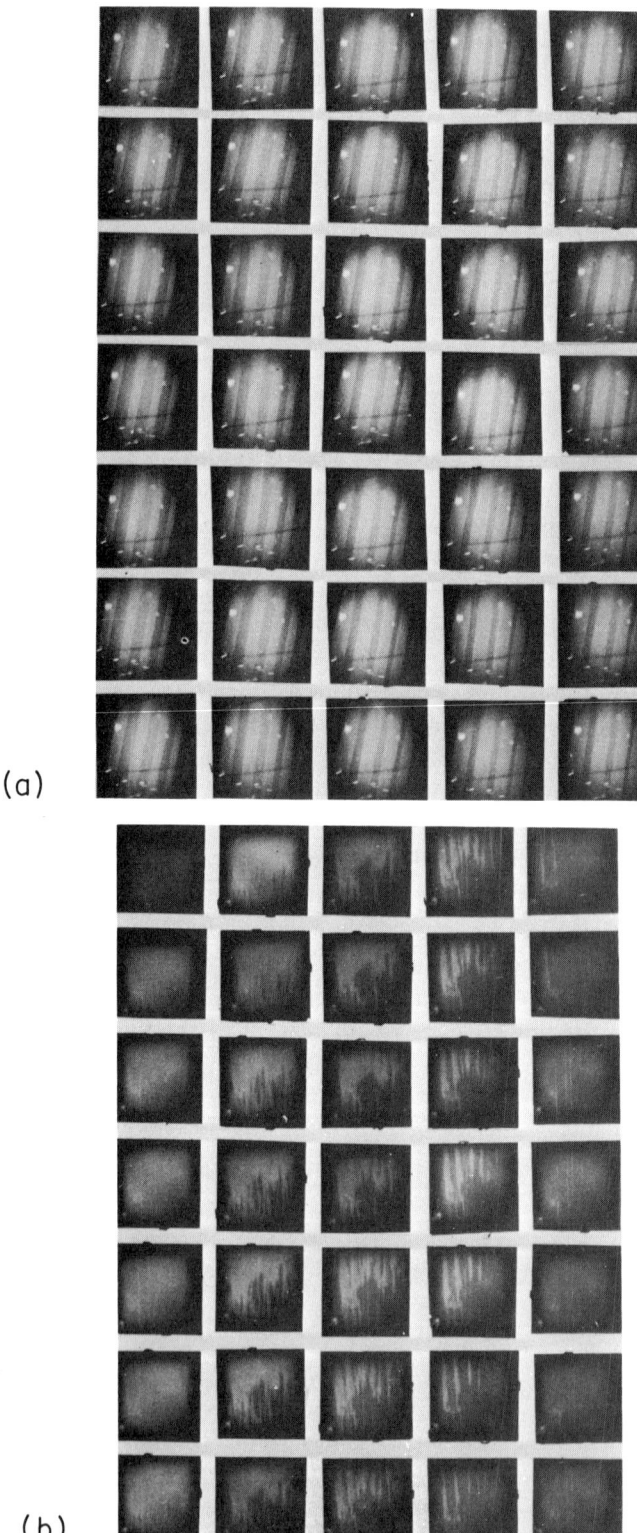

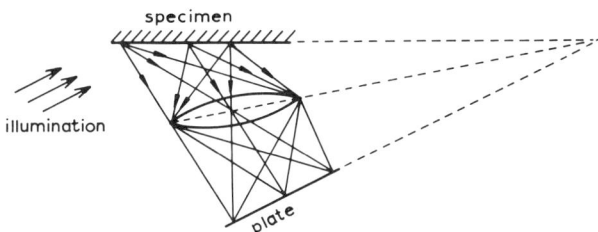

FIG. 29. Tilted lens arrangement by K. U. Stein and E. Feldtkeller [*Z. Angew. Phys.* **23**, 100 (1967)]. The specimen, lens, and film planes intersect in a common line.

image plane should intersect in a common line. This is clarified by Fig. 29, the camera lens acting as objective and further magnification being achieved photographically (resolution about 10^{-3} cm).

High resolution ($\sim 1.7 \times 10^{-4}$ cm) and moderate contrast is obtained by using a microscope in which the incident and reflected light beams are normal to the specimen surface, the angles of incidence of the rays depending on the power of the lens, which is used as both condenser and objective, and being controlled by an off-center aperture placed behind this lens, as by Kranz and Hubert[46] or Green and Prutton.[47] A conventional microscope may be modified for this and the whole field of view is, of course, focused, but the intensity is really low. The limiting resolution corresponds to the limits placed on the numerical aperture which can be used in practice.

The ellipticities introduced by metallic reflection and by the Kerr effect itself must affect the contrast. They may be reduced by using a quarter wave plate, but not eliminated since the ellipticities are inhomogeneous when the rays make a range of angles with the surface (particularly when large aperture lenses are used). For most studies quarter wave plates are not essential.

[46] J. Kranz and A. Hubert, *Z. Angew. Phys.* **15**, 220 (1963).
[47] A. Green and M. Prutton, *J. Sci. Instrum.* **39**, 244 (1962); A. Green, M. Prutton, and W. S. Carter, *ibid.* **40**, 490 (1963).

FIG. 28. Dynamic domain structures in a crystallite of (110) (001) textured 3% SiFe, frozen by high-speed photography using a xenon flash tube, at different stages in a magnetization cycle, starting upper left and in a vertical sequence: Kerr effect on a bloomed specimen with approximately 2×10^{-4} sec between frames. With 60-Hz triangular excitation, peak $H = 1.7$ Oe, the structure (a) resembles the 180° domains observed statically but with 60-Hz square wave and $H_p = 33.7$ Oe, (b) many more irregular domains form (G. L. Houze, Jr., *J. Appl. Phys.* **38**, 1089 (1967)].

13.6.5. Transverse Kerr Method

If one considers the specimen to be rotated in equipment aligned for the longitudinal Kerr method, the contrast (i.e., the rotation) gradually decreases to zero for a 90° rotation with no component of $\mathbf{M}_s$ in the plane of incidence. There remains, however, a difference in reflectivity from different domains due to the transverse Kerr effect, so long as the plane of polarization of the incident light is in the plane of incidence.

If the prepared surface of a specimen (or the surface of a film as grown) is illuminated by a laser or other intense polarized beam,[48] the modulation of the intensity of the reflected beam when the magnetization is cycled may be adequate for the measurement of magnetization loops. However, the inherent contrast is very low and can only be improved to the level obtained by the longitudinal method with uncoated specimens by incorporating phase correction by a second reflection from a surface of the same material,[49] with rather difficult alignment problems. Moreover the difference in intensities is expected to be independent of the reflectance so that coating the specimen does not enhance the contrast for the transverse effect. Thus, although Dove, for example,[49] obtained clear patterns on nickel–iron films these are probably better considered as a confirmation of the magnetooptic principles than as illustrations of a method having any special advantages.

13.7. Lorentz (Electron) Microscopy

The Lorentz force on an electron passing with velocity $\mathbf{v}$ through a material with induction $\mathbf{B}$ is (in dynes, with e in esu, etc.)

$$\mathbf{F} = -(e/c)\mathbf{v} \times \mathbf{B}. \tag{13.7.1}$$

In specimens which must be about 1000 Å or less thick for penetration by electron beams with acceleration potentials of $\sim$100 kV, the magnetization will lie in or near the surface, except for materials with high anisotropy fields. The angular deviation for a specimen with thickness d is then

$$\phi = \frac{4\pi e d M_s}{mcv} = \frac{4\pi d}{c}\left(\frac{e}{2Vm}\right)^{1/2} M_s \tag{13.7.2}$$

---

[48] A. Green and B. W. J. Thomas, *J. Sci. Instrum.* **43**, 399 (1966); R. Carey, E. D. Isaac, and B. W. J. Thomas, *J. Phys.* **D1**, 656 (1968).

[49] D. B. Dove, *J. Appl. Phys.* **34**, 2067 (1963).

which is around 10^{-4} rad for $V = 100$ kV and $d = 1000$ Å; M_s as for iron. This is adequate to give an effective magnetic contrast with suitable operation since, in order to control spherical aberration, electron microscopes always operate at a very low aperture, the optimum being $\sim 10^{-3}$ rad for 100-kV electrons. It will be recalled that contrast is obtained from crystalline specimens by cutting out diffracted beams using an aperture, in the back focal plane of the objective lens, which passes the direct beam only.

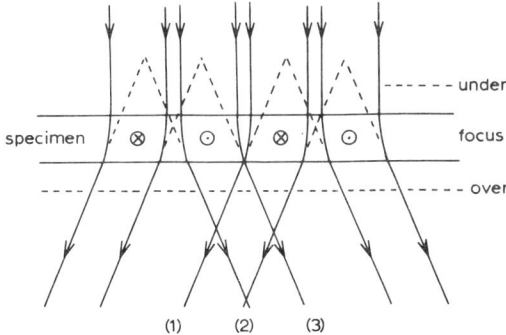

Fig. 30. Trajectories of electrons passing through a foil containing 180° domains. When over-focused, boundaries (1) and (3) appear dark (depletion) and (2) appears light (concentration). When under-focused more electrons appear to originate from walls (1) and (3) which appear light and so forth. The foil thickness is greatly exaggerated with respect to the domain spacing.

Replacing B by $4\pi M_s$, in the above treatment, amounts to the neglect of the contribution of stray fields in comparison with that of the magnetization. In principle, stray fields of the order of πM_s are to be expected in the neighborhood of discontinuities, but in practice it does seem that the effects observed can usually be ascribed wholly to the magnetization distribution and the stray fields are generally neglected.

The deflection of the electrons can be utilized in two ways. Referring to Fig. 30, it is apparent that no contrast at all is obtained when the specimen is focused by the objective lens, but the 180° domain walls appear as alternately dark and light lines on defocusing (i.e., focusing above or below the plane of the specimen) due to the overlap of the trajectories. For any one wall the contrast reverses on going through focus. This is known as the *Fresnel effect*, and is illustrated by Fig. 31, which shows 180 and 90° domain walls in a thin iron crystal with (100) orientation.[50]

[50] G. A. Jones, Private communication.

728   13. THE OBSERVATION OF MAGNETIC DOMAINS

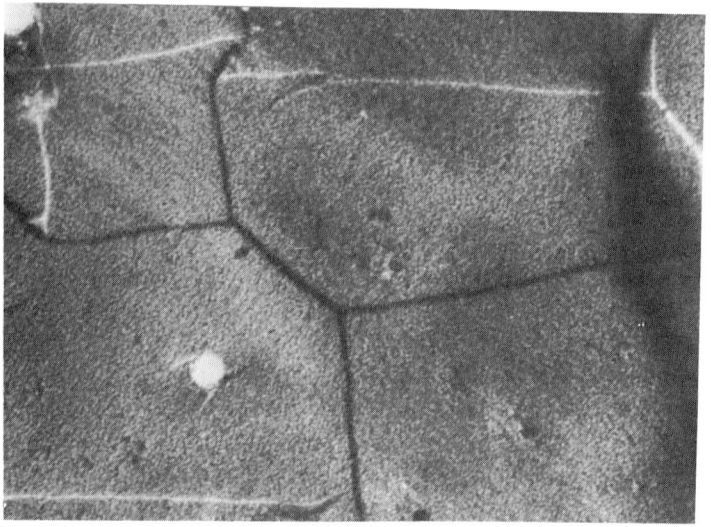

Fig. 31. Fresnel electron microscopy of an iron crystal with (100) surface, showing 90 and 180° walls (G. A. Jones).

In Fig. 30 the illumination is indicated to be parallel, and indeed this is the ideal for optimum resolution (minimum wall width in the image, as distinct from the actual wall width). The beam divergence should be less than the Lorentz deflection and with two condenser lenses 10^{-6} rad can in fact be achieved (cp 10^{-4}, above). However, for maximum brightness the condenser lenses are usually focused to give a strongly convergent beam, and the above condition thus corresponds to defocusing the condenser as compared to normal operation. This is quite a serious restriction when the difficulties of obtaining really thin specimens with good image brightness is recalled, but the condition is by no means critical and a reasonable compromise can be achieved. (In fact, in certain specimens the structures can scarcely be missed, and it is rather difficult to understand why the method was not discovered by accident during the study of iron specimens in the short specimen holders required for large-angle tilting.)

The above method was introduced by Hale et al.[51] and a second "Foucault" method was given independently by Fuller and Hale[52] and Boersch and Baith.[53] Figure 32 shows how an objective aperture may

[51] M. E. Hale, H. W. Fuller, and H. Rubinstein, J. Appl. Phys. **30**, 789 (1959).
[52] H. W. Fuller and M. E. Hale, J. Appl. Phys. **31**, 1699 (1960).
[53] H. Boersch and H. Raith, Naturwissenschaften **46**, 574 (1959).

13.7. LORENTZ (ELECTRON) MICROSCOPY

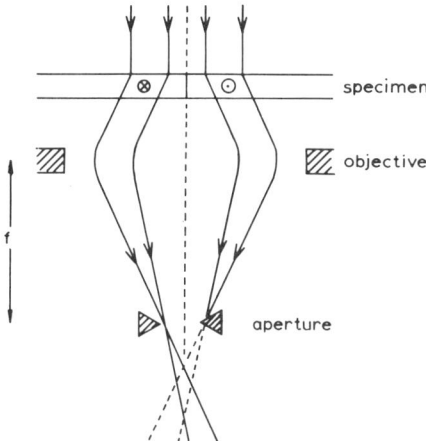

Fig. 32. The formation of two spots in the back focal plane of the objective lens, one of which may be cut out by displacing the objective aperture to give Foucault contrast.

be displaced to block one of the two beams produced by a set of 180° domains; if all four easy directions are occupied, four spots are produced. In practice this means that the microscope is put in the diffraction mode of operation, so that the diffraction spots themselves appear on the screen with a small splitting of the central spot as shown in Fig. 33a. (The

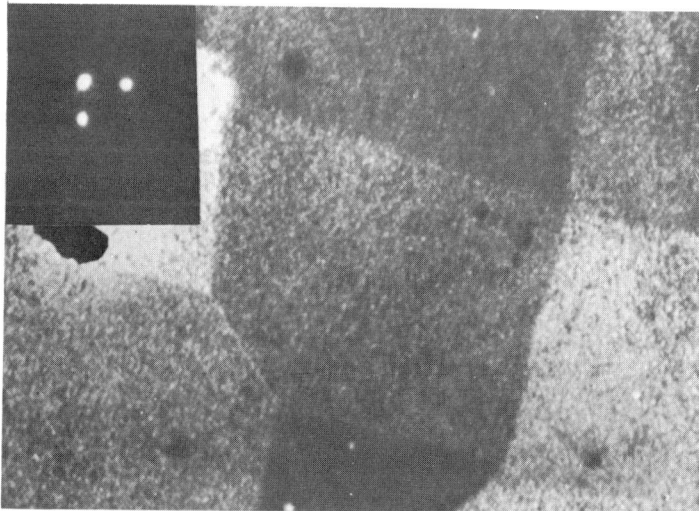

Fig. 33. (a) Image of spots, in diffraction mode, showing splitting corresponding to three directions of magnetization. (b) Domains in iron by Foucault contrast, with one spot obstructed and a second partly obstructed (G. A. Jones).

other crystal diffraction spots are also split). The objective aperture is then manipulated until it cuts out one of the two spots, and the imaging mode is resumed. Since the electrons from one set of domains are missing, these domains will appear dark in the final image, as in Fig. 33b.

In the Foucault method the specimen is, of course, in focus and this would appear to be the method of most general application. However, the great bulk of reported work appears to utilize the Fresnel technique. The latter gives by far the higher subjective visibility, partly because the walls appear as distinct lines rather than a change from one intensity to another, and in the presence of a multitude of contrast effects due to bend contours, thickness contours, defects, etc. this is particularly important (although contrast from some defects may be enhanced by defocusing). Moreover, although the resolution is higher for the Foucault method, the effective resolution (in electromagnetic microscopes) is limited for both methods by a most important practical feature not yet mentioned: the necessity to use a short specimen holder which raises the specimen well clear of the objective lens so as to remove it from the strong magnetic fields existing in the normal position. In the EM6 there is a field of about 5000 Oe normal to the specimen surface and this drops to $\sim$100 Oe in the "T.2 position." In view of the demagnetizing factors involved, this latter field has little effect. The strength of the objective lens must be reduced to focus on the modified specimen position and the resolution is reduced from $\sim$10 to $\sim$50 Å.

It is often difficult to achieve optimum contrast in the Foucault method since, unless the second condenser is correctly defocused, the back focal plane no longer coincides with the objective aperture and so either the spots or the aperture must appear blurred. (Also, the displacement of the aperture introduces some spherical aberration although this is probably inappreciable when compared with the effect of the specimen position.) It is not always easy to produce really clear domain patterns with the Foucault method in "difficult" specimens.

The splitting of the spots in the diffraction mode is clearly related to the thickness of the material through which the beam passes, and this underlies a method of estimating specimen thickness which is important when evaluating the observations.[54] The relation is

$$d = \frac{smcv}{8L\pi e M_s \sin \alpha}, \qquad (13.7.3)$$

[54] D. H. Warrington, J. M. Rodgers, and R. S. Tebble, *Phil. Mag.* **7**, 1783 (1962).

in which s is the spot separation, L is the camera length, and α is the half angle between magnetization vectors on either side of the wall: $\alpha = 90°$ for a 180° wall, for example. The effects from a single wall may be studied by isolating a small area crossed by the wall, as for selected area diffraction. If the aperture used is small enough for the wall itself to occupy a substantial proportion of the material in view, some information on the wall structure can be obtained.

Clearly the most appropriate fields of study by Lorentz microscopy are microscopic features such as the structure of the walls themselves and the interaction of walls with defects, which may themselves be imaged with high resolution. The possibility of imaging dislocations, for example, has lead to a great deal of searching for their interaction with domain walls. Much of this has been rather frustrating, but positive results have been recorded concerning dislocation clusters in FeCo.[55] It is essential to move the walls to study such interactions, and this can be achieved by using carefully designed coil systems, special specimen stages being described with references by Grundy and Tebble,[56] although tilting the specimen can also produce components of the objective lens field parallel to the specimen surface of about 1 Oe.

It is rarely that domain structures are too fine to be observable at all by optical methods, but it certainly may be the case that the spacing is too fine to be measured accurately other than by electron microscopy, as in very thin cobalt foils for example.[57] An example is shown in Fig. 34.

A further fascinating feature, which was first revealed by Lorentz microscopy, is magnetization ripple. This fine-scale variation of the magnetization direction occurs most notably in thin evaporated magnetic films which are very finely polycrystalline, as beautifully portrayed by Feldtkeller in Fig. 35,[58,59] but it can also exist within large crystallites. It is important in the present context to note that observations of ripple can be used to show the directions of the magnetization vectors within the domains, as illustrated by the figure.

Finally, although the images of domain walls do not bear any very obvious relation to the widths of the walls themselves, wall widths can be measured by rather complex methods involving interference effects. Systems of fringes are formed at the wall images if the beam is sufficiently

[55] M. J. Marcinowski and R. M. Poliak, *Acta Met.* **12**, 179 (1964).
[56] P. J. Grundy and R. S. Tebble, *Advan. Phys.* **17**, 153 (1968).
[57] J. Jakubovics, *Phil. Mag.* **10**, 277 (1964); **13**, 85 (1966); **14**, 881 (1966).
[58] E. Feldtkeller, *Z. Angew. Phys.* **17**, 121 (1964).
[59] E. Feldtkeller and W. Liesk, *Z. Angew. Phys.* **14**, 195 (1962).

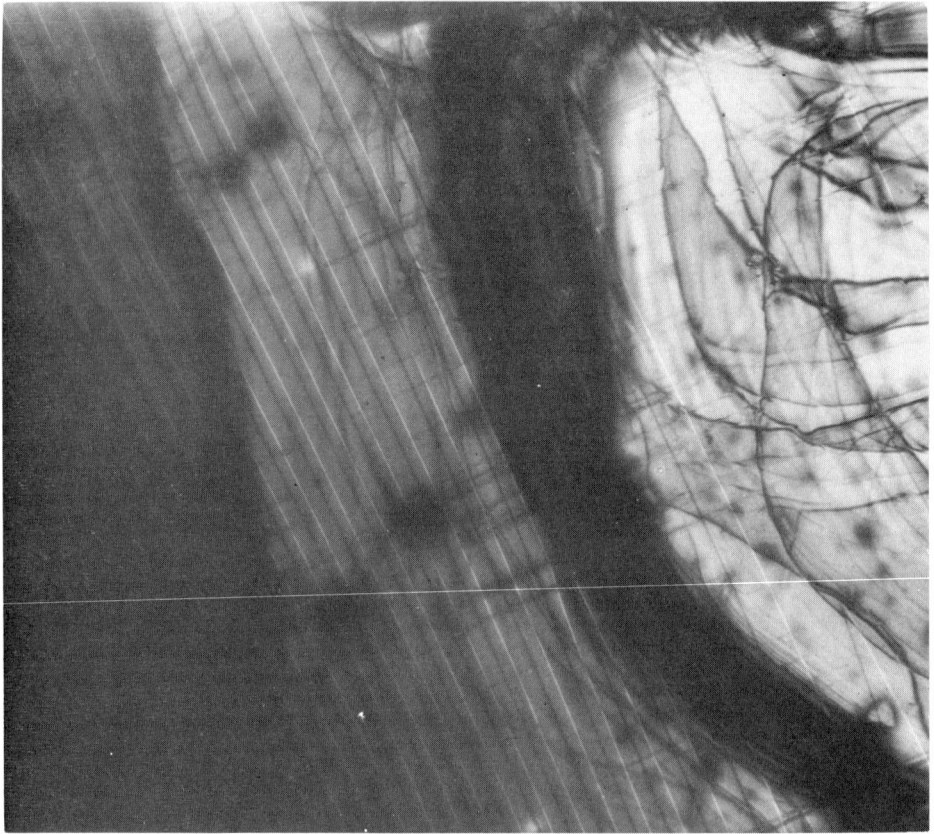

FIG. 34. Lorentz micrograph of 180° domains in a cobalt foil crossed by heavy contour lines. The closely spaced domains have a width of about 10^{-4} cm. There is a change of orientation in the center of the micrograph and the widely spaced walls occur in a crystallite with a surface close to the c axis. The moderately sharp images of slip lines show that it is not necessary to depart far from focus and walls may be seen in nominally focused images.

coherent due to effects analogous to biprism interference with light beams. Although of the greatest interest, the interference effects are only just beginning to be of value and reference should be made to the papers by Boersch et al.[60] and Hothersall,[61] and a useful review by Cohen.[62]

[60] H. Boersch, H. Hamisch, D. Wohlleben, and K. Grohmann, Z. Phys. **164**, 55 (1961); H. Boersch, H. Hamisch, K. Grohmann, and D. Wohlleben, ibid. **167**, 72 (1962); D. Wohlleben, Phys. Lett. **22**, 546 (1966); J. Appl. Phys. **38**, 3341 (1967).
[61] Hothersall, Phil. Mag. **20**, 89 (1969); **20**, 433 (1969).
[62] M. S. Cohen, J. Appl. Phys. **38**, 4966 (1967).

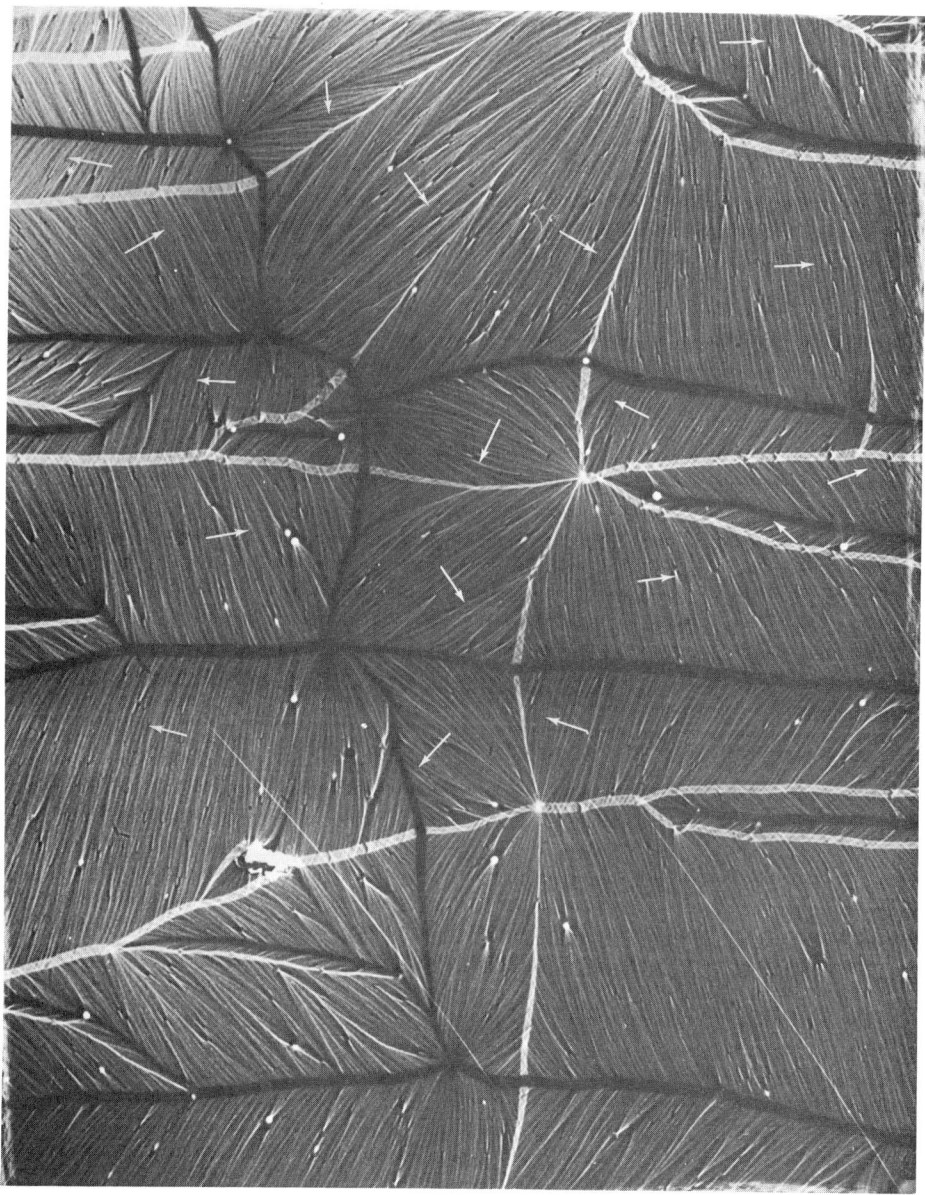

FIG. 35. Domain walls and magnetization ripple in a nickel–iron film. As indicated by the arrows the mean $\mathbf{M}_s$ is normal to the ripple structure [E. Feldtkeller, *Z. Angew. Phys.* **17**, 121 (1964)].

Problems of specimen preparation, as already discussed, cannot be separated from those of Lorentz microscopy and its interpretation. The method is most obviously suited to the study of thin films as such, whether polycrystalline or epitaxially grown single crystal, and sufficiently thin foils may otherwise be very difficult to prepare from some bulk materials. With films one is, in fact, directly studying the specimen of interest, whereas the applicability of results on artificially thinned foils to the properties and behavior of the parent material must always call for considerable thought. The increasing use of high-voltage electron microscopy will clearly be helpful in this respect.

13.8. Scanning Electron Microscopy

In recent years great efforts have been devoted to the visualization of domains by scanning electron microscopy, largely by Jakubovics and co-workers.[63] The contrast is due to the effect of the stray fields on the secondary electrons ejected from the specimen by the scanned primary electron beam. The technique can be carried out using a standard instrument with a metal plate with an off-center aperture fitted in part of the scintillator Faraday cage. Domains appear light or dark according to whether the associated stray field orientation is such as to scatter electrons into or away from the aperture.

The example shown in Fig. 36 has one particularly interesting feature. In some regions A the orientation of the 180° domain walls inside the cobalt crystal, with respect to that of the collector, is such as to give contrast which swamps that from the superficial (closure) domains. In regions such as B there is no contrast from the internal domains and the weaker and more complex stray field distribution from the superficial domains is effective. The shapes of internal domains can often be inferred simply from low-magnification powder patterns, particularly if the effects of surface structures are reduced by forming the pattern at some distance from the surface;[64] the more complex the charge distribution the more rapidly the field distribution evens out.

---

[63] D. C. Joy and J. P. Jakubovics, *Phil. Mag.* **17**, 145 (1968); D. C. Joy, E. M. Schulson, J. P. Jakubovics, and C. G. van Essen, *ibid.* **20**, 843 (1969); D. C. Joy and J. P. Jakubovics, *J. Phys. D.* **2**, 1367 (1969); J. R. Dorsey, *Proc. Nat. Conf. Electron Probe Microanal., 1st, Maryland* (1966); U. N. Vertsner, R. I. Lomunov, and Yu. V. Chentsov, *Bull. Acad. Sci. USSR. Phys. Ser.* **30**, 778 (1966); J. R. Banbury and W. C. Nixon, *J. Sci. Instrum.* **44**, 889 (1967); *J. Phys. E* **2**, 1055 (1969).

[64] B. Wyslocki, *Ann. Phys. Leipzig* **13**, 109 (1964).

13.8. SCANNING ELECTRON MICROSCOPY

FIG. 36. Scanning electron micrograph of the basal plane of cobalt (compare Fig. 9). [D. C. Joy and J. P. Jakubovics, *Phil. Mag.* **17**, 145 (1968)].

Apart from the indications of internal structures an advantage of this method is the ease of control of the specimen temperature over wide ranges. It is, however, scarcely practicable to apply fields to the specimen due to their influence on the primary beam and secondary electrons, and it is not known whether the resolution can ever be brought to anything like that of the instrument as used for topological studies. Jakubovics reports having obtained strong contrast with barium ferrite and cobalt ferrite and weak contrast even with haematite.

The method may be compared with that of Blackman and Grünbaum,[65] which is based on the study of the deflection of an electron beam passed close to the surface of a specimen.

---

[65] M. Blackman and E. Grünbaum, *Proc. Roy. Soc.* **A241**, 508 (1957).

13.9. X-Ray Method

X-ray diffraction topography was developed, notably by Lang,[66] to study dislocations and other defects in bulk crystals. In practice rather thin crystals are studied, but at the order of 0.1 mm these should reflect bulk properties, as opposed to the 1000 Å foils which transmit electron beams. Image formation (Fig. 37) is similar to dark-field electron microscopy. A Bragg-reflected beam would be of uniform intensity for a perfect crystal, but if the lattice is tilted (effectively by only a fraction of a second of arc) or distorted (by 10^{-5}), the reflected beam is affected to a detectable extent. On scanning the crystal and the cassette a direct image of imperfections is obtained as a projection of the crystal on a plane normal to the diffracted beam. The divergence of the incident beam must be less than the difference between the Bragg angles for the $K\alpha_1$ and $K\alpha_2$ wavelengths.

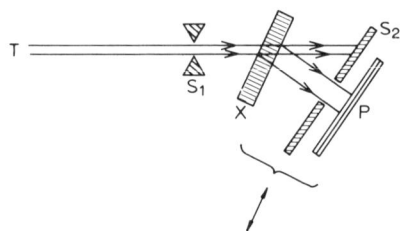

FIG. 37. Principle of X-ray topography equipment. S: slits, X: crystal, P: photographic place, T: fine-focus tube. The specimen and plate are moved slowly (total exposures may take many hours) parallel to the specimen surface.

The possibility of imaging domains by this method appears to have been discovered by Polcarova, while working with Lang,[67] although Merz[68] had already used a double crystal diffractometer in a reflection mode to detect domains in cobalt ferrite. (This applies only to materials with very high magnetostriction and does not appear to have been substantially pursued, the reflection mode losing many of the special advantages connected with the use of X-rays.) Although other factors have been suggested it seems generally adequate to interpret the effect as due to the magnetostrictive deformation of the lattice planes according

---

[66] A. R. Lang, *Acta Cryst.* **12**, 249 (1959).
[67] M. Polcarova and A. R. Lang, *Appl. Phys. Lett.* **1**, 13 (1962).
[68] K. M. Merz, *J. Appl. Phys.* **31**, 147 (1960).

to the direction of M_s. For example, silicon–iron has a cubic lattice above the Curie point but, with a positive magnetostriction, the lattice becomes elongated in the direction of M_s below the Curie point to give a tetragonal cell with $c/a = 1 + (3/2)\lambda_{100}$. Although $\lambda_{100} = 2.8 \times 10^{-5}$ for 3% Si, this is adequate to give contrast across a 90° domain wall where the lattice distortion itself rotates through 90°, but the lattices are equivalent on both sides of a 180° wall and no contrast is expected. In fact very weak contrast is sometimes obtained at 180° walls due to second-order effects (the interaction of the strain within the wall with the surrounding lattice, where the wall meets the specimen surface).

Usually the positions of 180° walls must be inferred on the principles of flux closure, as in Polcarova's micrograph in Fig. 38. Arrays of "fir-tree" closure structures help to indicate the wall positions.

The visibility of the 90° walls themselves varies according to the diffracted beam selected (i.e., the orientation of the specimen). For example, no contrast occurs if $(\mathbf{m}_2 - \mathbf{m}_1) \cdot \mathbf{g} = 0$, where $\mathbf{m}_1$ and $\mathbf{m}_2$ are the unit vectors parallel to $\mathbf{M}_s$ on each side of the wall and $\mathbf{g}$ is the diffraction vector. The wall becomes visible as a dark band if $(\mathbf{m}_2 - \mathbf{m}_1) \cdot \mathbf{g} > 0$, and as a bright band if $(\mathbf{m}_2 - \mathbf{m}_1) \cdot \mathbf{g} < 0$. Generally, selecting different diffracted beams leads to the formation of very different images, as shown in Fig. 39, and some study is required to derive the entire structure.

The simultaneous portrayal of lattice defects and domains should be of value, but, in fact, Polcarova reports that no interactions between the two have been definitely observed. To a certain extent this feature is a positive disadvantage since strong diffraction effects at defects obscure the domains in all but very good crystals. The results shown were obtained on silicon–iron crystals grown and annealed with the greatest care. Also there must be no surface deformation, and the silicon–iron crystals were electropolished. These difficulties are particularly severe so far as pure iron and nickel are concerned.

The other experimental features consist of the slit system, fine adjustment of the sample orientation, and the slow traverse mechanism. A fine focus tube is needed to give a powerful well-collimated X-ray beam. Suitable goniometers are now produced commercially, by Jarrell and Ash (USA), Rigaku Denki (Japan), and C.G.R. (France). The topographs are initially obtained as 1 : 1 images and magnification consists of photographic enlargement, calling for fine-grain emulsion such as Ilford Nuclear plates L4. These are sufficiently sensitive to X-rays, but need prolonged processing at low temperatures.

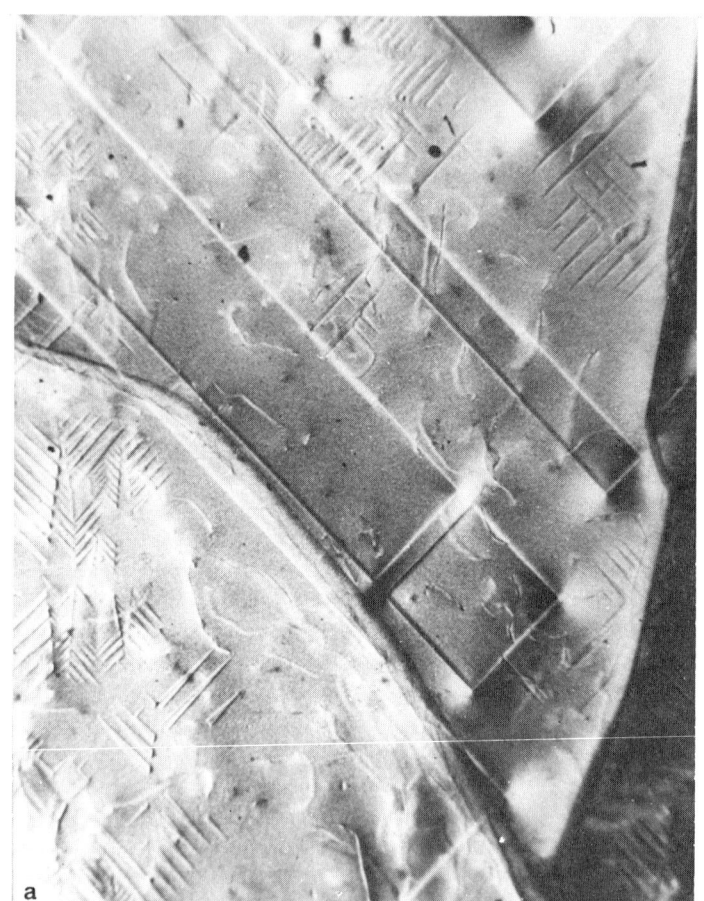
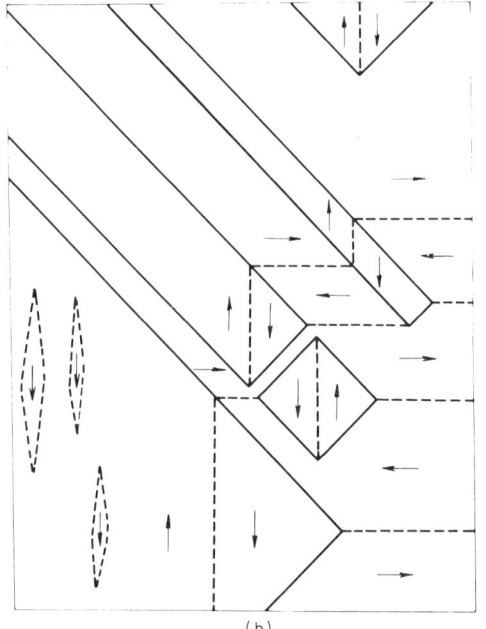

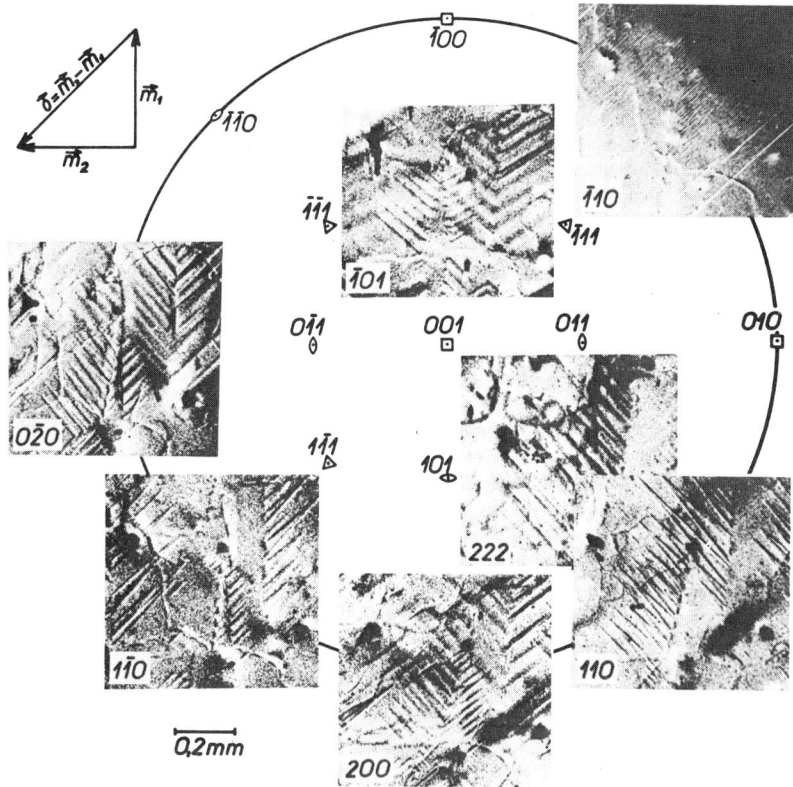

Fig. 39. X-ray topographs of a 3% SiFe crystal showing the variation in contrast when different reflections are used, as indicated by the positions of the images on the stereographic net. [M. Polcarova and J. Kaczer, presented at *Crystallogr. Congr.*, 7th, Moscow (1966)].

It is scarcely necessary to mention that the great advantage of this method is the portrayal of internal structures which might not intersect the surface at all, or would be obscured by closure structures at the surface. In several cases the patterns have been found to be quite different from those indicated by powder patterns or the Kerr method. The practical thickness for high resolution, about 0.1 mm, is very similar to that for optical studies in garnets by the Faraday effect.

Fig. 38. X-ray topograph of a 3% SiFe crystal, 120 μm thick with (001) surfaces; MoKα radiation; [200] reflection; (b) includes the positions of 180° walls which are not shown by the method but inferred on the grounds of flux closure and anisotropy (Polcarova).

13.10. Methods for Antiferromagnetic Domains

Antiferromagnetics are crystals in which the spins are ordered in an antiparallel, or rather more complex, manner to give zero net spontaneous magnetization. Clearly they cannot contain ferromagnetic domains, but the direction of ordering may only be constant over certain regions which are known as *antiferromagnetic domains*. For example, NiO is paramagnetic with cubic rock-salt structure above the Néel point of 523 K, while in the antiferromagnetically ordered state there is a slight distortion to rhombohedral. The original cubic cell is contracted along one of the [111] axes giving domains twinned on (100) or (110) planes. These twinning planes are called T walls. There are 12 possible T walls in NiO, i.e., 12 planes which satisfy the necessary condition for twinning by bisecting the (111) directions to give mirror images.[69]

Within a domain characterized by a certain contraction axis, i.e., by ordering within a (111) plane, there may be changes in the direction of the spins within the (111) plane giving S-walls of two different types: one type denoted $S_{\|}$ is parallel to the ferromagnetic (111) sheets while the second, $S_{\perp}$ walls, are perpendicular to these sheets and contain the rhombohedral axis.

Both S and T walls have finite energies and thicknesses.[70] T walls can be moved by applied stresses or fairly strong magnetic fields.[69] An extremely comprehensive and useful review of antiferromagnetic domain structures and methods of observation has been given by Farztdinov.[71]

Despite the absence of a spontaneous magnetization, optical methods are partly analogous to those for ferromagnetic domains; within each domain the axis of contraction corresponds to the optical axis of a uniaxial crystal. Thus NiO crystals exhibit birefringence and slight dichroism and T domains may be observed utilizing these effects (Fig. 40).[72] Gypsum half wave plates have been used in the 45° position. Surface deformation during specimen preparation introduces closely spaced structures, but after annealing the domains may have dimensions of several millimeters.

The contraction axis is not the axis along which the spins are ordered: the former is (111) and the antiferromagnetic axis is (110). Thus it

---

[69] G. A. Slack, *J. Appl. Phys.* **31**, 1571 (1960).

[70] T. Yamada, *J. Phys. Soc. Japan* **21**, 650 (1966).

[71] M. M. Farztdinov, *Sov. Phys. Usp.* **7**, 855 (1965) [*English transl.*: *Usp. Fiz. Nauk* **84**, 611, 1964].

[72] W. L. Roth, *Phys. Rev.* **111**, 772 (1958); *J. Appl. Phys.* **31**, 2000 (1960).

13.10. METHODS FOR ANTIFERROMAGNETIC DOMAINS

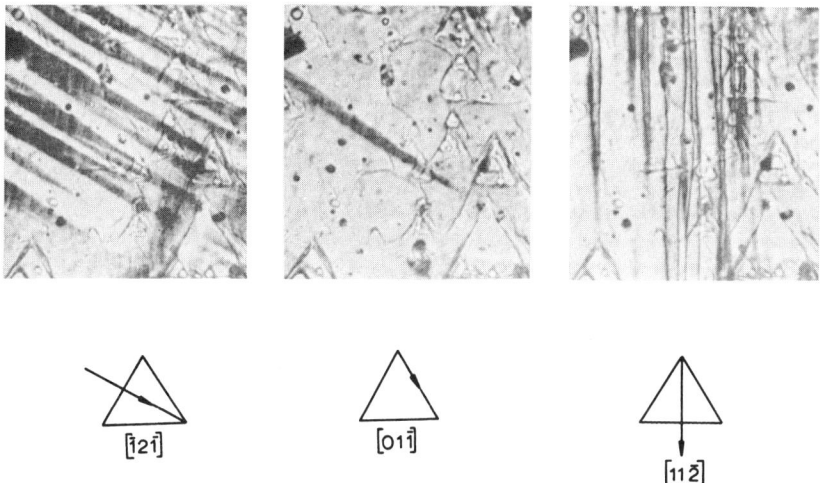

FIG. 40. Antiferromagnetic p domains observed by birefringence in a (111) cut crystal plate of NiO, carefully polished and annealed at 1600°C. The [1̄10] axis coincided with the polarizer axis. The p domains occur within crystallographic single domains and respond to moderate magnetic fields (10 kOe) as indicated [H. Kondoh and T. Takeda, J. Phys. Soc. Japan **19**, 2041 (1964)].

would appear that this method depends simply on birefringence in the normal optical sense, and not on magnetic birefringence as such. There is also an additional weak contraction along (110) axes, permitting the observation of S domains by very sensitive polarized light microscopy.[73]

A second optical method, that of specular reflection,[69] does not require the specimens to be transparent but conversely, of course, requires that the full structure be inferred from the surface observations. The angle between regions of the surface intersected by T domains is between 2.5' and 12' (again for NiO) and large domains in well-annealed crystals can be seen by the naked eye. Roth and Slack[74] polished and observed all the faces of a crystal in the form of a rectangular parallelopiped with the results indicated by Fig. 41.

X-ray diffraction topographs using the Berg–Barrett method[75,76] may

---

[73] H. Kondoh, J. Phys. Soc. Japan **17**, 1316 (1962); **18**, 595 (1963); H. Kondoh and T. Takeda, ibid. **19**, 2041 (1964).
[74] W. L. Roth and G. A. Slack, J. Appl. Phys. **31**, 352S (1960).
[75] W. Berg, Z. Kristallogr. **89**, 286 (1934).
[76] C. S. Barrett, Trans. AIME **161**, 15 (1945).

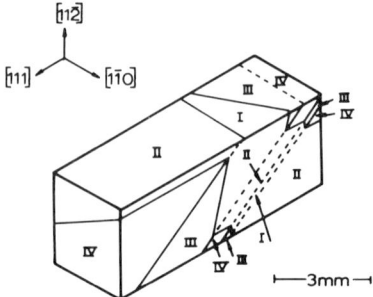

FIG. 41. The intersection of T domains with the surface of a crystal of NiO, as derived from specular reflection studies [W. L. Roth and G. H. Slaok, *J. Appl. Phys.* **31**, 3525 (1960)].

also be used for structures in antiferromagnetics.[77] A standard arrangement is suitable for T domains but inadequate, in view of the smaller distortion involved, for S domains. For the latter Yamada et al.[78] utilized the higher resolution of a double crystal system as shown in Fig. 42. The collimated beam is monochromatized by reflection from a (111) germanium surface and is 0.4 mm wide giving a topograph of a region 2 mm wide; for wider areas scanning was used as in the Lang method.

The main importance of this method appears to be the verification of the structures of S walls within the T domains, indicated by Kondoh's optical method. Results are shown in Fig. 43 (to be compared with Fig. 40), and good results have also been obtained for CoO by this method.[79]

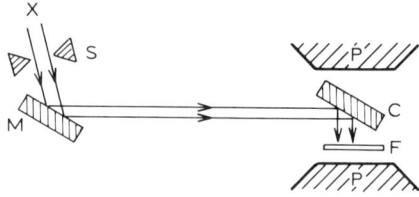

FIG. 42. Double crystal system for the Berg–Barrett method for antiferromagnetic domains. X: source, S: slit, M: monochromator, C: specimen crystal, F: film, P: pole pieces of magnet [T. Yamada, S. Saito, and Y. Shimomura, *J. Phys. Soc. Japan* **21**, 672 (1966)].

[77] S. Saito and Y. Shimomura, *J. Phys. Soc. Japan* **16**, 2351 (1961); S. Saito, *ibid.* **17**, 1287 (1962).
[78] T. Yamada, S. Saito, and Y. Shimomura, *J. Phys. Soc. Japan* **21**, 672 (1966).
[79] S. Saito, K. Nakahigashi, and Y. Shimomura, *J. Phys. Soc. Japan* **21**, 850 (1966).

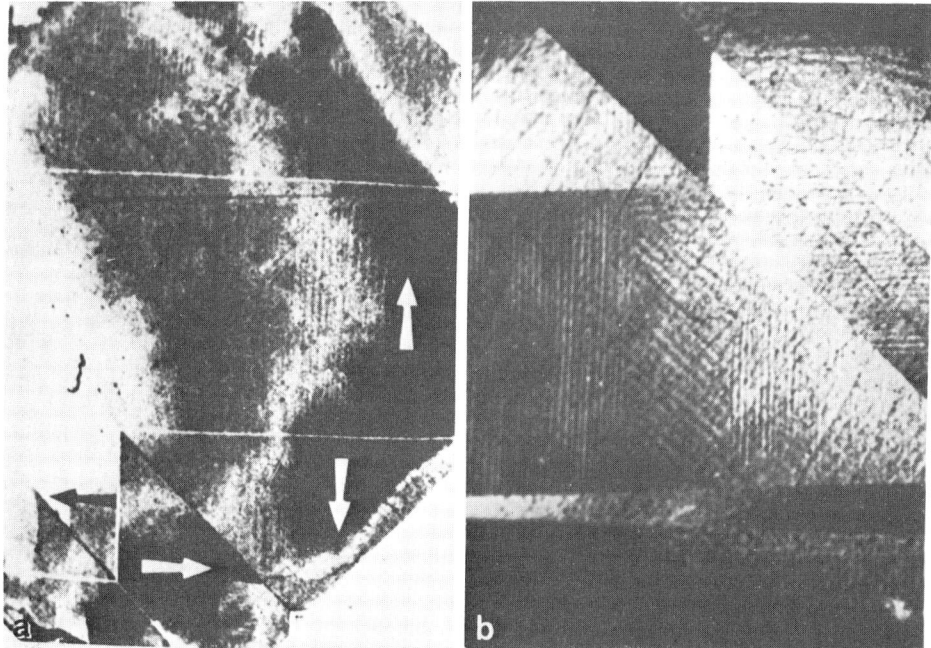

FIG. 43. Antiferromagnetic domains in NiO shown by (a) the X-ray method and (b) birefringence [T. Yamada, S. Saito, and Y. Shimomura, *J. Phys. Soc. Japan* **21**, 672 (1966)].

13.11. Further Methods

In a limited scope it has seemed best to give reasonably detailed accounts of the methods most often used at the expense of some others of considerable interest but less frequent application. However, no survey would be complete without mention of electron mirror microscopy as pioneered by Mayer[80] (noting that this preceded Lorentz electron microscopy) and of probe techniques. The latter are basically high-resolution magnetic field measurements, using minute Hall-effect probes[81] or vibrating Permalloy wires.[82] Part of the interest attaching to these methods is the possibility of applying the technology to other magnetic studies, such as detecting the presence of domains in storage systems. It can only be recommended that reference be made to the original papers.

[80] L. Mayer, *J. Appl. Phys.* **28**, 975 (1957); **30**, 252S (1959); **31**, 346 (1960); etc.
[81] B. Kostyshyn, J. E. Brophy, I. Oi, and D. D. Roshon, *J. Appl. Phys.* **31**, 772 (1960); B. Kostyshyn, H. Koehler, and T. C. Ku, *IBM J.* **5**, 326 (1961).
[82] J. Kaczer, *Czech. J. Phys.* **5**, 239 (1955); J. Kaczer and R. Gemperle, *ibid.* **6**, 173 (1956).

AUTHOR INDEX

A

Aalto, M. I., 513
Abel, W. R., 489, 503, 509
Abeles, B., 30
Abelès, F., 666, 669, 670(189)
Abragam, A., 513
Abraham, B. M., 509
Abramowitz, M., 46
Abrikosov, A. A., 550
Adams, C. M., Jr., 636
Adams, E. D., 492, 535, 536
Adde, R., 233, 246
Adkins, C. J., 239
Adler, J. G., 148, 165, 166, 167, 169, 170, 171, 194, 195(129), 196, 197, 210
Agee, F. J., Jr., 451
Agresti, D., 366
Aharonov, Y., 293
Ahlers, G. A., 59
Al-Bassam, T. S., 693
Albegova, I. Kh., 239(138), 240
Aleksandrov, L. N., 637
Aleksevskii, N. E., 25
Alers, G. A., 389, 397, 398, 402(68), 421
Alexander, J. H., 646
Alig, R. C., 396
Allen, F. G., 84, 97, 101, 112(50), 113
Allen, M. A., 542, 557, 560(4), 571(4), 573(4)
Allison, R., 82, 83(38)
Alworth, C. W., 592
Ambegaokar, V., 203, 247
Ambler, E., 504
Amick, J. A., 647
Amundsen, T., 26
Anderegg, M., 74
Anderson, A. C., 469, 489, 493, 494, 503, 505, 508, 509, 510, 586
Anderson D. E., 643
Anderson, G. S., 637
Anderson, J. C., 620, 622
Anderson, J. R., 34, 35, 40(16), 41(19), 50(19), 51, 52, 56(31, 33), 57, 64
Anderson, J. T., 216, 247, 251
Anderson, P. W., 127, 147, 152, 175, 200, 203(5), 212, 214(6), 218, 220, 221 (60), 256, 305
Andra, W., 692
Andreatch, P., Jr., 394, 399, 406, 412
Andres, K., 491
Andronikashvili, E. L., 451
Apker, L., 84
Appel, H., 521
Appelbaum, J. A., 164, 175
Arai, U., 181
Arams, F. R., 593
Archer, R. J., 674
Arenberg, D. L., 372
Arko, A. J., 20, 31
Arndt, D. G, 588, 592
Arnold, A. J. P. T., 261, 262(213)
Arnold, G. P., 466
Asakura, T, 307
Aschenbrenner, R. A., 319
Asher, W. R., 10
Ashley, J. R., 219, 229(63), 303(63), 304(63)
Aslamazov, L. G., 255, 270
Atkins, K. R., 445
Auracher, F., 221, 222, 270(79), 301
Averill, R. F., 238
Azbel, M. Ya., 1, 4(1), 9(1)

B

Bachman, C. H., 632
Bacon, D. D., 640
Bader, W.G., 13
Bahl, S. K., 621
Bailey, C. A., 508
Bakalyar, D., 538
Baker, A. D., 68
Baker, C., 68
Baker, G. L., 588, 592(105)
Bakker, J. W., 537
Banbury, J. R., 734
Banninger, U., 74
Banter, J. C., 670

Baratoff, A., 203, 258, 262(204), 552, 562 (33)
Bardeen, J., 203, 255, 262(195), 265, 268 (195), 429, 547, 550
Barnes, L. J., 239(139), 240, 241, 242 (139), 259(139)
Barnier, Y., 61
Barone, A., 254
Barrett, C. S., 741
Barron, T. H. K., 435, 440(145)
Bart, W. L., 663
Bartels, R. A., 412
Bartoli, A., 461, 462(35)
Bashara, N. M., 648
Basterfield, J., 686
Bateman, T. B., 393
Bates, L. F., 684, 692
Bauer, R., 83, 101(47), 105(47)
Bauer, W., 571, 591
Baumgardner, J. B., 333
Beams, J. W., 663, 664
Bearden, A. J., 318
Beasley, M. R., 296
Beattie, A. G., 395, 397, 398
Behrndt, K. H., 631, 653, 657(131)
Bell, R. L., 112, 113(100)
Bendall, R. G., 560, 573(60), 574, 591 (60)
Benedek, G. B., 365
Bennett, A. J., 140
Bennett, H. E., 656, 666(144)
Bennett, J. A., 658
Bennett, J. M., 656, 666(144)
Bent, M., 366
Bent, P. J., 466
Ben-Zvi, I., 588
Berckhemer, H., 621
Berg, W., 741
Berger, S. B., 648
Berger, W. G., 309, 310(12)
Berglund, C. N., 75, 76(22), 77(22), 78(22), 79(22), 92
Berglund, P. M., 513, 518, 526(99)
Bergmann, W. H., 688
Bergmark, T., 68, 99(9)
Berman, R., 534, 535
Bermon, S., 139, 148, 230
Bernstein, B., 379, 385(18), 435(18), 438(18), 440(18)

Bertin, E. P., 659
Bertolini, E., 346
Bethe, H. A., 583
Betjemann, A. G., 395
Betts, D. S., 510
Beyer, R. T., 374, 389, 430
Bianconi, A., 483, 484
Biquard, F., 585, 589, 590(98, 109)
Biquard, P., 409
Biran, A., 319
Birch, F., 434, 435
Birtwistle, J. K., 658
Bis, R. F., 632
Biscar, J. P., 342
Bishop, J. H., 533
Biswas, A. C., 247
Bitter, F., 683
Black, P. J., 350
Black, W. C., Jr., 489, 491, 492(17), 503, 537
Blackburn, H., 661
Blackburn, J. A., 212, 232, 258, 262(204)
Blackford, B. L., 140, 141
Blackman, M., 735
Blair, J. C., 658
Blake, C., 501
Blakely, J. M., 664
Blaney, T. G., 219, 302(65), 303(65), 304
Blaugher, R. D., 258
Blauth, E., 100
Blin-Stoyle, R. J., 518
Bloch, F., 200, 203(7), 263(7), 678
Blocher J. M., Jr., 645
Blodgett, A. J., 82
Blodgett, K. B., 651
Blount, E., 31
Blum, N. A., 361, 363
Blume, R. J., 403
Bömmel, H., 342
Bömmel, H. E., 426, 427, 428, 429, 431, 432
Boersch, H., 711, 728, 732
Boghosian, C., 511
Bohm, D., 293
Bohm, H. V., 395
Boiko, B. T., 621
Bol, M., 614
Brown, R., 659

AUTHOR INDEX

Bolef, D. I., 343, 398
Bonchev, Z. W., 347, 348(86)
Bondarenko, S. I., 261, 294
Bonse, W., 10
Booker, G. R., 635
Booth, R., 318
Bootsma, G. A., 674
Borgnis, F. E., 380
Borie, B., 663
Born, M, 413, 415, 417(92), 419
Bosacchi, B., 55
Bostock, J., 238
Bousquet, P., 666
Bowers, R., 27, 395
Bozorth, R. M., 505, 515
Bradley, C. C., 219, 302(65), 303(65), 304
Bradley, R. R., 646
Brady, B. G., 560, 573(60), 574, 591(60)
Brafman, H., 325, 340(45), 341(45)
Brailsford, A. D., 163, 164(62), 424
Brandt, B. L., 720, 721(40)
Breazeale, J. B., 663
Breazeale, M. A., 388, 389(31), 421
Breitling, G., 621
Brenner, A., 652
Brennert, G. F., 5, 19, 22(4), 28(4)
Brewer, W. D., 512
Bright, A. A., 12, 239(140), 240, 242(140)
Brinkman, W. F., 158, 159(32), 161(32), 164
Briscoe, C. V., 395, 444
Broadhurst, J. H., 318, 328, 337(55)
Brock, J. C. F., 535
Brockway, L. O., 622
Brodsky, M. H., 644
Brophy, J. E., 743
Brown, R. E., 615
Brown, W. F., Jr., 505, 683
Broyde, B., 646
Bruck, A., 252
Brugger, K., 377, 380(15), 383, 384(28), 385, 386, 387, 409(15), 413(25), 434, 435(28), 438, 440
Brundle, C. R., 68
Bruner, E. C., Jr., 83
Bruschi, L., 459, 461, 482
Brust, D., 76

Bruynseraede, Y., 542, 560, 562(58), 563(11), 572, 573(11, 58)
Bryukhanov, V. A., 351
Buckel, W., 621
Buckner, S A., 226, 228, 247
Buehler, E., 596
Buhrman, R. A., 250, 259
Bukshpan, S., 325, 341(48)
Burbank, R. D., 622
Burger, R. M., 107
Burgess, R. E., 283, 288(260)
Burns, J., 82, 83(38)
Burstein, E., 124, 151(1), 239
Busch, G., 74

C

Cabrera, B., 616, 617(14)
Cade, A. G., 445
Cady, W. G., 392
Calcott, T. A., 84, 109, 110
Cameron, J. A., 512, 525, 527, 528
Campagna, M., 74
Campbell, D. S., 660, 661
Campbell, I. A., 512, 528
Campbell, R. B., 648
Campi, M., 490
Carbotte, J. P., 217
Cardona, M., 552, 556, 557(48, 49), 562(33)
Careri, G., 444, 470, 478
Carey, R., 726
Carlson, R. V., 251
Carmeliet, J., 334
Carne, A., 560, 573(60), 574, 591(60)
Carrell, J. C., 345
Carter, W., 706, 721(31)
Carter, W. S., 725
Carver, T. R., 583, 584(96)
Cashion, J. K., 99
Casimir, H. B. G., 506
Cassell, K., 318
Castle, J. G.,Jr., 588
Catlin, A., 664
Cave, E. F., 646
Cedrone, N. P., 402
Ceperley, P. H., 570, 588
Cetas, T. C., 506, 509(50)
Chambers, E. E., 561, 573(62)
Chambers, R. G., 3, 26

Champion, A. R., 363, 364
Chandrasekhar, B. S., 165, 166, 167(69)
Chang, C. Y., 184
Chang, L. L., 165(91), 178, 179, 181, 191
Chang, Y. A., 412
Chang, Z. P., 416
Chase, C. E., 501
Chase, R. L., 319, 348(42)
Chen, F. S., 718
Chen, J. T., 222, 231, 247
Chen, T. T., 167, 169, 171, 194, 196, 197 (130)
Cheng, L., 220
Chentsov, Yu. V., 734
Chester, P. F., 600
Chick, B. B., 374, 389
Chikazumi, S., 684
Chittick, R. C., 646
Chivian, J. S., 158
Choi, S. I., 444
Chopra, K. L., 620, 621, 622, 631, 650, 653
Chow, C. K., 160
Christensen, N. E., 76
Christiansen, J., 309, 348(10)
Christiansen, P. V., 221, 257(72), 262 (72), 265(72)
Christoe, C. W., 364
Christopher, J. E., 171, 177, 650
Christy, R. W., 646, 648
Chu, T. L., 648
Citron, A., 564, 566, 570(67), 571, 591 (67)
Clairborne, L. T., 492
Clark, A. F., 21
Clark, T. D., 139, 148, 237, 243(122)
Clarke, J., 200, 216, 217, 218, 219(61), 220, 249, 254, 255, 260, 261, 268(191), 274, 283, 285(266), 290(19), 295, 298, 299
Clement, J. R., 490
Cline, J. E., 659
Clover, R. B., 511
Clow, H., 684
Coche, A., 346
Cochran, D. R. F., 316, 532
Cohen, D., 288, 613
Cohen, J., 363, 364
Cohen, M. A., 444
Cohen, M. G., 391

Cohen, M. S., 732
Cohen, R. L., 328, 336(57), 341, 344, 345 (76), 346
Coleman, R. V., 141, 171, 177, 192, 650
Coleridge, P. T., 52, 57
Collan, H. K., 513, 518, 526(99)
Collins, R. L., 309, 342, 345
Colvin, A. D., 406
Compton, J. P., 512, 525, 528
Compton, K. T., 67, 92(1)
Compton, W. D., 181
Condon, J. H., 8, 9, 27, 28, 34, 41
Conley, J. W., 183
Conrad, E. E., 655
Consadori, F., 258, 281
Cook, C. F., 223
Cook, J. R., 28
Cooke, A. H., 506
Coon, D. D., 229
Cooper, L. N., 429, 547
Cooper, P. V., 680
Cosgrove, J. G., 342
Cowen, J. A., 233, 235, 272(113)
Cracknell, A. P., 37
Craig, P. P., 310, 316, 361, 532
Craik, D. J., 680, 682, 684, 692, 697
Cranshaw, T. E., 325, 342(47), 343
Critchlow, P. R., 511
Crittenden, E. C., Jr., 658
Croxon, A. A. M., 52, 57(32)
Culler, G. J., 146
Cunsolo, S., 456, 457, 466, 478, 483
Cupp, J. D., 219, 228(64), 229(63, 66), 302(64), 303, 304(63)
Curran, D. R., 402
Cutler, L. S., 588
Czjzek, G., 309, 310(12)
Czorny, B. R., 646

D

Dabbs, J. W. T., 210
Dagman, E. I., 637
Dahlke, W. E., 191
Dahm, A. J., 239, 246, 445, 448
Dall' Oglio, G., 483
D'Amico, C., 107, 108(94)
Dammertz, G., 562, 571, 572(66), 590 (66), 591
Daniels, J. M., 509

AUTHOR INDEX

Daniels, W. B., 412
Danysz, J., 99
Dash, J. G., 316, 532
Datars, W. R., 16, 28
Davey, J. E., 621, 632
Davidenkov, N. N., 663
Davidse, P. D., 644
Davis, L. C., 163, 164(62), 181, 183
Davis, M., 646
Davis, R. S., 649
Dawson, J. W., 560, 562, 573(59, 65)
Day, E. P., 611
Daybell, M. D., 358
Dayem, A. H., 151, 218, 220, 221(60), 223, 232, 233, 234, 244, 256, 257(68)
Deaver, B. S., 611, 612, 613
Deaver, B. S., Jr., 202, 235, 236, 239 (142), 240, 242(142), 258, 260(142), 272(205), 281, 285(255)
Debrunner, P., 309, 363, 364, 365(107)
de Bruyn Ouboter, R., 261, 262(213), 294, 466
Debye, P., 409
Decker, D. L., 365, 492
DeGennes, P. G., 265, 547, 554, 555(35), 556, 561, 570(23)
de Groot, S. R., 518
de Haas, W. J., 33
de Klerk, D., 489, 490, 504(7), 506(7)
de Klerk, J., 424, 425
Dell'Oca, C. J., 645, 650(94)
Delyagin, N. N., 351
de Magistris, P., 462
Denbigh, P. N., 622
Denenstein, A., 218, 219, 220, 239, 246, 252, 273(62), 274
De Pasquali, G., 363, 365(107)
Derbenwick, G. F., 93, 107, 116, 117
DeReggi, A. S., 239(143), 240, 241
Derevyanchenko, A. S., 621
de Rosset, W. S., 409
Deruyter, H., 557
Deszi, I., 350
De Voe, J. R., 341
de Waard, H., 342
De Waele, A. Th. A. M., 294
Diamond, J. M., 499
Dick, G., 200, 299(201)
Dick, G. J., 542, 587(2), 592
Dicke, R. H., 545, 568(21), 587(21)

Dickey, J., 84
Diepers, H., 543, 564, 566(13), 571, 572 (13, 77, 78)
Dietrich, I., 139, 141, 142
Dietz, E., 101
Dignum, R., 374, 430
Dillon, J. F., 718
Di Nardo, A. J., 593
Di Stefano, T. H., 84, 85, 86, 87, 88(52), 94, 95(56), 105(59), 122(59)
Dixon, A. E., 16
Dmitrenko, I. M., 200, 230, 233, 239, 261, 294
Dmitriev, V. M., 200, 221, 257(70)
Dobbs, E. R., 395, 396
Dobrowolski, W., 530
Dodkin, A. P., 337
Doherty, P. E., 649
Doll, R., 202
Doniach, S., 77
Donnely, R. J., 480
Donnelly, R. J., 444, 471, 474, 475, 476, 480, 481
Donovan, T. M., 71, 76, 77(31)
Doran, J. C., 510
Dorsey, J. R., 734
Doty, C. T., 648
Douglass, D. H., Jr., 139
Douglass, R. L., 472, 473, 475, 481
Dove, D. B., 726
Drabble, J. R., 30, 412, 416
Dransfeld, K., 431, 432, 454, 483
Dreschel, W., 706, 721(30), 723
Drever, R., 365
Drexhage, K. H., 652, 659(128)
Drickamer, H. G., 363, 364, 365(107)
Druyvestein, W. F., 680
Du, H. Y., 163
Dubey, P. K., 161
DuBridge, L. A., 68
Duckett, S. W., 80
Duff, R. H., 622
Dugdale, J. S., 59
Duke, C. B., 124, 159, 164(2), 178, 179, 181, 182, 183, 184, 185, 186, 187
Dundon, J. M., 281
Dunkleburger, L. N., 294
Dunlap, B. D., 316
Dupré, F. K., 506
Durieux, M., 506

Durupt, P., 185
Dynes, R. C., 139, 149, 158, 159(32), 161 (32), 214, 257, 263(200), 294, 305

E

Eales, B., 658
Easaki, L., 139
Eastman, D. E., 76, 77(30), 99, 105(30), 109, 110, 121
Eck, R. E., 136, 229, 230, 231, 233(94), 234
Eckstein, Y., 509
Eckstrom, M. P., 594
Edelsack, E. A., 200(26), 201, 286
Edelstein, A. S., 139, 491, 492(18), 493 (18)
Eden, R. C., 69, 70, 71(10), 76, 79(10), 93, 96, 98(58), 111
Edge, C. K., 363, 365(107)
Edgecumbe, J., 643
Edmonds, D. T., 510
Edwards, D. O., 509
Edwards, G. J., 304
Edwards, H. H., 139
Edwards, L. R., 535
Eggleston, T. W., 588
Egorov, V. S., 25
Ehnholm, G. J., 357, 513, 518, 526(99)
Ehrenberg, W., 293
Ehrenreich, H., 432
Ehrman, J. R., 315, 366
Einspruch, N. G., 492
Einstein, A., 67
Eisenmenger, W., 151, 244
Ekin, J. W., 499
Ekstrom, L., 648
Elbaum, C., 374, 389
Eliashberg, G. M., 146, 149
Elleman, D. D., 611, 617(5)
Ellis, W. P., 670
Elmore, W. C., 684
Elsley, R. K., 236, 237
Endriz, J. G., 75, 102(20)
Enoch, R. D., 709
Erich, U., 510
Esaki, L., 178, 181, 185, 191
Esch, R. P., 656
Eschelbacher, H., 571, 591
Espinola, R. P., 405

Essmann, U., 693
Evans, M. J., 350
Evenson, K. M., 219, 228(64), 229(63, 66), 302(64), 303, 304(63)
Everett, G. E., 223
Eviatar, A., 307
Ewan, G. T., 346
Eylon, S., 328, 342

F

Faber, E. S., 646
Fack, H., 221, 260(75, 77), 270(75, 77), 274
Fahlman, A., 68, 99(9)
Fairbank, W. M., 202, 296, 543, 548, 561, 573, 575, 580, 581(88), 584(17, 84), 592
Fairstein, E., 348
Falicov, L. M., 9, 28
Fan, G. Y., 74, 704
Fane, R. W., 674
Fankuchen, I., 659
Farkas, V. D., 542, 557, 560(4), 571(4), 573(4)
Farnsworth, H. F., 107
Farrell, H. H., 566
Farztdinov, M. M., 740
Fate, W. A., 429, 430
Fawcett, E., 2, 4(3), 5, 7, 8, 15, 19, 21
Feder, J., 552
Federov F. I., 379, 380(22), 381(22), 399 (22)
Fedlmann, W. L., 159, 160(43), 167(43), 169(43), 170(43), 184(43), 197(43)
Feist, W. M., 647
Feldman, C., 621
Feldmann, W. L., 239(137), 240, 242 (137)
Feldtkeller, E., 723, 725, 731, 733
Ferentz, M., 521
Ferguson, W. F. C., 670
Ferrell, R. A., 214, 250, 443
Feucht, D. L., 185
Feuersanger, A. E., 632
Feynman, R. P., 203, 479
Fickett, F. R., 21
Field, B. F., 274
Fife, A. A., 238, 258, 281(206), 301
Finegan, J. D., 661, 662

Fink, J., 309
Finnegan, R. D., 215, 217
Finnegan, T. F., 220, 226, 228, 231, 239, 254, 274
Fiory, A. T., 538
Fischer, G., 556, 557(48)
Fischer, J. E., 71
Fischer, T. E., 74
Fiske, M. D., 214, 216, 229
Fitton, B., 74
Flanagan, T. P., 658
Flinn, P. A., 318, 342(33), 351
Floquet, J., 520, 530(101)
Florescu, V., 692
Fog Pedersen, G., 262, 263(217), 266(217), 272(217)
Ford, J., 388, 389(31)
Ford, P. J., 534
Forgacs, R. L., 281, 402
Forlani, F., 160
Forty, A. J., 101
Foster, E. H., 542, 560(8), 573(8), 574, 577, 584(8, 87)
Foster, F. G., 719
Foster, L. V., 719, 720
Foster, N. F., 632, 642, 643
Fournier, E., 31
Fowler, C. A., 720, 721
Fowler, R. H., 68, 72, 162
Fox, M., 682
Francombe, M. H., 639, 642
Francon, M., 657
Frankel, R. B., 363, 509
Frauenfelder, H., 363, 365(107)
Frederick, N. V., 282, 284(259)
Freedman, J. F., 663
Frei, E. H., 683
Freiser, M. J., 695, 704
Frenkel, R. B., 274
Fricke, J. L., 564, 566(67), 570(67), 571(67), 591(67)
Fried, B. D., 146
Frindt, R. F., 258, 281(206)
Fritzshe, H., 181
Fry, T. C., 68
Fryer, E. M., 720, 721
Fuchs, K., 413
Fujime, S., 621
Fulde, P., 139
Fuller, E. R., Jr., 415, 419

Fuller, H. W., 727, 728
Fulton, T. A., 214, 217, 257, 261, 263, 294, 305

G

Gabriel, J. R., 366
Gaerttner, M. R., 395, 396
Gaidukov, Yu. P., 30
Gaines, G. L., Jr., 651
Gaitanis, N., 342
Galkin, A. A., 171
Ganichev, D. A., 85
Garcia, N., 628, 629
Garrett, C. G. B., 504
Garrood, J. R., 689
Gasparovic, R. F., 136
Gatos, H. C., 642
Gaudefroy-Demonbvnes, A., 210
Gauster, W. B., 388
Gavaler, J. R., 258, 642
Geballe, T. M., 492
Geiger, A. L., 165, 166, 167
Gemperle, R., 743
George, J., 238
George, T. H., 107
Gerber, R., 655
Gerhardt, U., 101
Gerholm, T. R., 99
Gerlich, D., 420
Gerstenfeld, F., 318
Ghate, P. B., 415, 420
Giaever, I., 126, 131, 139, 148, 156, 159, 167, 168, 171, 177, 187, 191, 233, 239(141), 240, 242(141), 243, 252(141), 650
Giffard, R. P., 283, 284(265), 285(265), 286(265)
Gilad, P., 307
Gilchrist, J. LeG., 554, 555(37), 556, 557
Gill, D., 517
Gill, W. D., 644
Ginsberg, D. M., 139, 148
Ginzton, E. L., 577
Giovanielli, D. V., 12, 27
Gittleman, J. I., 556
Glaberson, W. I., 481
Glang, B., 178
Glang, R., 620, 647

Gobeli, G. W., 84, 97, 101, 104, 112(50), 113, 674
Göhre, H., 653
Gold, A. V., 8, 34, 35, 39, 40(16, 17), 41, 43, 47(7), 63(17), 64(17)
Goldman, A. M., 214, 216, 247, 251
Gollub, J. P., 296
Gomer, R., 452
Goodkind, J. M., 281, 284, 285(252), 286 (252, 261, 270), 615
Goodman, A. V., 122
Goodman, B. B., 555
Goodman, R. H., 319, 341(39)
Goodman, W. L., 281, 285(255)
Gopal, E. S. R., 512
Gordon, J. P., 223
Goree, W. S., 283, 284(267), 286(267), 611, 612, 613
Gorkov, L. D., 550
Gorlé, D., 542, 560(11), 562(58), 563 (11), 572(58), 573(11, 58), 588
Gorter, C. J., 506
Gould, G., 634, 636
Gould, R. W., 593, 594(125)
Grace, M. A., 518, 525
Graeffe, R., 252
Graham, R. L., 443
Grahman, C., 634, 636
Granato, A. V., 373, 389, 391, 397(6), 405, 409, 411, 412, 414, 415(95), 416 (88), 417, 418(94, 95), 420, 423, 424, 425, 426
Grant, P. M., 666, 667
Gray, P. V., 191
Green, A., 706, 721(31), 725, 726
Green, A. E., 435, 440(143)
Green, R. E., Jr., 380
Greenshpan, M., 325, 340(45), 341(45)
Gregers-Hansen, P. E., 220, 221(69), 257 (69, 71), 260(76), 262, 263(217), 266 (217), 270(76), 272(217)
Gregor, L. V., 645, 648
Gregory, W. D., 200(26), 201, 237, 238
Greiner, E., 29
Greiner, J. H., 252
Griffiths, C. H., 621
Griffiths, P. M., 684, 692
Grimes, C. C., 218, 219, 232, 233, 234, 300, 301, 302

Grimsrud, D. T., 462
Grissom, D., 592
Grobman, W. D., 121
Grohmann, K., 732
Grünbaum, E., 622, 634, 636, 735
Grundlach, K. H., 160
Grundy, P. J., 731
Gubankov, V. N., 304
Günther, K., 633
Guest, G. H., 318, 328, 337(55)
Guggenheim, H. J., 517
Guinau, T., 261, 262(213)
Gundlach, K. H., 163
Gunshor, R. L., 594
Gustafson, J. C., 511
Gustafsson, T., 101
Gustard, B., 692
Gutierrez, W. A., 621
Guyon, E., 210
Gygax, S., 238, 258, 281(206), 301, 537
Gylling, R. G., 513, 518, 526(99)

H

Hacskaylo M., 669
Haden, C. R., 588, 592, 593
Hadley, L. N., 671
Hafner, E., 588
Hafner, H., 100
Hafstrom, J. W., 141, 142
Hagenlocher, A. K., 632
Hagstrom, S. B. M., 99
Hagstrum, H. D., 107, 108(94)
Hahn, H., 542, 560(8), 562, 572, 573(8), 574, 577, 584(8, 87), 590(66), 591
Hahn, J., 348
Halama, H. J., 542, 560(8), 566, 572, 573 (8), 574, 577, 584(8, 87), 591(79)
Halbritter, J., 542, 543, 550, 551(15), 552, 557, 558, 559(50), 560, 561, 562, 565(57), 571(5, 50, 57), 572(66), 575 (54), 590(66)
Hale, M. E., 727, 728
Hall, H. E., 470
Hall, J. F., Jr, 670
Halperin, B. I., 247
Halpern, B., 452
Hamilton, C. A., 221, 224, 225, 226, 227, 228, 233, 269, 270(109)
Hamilton, W. O., 616, 617(14)

Hamisch, H., 732
Hamrin, K., 68, 99(9)
Haneman, D., 661
Hanna, S. S., 310, 343
Hansen, E. B., 221, 257(72), 262(72), 265(72)
Hansma, P. K., 167, 213, 272
Hanson, A. O., 591
Hanzel, D., 327, 329(52)
Haq, K. E., 642
Hardin, W. R., 492
Harding, J. T., 280, 281, 285(254), 286, 287, 288
Hardy, K. A., 309, 366
Hargrove, R. S., 342
Harley, R. T., 511
Harmmann, J. F., 82
Harrick, N. J., 668
Harris, L., 623, 633
Harris, L. A., 106
Harrison, W. A., 8, 418, 419
Hart, H. R., 131, 148
Hartman, P. L., 121
Hartman, T. E., 158, 160, 658
Hartmann-Boutron, F., 520, 530(100)
Hartwig, W. H., 585, 588, 592
Harvey, B. G., 347
Harvey, I. K., 274
Harwig, B, 315
Hass, G., 671
Hauser, J. J., 139, 640, 641
Hauser, J. L., 181
Hauser, M. G., 318
Hazony, Y., 325, 341(46)
Hearmon, R. F. S., 438
Heavens, O. S., 621, 622, 653, 657, 666, 670(147), 674(147)
Hebard, A. F., 296
Hebard, A. H., 611
Heberle, J., 343
Hedman, J., 68, 99(9)
Heidenreich, R. D., 622
Heiman, N. D., 309, 319(6)
Heinicke, W., 454
Heldmann, G., 160
Hempstead, C. F., 555
Henkels, W. H., 248, 250
Henshaw, D. G., 466

Henson, B. L., 453, 462
Herber, R. H., 325, 340(45), 341(45), 344, 346, 364
Hereford, F. L., 451, 452
Herman, F., 102
Hermstreet, R. A., 511
Herring, C., 30, 102
Hershkowitz, N., 309
Hetzler, M. C., 490
Heybey, O. W. G., 511
Hickmott, T. W., 650, 658
Hicks, J. A., 309
Hihara, T., 527, 530
Hikata, A., 389
Hiki, Y., 389, 405, 412, 417, 420
Hildebrandt, A. F., 611, 617(5)
Hill, A. E., 661
Hillman, P., 307
Hilsch, R., 621
Himmel, L., 412
Hindennach, P., 309, 348(10)
Hinds, J. J., 592
Hines, D. C., 51, 56(31), 57
Hinton, T., 380
Hirschkoff, E. C., 510, 533
Hoare, F. E., 489, 504(4)
Hodges, L., 34, 35, 40(17), 47(7), 63(17), 64
Hoenig, H. E., 284, 286(271)
Hörnfield, S., 55
Hoffman, G. R., 658, 661
Hoffman, R. W., 658, 660, 661, 662
Hoffswell, R. A., 591
Hogg, H. A., 542, 557, 560(4), 571(4), 573(4)
Holbrook, R. D., 409
Holder, J., 408, 409(80), 411, 416(88), 420
Holland, L., 620
Holliday, R. J., 531
Holly, S., 217, 221(57)
Holm, R., 160
Horner, H., 507, 508
Hothersall, D. C., 732
Houck, J. R., 395
Houze, G. L., Jr., 685, 725
Howard, D. G., 316
Howling, D. H., 531

Hoyt, E. W., 542, 557, 560(4), 571(4), 573(4)
Hsu, D., 395
Hsu, F. S. L., 596
Huang, K., 413, 415, 417(92), 419, 434, 435
Hubert, A., 693, 695, 725
Hubin, W. N., 139, 148
Hudak, J. J., 35, 41(19), 50(19)
Hudson, R. P., 504, 508, 509(61), 510
Huen, T., 101
Huff, R. W., 146
Hughes, A. L., 67, 68, 99
Hughey, L. R., 121
Huiskamp, W. J., 503, 509(39), 518
Hulbert, J. A., 59
Hull, G. W., Jr., 492
Hulm, J. K., 642
Hunter, W. R., 668
Huntington, H. B., 371
Huntley, D. J., 535
Huntzicker, J. J., 509
Hunzler, J. E., 492
Hurlbut, C. S., 103
Hurych, Z., 163
Hutchinson, R. I., 693, 695
Hutchinson, T. E., 622

I

Iignatiev, O. M., 171
Ikushima, A., 409, 433
Illarionova, N. V., 347
Ingalls, R., 363, 365(105)
Inn, E. C. Y., 82
Inouye, G. T., 296
Inukai, K., 163
Isaac, E. D., 726
Isaacson, R. A., 720, 721(40)
Isako, L. M., 337
Ishii, C., 255, 262(194), 268(194)
Ishiyaki, A., 357
Isin, A., 171, 177, 650
Itskevich, E. S., 30, 59
Ivanchenko, Yu. M., 231, 247, 248(158)
Ivanov, R. D., 643
Ives, H. E., 68, 100
Ivey, H F., 462
Iyengar, P. K., 318

J

Jaccarino, V., 517, 527
Jackson, G. N., 645
Jackson,, J. E., 194, 195(129), 197, 210
Jackson, L. C., 489, 504(4)
Jacobsohn, D. H., 319
Jaklevic, R. C., 164, 165, 166, 167, 257, 291(198), 292, 293
Jakubovics, J. P., 731, 734, 735
James, L. W., 96
Jan, J. P., 1
Janak, J. F., 76
Janocko, M. A.,258, 642
Janus, A. R., 217, 221(57)
Jaoul, O., 261
Jech, C., 110
Jeffries, C. D., 530
Jha, S. S., 247
Jiggins, A. H., 318
Jimenez, J. J., 589
Johansson, G., 68, 99(9)
Johnson, C. E., 525
Johnson, J. E., 632
Johnson, J. L., 255, 262(195), 268(195)
Johnson, R. T., 509, 536, 537
Johnson, W. J., 254
Johnson, W. L., 493, 586
Johnsrud, A. L., 68, 100
Jona, F., 191
Jones, B. K., 26
Jones, C. K., 258, 642
Jones, C. M., 564, 566(67), 570(67), 571(67), 591(67)
Jones, M. E., 689, 691
Jones, R. E., 644
Jones, R. V., 530
Jordan, E L., 646
Jordanov, A., 347, 348(86)
Jortner, J., 444
Joseph, A. S., 64
Josephson, B. D., 127, 199, 200, 201, 203, 204(8), 208(4), 238, 267, 268(3), 291(3, 4), 313, 468
Joy, D. C., 734, 735
Joyce, B. A., 645, 646
Judd, B. R., 518
Judge, D. L., 296
Jüngst, W., 571

K

Kaczer, J., 739, 743
Kaeser, R. S., 508, 509(61)
Kaganov, M. I., 1, 4(1), 9(1)
Kaindl, G., 336, 339(62), 345(62)
Kajimura, K., 251
Kalvius, G. M., 310, 344, 345(76), 346, 352, 357
Kalyamin, A. V., 350
Kambersky, V., 679
Kaminsky, M., 637
Kamm, G. N., 64
Kammerer, O. F., 566
Kamper, R. A., 200, 236, 282, 283(258), 288, 289, 537
Kane, E. O., 73, 76, 100, 101
Kankeleit, E., 326, 328(50), 333, 334, 336 (50), 338(50), 340(50), 348
Kanter, H., 246, 247, 301, 304
Kantor, H., 78
Kao, Y. H., 628, 629
Kaplan, L. H., 648
Kaplan, N., 517
Kappler, E., 10
Kappler, H. M., 318
Karlsson, S., 68, 99(9)
Katila, T. E., 357
Kawasaki, K., 409
Kay, E., 637, 644, 645
Keating, P. N., 419
Keen, B. E., 510
Keller, W. E., 532, 535
Kelly, R., 110
Kelly, T. M., 444
Kelsin, D., 327, 329(52)
Kemp, W. R. G., 586
Kerr, D. R., 650
Kerr, E. C., 466
Kestner, N. R., 444
Ketterson, 34
Ketterson, J. B., 39, 55
Khaikin, M. S., 233
Khalatnikov, J. M., 469, 550
Khan, I. H., 622
Khana, S. M., 148
Khutsishvili, G. R., 525
Kierstead, H. A., 499
Kim, D. J., 363
Kim, Y. B., 555, 610
Kim, Y. W., 222
Kinder, H., 243, 244
Kingery, W. D., 621
Kinser, J. P., 583
Kirk, R. S., 30
Kirk, W. P., 492
Kirschman, R. K., 247, 256, 257
Kislova, G. K., 643
Kistner, O. C., 313, 318, 319(35)
Kitchens, T. A., 363
Kittel, C., 35, 429, 513, 679
Klauder, J. R., 5, 6, 22, 28
Klein, H., 564, 566(67), 570(67), 571 (67), 591(67)
Klein, J. M., 210
Klein, M. L., 435, 440(145)
Kluth, E. O., 236
Knapp, R. A., 83
Kneisel, P., 542, 557, 558(50), 559, 560, 561(57), 562, 565(57), 571(5, 50, 57), 572(66), 573(10), 575(10), 590(66)
Knight, D. J. E., 304
Kodre, A., 327, 329(52)
Koehler, H., 743
Koehler, J. S., 423
Koehler, W. F., 671
Koenig, H. R., 644
Koi, Y., 527, 530
Koite, S., 31
Kolk, B., 315
Kondoh, H., 741
Kopf, L., 144, 145, 147, 148
Kopp, J., 534, 535(129)
Korostoff, E., 646
Kose, V. E., 219, 221, 228(64), 236, 260 (75, 77), 269(118), 270(75, 77), 272, 274, 302(64), 303(64)
Kosevich, A. M., 35
Kostikas, A., 342
Kostiner, E., 331
Kostyshyn, B., 743
Koyama, R., 83, 119, 121
Kraan, W. H., 294
Kranz, J., 706, 721(30), 723, 725
Krasnopolin, I. Ya., 233
Kreisman, P. J., 214
Kress, K. A., 98
Kreuzer, H. J., 170
Krikorian, E., 642

Kristianpoller, N., 83
Krolikowski, W. F., 71, 73, 78, 80, 102 (12), 109, 110, 118
Kroo, N., 171
Kruger, J., 657
Krusius, M., 513
Ku, T. C., 743
Kuchnir, M., 469
Kündig, W., 342, 366
Kugatt, C. E., 100
Kuhlman-Wilsdorf, D., 664
Kuhn, H., 652, 659(128)
Kulik, I. O., 200, 231, 251, 255, 262(193), 268(193)
Kullmann, D., 356
Kumagai, Y., 163
Kuntia, M., 215
Kuntze, M., 564, 566(67), 570(67), 571(67), 591(67)
Kunzler, J. E., 5, 6, 22(4), 28(4), 596
Kuper, C. G., 251
Kurkijärvi, J., 283
Kurti, N., 489, 504(4), 525
Kurtin, S., 161, 162
Kushida, T., 530
Kustom, R. L., 560, 573(60), 574, 591(60)
Kuz'man, R. N., 351

L

Labes, M. M., 621
Lagnado, I., 642
Lamb, J., 406
Lambe, J., 164, 165, 166, 167, 257, 291, 292, 293
Lambeck, M., 711, 715
Lambert R. M., 709
Landau, D. P., 507
Landau, L., 678
Landau, L. D., 378, 463, 464
Lander, J. J., 104
Landolt-Börnstein, 411
Lang, A. R., 736
Lang, G., 307
Langenberg, D., 200, 299(20)
Langenberg, D. N., 200, 218, 219, 220, 226, 228, 229, 230, 231, 234, 239, 246, 247, 252, 273, 274, 299(20)
Langmuir, I., 651

LaPeyre, G. J., 98
Laquer, H. L., 492
Larkin, A. I., 246, 255, 270
Larsen, P. K., 395
Larson, D. C., 621, 634, 636
Lass, J. S., 28
Laude, L. D., 74
Laugier, A., 185
Lavin, P. A., 693, 695
Lawless, K. R., 645, 649
Lawless, W. N., 538
Lawley, A., 664
Learn, A. J., 642
Leask, M. J. M., 509
Le Comber, P. G., 646
Lee, C. C., 31
Lee, P. A., 246
Lee, Y. K., 309, 356
Legendy, C. R., 26
Lehman, H. S., 644
Lehoczky, A., 395
Leibfried, G., 414, 419
Leich, D. A., 573
Lejeune, S., 334
Lekner, J., 468
Lengeler, H., 571
Leopold, L., 237, 238
Leroy, D., 542, 560, 562(58), 563(11), 572, 573(58), 573(11), 588
Leschenko, G. F., 304
Leslie, J. D., 217, 232
Letcher, S. V., 374, 430
Lever, R. F., 646
Levey, R. P., 364
Levgold, S., 535
Levine, J., 448
Levinsen, M. T., 220, 221, 257(69, 71), 260(76), 262,263(217), 266(217), 270(76), 272(217)
Levy, P. M., 507
Lewicki, G., 160, 161
Lewis, J. T., 395
Lewis, R. B., 583, 584(96)
Lichtensteiger, M., 642
Liesk, W., 731
Lifshitz, E., 678
Lifshitz, E. M., 378, 464
Lifshitz, I. M., 1, 4, 9, 35, 37
Likharev, K. K., 304
Lill, R. M., 560, 562, 573(59, 65)

Lim, C. S., 217
Linares, R. C., 648
Lindau, I., 99, 112
Lindberg, B., 68, 99(9)
Lindgren, I., 68, 99(9)
Lines, R. A. G., 512, 525, 527, 528
Lipson, S. G., 251
Lissberger, P. H, 665, 706, 707, 721(32), 723
Little, W. A., 200, 267, 299(20)
Logan, J. S., 644
Logan, R. A., 185, 186
Lomunov, R. I., 734
London, F., 202, 294, 469, 547, 554(24)
Longacre, A., Jr., 210, 232, 270, 302, 304
Longini, R. L., 185
Loram, J. W., 534
Losee, D. L., 174, 175(82), 176
Losee, P. L., 183, 184
Lounasmaa, O. V., 357, 490, 518, 526 (99), 536
Love, R. E., 426
Lowenheim, F. A., 645, 649
Lubbers, J., 503, 509(39)
Lucas, R., 409
Ludwig, W., 414, 419
Lücke, K., 423, 433
Lüders, K., 570
Lüthi, B., 433
Lukens, J. E., 283, 284, 285(268), 286
Lukirski, P., 85
Lundquist, S., 124, 151(1)
Lur'e, B. G., 350
Lynch, F. J., 333
Lythall, D. J., 171

M

McAshan, M. S., 561, 573(62), 590
McCaa, W. D., Jr., 594
McClintock, P.V. E., 452, 453
McConnell, R D., 556
McCormick, W. D., 470
McCrackin, F. L., 673
McCumber, D. E., 212, 217, 268, 269(33, 227), 270, 271, 272, 283
McDonald, D. G., 219, 228, 229, 302(64), 303, 304
MacDonald, D. K. C., 534
McDowell, W. P., 323
Macfarlane, J. C., 274
McGill, T. C., 161, 162
McGraw, R. B., 648
McIntyre, D. A., 682
Mackintosh, A. R., 8
McLachlan, D. S., 552
McMenamin, J. C., 220
McMillan, W. L., 136, 137(10), 139, 149, 150(10), 152, 159, 160(43), 164(10), 167(43), 169(43), 170(43), 184(43), 194, 195, 197(43)
McMillan, W. M., 139
McMullin, P. G., 328
MacRae, A. U., 109, 110
MacRae, D., Jr., 326, 328(51)
McSkimin, H. J., 371, 394, 399, 406, 407, 408, 409, 410, 412
MacVicar, M. L. A., 139, 141, 142, 143, 187, 192(16)
McWane, J. W., 294, 299(299)
McWhan, D. B., 181
Maddin, R., 10
Maddocks, F. S., 644
Mader, S., 633
Magill, P. J., 661
Mahan, G. D., 74, 100, 183
Maier, M. R., 336, 339(62), 345(62)
Maissel, L. I., 178, 620, 637, 644
Major, J. K., 309
Maki, K., 556
Malek, Z., 679
Manning, R. J., 451
Mapother, D. E., 537
Maraviglia, B., 482, 483, 484
March, R. H., 140, 141
Marcinowski, M. J., 731
Marcus, J. A., 22, 27, 30, 31, 34
Marcus, R. B., 622
Marcus, S. M., 164, 239
Margulies, S., 315, 366
Marius, J. A., 33
Marsh, D. M., 664
Marshall, R. F., 416, 417
Martens, H., 543, 564, 566(13), 571, 572 (13, 77, 78)
Martin, R. J., 223
Martin, R. M., 419
Martinet, A., 210
Mason, P. V., 593, 594(125)

Mason, W. P., 371, 374, 391, 392, 393, 397, 429, 430
Masuda, K., 449
Matcovich, T., 646
Mathews, W. N., Jr., 200(26), 201, 238
Mathis, R. D., 645
Matisoo, J., 200, 214, 215, 231, 251, 305
Matricon, J., 556
Matsuzaki J., 71
Mattern, P. L., 318
Matthews, J. W., 621, 622
Matthews, P. W., 510
Matthias, E., 316, 367, 520
Mattis, D. C., 550
Mattox, D. M., 637
Matzke, Hj., 110
Maxfield, B. W., 395, 396
Maxwell, E., 501, 543, 550(14), 574(14)
May, J. E., Jr., 401
Mayer, A., 318
Mayer, H., 620, 653
Mayer, L., 743
Mazak, R. A., 345
Mazei, F., 171
Mazzoldi, P., 478
Mead, C. A., 160, 161, 162, 183, 658
Mee, C. D., 695, 704
Megerle, K., 131, 139, 148, 156
Meiboom, S., 30
Meijer, P. H. E., 509
Meincke, P. P. M., 24
Melchert, F., 274
Mendoza, E., 504, 505(4)
Mercereau, J., 200, 299(20)
Mercereau, J. E., 200, 247, 256, 257, 258, 266(202), 273, 277, 279, 281, 284, 285 (248), 286(271), 291, 292, 293, 295, 299(20), 613
Meredith, D. J., 395
Merrill, J. R., 12, 27, 239(140), 240, 242(140)
Merz, K. M., 736
Merz, N., 564, 566(67), 570(67), 571 (67), 591(67)
Mesak, R. M., 230
Meservey, R., 139, 153, 154, 155, 267, 284
Mesnard, G., 185
Mess, K. W., 491, 492(18), 493(18), 503, 509(39)
Metzger, M., 371, 397(1)
Meyer, F., 674

Meyer, H., 511
Meyer, L., 454, 455, 465(32), 466, 468
Meyerhofer, D., 658
Meyerhoff, R. W., 566, 573
Mezei, F., 171
Mezhov-Deglin, L. P., 493
Michalski, M., 334, 342(61)
Migulin, V. V., 304
Mikkor, M., 182, 183(103), 184(103)
Miles, J. L., 139, 649
Miller, H., 571
Miller, J. G., 398
Miller, K. M., 55
Miller, P. B., 550
Miller, R. F., 622
Miller, R. J., 632
Millet, L. E., 365
Milliken, J. C., 16, 17
Minkova, A., 347, 348(86)
Mirskaya, G. G., 451
Mishory, J., 343
Mitrofanov, K. P., 347, 350
Mittag, K., 570
Mochel, J. M., 256
Modena, I., 462
Möller, H. S., 363
Mössbauer, R.L., 307, 316, 318
Mogab, C. J., 621
Molik, A., 327, 329(52)
Moll, J. L., 96
Molnar, B., 350
Monceau, P., 556
Montgomery, C. G., 545, 568(20, 21), 587(20, 21)
Montgomery, D. B., 601
Moon, J. R., 693, 695
Moon, P. B., 366
Moore, B. K., 499
Moore, S., 179
Moore, W. S., 221, 257(70)
Moraga, L., 634, 636
Moretti, A., 560, 562, 573(59, 65)
Morfeld, U., 309, 348(10)
Morignot, P. H., 542, 560(11), 562(58), 563(11), 572(58), 573(58), 588
Morris, K., 274
Morris, R. C., 141, 171, 177, 192, 650
Morrish, A. H., 529
Morrison, J., 104
Morrison, R., 495, 609, 612

Morse, A. L., 296
Morse, R. W., 428, 429, 430
Moss, F., 451, 462, 479, 480, 482
Moss, G. L., 622
Moyzis, J., 363, 364
Mühlschlegel, B., 203
Mueller, F. M., 37
Müller, J., 634, 636
Muir, A. H., Jr., 320
Mullen, L. O., 200, 236, 252
Mullendore, A. W., 637
Mundt, W. A., 356
Murbach, H. P., 661
Murnaghan, F. D., 377, 383(13), 435(13), 436(13), 437(13), 438(13)
Murphy, J. A., 430, 431
Murr, L. E., 645

N

Nadav, E., 325, 326(49)
Näbauer, M., 202
Nagasawa, H., 515
Nagle, D. E., 316, 318, 532
Nahman, N. S., 594
Naimon, E. R., 378, 414, 415, 418(16, 95)
Nakahigashi, K., 742
Narbut, T. P., 294
Nathan, P. S. P., 318
Neal, W. E. J., 674
Neeper, D A., 468
Neighbor, J. E., 294, 299(299)
Neppell, C. T., 511
Nesbitt, E. A., 640
Neugebauer, C. A., 664
Newbower, R. S., 294, 296, 299(299)
Newhouse, V. L., 200, 594
Nichols, M. H, 102
Nicol, J., 139
Niedermayer, R., 653
Niedzwiedz, S., 344
Nielsen, L., 503, 509(39)
Nielsen, P., 171, 173, 176(80)
Nielson, P., 174
Nilson, P. O., 101
Ninnaja, N., 160
Nisenoff, M., 255, 273, 281, 285(251)
Nistor, C., 331
Nixon, W. C., 734

Nobel, P. S., 318
Noll, W., 435
Nordberg, R., 68, 99(9)
Nordheim, L., 162
Nordling, C., 68, 99(9)
Nordman, J. E., 252
Noreika, A. J., 642
Northley, J. A., 444
Notarys, H. A., 256, 257, 258, 273
Novoseller, D. E., 220
Nussbaum, R. H., 318

O

Obenshain, F. E., 309
Oberley, L. W., 309
Ochici, S. I., 139, 141, 143, 187(16), 192(16)
Ochs, S. A., 658
Ochs, T., 396, 397
O'Connor, D. A., 318, 328, 337(55), 366
Odencrantz, F. K., 671
Oi, I., 743
O'Keeffe, D. J., 509
Oleson, J. R., 356
Olpin, A. R., 68, 100
Olson, D., 366
Olson, K. H., 622
Omar, M. H., 261, 262(213)
Onn, D. G., 449, 450
Ono, K., 357
Onodera, Y., 215
Orbach, R., 509
Oron, M., 636
Osborne, D. V., 470
Osborne, J. A., 677
Osmun, J. W., 181
O'Sullivan, W. J., 52, 56(33), 59
Otake, S., 31
Ouseph, P. J., 337
Ovchinnikov, Yu. N., 246, 255
Ovshinsky, S. R., 621
Owen, C. S., 214
Oxley, J. H., 645

P

Padovani, F., 182, 183
Pahor, J., 327, 329(52)
Pake, G. E., 506

Palatnik, L. S., 621
Palmai, E., 325, 326(49)
Panczyk, M. F., 535
Pankey, T., 621, 632
Panousis, P. T., 34, 47(7)
Panyushkin, V. N., 363, 365
Papadakis, E. P., 391, 401
Paré, V. K., 421
Parker, G.H., 183
Parker, W. H., 218, 219, 220(62), 246, 247, 248, 273, 274, 285
Parks R. D., 256
Pashley, D. W., 621
Passaglia, E., 657, 673
Passow, C., 558, 561(55), 594
Pasternak, M., 325, 341(48)
Paterson, J. L., 299
Patterson, D. O., 364
Paul, D. A. L., 443
Paul, W., 666, 667
Paulus, M., 687
Pauthenet, R., 61
Payne, R. T., 645
Pearson, W.B., 534
Pedersen, L., 220, 221(69), 257(69)
Peerson, J. J., 560, 562, 573(59, 65)
Penar, J. D., 490
Penley, J., 448
Perri, J. A., 650
Persson, B., 366
Peters, R. D., 421
Peterson, R. E., 510
Peterson, R. L., 236, 269(118), 272
Petley, B. W., 200, 274, 275
Petrosjan, V. I., 637
Pettit, G. D., 644
Pfeiffer, L., 309, 319(6)
Phillips, J. C., 76
Phillips, R. A., 39, 41, 43
Pickett, G. R., 513
Pickup, C. P., 586
Pickoszewski, J., 334, 342(61)
Pierce, D. T., 84, 85, 86, 87(52), 88(52), 95, 106
Pierce, J. M., 258, 272(205), 542, 543, 548, 549, 553, 554, 555, 558, 559(6, 36), 560(6), 573, 574(6), 580(36), 584(17, 36, 84), 615
Piosczyk, B., 564, 566(67), 570(67), 571(67), 591(67)

Pipkorn, D. N., 344, 363, 365(107)
Pippard, A. B., 8, 28, 35, 50(15), 255, 268(191)
Platt, A., 185
Pliskin, W. A., 644, 650, 653, 655, 656, 657(132)
Plotnikova, M. V., 350
Pogorelov, A. V., 37
Polcarova, M., 736, 739
Poliak, R. M., 731
Pollina, R. J., 433
Popov, G. V., 350
Poulsen, R. G., 55
Pound, R. V., 313, 365, 579
Povitskii, V. A., 337
Powell, C. F., 645
Powell, M. J. D., 509
Powell, R. J., 102, 107(82), 117, 122
Prange, R. E., 214
Pratt, W. P., 444, 475, 481(65)
Pratt, W. P., Jr., 526, 531
Preston, R. S., 323, 343
Priest, J. B., 661
Priestley, M. G., 47
Pritchard, J. P., Jr., 214
Probst, C., 483
Prothero, W. M., 615
Prothero, W. M., Jr., 284, 286(270)
Prutton, M., 706, 721(31), 725
Pulfrey, D. L., 645, 650
Puma, M., 235, 236, 239(142), 240, 242(142), 260(142)
Purcell, E. M., 99, 545, 568(21), 587(21)
Purrington, R. D., 367

Q

Qaim, S. M., 350
Queisser, H. J., 183
Quinell, E. H., 490
Quinn, J. J., 395

R

Rabinowitz, M., 557, 559, 565
Radebaugh, R., 289, 538
Raghaven, K. S., 664
Raimondi, D. L., 645
Raith, H., 728
Ramo, S., 545, 568(19)

Ramsey, J. B., 671
Randlett, M. R., 622, 631
Rahgeber, E., 571
Rayfield, G. W., 444, 448(9), 464, 476, 477
Raymond, J. P., 185
Read, D., 408, 409(80)
Read, T. A., 371, 397
Readey, D. W., 647
Rebarchik, F. N., 637
Rebka, G. A., Jr., 313
Rebuehr, M. T., 560, 562, 573(59, 65)
Recknagel, E., 309, 348(10)
Reed, W. A., 2, 4(3), 5, 7, 8, 9, 15, 19, 21, 22, 24, 28(4), 29, 30, 31
Reggev, Y., 325, 341(48)
Reif, F., 454, 455, 465(32), 466, 476, 477
Reivari, P., 343, 357
Reizman F., 656
Repici, D. J., 237, 238
Reynik, R. J., 346
Ricci, M. V., 483
Rice, S. A., 444
Richards, D. B., 535
Richards, J. L., 633
Richards, P. L., 200, 221, 222, 270(79), 299(20), 301, 302
Richardson, J. E., 319, 341(39)
Richardson, J. K., 318
Richardson, O. W., 67, 92(1)
Richardson, R. C., 538
Richter, H., 621
Riedel, E., 225, 226
Rieder, H., 560, 562(58), 572, 573(58)
Riegel, D., 309, 348(10)
Ries, R. P., 21, 499, 535, 537
Riesenman, R., 331
Rinef, G., 61
Risley, A. S., 219, 229(63), 303, 304(63)
Rives, J. E., 511
Roach, W. R., 491, 508, 509
Roberts, L. D., 364
Roberts, P. H., 444, 474, 475, 476, 480, 481(62)
Robichaux, J. E., Jr., 494, 510
Robinson, F. N. H., 509, 516
Rochlin, G I., 167, 213, 233, 239, 242 (135), 272(38)
Rockstad, H. K., 633
Roderick, R. L., 372

Rodgers, J. M., 730
Rodolakis, A. S., 632
Roellig. L. O., 444
Rogers, J., 148
Rogers, J. S., 148, 195, 210
Rogovin, D., 245, 246
Rojansky, V., 99
Rollason, R. M., 674
Romanov, Yu. F., 350
Ron, A., 251
Ron, M., 344
Rose, M. E., 518
Rose, R. M., 139, 141, 143, 187(16), 192 (16), 566
Rose-Innes, A. C., 586
Rosell, J. M., 171
Rosen, A., 296
Rosenbaum, R., 491, 492, 536
Rosenberg, H. M., 489, 504
Rosenberg, M., 692
Rosenblum, B., 556, 557
Rosenblum, S. S., 520
Rosenzweig, N., 521
Roshon, D. D., 743
Rosner, L. G., 643
Rossman, G. R., 284, 286(271)
Roth, W. L., 740, 741, 742
Rothwarf, A., 150
Rouard, P., 666
Rowell, J. M., 127, 135, 136, 137(10), 139, 144, 145, 147, 148, 149, 150(10), 158, 159, 160(43), 161, 164(10), 167 (43), 169(43), 170(43), 171, 172, 173, 175(8, 79), 176, 181, 186, 194, 195, 197, 210, 212, 239, 240, 242(137)
Rubin, L. G., 489, 504(6), 538
Rubinstein, H., 727, 728(51)
Rubinstein, M., 307
Ruby, S. L., 351, 366
Ruegg, F. C., 328, 341
Russell, D. C., 309, 366
Russer, P., 221, 260(74, 78), 270(78), 304
Ryden, W. D., 29

S

Saermark, K., 395
Saito, S., 742, 743
Salama, K., 389, 421

Salaün, M., 573, 574, 578
Salinger, G. L., 494
Samoilov, B. N., 525
Samson, J. A. R., 83, 120
Samuelsen, M. R., 223
Sanchez, J., 210
Sandell, R., 235, 236
Sanders, T. M., 444, 445, 448
Sandfort, R. L., 51, 56(31)
Santini, M., 459, 461, 478
Sapp, R. C., 510
Sarnot, S. L., 161
Sarwinski, R. E., 508, 509
Sasaki, W., 139, 141
Sato, H., 622
Satterthwaite, C. B., 21, 535
Sattler, K., 74
Sauer, C., 316
Sauer, W. E., 346
Sauro, J., 659
Sawatori, Y., 181
Sawatzky, E., 645
Sawicki, A., 334, 342(61)
Scalapino, D., 200, 299(20)
Scalapino, D. J., 146, 147, 164, 200, 214, 227, 229, 230, 231, 233(94), 234, 239, 245, 246, 250, 274, 299(20)
Scaramuzzi, F., 461, 462, 470
Schaller, H., 336, 339(62), 345(62)
Scheer, J. J., 112
Schemeckenbecher, A., 646
Schempp, A., 564, 566(67), 570(67), 571(67), 591(67)
Schermer, R. I., 526, 531
Scheule, D. E., 412
Schirber, J. E., 52, 56(33), 58, 59
Schlier, R. E., 107
Schmidt, O., 571, 572(78)
Schmunk, R. E., 412
Schneider, W., 367
Schoen, A. H., 318
Schoepe, W., 483
Schooley, J. F., 139
Schrader, H.-J., 221, 260(75), 270(75), 274
Schramm, C. W., 583
Schrey, F., 645
Schrieffer, J. R., 130, 146, 429, 547
Schrieffer, R., 203
Schroeder, H., 645, 650(91)

Schroeder, J. B., 648
Schroen, W., 214, 252, 653
Schulson, E. M., 734
Schulz, L. G., 668
Schulze, D., 570
Schuster, S., 664
Schwartz, B. B., 258, 262(204), 363
Schwartz, K. W., 467
Schwartz, N., 642, 659
Schwarz, H., 637
Schwettman, H. A., 543, 548, 561, 573 (17, 62), 584(17)
Schwidtal, K., 214, 215, 217, 252
Scott, A. C., 254
Scott, G. B., 52, 57(32)
Scott, R. B., 352
Scott, W. C., 268, 270
Scribner, R. A., 535, 536
Scruggs, D. M., 632
Scully, M. O., 246
Scurlock, R. G., 525
Seah, M. P., 101
Searle, C. L., 588
Sears, F. W., 409
Segal, D., 325, 341(48)
Segall, B., 72
Seib, D. H., 105, 106, 108
Seki, H., 391
Sella, C., 622
Sellmyer, D. J., 18
Septier, A., 573, 574, 589, 590(109)
Seto, J., 252
Shakespeare, E. C., 318, 328, 337(55)
Shal'nikov, A. I., 493
Shamov, A. I., 337
Shapiro, M. J., 387, 388(30), 389
Shapiro, S., 139, 217, 218, 219, 221 224, 225, 226, 232, 260, 262(59), 268(210), 270, 299, 300, 301, 302
Shapiro S. S., 139
Shaw, J. C., 55
Shaw, R. W., 426
Shaw, W., 552
Shay, J. L., 102
Shaylor, H. R., 318, 328, 337(55)
Sheard, F. W., 221, 257(70)
Shechter, H., 344
Shen, L. Y. L., 139, 171, 172, 173, 175 (79), 176
Sherrill, M. D., 139

Shewchun, J., 180, 181, 191
Shimomura, Y., 742, 743
Shinohara, M., 357
Shirley, D. A., 509, 512, 520, 527, 529 (113), 530(113)
Shoenberg, D., 33, 34, 41, 42
Shoshani, A., 319
Shpinel, V. S., 347, 350
Shtrikman, S., 307, 683
Shulze, D., 564, 566(67), 570(67), 571 (67), 591(67)
Siart, O., 564, 566(67), 570(67), 571 (67), 591(67)
Siday, R. E., 293
Sidlow, R., 560, 573(60), 574, 591(60)
Siegbahn, K., 68, 99, 346
Siegel, B. M., 623, 633
Siegmann, H. C., 74
Siegwarth, J. D., 289
Sierk, A. J., 591
Sievers, A. J., 236, 237
Silsbee, H. B., 397, 398(67)
Silver, A. H., 200, 233, 235, 236, 257, 258, 263, 265, 272(113), 277, 279 (219), 280, 284(219), 288, 291, 292, 293, 304, 537
Silver, M., 449, 450
Simmonds, M. B., 247, 248, 282, 283 (258), 285
Simmons, G., 411
Simmons, J. G., 159, 160(35, 36, 37, 38, 39, 40, 41, 42), 163, 658
Simopoulos, A., 342
Simpkins, R. A., 286
Simpson, J. A., 85, 100
Sites, J. R., 526
Sitton, D. M., 451, 462, 479, 480
Sjöström, C. J., 220, 221, 257(69, 72), 262(72), 265(72)
Sklyarevskii, V. V., 525
Skripkina, P. A., 637
Slack, G. A., 534, 535(129), 740, 741, 742
Slater, J. C., 545, 568(18, 22), 586
Slichter, C. P., 513
Sloope, B. W., 622
Slyusarev, V. A., 231
Smejtek, P., 449
Smith, C. S., 412
Smith, F. W., 552, 562(33)
Smith, H. A., 526

Smith, H. J., 217, 232
Smith, J. G., 593
Smith, N. V., 76, 77(26), 96, 100, 105, 108(87)
Smith, P., 139
Smith, P. H., 139, 649
Smith, T. I., 561, 573(62)
Sneed, R. J., 642
Soerensen, O. H., 223
Solomon, A. L., 632
Solyom, J., 177
Sommer, W. T., 444, 454
Soulen, R. J., 538
Soumpasis, D., 570
Southgate, P. D., 396
Southwell, W. H., 365
Spangler, G. E., 451, 452
Spanjaard, D., 520, 530(100)
Sparks, C. J., Jr., 663
Spear, W. E., 646
Spicer, W. E., 69, 70, 71, 73, 75, 76, 77, 78, 79(10, 22), 84, 92, 96, 102, 106, 111, 112, 113(100)
Spijkerman, J. J., 328, 341, 344, 347, 348 (85), 364
Spivak, G. V., 643
Spivey, S., 692
Springett, B. E., 444, 471
Springford, M., 55
Sprouse, G. D., 310
Standley, C. L., 644
Stark, R. W., 9, 24, 34, 48, 49(13), 50, 64, 467
Stebble, R., 731
Steckelmacher, W., 653, 655(133)
Steele, S. R., 647
Steffen, R. M., 367
Steger, J., 331
Stegum, I. A., 46
Stein, K. U., 723, 725
Stein, S. R., 541, 580, 585, 587(1), 589 (1)
Steinberg, H. L., 673
Steinberg, R. F., 632
Steinrisser, F., 181, 183
Stelzer, E. L., 648
Stepanov, E. P., 525
Stephen, M. J., 245, 246(148), 249, 610
Sterling, H. F., 646
Sterling, S. A., 302

Stern, R., 424, 425, 426
Stewart, A. T., 444
Stewart, W. C., 212, 268, 270
Steyert, W. A., 358, 363, 515, 518, 526, 528, 531
Stiles, J., 181
Stiles, P., 34
Stiles, P. J., 139, 178, 181(90)
Stockbridge, C. D., 655
Stokes, R. S., 239(143), 240, 241
Stolfa, D. L., 281, 285(252), 286(252)
Stoltz, O., 542, 557, 558(50), 559(50), 560, 561(57), 562, 565(57), 571(5, 50, 57), 572(66), 590(66)
Stone, D. R., 34, 35, 40(17), 41(19), 47(7), 50(19), 51, 56(31), 63(17), 64 (17)
Stone, J. L., 585, 588, 592
Stone, N. J., 509
Stoner, E. C., 677
Stoney, G. G., 662
Story, H. S., 661
Strässler, S., 243
Strait, S. F., 250, 259
Strathen, R. E. B., 412, 416
Stratton, R., 160,161(44), 182, 183
Straus, J., 194, 196, 197(130)
Strayer, D. M., 481
Strnad, A. R., 555
Stroberg, E., 631
Stromberg, R. E., 657, 673
Strong, J., 652
Strong, P. F., 139
Strongin, M., 566, 628, 629
Stroubek, Z., 181
Strube, H., 564, 566(67), 570, 571(67), 591(67)
Stuart, R. N., 80
Stünkel, D., 653
Stump, R., 443
Sudraud, P., 589
Suelzle, L. R., 591
Suhl, H., 556
Sullivan, D. B.,200, 236, 252, 269(118), 272, 274, 275(244), 284(244)
Sun, F. S., 571, 572(78)
Sun, R. K., 571, 572(77)
Sunyar, A. W., 313
Sutton, J., 139
Suzuki, K., 684

Suzuki, T., 389, 414, 415(95), 418
Suzuki, Y., 163
Svartholm, N., 99
Svidzinskii, A. V., 231
Svistunov, V. M., 230, 233, 239
Swanson, K. R., 347, 348(85)
Swartz, K. D., 412, 420
Sweet, J. N., 213, 223, 272(38)
Swenson, C. A., 59, 489, 499, 504(5), 506, 509(50)
Swihart, J. C., 552
Swinehart, R., 526, 531, 538
Symko, O. G., 493, 494, 510, 513, 517, 532(90), 533, 536
Sze, S. M., 184, 191
Szécsi, L., 558, 559, 560(52), 573(52)
Szentirmay, Zs., 171
Szymanski, A., 621

T

Tabor, W. J., 718
Taconis, K. W., 261, 262(213), 294, 466
Taft, E., 84
Taguchi, I., 217
Takayama, H., 251
Takeda, T., 741
Taloni, A., 661
Tanasoiu, C., 692
Tangherlini, F. R., 668
Tanner, D. J., 471, 472, 474
Tauber, R. N., 646
Taur, Y., 221, 222, 270(79), 301
Taylor, B. N., 136, 150, 200, 218, 219, 220(62), 229, 230, 231, 233(94), 234, 239, 273, 274
Taylor, M. T., 27
Taylor, R. D., 318, 363, 525, 532
Tebble, R. S., 682, 730
Tedman, C. S., Jr., 566
Tedrow, P. M., 139, 153, 154, 155, 267
te Kaat, E., 10
Templeton, I. M., 52, 57(32), 59, 534
Templeton, J. E., 512, 527
Tennart, W. E., 299
Ter Harmsel, H., 506
Terry, C., 491
Tesfandi, R. L., 181
Thelen, A., 665
Theuerer, H. C., 640, 641, 646

Thiene, P., 247, 280, 281, 285(254)
Thomas, B. W. J., 726
Thomas, D. E., 210
Thomas, G., 637
Thomas, J. F., Jr., 412, 414, 418, 420
Thomas, P., 183
Thomas, R. L., 395
Thompson, R., 517
Thomson, J. D., 364
Thorsen, A. C., 64
Thun, R., 671
Thurston, R. N., 374, 383, 384(28), 385, 386(28), 387, 388(30), 389, 434, 435, 437(26), 439, 440
Tiemann, J. J., 210
Tien, P. K., 223
Tietjen, J. J., 647
Tiller, C. O., 622
Tilley, D. R., 237
Tinkham, M., 296
Tinu, T., 331
Title, R. S., 644
Tittmann, B. R., 426, 427, 428
Todd, R. J., 222
Tolansky, S., 656
Tolhoek, H. A., 518
Tomasch, W. J., 152, 153
Tombrello, T. A., 573
Toots, J., 274
Toth, R. S., 622
Toupin, R. A., 379, 385(18), 435(18), 438(18), 439, 440(18)
Tourtellotte, H. A., 637
Townsend, P., 139
Toxen, A. M., 139
Träuble, H., 693
Traum, M. M., 96
Trautwein, A., 318
Travis, J. C., 344
Trela, W. J., 575, 580, 581(88), 592
Treves, D., 328, 342, 683, 706
Triftshäuser, W., 310
Trillat, J. J., 622
Trivisonno, J., 430, 431
Truell, R., 372, 373, 374, 391, 397(6), 424, 425
Truesdell, C., 379, 435, 439
Tschanz, J. F., 510
Tsui, D. C., 35, 40(18), 63(18), 64, 181, 182, 184, 191

Tsujimura, A., 527, 530
Tsuneto, T., 430
Tufte, O. N., 648
Turneaure, J. P., 541, 542, 543, 549, 550, 551, 552, 557, 558, 559(12), 560(3), 561, 562(3, 16), 563, 564, 565(16), 566, 569, 571, 572(7, 16), 573(3, 12, 29, 62), 585, 587(1), 589(1, 3), 591 (12)
Turner, D. W., 68
Turner, G., 395
Turner, J. A., 658
Turrell, B. G., 527
Tuzzolino, A. J., 82, 83(38)

U

Uddin, M. Z., 33
Uebbing, J. J., 107
Uehling, E. A., 397, 398
Ullman, F. G., 163
Ulrich, B. T., 296, 301
Ulrich, R. B., 236
Umkin, K. C., 85
Unvala, B. A., 635

V

Vaisnys, J. R., 30, 181
van Alphen, R. M., 33
Van Bladel, J., 583
Van Den Meijdenberg, C. J. N., 466
Vanderkooy, J., 34
van Dijk, H., 506
Van Duzer, T., 252, 302, 545, 568(19)
van Essen, C. G., 734
Vanfleet, H. B., 365
Van Hove, L., 144
van Kempen, H., 537
van Laar, J., 112
van Rijn, C., 506
Vant-Hull, L., 286
Vant-Hull, L. L., 533, 613
Van Vleck, J. H., 502
Varmazis, C. D., 566
Varshni, Y. P., 420
Varteresian, R., 210
Vashishta, P., 217

Vasicek, A., 672, 673
Vasilev, Y. V., 359
Vassell, W. C., 182, 183(103), 184(103)
Vaughan, R. W., 363, 364
Verderber, R. R., 658
Verleur, H. W., 667
Vermaak, J. S., 663
Vernet, G., 233, 246
Vernon, F. L., Jr., 246, 247, 301
Vertsner, U. N., 734
Ververkin, G. V., 359
Vetter, J. E., 564, 566(67), 570(67), 571(67), 591(67)
Viet, N. T., 542, 543(3), 560(3), 561, 562(3), 564, 566(3), 569, 571(3), 573(3), 578, 589, 590(109)
Vilches, O. E., 489, 537
Vincent, D. A., 281, 285(255)
Vinen, W. F., 470
Vinson, J. S., 451
Violet, C. E., 318, 344
Visscher, P. B., 28
Voderohe, R. H., 319
Vogel, H., 318
Voigt, W., 377, 415, 434, 435, 441
Vononov, F. F., 363
Vook, R. W., 663
Vortruba, J., 562, 572(66), 590(66)
Vystavskiy, A. N., 304

W

Wackerle, J., 443
Wagner, D. K., 499
Wagner, F., 336, 339(62), 345(62)
Waheed, A., 337
Wainfan, N., 659
Wajda, E. S., 647
Waldorf, D. L., 59
Waldram, J. R., 255, 268(191)
Walker, C. T., 511, 534, 535(129)
Walker, G. A., 663
Walker, G. B., 593
Walker, J. C., 309, 319(6), 356
Walker, L. R., 527
Walker, W. P., 664
Wallace, D. C., 383, 411, 434, 435, 437(27), 438(27), 440(27, 91), 441(27)
Wallace, W. D., 395, 396

Wallden, L., 101, 112
Walton, D., 490
Wang, H., 411
Wang, R. H., 273, 284, 286(271)
Warburton, R. J., 284, 285(268), 286
Warfield, G., 180, 181, 191
Warman, J., 212
Warner, A. W., 655
Warnick, A., 281
Warren, W.H., Jr., 13
Warrington, D. H., 730
Wasserman, H. J., 663
Watanabe, K., 82
Waterman, P. C., 379, 380(21), 396(21), 405
Wattamaniuk, W. J., 170
Waxman, A., 180, 181, 191
Webb, R. A., 283, 284(265), 285(265), 286(265)
Webb, W. W., 248, 250, 259, 283, 284, 285, 286, 296
Wehner, G. K., 107, 637
Weiser, K., 644
Weissman, I., 542, 549, 551, 552, 557, 563, 566, 571(7, 9), 572(7), 573(29), 580(29)
Wells, J. S., 219, 228(64), 229(66), 302(64), 303(64)
Wender, S. A., 309
Wernick, J. H., 596
Wert, C. A., 371, 397(1)
Werthamer, N. R., 223, 224, 225, 231(84), 232, 242, 243, 268(84), 270(106), 640, 641
Wertheim, G., 527
Wertheim, G. K., 328, 344
Weyer, G., 309, 348(10)
Weyhmann, W., 526, 531, 538
Whall, T. E., 534
Wheatley, J. C., 283, 284(265), 285(265), 286(265), 469, 489, 491, 494, 503, 509, 510, 533, 536, 537
Whinnery, J. R., 545, 568(19)
White, G. K., 489, 504(3)
White, P., 648
White, R. M., 433
White, W. C., 671
Whitton, J. L., 110,
Wiedemann, W., 356
Wiegand, J. J., 220, 256, 257(68)

Wiggins, J. W., 356
Wiik, T., 252
Wilcock, J. D., 661
Wilenzick, R. M., 309, 367
Wilkins, J. W., 146, 395
Wilkinson, A., 163
Wilks, J., 463, 535
Williams, A. R., 76
Williams, H. J., 719
Williams, J., 406
Williams, R., 122
Williams, R. L., 454
Willis, W. D., 281, 285(255)
Wilman, H., 661
Wilson, G. V. H., 525, 528
Wilson, I. G., 583
Wilson, P. B., 542, 543, 548, 557, 560(4), 561, 571(4), 573, 584(17, 84)
Wilson, P. J., 527
Wilson, T. R., 10
Windmiller, 34
Windmiller, L. R., 24, 34, 39, 47, 48, 49 (13), 50(13), 55, 64
Winsor H. V., 101
Winterling, G., 454
Wire, G. L., 390
Witt, F., 663
Witt, T. J., 274
Wohlleben, D., 732
Wolf, E. L., 174, 175(82), 176, 181, 183, 184
Wolf, I. W., 649
Wolf, W. P., 509, 510, 511
Wolfe, R., 30
Wood, E. A., 719
Wood, S. B., 210
Woodruff, T. O., 432
Woods, A. D. B., 466
Woody, D., 299
Woolam, J. A., 23
Woolf, M. A., 444, 448(9)
Wooten, F., 80, 101
Wu, T. M., 227
Wulff, J., 566
Wyatt, A. F. G., 139, 148, 171, 221, 257 (70)
Wyder, P., 537
Wyslocki, B., 734

Y

Yahia, J., 31
Yamada, T., 740, 742, 743
Yamashita, T., 215
Yanson I. K., 200, 230, 233, 239, 240, 242(136), 249
Yarnell, J. L., 466
Yates, M. J. L., 523
Yen, H. C., 542, 587(2), 592
Yonetani, T., 307
Yoshihiro, K., 139, 141, 251
Yoshioka, H., 217
Young, F. W., Jr., 10
Young, L., 645, 650
Young, R. C., 16, 17
Young, T., 661
Yu, A. Y. C., 184

Z

Zane, R., 329
Zanin, S. J., 653, 657(132)
Zavaritskii, N. V., 141, 142
Zawadowski, A., 177
Zeitman, S. A., 642
Zelevinskaya, V. I., 637
Zeller, H. R., 167, 168, 171, 177, 187, 191, 239(141), 240, 242(141), 243, 252 (141)
Zemel, J. N., 632
Zener, C., 432
Zeuthen Heidam, N., 223
Zeyfang, R., 663
Zharkov, G. F., 200
Zijlstra, H., 505
Zil'berman, L. A., 247, 248(158)
Zimmer, H., 304
Zimmerman, J. E., 200, 216, 217, 233, 235, 236, 247, 258, 263, 265, 269(118), 272, 276, 277, 279(219), 280, 281, 282, 284(219, 259), 285(247, 254), 286, 287, 288, 289, 291, 293, 295, 302, 537
Zimmerman, W., 444, 475, 481(65)
Zimmermann, W., Jr., 509
Zipfel, C., 444
Zvarich, S. I., 337
Zych, D. A., 295

SUBJECT INDEX

A

Acid-string saw, 10
Ac method, for energy distribution curves, 94
Acoustic cyclotron resonance, 430
Aerosol OT, in magnetic domain studies, 688–692
Alternating Josephson current, 235–238
Aluminum-aluminum oxide junctions, 173
Ammeter, absolute, 284
Amplitude factor, in de Haas-Van Alphen effect, 37–39
Angular energy distribution curves, 100–101
  see also Energy distribution curves
Angular fluid motion, in superfluid, 469–470
Anomalous skin effect, 546
Antiferromagnetic domains, methods for, 740–743
Auger process, in Mössbauer spectroscopy, 109, 310

B

Barrier(s)
  cadmium or zinc sulfide types, 191
  carbon, 192
  oxidation procedure for, 189–190
  preparation of, 188–189
  in tunnel junctions, 161–167
Barrier emission processes, in normal-metal tunneling, 167–169
BCS theory, 131, 137, 144–145, 153, 550
Bessel functions, 44, 55

Birch coefficients, 434
Birefringence
  defined, 715–716
  magnetic, 715–718
Birefringence method, in magnetic domain studies, 707–721
Birefringent crystals, 699
Bitter-type solenoid, in magnetoresistance measurements, 20, 595
Bloch energy, 137
Bloch functions, 75
Blume detection system, 403, 405
Bordoni peaks, Peierls barriers and, 422
Brewster angle of incidence, 704
Brillouin function, 502
Brillouin zone, 3
Brookhaven National Laboratory, 590
Brugger elastic constants, 383, 385–386, 414, 438, 440, 434
Bubble model, of negative ion and positronium atom, 443–445

C

Capacitors, thermometry and, 538
CARN technique, in thin films, 656
Cathodes, thermionic, 451
Cavities, materials for, 563–564
Cavity modes, tunnel junctions and, 229–233
Celacol film, 692
Cerous magnesium nitrate (CMN), 503–511
Cesium, in photoemission measurements, 110–112
Cesium ampoules, 113

Cesium channels, 113
Chemical vapor deposition, of thin films, 645–648
  see also Thin films
Circularly polarized light, 700
Clarke slug, 209, 260–262
  applications of, 296–299
Cleavage planes, in photoemission measurements, 103
Clebsch-Gordon coefficient, 521–522
CMN (cerous magnesium nitrate), in thermometry, 503–511
Cobalt, de Haas-Van Alphen effect, and, 61–63
Colloid technique, in magnetic domain studies, 687–692
Conducting pastes, in magnetoresistance studies, 13
Continuous wave composite oscillator systems, 397–398
Copper, magnetoresistance of, 7
Coulomb pseudopotential, 149
Coupling network, for superconducting resonators, 581–584
Cryogenic microwave techniques, 584
Crystals
  birefringent, 699
  light propagation in, 700–702
  small-amplitude waves in, 376–381
Cunsolo method, in ion drift velocity measurements, 456–458
Curie-Weiss law, 502–505, 517
Cyclotron mass measurements, in indium, 60
Czochralski technique, 57

D

Damping effects, in ultrasonic studies, 425–426
Dayem bridge, 255
Dc fields, superconductor shielding and, 612–617
Dc supercurrent, 212–217
  external noise and, 247–248
De Haas-Van Alphen effect, 9, 24, 33–34
  amplitude factor in, 37–39
  amplitude measurements in, 54
  Bessel functions in, 44, 55
  block diagram for, 47–48
  in cobalt, 61–63
  computer analysis in, 55–56
  Dingle temperature in, 55
  experimental methods for, 33–64
  Fermi surface and, 58–60
  in ferromagnetic metals, 61–64
  field dependence in, 54–55
  field modulation technique in, 44–56
  frequencies in, 35–37
  frequency measurements in, 51–52
  frequency selectivity in, 46–47
  impedance matching and filtering in, 49–50
  for lead, 56–60
  magnetocrystalline anisotropy in, 40–41
  modulation field in, 50–51
  for nickel, 63–64
  notation used in, 36
  oscillation frequency in, 37
  pickup coil for, 48–49
  preamplifier in, 50
  relative frequency measurements in, 53
  rotation oscillations in, 53
  spin-splitting factor in, 39
  temperature dependence in, 38, 54
  vector relations in, 40–44
Detector, regenerative, 302
Dewars, for Mössbauer spectroscopy, 352–356
dHvA, see De Haas-Van Alphen effect
Differential tunnel conductance, 126
Dingle temperature, 38, 54–59
Dislocation effects, in ultrasonic studies, 421–428
Doppler effect spectrometers, 316, 342
Drives
  constant-velocity, 318–319
  for Mössbauer spectrometers, 316–346
Drude theories, 68

E

EDC, see Energy distribution curve
e/h constant, Josephson effects and, 273–274
Elastic coefficients, wave equation and, 439–441
Elastic constants
  calculation of, 411–414
  central potentials and, 414–418

Elastic constants (contd.)
noncentral potentials in, 418–419
significance of, 410–411
static-lattice contributions to, 414–419
temperature dependence of, 419–421
ultrasonic measurements of, 410–421
Electric field gradient, in Mössbauer spectroscopy, 314
Electrode excitations, in normal-metal tunneling, 169–170
Electrodeposition, of thin films, 648–650
Electrode-semiconductor-electrode tunnel junctions, 187
Electrolytic saw, 10
Electromagnets, characteristics of, 595
Electron(s)
escape of at surface, 80–81
field emission of, 452–453
injection of from thermionic cathodes, 451–452
maximum kinetic energy of, 68
motion of in magnetic field, 2
open and closed orbits of, 2–3, 6
photoelectric injection of in liquid helium, 448–449
photoinjection of, 122
Electron-beam evaporation system, 635–636
Electron bombardment, in photoemission measurements, 104
Electron-electron scattering, in photoemission, 77–79
Electron microscopy
in magnetic domain studies, 726–734
replica method for, 692–693
scanning, 734–735
Electron multipliers, 99
Electron-phonon coupling constant, 149
Electron-phonon interaction, tunneling as probe of, 149–150
Electron-phonon scattering, 79
Elecron-plasmon scattering, 79
Electron tunneling, in solids, 123–197
see also Normal-metal tunneling; Semiconductor tunneling; Tunneling; Tunnel junction
Eliashberg equations, 149
Energy distribution curve, 69
ac method for obtaining, 92–97
block diagram of ac method in, 94
angular measurements for, 100–101
cesium channels and, 113–114
for copper, 71
future research in, 121–122
linear sweep method in, 97
lowered electron affinity and, 110–112
measurement of, 84–102
photoemission chambers and, 116
sample preparation in, 102
temperature and, 101–102
time differentiation method in, 97–98
weak structure in, 96
Energy gap, in superconductors, 135–140
Epitaxial growth, in thin films, 621
Epoxies, in magnetoresistance measurements, 13
Equivalent circuit models, of tunnel junctions, 268–273
Error signal, in Mössbauer spectroscopy, 326–327

F

Faraday effect, 704–705, 707–708
Faraday rotation, 703–704
Fast flyback operation, in Mössbauer spectroscopy, 340
FECO technique, in thin films, 656
Feedback velocity control, in Mössbauer spectroscopy, 326–331
Fermi energy, electron escape and, 80–81
Fermi level, 9
in copper, 72
photoemission analyzer and, 91
in semiconductor tunnel junctions, 180
Fermi surface, 6, 29
changes in, 58–60
of copper, 7
electron motion and, 2
Hall coefficient and, 4
longitudinal magnetoresistance and, 30
open and closed orbits on, 3
Fermi velocity, 550–551
Ferromagnetic nickel, spin-dependent tunneling in, 154
Field effect transistors, resistance thermometers and, 499–501
Field emission ion sources, 452–453
Field modulation, de Haas-Van Alphen effect in, 34, 44–56

SUBJECT INDEX

Field splitting, spin-dependent tunneling and, 154
Finite-amplitude waves, in solids, 387–389
Foucault method, in magnetic domain studies, 728–730
Fowler theory, 72–73
Fowler-Nordheim tunneling equation, 162
Free liquid surface, ions and, 481–483
Free-precession technique, in NMR measurements, 516
Frequency measurements, in de Haas-Van Alphen effect, 51–52
Fresnel effect, 727

G

Gallium arsenide
 cesiated, 110–111
 chemical cleaning of, 182
 contaminants of, 106–107
Galvanometer, Clarke slug as, 298
Gamma-ray anisotropy, 529
Gamma rays, recoil-energy loss in, 308
Gaseous discharges, ion injection in, 453–454
Gas flow coolers, for Mössbauer spectroscopy, 356
Gas photoionization, in ultraviolet photoemission measurements, 83
Germanium resistors, in thermometry, 491–492
Getter-sputtering system, 641–642
Glan-Thompson prisms, 706, 720
Glow-discharge sputtering, 638–642
Gold, electron-electron scattering in, 78
Gradiometers, 283–286
Gravimeter, 284

H

Hall coefficient, 4–5
 magnetoresistance and, 26
Hall effect, 1
 field dependence of, 4
 probes for, 743
Heat cleaning, in photoemission measurements, 105–107
Helium
 liquid, see Liquid helium
 turbulent superfluid, 479–481

Helium isotopes, 468–469
Helium-II
 thermodynamic properties of, 463–464
 "turbulent" state of, 479–481
Helium 3
 ion behavior in, 468–469
 melting curve for, 535–536
Helium 4, 469–470
Helmholtz free energy, 437
High-field magnetoresistance, measurement of in metals, 1–31
High magnetic fields, superconducting tunneling in, 153–156
High-Q superconductivity resonators
 design and fabrication of, 567–570
 frequency measurement and control in, 580–581
 frequency stability in, 585–586
 materials for, 570–575
 measurement techniques for, 575–580
Hinteregger discharge lamp, 121
Homodiodes, 185

I

Immersion spectrophotometry, 670–671
Impedance, surface, 544–550
Impedance matching, in de Haas-Van Alphen effect, 49–50
Impulsive fields, 34
Indium, cyclotron mass measurements in, 60–61
Induced torque technique, in magnetoresistance measurements, 27–29
Inductance, kinetic, 267
Inductance bridges, 284
Induction (helicon) method, in magnetoresistance measurements, 26–27
Inelastic tunneling, 163–164
Infrared lasers, frequency measurements for, 303–304
Interference experiments, in superconducting rings, 286–294
Interferometer, superconducting, 294–296
Ion(s)
 see also Negative ion; Positive ion
 capture of by quantized vortex lines, 470–474

Ion(s) *(contd.)*
  capture width for, 472–474
  escape of from quantized vortex lines, 474–476
  free liquid surface and, 481–483
  motions of in superfluid films, 483–484
  in quantum liquid studies, 443–484
  structure of in liquid helium, 443–446
  turbulent superfluid helium and, 479–481
  vortex ring creation with, 476–479
Ion bombardment cleaning, 107–110
Ionic drift velocity
  Cunsolo method in, 456–458
  measurement of, 454–463
  signal averages in, 458–461
Ion mobility
  along linear vortex lines, 481
  liquid helium microscopic excitations and, 463–469
  temperature dependence in, 465
Ion techniques, for studying microscopic quantum excitations, 469–481
Iron, hyperfine Fe^{57} structure in, 343–344
Isolation transformer, in magnetoresistance measurements, 22
Isotropic solid, longitudinal waves in, 375–376
I-V curve
  e/h constant and, 273–274
  for equivalent circuit model, 269–271
  external noise and, 247–248
  Josephson oscillation and, 242
  microwave-induced steps in, 217-225
  for point contacts, 259
  quasi-particle steps in, 222–225
  subharmonic structure in, 239–243
  supercurrent steps on, 217–222
  fo thin-film bridges, 257–258
  for tunnel junctions, 124, 193–194, 252–253
  for weak links, 209–212

J

Johnson noise, 288
Johnson-Nyquist relation, 245
Josephson current
  alternating or oscillating, 233–238
  frequency dependence of, 225–229

Josephson effect, 127, 199–200
  Dayem bridges and, 255–256
  e/h constant and, 273
  experiments on, 201–251
  first experimental evidence for, 212–213
  nature of, 201–208
  phonon generation by, 243–245
  supercurrent steps and, 217–222
  thermodynamic fluctuations and noise in, 245–251
  thin-film bridges and, 255–258
  volt and, 275
  voltage standard and, 274–275
Josephson frequency, 204, 301
  Riedel singularity and, 242
Josephson junction, 199
Josephson oscillation
  frequency modulation of, 246
  line width of, 247
  nature of, 265
  noise and, 247–249
  in thin-film bridge, 266
Josephson penetration depth, 205, 238
Josephson radiation, self-coupling of, 231
Joule-Thomson refrigerators, 356

K

Kanthal, 360
Kerr cell, 705
Kerr effects, 705–706
Kerr method
  longitudinal, 721–725
  polar, 718–721
  transverse, 725–726
Kinetic inductance, 267
Kondo effect, 530
Korringa equation, 512

L

Lagrangian strains, thermoelasticity and, 436
Landau-Ginsburg theory, 201, 208
Landau level, 9
Lande splitting factor, 39
Laser breakdown, ion injection in, 453–454

SUBJECT INDEX 773

Lead
  cleaning and plating of, 574
  dHvA studies in, 56–60
  for high-Q superconductivity
    resonators, 570, 573–575
LEED (low energy electron diffraction)
  pattern, 635
  in photoemission sample preparation,
    109–110
Light
  circularly polarized, 700
  propagation of in crystals, 700–702
Light ray, vibrations of, 699
Linear sweep method, in energy
  distribution curves, 97
Liquid helium
  ion production and structure in,
    443–454
  negative ions in, 450
  phonon region of, 464, 467–468
  photoelectric injection of electrons in,
    448–449
  roton region of, 464
  space-charge-limited diode and,
    461–462
  superconductivity magnet and, 606–607
Localized atomic moments, paramagnetic
  susceptibility of, 504-511
Lorentz (electron) microscopy, in
  magnetic domain studies, 726–734
Lorentz theory, 68

M

Magnet(s)
  in Mössbauer spectroscopy, 361–363
  superconducting, see Superconducting
    magnets
Magnet configuration, 362
Magnetic breakdown, galvanometric
  properties and, 9
Magnetic domains, 683–687
  antiferromagnetic domains and,
    740–743
  chemical polishing and, 686
  defined, 675
  Faraday rotation and, 703–704
  formation of, 676–683
  Hall probes in, 683
  Lorentz (electron) microscopy in,
    726–734
  magnetostatic effects in, 684
  observation of, 675–743
  optical and magnetooptical properties
    in, 697–726
  phase position in, 715
  powder pattern (colloid) technique in,
    687–692
  scanning electron microscopy in,
    734–735
  specialized techniques and pattern
    formation in, 692–697
  specimen preparation in, 684–687
  strain-free surfaces and, 685
  X-ray studies in, 736–739
Magnetic fields
  low temperatures and, 20
  in Mössbauer spectroscopy, 361–363
Magnetic films
  Faraday rotation in, 703–704
  powder patterns in, 695–697
  vapor condensation method for,
    693–694
Magnetic induction, equations for, 45
Magnetization, oscillations in, 33
Magnetization curves, computation of,
  681
Magnetization ripple, 731–733
Magnetoacoustic measurements, 430–431
Magnetocrystalline anisotropy, 40–41
Magnetometer, as magnetic flux meter,
  283, 286
Magnetometer circuit, superconducting
  rings and, 279–285
Magnetoresistance
  anomalous longitudinal, 30–31
  background, 6
  field dependence of, 4
  high-field, 1–31
  magnetic breakdown and, 8–9
  quadratic, 9
Magnetoresistance measurements
  ac method in, 23–25
  attaching leads in, 12
  boundary-value problem in, 26–27
  components in, 20
  data collection and processing in, 29
  de Haas-Van Alphen effect in, 24
  high-pressure, 29–30

Magnetoresistance measurements *(contd.)*
  induced torque method in, 27–29
  inductive (helium) method in, 26–27
  isolation transformer in, 22
  low-field effects in, 30
  measurement techniques in, 21
  noise in, 22
  pulsed field method in, 25–26
  pulsed magnets in, 20
  sample evaluation in, 15
  sample holders in, 15–20
  sample preparation and evaluation in, 10–15
  spark cutter in, 11
  special topics in, 29–31
Magnetoresistance rotation curve, 5
Magnetostatic effects, 684
Manganese ammonium Totton salt, 357
Manoxol OT, in magnetic domain studies, 688–692
Mattis-Bardeen theory, 550–553, 560
Meissner effect, 214, 555, 557, 609, 614
Metal(s)
  magnetoresistance of, 1–31
  phonon spectrum of, 146–148
  state of compensation of, 3
  ultrasonic attenuation in, 428–430
Metal-insulator-metal junctions, 147, 172, 174
Metal-insulator-semiconductor junctions, 147, 178–181
Metal-insulator-superconductor junction, 130
Metal-semiconductor tunnel junctions, 181–184
Microwave resonators, superconducting, *see* Superconducting microwave resonators
Minerals, cleavage planes of, 103
Molybdenum, for high-temperature furnaces, 359
Monochromators and light sources, in photoemission studies, 120–121
Mössbauer atom, 313
Mössbauer data, nature of, 311–312
Mössbauer effect
  experimental configurations in, 308
  susceptibility thermometer and, 532
Mössbauer experiments, radioactive sources of, 350–351
Mössbauer gamma-ray transition energy, 311
Mössbauer spectra
  Auger processes and, 310
  scattering geometry in, 309
Mössbauer spectrometers, 316–351
  accuracy of, 339
  calibration and standards for, 342–346
  constant-velocity systems and, 318–319
  drives and data collection in, 316–346
  electromagnetic transducers and, 331–337
  error signal in, 326–327
  fast flyback operation in, 340
  feedback velocity control in, 326–331
  gamma-ray detection in, 346–350
  modulator mode in, 320–323
  nuclear counter instrumentation for, 348–350
  preamplifiers and, 349
  resonance detectors in, 347
  scale factor for, 344
  scintillation counter and, 346–347
  sinusoidal drive in, 337–338
  sinusoidal wave form and, 339
  time mode in, 323–326
  transducers for, 331–337
  "tuning" of, 317
  zero-velocity point for, 344–345
Mössbauer spectroscopy
  auxiliary equipment for, 352–366
  constant acceleration in, 337–338
  cooled sources in, 356–357
  data analysis in, 366–367
  data handling in, 365–366
  defined, 307
  Dewars techniques in, 352–356
  electric field gradient in, 314
  experimental methods in, 307–368
  furnaces for, 359–361
  graphs in, 367–368
  high-pressure experiment in, 363–365
  introduction to, 307–316
  low-temperature techniques in, 352–359
  magnets and magnetic fields in, 361–363
  proportional feedback temperature control circuit for, 358
  quadrupole coupling in, 314
  refrigerators and flow coolers in, 356

Mössbauer spectroscopy (contd.)
  sources for, 350–351
  techniques used in, 309–310
  temperature control in, 358
  thermal leaks in, 354–355
  triangle generators for, 339
  ultralow temperatures in, 357
  wave form generators and, 337–342
Multichannel analyzers, 319–322
  see also Mössbauer spectrometers
  data handling in, 365–366
  "fast flyback" in, 324
  thermometry and, 522–523
  time mode in, 323
  triangle generator in, 339–340
  up-down multiscaling mode in, 325, 330
Multichannel scaling, 323–324
  sinusoidal wave form in, 339
Multimeters, 279–284

N

National Bureau of Standards, 343
Negative ion, bubble model for, 443–445
Nichrome, 360
Nickel
  de Haas-Van Alphen effect in, 63–64
  spin-dependent tunneling in, 154
NIM (nuclear instrument module) system, 349
Niobium
  as cavity material, 559, 563–566
  energy gaps in, 142
  for high-Q superconducting resonators, 570–573
  in point contacts, 258–259, 301–302
  Siemens process for, 571–573
  Stanford process for, 571–572
  in superconducting ring, 280
  tunneling in, 142
Niobium cavities, 559, 563–566
Niobium selenide, 258
NMR, see Nuclear magnetic resonance
Noise thermometer, 286–289
Normal metals, surface impedance of, 544–550
Normal-metal tunneling
  barrier excitations in, 167–169
  elastic, 157–161
  electrode excitations in, 169–170
  electron energy-momentum relationships in, 161
  high-voltage tunneling in, 161–163
  phenomena in, 159–160
  phonon interaction in, 170
  trapezoidal barrier in, 160–161
  tunneling conductance and, 159
  zero-bias anomalies in, 170–177
Nuclear counters, electronic instrumentation for, 348–350
Nuclear energy levels, electronic perturbation of, 307
Nuclear magnetic resonance (NMR)
  free precession technique in, 516
  measurements of, 513–518
  Mössbauer spectra and, 312–314
  resonance frequency in, 530
  specimens of, in de Haas-Van Alphen effect, 51–52
Nuclear moments, paramagnetic susceptibility of, 511–533
Nuclear orientation measurements, 518–533

O

Ohm's law, superconducting resonators and, 546
Optical density of states, 76
Optical excitation, 75–77
Oscillator systems, continuous-wave composite, 397–398

P

Pakadakis velocity-measurement system, 401
Paramagnetic susceptibility
  of localized atomic moments, 504–511
  of nuclear moments, 511–533
Parametric amplification, weak links for, 304
Particle tunneling, concept of, 123
  see also Electron tunneling; Tunneling
Peirels barriers, Bordoni peaks and, 422
Perturbation theory, optical excitation and, 75

Phase position, in magnetic domain studies, 715
Phonon
  generation and detection of, 150–151
  recombination of, 150–151
  region of, 467–468
  scattering of, 79–80
  spectrum of, with Van Hove singularities in, 146–147
Photoelectric emission, Fowler theory and, 72–73
Photoelectric quantum yield, in photoemission measurements, 81–84
Photoelectrons, escape depth of, 104
  see also Electron(s)
Photoelectron spectroscopy, 68, 99
Photoemission
  current-voltage curve in, 90
  direct transitions in, 76
  electron kinetic energy distribution in, 89–90
  energy distribution curves in, 70–71
  from metals, 67
  nondirect model of, 77
  quantum yield in, 81–84
  reverse, 88
  temperature-energy relationship in, 101–102
  three-step model of, 75–81
  transport and scattering events in, 77–80
  vacuum-chamber monochromator system and, 89
Photoemission analyzer, 84–89
  resolution in, 85–88
  zero potential in, 87
Photoemission chambers
  design of, 116–120
  high vacuum, 114–120
Photoemission measurements
  cesium channels or ampoules for, 113, 121
  cleavage in, 103–104
  contaminants in, 106–107
  current bridge circuit in, 93–94
  electron multipliers in, 99
  energy distribution curve measurements in, 84–102
  evaporation in, 104–105
  experimental techniques for, 67–122

film thickness in, 105
  future research in, 121–122
  heat cleaning in, 105–107
  ion bombardment cleaning in, 107–110
  LEED pattern in, 109–110
  lowered electron affinity in, 110–114
  monochromator and light sources in, 120–121
  photoelectric quantum yield and, 81–84
  resistance heating in, 104
  retarding field analyzer in, 84–89
  sample preparation in, 102–114
  vacuum pumps in, 114–116
Photoemission process, physics of, 74–75
Photoemission studies, at high photon energies, 121
  see also Photoemission measurements
Photon-assisted tunneling, 223
Photons, quantum yield in, 72
Phototube calibration, in photoemission measurements, 82–84
Physical properties, ultrasonic measurement of, 410–433
Piezoelectric effect, as transducer, 392
Plasmons, surface, 184
Platinum thermocouples, 359
P-n homodiode junction, 185–186
P-n tunnel junctions, 177, 184–187
Point contacts
  defined, 258–260
  in magnetometers, 281
  quantum interference effects in, 263
Point contact squid, 285
Polar effect, 705–706
Polarimetric measurements, in thin films, 671–674
Polarization
  Kerr component in, 721
  Kerr effects in, 705–706
Polarized light, 700
Polarizing field, 527
Polar Kerr method, 718–721
Positive ion, electrostriction model of, 445–446
Positronium atom, 443–445
Powder pattern (colloid) technique, in magnetic domain studies, 687–692, 695–697
Pulse-echo techniques, in ultrasonic studies, 371–372, 399–400

Pulsed field method
  de Haas-Van Alphen effect in, 34
  in magnetoresistance measurements, 25–26
Pulsed magnets, in magnetoresistance measurements, 20

Q

Quantized vortex lines
  capture of ions by, 470–474
  ion escape from, 474–476
Quantum liquids, ions in study of, 443–484
Quantum yield
  absolute, 81
  in photoemission measurements, 81–84
Quartz-crystal oscillator, in thin-film measurements, 654–655
Quartz transducers, 392
Quasi-particle steps, Josephson effect and, 222–225
Q values
  coupling networks and, 582–583
  or real resonator, 549
  for superconducting microwave resonators, 542–543

R

Radar technology, 371–372
Radioactive sources, for Mössbauer experiments, 350–351
Radioisotopes, as thermometers, 525
Real materials, ultrasonic study of, 421–433
Recoil-free gamma-ray resonance absorption, 307
  see also Mössbauer spectroscopy
Recombination phonons, 150–151
Regenerative detector, 302
Replica method, for electron microscopy, 692–693
Residual surface resistance, 553–560
Resistance bridge, for resistance thermometers, 499–501
Resistance thermometers, 489–501
  interference from, 495–496
  resistance bridge in, 499–500
  shielding of, 494–498

Resistors, in thermometry, 492–494
Resonance detectors, 347–348
Resonators
  high-Q, *see* High-Q superconducting resonators
  Q of, 549, 565, 568–570, 577–579
  superconducting microwave, *see* Superconducting microwave resonators
  ubiquity of, 541
Retarding field, defined, 85
Retarding field analyzer, 84–89, 99
  electronics of, 89–92
Rf field, superconducting surface impedance and, 560–566
RHEED (reflection high-energy electron diffraction) system, 635–636
Rhenium, energy gap anisotropy in, 143
Riedel singularity, 225–229
  Josephson frequency and, 242
Rotation oscillations, in de Haas-Van Alphen effect, 53
Roton gas, liquid helium and, 464
Roton system, 465–466

S

Sample holders, in magnetoresistance measurements, 15–20
Scanning electron microscopy, in magnetic domain studies, 734–735
Scattering, photoemission and, 77–78
Scattering geometry, in Mössbauer spectroscopy, 309
Schottky-barrier junctions, 172–174
Scintillation detectors, 346–347
Semiconductors
  cesiated, 112
  degenerate, 179–180
  in tunnel junctions, 177–187
  vacuum cleavage of, 181
Shubnikov-de Haas oscillations, 9, 23–24
Siemens process, for niobium, 571–573
Signal averagers, in ionic drift velocity measurements, 458–461
Single crystals, tunneling into, 140–141
Small-amplitude wave propagation, 384–387
SNS junctions, 253–255, 268–273

Sodium salicylate films, in reference phototube measurements, 82–83
Solids
 attenuation and dispersion of waves in, 389–391
 finite-amplitude waves in, 387–389
 finitely strained, 381–389
 properties of, 371–441
 small-amplitude waves in, 384–387
 stresss-train relationships in, 381
 ultrasonic studies of, 371–441
 ultrasonic waves in, 374–391
 wave equation for, 382–383
Space-charge-limited diode, 461–463
Spark cutting, 11
Spectrometer
 Mössbauer, see Mossbauer spectrometers
 photoemission measurements and, 99
 velocity scanning, 319–320, 342, 454–456
Spin-flip scattering, 175
Split-coil magnet, 362
Spot welding, in magnetoresistance measurements, 13
Sputtering
 in thin film preparation, 637–645
 triode, 642–643
Squid (superconducting quantum interference device), 281, 489, 532–535
Standing-wave resonances, in tunneling, 152
Stanford University High Energy Physics Laboratory, 590
Subharmonic structure, 212
Superconducting interferometers, 294–296
Superconducting quantum interference devices (squids), 281, 489, 532–535
Superconducting balloons, 616–617
Superconducting bladders, 615–616
Superconducting device technology, 595–618
Superconducting magnets, 595–608
 construction of, 597
 costs for, 607–608
 ductile alloys for, 598
 field measurement in, 602–603
 field uniformity in, 600–602
 high-field or compound type, 598
 laboratory arrangement for, 596
 liquid helium consumption and, 605–606
 materials of, 598–600
 operating procedure for, 606–607
 operation of in swept field, 604
 "persistent mode" operation of, 597–598
 power supply for, 603–604
 reversed field in, 605
 stabilization of, 600
 testing of, 599–600
Superconducting microwave resonators, 541–594
 applications of, 592–594
 breakdown of, 565–566
 coupling networks for, 581–584
 design and fabrication of, 567–570
 experiments using, 584–594
 frequency stability in, 585–587
 frequency standards referenced to, 588–590
 high-field applications of, 590–591
 high-Q, 567–584
 low loss in, 593
 material property studies of, 591–592
 oscillator and, 589
 Q values for, 542–543
 tuning of, 587–588
Superconducting rings, 275–289
 interference effects in, 289–296
 magnetometer circuits and multipliers in, 279–285
 noise limit in, 283
 quantized states of, 275–279
 quantum interference experiments in, 291–294
 with resistor, 286–289
 with single weak link, 285–289
 symmetrically biased, 290
Superconductor shielding, 609–618
Superconducting state, current flow in, 126
Superconducting surface impedance, rf field level and, 560–562
Superconducting surface reactance, 552–553
Superconducting surface resistance, 550–552

Superconducting tunnel characteristics, 126–128
Superconducting tunneling, in high magnetic fields, 153–156
Superconducting tunneling measurements, junctions for, 127
Superconductors
  coupling energy in, 203
  energy gap measurement in, 135–140
  Josephson effect in, 201–208
  microwave properties of, 544–566
  recombination lifetimes in, 150
  residual surface resistance in, 553–560
  surface impedance of, 544–550, 560–562
  surface reactance in, 552–553
  thermometry and, 536–537
  time-varying fields and, 610–611
  tunneling in, 126–156
  tunneling between normal metal and, 128–131
  two-fluid model of, 547–548
  weakly linked, 199–305
Superconductor shielding
  dc fields and, 612–617
  design of, 611–612
  time-varying fields and, 610–611
Supercurrent, oscillating, 233–238
Supercurrent steps, 217–222
  self-induced, 229–233
Supercurrent wave forms, for equivalent circuit, 272
Superelectrons, Josephson oscillator and, 265
Superfluid, angular fluid motion in, 469–470
Superfluid surfaces and films, studies of, 481–484
Surface impedance, of normal and superconducting metals, 544–550
Susceptibility thermometers, 502–533

T

Tantalum, tunneling measurements in, 142
Temperature, thermodynamic terminology for, 486–489
Texas, University of, 592

Thermal decomposition process, in thin films, 646
Thermal evaporation
  temperature and support materials in, 624–627
  of thin films, 622–637
Thermionic cathodes, electron injection from, 451–452
Thermocouples, 534–535
Thermodynamic temperature, terminology for, 486–489
Thermoelasticity theory, 433–438
Thermometers
  germanium resistors for, 491–492
  miscellaneous types of, 533–538
  nuclear orientation and, 488
  paramagnetic susceptibility and, 487
  radioisotopes as, 525–526
  resistance type, 489–501
  susceptibility type, 502–533
Thermometry
  capacitors in, 538
  cerous magnesium nitrate (CMN) in, 503–504
  energy gap and, 537
  high-temperature, 359–360
  multichannel analyzer and, 522–523
  nuclear magnetic resonance measurements and, 513–518
  nuclear orientation measurement in, 518–533
  superconductors and, 536–537
  thermocouples in, 534–535
  at ultralow temperatures, 485–539
Thin-film bridges, 255–258
  Josephson oscillation in, 266
  in magnetometers, 281
  quantum interference effects in, 263
  superconducting rings and, 278–279
Thin films, 619–674
  anodization of, 649–650
  chemical deposition of, 645–652
  chemical transfer processes in, 647
  electrical thickness measurements for, 658–659
  electrodeposition of, 648–650
  electron beam evaporation system in, 634–635
  ellipsometry for, 671–674
  epitaxial growth in, 621

SUBJECT INDEX

Thin films (contd.)
Getter-sputtering system in, 641–642
glow-discharge sputtering in, 638–642
immersion spectrophotometry for, 670–671
interferometer measurements of, 657
internal stresses in, 660–663
Langmuir-Blodgett technique in, 650–651
LEED (low-energy electron diffraction) system in, 635–636
low-pressure sputtering in, 642–645
measurements of, 652–674
mechanical techniques in, 660–665
Michelson interferometer and, 657
microbalance measurements in, 653–655
multilayer, 630–632
multiple-beam interferometric techniques in, 656
oxide layers and, 650
optical measurements of, 655–657, 665–674
polarimetric methods in, 671–674
preparation of, 620–652
quartz-crystal oscillation measurements of, 654
reflection and transmission measurements of, 666–669
refractory materials in, 633–637
resistance measurements of, 658
resistive heating in, 623–634
RHEED (reflection high-energy electron diffraction) system in, 635–636
single-crystal, 622
spectrophotometric methods in, 669–671
sputtering in, 637–645
stresses in, 663
stylus measurements of surface in, 659
substrate for, 620
surface contamination in, 620
tensile properties of, 663–665
thermal evaporation of, 622–637
thickness measurements in, 653–659
triode sputtering in, 642–643
vacuum evaporation system in, 629–630
X-ray absorption and reflection techniques for, 659

Thin-film squids, 285
Time differentiation method, for energy distribution curves, 98
Timing diagrams, in ultrasonic studies, 403–404
Tin-plated cavity, rf critical field of, 563
Transducers
bonding of, 393–395
electromagnetic, 331–337
in ultrasonic wave generation, 391–393
Transmission geometry, in gamma ray emission, 308–309
Triangle generator, 340
Triode sputtering, 642–643
Tunnel current, self field of, 215
Tunnel diodes, injection of hot electrons by, 449–450
Tunneling
see also Tunnel junction
artificial barriers in, 191–192
barrier preparation in, 188–189
contacts in, 192
and density of excited states, 144–149
electron, see Electron tunneling
energy gap determination in, 135–140
field splitting and, 154–155
Fowler-Nordheim equation for, 162
geometrical resonances in, 152–153
high-voltage, 161–163
inelastic, 163–164
I-V characteristic curve in, 135–136
junction fabrication in, 187–192
junction preparation in, 187–188
measurement circuits in, 193–197
normal-metal, 157–177
normal metal and superconductor, 128–131
phonon-assisted quasi-particle, 242
photon-assisted, 223
point contacts in, 258–260
as probe of electron-phonon interaction, 149–150
quasi-particle, 223
self-energy effects in, 183
in single crystals, 140–144
special topics on experimental techniques in, 187–197
spin-dependent, 154
superconducting energy gap and, 126
in superconductors, 126–156

SUBJECT INDEX

Tunneling *(contd.)*
 surface plasmons in, 184
 between two superconductors, 132–133
Tunneling behavior, deviation from ideal in, 133–135
Tunneling measurements, phenon spectra and, 144–149
Tunnel junctions
 see also Weak links
 barriers in, 254
 bias voltage across, 161–162
 cavity modes and, 229–233
 configuration of, 252–253
 construction of, 252
 contacts for, 192
 defined, 209
 equivalent circuit model of, 268–273
 fabrication of, 124–125, 252
 I-V characteristics of, 124, 193–194
 Josephson radiation in, 231, 234
 lead-copper-lead type, 254
 lead-lead oxide-lead type, 213, 217
 molecular excitations in barriers of, 165–167
 oxide barriers of, 161–164
 phonon intensity at, 243–245
 phonons emitted from, 243–245
 plasma oscillations in, 238–239
 p-n type, 184–187
 power spectrum for, 245–246
 quality of, 253
 semiconductors in, 133, 177–187
 SNS type, 254–255
 temperature dependence in, 216
 testing of, 125–126, 192–193
 thin-film bridges and, 255–258
 as weak links, 251–254

U

Ultralow temperatures, thermometry at, 485–539
Ultrasonic attenuation, in metals, 428–430
Ultrasonic measurements, 371–441
 attenuation in, 373
 magnetoacoustic measurements and, 430–431
Ultrasonic range, 371–372

Ultrasonic studies
 continuous-wave oscillator systems and, 397–399
 dislocation effects in, 421–428
 dispersion effect in, 424
 electronic effects in, 428–431
 losses in, 431–433
 measurement systems in, 400–410
 phonon viscosity and, 431–432
 pulse superposition in, 406–408
 sample preparation in, 396–397
 timing diagrams in, 403–404
 velocity and attenuation measurements in, 397–410
Ultrasonic waves
 attenuation and dispersion in, 389–391
 direct rf generation of, 395–396
 generation of by other means, 391–396
 losses in, 431–433
 physical property measurements of, 410–433
 in solids, 374–391
 thermoelastic losses in, 432–433
Ultraviolet photoemission, 67–122
 see also Photoemission
Up-down multiscaling, 325, 330

V

Vacuum pumps, in photoemission studies, 114–116
VAMFO technique, in thin films, 656
Van Hove critical points, 144
Vapor condensation method, for magnetic films, 693–695
Varian e-gun, 105
Velocity feedback control, in Mössbauer spectrometer, 326–331
Velocity-measurement systems
 block diagram of, 401
 drawbacks of, 409
Velocity scanning spectrometers, 319–320, 342–343
 in ionic drift velocity measurements, 454–456
Voight elastic coefficients, 434, 441
Volt, defined, 275
Voltage-tunable oscillator, 579
Vortex flow, ion techniques in, 469–470

Vortex rings
  creation of with ions, 476–479
  drift velocity vs. electric field in, 478

W

Wave equation, elastic coefficients in, 439–441
Wave form generators, in Mössbauer spectrometer, 337–342
Weak links (weakly linked superconductors), 209
  see also Tunnel junctions
  applications of, 273–305
  characteristics of, 251–273
  Clarke slug and, 260–261
  as computer elements, 305
  current-phase relations in, 260–268
  defined, 199
  detector applications for, 299–304
  digital devices for, 304–305
  experiments with, 199–305
  e/h measurements for, 273–274
  harmonic generation and, 299–304
  I-V characteristic curves for, 209–212
  Josephson effects and, 201–251
  kinetic inductance and, 267
  and microwave-induced steps on I-V curve, 217–225
  mixing applications of, 302–303
  parametric amplification of, 299–304
  phenomenological descriptions of, 262–268
  plasma resonance and, 238–239
  point contacts and, 258–260
  quasi-particle steps in, 222–225
  response of to ac signals, 299–301
  rf voltage across, 221
  Riedel singularity and, 225–229
  self-induced supercurrent steps and, 229–233
  in superconducting rings, 275–296
  supercurrent steps and, 217–225
  tunnel junctions as, 251–254
  types of, 251

X

X-ray diffraction topography, 736–739
X-ray photoelectron spectroscopy, 68, 99

Z

Zeeman splitting, in tunneling, 153–154
Zero-bias anomaly, 183
  "giant," 176
  in normal-metal tunneling, 170–177
  in p-n tunnel junctions, 186
Zero-crossing detector, 339
Zero-field mobility, in liquid helium, 465–466
Zero-velocity point, for Mössbauer spectrometers, 344–345